Load-Oriented Manufacturing Control

Hans-Peter Wiendahl

Load-Oriented Manufacturing Control

With 228 Figures, 14 Plates, and 52 Tables

Springer-Verlag Berlin Heidelberg GmbH

Hans-Peter Wiendahl, Univ.-Prof., Dr.-Ing., Dr.-Ing. h.c.
Institut für Fabrikanlagen,
Universität Hannover
Callinstrasse 36
D-30167 Hannover, Germany

ISBN 9978-3-642-63343-0

Library of Congress Cataloging-in-Publication Data
Wiendahl, Hans-Peter.
[Belastungsorientierte Fertigungssteuerung. English]
Load-oriented manufacturing control / H.-P. Wiendahl.
p. cm.
Includes bibliographical references and index.
ISBN 9978-3-642-63343-0 ISBN 978-3-642-57743-7 (eBook)
DOI 10.1007/978-3-642-57743-7
1. Production control. 2. Scheduling (Management) I. Title.
TS157.W54 1994 658.5-dc20 94-26685 CIP

Typesetting: Acso Trade Typesetting Ltd, Hong Kong

SPIN: 10019742 61/3020 - 5 4 3 2 1 0 - Printed on acid-free paper

Foreword

In the face of increasing international competition, production planning and control is becoming more and more important for manufacturers. Short delivery times, punctual delivery and low inventories are the most important objectives. In contrast, efficient utilization of production equipment is less significant than it was previously.

The use of data processing systems has now become very widespread in production planning and control. Numerous studies, however, have shown that existing manufacturing control systems can neither measure permanently nor have a direct influence on those objectives mentioned above. As a result, dispatching lists, although produced at considerable expense, very quickly become obsolete and are therefore no longer reliable. It is for this reason that foremen frequently have informal planning systems in addition to the official system, in order to be able to keep due dates despite numerous delays and changes.

In this situation, planners are very interested in new manufacturing control methods which are able to monitor specific manufacturing flow strategies in different planning conditions. One reason for this is the introduction of Computer Integrated Manufacturing (CIM), and of logistical concepts. One approach that has been much discussed in this respect is the Japanese Kanban system. It aims to obtain low inventories, short lead times and good schedule performance, and does so with an astonishingly little amount of control.

However, practical experience with the Kanban system has shown that it can only be used to a limited degree, since in many companies the production structure does not meet the system's necessary operational requirements.

This book presents a new, self-contained, logical approach to manufacturing control. It is mainly directed towards manufacturers with high-variant job-shop production. The book's central subject is a universal model of manufacturing flow that can be applied to conventional manufacturing as well as to automated manufacturing systems. A step-by-step guide for those readers with little mathematical knowledge of the subject is provided, and numerous examples are given to enable the method and its modules to be put into practice.

This book is based on many years of research work undertaken at the Institut für Fabrikanlagen at Hannover University, Germany. The work was started by the first director of the institute, Prof. Dr.-Ing. Hans Kettner, in 1972, and has been continued by the author since 1979. The research projects were funded by Federal and private German institutions including the Bundesministerium für Forschung und Technologie (Federal Ministry for Research and Technology). Since 1979, the manufacturing control method has seen numerous commercial software versions, and has been put into practice successfully in over 100 plants in Germany and throughout Europe.

I would like to express my gratitude to a number of people for their assistance in producing this book. For the original German-language edition I was particularly assisted by Dr.-Ing. W. Bechte, Dr.-Ing. W. Erdlenbruch, and Dr.-Ing. W. Buchmann. Herr Dipl.-Ing. H.G. von Wedemeyer helped in the revision of the numerous tables and the critical examination of the manuscript. I would like to thank Frau M. Bruns for her care in drawing the figures, and Frau I. Sommerfeld for laboriously reviewing and editing the manuscript.

The translation of this English edition was produced by Frau S. Heizmann, who was assisted by Frau I. Sommerfeld. The figures were produced by Frau M. Rohrsen and her colleagues. In the critical revision of the English manuscript, I was first aided by Prof. Dr. Fogarty, to whom I owe a great deal for his numerous suggestions. I am sorry to say that he passed away before he could finish the book. The advice given by Prof. Dr. John J. Kanet of Clemson University, South Carolina, was also a great help and enrichment to me.

Finally, I would like to honor my late colleague and predecessor, Herr Prof. Dr.-Ing. Hans Kettner, the father of the Hannover Funnel Model and of load-oriented order release, who laid the foundation for an extensive new approach to manufacturing control.

Hannover, February 1994 Hans-Peter Wiendahl

Contents

List of Abbreviations xiii

1 Introduction 1
1.1 Preface 1
1.2 Changes in the Manufacturing Environment 1
1.2.1 Productivity 3
1.2.2 Flexibility 3
1.2.3 Attractiveness of the Workplace 4
1.3 Shifting Objectives of Manufacturing Control 4
1.4 Scheduling in Practice 6
1.5 The Weak Points of Conventional Manufacturing Control 9
1.6 References 11

2 Conventional Production Scheduling and Control 13
2.1 Abstract 13
2.2 Survey 13
2.3 Lead Time Scheduling and Capacity Scheduling 16
2.3.1 Single Steps in Lead Time Scheduling 19
2.3.1.1 Determining Lead Times 19
2.3.1.2 Interoperation Time Reduction 20
2.3.1.3 Operation Overlapping 23
2.3.1.4 Operation Splitting 24
2.3.2 Single Steps in Capacity Scheduling 25
2.3.2.1 Finite Loading 25
2.3.2.2 Capacity Adjustment 25
2.3.2.3 Capacity Alignment 26
2.4 Evaluation of Conventional Lead Time Scheduling . . 28
2.5 Requirements for New Methods of Production Scheduling and Control 31
2.6 References 35

3 Lead Time – A Key Term in Manufacturing Control 37
3.1 Abstract 37
3.2 Introduction 37
3.3 Lead Time Components 41
3.4 Simple and Weighted Mean Lead Time at a Work Center 47

3.5 Order Lead Times 53
3.6 Statistical Evaluation of Work Center Lead Times . . 55
3.6.1 Absolute and Relative Frequency Distribution of Simple and Weighted Lead Time 55
3.6.2 Simple and Weighted Mean Operation Time . 59
3.6.3 Simple and Weighted Mean Operation Time Percentage 62
3.6.4 Median of Simple and Weighted Lead Time . . 64
3.6.5 Standard Deviation of Simple and Weighted Mean Lead Time 65
3.6.6 Coefficient of Variation of Simple and Weighted Lead Time 66
3.6.7 Median, Standard Deviation, and Coefficient of Variation of Simple and Weighted Order Time 66
3.7 Work Center Lead Time in the Shop Calendar 68
3.7.1 Transformation of Throughput Elements 68
3.7.2 Frequency Distribution and Statistical Evaluation in the Shop Calendar 69
3.8 Uncertainty of Measurement and Accuracy of Computed Lead Time Values 72
3.9 Examples of Authentic Lead Time Measurements . . . 75
3.9.1 Work Center Lead Times 75
3.9.2 Operation Lead Times 78
3.9.3 Order Lead Times 82
3.10 References 86

4 The Throughput Diagram – A General, Realistic Model of the Manufacturing Process 89
4.1 Abstract 89
4.2 Historical Evolution 89
4.3 Basic Form of the Throughput Diagram 93
4.4 The Work Center Throughput Diagram and its Basic Data 95
4.4.1 How to Produce a Work Center Throughput Diagram 95
4.4.2 Mean Inventory 101
4.4.3 Mean Time Data 103
4.4.3.1 Mean Range 103
4.4.3.2 Mean Advance Time 104
4.4.3.3 Weighted Mean Lead Time 106
4.4.4 Relationships between Mean Range, Mean Advance Time, and Weighted Mean Lead Time. 108
4.4.4.1 Inventory Trend Component of Lead Time 108
4.4.4.2 Sequence Component of Weighted Lead Time 111
4.4.4.3 Short- and Long-Term Relationships between Time Quantities 114
4.4.5 Mean Performance, Mean Capacity, and Mean Utilization 114

4.4.6 Relationships between Mean Inventory, Mean Performance, and Weighted Mean Lead Time . 116
4.4.7 Weighted Mean Lateness 117
4.5 Order Throughput Diagram 121
4.5.1 Extended Work Center Throughput Diagram . 121
4.5.2 How to Produce the Order Throughput Diagram 126
4.6 References 126

5 Analysis, Monitoring, and Diagnosis of the Manufacturing Flow 129
5.1 Abstract 129
5.2 Manufacturing Flow Monitoring 129
5.3 Manufacturing Flow Analysis 131
5.3.1 Procedure Outline 131
5.3.2 Forms of Evaluation and Representation 139
5.3.3 Representation of the Results 145
5.3.4 General Rules and Possibilities for Improving Manufacturing Flow 153
5.4 A Continuous Monitoring System for Manufacturing Flow 155
5.4.1 Objectives and Concept 155
5.4.2 Example of a Continuous Monitoring System . 158
5.4.3 Results and Use of a Continuous Monitoring System 167
5.5 Diagnosis of the Manufacturing Flow in the Throughput Diagram 175
5.5.1 Breaking down Inventories 175
5.5.2 Breaking down Lead Times 180
5.5.3 Measures to be Deduced from Monitoring and Diagnosis Results 186
5.6 Use of Graphics for the Representation of Throughput Diagrams and Key Data 188
5.6.1 Representation of the Results of Manufacturing Analyses 188
5.6.2 Graphics for Medium-Term Manufacturing Flow Monitoring 193
5.7 Implications for Manufacturing Control 195
5.8 References 200

6 Load-Oriented Order Release 203
6.1 Abstract 203
6.2 Fundamental Relationships 203
6.3 Procedure 206
6.4 Conversion of the Orders to be Loaded 212
6.5 Sample Demonstration of the Release Process 214
6.6 How to Choose the Load Limit and Loading Percentage Values 220
6.7 Interlinking Order Control with Manufacturing Control 231

6.8 Effects of the Load Limit and Time Limit Parameters in Simulation and in Practice 232
6.8.1 Simulation of Manufacturing Processes as an Aid in the Testing of Control Algorithms . . . 232
6.8.2 Effects of the Load Limit 245
6.8.3 Effects of the Time Limit 252
6.9 Controller Analogy of Load-Oriented Order Release 254
6.10 Priority Rules and Order Sequencing in Load-Oriented Order Release 256
6.11 References 261

7 Schedule-Oriented Capacity Planning and Control 263
7.1 Abstract 263
7.2 The Problem and the Method 263
7.3 Defining the Load Centers 266
7.4 Schedule-Oriented Capacity Planning Procedure . . 269
7.5 References 277

8 Implementation of Load-Oriented Manufacturing Control . 279
8.1 Abstract 279
8.2 Prerequisites 279
8.2.1 Influence of the Lot Size on Inventories and Mean Lead Time 279
8.2.2 Orders Must Have a Due Date 286
8.2.3 An Operation Sheet with Standard Times Must Exist 288
8.2.4 Material, Tooling, Fixtures and NC Programs Must Be Available 288
8.2.5 Machine Availability and Personnel Capacity Must Be Known 288
8.2.6 Operation Feedback Must Be Complete and Sufficiently Accurate 291
8.3 Program Modules of Load-Oriented Manufacturing Control 292
8.3.1 Overview 292
8.3.2 Capacity Planning 292
8.3.3 Release Planning 295
8.3.4 Sequencing 296
8.3.5 Calculation of Monitored Data 299
8.4 System-to-User Interfaces and Hardware Configuration 301
8.5 Implementation Strategies 305
8.6 Load-Oriented Manufacturing Control in Automated Production Sites 308
8.6.1 Controlling Flexible Manufacturing Systems . 308
8.6.2 Integration into CIM Concepts 310
8.7 Effects of Load-Oriented Manufacturing Control . . 316
8.7.1 Cost Effectiveness 316
8.7.2 Qualification and Motivation 318
8.8 References 319

9 Comparison of Load-Oriented Manufacturing Control with Other Methods 323
9.1 Abstract 323
9.2 Overview 323
9.3 The Kanban Principle 328
9.4 The Job-Progress Number System 331
9.5 Finite Control with a Graphic Control Unit (Electronic Leitstand) 333
9.6 Queueing Models 335
9.7 The OPT System 337
9.8 References 341

10 Summary 343

Appendix A 347

Appendix B 355

Subject Index 365

List of Abbreviations

AB	Basis Area
AC	Control Area
AF	Flow Area
AFI	Final Inventory Area
AFIA	Final Inventory Additional Area
AH	Anticipation Horizon
AI	Inventory Area
AII	Initial Inventory Area
AIIA	Initial Inventory Additional Area
AINSD	Negative Schedule Deviation Area of Input
AIPSD	Positive Schedule Deviation Area of Input
AL	Lot Area
ALT	Lead Time Area
AONSD	Negative Schedule Deviation Area of Output
AOPSD	Positive Schedule Deviation Area of Output
AP	Position Area
ATA	Advance Time Area
BF	Final Basis Inventory
BOM	Bill of Material
C	Capacity
CC	Container Content
CDAY	Daily Capacity
CF	Conversion Factor
CIM	Computer Integrated Manufacturing
CV	Coefficient of Variation
DD	Daily Demand
DEL	Delay
ENT	Entry
EOQ	Economic Order Quantities
ER	Efficiency Rate
FAS	Flexible Assembly System
FIFO	First in – First out
FITC	Final Inventory Trend Component
FMS	Flexible Manufacturing System
Hrs	Hours
IB	Basis Inventory

IBI	Initial Basis Inventory
IC	Control Inventory
IF	Final Inventory
IFL	Flow Inventory
II	Initial Inventory
IITC	Initial Inventory Trend Component
IL	Lot Inventory
ILO	Leftover Inventory
INP	Input
IS	Scheduled Input
ITC	Inventory Trend Component
LC	Load Center
LHC	Labor Hours Content
LIFO	Last in – First out
LL	Load Limit
LOT	Longest Operation Time Rule
LPG	Loading Percentage
MED	Median (Value)
MRPII	Manufacturing Resource Planning
NK	Number of Kanbans
NSD	Negative Schedule Deviation
OLO	Lateness of the Outputs/Order Lateness
OLOP	Positive Weighted Output Lateness
OP	Operation
OPO	Number of Operations per Order
OS	Scheduled Output
OUT	Output
P	Schedule Period/Reference Period
PDC	Production Data Collecting
PDP	Production Data Processing
PE	Performance
P_{inp}	Input Probability for an Order
POS	Weighted Position
P_{out}	Output Probability for an Order
PPC	Production Planning and Control
PSD	Positive Schedule Deviation
Q	Quantity/Lot Size
R	Range
RAT	Rated
RB	Basis Range
RC	Control Range
REL	Release
RF	Flow Range
RFI	Range of Final Inventory
RII	Range of Initial Inventory
RL	Lot Range
RPR	Relevant Production Rate
SCD	Shop Calendar Day(s)
SD	Weighted Schedule Deviation
SOT	Shortest Operation Time Rule

STA	Standard Deviation
TA	Advance Time
TB	Buffer Time
TIO	Interoperation Time
TLG	Top Value of a Lead Time Group
TL	Lead Time
TL_m	Simple Mean Lead Time
TLM	Weighted Mean Lead Time
TL_{mw}	Weighted Mean Lead Time
TLSC	Sequence Component of Lead Time
TLV	Coefficient of Variation of Simple Lead Time
TO	Order Time
tOC	Order Completion Date
TOL	Order Lead Time
TO_m	Simple Mean Order Time
TO_{mw}	Weighted Mean Order Time
TOP	Operation Time
TOP_m	Mean Operation Time
TOP_{mw}	Weighted Mean Operation Time
TOPP	Operation Time Percentage
TOPS	Standard Deviation of Operation Time
TOP_w	Weighted Operation Time
tOR	Order Release Date
TO_S	Standard Deviation of Simple Order Time
TO_V	Coefficient of Variation of Simple Order Time
TO_{vw}	Coefficient of Variation of Weighted Order Time
tPB	Process Begin
tPE	Process End Date
tPEU	Process End at Upstream Work Center (Input Date)
TPO	Process Time per Order
TPU	Process Time per Unit
TS	Setup Time
tSB	Setup Begin
TT	Transit Time
TU	Time per Unit
TW	Waiting Time
TWA	Waiting Time After Processing
TWB	Waiting Time Before Processing
U	Utilization
VAC	Coefficient of Variation
WC	Work Center
WD	Working Days
Wks	Weeks

Chapter 1
Introduction

1.1 Preface

Many advances have been made in manufacturing management in the last thirty or forty years. These improvements originated in many different areas, and included such concepts as Manufacturing Resource Planning (MRP II), Statistical Process Control (SPC), Group Technology (GT), the Just-In-Time Approach (JIT), the Theory of Constraints (TOC), and Total Quality Management (TQM). These advances usually built on earlier concepts, utilizing their strengths while introducing new ideas to meet unsolved challenges. This work follows that pattern. The first two chapters discuss the present status of manufacturing management and the challenges that exist. The chapters which follow describe an approach that has overcome many of these challenges in a broad range of manufacturing settings.

1.2 Changes in the Manufacturing Environment

Traditionally, the competitive strength of a manufacturing company has been determined by both the performance and reliability of its products. However, short and reliable delivery times have now become equally important for corporate success. In the automotive, mechanical engineering and electrical industries, for instance, since the early 1980s customer pressure has frequently dictated that delivery times be cut in half. This has taken place in spite of heavy pressure to reduce production costs.

Many companies have realized that they can no longer meet this challenge by simply moving production sites into, so called, 'low-wage' countries. To be competitive, improved production management must become a strategic objective equal in priority to product innovation. This is true in both capital and labor intensive environments, and in relatively high and low labor rate environments.

Individual, isolated measures such as overhead analyses, inventory reduction programs, or the use of a few highly automated processing machines will not usually produce substantial improvements in a company's performance. Individual changes must be planned and implemented on the basis of an integrated corporate strategy

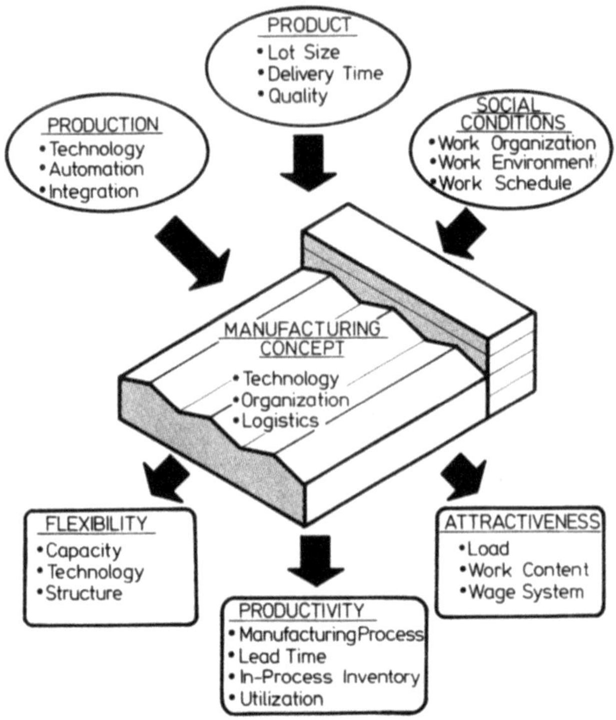

Fig. 1.1. Factors and objectives of a progressive manufacturing company

– including the manufacturing strategy [1–5, 22–26]. The first steps must be an analysis of the relevant factors affecting production, and a definition of corporate objectives. This provides a blueprint upon which to carry out and control multi-plant and coordinated measures on the basis of a comprehensive evaluation of the manufacturing process and its relationship to the marketing strategy. Figure 1.1 shows the relevant factors affecting production, and the corporate objectives within this context.

The market situation and competition are largely responsible for the demands placed on a product, thus the company's master production schedule is a direct result of these demands. In general, customer demand today is characterized by increasing product variety (and thus smaller lot sizes), and by shorter delivery times and higher quality standards. Moreover, due to shorter new product design cycles, product life cycles are shorter. As a result, several product generations frequently have to be manufactured simultaneously in different lot sizes, often using different production processes.

Companies are also being affected by developments in production engineering. Most spectacular is the rapid expansion of new technologies (e.g. the use of lasers in cutting, welding, marking, measuring, etc.), and the increasing automation of the manufacturing processes. This is particularly due to advances in electronic process control systems as well as in numerical and graphical data processing. In addition, there is an increasing trend towards integrating the flow of information and materials, characterized by the Flexible Manufacturing Systems (FMS), Computer Integrated Manufacturing (CIM) [6] and logistics concepts [7].

Finally, changes in factory working conditions have also affected manufacturing. These changes include a shorter working week, new regulations concerning breaks and shifts, and new organizational methods such as detaching workers from time-phased assembly-line processes and forming autonomous work groups. When designing and operating factories, the working environment now has to meet new requirements that limit the physical and mental strain placed on workers. This results, in particular, from intensive efforts in the field of job humanization. A modern manufacturing approach must therefore possess a number of essential defining attributes: technology, logistics – i.e. the overall spatial-temporal manufacturing process – and a structural and managerial organization consistent with productivity, flexibility and attractiveness objectives.

1.2.1 Productivity

There is no doubt that long-term profit based on productivity ranks as the number one goal in manufacturing industry. However, besides the traditional aim of cost-efficient techniques for design, planning, processing, storage, transport and handling, there are additional objectives which follow directly from those factors previously mentioned. These additional objectives – shorter manufacturing lead times, lower inventories and improved throughput – call for new approaches.

First, customers demand shorter lead times. The strategic aim must be to achieve a lead time (i.e. the time span from customer order entry to product delivery) that is shorter than, or at least equal to, the requested delivery time. Low inventories – the second objective – have two effects on manufacturing companies: one is that the working capital thus released can be invested in modern production equipment, thereby reducing entrepreneurial risks; the other effect is that lower inventories mean that less space is used for storage, a more transparent manufacturing flow results, and there are lower scrap risk and shorter lead times because of the shorter queues at work centers. The final objective, increased throughput, expresses that either simple or complex, strongly interlinked systems such as FMSs or Flexible Assembly Systems (FAS), if misdesigned or operated improperly, can result in reduced output. This can, in turn, have a catastrophic effect on delivery performance and thus profit. Therefore the aim is to create low-fault and fault-tolerant production structures which can, due to a corresponding buffer layout, compensate for minor breaks in production, and which are easy to maintain or self-correcting, even in the case of large scale breakdowns.

1.2.2 Flexibility

Flexibility has become an important objective for most companies. Often, the master production schedule changes very quickly with respect to quantity and product mix. To react flexibly to changes, it must be possible to adapt the capacity of many resources, for example by taking smaller, product-specific production units in or out of the manufacturing process or by adapting the use of automatic multi-purpose equipment.

The implementation of new technologies must be made possible by appropriately organizing both the interfaces of the parts flow and of the control information. Finally, equipment should be installed so that, if necessary, it can be inexpensively rearranged, thus ensuring much greater flexibility. The same flexibility demands of course apply to scheduling and control.

1.2.3 Attractiveness of the Workplace

Finally, when designing or redesigning facilities attention must be paid to the attractiveness of the workplace. Not only should physical strain on workers be reduced by using machines and robots, but the mental strain on operators resulting from new information and communication technologies used in control and direction must also be taken into account. Humane computer programs are, according to 'software ergonomy', designed in such a way that it is not the machine but the operators who determine the pace of work, with the computer only assisting. Work content also has to be analyzed. Workers involved in the manufacturing process must be able to relate to the results of their work without feeling overstrained by an extremely broad range of activities. A certain amount of routine is essential, but it should not be an obligation. Finally, new work structures have a considerable effect on the form of payment. Apart from the output, attention must also be paid to product quality and machine utilization as factors affecting payment. New objectives and requirements for the design and redesign of job shops, and for manufacturing control, therefore follow from these conditions.

1.3 Shifting Objectives of Manufacturing Control

Conflicting objectives have long existed in manufacturing companies. Marketing and customers desire short delivery times and punctual delivery; manufacturing desires level loads, and has traditionally desired a high utilization of production equipment; and both finance and manufacturing desire the lowest possible inventories of raw materials, semi-completed and completed products [8].

However, the increased emphasis on speedy and punctual delivery has shifted the weighing of these objectives (Fig. 1.2). Previously, utilization of both machinery and manpower were most important. Today, more emphasis is placed on meeting customer delivery requirements, maximizing throughput, and reducing inventories, all at the same time.

Short lead times, and thus short delivery times, not only increase the competitive strength of a firm, since the customer can decide on final specifications at much shorter notice, but they also reduce the risk of having to redesign components or products. Good schedule performance is also needed to be competitive. For the company, detail scheduling has the added effect that work can be carried out much more smoothly and thus more cost-effectively, as well planned and executed processes are not constantly interrupted by having to deal with emergency order fulfilment.

There are various reasons for having low inventories. On the one hand, for many companies lower inventories free working capital, thus reducing the percentage of a company's borrowed capital and freeing capital for the purchase of modern produc-

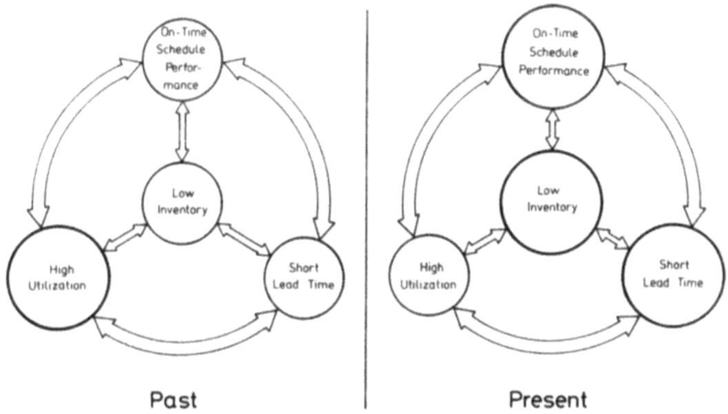

Fig. 1.2. Manufacturing control objectives – shift in emphasis within the past decade

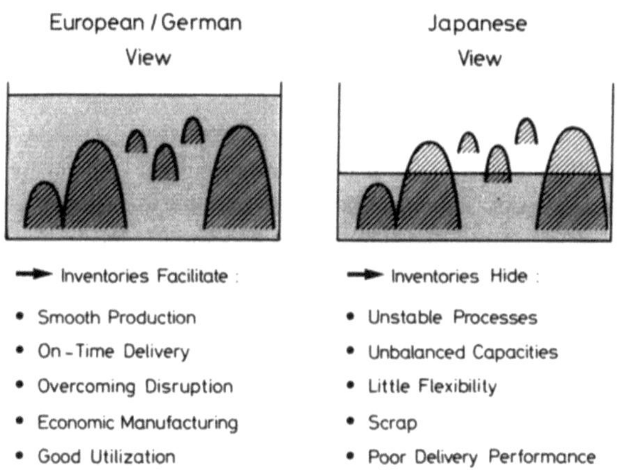

Fig. 1.3. Effects of inventories (After Siemens AG)

tion equipment. One machine-tool company, for example, reduced inventories by nearly 70% within eight years through systematic methods while still maintaining their manufacturing output. On the other hand, many firms have come to the conclusion – above all due to the success of Japanese industry – that high inventories conceal many shortcomings in a company, such as excessively long setups, unreliable processes, and poor product quality that in reality cause the long lead times. The latter delay results from long queues which form at the individual work centers concerned.

Figure 1.3 clearly contrasts the Japanese and the traditional Western views on Work-In-Process (WIP) by way of comparison with the liquid level in a container (representing the company). Lowering the WIP brings to light those problems mentioned above, as well as the lack of flexibility on the part of the workforce to work at varying work centers and at different times. This calls for a gradual elimination of such deficiencies.

1.4 Scheduling in Practice

As many companies do not have reliable information on their rate of late order fulfilment, they often believe they are in a much healthier position than is actually the case [9]. One typical example is a well known metalworking company where an investigation by the Institut für Fabrikanlagen of Hannover University, Germany produced unexpected results. Contrary to the organization's allowance of five workdays per operation, the average flow time per operation was 8.5 days. The mean order flow time was 80.1 days instead of an allowance of 55.4 days, and the average tardiness was 13 days, including the late start of orders.

Figure 1.4 shows the order flow time distribution of the 6758 orders investigated, and Figure 1.5 shows their lateness distribution. The mean tardiness of 13 days is, on average, 16% of the mean flow time and 23% of the allowance. Another interesting finding in addition to these mean values was the variance in delivery dates, which is also made clear in Figure 1.5. The company had assumed an allowed tolerance of ±5 days for due date deviation, but only approximately 15% of the orders were delivered within this range. More than 70% of orders were completed behind schedule.

Thus, in addition to the average order taking 25 days longer than planned, many better than average orders took 10 to 15 days longer than planned.

The outcome of this investigation was completely unexpected by the operations manager. He had thought that the schedule performance was, on the whole, satisfactory. As a by-product of the investigation, it was found that feedback data did not exist for at least one operation for 36% of orders, and that the recorded work center numbers were incorrect for 23% of operations.

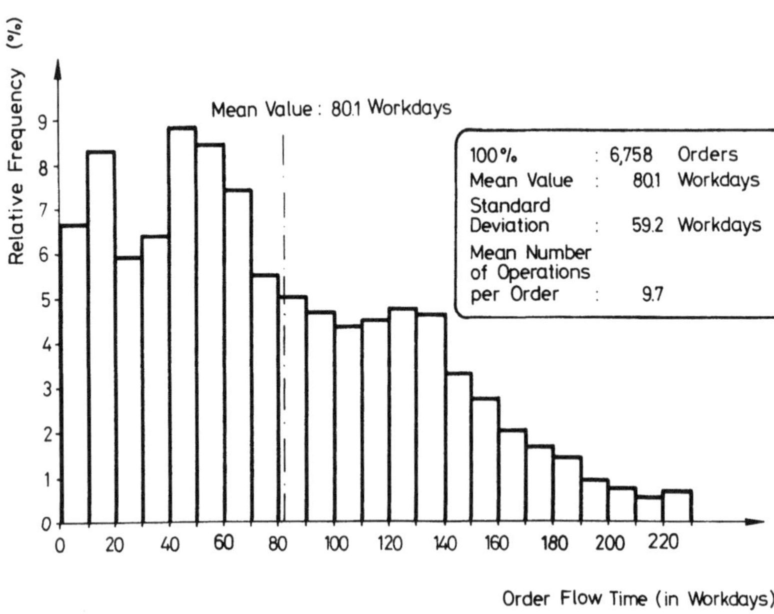

Fig. 1.4. Flow time distribution of shop orders (example from a job shop)

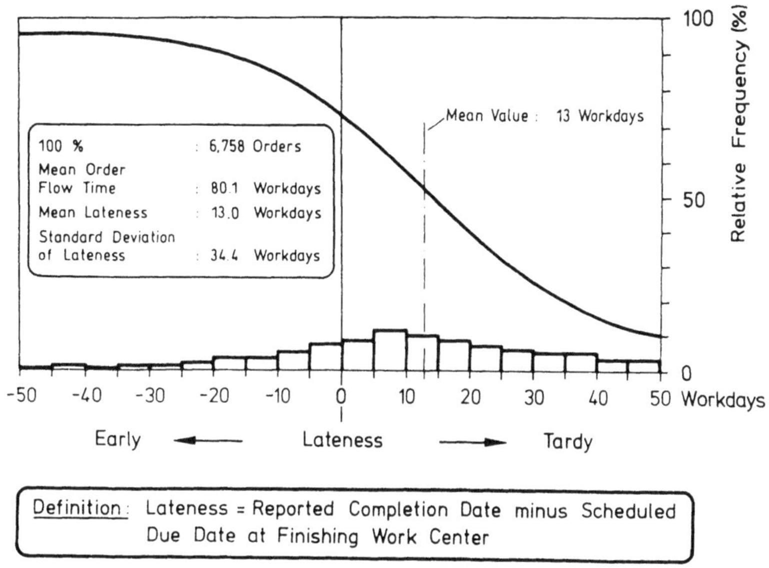

Fig. 1.5. Lateness of shop orders (example from a metal working production)

Such conditions are not unusual. Figure 1.6 shows the mean flow times per operation taken from six extensive investigations at companies from different branches of industry in Germany. The values lie between 7 and 16 workdays, the lower values from orders with 10–12 operations, and the higher values coming from orders with 4–6 operations. It is further noted that the processing time was less than 15% of the total manufacturing flow time [10], thus the jobs were waiting, being moved, or in a queue for over 85% of their production time.

In view of such ignorance of the actual schedule performance of manufacturing, it is not surprising that there are many similar 'shop-related experiences'. In 1973, some were described (a little sarcastically) by G. W. Plossl, one of the most renowned authors in the field of production planning, but similar situations are just as real today [11]. In 1979, B. Kivenko [12] quite rightly called them 'myths', and complemented them with the corresponding counter-arguments (Table 1.1). Both authors are correct in pointing out the problem of excessive inventories, which both decrease clarity and increase flow times. This is caused by too much emphasis laid on the utilization objective, and by the lack of a rationally justified 'correct' inventory level.

Ignorance of the relationship between the objectives and the actual performance results in a vicious cycle of mistakes in manufacturing control (Fig. 1.7) [13], which has its beginnings in the false conclusion that poor scheduling performance is due to planned lead times being too short. If one increases the allowances used in the MRP system and for the purpose of backward scheduling, the orders reach the shop earlier; the queues and thus the WIP increase. On average this means longer waiting times, and thus longer flow times for orders coupled with a greater variance in flow time. The result is that order lateness deteriorates instead of improving. Only rush orders and expediting can ensure that the most important orders reach the assembly department on time. The typical order takes longer, and the normal reaction is therefore to increase the allowance again. This cycle becomes a vicious spiral of

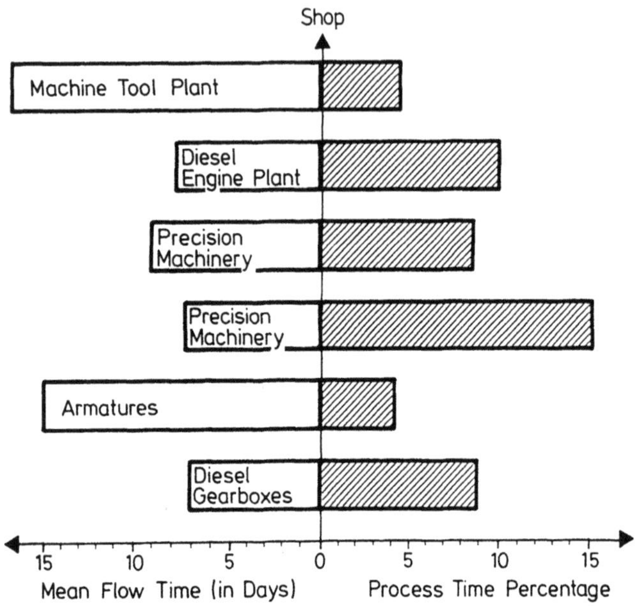

Fig. 1.6. Actual values of mean flow time and mean process time per operation (taken from investigations in six plants). □: mean flow time per operation (standardized to 8 hour workday); ▨: process time percentage $= \frac{\text{Mean process time} \times 100}{\text{Mean flow time}}$. (Basis: 122 000 operations (min. 1000, max. 62 000 per investigation)

Table 1.1. Six myths of production control (After Plossl [11] and Kivenko [12])

1. **To get more production from a factory, you must put more orders into it**
Unfortunately, putting more work into an overloaded plant makes it even more difficult to get the right things out on time.
2. **To get important jobs completed on time, ensure they are started as soon as possible**
Unfortunately, like releasing more orders, this approach just adds to work-in-process, and makes it more difficult to get specific jobs out on time.
3. **If the planned lead time is not long enough, increase it**
You cannot close the gap between planned and actual lead times simply by changing the planned figures. You must obtain more capacity, and work off the excesses causing the delay.
4. **Stage sets of parts further in advance of assembly if you need more time to expedite the shortages**
This will guarantee that you will have more shortages to expedite. It reduces the flexibility of inventories, puts more work into the plant to compete with the true shortages, and aggravates record accuracy – usually the basic problem that makes staging necessary in the first place.
5. **If you are short of several items made on one machine, get out of trouble quicker by cutting all the lot sizes**
This is fine, except that you will really be getting into deeper trouble if your problem is lack of machine capacity to produce all you need of everything it makes.
6. **If a little expediting is good, a little more will be better**
This approach attempts to overpower the problems resulting from poor planning and control, and becomes self-defeating as soon as expediters begin to compete for capacity.

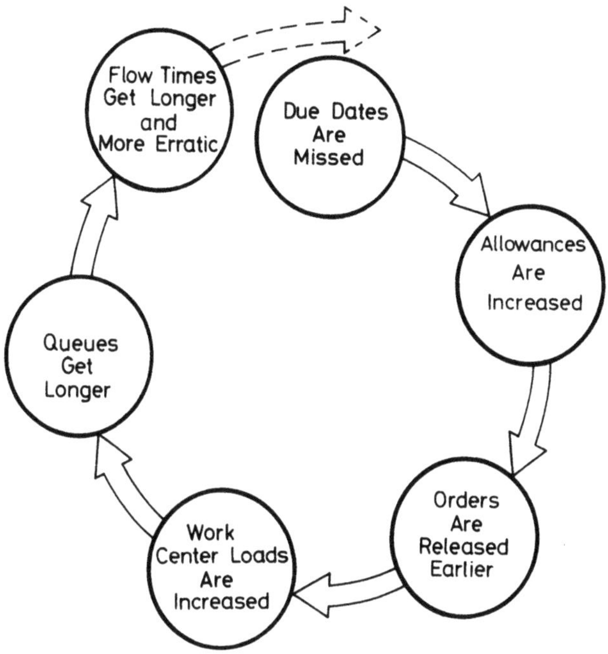

Fig. 1.7. The vicious manufacturing cycle (After Plossl [8])

events, one which can only be stabilized by having flow time levels and variances that are much higher than required.

1.5 The Weak Points of Conventional Manufacturing Control

Advances in manufacturing planning have brought about substantial improvements in manufacturing performance [14–18]. Many problems remain, however, even in organizations that have implemented planning concepts and techniques. A general lack of confidence, especially in manufacturing control systems, is often to be found. The general criticism can be summarized in three points (Table 1.2):

1. Despite the extensive use of data processing for planning, feedback and production activity control systems, and considerable computing cost for capacity schedul-

Table 1.2. The weak points of conventional manufacturing control systems

- Despite costly systems and data processing, often poor actual-to-plan correspondence.
- Inadequate reporting of performance data and little actual control of lead time, schedule performance, and inventory.
- Often little decision-making responsibility of planning and operating personnel.

ing and finite scheduling, the coincidence between the 'optimum' manufacturing process given by the computer and the actual facts is still very small. Several authors have observed that "as a rule, the weekly plan becomes obsolete due to rush orders, technological surprises and defects" [19], and that "finite scheduling must, as a rule, be revised after a very short time" [20]. Thus, expediting with 'stock-chasers' to 'bypass the system' is still common practice.

2. Conventional manufacturing control systems do not give directly usable information or indications with which one can selectively influence flow time, lateness, inventory, and throughput objectives. In addition, manufacturing control systems that are marketed with rational justifications and which at first seem clearly self-evident, can often not even measure the values they claim to be able to control, and do not possess appropriate control parameters.

3. These increasingly intricate systems leave less and less room for autonomous decision-making from the 'man on the job'. When there is insufficient information and training for personnel involved in and affected by planning, and the scheduling specialists are not fully qualified, it is possible that the recently installed, costly system will be avoided altogether. In these cases personnel will revert to an informal planning system, with foremen keeping the database and the scheduling rules in their notebooks.

It has frequently been thought that the poor performance of manufacturing control systems is due to insufficient lead time allowance and poor feedback. This accounts for the widespread use of on-line data collection systems. In many factories, however, the bad situation only became clearly visible with the implementation of these systems, and other solutions for manufacturing control had to be found.

The great success of Japanese industry and reports on the Kanban technique (in Japanese, Kanban means card, billboard, sign) at the end of the 1970s highlighted the fact that the conceptual approach fundamental to many systems in the West was completely wrong (Chapt. 9 gives an analysis of the Kanban technique). Starting in about 1980, numerous companies in the Federal Republic of Germany introduced the Kanban technique – in some cases with considerable success [21]. However, some companies had to acknowledge that they did not or could not meet the necessary requirements concerning product and production structure, therefore this solution was not viable, particulary for companies with job-shop production, and small batch or high-variant batch production.

It is also remarkable that most publications on the Kanban technique and its further development do not provide a self-contained theory of the manufacturing process, but rather prescribe individual actions to be taken.

We recognize that neither 'conventional' manufacturing control systems nor newer solutions such as the Kanban technique are founded on a feasible, generally applicable model of the manufacturing process, from which single techniques could be logically deduced, and from which the prerequisites for, and limits to their use can be predicted.

Before we present a new approach which better accounts for the real, statistical nature of the manufacturing process, the procedures commonly used today are analyzed so as to highlight the reasons why they have failed in practical applications. One important finding is that existing assumptions about the structure of lead times are too simple.

1.6 References

[1] Hall, RW. *Attaining Manufacturing Excellence*. Dow Jones-Irwin, Homewood, IL, 1987.

[2] Hall, RW. *Zero Inventories*. Dow-Jones-Irwin, Homewood, IL, 1983.

[3] Hill, T. *Manufacturing Strategies: Text and Cases*. Addison-Wesley, Wokingham, UK, 1989.

[4] Schonberger, RJ. *Japanese Manufacturing Techniques: Nine Hidden Lessons in Simplicity*. Free Press, New York, 1982.

[5] Shingo, S. *Non-Stock Production: The Shingo System for Continuous Improvement*. Productivity Press, Cambridge, MA, 1988.

[6] Scheer, A-W. *CIM (Computer-Aided Manufacturing) – Der computergesteuerte Industriebetrieb (CIM – The Computer-Controlled Factory)*, 4th edn. Springer-Verlag, Berlin, 1990.

[7] Jünemann, R. *Materialfluß und Logistik. Systemtechnische Grundlagen mit Praxisbeispielen (Material Flow and Logistics. Fundamentals and Practical Examples)*. Springer-Verlag, Berlin, 1989.

[8] Plossl, GW. *Production and Inventory Control – Principles and Techniques*, 2nd edn. Prentice-Hall, Englewood Cliffs, NJ, 1985.

[9] Plossl, GW. *Production and Inventory Control – Applications*. George Plossl Educational Services, Atlanta, GA, 1983.

[10] Melnyk, SA, Carter, PL. *Production Activity Control – A Practical Guide*. Dow Jones-Irwin, Homewood, IL, 1987.

[11] Plossl, GW. *Manufacturing Control – The Last Frontier for Profits*. Reston, NY, 1973.

[12] Kivenko, B. Reducing Work-In-Progress Inventory. *Prod. Eng.* **26** (3), 1979: 48–50.

[13] Mather, H, Plossl, GW. Priority Fixation versus Throughput Planning. *APICS Intern. Conf.*, Cleveland, OH, 1977.

[14] Orlicky, J. *Material Requirements Planning*. McGraw-Hill, New York, 1975.

[15] Wight, OW. *MRP II: Unlocking American Productivity Potential*. CBI Publishing, Boston, MA, 1982.

[16] Goldratt, EM, Cox, J. *The Goal: A Process of Ongoing Improvement*. North River Press, Croton-on-Hudson, NY, 1986.

[17] Tincher, M, Buker, DW. Parallel Paths: A New Succsessful Path to Class A MRP II. *31st Annual International Conference Proceedings*, 1–3. American Production and Inventory Control Society, Falls Church, VA, 1989.

[18] Fogarty, DW, Blackstone Jr., JH, Hoffman, TR. *Production and Inventory Management*, 2nd edn. South-Western Publishing, Cincinnati, OH, 1991.

[19] Gerlach, H-H, Vortherms, B. *Probleme beim Einsatz von EDV-Systemen in der Produktionssteuerung (Problems with the Application of Computer Systems in Production Control)*. Werkstattstechnik **67** (10), 1977: 629–634.

[20] Kettner, H, Bechte, W. Neue Wege der Fertigungssteuerung durch belastungsorientierte Auftragsfreigabe (Recent Production Control Methods by Use of Load-oriented Order Release). *Zeitschrift des Vereins Deutscher Ingenieure* **123** (11) 1981: 459–466.

[21] Wildemann, H et al. *Flexible Werkstattsteuerung durch Integration von KANBAN-Prinzipien (Flexible Shop-Floor Control by Integrated Kanban Principles)*. CW-Publikationen, München, Germany, 1984.

[22] Hayes, RH, Wheelwright, SC. *Restoring our Competitive Edge: Competing Through Manufacturing*. Wiley, Chichester, UK, 1984.

[23] Buffa, ES. *Meeting the Competitive Challenge*. Dow Jones-Irwin, Homewood, IL, 1984.

[24] Skinner, W. *Manufacturing in the Corporate Strategy*. Wiley, Chichester, UK, 1978.

[25] Krajewski, LJ, Ritzmann, LP. *Operations Management, Strategy and Analysis*, 2nd edn. Addison-Wesley, Wokingham, UK, 1990.

[26] Hill, T. *Manufacturing Strategy, Text and Cases*. Richard Irwin, Homewood, IL, 1989.

Chapter 2

Conventional Production Scheduling and Control

2.1 Abstract

After a brief description of the well known MRP II concept, this chapter concentrates on the way in which allowances for shop orders are determined in backward and capacity scheduling processes. It then analyses the weak points of conventional procedures and develops a set of requirements for a better method of order scheduling.

2.2 Survey

The common process of production scheduling and control can be characterized by the features outlined below.

The load resulting from the order stock and the available capacity can obviously be determined less precisely the further into the future is the time for which a statement concerning deadlines is to be made. For this reason, planning using stages of increasing precision is common practice. These stages are referred to as long-, medium- and short-range planning [1]. Figure 2.1 shows these stages and their inter-relationships [2].

Long-range planning characteristically takes place at the level of the products, either based on customer orders or on sales forecasts. This is because data on quantities and due dates are frequently still imprecise at this stage, and there may also be no exact data on the structure of the products and the routing.

Shop drawings and bills of material (BOM) are the prerequisites for medium-term planning. The BOM are used to determine the required quantities of components and parts, as well as the planned due date in the course of job-planning. For this, the dependent requirements of manufactured and purchased parts calculated from the BOM which determine the lead time of the next operation are offset with respect to the due date (called 'back shifting'). The gross requirements thus obtained are compared with the stock-in-hand for each period, thus yielding the net requirements. If the same item is required for several orders during the same period, these

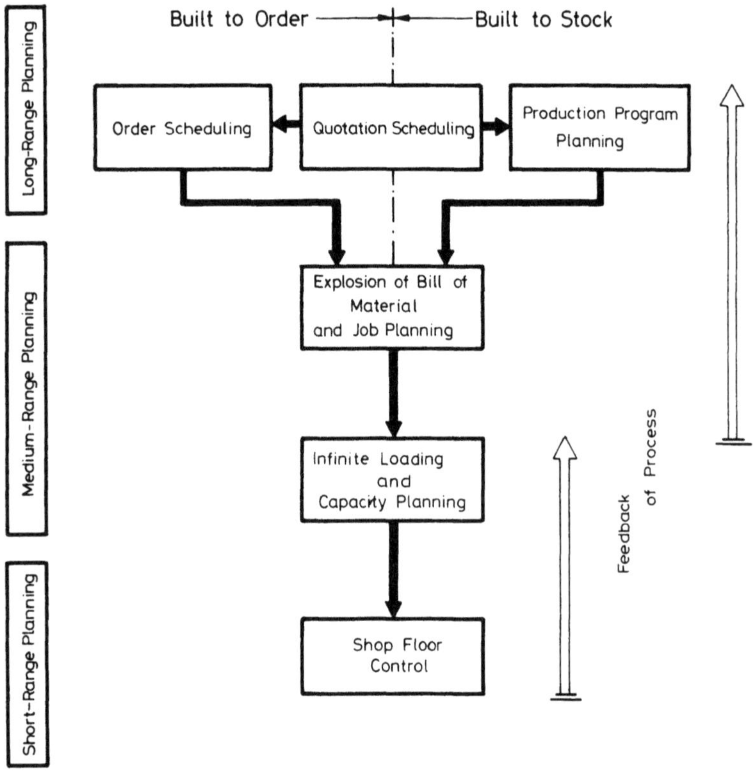

Fig. 2.1. Production planning and control framework (After Brankamp [2])

orders are combined, following the imperative of minimizing costs, to obtain so called 'economic orders' (economic order quantities (EOQ) or economic lot sizes, respectively) [1]. For each part to be manufactured, the operations are scheduled based on the routing sheets, the order loads are allotted to the capacity centers, and the leveling of peak and valley loads (capacity planning) is carried out. This is done as a second step in medium-range planning.

In short-term planning, the scheduling of individual jobs in individual capacity centers takes place. At this stage, both costs and schedule performance can be of importance, for instance when sequencing orders at one work center according to minimum total setup time. Dispatching is the next step in processing orders and is referred to, together with the feedback system, as the *production activity control system.*

In addition to this three-stage concept there is another characteristic model in production scheduling and control: a control loop to be followed at regular intervals (a planning cycle) or in individual cases. Figure 2.2 illustrates this concept using Brankamp's frequently quoted model [2]. The scheduling – carried out by an employee, a computer program, or a combination of both – acts as a controller, fixing the planned status and thus leading in the control line (i.e. in production) to the actual process. This process is monitored by feedback records so as to minimize disruptive factors. This process as a whole is called 'production scheduling and control'; its most important component is production scheduling. The closed-loop

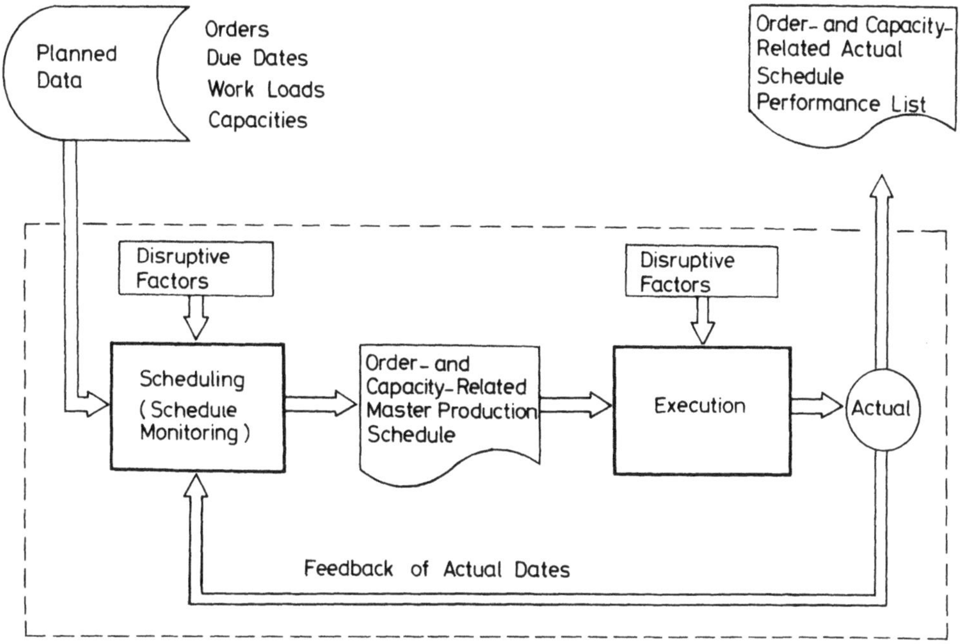

Fig. 2.2. Schedule control within manufacturing control (After Brankamp [2])

scheduling and control cycle responsible for a continuous flow of manufacturing processes is an internal part of this system.

This scheduling of orders within the given margin results in the order-related schedule (when and where the individual operations of an order are to be processed), and the capacity-related schedule (what operations of which orders are to be processed in what sequence at a work center within a given time bucket?). This means that the same plan can be described in two different ways.

The third characteristic feature of production scheduling and control is the increasingly fine distinction between functions. This has been the inevitable consequence of the growing formalization of these techniques due to the use of computers in this field.

Figure 2.3 shows a production planning and control model (cited after Hackstein [3]) based on a three-stage structure [4]. This model also has an increasingly sophisticated structure of production scheduling and control, concerning the planning horizon and level of detail, from top to bottom. No explicit distinction is made between long-range, medium-range and short range planning stages. The basic functions are, however, additionally divided into two more substages, which can be further subdivided into the levels of flow charts and program modules.

Standard programs are available for many functions; with so-called modular programs, customized combinations are possible [5]. Depending on the product range and type of production, the structure described can take different forms in practical applications, and additional functions such as schedule monitoring in quality inspection can be included. On the whole, the function structure described can be considered as being representative of manufacturing practice.

Another aspect of conventional manufacturing control that is particularly impor-

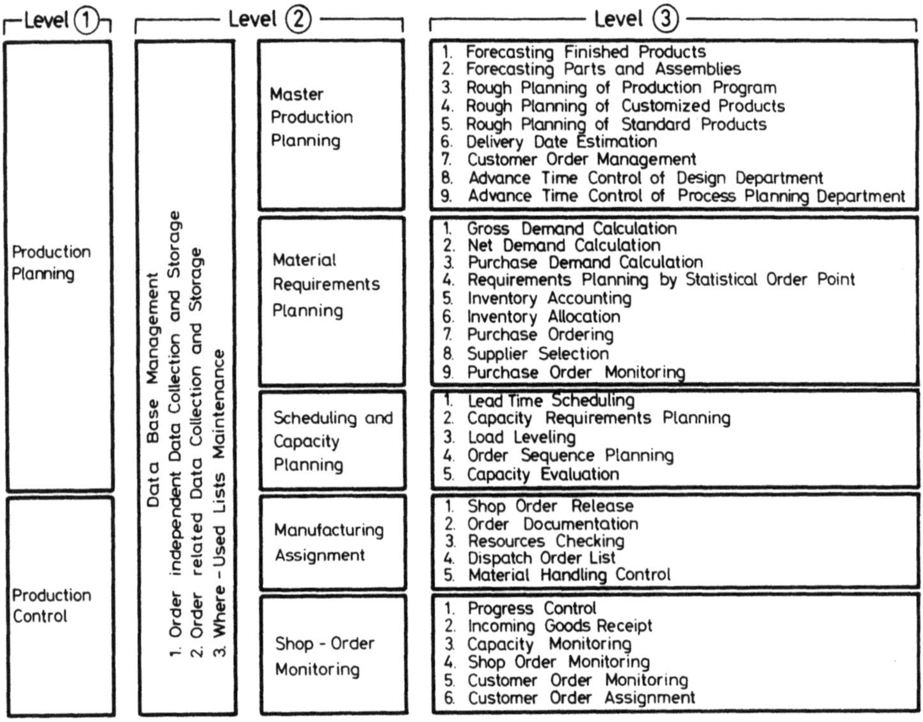

Fig. 2.3. Production planning and control functions (After Hackstein [3])

tant in the context of this book is the way in which lead time scheduling and capacity alignment are carried out, as well as the underlying model of the manufacturing process. In the following sections, this is dealt with in more detail, since it is the key to many problems associated with finite control.

2.3 Lead Time Scheduling and Capacity Scheduling

The crucial problem in production scheduling and control is the question of when an order has to be started given a certain due date, and when it will arrive at individual work centers. The problem is tackled in two separate steps at each planning stage. First, the manufacturing steps of the individual order are put in a chronological sequence (Fig. 2.4) [6].

The time needed for the individual steps is either estimated from preceding orders, given by standard routing sheets, or calculated anew for the case in question. If a simple bar chart of the type shown in Fig. 2.4 does not suffice to represent the temporal interlinkage, it is replaced by network plans. In either case, the result of the calculation is the expected lead time (or allowance) [7] without taking competing orders at the individual capacity centers into consideration. The process is therefore called *lead time scheduling*. It determines when an order (assuming the required capacities are available at any place and in any quantity demanded) must be started

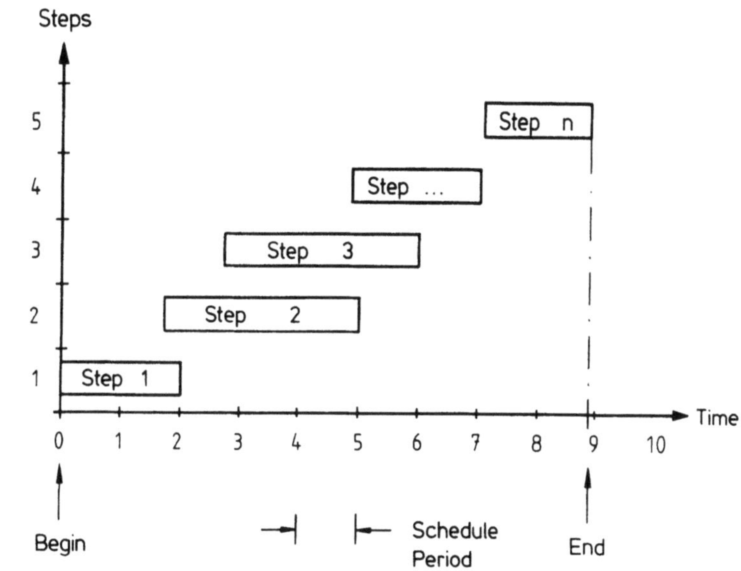

Fig. 2.4. Infinite loading (backward scheduling)

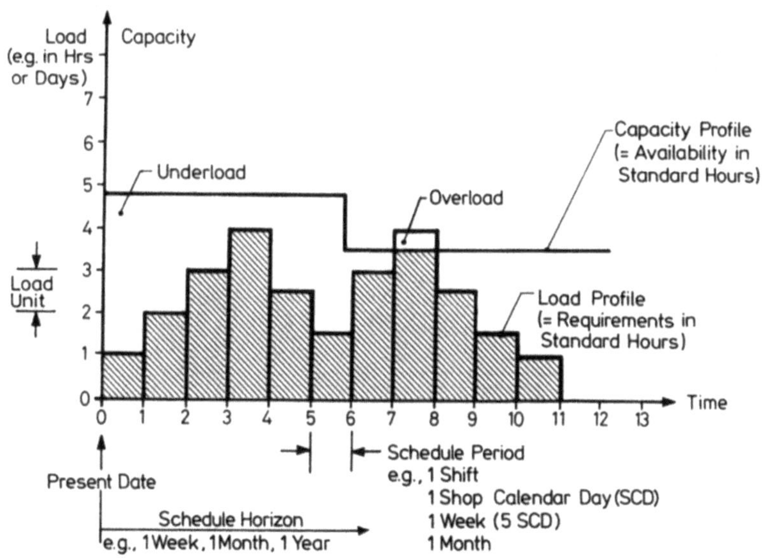

Fig. 2.5. Infinite loading and finite loading principle

to meet a known due date, or when completion is to be expected given a known start date. If the desired aim cannot be reached, i.e. if the scheduled start date lies in the past or if the scheduled due date is later than the planned due date, special actions are generally recommended, such as overlapping successive operations, splitting operations by alternate processing, or interoperation time reduction [8].

Once lead time scheduling has been carried out, loading follows (Fig. 2.5) [6]. In this process, the forseeable future – the planning horizon – is divided into equal time buckets – the schedule periods.

The load values obtained from lead time scheduling are then added onto the loading account of the relevant load center at the appropriate time. In this process already released orders have to be taken into account. Thus for each load center, a load profile is found for comparison with the capacity profile stating the available capacity.

If the available and required capacities differ over a long period of time, then the capacity must be adjusted, for example with overtime, shorter working weeks, multiple shifts, or subcontracting. Short-term fluctuations are counter-balanced, if possible, by shifting single operations in time without, however, jeopardizing the whole orders's due date. Capacity loading performed in this way is called *capacity scheduling.*

From this the common two-tier scheduling method shown in Fig. 2.6 results (after Mertens [8]). As experience shows, the automatic capacity leveling and availability check are, however, rarely carried out in practice because of a lack of reliability resulting from frequent changes in standards and breaks in the process.

The reasons cited above certainly explain the failure of capacity scheduling. On the other hand, though, we should ask whether it is the method itself that causes

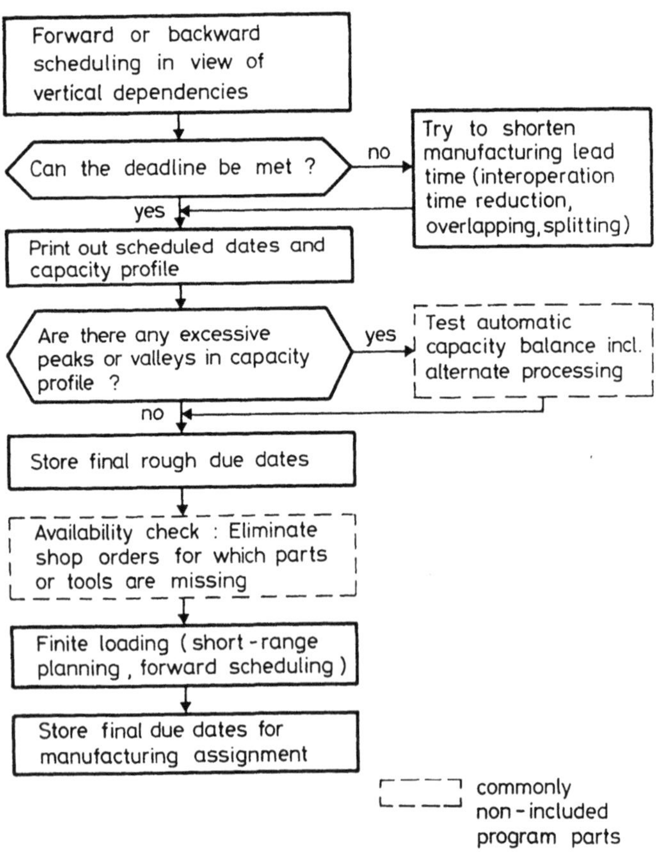

Fig. 2.6. Flow Chart for a Two-Tier Scheduling Method (After Mertens [8])

the problems, since after all, a scheduling method should adapt to the facts of manufacturing life, and not *vice versa*.

Following this train of thought, let us look more closely at lead time scheduling at the level of single operations.

2.3.1 Single Steps in Lead Time Scheduling

2.3.1.1 Determining Lead Times

In order lead time scheduling, the aggregate lead time is composed of the lead times of each of the single operations. Each operation lead time consists of two components: the operation time, and the interoperation time between the upstream work center and the work center in question. For the very first operation, the interoperation time is the time between the release of material and its arrival at the gateway work center. In backward scheduling – with a known due date and a start date to be determined – the calculation is performed as shown in Fig. 2.7 upper part.

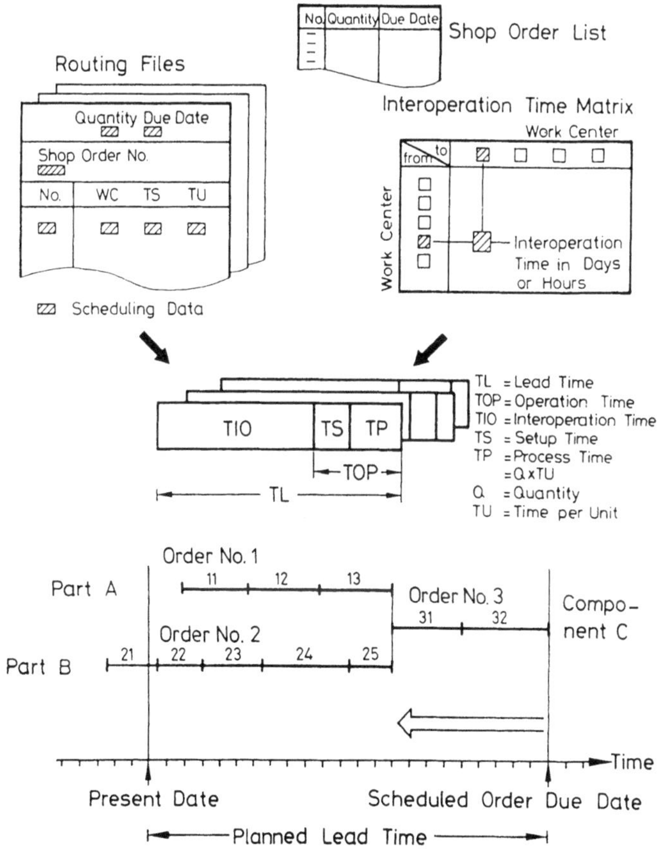

Fig. 2.7. Backward scheduling for related orders

The shop order list is arranged in the same sequence as the due dates. The so called 'order time' is then computed from the setup time *TS* plus the process time *TP*. The process time is the product of quantity *Q* and time per unit *TU*, these two values being specified in the operation sheet. The interoperation time is usually documented in a so called 'interoperation time matrix,' and specifies how many hours or days the job normally takes to work its way from one work center to another. The order time, given in standard hours, must be divided by the daily capacity to obtain the operation time. For example: an operation with an eight hour standard time takes 1 day with single-shift work, but only 0.5 days with two-shift work (eight hour shifts presupposed). The sum of the operation time and the interoperation time is called the *operation lead time*, and is used for lead time scheduling.

Figure 2.7 lower part shows the backward scheduling of three related orders. Let Order 1 and Order 2 be the manufacturing of Parts A and B, and Order 3 their assembly forming Component C. Moving backwards in time from the scheduled due date, the throughput elements for the individual suborders are mapped onto the time axis in their logical sequence, derived from the product structure in the BOM. In this example, the planned start date for Part B lies in the past, i.e. the due date cannot usually be met.

2.3.1.2 Interoperation Time Reduction

This is where conventional scheduling takes into consideration the fact that interoperation times are given as standard values, which means that they can be shorter in individual cases. To this end, interoperation time is divided into several components [7], some being reducible and others not. Mertens [8] lists the following interoperation time components.

1. Average queue time before an operation can be started.
2. Process-dependent waiting time before processing (e.g. for marking).
3. Process-dependent waiting time after processing (e.g. for cooling).
4. Queuing time before inspection.
5. Inspection time.
6. Queuing time before transportation.
7. Transit time to downstream work center.

Figure 2.8 shows an example of the structural components of work center lead time which was taken from the software documentation of a widespread production scheduling and control program [9]. In this case there is the possibility of overlapping the preparation time T_2 and the setup time *TS*. Queue time T_1, waiting time T_4, and transit time T_5 can be reduced in case of delay.

The individual operation allowances can be determined in several ways. One simple possibility is to give the interoperation time as a multiple of the considered operation's actual operation time. This is generally recommended if the waiting time before processing and the inspection time are the main components, and if they are at least approximately a function of the operation time. If waiting times are independent of the operation time, and if operations require a special work center, the interoperation time should be stored as a constant in the work center file. If, however, transit times prevail, an interoperation time matrix, as suggested in Fig. 2.7, is recommended. Another possibility is to calculate each single component.

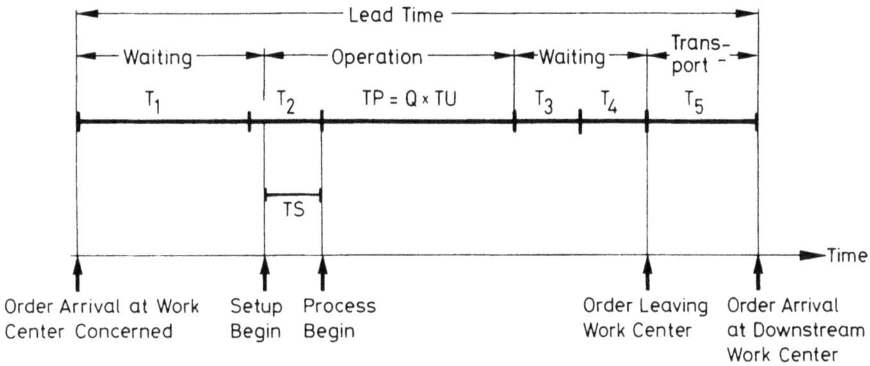

Fig. 2.8. Components of work center lead time. T_1: queue time*; T_2: preparation time (e.g. marking cleaning, inspection); *TS*: setup time; *TP*: process time = Time per unit *TU* × Lot size *Q*; T_3: Post-operation time (e.g. cooling, deburring, inspection); T_4: wait time (for transport)*; T_5: transit time; *: reducible in case of delay (After IBM [9])

Table 2.1. Handling of interoperation times. O: simulated; E: empirical value (After IBM [9])

Interoperation time component	Calculated from				Used by				Reducible within			
	Parameter	Work center	Operation time	Work center layout	Order scheduling	Order release planning	Order scheduling	Final assembly scheduling	Order scheduling	Order release planning	Order scheduling	Final assembly scheduling
Queue time (T_1)	×	×			×	×	O	E	×	×		
Preparation time (T_2)		×	×		×	×	×	×				
Postoperation time (T_3)		×	×		×	×	×	×				
Wait time (T_4)	×	×			×	×	×	×	×	×	×	×
Transit time (T_5)	×	×		×	×	×	×	×	×	×	×	×

As an example, Table 2.1 shows the possibilities provided by IBM's CAPOSS-E software [9].

As an added method, the queuing time can be determined before processing by simulating the queue in front of each work center. This yields a theoretical mean queuing time, assuming that processing is performed without breaks or faults. These values are determined periodically and, taking values from past operations into account, smoothed exponentially to find average historical values for each work center [21]. Explicit warnings are, however, given not to update these values automatically, and not to replace them by those values for the queue time defined by the user, since the scheduled values could easily slip out of the user's control [9].

It remains to be said that any interoperation allowance must be determined 'by the user'. Neither the relevant literature nor most software manuals for such programs give plausible and easy-to-follow instructions.

Interoperation allowance reduction in the case of delay seems to be just as complicated to carry out as interoperation allowance determination. One commonly used procedure is the uniform reduction of all operations involved by a constant factor, the size of which is limited by a maximum reduction factor. Another possibility is to cut single interoperation time components by different factors either in a given sequence or all at one time. Interoperation time reduction is sometimes thought desirable even if no delays are expected. This is the case when production is very expensive, for example because of high material value or because of large added value due to high processing costs.

Unfortunately, no advice is given on how to compensate for reductions, because when some orders are put through faster, others must necessarily wait longer.

Returning to the example of the related orders in Fig. 2.7, let us assume that backward scheduling has produced a start date that lies in the past. This is due to the fact that Suborders 2 and 3 last longer than the planned lead time, thus they are the critical orders. Figure 2.9a starts with this situation. Let the subsequent lead time reductions lead to a scheduled start date that now lies in the future (Fig. 2.9B). Now, however, Order 1 is subject to an unnecessarily long waiting time; this is eliminated by shifting Order 1 in such a way that it links up closely with Order 3 (Fig. 2.9C). One can see even from this simple example how complex a calculation

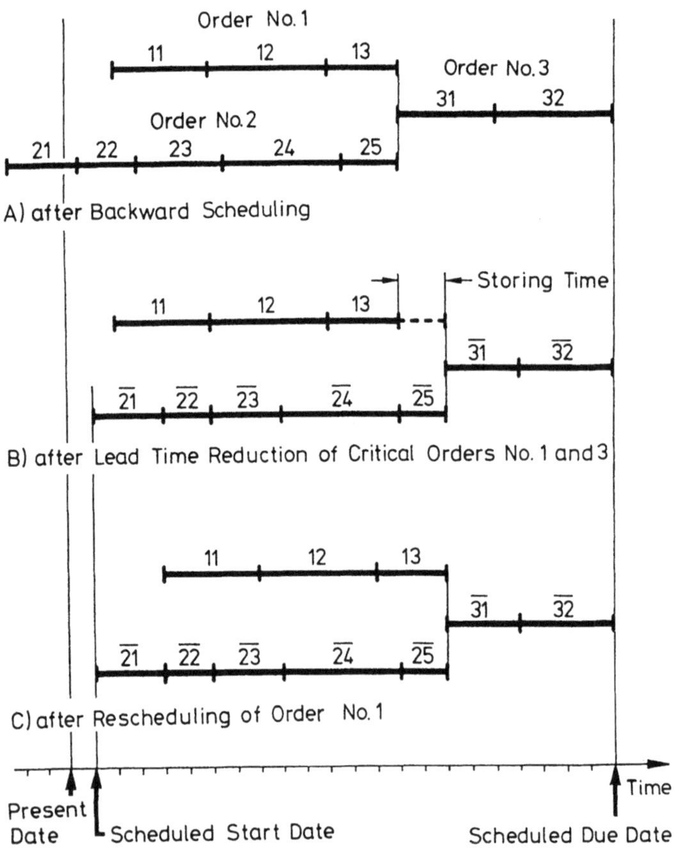

Fig. 2.9. Scheduling steps for related orders [After Mertens]

becomes if one considers that every date must be read, changed, checked for permissible limits, and rewritten over and again for all operations involved.

2.3.1.3 Operation Overlapping

For some parts, even with the new allowance after interoperation time reduction, one will still not be able to meet the planned due date. Instead of simply shifting the whole order, which would cause an avalanche of more calculations, one can now take other measures, depending on the importance of the order, which, however, causes additional costs compared with the original procedure that was planned to minimize costs. The measures in question are the overlapping and splitting of orders, which were briefly mentioned above. The basic idea behind overlapping is not to process all parts of an order at one work center and then transport them on to the next work center, but to transport smaller quantities, when completed, to the downstream work center (possibly inspecting them first), and start the next operation immediately [1, 7]. Figure 2.10 makes this clear by showing an order with two successive operations.

The lot originally consisting of 15 items is split into three sublots containing five items each. Once the first sublot has finished at Work Center A, the items are transported to Work Center B during the interoperation time TIO_1. Work Center B is set up during its setup time TS_B, and then starts processing Sublot 1. In the same way, Sublots 2 and 3 are processed as soon as they are ready. Since, however, the process times of two successive operations are rarely the same, interoperation times differ. In our example they become longer, since the process time of B is longer than that of A. The result is that, with equal transit times, there are waiting times for Sublots 2 and 3. If the process time is shorter, though, even idle times at Work

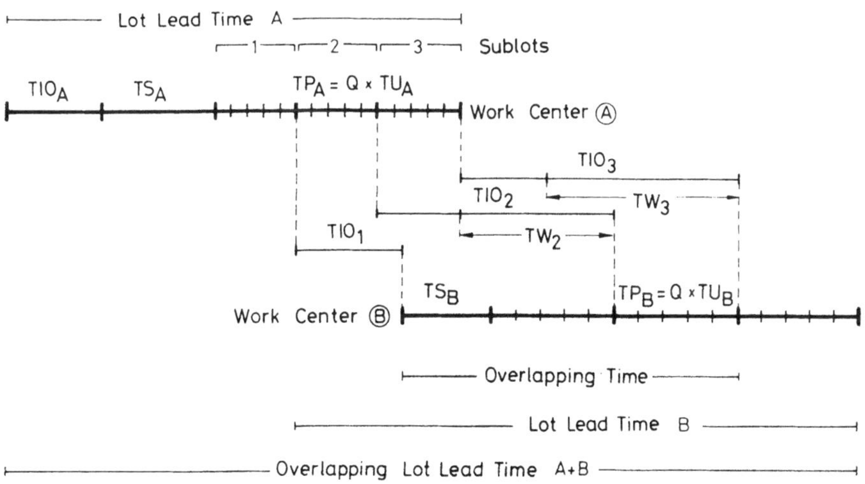

Fig. 2.10. Overlapping of operations. *TIO*: interoperation time; *TS*: setup time; *TS*: process time; *TU*: time per unit; *TW*: wait time sublot

Center B can occur, for example if Sublot 1 is finished at Work Center B before Sublot 2 arrives. Because the time differences add up, there must necessarily be idle times at Work Center B with a larger number of sublots.

In practice, such overlapping is only carried out when a considerable amount of time can be saved. Shortening interoperation time in this way not only causes additional costs for transporting the sublots, but requires complex scheduling and much more accurate coordination between those work centers involved than is usually the case. Finally, it requires substantial additional effort in dispatching and schedule monitoring.

Again, we should ask whether it is the symptoms instead of the disease which are being remedied. Long operation times are frequently due to so called 'economic lot size' calculation. For this, only reorder costs and capital tied-up are taken into account under very simplified conditions. Neither the additional costs described above nor the effects on lead time are considered in conventional lot sizing, so this method is increasingly being questioned today. Instead of performing dubious lot sizing and 'ironing out' the consequences at great expense, companies should direct their efforts towards setup time reduction [10] (see Sect. 8.2.1).

2.3.1.4 *Operation Splitting*

When splitting operations as an alternative to overlapping, the lot is divided up between two or more work centers, instead of being processed entirely at a single work center [1, 7]. Figure 2.11A shows the lead time of an unsplit lot containing 15 items at Work Place N. One distinguishes between quantity splitting and time splitting. In Fig. 2.11B a quantity split has been carried out forming three equal sublots (five items each), which are distributed over work centers N, M, and P.

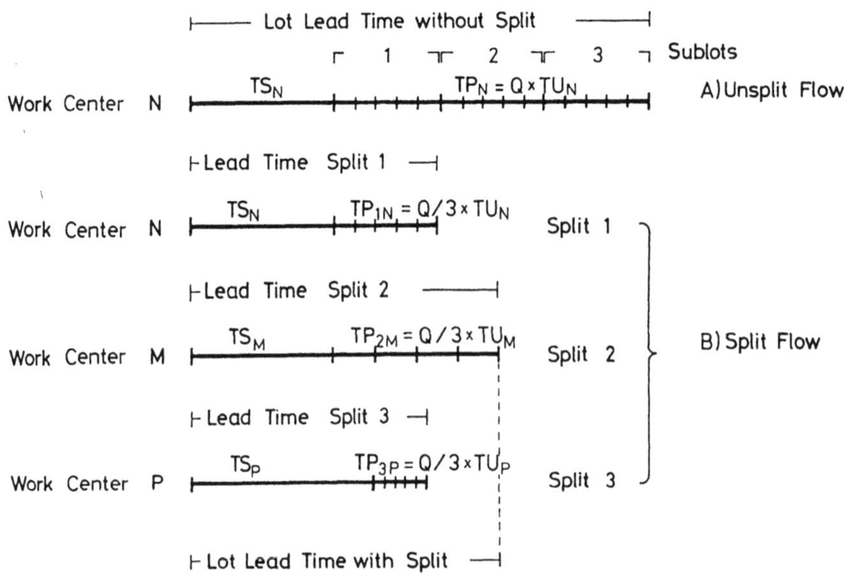

Fig. 2.11. Splitting of operations

Sometimes the setup time (*TS*) and the time per unit (*TU*) are equal at the machines, but this is not necessarily the case, especially in small shops where there is frequently only one of each type of machine. In this case there are different lead times. The lead times can be leveled out when performing a time split, where the process time is split up in such a way that it is about equal at all the work centers.

The price for the shorter lead time thus obtained is the additional setup times at work centers M and P, which not only causes an increase in costs but also reduces the available capacities at these centers. Furthermore, the sublots necessitate additional paperwork and, in the case of fundamentally different machine types, even new operation sheets, and possibly fixtures, tooling and NC programs. As with order overlapping, scheduling must be very precise here so as to really obtain the hoped for time gain. Thus, operation splitting requires very sophisticated planning and monitoring, is costly to carry out, and in practical application should, therefore, be restricted to special cases. Also, operation splitting only attempts – despite considerable effort – to solve those problems caused by overly high setup times and inappropriate lot sizing.

If scheduling shows that despite all economically justifiable special measures the requested time allowance cannot be reached even presuming unlimited capacity, then there generally has to be a due date offset for the order and a new lead time calculation because of the effects on other orders, depending on the extent to which orders are interlinked. In practice, however, this is frequently omitted, either because one hopes to 'put the order through somehow', because the rescheduling costs do not seem justifiable, or because frequent shifting of orders has shattered the credibility of the whole scheduling system to such an extent that people in the shop no longer take the dates seriously.

The results of the scheduling process, with all their uncertainties concerning inter-operation times and the equally uncertain and complex planned measures for lead time reduction in critical orders, now enter the next important phase of production scheduling – capacity scheduling.

2.3.2 Single Steps in Capacity Scheduling

2.3.2.1 Finite Loading

As a first step in capacity scheduling, the single operations with their work content are loaded onto the loading account of the respective work centers, corresponding to their position on the time axis. The capacity-related schedule is produced from the order-related schedule (Fig. 2.12) [2]. A work center consisting of technically similar machines forms a load center; the time unit is normally one working day, and the planning period is commonly one week. Orders with a time buffer – such as those which are not on a critical path in a schedule network – are scheduled with the earliest start date.

2.3.2.2 Capacity Adjustment

The result of this is the so called 'load profile', as opposed to the capacity profile shown in Fig. 2.5. Some typical situations are outlined in Fig. 2.13 [6].

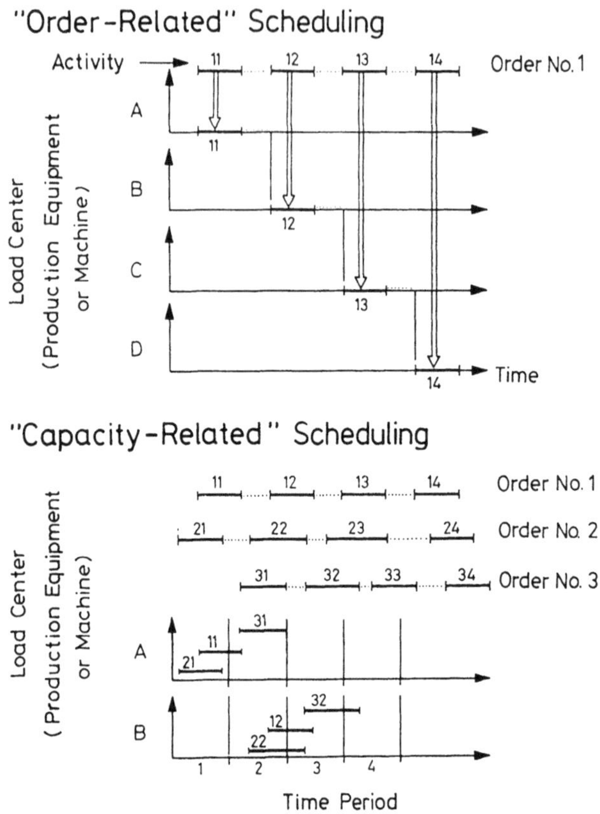

Fig. 2.12. Order-related and capacity-related scheduling (After Brankamp [2])

In Case A the load varies about a normal capacity value for short periods. In this case one can try to smooth the load profile by operations offset within the limits of their buffers, or by alternate processing, in order to avoid both idle times at the work centers or overloads. In Case B one can see a medium-term reduction in requirements. In this case capacity can be reduced, for example by cutting out a shift or disusing a machine. In Case C there is, in total, too little capacity. Capacity must be increased, for example by overtime; the remaining fluctuations can be handled as in Case A. Case D does not currently require any corrective actions.

2.3.2.3 *Capacity Alignment*

For capacity alignment there are several possibilities, as shown in Fig. 2.14 [2]. Overload is either eliminated by subcontracting or by alternate processing, presuming that work centers, fixtures, tooling, and working papers are available. With alternate routing (Case B), on the other hand, one tries to pull ahead individual operations or, on the contrary, to load operations that are behind schedule within the limits of their buffers.

Several authors have examined the problem of capacity alignment [8, 1, 7, 11].

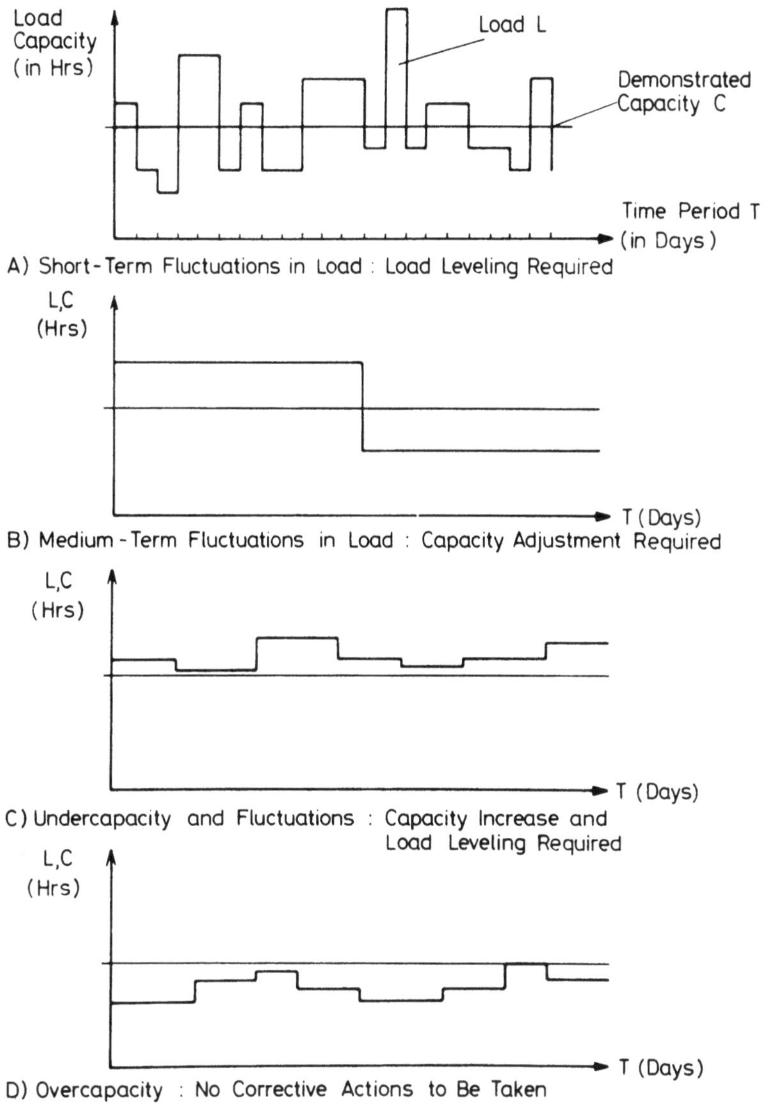

Fig. 2.13. Capacity adjustment and finite loading after scheduling

For example, Brankamp [2] has developed a method which, by use of a relative lateness value, recognizes orders according to their alignment priority, and shifts them until the difference between the load profile and the calculated capacity profile is within an upper or lower tolerance limit determined by the user.

All the alignment methods, however, proceed on the assumption that the individual operations – breaks and changes in production excluded – arrive at the work center at the time determined by scheduling. The question again is, considering the uncertainties in lead time scheduling already mentioned, whether it is sensible to go to such lengths.

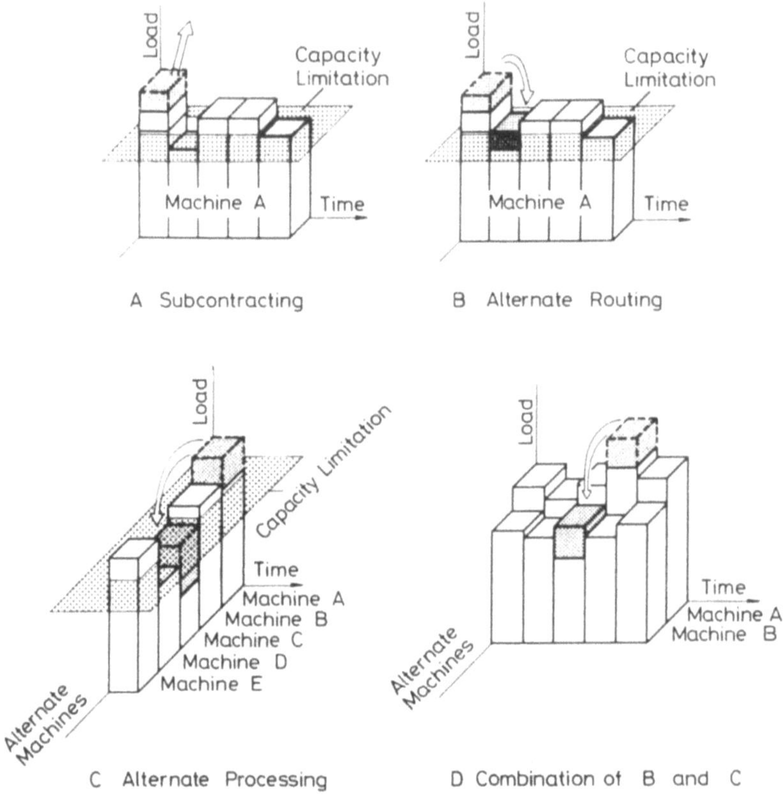

Fig. 2.14. Capacity alignment by load modification (After Brankamp [2])

2.4 Evaluation of Conventional Lead Time Scheduling

The results of conventional allowance and capacity scheduling can be understood with the help of Fig. 2.15 [12]. The individual operations are given in an 'optimized' sequence for each work center after complicated calculations. What has not been sufficiently considered so far is the fact that the operations are now, so to speak, 'frozen' in a complete sequence on the time axis. Prior to that, each operation element could be shifted within a flexibility margin as large as a few days, which was relatively large compared with its operation time. After capacity scheduling this margin is only at the scale of one hour or less. Thus the buffer, which is supposed to be able to compensate for unavoidable breaks in production, is no longer available at the beginning of the process.

The consequences of this mistaken approach are as logical as they are unpleasant: whenever there is even a comparatively small delay in one of the continuously arranged operations at one work center – for instance, due to personnel absenteeism or to the breakdown of machines or fixtures – all subsequent operations will also

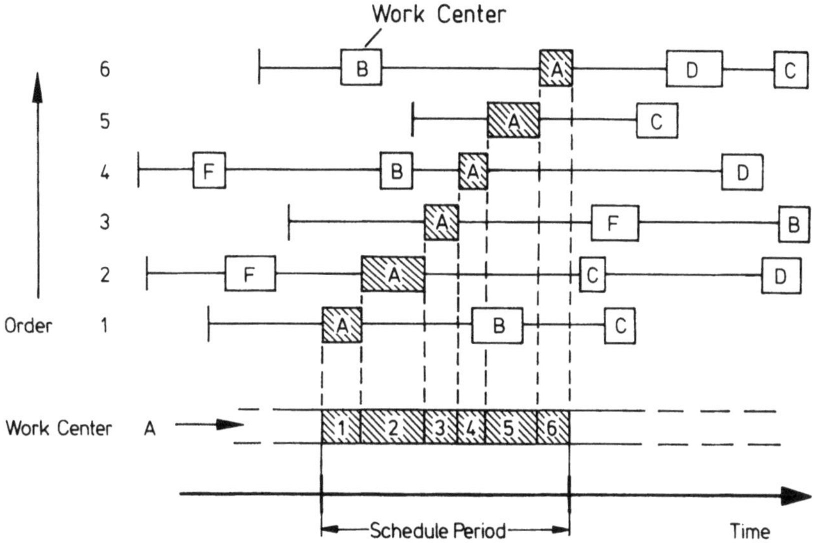

Fig. 2.15. Finite scheduling diagram (After Bechte [12])

suffer delays. If one or several of these orders are, by chance, arranged in such a way that interoperation allowance has already been reduced to its minimum permitted value, the consequence is due date delays for these orders which, in turn, results in increased priority and thus in order sequence transpositions at the downstream work centers. The whole complex balanced schedule becomes mixed up. On the situation in the shops, Mertens notes: "In many shops alterations occur at such a pace that the optima determined at great expense are only valid for a few hours" [8].

This is why manufacturers now rarely use conventional capacity alignment, with its complicated and questionable optimization procedures. One widespread alternative is to use the loading list determined in order scheduling as a recommendation for 'on-the-job' finite scheduling. The final order sequence is determined by the foreman, or by a dispatcher in agreement with the foreman, sequencing and allocations being defined according to the criteria of the technical feasibility of jobs on individual machines, the qualification of personnel, order priority, and setup time minimization. Before this can be done, the availability of tooling, fixtures and working papers, including NC programs, of course has to be checked.

How is it possible that manufacturers are still able to deliver orders in time in spite of the obvious and serious deficiencies in the present methods of production scheduling? A typical lead time distribution may help in answering this question. Figure 2.16 shows such a distribution at an individual work center taken from an investigation in a precision machining factory [13].

The orders processed at a lathe work center within 16 weeks have been grouped together according to the lengths of their historical lead times (from leaving the upstream work center to leaving the work center in question), multiplied by their work content (i.e. their operation time calculated from the operation sheets). Thus the percentage of orders per group has been determined. (This type of lead time refers to one unit of order time; it is also called the *weighted flow time*, and is defined and calculated exactly in the following chapter.)

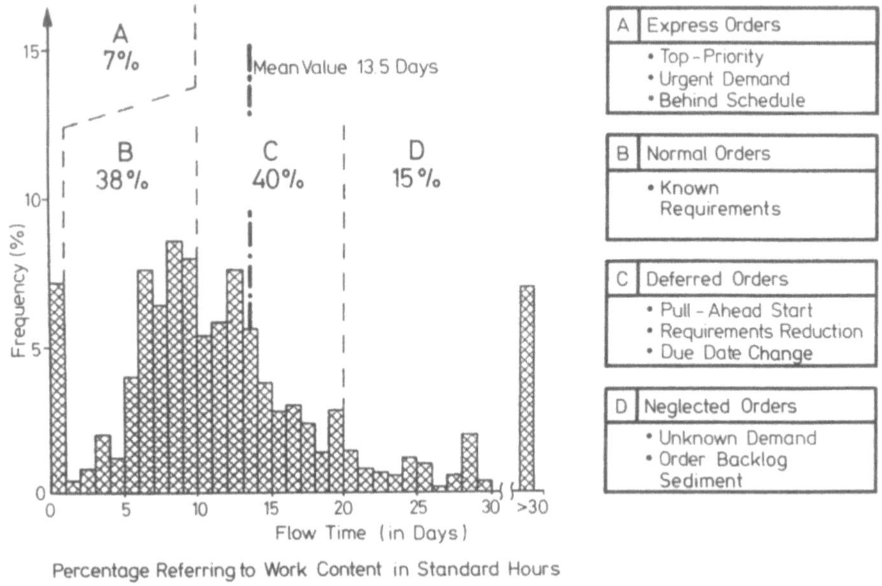

Fig. 2.16. Example distribution of flow time for a lathe work center (After Bechte [12])

First, one notices a wide variance in flow time with a very high mean value of 13.5 working days. If the orders are divided into four groups of declining priority (as proposed by Orlicky [14]), the following can be stated: About 7% of the orders (in order hours) passed the work center at top priority (up to one day). They represent a high-priority demand and are already suffering a delay. If one arbitrarily assumes 10 days to be the limit for so called 'normal orders', this group contains about 38% of the orders. The completion of these orders is required in the near future. The rest of the orders – up to 20 days – are either unimportant orders that were started ahead of schedule because of uncertainties in scheduling, or are so called 'filling' orders. They may also be orders where an 'economic lot' was split up due to a top priority requirement, the rest of the lot no longer being as urgent. They may also be orders for which, for whatever reason, the due date of the lot has been deferred. Finally, there are the neglected orders, the demand for which is unclear, for instance because there has been a technical alteration, or because the quality assurance department has stopped their further processing, or because orders have been altered or canceled.

Thus only the really important orders are put through in about the time assumed in scheduling. The rest of the orders are not regarded as important during schedule monitoring, and are therefore ignored when it comes to judging the schedule performance of the company. Since the common scheduling methods do not provide any objective, statistically provable data on real lead times, schedule performance and inventory tie-ups (only the utilization of production facilities is frequently given to the second decimal point), many companies falsely believe that their system cannot be that bad. 'We always get important orders out of the shop on time' is a frequent answer to questions regarding production scheduling and control.

The real situation is quite different:

- Lead times of orders and operations suffer wide variances
- Scheduled lead times are only achieved for important orders. The rest of the orders cause high inventories and blockages in the shop.
- The common methods of production scheduling and control do not account for the fact that flow time is a variable determined by statistical laws.
- The variables flow time, inventory, utilization, and scheduling performance, which such methods are said to be able to influence, are not even measured (except, perhaps, for utilization). In none of the individual steps of allowance and capacity scheduling is there an overall check to determine whether, and to what extent, these variables improve or deteriorate during the scheduling periods. Instead, there is an 'overall optimizing' of the objectives of schedule performance and utilization, which oppose each other, an operation which can neither be reconstructed nor verified.
- No rules or directions for dimensioning allowances are given in any of the relevant literature or method descriptions.

In conclusion, it can be observed that there is no realistic and self-contained theory of the manufacturing flow process that clearly describes the obvious connections between the objectives, and thus gives a basis for new methods (or the classification and evaluation of known methods) for production scheduling and control.

2.5 Requirements for New Methods of Production Scheduling and Control

To overcome practical difficulties, a growing number of computer firms, software houses, and management consultants are offering a wide variety of more or less standard solutions for production scheduling and control. Generally, the following developments can be observed [15]:

- Use of workstations in dialogue mode in or near the shop
- Possibility of integration with mainframe or subordinate computers
- Increasing availability of standard software for the processing and protection of mass data
- Integration of evaluation and representation routines, particularly for graphic representation (diagrams)
- Low-cost databank systems.

In addition to these systems, a rapid development of production data collecting and processing systems (PDR-PDP systems) can be seen. The basic idea is to record the start and end dates of processing and transporting operations as realistically as possible – up to automatic recording directly from the controller of a fully automatic machine [16]. These data are then collected and processed to obtain progress reports and key data [17]. In job shop production, the scheduling part is often dropped completely, and production progress is instead shown on planning boards. These control units are presently being replaced by color graphics displays with dialogue capabilities, so-called Leitstands [18].

Since the early 1980s, an increased effort to include manufacturing control in an overall integrated model of manufacturing companies has been evident. This approach was discussed in the mid-1960s under the heading 'integrated data processing', but was unsuccessful because of the enormous difficulties in joining different hardware and software systems. Today, this integrational approach is being rediscovered under the various headings of Computer Integrated Manufacturing (CIM), Logistics, and Just-In-Time (JIT) production, and, to an increasing extent, implemented in practice. While CIM emphasizes the computerization and integration of the design functions (Computer Aided Design – CAD), operations planning (Computer Aided Planning – CAP), manufacturing (Computer Aided Manufacturing – CAM), and quality assurance (Computer Aided Quality – CAQ) in a computerized factory, the Logistics and JIT approaches examine the complete material flow and the corresponding information flow between the sales market and the supply market; the aim is that the company involved meet the market needs in the shortest possible time at the right date and at the right place, with the lowest possible costs. This can even mean on-line connections with suppliers on the one hand, and with important customers on the other hand. The remarkable aspect of these solutions is that the control methods themselves are not questioned in most cases, and even in the more recent literature no possibilities are seen for further development – with a few exceptions. One exception is the Japanese Kanban system [19], and another is the Optimized Production Technology (OPT) system which has received widespread publicity above all in the United States [20]. Chapter 9 deals with these and other methods, but it should be mentioned here that none of the publications on these two methods have proposed a model with which the respective rules and formulas could be made understandable or the results reproducible.

One remarkable exception is the opinion on manufacturing control held by G.W. Plossl (USA) [21, 22]. He emphasizes the importance of input and output monitoring for inventory and lead time control, and shows this in a quite descriptive model (Fig. 2.17): the order input rate must be defined and controlled corresponding to the output rate. The output rate is, in turn, influenced by the capacity, which is also variable. The amount of WIP is determined by the input and output rates, which in turn determines the manufacturing lead time. Plossl actually expresses the essential relationships, but has not yet developed them into a formal model.

Therefore we shall first set up requirements for a manufacturing control system that will also hold in the future. These requirements are based on former publications [15, 24]. They can be divided into three categories (as shown in Table 2.2); the trends in hardware and software described are included for the sake of completeness.

The main requirements for the principles must be the following (Table 2.2):

1. They must represent and deal with a manufacturing process model in the computer as a stochastic process.
2. Further, it is necessary that, depending on the market situation and the manufacturing level involved (cost center or work center), there is the possibility of using different control strategies. For example, with a good order situation one would prefer to emphasize the utilization of capacities, whereas with few orders, short lead times can be of importance. The method should not try to pull orders ahead to underloaded work centers, because this means a greater load on bottleneck capacities.
3. Measurement and monitoring of the four main variables (flow time, inventory, scheduling performance, and utilization), as well as the diagnosis of deviations, is another important requirement.

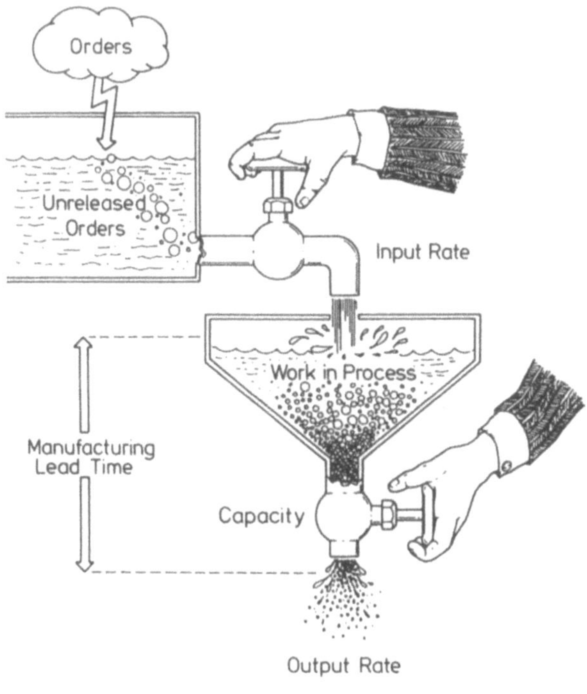

Fig. 2.17. Typical situation in a factory (After Plossl [21])

Table 2.2. Standards for production scheduling and control systems

Hardware/software	Principles	User requests
• Decentralized and modular extendable	• Model design according to stochastic process nature	• Provide decision and planning flexibility
• Integrable	• Alternative control strategies	• Extend information exchange beyond production equipment
• Ergonomic equipment and transmission of information	• Monitoring and diagnostic system of objectives by key data and diagrams	• Promote process transparency
• Dialog-oriented systems management	• Feedback accuracy review	• Differentiate personnel- and process-based data
• Data-bank-oriented		

4. Finally, in a control loop it is necessary to check on the quality of the feedback data. If these basic principles are not followed, any production planning and control system will fail in practice because of its inability to match the requirements of modern production.

As well as the requirements placed on techniques, the role of the user has to be remembered; this means that requirements for applications must be defined [24].

First, there must be room for decision-making and planning. The very stochastic nature of the processes renders planning to the minute or to the hour senseless. Instead, the personnel involved in and affected by planning must be included in the planning and control of the manufacturing process within the framework of defined work volumes and planning scopes. For this we need a clear representation of the process, the above-mentioned support from alternative planning methods, and user-friendly monitoring and diagnosis aids; on the other hand, we need a deliberate effort not to perform each and every activity with the system. Personal contact (meetings, telephone calls, etc.) is still necessary: too much formalism will suffocate the system. Finally, separation of personnel data – used for waging purposes – from process-based data, used for order monitoring, as early as their recording is desirable and, with today's technology, feasible.

The integration of scheduling in quality assurance [25] is a good place to start when trying to come to terms with this way of thinking. Figure 2.18 shows the quality–time–cost–cycle according to Geiger [26], which is based on the quality cycle (an important concept in quality management). The QTC-cycle reflects the basic sequence of planning, execution, and utilization of the output (which can be a physical product or a service). Planning, control, and monitoring of the functional product characteristics can be considered the state-of-the-art for ambitious products, and this approach is increasingly being used in industrial calculations as so

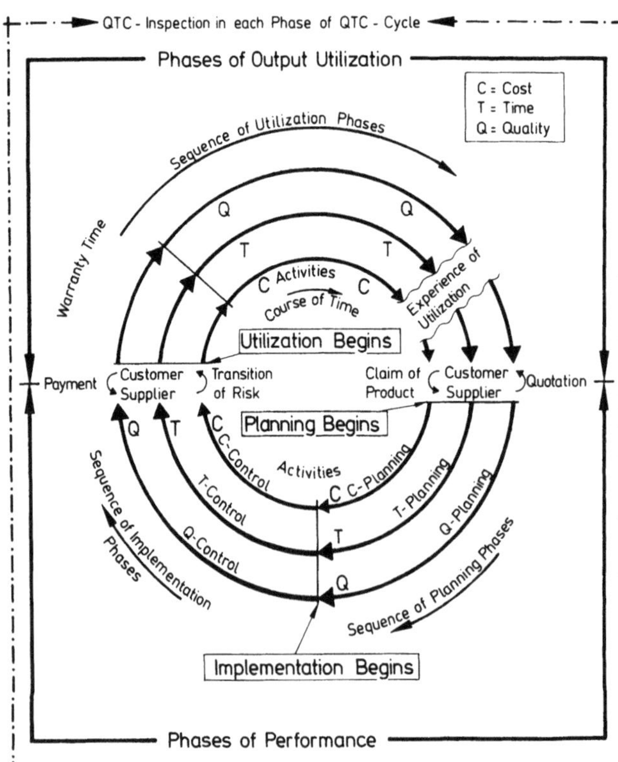

Fig. 2.18. Quality–time–cost–cycle (Adapted from Geiger [26])

called 'cost controlling', but time has not yet become a commonly accepted feature of a quality cycle in the theory and practice of production scheduling and control.

All the prerequisites for accepting time as a quality cycle feature have been met: dates can also be scheduled, controlled, and monitored. Time also obeys statistic laws that have to be recognized and used, thus it cannot be sensible to plan dates to an exact point. Instead, they should lie within a tolerance range to be defined with a certain frequency distribution (the tolerance range is given by the market). The production scheduling and control system must then, together with the production activity control system, ensure that the tolerance limits are kept. Dates thus become logistical quality characteristics.

Many findings of quality management can therefore be used for production scheduling and control, from statistical techniques for planning and monitoring, to the analysis and diagnosis of deviations, and to the organization of quality management and the motivation for quality. Not until the same feelings of responsibility and of pride in good results develop in a company for schedule performance as they do for product quality and costs can there be perceptible progress in this area.

The essence of all the demands placed on modern production scheduling and control is the demand for realistic modeling of manufacturing processes, and for a logical and understandable concrete representation of the interdependencies between the four basic objectives of manufacturing control.

It has already been mentioned that, for this, the lead time has to be examined more closely, and this is done in the following chapter.

2.6 References

[1] Fogarty, DW, Blackstone Jr., JH, Hoffman, TR. *Production and Inventory Management*, 2nd edn. South-Western Publishing, Cincinnati, OH, 1991.

[2] Brankamp, K. *Ein Terminplanungssystem für Unternehmen der Einzel- und Serienfertigung* (*A Scheduling System for Job Shop and Batch Manufacturing*), 2nd edn. de Gruyter Würzburg, Wien, 1973 (Based on the authors's doctoral thesis, TH Aachen, 1967).

[3] Hackstein, R. *Produktionsplanung und -steuerung* (*PPS*) – *Ein Handbuch für die Betriebspraxis* (*Production Planning and Control* (*PPC*) – *A Handbook for Manufacturing*). VDI-Verlag, Düsseldorf, 1984.

[4] Schomburg, E. Entwicklung eines betriebstypologischen Instrumentariums zur systematischen Entwicklung der Anforderungen an EDV-gestützte Produktionsplanungs- und -steuerungssysteme im Maschinenbau (Shop-specific Qualifications for the Development of Standards for Computer-Aided Production Planning and Control Systems in Mechanical Engineering). Dissertation, TH Aachen, 1980 (doctoral thesis).

[5] Grupp, B. *Modularprogramm für die Fertigungsindustrie* (*A Modular Program for Manufacturing Industry*). Springer-Verlag Berlin/New York, 1973.

[6] Wiendahl, H-P. *Betriebsorganisation für Ingenieure* (*Abschnitt* 6: *Produktionssteuerung*) (*Manufacturing Management for Industrial Engineers* (Chapter 6: Production Control)). 3rd edn. Hanser München/Wien, 1989.

[7] Melnyk, SA, Carter, PL. *Production Activity Control. A Practical Guide*. Dow-Jones-Irwin, Homewood, IL, 1985.

[8] Mertens, P. *Industrielle Datenverarbeitung. Teil 1: Administrations- und Dispositionssysteme* (*Data Processing in Industry, Part 1: Administration and Planning Systems*), 5th edn. Gabler, Wiesbaden, 1983.

[9] IBM (ed.). *Capacity Planning and Operation Sequencing. System Extended* (*CAPOSS – E*) – *Programmbeschreibung* (description of the program), *Vol 1: Planungssystem* (planning system). IBM GmbH, Germany, 1981.

[10] Nyhuis, F. Rüstzeitanalyse – Voraussetzung für eine systematische Rüstzeitreduzierung (Setup Time

Analysis – A Prerequisite for Systematic Setup Time Reduction). Vortrag zum Fachseminar "Statistisch orientierte Fertigungssteuerung" des Instituts für Fabrikanlagen der Universität Hannover (seminar paper). Institut für Fabrikanlagen (IFA), Hannover, 1984.

[11] Melnyk, SA, Carter, PL, Ditts, DM, Lyth, DM. *Shop Floor Control.* Dow Jones-Irwin, Homewood, IL, 1985.

[12] Bechte, W. Belastungsorientierte Auftragsfreigabe – eine neue Methode in der Fertigungssteuerung (Load-oriented Order Release – A New Production Control Technique). Vortrag zum Seminar "Neue Wege der Fertigungssteuerung" des Instituts für Fabrikanlagen der Universität Hannover am 16./17.03.1982 in Hannover (seminar paper). IFA, Hannover, 1982.

[13] Kettner, H, Bechte, W. Neue Wege der Fertigungssteuerung durch belastungsorientierte Auftragsfreigabe (Recent Production Control Methods by Use of Load-oriented Order Release). Zeitschrift des *Vereins Deutscher Ingenieure,* **123** (11): 1981, 459–466.

[14] Orlicky, J. *Materials Requirement Planning.* New York, 1975.

[15] Wiendahl, H-P. Stand und Entwicklungstendenzen in der Fertigungssteuerung (Production Control – State of the Art and Trends). *Industrie Anzeiger,* **103** (104): 1981, 40–44.

[16] Brankamp, K, Bongartz, B. *Der moderne Stanzbetrieb – Vom Sensormonitoring zur Geisterschicht (The Progressive Punching Shop – From Sensor Monitoring to the Man-less Shift).* VDI-Verlag, Düsseldorf, 1986.

[17] Virnich, M et al. Dezentrale BDE und Werkstattsteuerung (Decentralized Production Data Collecting and Shop Floor Control). Industrie Anzeiger **105** (23): 1983, 66–69.

[18] Kanet, JJ, Sridharan, V. The Electronic Leitstand: A New Tool for Shop Scheduling. *Manuf. Rev.*, **3** (3): 1990, 161–170.

[19] Wildemann, H. *Flexible Werkstattsteuerung durch Integration von KANBAN-Prinzipien (Flexible Shop Floor Control by Use of Integrated Kanban Principles).* CW Publikationen, München, 1984.

[20] Fox, RE. OPT vs. MRP. Thoughtware vs. Software, Part I and II. *Inventories & Prod. Mag.*, Vol 3 November/December, 1983, Vol 4, January/February, 1984.

[21] Plossl, GW. *Production and Inventory Control – Principles and Techniques,* 2nd edn. Prentice-Hall, Englewood Cliffs, NJ, 1985.

[22] Plossl, GW. *Production and Inventory Control – Applications.* Prentice-Hall, Englewood Cliffs, NJ, 1983.

[23] Scheer, AW. Trends zur Computerleistung am Arbeitsplatz im Bereich der Produktionsplanung- und -steuerung (Trends towards Computer Application in Production Planning and Control). In: *Tagungsband zur Ausschuß für Wirtschaftliche Fertigung Bad Soden Tagung PPS '80,* 12–14 November 1980, pp 3–23.

[24] Fröhner, K-D. Stand und Entwicklungsmöglichkeiten der Fertigungssteuerung unter Berücksichtigung personenbezogener Auswirkungen (State of the Art and Development in Production Control in View of Effects on the Personnel). Kernforschungszentrum Karlsruhe (Projektträger Datenverarbeitung-Bericht Kernforschungszentrum Karlsruhe/Projektträger Datenverarbeitung 205), 1981.

[25] Masing, M. *Handbuch der Qualitätssicherung (Handbook of Quality Assurance),* 2nd edn. Hanser München/Wien, 1989.

[26] Geiger, W. Der Qualitäts-Termin-Kosten-Kreis (QTK-Kreis) – Ein Modell für das Zusammenwirken aller Tätigkeiten im Unternehmen (Quality–Time–Cost–Cycle (Q–T–C-Cycle) – A Model for Manufacturing Activities Interaction). *Zeitschrift des Vereins Deutscher Ingenieure,* **125** (9): 1983, 313–37.

Chapter 3

Lead Time – A Key Term in Manufacturing Control

3.1 Abstract

Work center lead time and order lead time are the basic elements for the systematics of production planning and control. A precise definition of these terms is therefore given first, and the difference between simple and weighted values is explained. The formulas and tables necessary to compute statistical values for the evaluation of recorded or simulated samples are then explained, as is their graphical representation. The chapter closes with examples of authentic flow time measurements performed in industrial firms in Germany. Thus, readers will be able to carry out such evaluations in real shops by themselves.

3.2 Introduction

Planning, measuring, and monitoring lead time is one of the essential tasks of production scheduling and control. As opposed to the commonly accepted composition of order time as defined by REFA (Verband für Arbeitsstudien und Betriebsorganisation e.V., West Germany) [1], it has not yet been possible to find an equally accepted classification of manufacturing lead time, despite numerous attempts. Continual or at least periodical measuring of actual lead times is not common practice either, only being performed on the occasion of scientific investigations or problem solving studies.

As early as 1963, Tully [2], from his experience as a manager of a machine tool factory, wrote the following on this subject.

> The actual lead time of an operation is normally 5 workdays. With high-precision parts, with large standard time per operation, with a large number of operations (up to about 20), and with long inspection times for single operations, an even longer lead time is frequently needed. The waiting time component of manufacturing lead time, i.e. the time from the issue of raw materials to the finished goods' arrival at stock, is about 90 per cent to 95 per cent.
>
> A time study in one plant yielded the following percentages.

Waiting time before processing	75%
Process time	6%

Waiting between processing and transport to inspection	7%
Waiting time before inspection	7%
Inspection time and transport to downstream work center	2%
Breaks in processing	3%

This is a poor result and is partly determined by the type of production. These numbers suggest that we should examine more closely the problem of lead time reduction.

This problem, however, only became an object of large scale investigations nearly ten years later. In 1972, Hackstein reported on the first results of a lead time study carried out in three companies by his institute [3]; a more detailed report by Stommel and Kunz [4] was published in 1973. Figure 3.1 shows a frequently cited diagram from this research report which proves the statements by Tully, and which is still valid for many factories today. This report also confirmed the fact that the

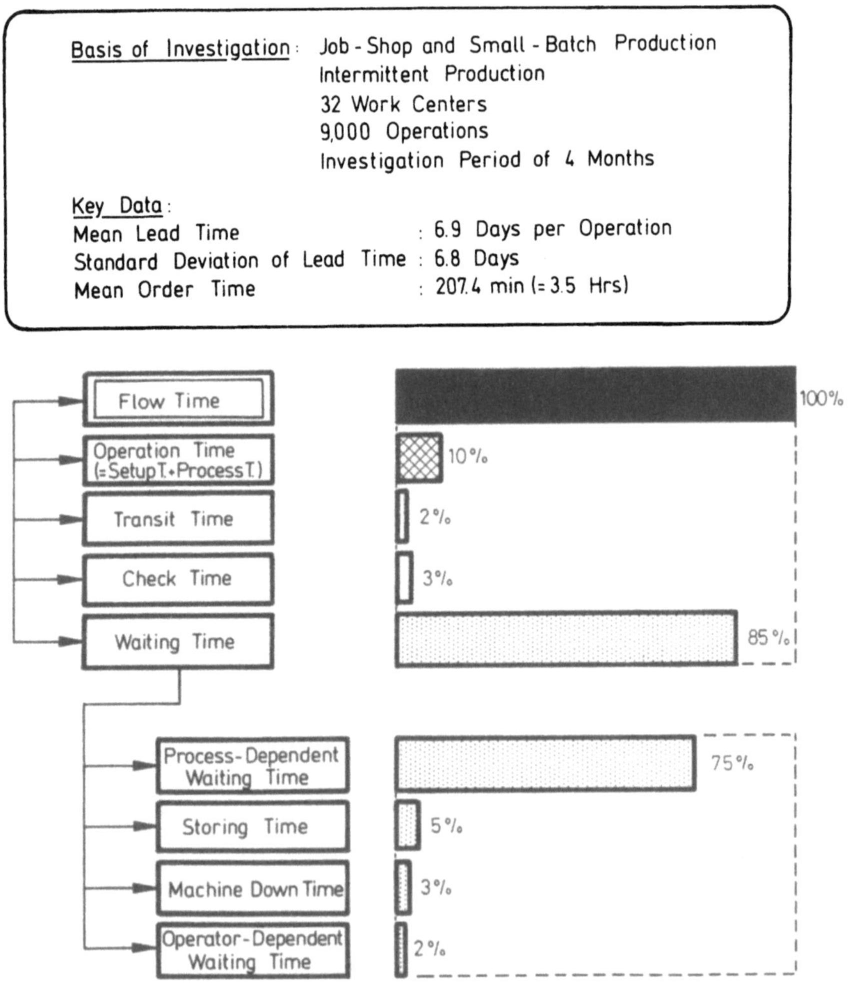

Fig. 3.1. Components of flow time in a metalworking plant with job-shop and series production (After Stommel and Kunz [4])

process time component of manufacturing lead time is usually under 10% and that, consequently, measures to reduce lead time must aim to reduce waiting time before processing.

This research report also shows that actual lead times (also called flow times) at the individual work centers vary between wide limits. Figure 3.2 shows the actual mean work center lead times and their standard deviations for the 32 work centers of the same factory as in Fig. 3.1. The actual mean value of all work center lead times is about 5 working days, but there are also considerably lower and higher individual values, such as 1.2 days at Work Center 7/72 (lapping), and 16.9 days at Work

Work Center	Description	Mean Flow Time (Days)	Standard Deviation of Mean Flow Time (Days)
1/02	Identification	1.70	1.89
1/10	Lathes	4.01	5.22
1/11	Lathes	4.51	4.26
1/13	Lathes	2.33	1.80
1/20	Milling Machines	6.38	5.31
1/23	Milling Machines	5.30	6.97
1/36	Drilling Machines	2.36	2.31
1/40	Planing Machines	4.51	6.06
1/47	Broaching Machines	10.00	7.92
1/81	Sawing	3.27	3.75
1/89	Sawing	3.42	4.56
3/01	Tool Making	6.31	8.72
3/34	Drilling Machines	4.36	4.47
3/36	Drilling Machines	5.78	5.34
3/82	Separating	6.58	10.84
3/92	Signing	1.74	2.52
3/94	Engraving	4.15	4.63
3/96	Pressing	2.45	2.41
4/02	Marking	1.97	2.29
5/00	Hardening Room	3.24	1.74
7/50	Grinding Machines	2.54	2.59
7/51	Grinding Machines	3.31	8.40
7/54	Grinding Machines	9.18	3.17
7/61	Grinding Machines	6.60	8.36
7/62	Grinding Machines	5.64	8.39
7/65	Grinding Machines	4.98	5.27
7/66	Grinding Machines	2.43	3.02
7/70	Honing	1.75	1.58
7/72	Lapping	1.20	.63
99/10	Lathes	9.10	5.57
99/13	Lathes	6.93	5.25
99/16	Lathes	16.94	13.64
Σ 32	Mean Value	4.84	4.97

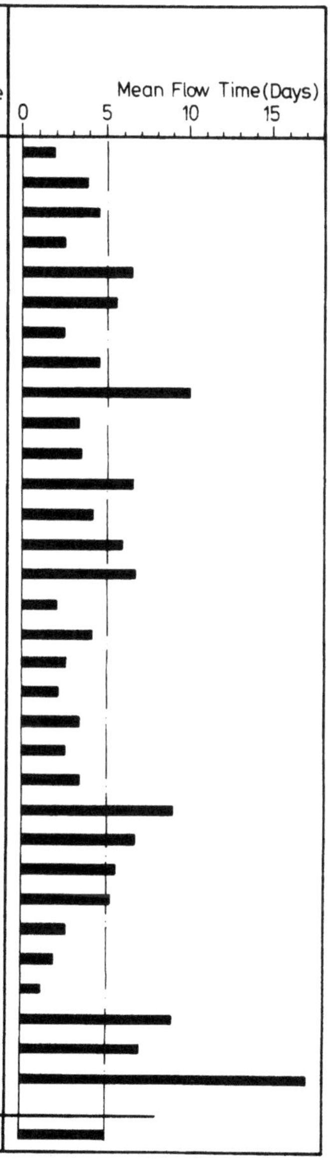

Fig. 3.2. Mean flow time of different work centers (After Stommel and Kunz [4])

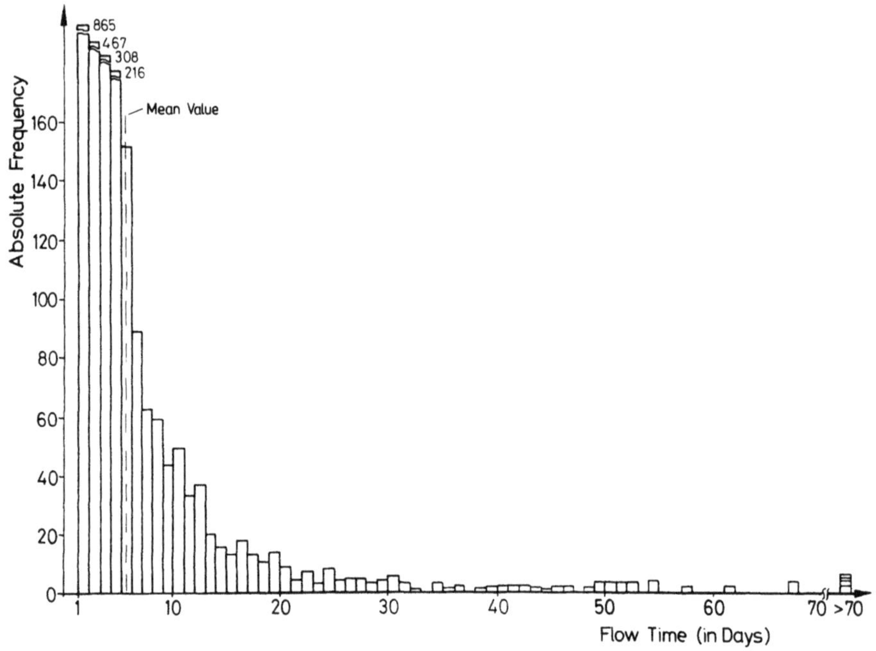

Fig. 3.3. Frequency distribution of actual operation flow times in a tool shop. Total = 2600 operations = 100%; mean value: 5 days (After Hackstein [3])

Center 99/16 (lathes). It is further remarkable that the standard deviations of the individual values and the mean value of all work centers are, with a few exceptions, in the same order of magnitude as the mean values themselves.

Finally, the frequency distribution is also of interest in this investigation. Figure 3.3 shows the results of an analysis of actual lead times per operation (for the same company as in Figs. 3.1 and 3.2 [4]). Besides the large variance in lead times, it can also be seen that the frequencies in the single groups are not subject to a normal distribution, but rather to an exponential distribution. The mean value of this distribution is about five days.

Other authors have also examined the problems of lead time analysis [5–7].

From 1971 on, Kettner [8] also worked on fundamental studies concerning lead times at the Institut für Fabrikanlagen (IFA) at the University of Hannover, Germany. The project, carried out with a large number of manufacturing companies served to analyze more closely the actual lead times and inventories in the shops [9]. From this, a general method of data recording and a computer-aided evaluation system, DUBAF (Durchlaufzeit- und Bestandsanalyse in der Fertigung = Manufacturing Lead Time and Inventory Analysis), were produced. DUBAF is continuously being advanced and is, at present, in constant use in practice [11, 12].

During the IFA investigations it became absolutely necessary first of all to give a clear and uniform definition of manufacturing lead time and its components, in order, first, to standardize the methods of data recording and enable the data evaluation to be programmed, and, second, to permit a comparison between different factories and job shops. Based on these investigations, work center and order lead time and their components will be defined in the following.

3.3 Lead Time Components

Using the term 'lead time', we have to distinguish between planned and actual values. Kanet [24] defines planned lead time as allowance and actual lead time as flow time (he refers to Conway et al. [25]). This book will also use these terms. In the following basic definitions, however, the term 'lead time' will be used until it becomes necessary to use allowance, resp. flow time.

In considering the throughput of a shop order, it is common practice – as shown in Chapt. 2 – to plot the individual orders with their operations over the time axis. The time from the issue of materials to parts' or components' arrival at an intermediate store or at the assembly department is commonly called the *order lead time.* This is a rather inexact term, for it should rather be called *order lead time period* or *time interval* or *time span.* The time needed for one operation is the smallest unit: it is called the *operation lead time.* Further classification into lead time components and their definition varies both in the literature and in their practical application.

Figure 3.4 shows a slightly altered proposal by Heinemeyer, which will also be the basis for the following statements [13]. Heinemeyer distinguishes between three levels. On the order level there are the separate operations OP_1 to OP_k. Each operation is split into five more components on the operational level:

- Waiting after processing
- Transportation
- Waiting before processing (Queuing)
- Setup and
- Processing

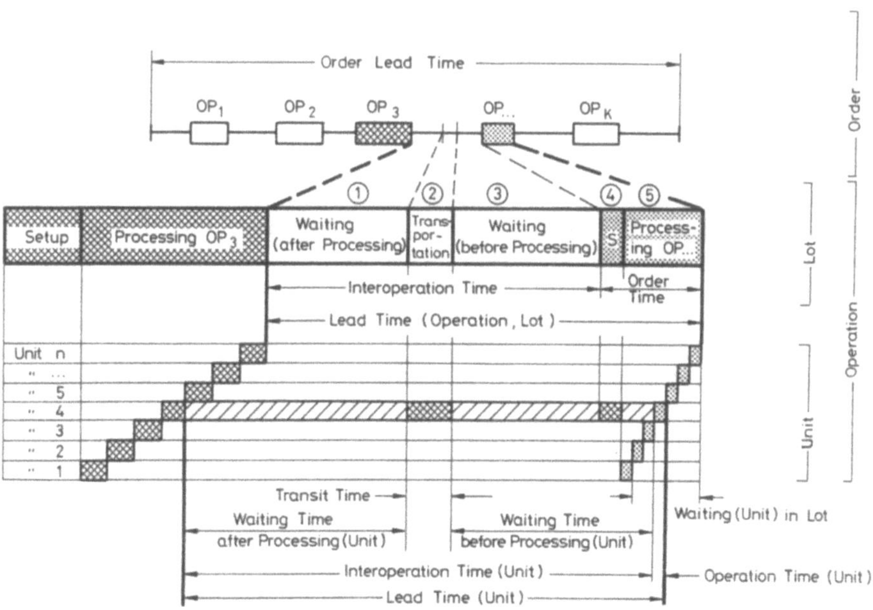

Fig. 3.4. Components of lot and order lead time

In contrast to other models (e.g. [4]), waiting after processing at the upstream work center and transit time are assigned to the downstream work center. With the very first operation, this starts with the order release; for the other operations it starts with the process end at the previous work center. A manufacturing order generally consists of several units, the *manufacturing lot*. The throughput of these units 1 to *n* is examined at the level of the single units, i.e. manufactured parts. We can see that there is an additional waiting time for each part within the process time of the lot, the so-called *lot waiting time*.

In job shop production, the single lots are usually transported together, processed in succession at one work center, and controlled accordingly. For our further considerations, it is therefore sufficient to look at the whole lot on the operational level. Only in batch production is a discussion of the throughput of single parts or sublots necessary, since the process time is longer than the interoperation time, necessitating a coordination of the overlapping successive operations.

In Figure 3.5A, order throughput is shown on the time axis. An operation ends at *tPE* (process end) and starts at *tPEU* (process end at upstream work center). The date *tPEU* does not characterize the physical arrival of an order at a work center, but signifies the date of the ready note at the upstream work center. The same applies to *tPE*. In this way, all lead time components can be measured completely if the release date is known. The graphical representation of the lead time at a work center is the basic element of the following considerations, and is called the *throughput element* [13].

Figure 3.5B shows the throughput element of Operation 3 with its components, as discussed above. Since it has, for the moment, only one dimension – namely time – it is called the *one-dimensional throughput element* [18].

The operation time (*TOP*) is small compared to the lead time (*TL*), typically between 2% and 10%. Therefore, recording of the process begin (*tPB*) or setup begin (*tSB*) can be omitted for analytical and control purposes (it suffices to calculate this value from the order time). Thus the throughput element can be simplified to form the reduced representation in Fig. 3.5C. For this calculation, only the process end at the upstream work center (*tPEU*), the process end at the work center concerned (*tPE*), and the operation time (*TOP*) resulting from the order time are needed.

The calculation of the reduced throughput element is first discussed using Fig. 3.6 [14]. The actual lead time (*TL*) is calculated from the difference between the reported dates *tPEU* and *tPE*. To calculate the interoperation time (*TIO*), the operation time (*TOP*) must be known. It is calculated from the order time (*TO*) multiplied by the efficiency rate (*ER*), and divided by the daily capacity (*CDAY*) of the work center concerned [15]. If, for instance, the order time (in standard hours) is 12 hours and the efficiency rate is 120%, then the work center will be used for $12/1.2 = 10$ hours for this order. If, at this work center, the working day is eight hours long, the operation time will be $10/8 = 1.25$ days. The order time of a lot itself is composed of setup time (*TS*) and process time per order (*TPO*). *TPO* is the product of process time per unit (*TPU*) and lot size (*Q*).

The interoperation time consists of the components' transit time (*TT*) and the waiting times (*TWA* and *TWB*). The transit time is only important when transport organization is poor [14]. Often, waiting times also include quality control checks or other activities at this work center, for which there is no time allowed in the operation sheet, e.g. for marking, cleaning, or deburring the parts. Breakdowns also cause waiting times, but they are only important with highly automated manufacturing or assembly facilities working with cycle times of a few minutes or even

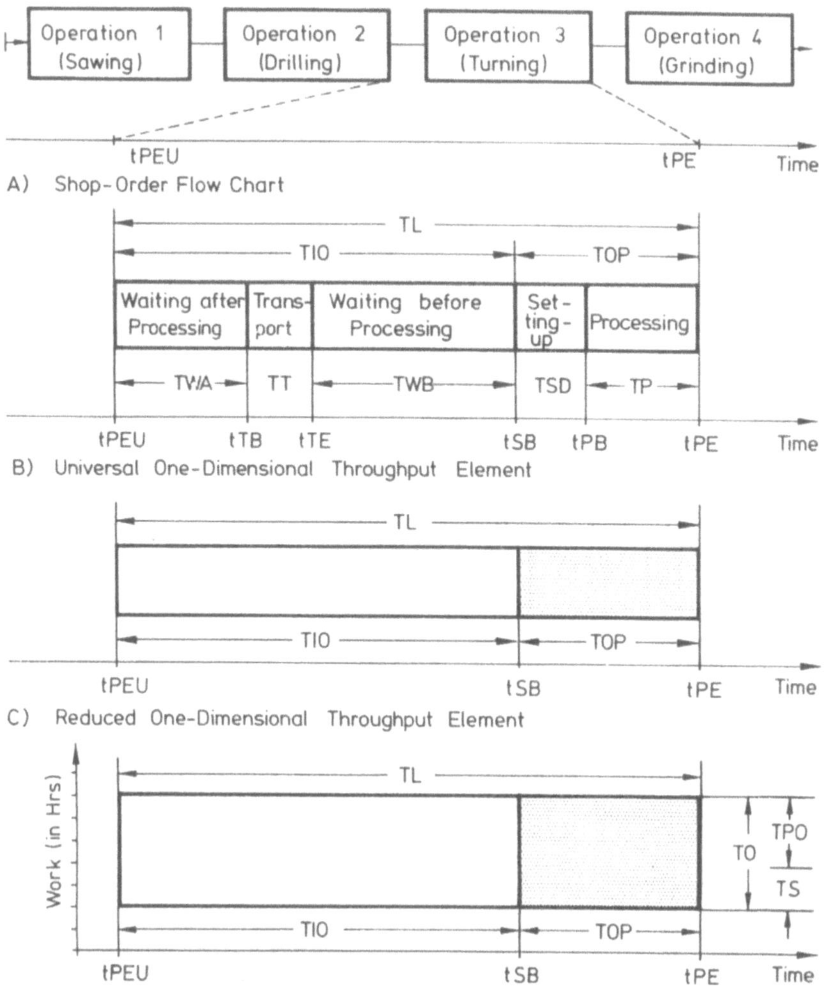

D) Reduced Two-Dimensional Throughput Element

tPEU : Process End at Upstream Work Center
tTB : Transport Begin
tTE : Transport End
tSB : Setup Begin
tPB : Process Begin
tPE : Process End
TO : Order Time
TS : Setup Time
TPO : Process Time per Order

TL = tPE - tPEU : Lead Time
TIO = tSB - tPEU : Interoperation Time
TOP = tPE - tSB : Operation Time
TSD = tPB - tSB : Setup Time (Duration)
TP = tPE - tPB : Process Time
TWB = tSB - tTE : Waiting before Process
TT = tTE - tTB : Transit Time
TWA = tTB - tPEU : Waiting after Process

Fig. 3.5. One- and two-dimensional throughput elements (universal and reduced representation)

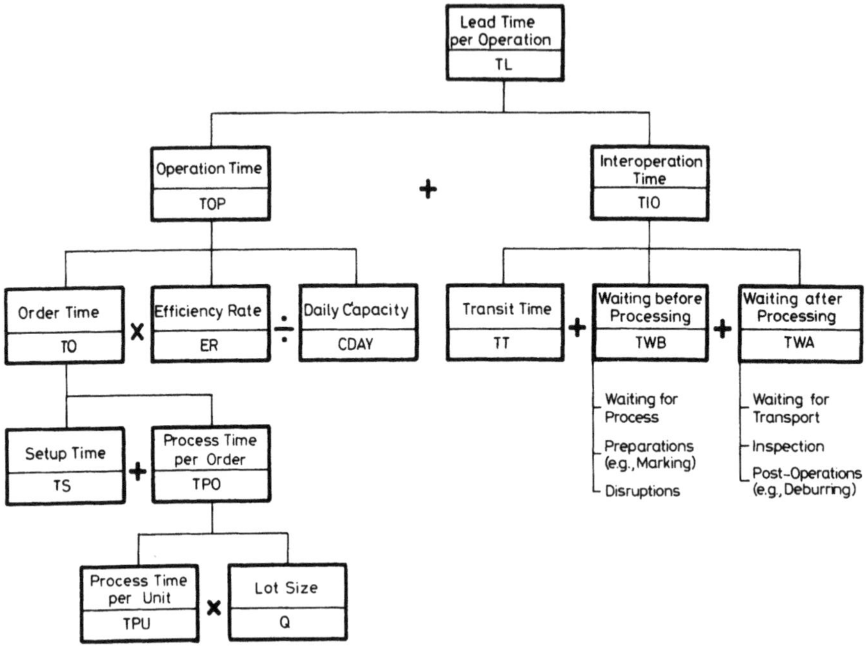

Fig. 3.6. Components of work center lead time

seconds. The predominant component of interoperation time is the waiting time in the queue before the work center [16]. Only in special cases, as in special studies, are these components determined and evaluated by recording the individual times.

Therefore, the reduced throughput element, as shown in Fig. 3.5C, will first be the basis of the following. It consists only of the components operation time (*TOP*) and interoperation time (*TIO*).

The date of actual output at a work center is determined in practice as shown in Fig. 3.7. The date of reported output (here Order 4718 at Work Center A, at 7.18 on Day 282) is recorded twice. First, it is the output date of Order 4718 at this work center; and second, it is the input date of the same order at downstream Work Center B. When Order 4718 leaves Work Center B, the date of reported output (here 13.24, Day 285) at Work Center B is also the date of reported input at Work Center C, according to the operation sheet.

The actual lead time (*TL*) for a throughput element is then calculated using the recorded dates.

$$TL = tPEU - tPE \tag{3.1}$$

where:

$tPEU$ = date of reported process end at upstream work center
tPE = date of reported process end at work center concerned.

The values for our example, Order 4718 at Work Center B, are then:

$$\begin{aligned} TL &= \text{day } 285, 13.24 - \text{Day } 282, 7.18 \\ &= (285 \text{ days} + 13 \text{ hrs} + 24 \text{ min}) - (282 \text{ days} + 7 \text{ hrs} + 18 \text{ min}) \\ &= 3 \text{ days} + 6 \text{ hrs} + 6 \text{ min} \end{aligned}$$

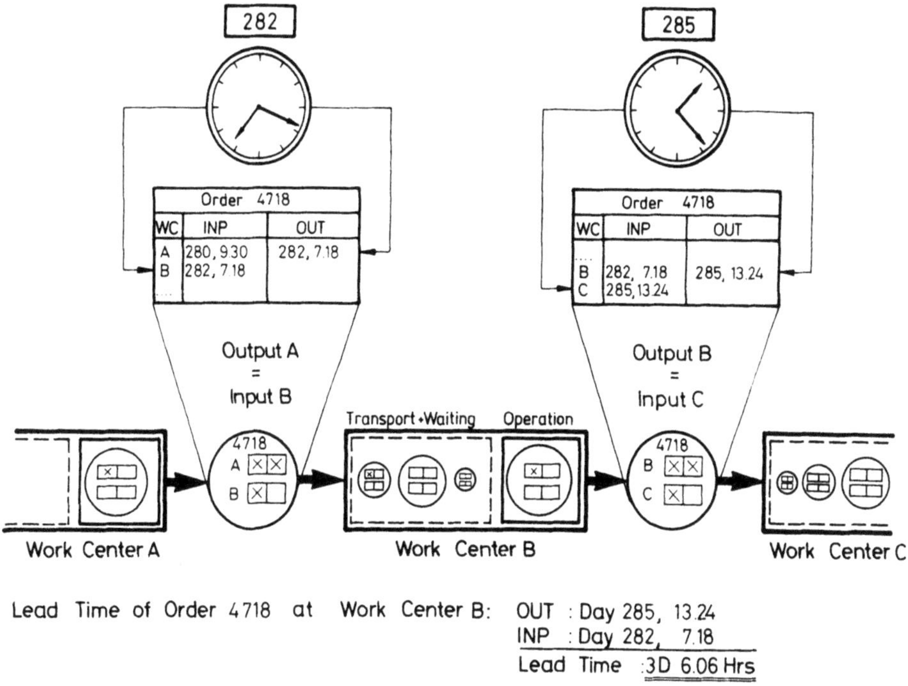

Fig. 3.7. Measuring work center lead time

The conversion of the process time into workdays obviously depends on the number of working hours per day. The following equation expresses the relationship.

$$TL = TL \text{ days} + \frac{TL \text{ minutes}}{CDAY} \tag{3.2}$$

where:

TL = lead time
$CDAY$ = daily capacity in working minutes per day

In our example, the value is, with eight hours per day:

$$TL_1 = 3 \text{ days} + \frac{6 \text{ hrs} + 6 \text{ min}}{8 \text{ hrs/day}} = 3 + \frac{366 \text{ min} \times \text{day}}{480 \text{ min}} = 3.7 \text{ days}$$

and with 16 hours per day (two-shift work):

$$TL_2 = 3 \text{ days} + \frac{366 \text{ min} \times \text{day}}{16 \times 60 \text{ min}} = 3.35 \text{ days}$$

In this fictitious example, the whole lead time days are not affected by the number of working hours per day, since calendar days and hour of the day were given as reported dates. One could assume that in practice, with two-shift work, lead time will, as a whole, become shorter resulting in different reported output dates. However, this does not necessarily have to be the case, since lead time is primarily determined by waiting time before processing which, besides being dependent on capacity, depends on the quantity of work waiting for processing at a work center. (These relationships will be dealt with more closely below.) Calculation of operation time (TOP) obviously depends on the daily capacity ($CDAY$) with which an order with the order time TO is processed. For instance, an order with a standard time of 20 hours needs, with an efficiency rate of 1.0 and single-shift work (8 hrs/day), 2.5 days for processing. With two-shift work (16 hrs/day), this is 1.25 days, and with three shift work, 0.83 days.

For the operation time (TOP) at a work center and thus for one operation of an order, it follows that

$$TOP = \frac{TO}{CDAY \times ER} \tag{3.3}$$

where:

TO = order time in standard hours
$CDAY$ = daily capacity in standard hours per day
ER = efficiency rate at work center concerned (= actual labor hours/standard hours).

As already mentioned, the operation time (TOP) is small compared to the overall lead time (TL). Therefore, one can easily assume that the efficiency rate is equal to one, thus saving unnecessary data recording and computing time.

Operation time can more easily be calculated as:

$$TOP = \frac{TO}{CDAY} \tag{3.4}$$

where:.

TO = order time in standard hours (= setup time + lot size × time per unit)
$CDAY$ = capacity in standard hours per day.

When TOP is known, setup begin (tSB) and interoperation time (TIO) can also be determined. tSB is determined as follows:

$$tSB = tPE - TOP \tag{3.5}$$

where:

tPE = date of reported process end
TOP = operation time.

Interoperation time (TIO) is:

$$TIO = TL - TOP \tag{3.6}$$

where:

TL = operation lead time
TOP = operation time.

Using the definition of and the equations for the throughput element, the mean lead time can now be defined. Here, as well, we shall first look at only one work center.

3.4 Simple and Weighted Mean Lead Time at a Work Center

Mean lead times are the object of measurements, for example for problem solving studies, as well as a prerequisite for planning purposes, e.g. for scheduling. Mean lead times can refer to work centers, operations, and orders.

Normally, mean lead time values are determined by calculating the arithmetic mean of the individual lead times of all orders reported leaving a work center within a certain period (a reference or investigation period). This is called the *simple mean lead time* (TL_m). The equation for this is:

$$TL_m = \frac{1}{n}(TL_1 + TL_2 + \cdots + TL_n)$$

$$TL_m = \frac{\sum_{i=1}^{n} TL_i}{n} \tag{3.7}$$

where:

TL_m = simple mean lead time
TL_i = lead time of order i ($= tPE_i - tPEU_i$)
n = number of orders considered.

This calculates how long, on average, an order (i.e. a lot) stays at a work center.

One can also use a funnel model to represent this process (Fig. 3.8). The model has been developed in this form by Kettner and Bechte [17], and is very important for understanding the manufacturing process.

The arriving lots form an inventory of waiting lots; they all want to pass the funnel. The funnel exit gauge corresponds to the capacity, which is variable within certain limits. Obviously the mean lead time for an arriving lot is greater the larger is the number of waiting lots. It is smaller the larger is the utilized capacity, i.e. the output of the work center.

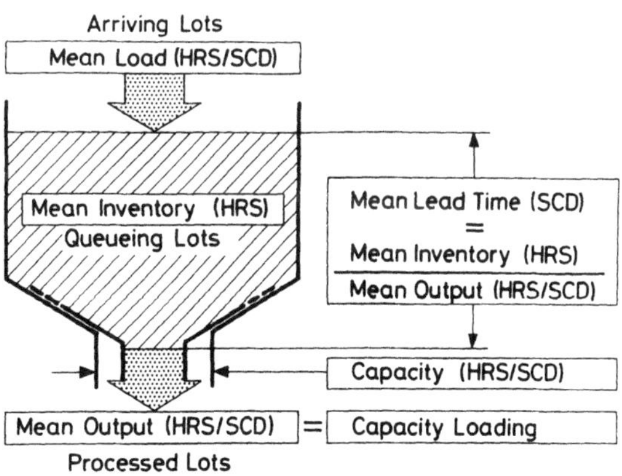

Fig. 3.8. The funnel model of a work center. SCD: shop calendar day (After Kettner and Bechte [17])

In continuous production, capacity and thus output are given in parts per hour or day. This, however, is not possible in job shop production because of the wide variations in order time. Capacity is therefore given in hours per day. With hours (hrs), we mean order time (in standard hours); 1 day is defined as 1 working day, frequently called a shop calendar day (SCD).

When capacity and output are given in hours, it makes sense to also give lot inventories in hours instead of in items. From this it follows that each single lead time must be weighted by the order time.

The one-dimensional throughput element thus enters the second dimension. Bechte has coined the term *two-dimensional throughput element* [18]. If this second dimension is the work content in hours, we can speak of the work-related two-dimensional throughput element [19] (cf. Fig. 3.5D). If this dimension is the quantity of items in an order, it is called the *quantity-related two-dimensional throughput element*. Figure 3.9 compares the three concepts.

In the following, we use the work-related two-dimensional model. The corresponding work center lead time will be called the *weighted lead time* as opposed to *simple lead time*. It has the dimension days × hours. The mean value of weighted lead time is correspondingly called the *weighted mean lead time*.

As experience has shown that it is very difficult to understand exactly what is

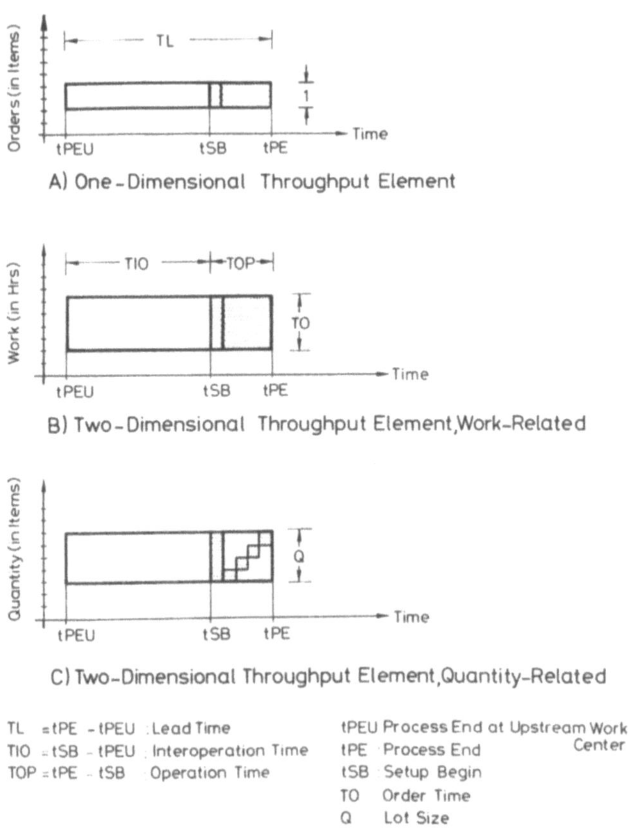

Fig. 3.9. Work-related versus quantity-related throughput element

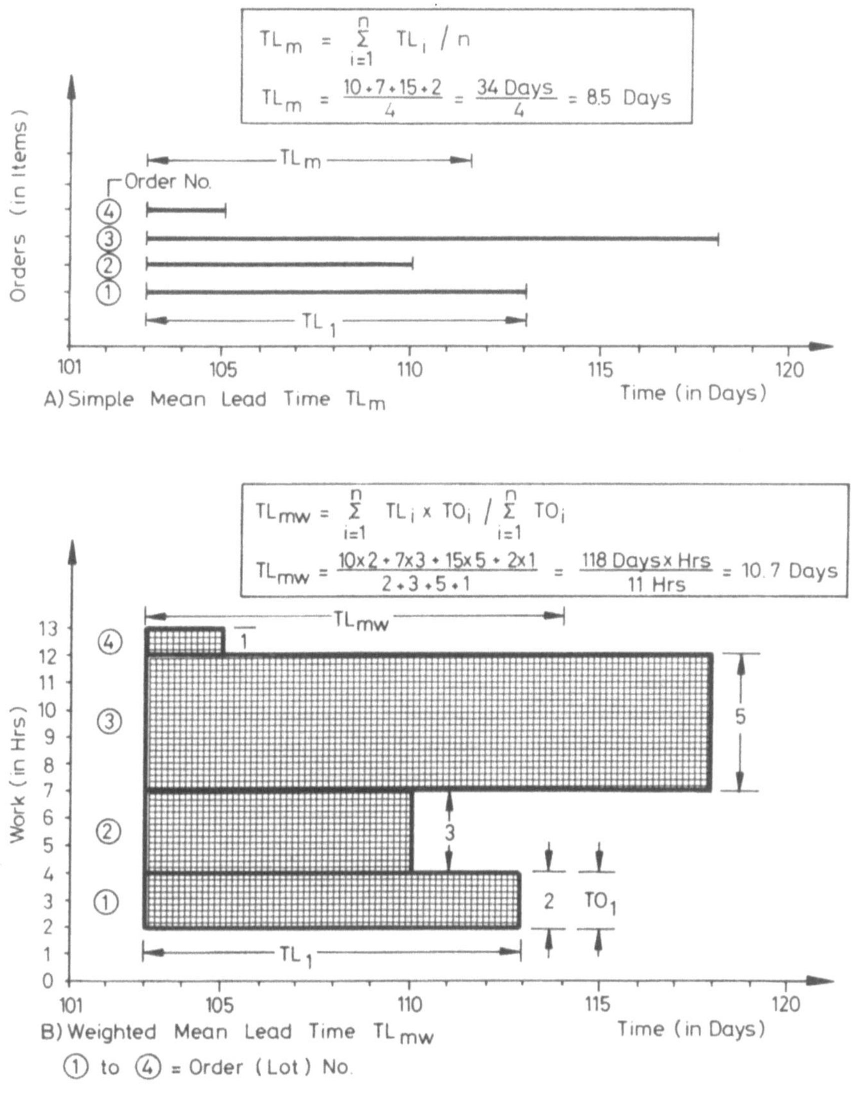

Fig. 3.10. Simple mean and weighted mean lead time at a work center. Variant A: weighted lead time greater than simple lead time

meant by this weighted mean lead time, it is now explained with a simple example. In Fig. 3.10A the lead times of four orders are plotted on the calendar time axis. The mean value for TL is 8.5 days.

With weighted lead time, however, it is not the number of orders but the work (in hours) put through the work center that is of interest here. So each lead time is multiplied (weighted) by the work content, i.e. order time (TO), thus producing an area which represents the weighted lead time of this operation.

To find the mean value, one consequently has to divide the sum of these areas by the number of processed hours:

$$TL_{mw} = \frac{TL_1 \times TO_1 + TL_2 \times TO_2 + \cdots + TL_n \times TO_n}{TO_1 + TO_2 + \cdots + TO_n}$$

$$TL_{mw} = \frac{\sum_{i=1}^{n} TL_i \times TO_i}{\sum_{i=1}^{n} TO_i} \tag{3.8}$$

where:

TL_{mw} = weighted mean lead time
TL_i = lead time of Order i
TO_i = order time of Order i.

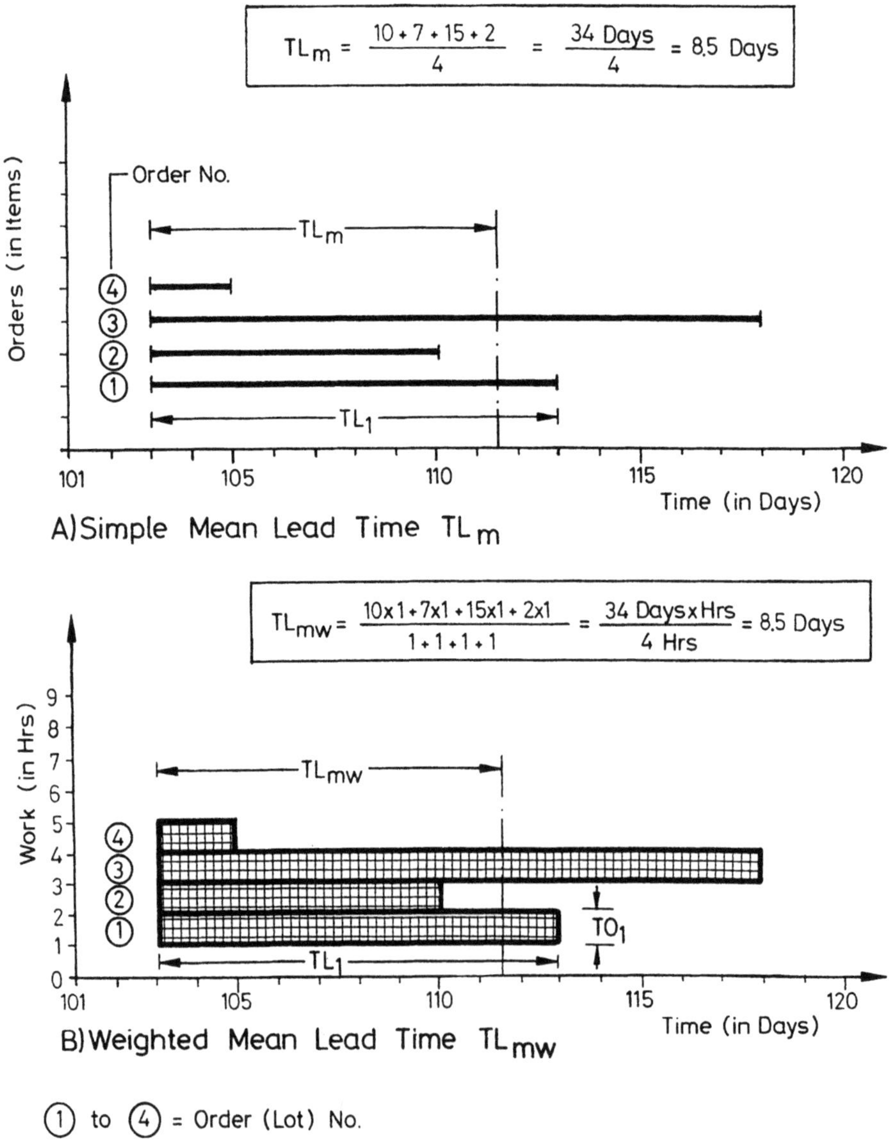

Fig. 3.11. Simple mean and weighted mean lead time at a work center. Variant B1: $TL_m = TL_{mw}$ (TO = constant)

Figure 3.10B shows the lead time areas of Orders 1–4 (from Fig. 3.10A), weighted by the order time, and the calculation of the weighted mean lead time. The value of TL_{mw} (10.7 days) is more than two days greater than the simple mean value TL_m (8.5 days).

The weighted mean lead time states how long, on average, a unit of work (e.g. 1 hour) takes to pass through the work center in question.

Weighted mean lead time can be greater than, equal to, or less than simple mean lead time. The first case has been shown in Fig. 3.10.

Equal simple and weighted mean lead times are possible in two cases: either all order times (TO) or all lead times (TL) are equal.

Figure 3.11 shows the first of these possibilities. Let the order time be 1 hour for

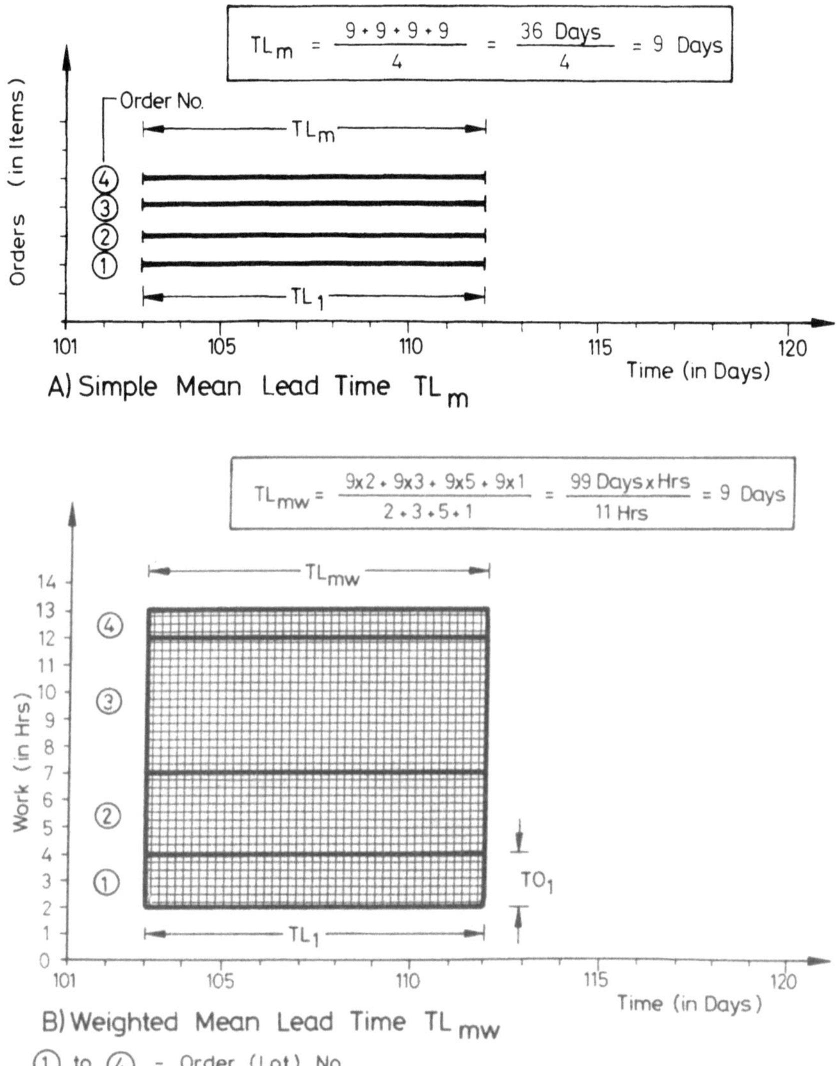

Fig. 3.12. Simple mean and weighted mean lead time at a work center. Variant B2: $TL_m = TL_{mw}$ (TL = constant)

every order. Since variance in order time is zero, simple and weighted mean lead time values are equal. On average, one lot is processed by the work center in the same time as 1 hour of work.

In such a case, inventory can also be given in items, and output and capacity can be given in items per day, week, or month. Lead time variance is, in this case, only determined by interoperation time (*TIO*) variance, since operation time (*TOP*) is always the same, given constant capacity.

If lead times are equal, but order times different, TL_m and TL_{mw} are also equal (Fig. 3.12). This case is, however, rather hypothetical, and will rarely be found in practice.

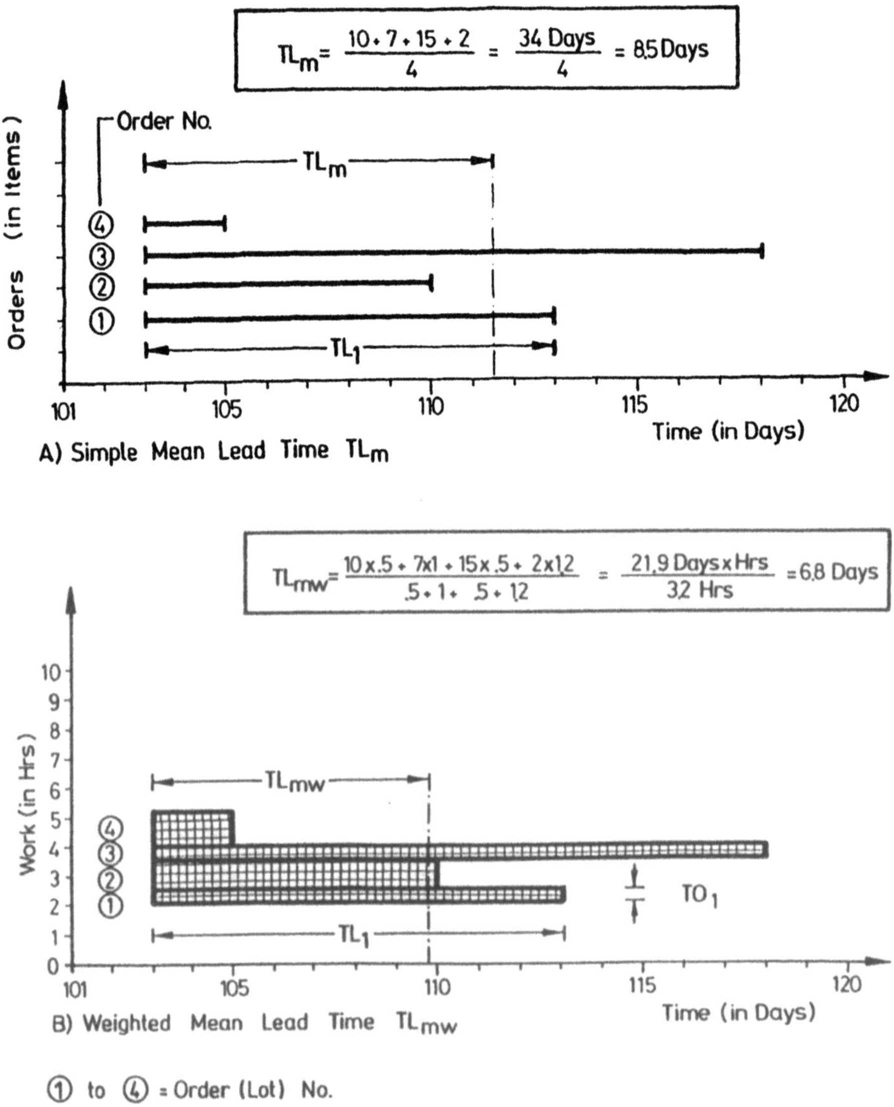

Fig. 3.13. Simple mean and weighted mean lead time at a work center. Variant C: weighted lead time less than simple lead time

Thirdly, weighted mean lead time can be less than simple mean lead time. Figure 3.13 shows such an example. This can be observed when orders with a large work content are, on average, processed faster than those with little work content. This is the case when, for example, the priority rule LOT (Longest Operation Time) is applied.

3.5 Order Lead Times

After having defined the throughput element, we can now also define order lead time (TOL) (Fig. 3.14) [14]. It consists of the sum of the individual operation lead times at all work centers passed by this order. For the first throughput element, there is no upstream work center. One therefore sets the order start date – order release date (tOR) – to be the start date of the first operation. It is defined as the date on which the supply of raw materials is reported. The material provided enters the inventory of the first work center at the date tOR. The order is completed with the order completion date tOC, which is also the process end date tPE of the last operation.

Now the order lead time (period) can be calculated as follows:

$$TOL = tOC - tOR \tag{3.9}$$

where:

tOC = order completion date
tOR = order release date.

As was shown for work center lead time (TL), the daily capacity, i.e. the number of working hours per day, must also be taken into account in this case (cf. Eqn. (3.2)). Order lead time (TOL) can also be calculated from the sum of the operation lead times:

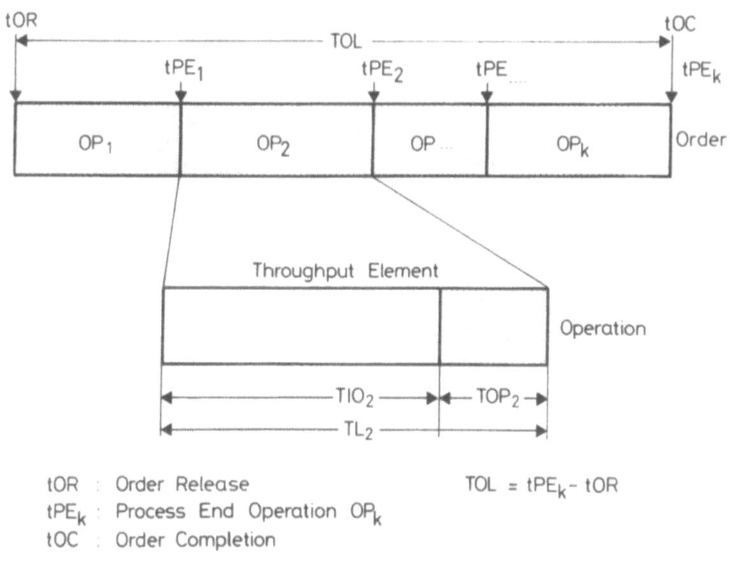

Fig. 3.14. Components of order lead time: one-dimensional throughput elements

$$TOL = TL_1 + TL_2 + \cdots + TL_k$$

$$TOL = \sum_{i=1}^{k} TL_i \tag{3.10}$$

where:

TL_i = lead time of operation i
k = number of operations.

When calculating mean order lead time, one must, as with work center lead time, distinguish between simple and weighted mean values.

The simple mean order lead time (TOL_m) of n orders is:

$$TOL_m = \frac{TOL_1 + TOL_2 + \cdots + TOL_n}{n}$$

$$TOL_m = \frac{\sum_{i=1}^{n} TOL_i}{n} \tag{3.11}$$

$$TOL_m = \frac{\sum_{i=1}^{n} \sum_{j=1}^{k_i} TL_{i,j}}{n} \tag{3.12}$$

where:

$TL_{i,j}$ = lead time of operation j of order i
k_i = number of operations of order i
n = number of orders
TOL_i = lead time of order i.

If one defines the mean number of operations per order as:

$$OPO_m = \frac{k_1 + k_2 + \cdots + k_n}{n}$$

$$OPO_m = \frac{\sum_{i=1}^{n} i \sum_{j=1}^{k_i} j}{n} \tag{3.13}$$

where:

j = operation j of order i
k_i = number of operations of order i
n = number of orders concerned

and the mean operation lead times of the orders in question as:

$$TL_m = \frac{\sum_{i=1}^{n} \sum_{j=1}^{k_i} TL_{i,j}}{\sum_{i=1}^{n} k_i} \tag{3.14}$$

where:

$TL_{i,j}$ = lead time of operation j of order i
k_i = number of operations of order i
n = number of orders concerned.

then TOL_m can be defined as:

$$TOL_m = TL_m \times OPO_m \tag{3.15}$$

According to the definition of weighted lead time, $TL_w = TL \times TO$ (cf. Fig. 3.9), weighted mean order lead time (TOL_{mw}) is:

$$TOL_{mw} = \frac{\sum_{i=1}^{n} \sum_{j=1}^{k_j} (TL_{i,j} \times TO_{i,j})}{\sum_{i=1}^{n} \sum_{j=1}^{k_j} TO_{i,j}} \tag{3.16}$$

where:

$TL_{i,j}$ = lead time of operation j of order i
$TO_{i,j}$ = order time of operation j of order i.

These simple definitions allow the throughput behavior of work centers and orders to be described based on one common element, namely the throughput element.

3.6 Statistical Evaluation of Work Center Lead Times

3.6.1 Absolute and Relative Frequency Distribution of Simple and Weighted Lead Time

Measured values are commonly evaluated by classifying them and then determining frequency distributions and statistical key data. The values essential for lead time are the simple and the weighted mean values TL_m and TL_{mw}.

The following example shows how such an evaluation is carried out. Table 3.1 contains, in columns 1–3, a list of authentic data for one work center [19]. Column 1 contains the order number, Column 2 the input date ($tPEU$), i.e. the output date of the upstream work center, given as shop calendar day (SCD), Column 3 is the output date (tPE) of the work center concerned (as SCD), and Column 4 is the standard order time (TO) in standard hours taken from the routings. In this example, feedback data are only given to the day, which is common practice.

The first step in the calculation is to determine the lead time of each order by subtracting the input date from the output date. The result is given in Column 5 as the actual lead time (TL) in SCDs. From the sum of all lead time values divided by the number of orders, one obtains the actual simple mean lead time (TL_m) (284/20 = 14.2 days). As a by-product, simple mean order time (TO_m) is 146 hours/20 = 7.3 hours. (The meaning of Column 6 is explained later.)

Now lead time groups are created; here steps of 1 day are the most practical (Column 7). Using the values from Column 5, the orders are now classified into individual groups with the same value of TL (where applicable). For instance, Group 2 contains Order 125 and Order 145. Therefore, Column 8 states two occurrences,

Table 3.1. Sample calculation for simple and weighted actual lead time distribution, and for mean weighted lead times by setting up groups (in calendar days)

A) Table of Initial Values

Order Number (-)	Input tPEU (SCD)	Output tPE (SCD)	Order Time TO (Hrs)	Lead Time TL (SCD)	TOxTO (HrsxHrs)
1	2	3	4	5	6
115	181	206	0.5	25	0.3
119	194	207	11.4	13	130.0
110	170	208	15.4	38	237.2
125	206	208	3.8	2	14.4
120	194	209	6.8	15	46.2
124	205	212	9.6	7	92.2
118	195	213	7.7	18	59.3
127	208	213	5.3	5	28.1
121	199	214	8.8	15	77.4
126	208	219	2.6	11	6.8
131	213	220	2.1	7	4.4
116	187	220	7.4	33	54.8
108	202	222	13.8	20	190.4
117	198	222	4.2	24	17.6
135	219	226	6.9	7	47.6
132	215	226	13.8	11	190.4
140	222	227	5.1	5	26.0
123	207	229	13.6	22	185.0
145	227	229	5.1	2	26.0
142	226	230	2.1	4	4.4
20	< TOTAL >		146.0	284	1,438.5

B) Calculation of Group Values

Lead Time Group (Days)	Absolute Occurences (-)	Relative Occurences (%)	Absolute Work (Hrs)	Relative Work (%)	Group Value x Occurence (Days)	Group Value x Work (DaysxHrs
7	8	9	10	11	12	13
2	2	10	8.9	6.1	4	17.8
4	1	5	2.1	1.4	4	8.4
5	2	10	10.4	7.1	10	52.0
7	3	15	18.6	12.7	21	130.2
11	2	10	16.4	11.2	22	180.4
13	1	5	11.4	7.8	13	148.2
15	2	10	15.6	10.7	30	234.0
18	1	5	7.7	5.3	18	138.6
20	1	5	13.8	9.5	20	276.0
22	1	5	13.6	9.3	22	299.2
24	1	5	4.2	2.9	24	100.8
25	1	5	0.5	0.3	25	12.5
33	1	5	7.4	5.1	33	244.2
38	1	5	15.4	10.5	38	585.2
TOTAL >	20	100	146.0	100.0	284	2,427.5

Simple Mean Order Time: TOm = 146 Hrs/20 = 7.3 Hrs

Weighted Mean Order Time: TOmw = (1,438.5 Hrs x Hrs)/146 Hrs = 9.9 Hrs

Simple Mean Lead Time: TLm = 284 Days/20 = 14.2 Days

Weighted Mean Lead Time: TLmw = (2,427.5 Days x Hrs)/146 Hrs = 16.6 Days

Simple Mean Operation Time Percentage: *) TOPPm = 7.3 Hrs/(8 Hrs/Day x 14.2 Days) = 6.4 %

Weighted Mean Operation Time Percentage: *) TOPPmw = 9.9 Hrs/(8 Hrs/Day x 16.6 Days) = 7.5 %

*) With: Daily Capacity CDAY = 8 Hrs/Day

which amount to 10% of the total of 20 orders. The percentage is given in Column 9 (relative frequency of occurrences). Column 10 gives the total of the order times in the individual groups, which means for Group 2: 3.8 hrs (Order 125) plus 5.1 hrs (Order 145) = 8.9 hrs. These values can again be related to the total of all order times (146.0 hrs), thus giving the relative work content (Column 11). Column 12 gives the lead time per group, which is the product of Columns 7 and 8. The total of Column 12 is, as expected, 284 days, which is known as the sum of all lead times from Column 5. If one divides this sum by the number of all orders (total of Column 8), this equals the actual simple mean lead time, which is known to be 14.2 days.

To calculate the actual weighted mean lead time, the weighted lead times of each group are now determined in Column 13. The group value (lead time in days) must be multiplied by the total order time for this group (Column 7 × Column 10).

By dividing the total of Column 13 by the total of Column 10, one obtains the actual weighted mean lead time:

$$TL_{mw} = 2427.5 \text{ days} \times \text{hrs}/146 \text{ hrs} = 16.6 \text{ day}$$

This is greater than the actual simple mean lead time (TL_m), which is 14.2 days. The difference is not very great in this example. However, when there are many work centers in a shop, the differences are considerably greater.

One can now plot the values in a frequency diagram. Figure 3.15 shows the frequency distribution of the order quantities and order times in hours, using their absolute values (Columns 8 and 10 from Table 3.1). Actual simple and weighted mean lead times are also given. It is advisable to add the number of measured values and the totals of the actual lead times and the order times to the diagram, as well as the type and number of the work center and the registration period.

Relative frequency values express more than absolute frequency values (Columns

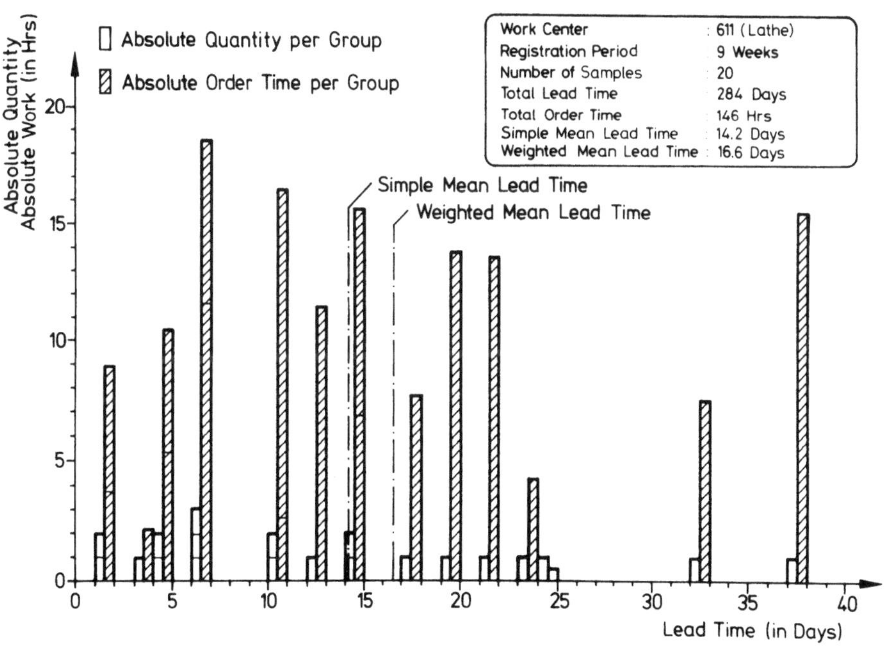

Fig. 3.15. Absolute frequency distribution of actual lead time in calendar days for a lathe work center (quantity and order time)

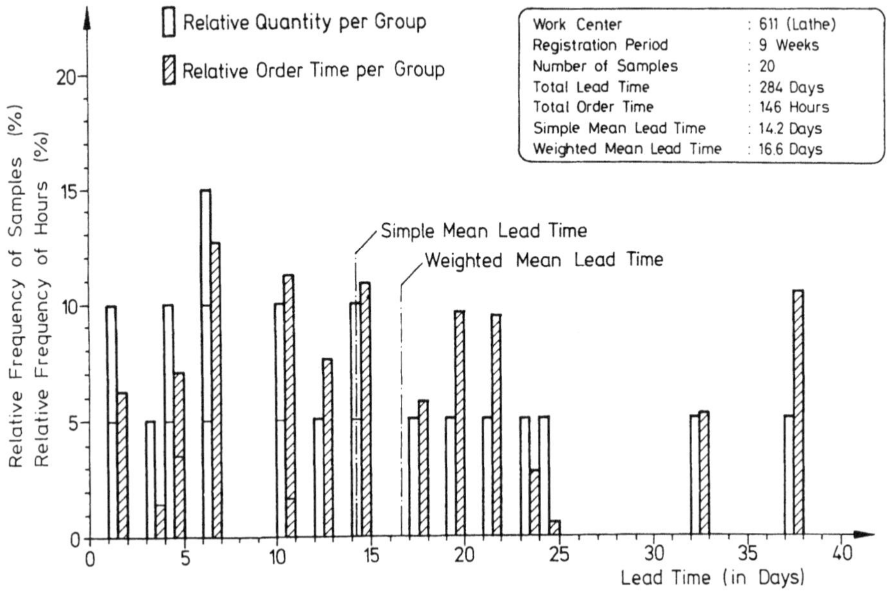

Fig. 3.16. Relative frequency distribution of actual lead time in calendar days for a lathe work center (quantity and order time)

9 and 11 in Table 3.1). Figure 3.16 gives the respective values. The comparison immediately makes clear how quantity and order time relate to each other in each group.

The following equations for calculating simple and weighted mean lead time when evaluated in lead time groups can then be formulated: for simple mean lead time:

$$TL_m = \frac{1}{n}(n_1 \times TLG_1 + n_2 \times TLG_2 + \cdots + n_k \times TLG_k)$$

$$TL_m = \frac{\sum_{i=1}^{k} n_i \times TLG_i}{n} \tag{3.17}$$

where:

k = number of lead time groups
TLG_i = top value of group i
n_i = number of values in group i
n = number of measured values.

For weighted mean lead time (cf. Eqn. (3.8)):

$$TL_{mw} = \frac{TLG_1 \times TOT_1 + TLG_2 \times TOT_2 + \cdots + TLG_k \times TOT_k}{TOT_1 + TOT_2 + \cdots + TOT_k}$$

$$TL_{mw} = \frac{\sum_{i=1}^{k} (TLG_i \times TOT_i)}{\sum_{i=1}^{k} TOT_i} \tag{3.18}$$

where:

k = number of lead time groups
TLG_i = top value of group i
TOT_i = total of order times in group i.

3.6.2 Simple and Weighted Mean Operation Time

Operation time was defined in Sect. 3.3 (Eqn. (3.3)) as:

$$TOP = \frac{TO}{CDAY \times ER}$$

where:

TO = order time
$CDAY$ = daily capacity
ER = efficiency rate.

As was shown in Eqn. (3.4), this equation can be simplified as:

$$TOP = \frac{TO}{CDAY}$$

In Fig. 3.17A, this case is demonstrated with a simple example, the values for the lead times being the same as in Fig. 3.10. The general equation for simple mean operation time at a work center is:

$$TOP_m = \frac{1}{n}(TOP_1 + TOP_2 + \cdots + TOP_n)$$

$$TOP_m = \frac{1}{n}\sum_{i=1}^{n} TOP_i \tag{3.19a}$$

where:

TOP_i = individual operation time value
n = number of measured values.

With $TOP = \frac{TO}{CDAY}$ (Eqn. (3.4)) and C_{DAY} being constant:

$$TOP_m = \frac{1}{n}\left(\frac{TO_1}{CDAY} + \frac{TO_2}{CDAY} + \cdots + \frac{TO_n}{CDAY}\right)$$

$$TOP_m = \frac{1}{n \times CDAY} \times \sum_{i=1}^{n} TO_i \tag{3.19b}$$

where:

TO_i = individual order time value
n = number of measured values
$CDAY$ = daily capacity.

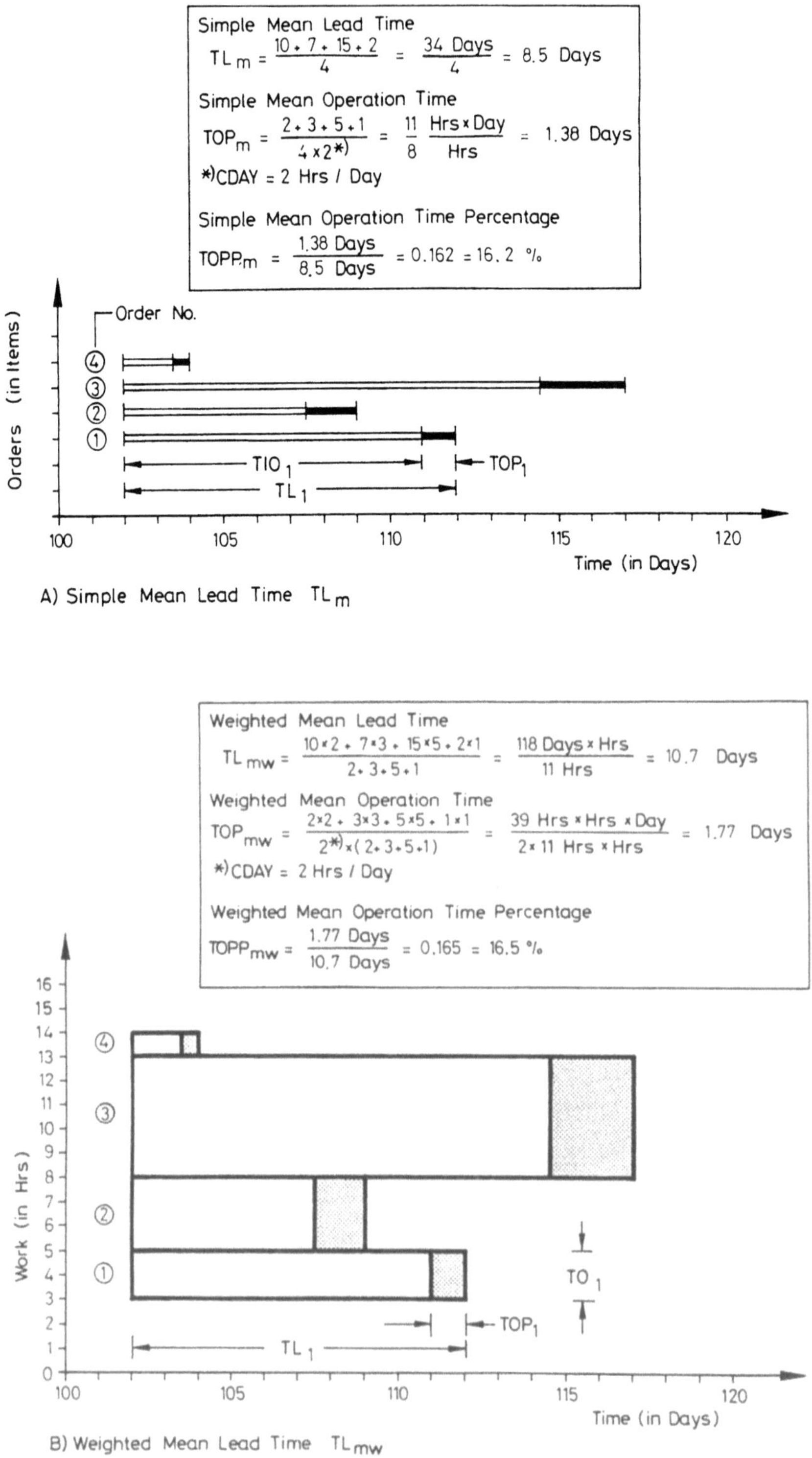

Fig. 3.17. Calculation of simple and weighted mean actual lead time at a work center

In Fig. 3.17A, TOP_m equals 1.38 days. For the values in Table 3.1, assuming a daily capacity of 8 hrs/day:

$$TOP_m = \frac{146 \text{ hrs} \times \text{day}}{20 \times 8 \text{ hrs}} = 0.9 \text{ days}$$

As with weighted mean lead time (TL_{mw}), one can also define a weighted mean operation time (TOP_{mw}).

The weighted operation time (TOP_w) of an operation, given an efficiency rate of 1, is calculated as:

$$TOP_w = \frac{TO}{CDAY} \times TO \tag{3.20}$$

where:

TO = order time
$CDAY$ = daily capacity.

The weighted mean operation time (TOP_{mw}) is usually given as:

$$TOP_{mw} = \frac{TOP_1 \times TO_1 + TOP_2 \times TO_2 + \cdots + TOP_n \times TO_n}{TO_1 + TO_2 + \cdots + TO_n}$$

$$TOP_{mw} = \frac{\sum_{i=1}^{n} TOP_i \times TO_i}{\sum_{i=1}^{n} TO_i} \tag{3.21}$$

where:

TOP_i = operation time of operation i
TO_i = order time of operation i
n = number of measured values.

Using the equation:

$$TO_P = \frac{TO}{CDAY}$$

Eqn. (3.21) can be converted into:

$$TOP_{mw} = \frac{TO_1 \times TO_1 + TO_2 \times TO_2 + \cdots + TO_n \times TO_n}{CDAY(TO_1 + TO_2 + \cdots + TO_n)}$$

$$TOP_{mw} = \frac{\sum_{i=1}^{n} (TO_i)^2}{CDAY \times \sum_{i=1}^{n} TO_i} \tag{3.22}$$

where:

TO_i = order time of operation i
n = number of measured values
$CDAY$ = daily capacity.

According to this equation, the weighted mean operation time from the example in Fig. 3.17B equals 1.77 days. As opposed to lead time, weighted mean operation time is, with constant daily capacity, equal to or greater than, but never less than, the simple mean value, and only depends on the simple mean value of operation time and on the variance in the individual values.

In the example in Table 3.1, weighted mean order time (TO_{mw}) was first determined to find out TOP_{mw}. Column 6 contains the value of TO (Column 4) multiplied by itself. The total of these values divided by the total of the non-weighted order times equals a TO_{mw} of 9.9 hours. With $CDAY = 8$ hrs/day, weighted mean operation time (TOP_{mw}) then equals 9.9 / 8 = 1.2 days.

3.6.3 Simple and Weighted Mean Operation Time Percentage

Another key factor in the evaluation of lead times is the ratio of lead time to operation time, the so called operation time percentage ($TOPP$). It is defined as:

$$TOPP = \frac{TOP}{TL} \tag{3.23}$$

where:

TOP = operation time
TL = lead time.

To obtain a dimensionless number, TOP must be given in the same dimension as TL, e.g. in days. If TOP has not been provided in days, it must be converted from process time. The simple mean operation time percentage ($TOPP_m$) is then:

$$TOPP_m = \frac{(TOP_1 + TOP_2 + \cdots + TOP_n) \times \frac{1}{n}}{(TL_1 + TL_2 + \cdots + TL_n) \times \frac{1}{n}}$$

$$TOPP_m = \frac{\sum_{i=1}^{n} TOP_i}{\sum_{i=1}^{n} TL_i} \tag{3.24}$$

$$TOPP_m = \frac{TOP_m}{TL_m} \tag{3.25}$$

where:

TOP_i = individual operation time value
TL_i = individual lead time value
n = number of measured values.

According to the definition of weighted mean operation time, the weighted mean operation time percentage ($TOPP_{mw}$) can be defined as:

$$TOPP_{\mathrm{mw}} = \frac{\frac{1}{n}(TOP_1 \times TO_1 + TOP_2 \times TO_2 + \cdots + TOP_n \times TO_n)}{\frac{1}{n}(TL_1 \times TO_1 + TL_2 \times TO_2 + \cdots + TL_n \times TO_n)}$$

$$= \frac{\sum_{i=1}^{n} (TOP_i \times TO_i)}{\sum_{i=1}^{n} (TL_i \times TO_i)}$$

If one reduces $\sum TO_i$ in this expression then, with Eqns. (3.8) and (3.21),

$$TOPP_{\mathrm{mw}} = \frac{TOP_{\mathrm{mw}}}{TL_{\mathrm{mw}}} \tag{3.26}$$

where:

TOP_i = individual operation time value
TO_i = individual order time value
TL_i = individual lead time value
n = number of measured values.

For the example in Fig. 3.17 it follows that:

$$TOPP_{\mathrm{m}} = 1.38/8.5 = 0.162 = 16.2\% \text{ and}$$
$$TOPP_{\mathrm{mw}} = 1.77/10.7 = 0.165 = 16.5\%.$$

For the example in Table 3.1, $TOPP_{\mathrm{m}}$ and $TOPP_{\mathrm{mw}}$ for the 20 values are then 6.4% and 7.5%, respectively.

As will be shown in Sect. 3.7, lead time components do not tell us very much when they are related to calendar days, since it is obviously pure chance whether a period of operation time and of interoperation time is interrupted by non-productive days (such as Sundays, holidays), or not. Depending on how the non-productive days are distributed, TL and TOP, and thus $TOPP$, can have totally different values.

We must further point out a peculiarity in the calculation of mean lead time components for the values in Table 3.1. The reason why the operation time (TOP) can not be calculated as the difference between process end (tPE) and setup begin (tSB), but from order time (TO) and daily capacity ($CDAY$), was given in Sect. 3.3. This means, however, that idle times during non-productive days, which can occur during interoperation time as well as when setting up and processing, are neglected when determining order time and thus lead time. The values calculated in Table 3.1 for simple and weighted operation times are therefore 'artificial' values which do not, in this form, exist in practice, whereas the lead time value does exist in reality, because it is based on the difference between two measured dates. This is another important reason why simple and weighted operation time percentages, which refer to calendar days, should not be used in this form. If one still wants to use them, one must either calculate them on the basis of the measured dates $tPEU$ and tSB, or determine the real setup begin date from the shop calendar and from the pure operation time calculated from the order time.

3.6.4 Median of Simple and Weighted Lead Time

In frequency distribution studies it is common to determine, besides the mean value, three more statistical values known as median, standard deviation, and coefficient of variation.

The median (middle value) is the value which lies in the middle of the distribution. It is defined as:

$$x(n/2 + 1/2), \quad \text{if } n \text{ is an odd number}$$

$$\tfrac{1}{2}(x(n/2) + x(n/2 + 1)) \quad \text{if } n \text{ is an even number} \tag{3.27}$$

To calculate the actual median of simple lead time (TL_{med}), the values are normally arranged by increasing order of magnitude. As an example, the measured values from Table 3.1 have been arranged by increasing lead time and copied into Table 3.2 (Columns 1 to 3).

Since there is an even number of values, it follows from Eqn. (3.27) that:

$$TL_{\text{med}} = 1/2\,(TL_{10} + TL_{11})$$

From the total in Column 4 in Table 3.2, we can see that the 10th value has the lead time of 11 days and the 11th value 13 days. From this it follows that:

$$TL_{\text{med}} = 1/2\,(11 \text{ days} + 13 \text{ days}) = 12 \text{ days}$$

Table 3.2. Sample calculation for the median and the standard deviation of simple and weighted actual lead times (in calendar days)

Order Number (-)	Lead Time TL (Days)	Order TIME TO (Hrs)	Cumulative Value (-)	Cumulative Order Time TO (Hrs)	(TLm-TLi)² (Days²)	(TLmw-TLi)² x TO (Days² x Hrs)
1	2	3	4	5	6	7
125	2	3.8	1	3.8	148.8	810.0
145	2	5.1	2	8.9	148.8	1,087.1
142	4	2.1	3	14.0	104.0	333.4
127	5	5.3	4	16.3	84.6	713.2
140	5	5.1	5	21.4	84.6	686.3
124	7	9.6	6	31.0	51.8	884.7
131	7	2.1	7	33.1	51.8	193.5
135	7	6.9	8	40.0	51.8	635.9
126	11	2.6	9	42.6	10.2	81.5
132	11	13.8	10	56.4	10.2	432.8
119	13	11.4	11	67.8	1.4	147.7
120	15	6.8	12	74.6	0.6	17.4
121	15	8.8	13	83.4	0.6	22.5
118	18	7.7	14	91.1	14.4	15.1
108	20	13.8	15	104.9	33.6	159.5
123	22	13.6	16	118.5	60.8	396.6
117	24	4.2	17	122.7	96.0	230.0
115	25	0.5	18	123.2	116.6	35.3
116	33	7.4	19	130.6	353.4	1,990.3
110	38	15.4	20	146.0	566.4	7,052.6
TOTAL >	284	146.0	20	146.0	1,990.4	15,925.4

Median of Simple Lead Time: TLmed = (11 Days + 13 Days)/2 =12.0 Days

Median of Weighted Lead Time: TLmedw = (15 Days + 15 Days)/2 = 15.0 Days

Standard Deviation of Simple Lead Time: TLS $= \sqrt{1{,}990.4 \text{ Days}^2/20} = 10.0$ Days

Standard Deviation of Weighted Lead Time: TLSw $= \sqrt{(15{,}925.4 \text{ Days}^2 \times \text{Hrs})/146.0 \text{ Hrs}} = 10.4$ Days

From the difference between this and the arithmetic mean (14.2 days), we can see that the distribution is not symmetrical.

To determine the median of actual weighted lead time (TL_{medw}), one has to find the value which lies in the middle of the distribution of weighted lead time. For this, the overall order time was calculated in Column 5 of Table 3.2. This means that each value in the column is the sum of all values above it. Proceeding according to Eqn. (3.27), we must make n not the number of orders but the number of hours. Since $n = 146$ (even), $TL_{medw} = 1/2\,(TL_{73} + TL_{74})$.

The 73rd order hour lies – as can be seen in Column 5 of Table 3.2 – in the lead time group of up to 15 days, as does the 74th hour. For the median of weighted lead time it follows that.

$$TL_{medw} = 1/2\,(15 \text{ days} + 15 \text{ days}) = 15 \text{ days}$$

3.6.5 Standard Deviation of Simple and Weighted Mean Lead Time

The standard deviation of a series of measurements is defined as:

$$S = \sqrt{\frac{1}{n}\sum_{i=1}^{n}(\overline{X} - X_i)^2} \tag{3.28}$$

where:

n = number of measured values
X = arithmetic mean of all values
X_i = individual value.

The standard deviation of actual simple lead time is then calculated as:

$$TLS = \sqrt{\frac{1}{n}\sum_{i=1}^{n}(TL_m - TL_i)^2} \tag{3.29}$$

When determining actual weighted lead time, each value must, according to the definition, be multiplied by the order time, n being the number of operations. For the standard deviation of weighted lead time it follows that:

$$TLS_w = \sqrt{\frac{\sum_{i=1}^{n}(TL_{mw} - TL_i)^2 \times TO_i}{\sum_{i=1}^{n} TO_i}} \tag{3.30}$$

where:

TL_i = simple lead time of order i
TO_i = order time of order i
TL_m = simple mean lead time
TL_{mw} = weighted mean lead time
n = number of measured values.

Columns 6 and 7 of Table 3.2 give the individual values of $(TL_m - TL_i)^2$ and $(TL_{mw} - TL_i)^2 \times TO_i$, respectively. According to Eqns. (3.29) and (3.30), this results in standard deviations TLS of 10.0 days and TLS_w of 10.4 days, respectively.

3.6.6 Coefficient of Variation of Simple and Weighted Lead Time

The third important value in a distribution is the coefficient of variation. For a series of measurements, it is defined as:

$$V = \frac{S}{\overline{X}} \tag{3.31}$$

where:

S = standard deviation of values 1 to n
$\overline{X}$ = arithmetic mean of values 1 to n.

The coefficient of variation of actual simple lead time

$$TLV = \frac{TLS}{TL_m} \tag{3.32a}$$

where:

TLS = standard deviation of simple lead time
TL_m = simple mean lead time.

and the coefficient of variation of actual weighted lead time

$$TLV_w = \frac{TLS_w}{TL_{mw}} \tag{3.32b}$$

where:

TLS_w = standard deviation of weighted lead time
TL_{mw} = weighted mean lead time.

With the values from Table 3.2 these will be.

$$TLV = 10.0/14.2 = 0.70$$

$$TLV_w = 10.4/16.6 = 0.63$$

3.6.7 Median, Standard Deviation, and Coefficient of Variation of Simple and Weighted Order Time

As with lead time, these values can also be determined for simple and weighted order time. Table 3.3 shows the respective values of the series of measurements from Table 3.1. This time, they are arranged by increasing order time (TO).

The median is determined according to Eqn. (3.27):

$$TO_{med} = 1/2\,(TO_{10} + TO_{11}) = 1/2\,(6.8 \text{ hrs} + 6.9 \text{ hrs}) = 6.85 \text{ hrs}$$

For the median of weighted order time we must find the value of TO which corresponds to order time Hours 73 and 74. From Column 4 (total order time), we can see that both values correspond to order no. 124, i.e. TO of 9.6 hrs.

$$TO_{medw} = 1/2\,(T_{O73} + T_{O74}) = 1/2\,(9.6 \text{ hrs} + 9.6 \text{ hrs}) = 9.6 \text{ hrs}$$

For the standard deviation of the simple order time (according to Eqn. (3.28)):

Table 3.3. Sample calculation for median, standard deviation, and variation coefficient of simple and weighted order times

Order Number (-)	Order Time TO (Hrs)	Cumulative Value (-)	Cumulative Order (Hrs)	TO x TO (Hrs²)	(TOm-TOi)² (Hrs²)	(TOmw-TOi)² x TO (Hrs²)
1	2	3	4	5	6	7
115	.5	1	.5	.3	43.7	43.7
131	2.1	2	2.6	4.4	27.0	126.2
142	2.1	3	4.7	4.4	27.0	126.2
126	2.6	4	7.3	6.8	22.1	136.8
125	3.8	5	11.1	14.4	12.3	139.2
117	4.2	6	15.3	17.6	9.6	134.2
140	5.1	7	20.4	26.0	4.8	115.2
145	5.1	8	25.5	26.0	4.8	115.2
127	5.3	9	30.8	28.1	4.0	109.9
120	6.8	10	37.6	46.2	.3	63.4
135	6.9	11	44.5	47.6	.2	60.2
116	7.4	12	51.9	54.8	.0	44.5
118	7.7	13	59.6	59.3	.2	35.7
121	8.8	14	68.4	77.4	2.3	9.8
124	9.6	15	78.0	92.2	5.3	0.6
119	11.4	16	89.4	130.0	16.8	27.3
123	13.6	17	103.0	185.0	39.7	191.0
108	13.8	18	116.8	190.4	42.3	215.0
132	13.8	19	130.6	190.4	42.3	215.0
110	15.4	20	146.0	237.2	65.6	473.9
TOTAL >	146.0	20	146.0	1,438.5	372.8	2,382.9

Median of Simple Order Time:	TOmed	= (6.8 Hrs + 6.9 Hrs)/2 = 6.85 Hrs
Median of Weighted Order Time:	TOmedw	= (9.6 Hrs + 9.6 Hrs)/2 = 9.6 Hrs
Simple Mean Order Time:	TOm	= 146 Hrs/20 = 7.3 Hrs
Weighted Mean Order Time	TOmw	= 1,438.5 Hrs²/146 Hrs = 9.9 Hrs
Standard Deviation of Simple Order Time:	TOs	= $\sqrt{372.8\ \text{Hrs}^2/20}$ = 4.3 Hrs
Standard Deviation of Weighted Order Time:	TOsw	= $\sqrt{2{,}383.5\ \text{Hrs}/146\ \text{Hrs}}$ = 4.0 Hrs
Coefficient of Variation of Simple Order Time:	TOv	= 4.3 Hrs/7.3 Hrs = .58 Hrs
Coefficient of Variation of Weighted Order Time:	TOvw	= 4.0 Hrs/9.9 Hrs = .40 Hrs

$$TOS = \sqrt{\frac{1}{n}\sum_{i=1}^{n}(TO_{\mathrm{m}} - TO_i)^2} \tag{3.33}$$

and for the standard deviation of weighted order time:

$$TOS_{\mathrm{w}} = \sqrt{\frac{\sum_{i=1}^{n}(TO_{\mathrm{mw}} - TO_i)^2 \times TO_i}{\sum_{i=1}^{n} TO_i}} \tag{3.34}$$

where:

TO_i = order time of order i
TO_{m} = simple mean order time
TO_{mw} = weighted mean order time.

Table 3.3 contains the calculations necessary for TOS (Column 6) and TOS_{w} (Column 7), which produces values of 4.3 hrs for the standard deviation of mean order times and 4.0 hrs for the standard deviation of weighted order time.

The coefficient of variation of simple and weighted operation time (according to Eqn. (3.31)) is:

$$TOV = \frac{TO_S}{TO_m} \tag{3.35}$$

and

$$TOV_w = \frac{TOS_w}{TO_{mw}} \tag{3.36}$$

where:

TOS = standard deviation of simple order time
TOS_w = standard deviation of weighted order time
TO_m = simple mean order time
TO_{mw} = weighted mean order time.

The corresponding values are given in Table 3.3: $TOV = 0.58$ and $TOV_w = 0.40$.

3.7 Work Center Lead Time in the Shop Calendar

3.7.1 Transformation of Throughput Elements

When discussing operation time and operation time percentage, it becomes clear that besides calendar time, working days (commonly called shop calendar days [20]) are also important. When calculating actual lead time in calendar days it is mere chance whether or not there are non-productive days such as Sundays contained in the lead time of an order. Thus, mean lead time values are of course arbitrary, which has particularly undesirable results when several of these days come in succession, such as Christmas holidays or a holiday plus a weekend.

For this reason, lead times are usually calculated not in calendar days, but in shop calendar days (SCD). There are, however, some cases where lead time values in calendar days can also be informative; namely, when one wants to measure the capital tied up in the orders. This is because interest is charged for calendar days.

If one wishes to relate lead times to workdays, each throughput element must be transformed from the common calendar into the calendar of the individual shop. Figure 3.18 shows a simple example of this transformation.

Case A shows a reduced two-dimensional throughput element over a 7-days-per-week calendar. Each weekday has a calendar day number. We can see that (by chance) there are seven non-productive days in the lead time period of this element. If we now 'cut out' these days we get a shortened throughput element (B) with 13 shop calendar days (SCD) instead of 20 days (A). Of course, the dates $tPEU$, tSB, and tPE must always fall on a shop calendar day.

This is another problem in calculating lead time in calendar days. If one calculated the operation time to be the difference between tPE and tSB, this would equal five days. The real processing of the operation is, however, performed in only three days, as can be seen from the dark areas representing TOP_a and TOP_b. If one then calculated tSB to be the difference between tPE and TOP, setup begin would be on

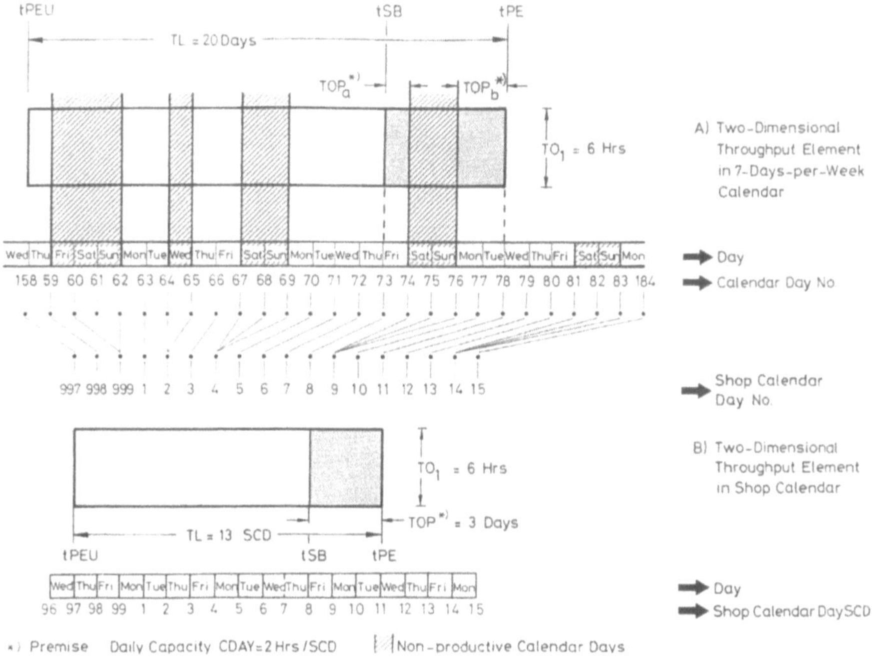

Fig. 3.18. Transformation of a two-dimensional throughput element from the 7-Days per-Week calendar (A) into the shop calendar (B)

Sunday morning, which is impossible. This problem disappears when the throughput element is transformed according to the shop calendar. Interoperation time and operation time are now in continuous succession. They can only be interrupted by breaks in production, which is not accounted for in normal lead time studies. Only when the average machine down time percentage of manufacturing lead time reaches the order of magnitude of the average operation time percentage, does one have to account for disruptions. This can be the case with highly automated facilities, but this case will be neglected for the time being.

3.7.2 Frequency Distribution and Statistical Evaluation in the Shop Calendar

The feedback data contained in Table 3.1 will now be converted to suit the shop calendar, and statistically evaluated. For this we need a shop calendar. Figure 3.19 shows the months January to March of a shop calendar for an arbitrary year. It gives calendar days, weekdays, a calendar day number created especially for this example, the workdays per month, the serial number of each week within the year and, most importantly, the serial number of the workday, called the shop calendar day (SCD). We start counting with number one, which is the first workday of the year (here: Monday Jan 4th, Calendar Day 163).

01 January					02 February					03 March				
Calendar Day	Calendar Day No.	Workdays	Week	Monthly Serial No.	Calendar Day	Calendar Day No.	Workdays	Week	Monthly Serial No.	Calendar Day	Calendar Day No.	Workdays	Week	Monthly Serial No.
1 Fri	160	New Year's Day			1 Mon	191	1	5	20	1 Mon	219	1	9	40
2 Sat	161				2 Tue	192	2		21	2 Tue	220	2		41
3 Sun	162				3 Wed	193	3		22	3 Wed	221	3		42
4 Mon	163	1	1	1	4 Thu	194	4		23	4 Thu	222	4		43
5 Tue	164	2		2	5 Fri	195	5		24	5 Fri	223	5		44
6 Wed	165	Local Holiday			6 Sat	196				6 Sat	224			
7 Thu	166	3		3	7 Sun	197				7 Sun	225			
8 Fri	167	4		4	8 Mon	198	6	6	25	8 Mon	226	6	10	45
9 Sat	168				9 Tue	199	7		26	9 Tue	227	7		46
10 Sun	169				10 Wed	200	8		27	10 Wed	228	8		47
11 Mon	170	5	2	5	11 Thu	201	9		28	11 Thu	229	9		48
12 Tue	171	6		6	12 Fri	202	10		29	12 Fri	230	10		49
13 Wed	172	7		7	13 Sat	203				13 Sat	231			
14 Thu	173	8		8	14 Sun	204				14 Sun	232			
15 Fri	174	9		9	15 Mon	205	11	7	30	15 Mon	233	11	11	50
16 Sat	175				16 Tue	206	12		31	16 Tue	234	12		51
17 Sun	176				17 Wed	207	13		32	17 Wed	235	13		52
18 Mon	177	10	3	10	18 Thu	208	14		33	18 Thu	236	14		53
19 Tue	178	11		11	19 Fri	209	15		34	19 Fri	237	15		54
20 Wed	179	12		12	20 Sat	210				20 Sat	238			
21 Thu	180	13		13	21 Sun	211				21 Sun	239			
22 Fri	181	14		14	22 Mon	212	16	8	35	22 Mon	240	16	12	55
23 Sat	182				23 Tue	213	17		36	23 Tue	241	17		56
24 Sun	183				24 Wed	214	18		37	24 Wed	242	18		57
25 Mon	184	15	4	15	25 Thu	215	19		38	25 Thu	243	19		58
26 Tue	185	16		16	26 Fri	216	20		39	26 Fri	244	20		59
27 Wed	186	17		17	27 Sat	217				27 Sat	245			
28 Thu	187	18		18	28 Sun	218				28 Sun	246			
29 Fri	188	19		19						29 Mon	247	21	13	60
30 Sat	189									30 Tue	248	22		61
31 Sun	190									31 Wed	249	23		62

Fig. 3.19. Extract from a shop calendar (calendar days versus shop calendar days)

Table 3.4 assigns the input and output dates from the example in Table 3.1 with the SCDs from Fig. 3.19. From this, the actual simple and weighted mean lead times have been calculated as in Table 3.1 which, as expected, results in lower values, namely a TL_m of 10.1 SCD as against 14.2 days, and a TL_{mw} of 11.8 SCD as against 16.6 days.

Compared with a calculation in calendar days (cf. Table 3.1), the simple and weighted operation time percentages are higher in this example: $TOPP_m = 9.0\%$ (6.4%) and $TOPP_{mw} = 10.5\%$ (7.5%), since the mean lead time is shorter in the shop calendar.

As in Figs. 3.15 and 3.16, we can now develop diagrams from Table 3.4 for the absolute and relative frequency distributions of orders and their operation times per group. Figure 3.20 shows only the relative frequency diagram. Compared to Figure 3.16, it has fewer lead time groups and a different distribution.

Calculation of the statistical values median, standard deviation, and coefficient of variation is basically the same. The equations shown in Sect. 3.6 can be used.

The results are given in Table 3.5. They are also contained in Fig. 3.20 to allow a comparison with the respective values in calendar days.

Table 3.4. Sample calculation of simple and weighted actual lead times and their relative frequency distribution by setting up groups (in SCDs)

A) Table of Initial Values

Order Number (-)	Input tPEU (SCD)	Output tPE (SCD)	Order Time TO (Hrs)	Lead Time TL (SCD)
1	2	3	4	5
115	14	31	.5	17
119	23	32	11.4	9
110	5	33	15.4	28
125	31	33	3.8	2
120	23	34	6.8	11
124	30	35	9.6	5
118	24	36	7.7	12
127	33	36	5.3	3
121	26	37	8.8	11
126	33	40	2.6	7
131	36	41	2.1	5
116	18	41	7.4	23
108	29	43	13.8	14
117	25	43	4.2	18
135	40	45	6.9	5
132	38	45	13.8	7
140	43	46	5.1	3
123	32	48	13.6	16
145	46	48	5.1	2
142	45	49	2.1	4
20	< TOTAL >		146.0	202

B) Calculation of Group Values

Lead Time Group (SCD)	Absolute Occurrence (-)	Relative Occurrence (%)	Absolute Work (Hrs)	Relative Work (-)	Group Value x Work (SCD x Hrs)
6	7	8	9	10	11
2	2	10.0	8.9	6.1	17.8
3	2	10.0	10.4	7.1	31.2
4	1	5.0	2.1	1.4	8.4
5	3	15.0	18.6	12.7	93.0
7	2	10.0	16.4	11.2	114.8
9	1	5.0	11.4	7.8	102.6
11	2	10.0	15.6	10.7	171.6
12	1	5.0	7.7	5.3	92.4
14	1	5.0	13.8	9.5	193.2
16	1	5.0	13.6	9.3	217.6
17	1	5.0	.5	.3	8.5
18	1	5.0	4.2	2.9	75.6
23	1	5.0	7.5	5.1	170.2
28	1	5.0	15.4	10.5	431.2
TOTAL >	20	100.0	146.0	100.0	1,728.1

Simple Mean Lead Time: TLm = 202 SCD/20 = 10.1 SCD

Weighted Mean Lead Time: TLmw = (1,728.1 SCD x Hrs)/146 Hrs = 11.8 SCD

Simple Mean Operation Time Percentage: *) TOPPm = 7.3 Hrs/(8 Hrs/SCD x 10.1 SCD) = 9.0 %

Weighted Mean Operation Time Percentage: *) TOPPmw = 9.9 Hrs/(8 Hrs/SCD x 11.8 SCD) = 10.5 %

*) With: Daily Capacity CDAY = 8 Hrs/Day

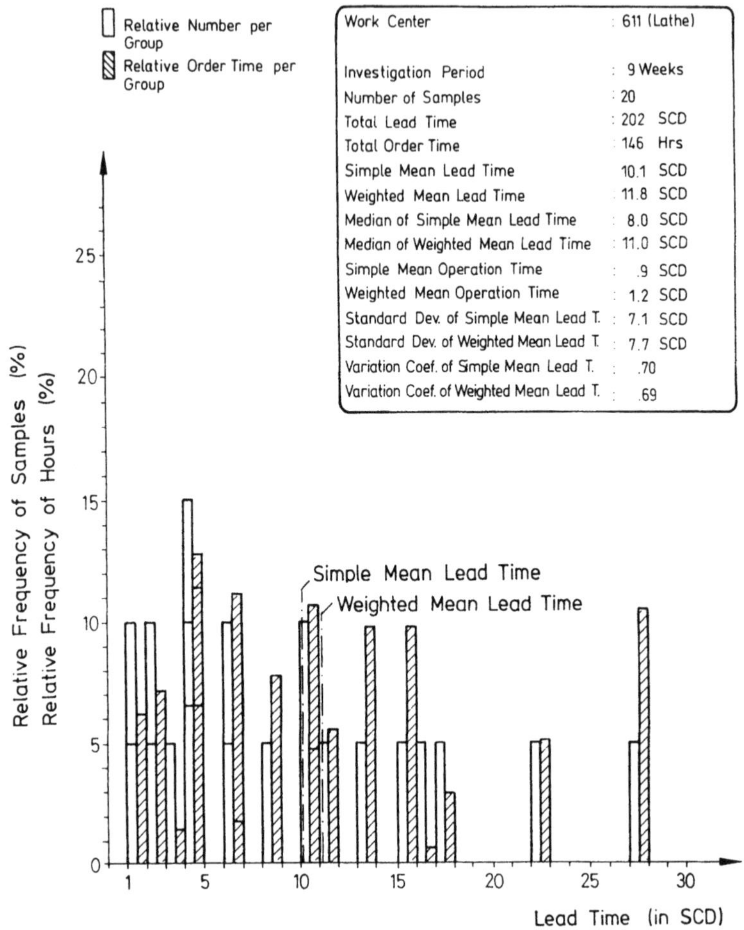

Fig. 3.20. Relative frequency distribution of actual lead time at a work center in numbers of operations and in order time (in shop calendar time brackets)

3.8 Uncertainty of Measurement and Accuracy of Computed Lead Time Values

To evaluate measured values, one must first make an accuracy check before interpreting and further using them. With actual lead times, it is useful to know how the values on which statistical evaluations are based have been calculated.

Figure 3.21 (top) shows the throughput of a few orders at one work center. The defined output dates are not real physical movements, but reported dates of the upstream work center which finishes the previous operation. These dates are, within certain limits, left to the judgement of the reporting person (be it the worker at the machine, the foreman, or the dispatcher). Consequently, a date recorded in this way probably deviates from the actual date of operation end. This deviation is mainly influenced by the individual wage system, and by the corresponding employment agreement.

Table 3.5. Sample calculation for the median and the standard deviation of simple and weighted actual lead times (in SCDs)

Order Number (-)	Lead Time TL (SCD)	Order TIME TO (Hrs)	Cumulative Value (-)	Cumulative Order Time TO (Hrs)	$(TLm-TLi)^2$ (SCD^2)	$(TLmw-TLi)^2 x TO$ (SCD^2 x Hrs)
1	2	3	4	5	6	7
125	2	3.8	1	3.8	65.6	365.0
145	2	5.1	2	8.9	65.6	489.8
140	3	5.1	3	14.0	50.4	394.9
127	3	5.3	4	19.3	50.4	410.4
142	4	2.1	5	21.4	37.2	127.8
131	5	2.1	6	23.5	26.0	97.1
124	5	9.6	7	33.1	26.0	443.9
135	5	6.9	8	40.0	26.0	319.1
132	7	13.8	9	53.8	9.6	318.0
126	7	2.6	10	56.4	9.6	59.9
119	9	11.4	11	67.8	1.2	89.4
120	11	6.8	12	74.6	.8	4.4
121	11	8.8	13	83.4	.8	5.6
118	12	7.7	14	91.1	3.6	.3
108	14	13.8	15	104.9	15.2	66.8
123	16	13.6	16	118.5	34.8	239.9
115	17	.5	17	119.0	47.6	13.5
117	18	4.2	18	123.2	62.4	161.4
116	23	7.4	19	130.6	166.4	928.3
110	28	15.4	20	146.0	320.4	4,041.6
TOTAL >	202	146.0	20	146.0	1,019.6	8,577.1

Median of Simple Lead Time: TLmed = (7 SCD + 9 SCD)/2 = 8.0 SCD

Median of Weighted Lead Time: TLmedw = (11 SCD + 11 SCD)/2 = 11.0 SCD

Standard Deviation of Simple Lead Time: $TLS = \sqrt{1{,}019.6\ SCD^2/20} = 7.1\ SCD$

Standard Deviation of Weighted Lead Time: $TLSw = \sqrt{(8{,}577.1\ SCD^2 \times Hrs)/146\ Hrs} = 7.7\ SCD$

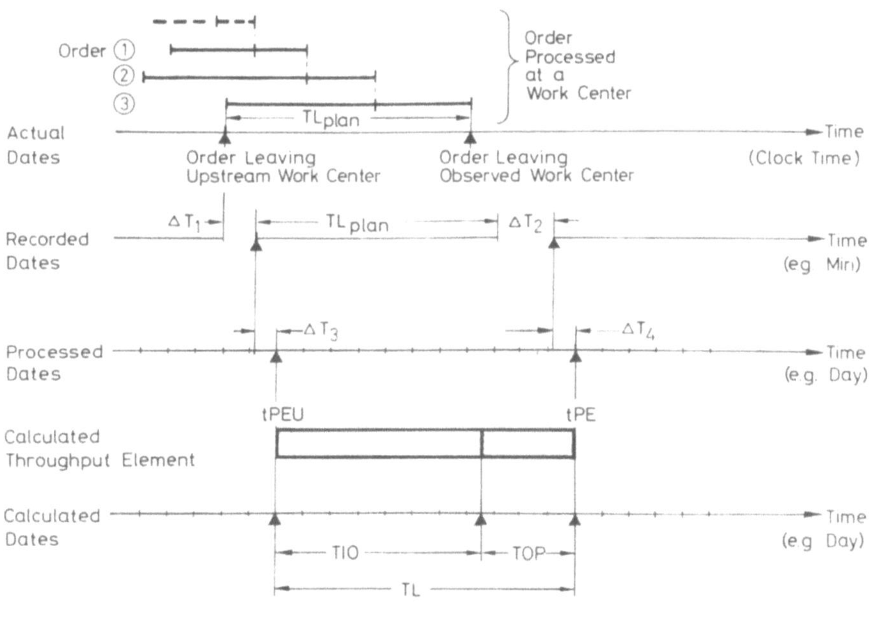

Fig. 3.21. Influential factors on accuracy in throughput-element calculation

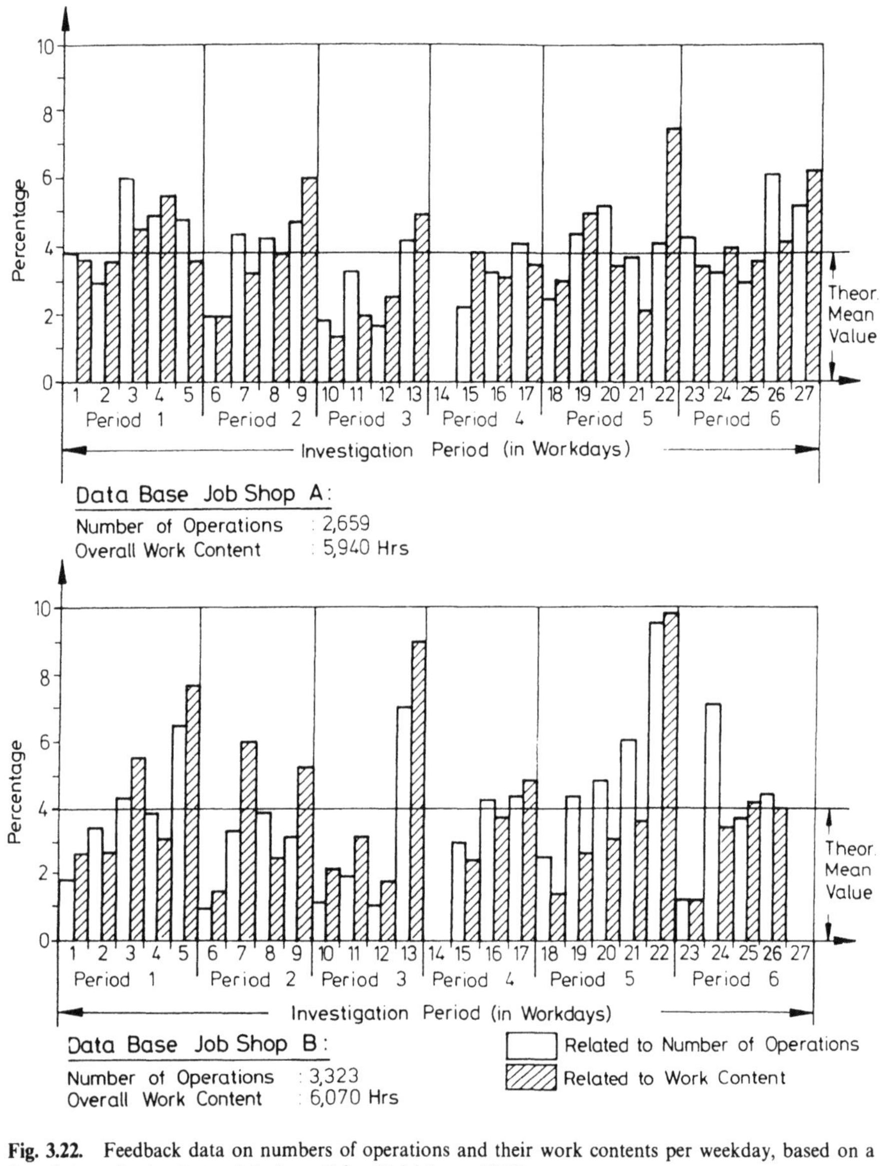

Fig. 3.22. Feedback data on numbers of operations and their work contents per weekday, based on a 6-week investigation in two job shops (After Holzkämper [21])

Figure 3.22 shows interesting results of an investigation in two job shops of the same mechanical engineering company [21]. The work reported each day was plotted in the number of orders and in the hours of work. In Job Shop A, feedback of dates and quantities of work (numbers of hours) were strictly separated. It was, within a certain scope, left to the worker when to charge the standard hours of his reported orders. In Job Shop B, however, standard hours were booked onto the wages account together with the feedback of the dates. This causes the typical 'lump wages effect': as compared with the theoretical mean value (100%/26 workdays = 3.8% in Case A, 100%/25 workdays = 4% in Case B), hours were reported with unusual cumulation at the weekends. One consequence of such feedback inaccuracy is an increase in mean lead time by about one or two days per operation, which means an increase in order lead time by one or two weeks, given a mean of five operations per order.

When using feedback dates, one must consequently take into account more or less large deviations from actual dates. As Fig. 3.21 indicates, the dates are, however, frequently altered in a second respect.

For lead time scheduling purposes, the reported dates are, as a rule, not given to the minute, but, depending on individual shop practice, rounded up to full hours or days, and then used in this form. Accuracy to the day is common practice, which means that all feedback data occurring during one day are given the feedback date of that day, irrespective of the hour and minute. This is also the case with the feedback data in our example in Table 3.1. This rounding up causes another deviation which, given arbitrarily distributed dates of reported outputs, is on average one half of a working day. Both processed dates are, for the reasons shown above, rounded into the future.

In calculating lead time, the deviations resulting from feedback inaccuracy have the consequence that, with each operation, the waiting time TL_{plan} starts by ΔT_1 later and ends correspondingly later at the next work center. The delay in feedback by ΔT_2 thus causes an increase in lead time by ΔT_2, while the rounding errors ΔT_3 and ΔT_4 cancel each other out. Feedback accuracy is therefore of significant importance for accuracy in lead time calculation.

3.9 Examples of Authentic Lead Time Measurements

In the following we present some results of flow time studies from analyses carried out by the Institut für Fabrikanlagen. The data were in all cases collected from job shop feedback information, according to the method suggested in Sect. 3.3, checked for various errors, and then evaluated using the DUBAF system (Durchlaufzeit- und Bestandsanalyse in der Fertigung = Manufacturing Lead Time and Inventory Analysis). (DUBAF is dealt with in Sect. 5.3.) The data come from precision parts manufacturing [18].

3.9.1 Work Center Lead Times

Table 3.6 shows part of the list of single work places which are hierarchically classified into job shops, cost centers, and work centers. Serial numbers 1 to 50 were allotted to the work centers examined. There were in total 90 single work places, mostly lathes as well as some milling and drilling machines.

Figure 3.23 summarizes the evaluation of actual lead times at these 50 work centers during a period of 16 weeks [18]. In Figs. 3.23A and B, the single work places were arranged according to the simple and weighted actual mean lead times respectively, rounded up to whole days. In Fig. 3.23C, they are ordered according to the quotient of weighted and simple actual mean lead time. The actual mean values of 12.4 workdays (simple) and 14.2 workdays (weighted) are more than double the scheduled lead time value of five days thought common in this shop. Only five of the

Table 3.6. List of work centers in a precision engineering shop (After Bechte [19])

Field of Investigation
50 Work Centers
90 Single Work Centers

Coding (Sample)

30 631337 Automatic Lathe (4)

- 30 — Serial No.
- 631337 — Code No.
- Automatic Lathe — Description
- (4) — Number of Single Work Centers

Note:
For the Job Shops 1, 2, and 7, only those work centers are listed which are really loaded by the orders in question. There are actually more work centers and multiple single work centers. Work centers with a serial no. are objects of the work center related analysis. The others are only loaded now and then, except for Work Center 552001, which processes nearly all orders, but has neither standard times nor capacity limits.

JOB SHOP 1

Cost Center 105

	105001	Pre-Assembly
	105091	Deburring
	105230	Milling Machine
	105281	Engraving Machine
	105310	Lathe
	105354	Automatic Lathe
	105430	Bench Drilling Machine (2)
	105470	Geng Drilling Machine (2)
	105691	Lapping Machine

JOB SHOP 2

Cost Center 512

	512093	Deburring
	512550	Ultrasonic Welding
	512831	Injection Molding Machine
	512832	Injection Molding Machine

Cost Center 520

	520132	Press
	520526	Heat Treatment
	520708	Surface Treatment

Cost Center 530

	530704	Sand-Blasting
	530707	Slide Grinding

Cost Center 540

	540001	Assembly

Cost Center 551

	551001	Painting
	551002	Painting

Cost Center 552

	552001	Degreasing

Cost Center 560

1	560001	Electroplating (2)
2	560036	Electroplating (3)
3	560711	Electroplating Machine (1)

JOB SHOP 6

Cost Center 611

4	611210	Milling Machine (1)
5	611220	Milling Machine (1)
6	611300	Center Lathe (1)
7	611410	Drilling Machine (1)
8	611420	Drilling Machine (4)
9	611425	Countersinking Machine (1)
10	611430	Drilling Machine (2)

Cost Center 614

11	614380	NC Chucking Lathe (1)
12	614385	NC Chucking Lathe (1)
13	614388	NC Swiss-Type Automatic Lathe (1)
14	614389	NC Swiss-Type Automatic Lathe (1)
15	614460	NC Drilling Machine (2)
16	614461	NC Drilling Machine (1)
17	614462	NC Drilling Machine (1)

Cost Center 621

18	621371	Semi-Autom. Chucking L. (6)
19	621374	Semi-Autom. Chucking L. (3)
20	621376	Semi-Autom. Chucking L. (1)
21	621378	Semi-Autom. Chucking L. (1)

Cost Center 622

22	622250	Machine Saw (1)
23	622332	Automatic Bar Lathe (2)
24	622333	Automatic Bar Lathe (1)
25	622341	Autom. Chucking Lathe (2)
26	622342	Autom. Chucking Lathe (1)

Cost Center 631

27	631326	Automatic Lathe (1)
28	631328	Automatic Lathe (2)
29	631335	Automatic Lathe (1)
30	631337	Automatic Lathe (4)
31	631351	Automatic Lathe (1)
32	631352	Swiss-Type Automatic Lathe (1)
33	631353	Swiss-Type Automatic Lathe (2)
34	631357	Automatic Lathe (5)
35	631358	Automatic Lathe (1)
36	631361	Automatic Lathe (1)
37	631362	Automatic Lathe (2)

Cost Center 641

38	641092	Deburring (4)
39	641221	Milling Machine (1)
40	641230	Milling Machine (2)
41	641310	Finishing Lathe (2)
42	641331	Finishing Lathe (7)
43	641312	Finishing Lathe (1)
44	641313	Finishing Lathe (1)
45	641314	Finishing Lathe (1)
	641315	Automatic Bar Machine (–)
46	641330	Automatic Bar Machine (3)
47	641390	Special-purpose Lathe (1)
48	641392	Special-purpose Lathe (1)
49	641431	Spec.-p. Drilling Mach. (1)
50	641600	Grinding Machine (1)

JOB SHOP 7

Cost Center 705

705022	Flame Cutting
705095	Deburring
705100	Cutting
705180	Edging
705256	Sawing

Cost Center 710

710026	Seal Inspection
710116	Drawing Press
710130	Eccentric Press
710141	Eccentric Press
710170	Screw-Threading Machine
710310	Finishing Lathe
710520	Reheating Furnace
710525	Continuous Heating Furn.

Cost Center 711

711190	NC-Nibbling Machine

Cost Center 720

720040	Electric Welding
720051	Metalworking Shop
720500	Spot Welding

Cost Center 730

730050	Metal-Working
730054	Metal-Working
730055	Metal-Working
730185	Push Bench
730530	Induction Soldering
730532	Flame-Brazing

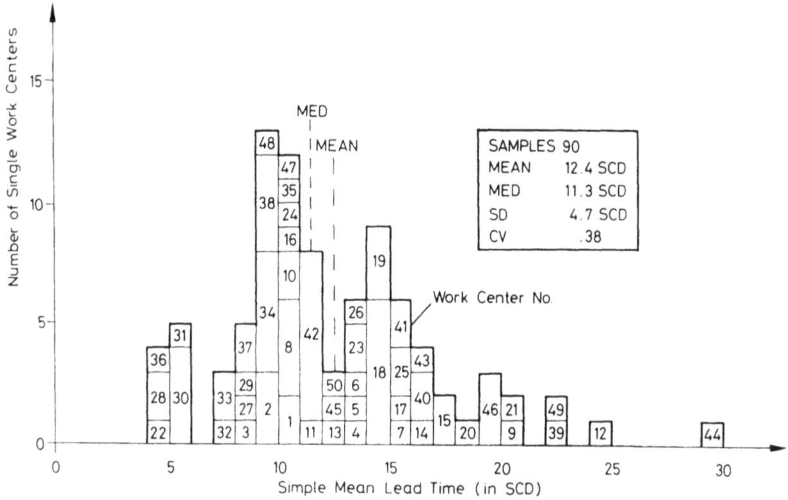

A) CLASSIFICATION OF SINGLE WORK CENTERS ACCORDING TO SIMPLE ACTUAL MEAN LEAD TIME

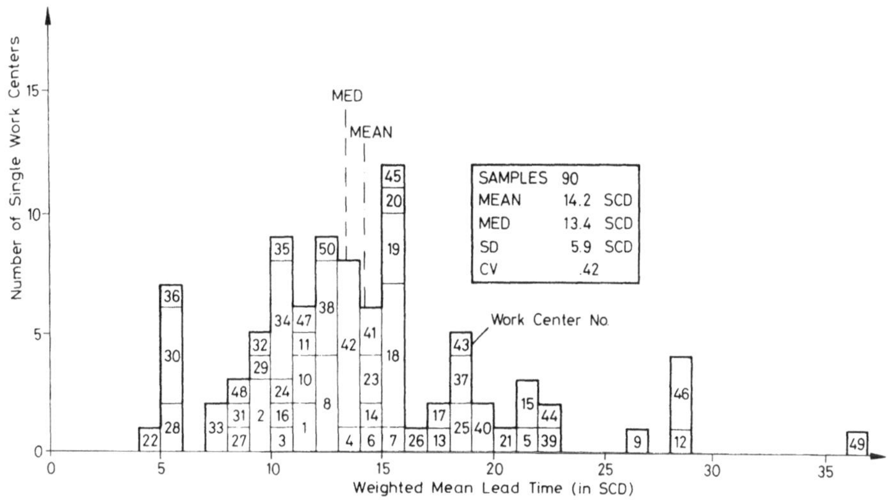

B) CLASSIFICATION OF SINGLE WORK CENTERS ACCORDING TO WEIGHTED ACTUAL MEAN LEAD TIME

Fig. 3.23. Simple and weighted mean actual lead times from 50 work centers (After Bechte [19])

continued

90 single work places really had a simple actual mean lead time corresponding to this value; some have 20 or more days of mean flow time (for the definition of "flow time" see Sect. 3.3). The weighted mean flow time is mostly larger than the simple mean flow time: on average 14.2 – 12.4 = 1.8 workdays. With some machines, however, weighted flow times are less than simple flow times, as can be seen in Fig. 3.23C, e.g. for no. 44 (finishing lathe), no. 20 (semiautomatic chucking lathe), and no. 48 (special-purpose lathe). For all work places, the variation coefficients of simple and weighted mean flow times were determined and compared with each other (Fig. 3.24).

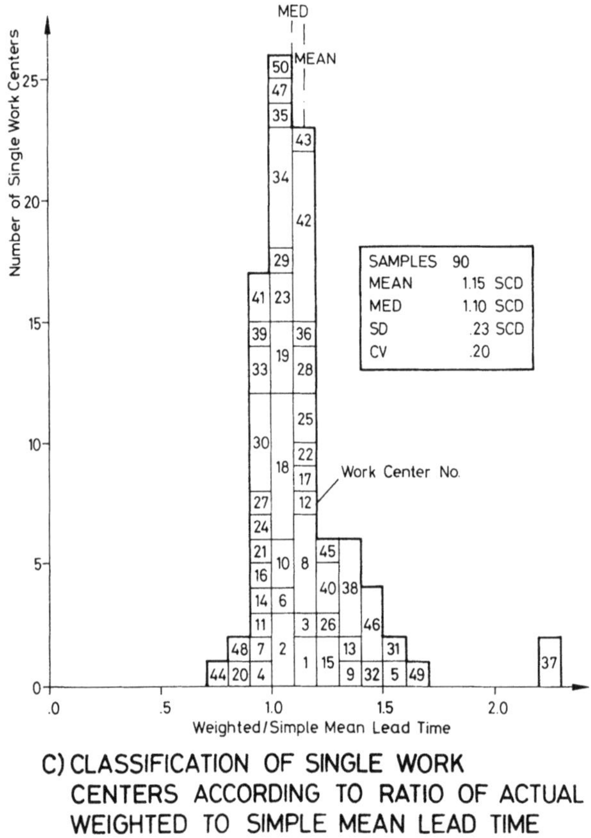

Fig. 3.23 (*continued*)

This has shown that with most of the work centers, the standard deviations and the mean flow times lie close to each other (i.e. their ratio, which is the coefficient of variation, is about 1, but there are also exceptions.

In general, we can conclude from this that the variance in flow times is greater the higher their mean values. This means that in practice, the longer the flow times the more difficult schedule performance becomes.

3.9.2 Operation Lead Times

In the investigation described, a total of 11 269 operations were measured and, among other things, arranged by their simple and weighted flow times. Figure 3.25 shows the result in a frequency distribution [18]. About 40% of all operations (measured in numbers of operations), but only 22% of the work hours have a flow time of up to five days, which accounts for the high mean values of 9.1 and 13.6 workdays for simple and weighted mean flow times, respectively.

Figure 3.26 shows the time fluctuations to which mean flow time is subject. It shows the weekly mean weighted flow time values and the familiar mean value of all operation flow times. One can see that the weekly values deviate from the medium-term mean value by about one day upwards or downwards [18].

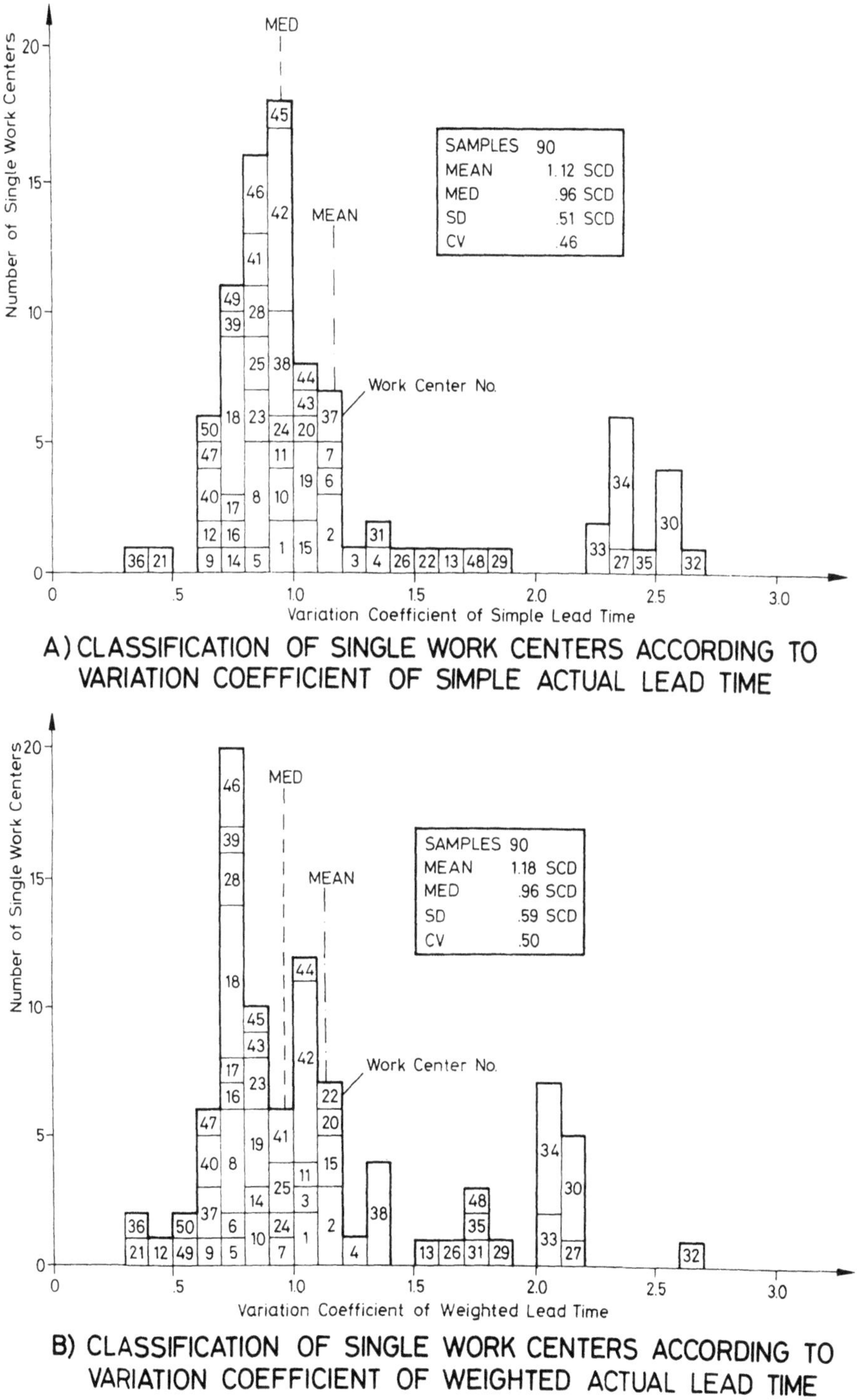

Fig. 3.24. Variation coefficients of the simple and weighted actual lead times from 50 work centers (After Bechte [19])

continued

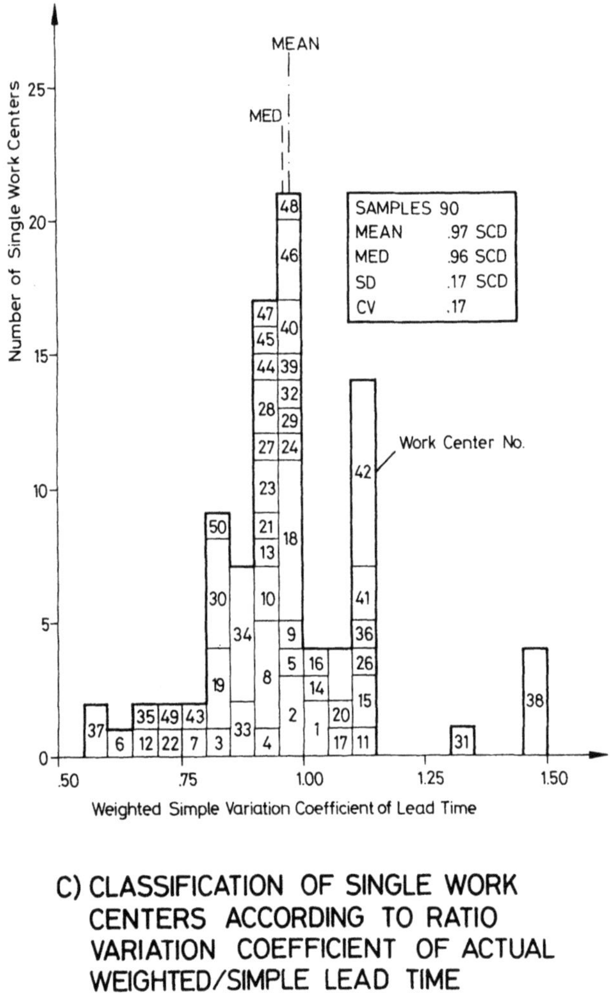

Fig. 3.24 (*continued*)

When looking at the flow time values, one is inclined to think that these are exceptions. This is, however, not the case, as was found in an investigation carried out using the method described above over several years in seven precision or mechanical engineering companies performing small-batch and job shop production (Fig. 3.27). All operations in the respective shops during times of two to twelve months were evaluated, and the simple and weighted mean flow times compared.

Except for Case 7, the simple mean flow time per operation is about 9 workdays, the weighted mean flow time is, except for Cases 1 and 7, about 14 workdays. The extraordinarily high weighted flow time in Case 1 is due to a larger number of operations with a very great work content and long waiting times, whereas the small values in Case 7 can be explained as being the result of comparably good sequencing. The operation time percentage is interesting in Case 7: the simple mean order time per operation was 6.4 hours, and the weighted mean order time was 37.1 hours. Since, in this case, all orders were processed in two-shift work, the values for simple

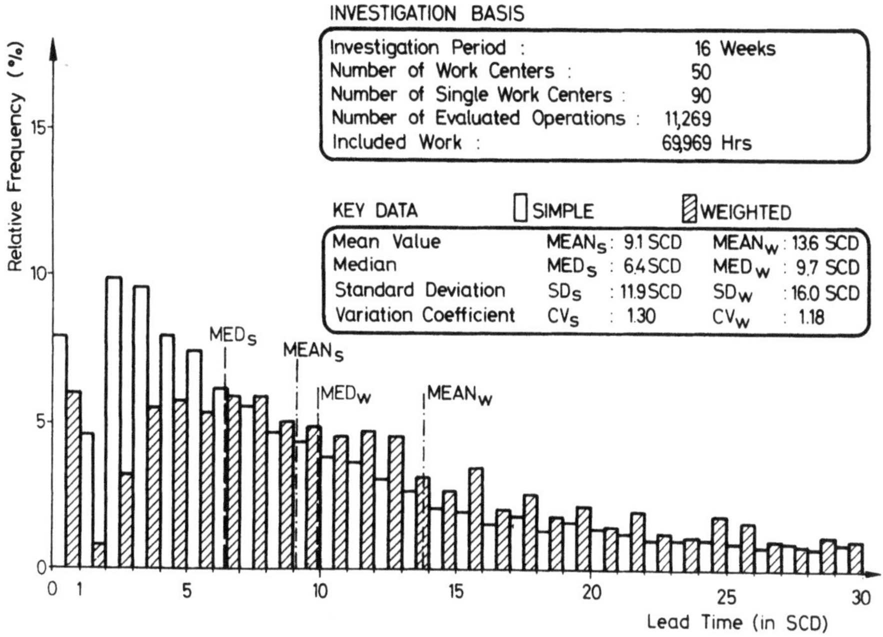

Fig. 3.25. Example distribution of simple and weighted mean actual lead time per operation in a mechanical manufacturing plant (After Bechte [19])

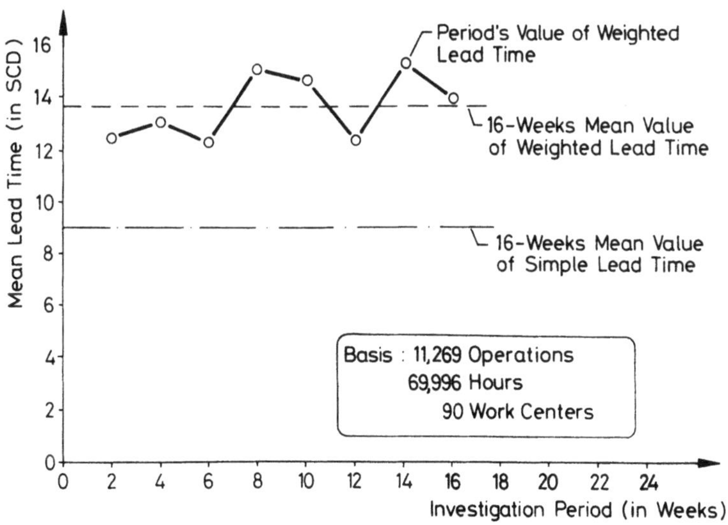

Fig. 3.26. Weighted mean actual lead time per operation in a mechanical manufacturing plant over time (After Bechte [19])

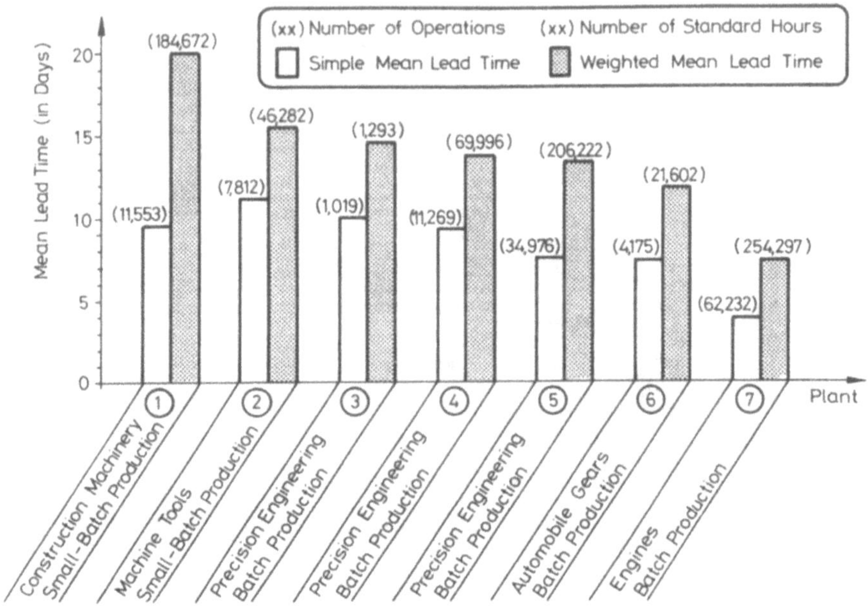

Fig. 3.27. Actual mean lead times per operation (taken from investigations in seven plants)

and weighted operation time were, with an efficiency rate of 1.0, 6.4/16 = 0.4 days and 37.1/16 = 2.3 workdays, respectively.

Consequently, the values for the simple and weighted operation time percentages are 0.4/4 = 10% and 2.3/7.4 = 31%, respectively. This very example shows that weighted flow time values give a different image of the production process than simple flow times.

3.9.3 Order Lead Times

Finally, we want to examine the order flow times over a representative period in the company from which the examples in Table 3.6 were taken. Within the investigation period of 16 weeks, 3261 shop orders were processed completely.

Figure 3.28 shows the single and the overall frequency values of operations per order, the mean value being 4.4 operations per order. Most orders have three or four operations.

The work content of the orders is shown in Fig. 3.29, the distribution diagram of simple and weighted order time. We can see a wide variance of simple and weighted order times and the great difference between their mean values [18].

Another descriptive representation is given by the Pareto distribution of order time (Fig. 3.30) [18]. For such a curve, one plots the respective cumulative percentage of work content over the percentage of the number of orders. (A detailed description on how to plot a Pareto curve is given in Sect. 6.6.) It shows, for example, that 50% of the orders have only about 15% of the overall work content, while 50% of the work content is processed in only about 15% of the orders. The median and

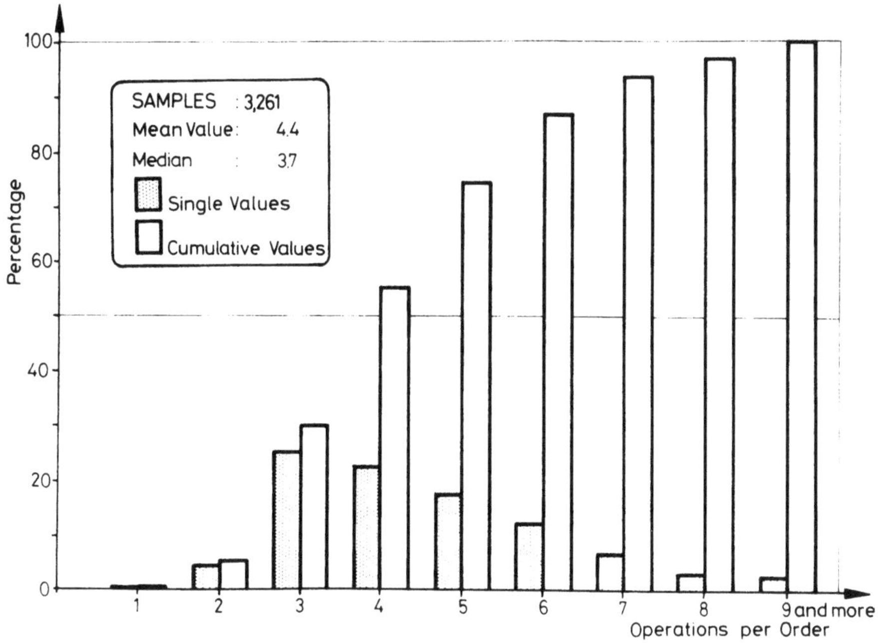

Fig. 3.28. Frequency distribution of the number of operations per order (data taken from a precision engineering company) (After Bechte [19])

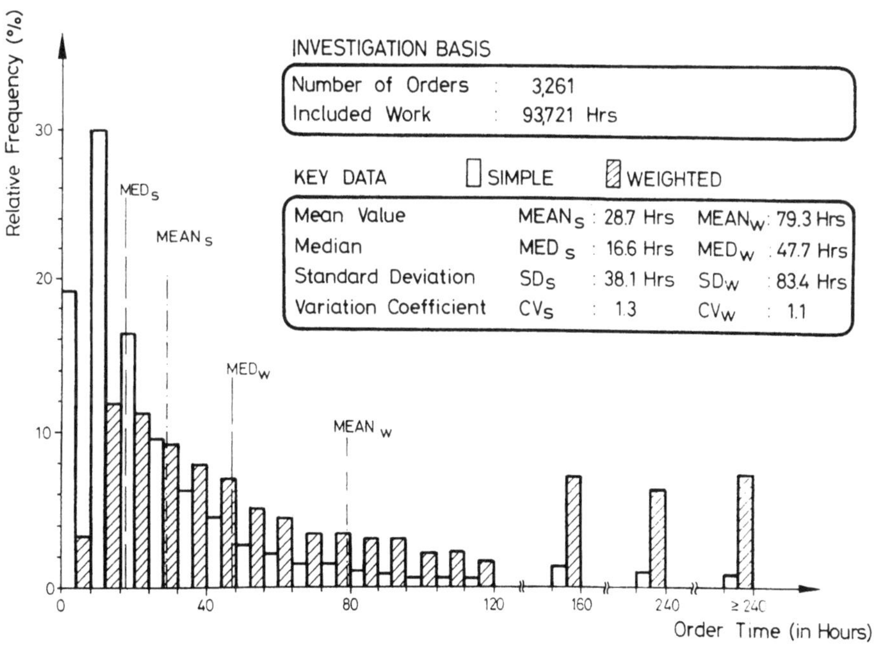

Fig. 3.29. Frequency distribution of simple and weighted order time (data taken from a precision engineering company) (After Bechte [19])

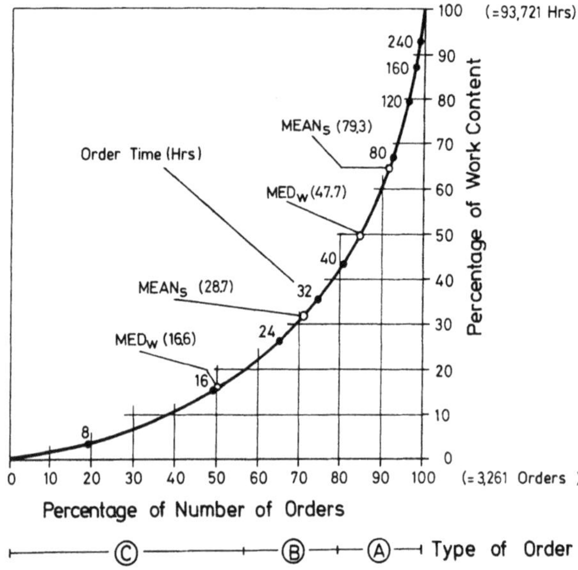

Fig. 3.30. Pareto curve of order time (data taken from a precision engineering company) (After Bechte [19])

mean values of simple and weighted order times can also be entered and interpreted here. The Pareto curve can also be used for the better known ABC analysis. 'A' orders are those 20% which contain about 58% of the work. On the other hand, those 58% of the orders which contain only 20% of the work can be called 'C' orders. The orders between them could be classified as 'B' orders.

Figure 3.31 shows the interesting frequency distribution of simple and weighted order flow times for all 2269 evaluated orders. If one calculates from the values of mean simple and weighted flow times (41 and 52.5 workdays, respectively), the values of flow time per operation, this equals, given a mean of 4.4 operations per order (see Fig. 3.28), 9.3 and 11.9 workdays respectively, of simple and weighted mean flow time per operation. The fact that these values do not correspond with the operation flow time values (Fig. 3.25), is due to the different statistical bases being examined. The 61 815 hours contained in the orders only come from the operations of completed orders, whereas for the examination of operations, all operations of the investigation period are counted, which was 69 996 hours [14, 18].

Finally, it can be of interest to evaluate order flow time as a function of the number of operations [18]. Figure 3.32 shows the evaluation of unweighted order lead time. We can see an almost proportional increase in order lead time with the number of operations, the mean value being about 10.8 workdays per operation.

Our findings in Chapter 3 prove the statement made in Chapter 1, namely that lead times in job shop production have large mean values of about 10 days per operation, with large variances, and that there is a very poor proportion of operation time to lead time, which is 1:10.

From this it follows that throughput control at the work centers is the key to efficient scheduling and scheduling control. For this we need a throughput model which gives a realistic image of the actual process. This is the subject of the next chapter.

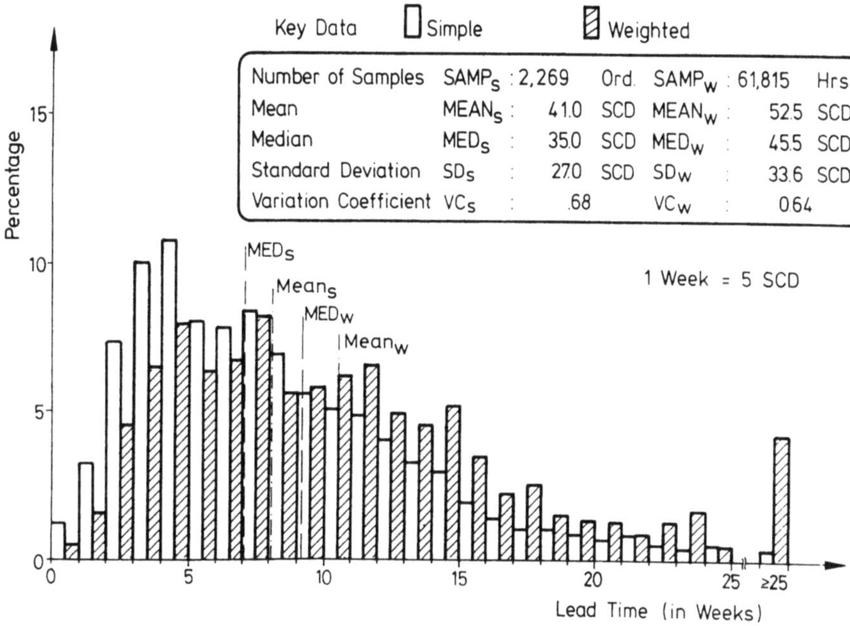

Fig. 3.31. Distribution of simple and weighted order lead time (data taken from a precision engineering company) (After Bechte [19])

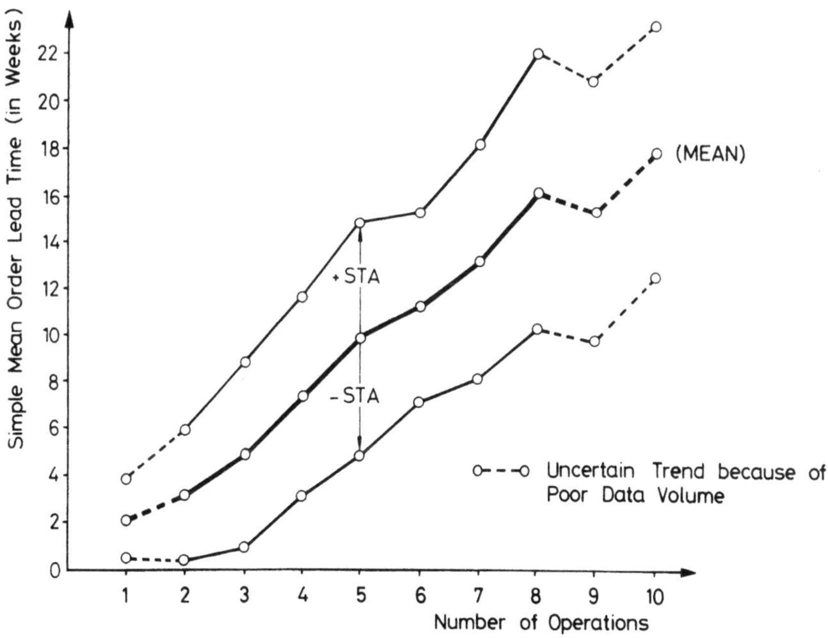

Fig. 3.32. Relationship between simple actual order lead time and number of operations per order (data taken from a precision engineering company) (After Bechte [19])

3.10 References

[1] REFA (ed.). *Methodenlehre des Arbeitsstudiums, Teil 2: Datenermittlung* (*Work Study Theories, Part 2: Data Recording*), 2nd edn. Hanser, München, 1971.

[2] Tully, H. Fertigungssteuerung im Werkzeugmaschinenbau (Production Control in Machine Tool Manufacturing). *Werkstattstechnik* **53** (9): 1963, 435–442.

[3] Hackstein, R. Problematik der Terminplanung im Industriebetrieb (Scheduling Problems in Manufacturing Companies). *Zeitschrift für Wirtschaftliche Fertigung*, **67** (1): 1972, 43–45.

[4] Stommel, HJ, Kunz, D. Untersuchungen über Durchlaufzeiten in Betrieben der metallverarbeitenden Industrie mit Einzel- und Kleinserienfertigung (Lead Time Studies in Metalworking Shops with Job-Shop and Small-Batches Production). Forschungsbericht Nr. 2355 des Landes Nordrhein-Westfalen (research report). Opladen, 1973.

[5] Loos, V. Maßnahmen zur Verkürzung der Durchlaufzeit in Betrieben mit Einzel- und Kleinserienfertigung (Lead Time Reduction in Job-Shop and Small-Batches Production). *Zeitschrift des Vereins Deutscher Ingenieure*, **119** (19): 1977, 913–921.

[6] Nadzeyka, H, Schnabel, B. Untersuchungen über die Fertigungsdurchlaufzeiten in der Maschinenbauindustrie (Lead Time Studies in Mechanical Engineering Companies). REFA-Nachrichten, **28** (5): 1975, 267–271.

[7] Weber, N, Stader, H. Kenngrößen zur Beurteilung der Qualität von Terminplanung und Fertigungssteuerung (Key Factors for Scheduling and Production Control Evaluation). *Industrie Anzeiger*, **91** (50): 1969, 1165–1166.

[8] Heinemeyer, W. Durchlaufzeitanalyse als Grundlage für eine systematische Rationalisierung im Maschinenbau (Lead Time Analysis as a Basis for Systematic Rationalization in Mechanical Engineering). *Metallwissenschaft und Technik*, **26** (6): 1972, 603–612.

[9] Gottwald, H, Heinemeyer, W, Kreutzfeldt, H-F, Wegner, N. Fertigungsdurchlaufzeit und Kapitalbindung (Manufacturing Lead Time and Capital Tie-up). Repräsentative Erhebung des Arbeitsausschusses Fertigungswirtschaft (AFW) der Deutschen Gesellschaft für Betriebswirtschaft (DGfB). Teil 1: Datenerfassung, Teil 2: Auswertung (als Manuskript veröffentlicht) (DGfB paper, part 1, part 2). Institut für Fabrikanlagen (IFA), Technische Universität Hannover, 1973.

[10] Kettner, H. (ed.). *Neue Wege der Bestandsanalyse im Fertigungsbereich* (*Recent Inventory Analysis Methods in Manufacturing*). IFA, Hannover, 1976.

[11] Autorenkollektiv (several authors). Neue Wege der Fertigungssteuerung (Recent Manufacturing Control Methods). Dokumentation zum Seminar 'Neue Wege der Fertigungssteuerung' des Instituts für Fabrikanlagen der Universität Hannover am 16./17.3.1982 in Hannover (seminar paper). IFA, Hannover, 1982.

[12] Nyhuis, F. Fertigungsablaufanalyse benötigt für gute Ergebnisse systematische Planung (Systematic Planning as a Prerequisite for Manufacturing Process Analysis). *Maschinenmarkt*, **90** (20): 1984, 442–445.

[13] Heinemeyer, W. Die Analyse der Fertigungsdurchlaufzeit im Industriebetrieb (Lead Time Analysis in Manufacturing Companies). Dissertation Technische Universität Hannover, 1974 (doctoral thesis).

[14] Erdlenbruch, B. Grundlagen neuer Auftragssteuerungsverfahren für die Werkstattfertigung (Principles of Recent Order Control Techniques in Job-Shop Production). Dissertation Universität Hannover, 1984 (veröffentlicht in: Fortschritt-Berichte der VDI Zeitschriften, Reihe 2, Nr. 71, Düsseldorf, 1984) (doctoral thesis, published in the above VDI report).

[15] Fogarty, DW, Blackstone, Jr., JH, Hoffman, TR. *Production and Inventory Management*, 2nd edn. South-Western Publishing Co, Cincinnati, OH, 1991.

[16] Plossl, GW. *Production and Inventory Control – Principles and Techniques*, 2nd edn. Prentice-Hall, Englewood Cliffs, NJ, 1985.

[17] Kettner, H, Bechte, W. Neue Wege der Fertigungssteuerung durch belastungsorientierte Auftragsfreigabe (Recent Manufacturing Control Methods by Use of Load-oriented Order Release). *Zeitschrift des Vereins Deutscher Ingenieure*, **123** (11): 1981, 459–466.

[18] Bechte, W. Steuerung der Durchlaufzeit durch belastungsorientierte Auftragsfreigabe bei Werkstattfertigung (Lead Time Control in Job-Shop Production by Load-oriented Order Release). Dissertation Universität Hannover, 1980 (veröffentlicht in: Fortschritt-Berichte der VDI-Zeitschriften, Reihe 2, Nr. 70, Düsseldorf, 1984) (doctoral thesis, published in the above VDI report).

[19] Bechte, W. Arbeitsinhalt-Zeit-Funktionen – ein Kontrollinstrument für die Fertigungssteuerung (Relationships between Work Content and Time – A Means for Monitoring in Manufacturing Control). *Fertigung und Betrieb/Industrial Engineering*, **32** (2): 1983, 7–14.

[20] American Production and Inventory Control Society (APICS). *The Official Dictionary of Production and Inventory Management Terminology and Phrases.* APICS, Falls Church, VA, 1987.

[21] Holzkämper, R. Voraussetzungen für die Realisierung eines Kontroll- und Diagnosesystems zur organisierten Fertigungsablaufüberwachung (Prerequisites for the Acquirement of a Monitoring and Diagnosis System for Manufacturing Process Monitoring). *Zeitschrift für Wirtschaftliche Fertigung,* **80** (6): 1985, 238–243.

[22] Wiendahl, H-P. Grundlagen neuer Verfahren der Fertigungssteuerung (Fundamental Principles of Recent Manufacturing Control Techniques). In: Dokumentation zum Fachseminar 'Statistisch orientierte Fertigungssteuerung' des Instituts für Fabrikanlagen der Universität Hannover, 14–15 May 1984, Hannover (seminar paper), pp 1–19.

[23] Lorenz, W. Differenzierte Durchlaufzeitanalyse im Fertigungsbereich – Ein Verfahren zur Durchlaufzeitkontrolle (Sophisticated Lead Time Analysis in Manufacturing – A Lead Time Monitoring Method). Arbeitsvorbereitung, **20** (5): 1983, 144–149.

[24] Kanet, JJ. Towards Understanding Lead Times in MRP Systems. *Prod. & Inventory Manage.* Third Quarter, 1982, 1–14.

[25] Conway, RW., Maxwell, WL., Miller, LW. *Theory of Scheduling.* Addison-Wesley, Reading, MA, 1967.

Chapter 4

The Throughput Diagram – A General, Realistic Model of the Manufacturing Process

4.1 Abstract

This chapter develops the throughput diagram as the basis for describing the input and output events of a work center. It is shown that the four objectives of production control – inventory, lead time, utilization and lateness – can be graphically represented over time in this diagram. Furthermore, the basic relationship between mean lead time, mean inventory and mean utilization is derived from this diagram. Finally, the flow of orders through a whole shop is also represented by means of the throughput diagram.

4.2 Historical Evolution

With increasing product variety, an intensifying division of labor, and with the growing size of manufacturing companies, a clear representation of the manufacturing process becomes increasingly important. It has, however, become apparent that a merely numerical representation in lists, tables and updated key data can no longer meet today's requirements, despite – or rather because of – extensive use of electronic data storage and processing. On the contrary, encouraged by decreasing prices in graphic data processing, there is a growing demand for graphic representation of manufacturing processes, exemplified by the state-of-the-art displays in control units of power stations and chemical plants.

It is, however, not sufficient to represent the physical manufacturing process synchronously, as is often the case in complex automated manufacturing systems. In such systems, a simplified layout of the plant is displayed on a screen. In this layout the operator can distinguish between certain states of individual pieces of equipment – for instance, whether a machine is in use or standing idle, whether a transport vehicle is moving or waiting, loaded or empty, etc. However, such representations only offer momentary information; no information is provided on past activity or the planned processing of an order, or on the previous or planned loading status of

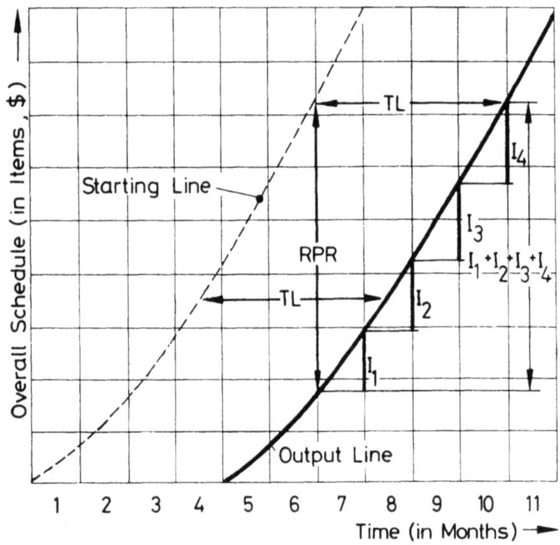

Fig. 4.1. Lead time and relevant production rate in a variable production program. RPR: relevant production rate (in items, $); I_1, I_2 ...: monthly schedule (items/month, $/month); *TL*: lead time of product (days, months). RRR indicates the number or value of all in-process items at a certain point of time (After Schmitz [1])

a machine. Although these data are available in a production planning and control system, they are largely provided in the form of lists.

Schmitz [1] was one of the first to try and represent order throughput graphically. In 1961 he attempted to describe the interdependencies between the production plans and plant utilization on the basis of his experience at the Carl Zeiss factory in Jena, Germany. The fundamental idea was to plot the manufacturing progress at a machine cumulatively over time using the starting and completion dates. Figure 4.1 shows the basic graph for this. We can see the output line corresponding with the production schedule and the starting line determined from the lead time (*TL*) of the individual product, as well as the inventory trend (referred to here as the *Relevant Production Rate* – RPR) of in-process items, resulting from the vertical distance between the starting and output lines. With this, Schmitz was able to provide a general representation of the manufacturing process for entire products, which analytically connects and describes the fundamental objectives of lead time, inventory, and performance.

In 1967, Conway et al. [2] proposed plotting of the cumulative number of arrivals and the cumulative number of job completions in a shop against time, and to combine them in the same graph (Fig. 4.2A). They also take note of the fact that jobs are not always completed in arrival order, which makes the graph appear like that shown in Fig. 4.2B. They conclude that for a steady-state process and with a large number of jobs, it results that the mean flow time through such a system equals the product of the mean number of jobs in the system and the mean time between arrivals. The authors also refer to Little [3], who published a formal proof of this relationship in a queueing context. This relationship applies to an entire plant, to a department within this plant, and to an individual machine. It also applies to the

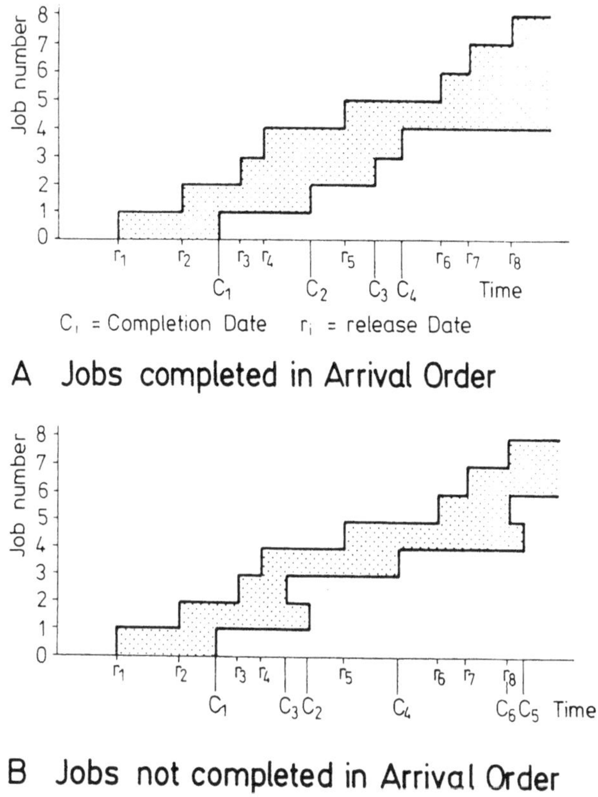

Fig. 4.2. Cumulative plot of arrivals and completions of jobs against time in a job shop (After Conway et al. [2])

complete product mix of a company, or to just one product line, or even to an individual product.

In't Veld [4] presented similar ideas in 1971. On the basis of his experiences at the plants of aircraft manufacturer Vocker (Netherlands), he proposed a representation of the same facts in a so called 'Z-diagram'. Figure 4.3 shows such a Z-diagram, which consists of three single graphs. The centre graph shows the input and output curves, as in Schmitz's and Conway's diagrams; the vertical and horizontal distances describe the inventory trend and lead time, respectively. The top graph shows the lead time trend with the planned value and the alarm limit, and the bottom graph shows the input and output trends of production units. The bottom graph was produced by calculating the changes in the input and output curves for each time bucket.

Since the three graphs together form a Z, the diagram as a whole was called a Z-diagram. In't Veld pointed out that the units can be given in items (number of products) or in hours of work, or – when considering a design department, for example – as the number of drawings; and he stated that such diagrams can be useful for managers as a monitoring and diagnostic system [4].

In the course of the flow time investigations proposed by Kettner – the first di-

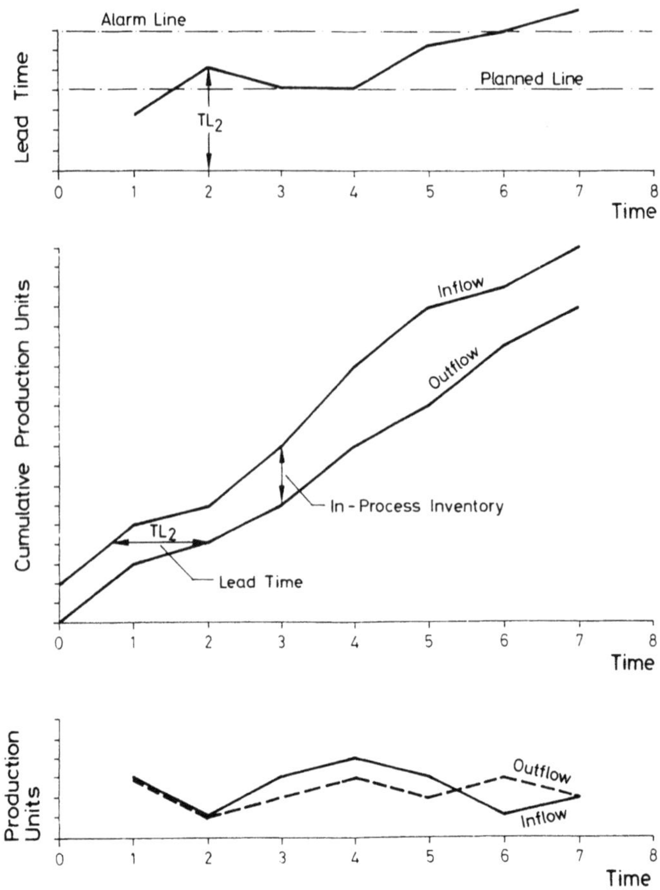

Fig. 4.3. Z-diagram of inflow, outflow and lead time (After In't Veld J. [4])

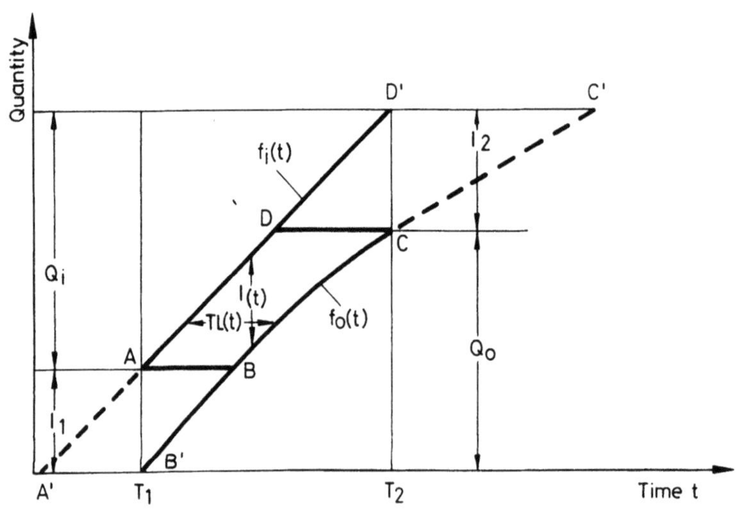

Fig. 4.4. Generalized representation of the throughput element. Q_i: input quntity at T_1, T_2; Q_0: output quantity at T_1, T_2; I_1: initial inventory at T_1; I_2: final inventory at T_2; $T_L(t)$: lead time as function of time t; I_t: inventory as function of time t; $f_i(t)$: function of input quantity; f(t): function of output quantity

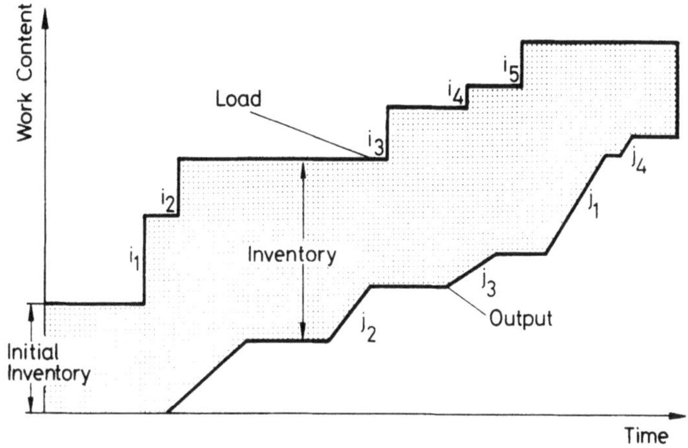

Fig. 4.5. Trend of load and output curves (After Kreutzfeldt [6])

rector of the Institut für Fabrikanlagen at the University of Hannover – Heinemeyer took up Schmitz's ideas and, in 1974, suggested the representation shown in Fig. 4.4, which permits an investigation of the order throughput at a work center period by period [4]. Heinemeyer [5] referred to the time between the output of an order at the upstream work center and its output at the work center concerned as the *throughput element* (for more detailed information see Chapt. 3).

What is of interest here is the definition of initial and final inventories (I_1 and I_2), as well as the essentially arbitrary trends of the input and output curves.

In 1977, in the course of his work at the same institute, Kreutzfeldt extended the quantity unit in the Z-diagram to the more general unit work content (Fig. 4.5) [6]. His successor, W. Bechte, used the diagram (called the work-content-time-function) as the basis for a complete description of the manufacturing process, from which he – following the ideas of Jendralski [11] – developed the method of load-oriented order release [7]. With the concept throughput diagram, Erdlenbruch extended Bechte's work-content-time-function to include the representation of complete order throughputs [8], whereas his colleague Lorenz used it as a basis for an extended queueing model [9].

The throughput diagram has now become the most important means for describing the complex manufacturing process in job-shop production, and is especially suitable for visualizing the process in view of the demand for transparency. It can also be used for planning purposes, and is thus a general model of the manufacturing process that can be dealt with by mathematical means.

Because of its fundamental importance for manufacturing control, the throughput diagram is here developed step-by-step, building on the last three publications mentioned [7–9].

4.3 Basic Form of the Throughput Diagram

A work center – be it a single work place, a group of work places, a job shop, or a total shop – can be compared with a funnel, at which orders arrive (input), wait for processing (inventory), and leave the system (output). Figure 4.6 (left) shows such a

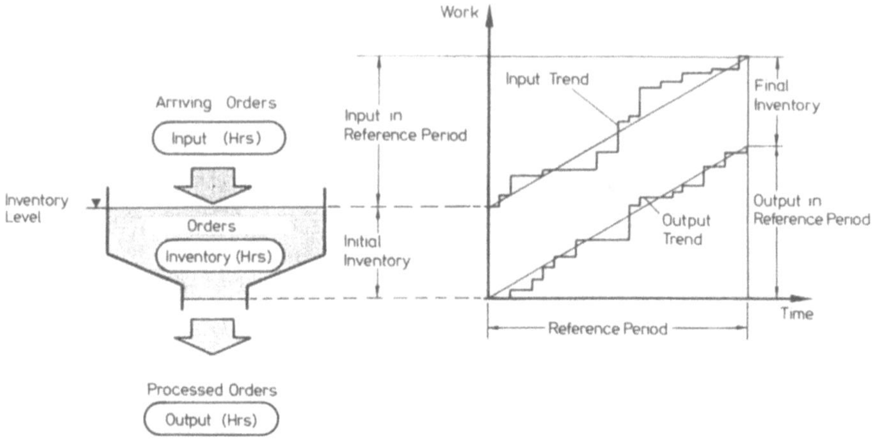

Fig. 4.6. Derivation of the throughput diagram from the funnel model (After Kettner and Bechte [7])

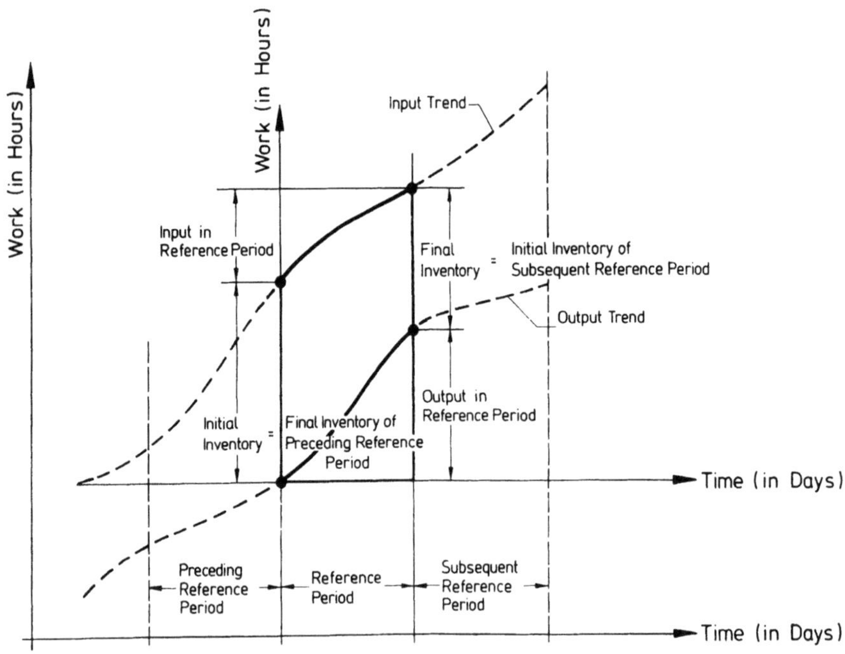

Fig. 4.7. Principle of the throughput diagram of a work center

funnel. When the work center is observed over a longer period of time (reference period), the results can be plotted as curves.

In the graph on the right we can see an input trend and an output trend. For the input trend, one must determine the quantity of work waiting at the work center at the beginning of the reference period (initial inventory). From this point, the curve is plotted by adding the inflowing work with its work content in standard hours at the respective input times, until the end of the reference period. This produces the input trend. Accordingly, the output trend is plotted by adding the completed orders with their work content at the respective reported output times, starting at the origin of the coordinate system. Since both curves together describe the throughput of orders through the system, this representation is, according to Erdlenbruch [8], called the *throughput diagram.*

At the end of the reference period there is also a certain inventory, called the *final inventory* in Fig. 4.6. If this final inventory is understood to be the initial inventory of the next reference period, the throughput diagram turns out to be a snapshot of the continuous description of a work center.

Figure 4.7 outlines this with arbitrary input and output curves plotted over time. Assume the coordinate system is set up anew at the end of each reference period; the input and output curves then 'move' as if seen through a window. The total of all inflowing order hours recorded during the reference period is called the *input*, and the total of all outflowing orders is called the *output*.

The type of representation shown in Fig. 4.7 does not tell us whether the processed work is the work content of single operations of a shop order, of whole orders, or of entire modules or products. Nor do we know whether the work center in question is a single work place, a group of work places, one job shop, or a total shop. In Chapter 3, however, we found out that the throughput element (i.e. the work center lead time) is the fundamental element of scheduling. Correspondingly, the description of one work center with the operations carried out for the individual orders, represents the fundamental throughput diagram on which all further models of the manufacturing process are based. Therefore, the throughput diagram is first presented as a work center throughput diagram.

4.4 The Work Center Throughput Diagram and its Basic Data

4.4.1 How to Produce a Work Center Throughput Diagram

As has already been shown, the following definition of flow time is useful for a throughput evaluation of operations: the flow time of an operation at an individual work center is calculated as the difference between the reported date of output of the operation at the work center concerned, and that of the previous operation at the upstream work center. The definitions of the input and output curves in the throughput diagram are also based on this figure. Because the throughput diagram is valid for both planned and actual lead times, the term *lead time* is used in this chapter to represent both allowance and flow time.

For plotting a throughput diagram we can make use of feedback data supplied by the production data collecting (PDC) systems that are increasingly being used today. These data from various work centers are collected by a central system and can be represented, among other things, in the form of a list of events for each work center, which can be considered as the protocol of events at the work center. Figure 4.8 (left) shows such a list of events at a work center in a simplified form. Besides the dates of events (given here in Column 1 in calendar days, i.e. accuracy to the day), the list contains the order number (Column 2), the order time in standard hours (Column 3), and the type of event, which is either input (Column 4) or output (Column 5). We now want to plot the throughput diagram for the reference period Calendar Days 61 to 67 – i.e. for one week. The reference period is the 'window' through which the process at the work center is observed.

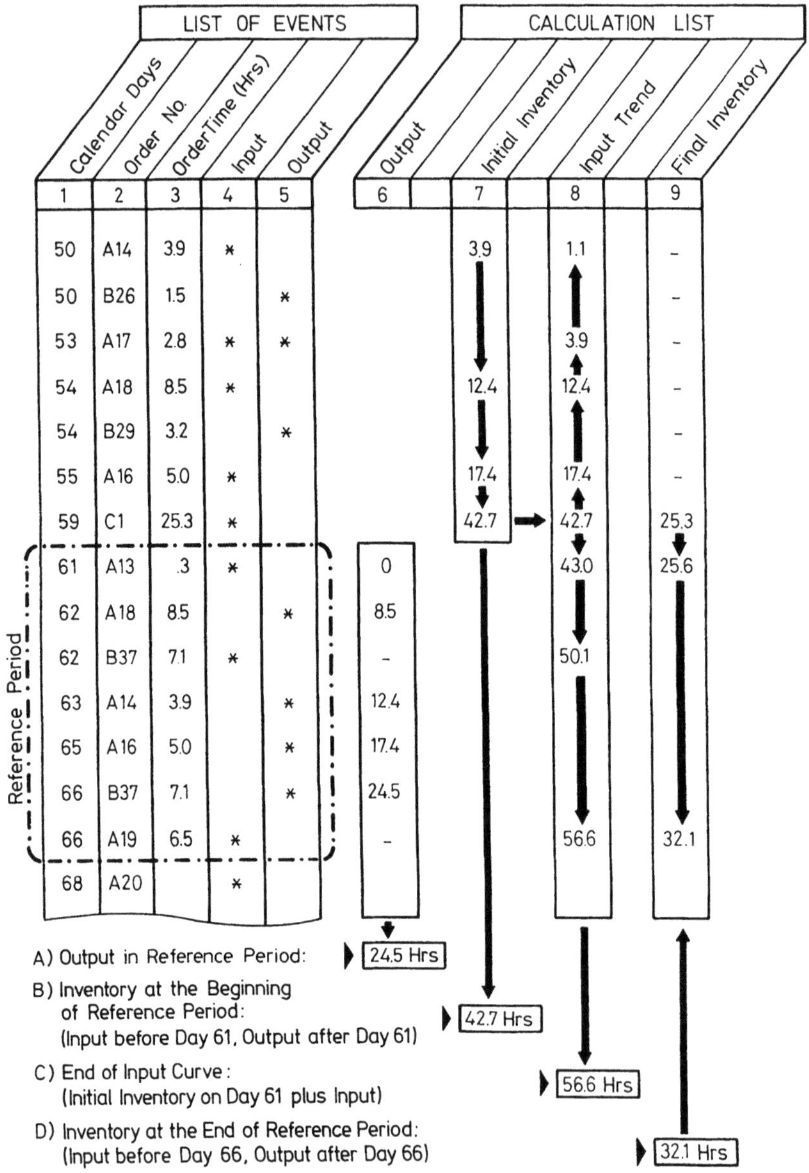

Fig. 4.8. Calculation of output, input, initial inventory, and final inventory of a work center, based on a list of events

It is advisable to start with the output curve. For this, one adds up the output order hours, as in Column 6, from the beginning to the end of the reference period; in our example, the sum is 24.5 hours. (Note that there were no events during Days 64 and 67.) In Fig. 4.9, the output curve has been plotted between points 1 and 2, according to the output dates of Orders A18, A14, A16, and B37. Since all the feedback data for one day are related to the end of the day, the diagram starts at the end of Day 60, which is at the same time the beginning of Day 61.

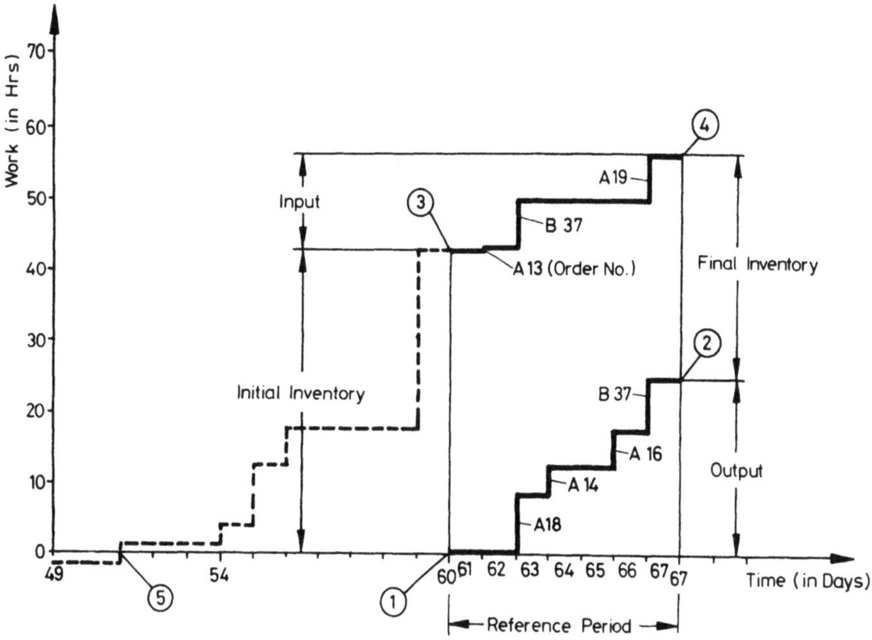

Fig. 4.9. Construction of a throughput diagram from the input and output trends. 1: Beginning of output curve; 2: end of output curve; 3: beginning of input curve (≅ initial input); 4: end of input curve (≅ final inventory + output); 5: beginning of dummy output curve (≅ zero point)

To find the starting point of the input curve we must now determine the initial inventory. For this we add up the hours of all the orders that came in before the beginning of the reference period but had not been completed before this date, as shown in Column 7 of Fig. 4.8. This applies for Orders A14, A18, A16, and C1 with a total of 42.7 hours. In practice, one generally uses the data from order planning, which normally show orders that were allotted to specific work centers on a certain day – here the beginning of the reference period. In this way, one avoids having to trace data on the input lists (if these exist at all) over a very long period. If the initial inventory cannot be provided by the order planning system, one may have to determine it by physically counting the orders to be processed.

Proceeding from this initial inventory, the input curve is now plotted by summing up the orders that came in during the reference period (i.e. Orders A13, B37, and A19) so that the end point of the input curve is now 56.6 hours, as shown in Column 8 in Fig. 4.8 (Point 4 in Fig. 4.9). Input is calculated as the total of the order hours of these orders: 0.3 + 7.1 + 6.5 = 13.9 hours.

The final inventory at Day 67 is then calculated as the difference between the end values of the input and output curve. In our example, this is 56.6 – 24.5 = 32.1 hours. The final inventory can, however, be determined as initial inventory has been, namely by adding all the order hours which came in before Day 67 and were not completed until after Day 67. As expected, the result of this calculation is 32.1 hours (Column 9 in Fig. 4.8).

It is interesting to complete the throughput diagram by giving the input trend of the orders which came in before the beginning of the reference period. For this, one follows the input curve backward, using the values in Column 8. We can see that the input curve crosses the abscissa on Day 50 and takes 1.1 hours as its first value. If we subtract from this the work content of A14 (3.9 hours), which flowed in during this day, we get −2.8 hours as the second value.

The fact that the input curve does not start with the value of a whole order can be explained as follows: at the beginning of the reference period, the input curve has the value of the initial inventory, which only includes the orders that were really contained in the inventory on Day 61. Order A17, for instance, is not contained in the initial inventory, since it came in and was also completed before this day. It is, however, contained in the input curve, as this describes the 'history' of this work center, i.e. the chronological sequence of input, independent of any reference period. If one were to plot the input curve of all orders contained in the initial inventory, its course would thus be different from that of the 'historical' curve. This is why an order input can accidentially be 'cut in two'. This phenomenon will be treated in some more detail as discussion of the throughput diagram proceeds.

For further discussion of the throughput diagram, this fictitious example is, however, too simple. Therefore, Table 4.1, Columns 1 to 4, gives an excerpt taken from the feedback records of an authentic work center (a lathe). The dates were ordered according to their input dates, and given to the day. The order times were taken from the working papers; this means that it is a matter of standard hours, not the hours actually needed. As has already been explained, standard hours are fully sufficient to describe the throughput processes.

In this example, the reference period is to be 28 days (or four weeks) from Day 205 to Day 232. From Column 2 we can see that on some days there are no inputs, e.g. on Days 231 and 232. Nor are there outputs every day: Days 205, 231, and 232 have none.

The first step for our throughput diagram was to determine the inventory for the first day of the reference period. The value is given in Column 5: 95.8 hours. Since the feedback data are grouped by input dates, determining the input curve was simple; total input in the reference period was 184.9 hours (Column 6). As the starting point of the curve on Day 205 is 95.8 hours, the end point is 95.8 hours plus 184.9 hours = 280.7 hours. We now plot the individual inputs (from Column 7) backwards from this point to produce the input curve. As expected, it takes the value 95.8 hours at the beginning of the reference period (beginning of Day 205 = end of Day 204) and crosses the time axis at Day 178 with Order 113 (zero point of input curve).

For the output curve, we need all the outputs reported in the reference period, 146.0 hours in total (Column 8). The output values are now listed in order of the output dates (it is the same list as was used as Table 3.1 in Chapt. 3). After having also plotted these values, we obtain Fig. 4.10, which is an accurate representation of the input and output trends [5].

The figure also contains several values with their abbreviations, which are used in the following. We already know initial inventory (II), input (INP), final inventory (IF), output (OUT), and the reference period (P). New values in this throughput diagram are inventory (I), range (R), advance time (TA), and lead time (TL) [7].

The inventory (I) of work at this work center can be determined by drawing a vertical line (parallel to the ordinate axis) in the diagram at an arbitrary time T. The distance between the resulting intersections of this line with the input and output curves represents the inventory at the work center at the respective time, measured in hours of work. These can either be standard hours or actual hours, depending on the available data. The quantity thus determined is called the inventory $I(T)$ and

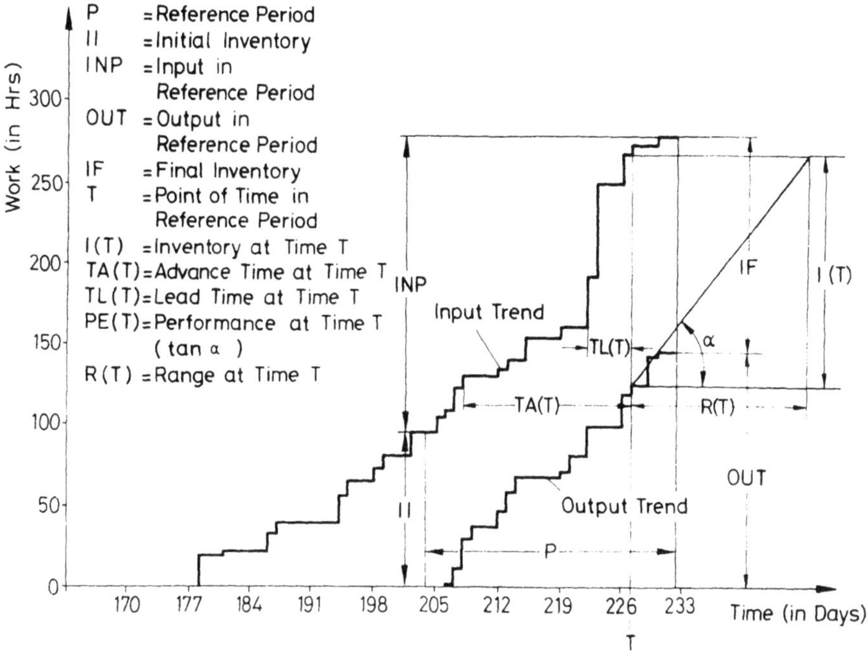

Fig. 4.10. General form of the throughput diagram (After Bechte [7])

varies with time [7]. It can be compared with the balance of a bank account: each change on the debit or credit side leads to a change in the balance. A negative balance, i.e. a negative inventory, however, cannot possibly occur in a real throughput diagram. In our example the time T was Day 227; the difference between the values of the input and output curves thus equals the inventory at this time. Table 4.1, Column 7, Line 37, gives the input value 275.6 hours for this day. From Column 8 we obtain the output value for this day by subtracting the output hours reported between Days 232 and 227 from the output OUT, that is, Orders 142, 145, and 123. From this, the value of the output curve for Day 227 is: $146 - (2.1 + 5.1 + 13.6) = 125.2$ hours. The inventory for this day is then $275.6 - 125.2 = 150.4$ hours.

The second new value in the diagram, range (R), is used in store management to express how long the stock of an individual item will last, given a known rate of consumption. It is defined as the ratio of the available inventory to expected consumption. If one applies this definition to a work center, 'stock' corresponds to the described vertical distance between the curves, and 'consumption' is the current output rate, or performance, of the work center. The range $R(T)$ describes the time the work center would need to process the inventory $I(T)$ with the performance $PE(T)$. The range $R(T)$ at a work center is defined as the ratio of the inventory $I(T)$, to the performance $PE(T)$ ((T) = current values)

$$R(T) = \frac{I(T)}{PE(T)} \tag{4.1}$$

In our example, the capacity of the work center concerned was 8 hours per day. Let us assume that the work center was loaded to its capacity at time T, and that the capacity was equal to the performance $PE(T)$, i.e. to the actual work hours pro-

Table 4.1. Calculation of the input and output curves of a work center, based on the feedback records of the reference period (Day 205–Day 232) in the order of the input dates

A) Data Base | B) Inventory Input and Output Trend

Line Number	Order Number (-)	Input Date tPEU (Day)	Output Date tPE (Day)	Order Time TO (Hrs)	Inventory at the End of SCD 204 (Hrs)	Input from SCD 205 to 232 (Hrs)	Cumulative Overall Input (Hrs)	Output between SCD 205 to 232 (Hrs)	Inventory at the End of SCD 232 (Hrs)
	1	2	3	4	5	6	7	8***)	9
1	103	81	246	3.3	3.3		-208.4		3.3
2	104	91	340	13.1	16.4		-195.3		16.4
3	102	151	198	26.4			-168.9		
4	101	156	156	18.8			-150.1		
5	105	157	181	17.8			-132.3		
6	106	157	160	2.1			-130.2		
7	107	157	187	34.9			- 95.3		
8	109	166	180	70.0			- 25.3		
9	110	170	208	15.4	31.8		- 9.9	15.4	
10	113	178	198	30.3			20.4*)		
11	115	181	206	.5	32.3		20.9	.5	
12	114	186	201	11.4			32.3		
13	116	187	220	7.4	39.7		39.7	7.4	
14	119	194	207	11.4	51.1		51.1	11.4	
15	120	194	209	6.8	57.9		57.9	6.8	
16	118	195	213	7.7	65.6		65.6	7.7	
17	112	198	235	3.4	69.0		69.0		19.8
18	117	198	222	4.2	73.2		73.2	4.2	
19	121	199	214	8.8	82.0		82.0	8.8	
20	108	202	222	13.8	95.8		95.8**)	13.8	
21	124	205	212	9.6		9.6	105.4	9.6	
22	125	206	208	3.8		13.4	109.2	3.8	
23	123	207	229	13.6		27.0	122.8	13.6	
24	126	208	219	2.6		29.6	125.4	2.6	
25	127	208	213	5.3		34.9	130.7	5.3	
26	122	212	233	3.8		38.7	134.5		23.6
27	131	213	220	2.1		40.8	136.6	2.1	
28	128	213	258	4.7		45.5	141.3		28.3
29	132	215	226	13.8		59.3	155.1	13.8	
30	135	219	226	6.9		66.2	162.0	6.9	
31	140	222	227	5.1		71.3	167.1	5.1	
32	136	222	234	26.9		98.2	194.0		55.2
33	141	223	247	57.4		155.6	251.4		112.6
34	142	226	230	2.1		157.7	253.5	2.1	
35	129	226	250	14.1		171.8	267.6		126.7
36	143	226	236	2.9		174.7	270.5		129.6
37	145	227	229	5.1		179.8	275.6	5.1	
38	134	230	298	5.1		184.9	280.7		134.7
				TOTAL >	95.8	184.9	280.7	146.0	134.7

↕ Reference Period

*) Zero Point of Input Curve
**) Beginning of Input Curve in Reference Period
***) In order to plot the Output Curve, TO Values (Column 4) have to be rearranged according to Output Dates (Column 3)

cessed per day. Then the range $R(T)$ is 150.4 hours/8 hours/day = 18.8 days. To illustrate this, Fig. 4.10 shows a triangle with the sides $I(T)$ and $R(T)$ and the hypotenuse at the angle α, where $\tan\alpha = PE(T)$.

Two more values result from drawing horizontal lines through the throughput diagram. The lead time $TL(T)$ tells us what actual lead time was needed for the last lot processed at the time T; the advance time $TA(T)$ is the time difference between the input curve and the output curve at the time T. The reason for the difference between the two values is the transposition of input and output sequence. If the orders were processed in the same sequence as they arrived, i.e. by following the priority rule FIFO (first in, first out), the input and output curves would be identical and the values $TL(T)$ and $TA(T)$ would be equal for every time T.

For the case examined, we can see from Table 4.1, Column 8, Line 31, that Order

140 was processed on Day 227 and that its input date was Day 222. The lead time $TL(T)$ for Day 227 was therefore five days.

In contrast, the advance time $TA(T)$ can only be calculated in a roundabout way: we have to find the date on which the input curve has the same value as the output curve at time T. In this case, the value – as discovered when determining the inventory – is 125.2 hours. Table 4.1, Column 7, Line 24, shows that the value of the input curve is just over this value (125.4 hours) on Day 208 for Order 126. Thus, the advance time $TA(T)$ with T = Day 227, is $227 - 208 = 19$ days.

In practice, a daily measurement of inventory, range, lead time and advance time is of little use, since there are great variances in the values because of the great unevenness of the input and output curves (due to the variance of order times), the irregular intervals between input and output events, and the arbitrary transpositions in the sequence of orders. Moreover, no analytical relationship can be determined between them. Instead, discontinuous processes should be studied period by period, and mean values should be determined for these quantities. Such mean values can, first, be used for process monitoring and, second, as scheduled values for controlling the following periods. Therefore, in the following we first define the mean inventory and then the three mean, period-related time-function quantities range, advance time, and lead time in the work center throughput diagram, and show their interrelationships. As the period length we have chosen a time interval of four weeks, because other monitoring data in business management such as order entry, turnover, and production efficiency, are commonly determined in the same monthly rhythm. The period length chosen here, however, is only an example and can also, of course, have other values, e.g. one, two or three weeks.

4.4.2 Mean Inventory

The mean inventory in a period is calculated from the total of the individual inventories for each day of the period P, divided by the number of days in the period P. Table 4.2 shows the calculation of the inventory trend from inputs and outputs in the reference period (Days 205 to 232), and the calculation of the mean inventory. Note that all input and output events for one day are reported at the end of that day. This means that the corresponding changes in inventory do not take effect until the next day. In our example, the inventory is 95.8 hours at the beginning of the reference period, which is the end of Day 204 or the beginning of Day 205. This inventory exists for the whole of Day 205, thus resulting in an inventory tie-up, which can also be interpreted as the inventory area, of 95.8 hours × day. Input during Day 205 amounts to 9.6 hours, which leads to an inventory of 105.4 hours at the end of Day 205. This inventory is valid for the whole of next day; the inventory area for Day 206 is therefore 105.4 hours × day. The inputs and outputs during Day 206 lead to an inventory of 108.7 hours at the end of the day, which determine the inventory area on Day 207, etc.

The mean inventory (I_m) can thus be described as.

$$I_m = \frac{AI}{P} \tag{4.2}$$

where:

AI = inventory area (= total inventory) in reference period
P = reference period in days.

Table 4.2. Calculation of inventory trend, mean inventory, and mean range at a work center (in hours or calendar days)

Day (No.)	Input INP (Hrs)	Output OUT (Hrs)	Inventory Area AI (Hrs x Days)	Inventory at the End of Day I (Hrs)
1	2	3	4	5
			Initial Inventory	95.8
205	9.6	.0	95.8	105.4
206	3.8	.5	105.4	108.7
207	13.6	11.4	108.7	110.9
208	7.9	19.2	110.9	99.6
209	.0	6.8	99.6	92.8
210	.0	.0	92,8	92.8
211	.0	.0	92.8	92.8
212	3.8	9.6	92.8	87.0
213	6.8	13.0	87.0	80.8
214	.0	8.8	80.8	72.0
215	13.8	.0	72.0	85.8
216	.0	.0	85.8	85.8
217	.0	.0	85.8	85.8
218	.0	.0	85.8	85.8
219	6.9	2.6	85.8	90.1
220	.0	9.5	90.1	80.6
221	.0	.0	80.6	80.6
222	32.0	18.0	80.6	94.6
223	57.4	.0	94.6	152.0
224	.0	.0	152.0	152.0
225	.0	.0	152.0	152.0
226	19.1	20.7	152.0	150.4
227	5.1	5.1	150.4	150.4
228	.0	.0	150.4	150.4
229	.0	18.7	150.4	131.7
230	5.1	2.1	131.7	134.7
231	.0	.0	134.7	134.7
232	.0	.0	134.7	134.7
			Final Inventory	134.7
28	184.9	146.0	3,036.0	< TOTAL

Mean Inventory Im = AI/P = (3,036.0 Hrs x Days)/28 Days = 108.4 Hrs

Mean Range Rm = AI/OUT = (3,036.0 Hrs x Days)/146 Hrs = 20.8 Days

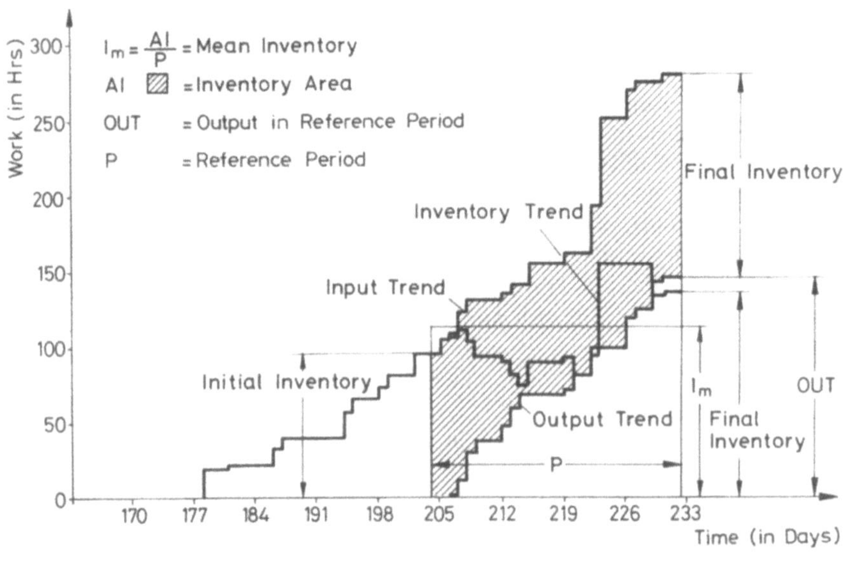

Fig. 4.11. Inventory trend and mean inventory in the throughput diagram (After Bechte [7])

If we now plot this into a throughput diagram (Fig. 4.11), the calculated total inventory area corresponds exactly to the hatched area, which is consequently called the inventory area (AI). Besides the inventory trend, the throughput diagram also gives the mean inventory I_m, which was calculated from Table 4.2: 3036.0/28 = 108.4 hours. The unit for the time axis in Fig. 4.11 is 1 day, since the values in Table 4.2 are given in whole calendar days. If changes in inventories are reported by the hour, or even by the minute, the time axis must have the corresponding unit. In practice, the unit shop calendar day (SCD) is used in most cases.

4.4.3 Mean Time Data

4.4.3.1 Mean Range

The mean range (R_m) in a period P is, according to the definition of the current range, calculated as:

$$R_m = \frac{I_m}{PE_m} \tag{4.3}$$

where:

I_m = mean inventory in reference period P
PE_m = OUT/P = mean performance in reference period P
OUT = output in reference period P
P = reference period (= length of current reference period in days).

If this is plotted into a throughput diagram, the following relationships result (Fig. 4.12). The mean inventory in the period P is the total of the individual inventories for each day of the period P, divided by the total of the days contained in period

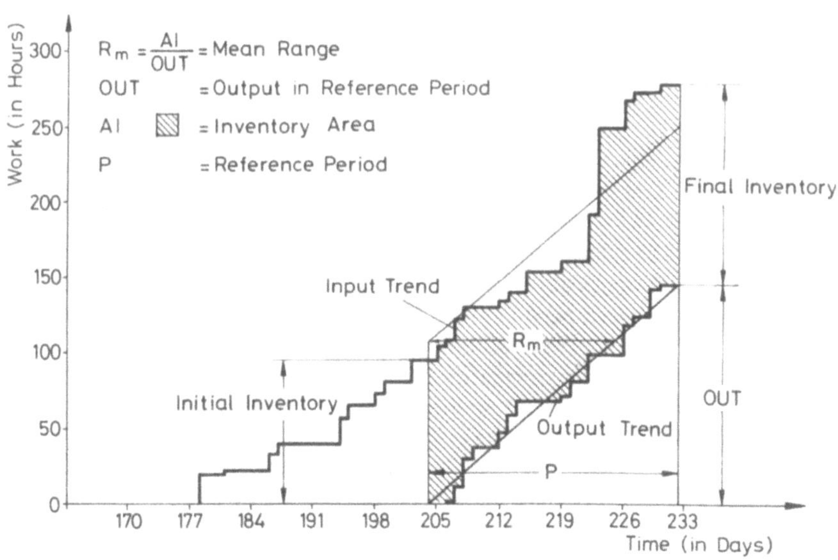

Fig. 4.12. Mean range in the throughput diagram (After Bechte [7])

P. The total of individual inventories corresponds exactly to the hatched area AI known from Fig. 4.11.

The mean performance (PE_m), or output rate, is the total of the outputs (OUT) in the period P, divided by the number of days in period P. From this follows:

$$R_m = \frac{AI}{OUT} \tag{4.4}$$

where:

AI = inventory area of reference period P
OUT = output in reference period P.

Using the values from Table 4.2, we can calculate R_m as 3,036.0 hours × days, divided by 146 hours, which equals 20.8 days.

4.4.3.2 *Mean Advance Time*

Advance time has been defined as the time difference between the input and output curves. If we want to determine the mean advance time (TA_m) in a period P, we must cumulate the individual values $TA(T)$ in the reference period and then find the mean value.

Table 4.3 shows the necessary calculations. The single calculation steps are illustrated. First, the auxiliary area $A1$ is calculated from the section of the input curve which runs from zero to the value OUT of the output curve. The values in Table 4.3A, which was set up for this calculation, were based on Column 7 of Table 4.1. To determine the advance time, the input curve must have the same value as the output curve at the date concerned, as shown above. At the beginning of the reference period (Day 205), the initial value of the output curve is zero by definition. The crossover point of the input curve is – as shown in Table 4.1, Column 7, Line 20 – on Day 178, with Order 113. Since only part of the order, namely 20.4 of 30.3 hours, is on the positive branch of the input curve, Order 113 is 'cut in two'; this means that only this portion is used to calculate the area $A1$ (cf. Table 4.3, Column 5). Then all orders, arranged according to their input dates, are listed until a cumulated input of 146 order hours is reached, which corresponds to the output reported over the reference period. The last individual value represents only part of Order 132, namely 4.7 of 13.8 hours, since only this portion lies within the 146 hours of order time concerned. Thus, Order 132 is also 'cut in two'. The rest of the calculation can be seen from Table 4.3A.

Table 4.3B for the calculation of the auxiliary area $A2$ was produced in the same manner from the output curve. The individual values were based on Column 8 of Table 4.1, and were rearranged in order of increasing output dates. The calculation results in a mean advance time value (TA_m) of 20.9 days.

Figure 4.13 illustrates the calculation. The cumulated advance time values correspond with the hatched advance time area ATA. The mean value TA_m is obtained by dividing this advance time area by the output OUT. The area of the parallelogram with the altitude OUT and the base TA_m corresponds to the hatched area ATA.

As already been shown in Fig. 4.13, mean advance time is then

$$TA_m = \frac{ATA}{OUT} \tag{4.5}$$

Table 4.3. Calculation of mean advance time at a work center, based on the outputs in the reference period and the inputs up to the output level in the reference period (in calendar days)

A) In Order of Input Dates

Order Number (-)	Input Date tPEU (Day)	Order Time TO (Hrs)	End of Reference Period Minus Input Date (Days)	Advance Time Area A1 (Hrs x Days)
1	2	3	4	5
113	178	20.4*)	54	1,101.6
115	181	.5	51	25.5
114	186	11.4	46	524.4
116	187	7.4	45	333.0
119	194	11.4	38	433.2
120	194	6.8	38	258.4
118	195	7.7	37	284.9
112	198	3.4	34	115.6
117	198	4.2	34	142.8
121	199	8.8	33	290.4
108	202	13.8	30	414.0
124	205	9.6	27	259.2
125	206	3.8	26	98.8
123	207	13.6	25	340.0
126	208	2.6	24	62.4
127	208	5.3	24	127.2
122	212	3.8	20	76.0
131	213	2.1	19	39.9
128	213	4.7	19	89.3
132	215	4.7**)	17	79.9
	TOTAL >	146.0		5,096.5

*) Order 113: Only 20.4 of Total 30.3 Hours are relevant for Reference Period Day 205 to 232
**) Order 132 : Only 4.7 of Total 13.8 Hours are relevant for Reference Period Day 205 to 232

B) In Order of Output Dates

Order Number (-)	Output Date tPE (Day)	Order Time TO (Hrs)	End of Refer Period Minus Output Date (Days)	Advance Time Area A2 (Hrs x Days)
6	7	8	9	10
115	206	.5	26	13.0
119	207	11.4	25	285.0
110	208	15.4	24	396.6
125	208	3.8	24	91.2
120	209	6.8	23	156.4
124	212	9.6	20	192.0
118	213	7.7	19	146.3
127	213	5.3	19	100.7
121	214	8.8	18	158.4
126	219	2.6	13	33.8
131	220	2.1	12	25.2
116	220	7.4	12	88.8
108	222	13.8	10	138.0
117	222	4.2	10	42.0
135	226	6.9	6	41.4
132	226	13.8	6	82.8
140	227	5.1	5	25.5
123	229	13.6	3	40.8
145	229	5.1	3	15.3
142	230	2.1	2	4.2
	TOTAL >	146.0		2,050.4

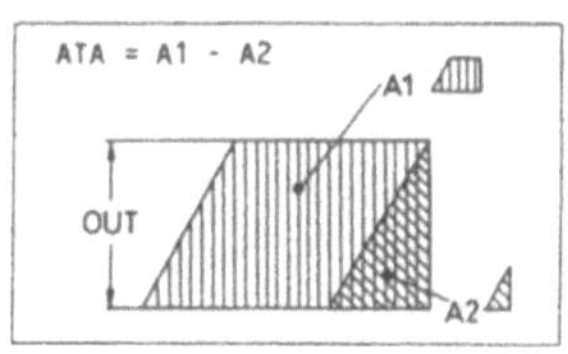

Sketch for Calculation of ATA

Mean Advance Time TAm = ATA/OUT = (A1 - A2)/OUT
= (5,096.5 - 2,050.4)/146
= (3,046.1 Hrs x Days)/146 Hrs = 20.9 Days
= 20.9 Days

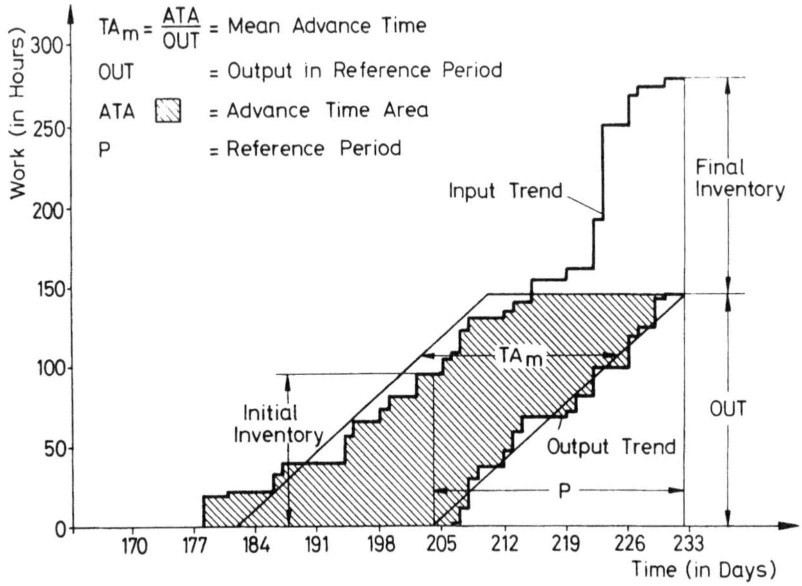

Fig. 4.13. Mean advance time in the throughput diagram (After Bechte [7])

where:

ATA = advance time area in reference period P
OUT = outputs in reference period P.

4.4.3.3 Weighted Mean Lead Time

The actual mean lead time (or flow time) expresses how long, on average, the work contents of the orders processed in one period have been at an individual work center. The concept of mean lead time was already discussed in Sect. 3.4, and the reasons why calculations of the lead times of operations at a work center must be based on the weighted lead time were shown using the funnel model. This very important concept of weighted lead time is now illustrated with the throughput diagram.

Table 4.4 shows the orders leaving the work center in the reference period with their weighted lead times (Column 6) and the sum of the weighted lead times. Since the orders in Table 4.4 are the same as those in Table 3.1, the values of the simple and weighted mean lead times, 14.2 days and 16.6 days respectively, are the same, as expected.

The weighted lead time is illustrated in Fig. 4.14. In addition to the input and output curves, the weighted throughput element is plotted for each order. As shown *in* Fig. 3.9, it consists of a rectangle, the length and altitude of which correspond to the value output date minus input date and the value of the order time (TO) respectively.

For the weighted mean lead time (TLM) it follows that

Table 4.4. Calculation of the weighted mean lead time of a work center, based on the feedback records in the order of the output dates (in calendar days)

Order Number (-)	Input Date tPEU (Day)	Output Date tPE (Day)	Order Time TO (Hrs)	Lead Time TL (Days)	Lead Time Area ALT (Hrs x Days)
1	2	3	4	5	6
115	181	206	.5	25	12.5
119	194	207	11.4	13	148.2
110	170	208	15.4	38	585.2
125	206	208	3.8	2	7.6
120	194	209	6.8	15	102.0
124	205	212	9.6	7	67.2
118	195	213	7.7	18	138.6
127	208	213	5.3	5	26.5
121	199	214	8.8	15	132.0
126	208	219	2.6	11	28.6
131	213	220	2.1	7	14.7
116	187	220	7.4	33	244.2
108	202	222	13.8	20	276.0
117	198	222	4.2	34	100.8
135	219	226	6.9	7	48.3
132	215	226	13.8	11	151.8
140	222	227	5.1	5	25.5
123	207	229	13.6	22	299.2
145	227	229	5.1	2	10.2
142	226	230	2.1	4	8.4
20	< TOTAL >		146.0	284	2,427.5

Weighted Mean Lead Time TLM = ALT/OUT
= (2,427.5 Hrs x Days)/146 Hrs ≈ 16.6 Days

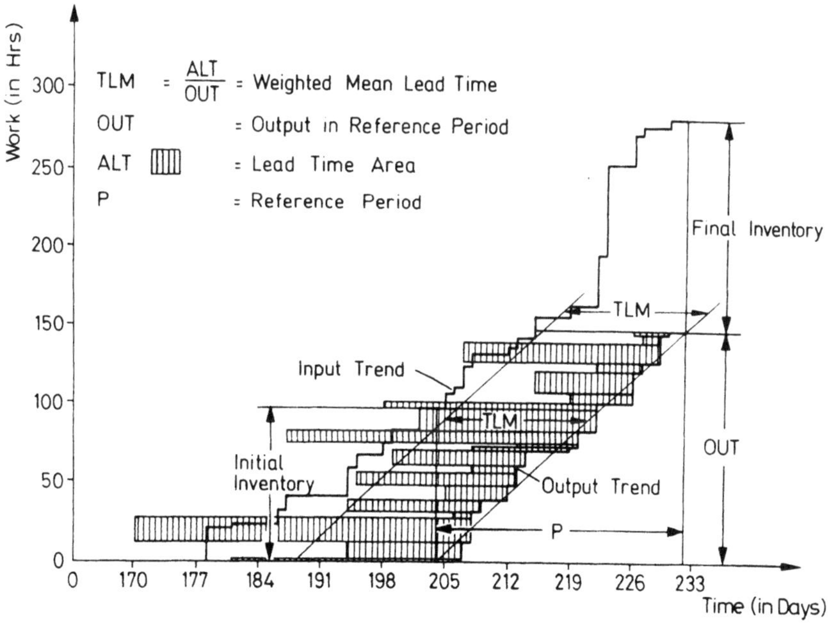

Fig. 4.14. Weighted mean lead time in the throughput diagram (After Bechte [7])

$$TLM = \frac{ALT}{OUT} \tag{4.6}$$

where:

ALT = lead time area (= total weighted lead time)
OUT = output in reference period P.

Although the abbreviation TL_{mw} was used for the weighted mean lead time in Chapt. 3, we use TLM from now on to emphasize the difference between the statistical counting of lead times applied up to now and the calculation of weighted lead time from the throughput diagram. In the following, we only use TLM, except for references to Chapt. 3.

The parallelogram with the base TLM and the altitude OUT thus represents the lead time area ALT in the reference period P.

There is another important message which can be deduced from Fig. 4.14: the dispatching procedure for the orders at this work center. The orders were obviously not processed according to the sequence of arrival (FIFO), since that would mean that all throughput elements would lie exactly between the input and output curves, and that their sum would correspond to the advance time area (ATA) in Fig. 4.13. In our case, one cannot distinguish any dispatching rule, which is common practice.

However, the figure also tells us that the usual mean value of the single lead times does not express very much, since the number of orders does not have any relationship to the work content. Consequently, all further considerations will be based on the weighted mean lead time, called TLM.

4.4.4 Relationships between Mean Range, Mean Advance Time, and Weighted Mean Lead Time

4.4.4.1 *Inventory Trend Component of Lead Time*

An obvious first step is to set up an analytical relationship between the mean range (R_m) and the mean advance time (TA_m), since both are based on a common inventory area between the input and output curves. Figure 4.15 was produced by combining Figs. 4.12 and 4.13. It allows us to deduce directly the following relationship, which uses Bechte's [7] concepts of initial inventory area (AII) and final inventory area (AFI):

$$AII + AI = ATA + AFI \tag{4.7}$$

A test calculation for our sample work center produces the following:

$$AII = 1241.4 \text{ hrs} \times \text{days (see Table 4.5)}$$

$$AI = 3036.0 \text{ hrs} \times \text{days (see Table 4.2)}$$

$$ATA = 3046.1 \text{ hrs} \times \text{days (see Table 4.3)}$$

$$AFI = 1231.3 \text{ hrs} \times \text{days (see Table 4.5.)}$$

$$1{,}241.4 + 3{,}036.0 = 3{,}046.1 + 1{,}231.3$$

$$4{,}277.4 = 4{,}277.4$$

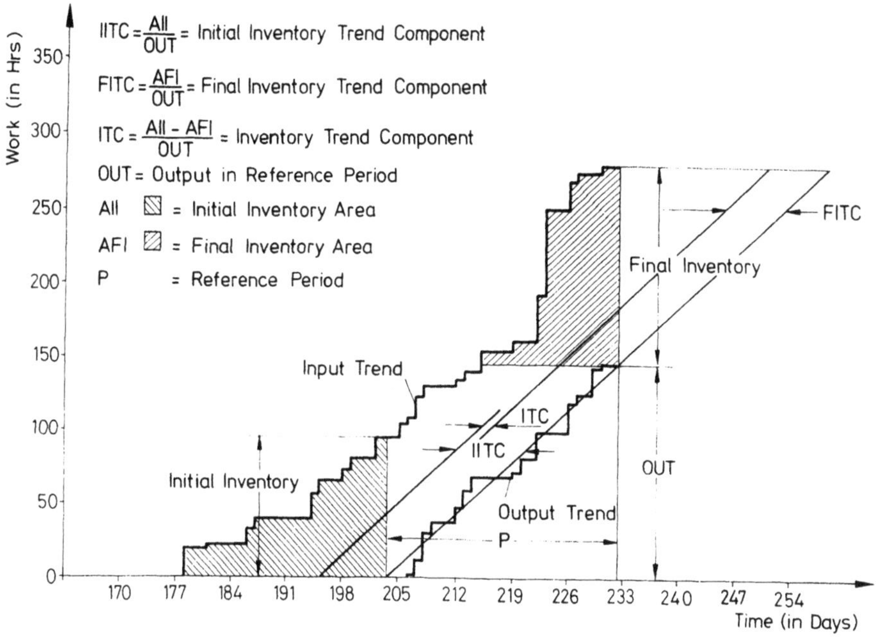

Fig. 4.15. Inventory trend component in the throughput diagram (After Bechte [7])

Furthermore, it follows from Eqn. (4.7) that

$$ATA = AI + (AII - AFI).$$

With the equations

$$R_m = \frac{AI}{OUT} \quad \text{(Eqn. 4.4)} \quad \text{and} \quad TA_m = \frac{ATA}{OUT} \quad \text{(Eqn. 4.5)}$$

where:

AI = inventory area in reference period P
ATA = advance time area in reference period P
OUT = output in reference period P,

and a so-called range of initial inventory defined as

$$RII = \frac{AII}{OUT}$$

and a range of final inventory defined as

$$RFI = \frac{AFI}{OUT}$$

we can define an inventory trend component as:

$$ITC = \frac{AII - AFI}{OUT} \tag{4.8}$$

The inventory trend component tells us how the inventory range changes between the initial and the final inventories due to changes in inventory during the period P. Table 4.5 gives the calculation of our sample work center. Thus the inventory trend component corresponds exactly to the difference between the mean advance time and the mean range:

$$TA_{\mathrm{m}} = R_{\mathrm{m}} + ITC \tag{4.9}$$

This can be proved with a sample calculation:

Table 4.5. Calculation of the inventory trend component (in calendar days)

A) Initial Inventory Area AII (Historical Input Curve from Zero to Initial Inventory)

Order Number (-)	Input Date tPEU (Day)	Order Time TO (Hrs)	Time up to Day 204 (Day)	Initial Inventory Area AII (Day x Hrs)
1	2	3	4	5
113	178	20.4*)	26	530.4
115	181	.5	23	11.5
114	186	11.4	18	205.2
116	187	7.4	17	125.8
119	194	11.4	10	114.0
120	194	6.8	10	68.0
118	195	7.7	9	69.3
112	198	3.4	6	20.4
117	198	4.2	6	25.2
121	199	8.8	5	44.0
108	202	13.8	2	27.6
11	< TOTAL >	95.8		1,241.4

*) 20.4 Hrs out of 30.3 Hrs

B) Final Inventory Area AFI (Historical Input Curve from Output (Day 232) up to Final Inventory)

Order Number (-)	Input Date tPEU (Day)	Order Time TO (Hrs)	Time up to Day 232 (Day)	Final Inventory Area AFI (Day x Hrs)
1	2	3	4	5
132	215	9.1*)	17	154.7
135	219	6.9	13	89.7
140	222	5.1	10	51.0
136	222	26.9	10	269.0
141	223	57.4	9	516.6
142	226	2.1	6	12.6
129	226	14.1	6	84.6
143	226	2.9	6	17.4
145	227	5.1	5	25.5
134	230	5.1	2	10.2
10	< TOTAL >	134.7		1,231.3

*) 9.1 Hrs out of 13.8 Hrs

Range of Initial Inventory RII =
AII/OUT = (1,231.4 Hrs x Days)/146 Hrs = 8.5 Days

Range of Final Inventory RFI =
AFI/OUT = (1,231.3 Hrs x Days)/146 Hrs = 8.4 Days

Inventory Trend Component ITC =
(AII - AFI)/OUT = (1,241.4 - 1,231.3 Hrs x Days)/146 Hrs = .1 Days

$$TA_m = \frac{ATA}{OUT} = \frac{3046.1 \text{ hrs} \times \text{days}}{146 \text{ hrs}} = 20.9 \text{ days}$$

$$R_m = \frac{AI}{OUT} = \frac{3036.9 \text{ hrs} \times \text{days}}{146 \text{ hrs}} = 20.8 \text{ days}$$

$$ITC = \frac{AII - AFI}{OUT} = \frac{1241.4 - 1231.3 \text{ hrs} \times \text{days}}{146 \text{ hrs}} = 0.1 \text{ days}$$

20.9 days = 20.8 days + 0.1 day.

Despite a considerable difference between the initial and final inventories, the inventory trend component in our example is very small, because the initial and final inventory areas happen to have nearly the same values.

Figure 4.15 illustrates *RII*, *RFI*, and *ITC*. If the initial and final inventory areas are of equal size, mean advance time and mean range are also equal.

4.4.4.2 Sequence Component of Weighted Lead Time

The calculation for determining the relationship between advance time and range is based on the inventory areas, and the ratio between the inventory and the output in the reference period. To interrelate the weighted lead time and the advance time (and thus the range), however, we must compare the throughput elements in the different sections of the input and output curves and relate them to the output in the reference period. We must therefore determine the analytical relationship between the advance time area (Fig. 4.13) and the lead time area (Fig. 4.14). Let us start with the following consideration: at the beginning of the reference period there is an initial inventory with known throughput elements. By assigning the throughput elements to the respective orders in the initial inventory and ordering these according to their input dates, one obtains a second input curve which is normally different from the input curve produced by updating the input events. The difference is due to the fact that there are orders contained in the 'historical' input curve that were processed before the beginning of the reference period.

Figure 4.16 shows the throughput elements of the orders contained in the initial inventory, which lie partly inside and partly outside the 'historical' input curve (see Table 4.6A). The difference between the initial inventory area (*AII*) and the area of the throughput elements of the inital inventory is defined as the initial inventory additional area (*AIIA*) [7].

The area enclosed by this new input curve and the output curve must correspond to the sum of all the throughput elements in the reference period.

For this, all the throughput elements were plotted in Fig. 4.17. We can see the throughput elements derived from the output curve with their area *ALT* as well as the throughput elements contained in the final inventory, which also deviate at times from the 'historical' input curve (see Table 4.6B). The difference between the final inventory area (*AFI*) and the area of the throughput elements in the final inventory is defined as the final inventory additional area (*AFIA*). While the initial and final inventory areas represent the 'minimum age' of the inventory (when the FIFO rule is applied), they add up to an 'effective age' of the inventory together with the initial and final inventory additional areas. The latter are fictitious areas, which cannot be

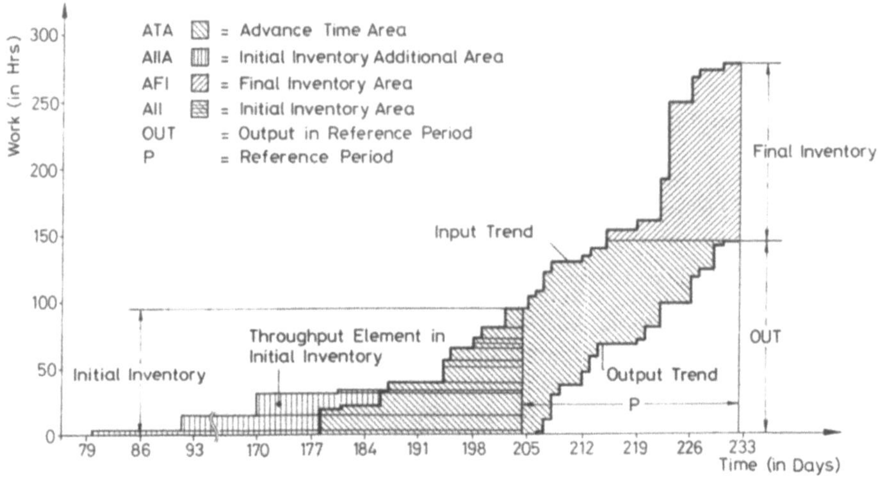

Fig. 4.16. Initial inventory additional area in the throughput diagram (After Bechte [7])

found directly in Figs. 4.10–4.15. The additional areas are always zero when the FIFO rule is applied.

The following area equation can, however, be derived from a comparison of Figs. 4.16 and 4.17:

$$AIIA + ATA + AFI = ALT + AFI + AFIA \tag{4.10}$$

It follows that $ALT = ATA + (AIIA - AFIA)$. With Eqn. (4.5), $TAm = \dfrac{ATA}{OUT}$ and Eqn. (4.6), $TLM = \dfrac{ALT}{OUT}$, and the following definition of a so called *sequence component of lead time* ($TLSC$)

$$TLSC = \frac{AIIA - AFIA}{OUT} \tag{4.11}$$

it follows that this sequence component describes the change in lead time between the initial and final inventories due to the respective dispatching procedure and thus corresponds exactly to the difference between the mean advance time and the mean lead time

$$TLM = TAm + TLSC \tag{4.12}$$

In Table 4.6 the sequence component is calculated from the differences between the single areas in the throughput diagram, as is the course of the input function of the initial and final inventories. In this case, a negative sequence component value of −4.2 days was found, which indicates that in contrast to the FIFO rule, lots with later input dates were largely processed. This deduction can be proved with a sample calculation, since TLM had equalled 16.6 days and TAm 20.9 days. (The difference of 0.1 day is due to rounding errors.)

Table 4.6. Calculation of the sequence component (in calendar days)

A) Initial Inventory Additional Area

A1) Lead Time Area of Initial Inventory

Order Number	Input Date tPEU	Order Time TO	Cumul. Input up to Day 204	Time up to Day 204	Lead Time Area of Initial Inventory ALTI
(-)	(Day)	(Hrs)	(Hrs)	(Day)	(HrsxDay)
1	2	3	4	5	6
103	81	3.3	3.3	123	405.9
104	91	13.1	16.4	113	1,480.3
110	170	15.4	31.8	34	523.6
115	181	0.5	32.3	23	11.5
116	187	7.4	39.7	17	125.8
119	194	11.4	51.1	10	114.0
120	194	6.8	57.9	10	68.0
118	195	7.7	65.6	9	69.3
112	198	3.4	69.0	6	20.4
117	198	4.2	73.2	6	25.2
121	199	8.8	82.0	5	44.0
108	202	13.8	95.8	2	27.6
12	<TOTAL>	95.8	95.8	TOTAL >	2,915.6

A2) Initial Inventory Area

Order Number	Input Date tPEU	Order Time TO	Time up to Day 204	Initial Inventory Area AII
(-)	(Day)	(Hrs)	(Day)	(HrsxDay)
1	2	3	4	5
113	178	20.4*)	26	530.4
115	181	.5	23	11.5
114	186	11.4	18	205.2
116	187	7.4	17	125.8
119	194	11.4	10	114.0
120	194	6.8	10	68.0
118	195	7.7	9	69.3
112	198	3.4	6	20.4
117	198	4.2	6	25.2
121	199	8.8	5	44.0
108	202	13.8	2	27.6
11	<TOTAL>	95.8	TOTAL >	1,241.4

*) 20.4 Hrs out of 30.3 Hrs

Initial Inventory Additional Area AIIA = ALTI - AII
= 2,915.6 - 1,241.4 = 1,674.2 Hrs x Days

B) Final Inventory Additional Area

B1) Lead Time Area of Final Inventory

Order Number	Input Date tPEU	Order Time TO	Cumul. Input up to Day 232	Time up to Day 232	Lead Time Area of Final Inventory ALTF
(-)	(Day)	(Hrs)	(Hrs)	(Day)	(HrsxDay)
1	2	3	4	5	6
103	81	3.3	149.3	151	498.3
104	91	13.1	162.4	141	1,847.1
112	198	3.4	165.8	34	115.6
122	212	3.8	169.6	20	76.0
128	213	4.7	174.3	19	89.3
136	222	26.9	201.2	10	269.0
141	223	57.4	258.6	9	516.6
129	226	14.1	272.7	6	84.6
143	226	2.9	275.6	6	17.4
134	230	5.1	280.7	2	10.2
10	<TOTAL>	134.7	280.7	TOTAL >	3,524.1

B2) Final Inventory Additional Area

Order Number	Input Date tPEU	Order Time TO	Time up to Day 232	Final Inventory Area AFI
(-)	(Day)	(Hrs)	(Day)	(HrsxDay)
1	2	3	4	5
132	215	9.1*)	17	154.7
135	219	6.9	13	89.7
136	222	26.9	10	269.0
140	222	5.1	10	51.0
141	223	57.4	9	516.6
142	226	2.1	6	12.6
143	226	2.9	6	17.4
129	226	14.1	6	84.6
145	227	5.1	5	25.5
134	230	5.1	2	10.2
10	<TOTAL>	134.7		1,231.3

*) 9.1 Hrs out of 13.8 Hrs

Final Inventory Additional Area AFIA = ALTF - AFI
= 3,524.1 - 1,231.3 = 2,292.8 Hrs x Days

Sequence Component SC = (AIIA - AFIA)/OUT
= (1,674.2 - 2,292.8)/146 = (-618.6 Hrs x Days)/146 Hrs = -4.2 Days

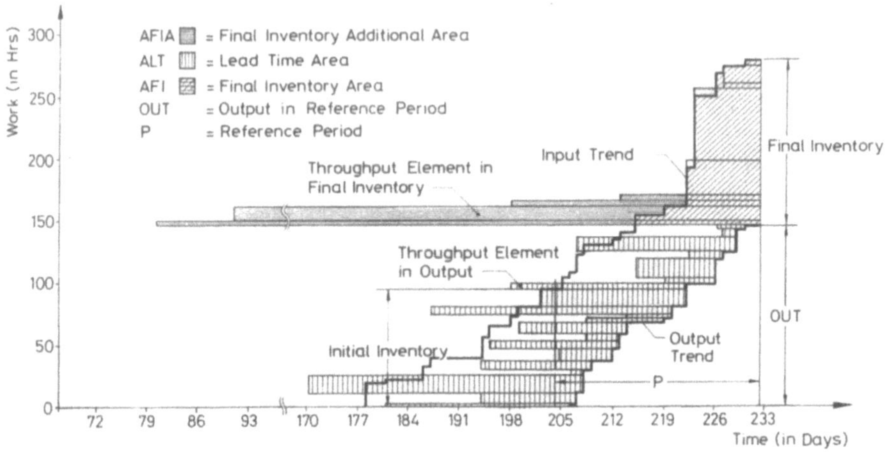

Fig. 4.17. Final inventory additional area in the throughput diagram (After Bechte [7])

4.4.4.3 *Short- and Long-Term Relationships between Time Quantities*

Mean range, mean advance time and mean lead time are related to the output trend, which starts at the zero point of the reference period and ends with the value *OUT*.

They have the following interrelationships:

$$\text{Eqn. (4.12):} \quad TLM = TA_m + TLSC$$
$$\text{Eqn. (4.9):} \quad TAm = R_m + ITC$$

It follows that

$$TLM = R_m + TLSC + ITC \tag{4.13}$$

The weighted mean lead time at a work center thus corresponds to the mean range, adjusted for the changes resulting from the different initial and final inventories. These are changes in the sequence (*TLSC*) of the orders in the initial and final inventories, and changes in the inventory (*ITC*) in the initial and final inventories.

Seen over longer periods of time, the sequence and inventory trend components disappear, since the variations in the single periods cancel each other out. Thus, it follows for stretches of time which are considerably greater than *TLM*:

$$TLM = TAm = Rm$$

As will be shown in Chapt. 5, considerable deviations between lead time, range and advance time can occur, which indicates uneven input and output trends as well as a deviation from the FIFO rule. These data can be used to introduce measures for improving the manufacturing process.

4.4.5 Mean Performance, Mean Capacity, and Mean Utilization

To find the performance and capacity curves, we only need the output curve. Figure 4.18 shows the output curve plotted with data from Table 4.3B.

In this case, 146 output hours were reported in the reference period P. If we define performance as work per unit time – as is common practice – mean performance (PE_m) in the reference period P is then defined as

$$PE_m = \frac{OUT}{P} \tag{4.14}$$

where:

OUT = outputs in reference period P
P = reference period in days.

In our case, PE_m is:

$$PE_m = 146/28 = 5.2 \text{ hours per day.}$$

This value is, however, not very useful, since P also contains non-working days. Mean performance should therefore be related to workdays, or shop calendar days (SCD). From Fig. 4.18, we can see that there were 20 workdays in the reference

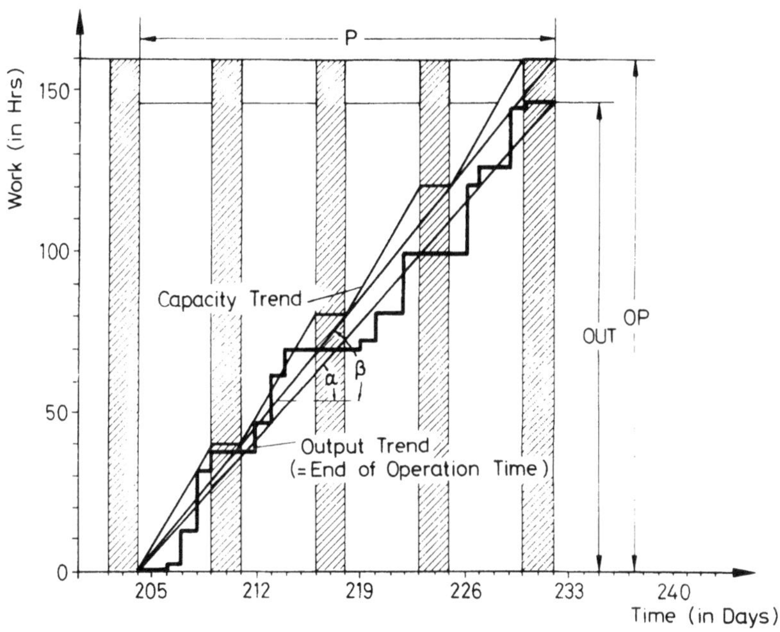

Non-productive Calendar Days

$\tan\alpha = PE_m = \frac{OUT}{P}$ = Mean Performance within Reference Period

$\tan\beta = C_m = \frac{OP}{P}$ = Mean Capacity within Reference

OUT = Actual Output in Reference Period

OP = Potential Output in Reference Period

$L_m = \frac{PE_m}{C_m} = \frac{OUT}{OP}$ = Mean Utilization within Reference Period

P = Reference Period

Fig. 4.18. Performance, capacity, and utilization in the throughput diagram period, which gives us a PE_m value of 146/20 = 7.3 hours per workday. PE_m can be interpreted as a straight line with the angle of inclination α. It was also included in Fig. 4.18.

Besides the work reported, the potential output (i.e. the capacity) can be plotted in the same diagram. In our example, we assume that the work center has a capacity of 8 hours per day (for a single shift), giving us the capacity curve plotted in Fig. 4.18, which reaches the potential output value OP at the end of the reference period.

The mean capacity (C_m) in the reference period can then be defined as:

$$C_m = \frac{OP}{P} \tag{4.15}$$

where:

OP = scheduled output in reference period P
P = reference period in days.

Mean capacity should also be related to workdays which, of course, results in this case in the value of the daily capacity (C_m = 8 hours per workday). The potential output (OP) is $20 \times 8 = 160$ hours. The mean capacity can be interpreted as a straight line with the angle of inclination β.

Mean utilization (U_m) of a work center in the reference period is:

$$U_m = \frac{PE_m}{C_m} = \frac{OUT}{OP} \tag{4.16}$$

In our case, this value is:

$$U_m = 7.3/8 = 0.913, \text{ or } 91.3\%$$

This calculation was based on the assumption that the hours actually processed correspond to the order time (TO). If this is not the case, as, for example, with piece-work wages, the reported hours must be used as the ouput value.

4.4.6 Relationships between Mean Inventory, Mean Performance, and Weighted Mean Lead Time

With the range (R_m) defined as

$$\text{(Eqn. 4.4)} \quad R_m = \frac{AI}{OUT}$$

and mean inventory (I_m) defined as

$$\text{(Eqn. 4.2)} \quad I_m = \frac{AI}{P}$$

and mean performance (PE_m) as

$$\text{(Eqn. 4.14)} \quad PE_m = \frac{OUT}{P}$$

it follows directly that

$$R_{\mathrm{m}} = \frac{AI}{OUT} = \frac{I_{\mathrm{m}} \times P}{PE_{\mathrm{m}} \times P}$$

$$R_{\mathrm{m}} = \frac{I_{\mathrm{m}}}{PE_{\mathrm{m}}} \tag{4.17}$$

where:

AI = inventory area in reference period P
OUT = outputs in reference period P
P = reference period.

Since, over longer periods of time, $R_{\mathrm{m}} = TA_{\mathrm{m}} = TLM$, it follows that

$$TLM = \frac{I_{\mathrm{m}}}{PE_{\mathrm{m}}} \tag{4.18}$$

As Bechte proved in empirical studies, this equation is also valid in practice [7], which means that the mean lead time of a work center over a longer period of time can generally be described with the so called *funnel* formula:

$$\text{Weighted Mean Lead Time} = \frac{\text{Mean Inventory}}{\text{Mean Performance}}$$

The relationship, deduced only qualitatively with the funnel in Fig. 3.8, can thus also be proved mathematically. Conway et al. treated this problem analytically [2]. Starting from the general queuing formula $\bar{F} = \bar{N}/\lambda$ (where $\bar{F}$ is the mean flow time, $\bar{N}$ the average number of jobs in the system and λ the mean arrival rate), they prove that this relationship can also be expressed in terms of average utilization and average amount of work per job. From this they derive (p. 20) the following formula for the average flow time $\bar{F}$:

$$\bar{F} = \frac{\bar{N} \times \bar{p}}{m \times \bar{U}} \tag{4.19}$$

where:

$\bar{N}$ = average number of jobs in the system
$\bar{p}$ = average work content of the jobs in the system
m = number of machines in the system
$\bar{U}$ = average utilization of the machines.

Although it seems to be the same formula as that for calculating weighted mean lead time, it does not produce the same values. Rather, it gives the values of the simple mean lead time.

4.4.7 Weighted Mean Lateness

Whereas inventory, lead time, and utilization are quite tangible concepts, measurement and calculation of lateness frequently fail owing to the lack of a precise definition. As the concepts explained above were related to the processes at a work center and represented in the work center throughput diagram, we first explain lateness also in this way.

Table 4.7. Calculation of the lateness key data of the work center output function (in calendar days)

Order Number	Actual Completion Date tPEa	Planned Completion Date tPEr	Order Time TO	Order Lateness Output OLO	Weighted Lateness Output OLO x TO	Simple Quadratic Deviation (OLOm-OLO)²	Weighted Quadratic Deviation (OLOmw-OLO)²xTO
(-)	(Day)	(Day)	(Hrs)	(Days)	(Days x Hrs)	(Days²)	(Days² x Hrs)
1	2	3	4	5	6	7	8
115	206	191	0.5	15	7.5	112.4	25.2
119	207	204	11.4	3	34.2	2.0	273.7
110	208	180	15.4	28	431.2	557.0	6,221.8
125	208	216	3.8	- 8	- 30.4	153.8	960.7
120	209	204	6.8	5	34.0	.4	57.2
124	212	215	9.6	- 3	- 28.8	54.8	1,140.6
118	213	212	7.7	1	7.7	11.6	366.6
127	213	218	5.3	- 5	- 26.5	88.4	882.0
121	214	214	8.8	0	.0	19.4	549.2
126	219	218	2.6	1	2.6	11.6	123.8
131	220	223	2.1	- 3	- 6.3	54.8	249.5
116	220	202	7.4	18	133.2	185.0	754.9
108	222	187	13.8	35	483.0	936.4	10,134.9
117	222	212	4.2	10	42.0	31.4	18.5
135	226	229	6.9	- 3	- 20.7	54.8	819.8
132	226	225	13.8	1	13.8	11.6	657.0
140	227	232	5.1	- 5	- 25.5	88.4	848.7
123	229	218	13.6	11	149.6	43.6	130.7
145	229	237	5.1	- 8	- 40.8	153.8	1,289.3
142	230	236	2.1	- 6	- 12.6	108.2	405.7
20	< TOTAL >		146.0	87	1,147.2	2,679.4	25,909.8

Rated Output In Order of Rated Due Dates	Rated Output Cumulative Order Time
(Day)	(Hrs)
9	10
180	15.4
187	29.2
191	29.7
202	37.1
204	43.9
204	55.3
212	63.0
212	67.2
214	76.0
215	85.6
216	89.4
218	94.7
218	97.3
218	110.9
223	113.0
225	126.8
229	133.7
232	138.8
236	140.9
237	146.0
TOTAL >	146.0

Total Negative Weighted Lateness (Earliness) = 191.6 Days x Hrs

Total Positive Weighted Lateness (Tardiness) = 1,338.8 Days x Hrs

Simple Mean Output Lateness OLOm = 87 Days/20 = 4.4 Days

Weighted Mean Output Lateness OLOmw = (1,147.2 Days x Hrs)/146 Hrs = 7.9 Days

Negative Weighted Mean Output Lateness OLONmw = (191.6 Days x Hrs)/146 Hrs = 1.3 Days

Positive Weighted Mean Output Lateness OLOPmw = (1,328.8 Days x Hrs)/146 Hrs = 9.2 Days

Simple Standard Deviation of Output Lateness OLOS = $\sqrt{2{,}679.4 \text{ Days}^2/20}$ = 11.6 Days

Weighted Standard Deviation of Output Latneness OLOSw = $\sqrt{(25{,}909.8 \text{ Days}^2 \times \text{Hrs})/146 \text{ Hrs}}$ = 13.3 Days

Let us assume that each order's scheduled output date is known before its arrival at the work center concerned. After the operation is performed, the actual output date is reported. For each actual date in a reference period, the scheduled date can be found and a schedule deviation calculated.

The basic data are given in Table 4.7, Columns 1 to 4. From these, the actual output curve can be plotted (see Fig. 4.19). In Column 5, the order lateness (OLO) in days was determined for each order. Multiplication of these values by the order times TO results in a positive or negative schedule deviation (order lateness) area for each order (Column 6). A positive value indicates tardiness, a negative value earliness of the order completion. If actual and scheduled output dates coincide, the schedule deviation area is, of course, zero. The areas are plotted in Fig. 4.19. As in lead time calculations, a mean value of schedule deviation can be determined for the orders processed in period P. It seems convenient to define a weighted mean lateness, which could be directly compared with the weighted mean lead time.

The weighted mean lateness of the outputs (OLO_{mw}) can then be defined as

$$OLO_{mw} = \frac{AOPSD - AONSD}{OUT} \tag{4.20}$$

where:

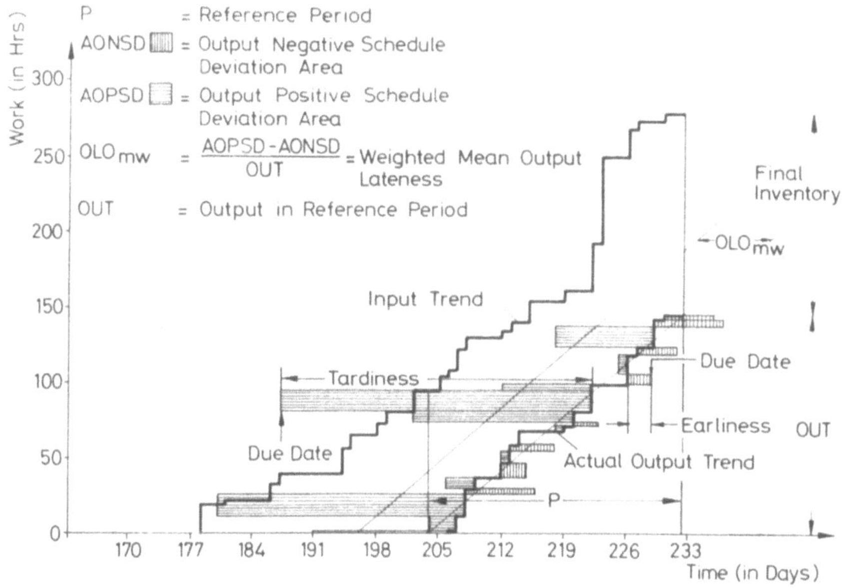

Fig. 4.19. Lateness of outputs in the throughput diagram

$AOPSD$ = positive schedule deviation area of output (order completed behind schedule)

$AONSD$ = negative schedule deviation area of output (order completed ahead of schedule)

OUT = outputs in reference period P.

In our example, this produces a weighted mean lateness (OLO_{mw}) of 1147.2 days × hours/146 hours = 7.9 days.

The positive and negative mean schedule deviations should be calculated as well as the standard deviation, since a mean value of zero could lead to the false assumption that there is no lateness at the work center in question. In Table 4.7, the negative weighted mean output lateness ($OLON_{mw}$) is calculated as 191.6/146 = 1.3 days, and the positive weighted mean output lateness ($OLOP_{mw}$) as 1338.8/146 = 9.2 days.

The value of the simple mean output lateness OLO_m (4.4 days) gives a much too positive reflection of the actual scheduling performance. The reason for this can be seen in Fig. 4.19: some orders with a large work content (nos. 115, 119, 125, and 117) also have large schedule deviations, and therefore affect the weighted mean value. The weighted standard deviation of schedule deviation also indicates great variances.

As defined above, the lateness is only calculated from the difference between scheduled and actual dates of the output curve. This is, however, not an objective statement on the lateness of a work center, since it might be that schedule deviations could not be avoided because the order input into the work center have already taken place too late.

Consequently, we must also calculate the schedule deviation of the input curve. As the input dates of a work center, by definition, correspond to the output dates of the upstream work center (or to the release date for the gateway work center), the weighted mean input lateness can be calculated according to the method shown in Fig. 4.19. As was shown in Fig. 4.16, there is an initial inventory at the beginning of the reference period from which the actual input curve can be determined. For these orders weighted lateness can be calculated. For the incoming orders up to the output level (OUT) a lateness can also be calculated. The weighted mean lateness of these two groups of orders is the weighted mean lateness of the input which can be compared with the weighted mean output lateness. It follows for the weighted mean input lateness that

$$OLI_{mw} = \frac{AIPSD - AINSD}{OUT} \tag{4.21}$$

where:

$AIPSD$ = positive schedule deviation area of input (order started ahead of schedule)
$AINSD$ = negative schedule deviation area of input (order started behind schedule)
OUT = actual input up to output level = output in reference period P.

Now we can determine a relative weighted mean lateness (OL_{mwr}) as

$$OL_{mwr} = OLO_{mw} - OLI_{mw} \tag{4.22}$$

This value tells us whether, on average, the work center concerned accelerated or delayed the processing of the orders compared with the scheduled lead time.

We now complete our description of the order throughput at a work center. The key data introduced permit a comprehensive representation of the processes. In Sect. 3.7.1 we pointed out the alternative of calculating lead time values in calendar days or in shop calendar days. Giving key data in calendar days is of use when determining the capital tie-up or the lateness, for example. For production planning and control monitoring purposes, however, key data in shop calendar days are more useful. Therefore, Tables 4.1–4.7 are recalculated in shop calendar days in Appendix A (Tables A1–A7).

Table 4.8 briefly compares the key data in calendar and shop calendar days.

By comparing the respective values of mean advance time, we can check whether this is reasonable. Since the advance time in the throughput diagram corresponds to the horizontal distance between the input and output curves in days or shop calendar days, respectively, it must be valid for longer periods of time that

$$\frac{TA_{mSCD}}{5\ \text{SCD}} = \frac{TA_{mday}}{7\ \text{days}}$$

$$TA_{mSCD} = TA_{mday} \times \frac{5\ \text{SCD}}{7\ \text{days}}$$

With the values from Table 4.8, it follows that

$$14.9 = 20.9 \times \frac{5}{7}$$

$$14.9 = 14.9\ \text{SCD}$$

Table 4.8. Key data of a work center – calculations in calendar days versus shop calendar days (individual values taken from Tables 4.2–4.6 and Appendix Tables A2–A6)

Key Data		by Calendar Days (Days)	by Shop Calendar Days (SCD)
Mean Range	Rm	20.8	14.4
+ Inventory Trend Component	ITC	.1	.4
= Mean Advance Time	TAm	20.9	14.9*)
+ Sequence Component	SC	-4.2	-3.0
= Mean Lead Time	TLm	16.6*)	11.8

*) Deviations behind Decimal Point Result from Rounding in Individual Values Calculation

Reference Period: Day 205 - 232 or SCD 30 - 49
Corresponding to 28 Days or 20 SCD

Which representation is chosen will depend on the individual application. When looking at processes on the work center level, the representation in shop calendar days will often be chosen.

In the next section we expand the throughput diagram, which so far has only described one work center, to cover the throughput of entire manufacturing orders. Here, the representation in calendar days is useful.

4.5 Order Throughput Diagram

4.5.1 Extended Work Center Throughput Diagram

To develop an order throughput diagram we must find the transition from the throughput of a single work center to the throughput of the entire manufacturing order.

Figure 4.20A shows an order with sawing, turning, drilling, and grinding operations in their time sequence and with their work content. Each of the rectangles corresponds to a weighted throughput element. If one separates the single elements of the operational sequence in a second step, one can determine, in addition to the known input and output dates for each single operation, when the corresponding order was entered (i.e. generated in the material requirement planning), when it was released, and when it was completed. The usual operation throughput element becomes an extended (weighted) throughput element, which can be assigned to the corresponding work center (Fig. 4.20B). The (weighted) order throughput element is then a rectangle which consists of the total lead time of the order and the total order time of the operations contained in it (Fig. 4.20C). Unlike the element shown in Fig. 3.14, it also includes the time between order entry and order release.

Within the extended operation throughput element, we distinguish between different portions of the total lead time, as shown in Fig. 4.21, here called the dispatch lead time, indirect lead time, direct lead time, and post lead time [10].

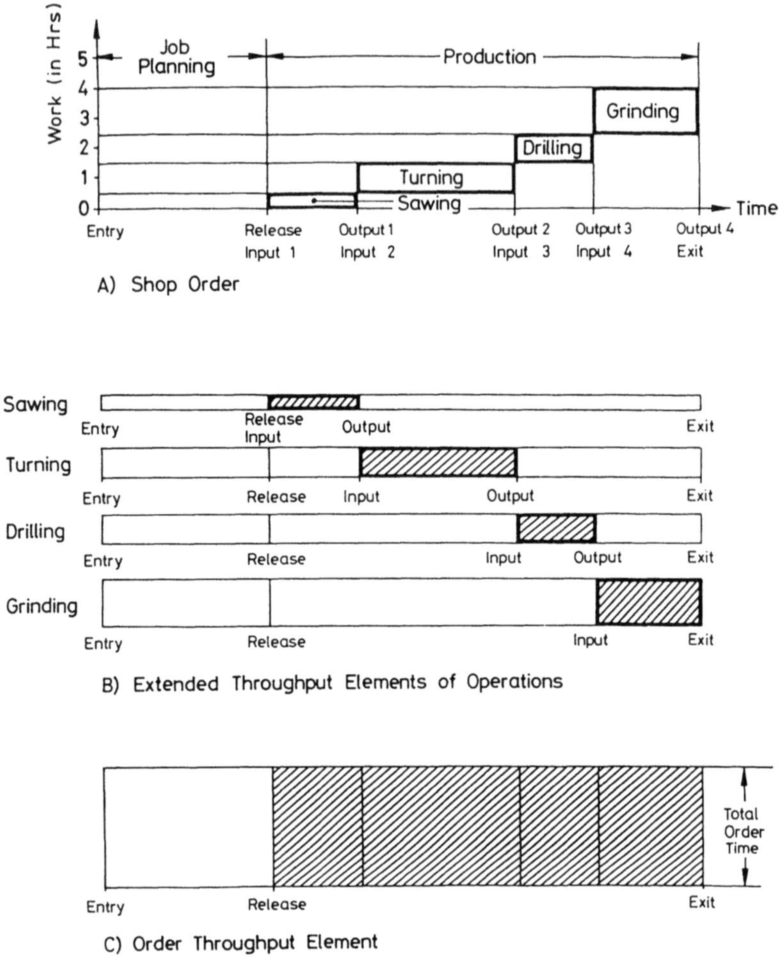

Fig. 4.20. Operation throughput elements and order throughput elements of a shop order

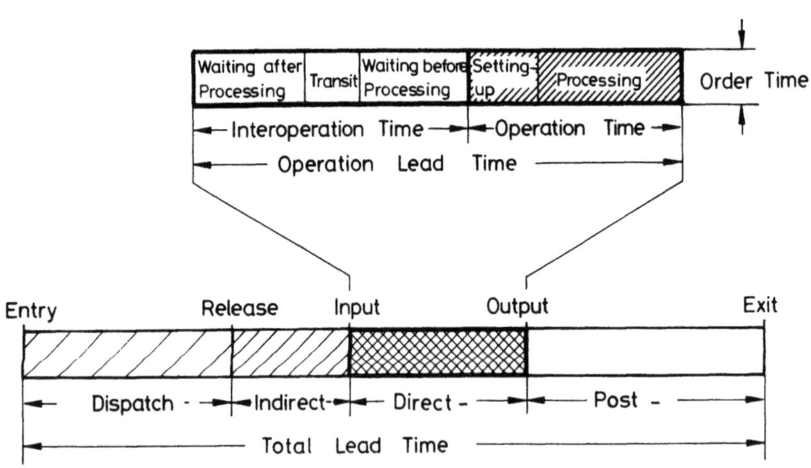

Fig. 4.21. Definition of lead time components within the overall order lead time (After Bechte [7])

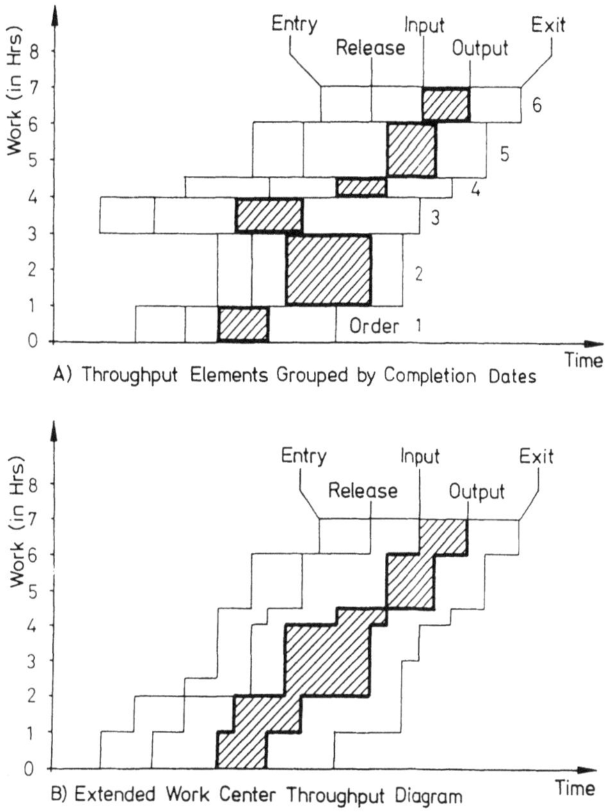

Fig. 4.22. Derivation of the extended work center throughput diagram (After Bechte [7])

If the orders passing through a work center are ordered according to increasing exit dates, an extended throughput diagram is produced (Fig. 4.22B). Since, as a rule, order sequence transpositions occur in the queue, the individual events entry, release, input, output, and exit must be arranged according to increasing time so as to produce the final throughput diagram. From this results the so called *extended work center throughput* diagram (Fig. 4.22B) [8, 10]. The usual input and output curves are marked with bold lines. This throughput diagram makes clear how long before the orders reached the work center concerned they were dispatched and released, and how long it took the downstream work centers to complete the order.

The extended throughput diagram can also be represented in periods; here it is the exit curve, not the output curve which starts in the zero point (Fig. 4.23). Individual inventory components can be defined within the inventory, corresponding to the components of lead time [7, 8, 10]. The dispatch inventory tells us how many working hours are planned for this work center at a certain time. The indirect inventory shows how much work released for this work center is still being processed by upstream work centers. Direct inventory is the work in process (WIP) at this work center, and post-inventory represents the work being processed at the downstream work centers. For the gateway work center, the release curve is identical to the input curve, for the final work center, the output curve is the same as the exit curve.

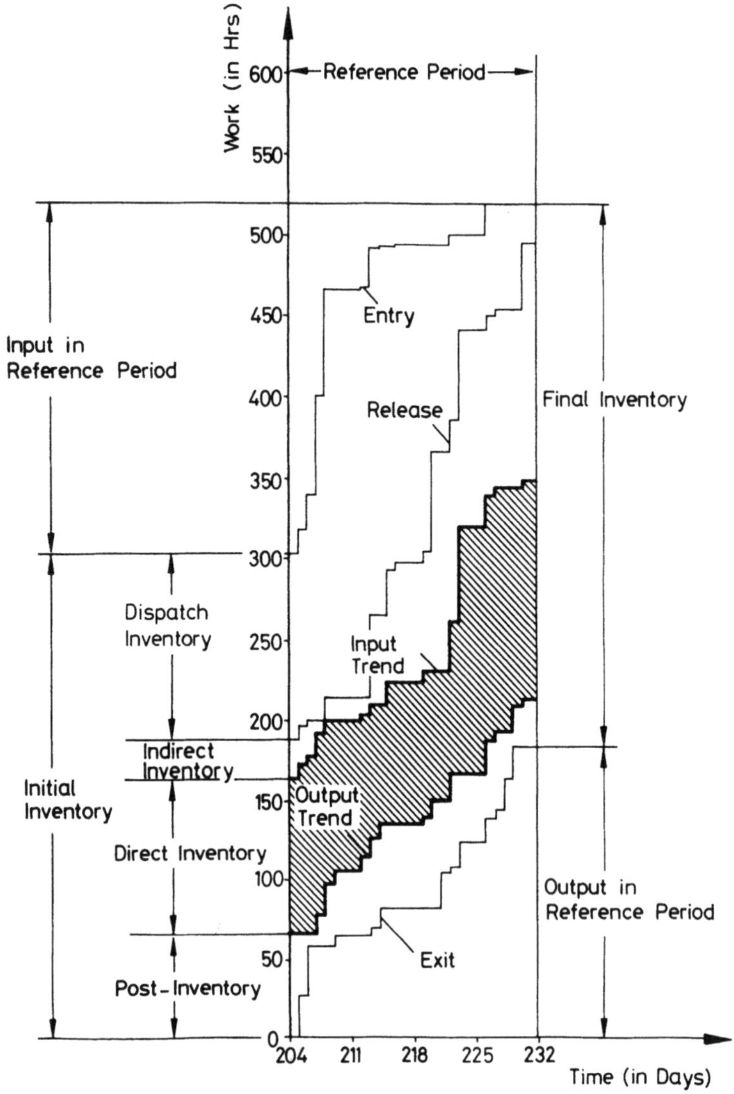

Fig. 4.23. Example of an extended work center throughput diagram (After Bechte [7])

We do not need to show the calculation tables for the extended throughput diagram, since they are analogous to those for the work center throughput diagram. The mean values for the individual inventories and components of lead time are also calculated as described above. The extended throughput diagram gives a good idea of where the work center concerned is located in the flow of the orders it processes.

This can also be characterized by a key value, called the *weighted mean position* of a work center [8], which can be used within a monitoring system (see Sect. 5.4.2). For each order processed by the work center, the work center's position in the operational sequence of the order is plotted in a diagram.

Figure 4.24 shows the positions of our sample work center, weighted with the work contents, from which the weighted mean position (POS_{mw}) can be calculated:

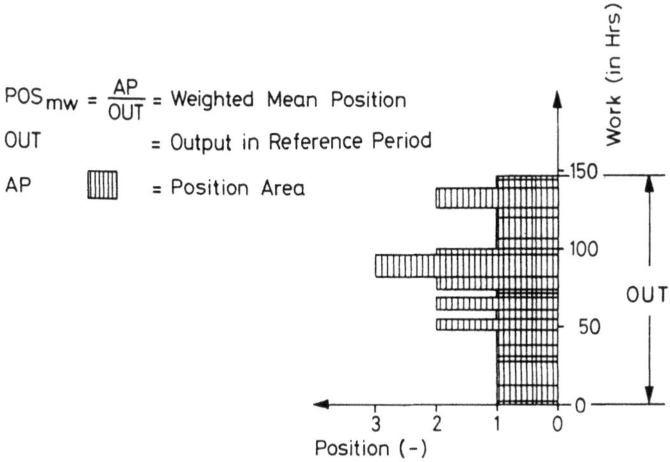

Fig. 4.24. Weighted mean position of a work center

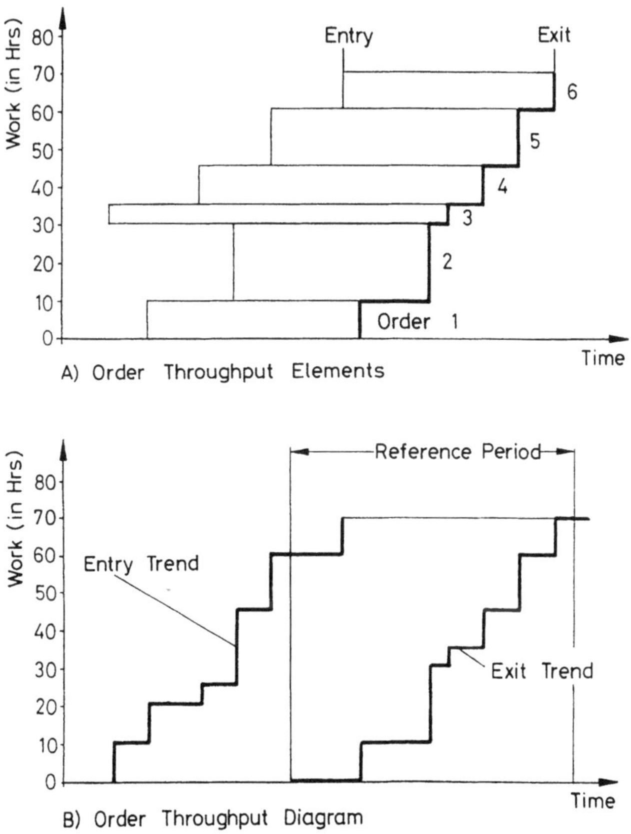

Fig. 4.25. Derivation of the order throughput diagram

$$POS_{mw} = \frac{AP}{OUT} \tag{4.23}$$

where:

AP = position area of outputs in reference period
OUT = outputs in reference period P.

In our example, the mean weighted position of the work center is $POS_{mw} = 215.3$ hrs/146 hrs = 1.5. This means that the work center is used at the beginning of most orders.

4.5.2 How to Produce the Order Throughput Diagram

From the extended work center throughput diagram, the order throughput diagram is developed by ordering the weighted order throughput elements according to increasing exit dates (Fig. 4.25A). For plotting the entry curve, the entry dates of the individual orders must be reordered in increasing sequence, because of the frequent sequence transpositions (Fig. 4.25B). The resulting order throughput diagram can be analyzed with the same methods as the work center throughput diagram (which we do not deal with in detail here).

To conclude, it has been shown that it is possible to clearly define the four basic objectives of manufacturing control – inventory, flow time, utilization, and lateness – for the individual work center and for the order, to analyze their relationships with each other, and thus to create a universal model for the representation of a manufacturing process.

The different throughput diagrams and their key data can be used for two purposes: continuous monitoring and diagnosis of the manufacturing process, which will be dealt with in Chapt. 5, and the scheduling of the order throughput, which will be described in Chapts. 6 and 7.

4.6 References

[1] Schmitz, P-G. Analytische Beziehungen zwischen Produktionsverlauf, Belastung und unvollendeter Produktion in Maschinenbaubetrieben (Manufacturing Process, Loading, and In-Process Inventory – Their Analytical Relationships in Mechanical Engineering Companies). VEB Verlag Technik, Berlin, 1961.

[2] Conway, RW, Maxwell, WL, Miller, LW. *Theory of Scheduling*. Addison-Wesley, Reading, MA, 1967.

[3] Little, JDC. A Proof for the Queuing Formula $L = \lambda W$. *Operat. Res.*, **9** (3): 1961.

[4] In't Veld J. Bedrijfsinformatie (Industrial Science). Wissenschaftliche Beilage. TED (Netherlands), February 1971.

[5] Heinemeyer, W. Die Analyse der Fertigungsdurchlaufzeit im Industriebetrieb (Lead Time Analysis in Manufacturing Companies). Dissertation Technische Universität Hannover, 1974 (doctoral thesis).

[6] Kreutzfeldt, H-F. Analyse der Einflußgrößen auf die Terminplanung bei Werkstättenfertigung (Analysis of Relevant Factors of Scheduling in Job-Shop Production). Dissertation Technische Universität Hannover, 1977 (doctoral thesis).

[7] Bechte, W. Steuerung der Durchlaufzeit durch belastungsorientierte Auftragsfreigabe bei Werkstattfertigung (Lead Time Control in Job-Shop Production by Load-oriented Order Release). Disserta-

tion Universität Hannover, 1980 (veröffentlicht in: Fortschritt-Berichte der VDI-Zeitschriften, Reihe 2, Nr. 70, Düsseldorf, 1984) (doctoral thesis, published in above VDI report).

[8] Erdlenbruch, B. Grundlagen neuer Auftragssteuerungsverfahren für die Werkstattfertigung (Principles of Recent Order Control Techniques in Job-Shop Production). Dissertation Universität Hannover, 1984 (veröffentlicht in: Fortschritt-Berichte der VDI-Zeitschriften, Reihe 2, Nr. 71, Düsseldorf, 1984) (doctoral thesis, published in above VDI report).

[9] Lorenz, W. Entwicklung eines arbeitsstundenorientierten Warteschlangenmodells zur Prozeßabbildung der Werkstattfertigung (Working-hour-related Queueing Theory as a Representation of the Manufacturing Process in Job-Shop Production). Dissertation Universität Hannover, 1984 (veröffentlicht in: Fortschritt Berichte der VDI-Zeitschriften, Reihe 2, Nr. 72, Düsseldorf, 1984) (doctoral thesis, published in above VDI report).

[10] Bechte, W. Aufbau und Realisierung des Kontroll- und Planungssystems zur Fertigungssteuerung KPSF (Design and realization of the control and planning system for job shop control KPSF). In: Wiendahl, H-P. (ed.). *Anwendung der Belastungsorientierten Fertigungssteuerung (Application of Load-oriented Manufacturing Control)*, Hanser, München, 1991, pp 117–152.

[11] Jendralski, H. Kapazitätsterminierung zur Bestandsregelung in der Werkstattfertigung (Capacity Scheduling for Inventory Control in Job-Shop Production). Dissertation Technische Universität Hannover, 1978 (doctoral thesis).

Chapter 5

Analysis, Monitoring, and Diagnosis of the Manufacturing Flow

5.1 Abstract

Monitoring is continuous observation of the flow of orders through the manufacturing process by means of key figures and graphics. This chapter describes first how to organize and analyze a work shop, then explains the concept, and shows authentic results of such a continuous monitoring system. The outputs of such a system can be interpreted with a diagnostic system, of which the basic theory is given. Examples of color graphics taken from experimental and commercial systems illustrate possible outputs of monitoring systems. General rules on how to improve the manufacturing flow, and how existing production planning systems can be better used, are also given.

5.2 Manufacturing Flow Monitoring

Continuous monitoring of the production processes is a matter of course in a progressive company. However, the processes monitored are primarily those which can be expressed in terms of money. Order entry, turnover, investments, purchases, for example, are summed up at the end of the month and compared with the annual master production plan. For the shop, the main point to compare is the planned production hours with the actually processed production hours, and the resulting costs. As already shown in Chapt. 2, it is not yet common practice to monitor the equally important key data for inventories, lateness, and flow times, and compare them with the planned values. These conditions call for a system for organizational manufacturing flow monitoring in addition to existing cost and quality monitoring systems. This could be referred to as a 'continuous logistic quality review' of manufacturing.

Such a system must be based on a model which, in describing the manufacturing flow, calculates and uses the above-defined key data and their logical relationships. The model must also make clear how widely the actual status differs from a 'realistic

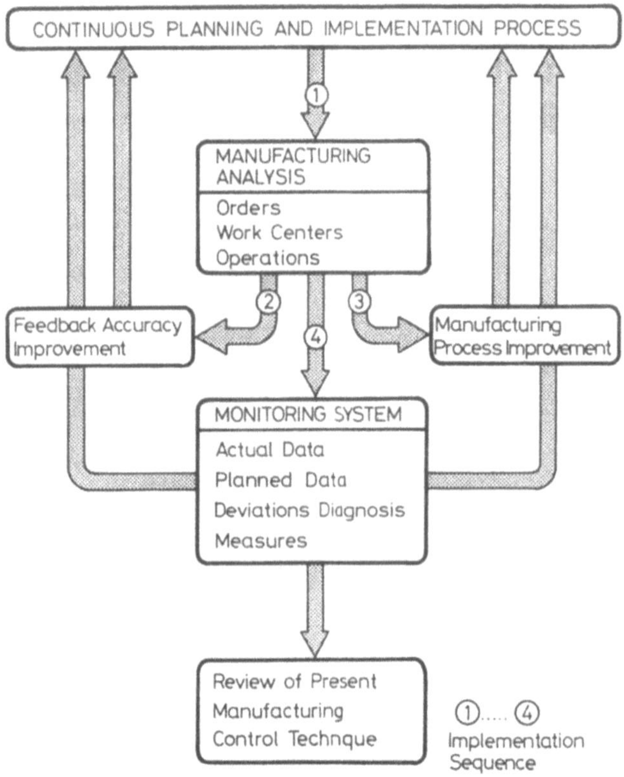

Fig. 5.1. Steps towards an implementation of manufacturing process monitoring

ideal status'. Only then is it possible to find selective manufacturing control measures. The throughput diagram in its different forms and with its key data, as developed above, meets these requirements, and will thus be the basis of our further considerations.

It is, however, highly recommended to develop and introduce such a system via the steps shown in Fig. 5.1. First, a manufacturing flow analysis must be carried out based on a certain data collection period, analyzing the throughput behavior of orders, work centers, and operations. As experience shows, this leads to first improvements in both feedback data quality and the manufacturing flow. Only after those weak points have been removed can a monitoring system be developed and introduced, which, on the one hand, is adapted to the needs of individual users and, on the other hand, can quite easily be modified or completed. The use of this permanent monitoring system during a certain period will automatically result in increased improvements. At this point, at the latest, the question will arise as to whether the existing manufacturing control method meets the new requirements, or whether it, too, should be improved.

Another characteristic of such analyses and monitoring systems must be that they provide a graphic representation of the manufacturing flow and its essential key data, since conventional printed lists and tightly packed screen displays have a rather discouraging effect on the user. The hardware and software for good graphic representation – for example color graphics monitors – are constantly improving (see Sect. 5.6).

The procedure of a manufacturing flow analysis is described below.

5.3 Manufacturing Flow Analysis

5.3.1 Procedure Outline

An analysis of the manufacturing flow in a company is generally initiated by unsatisfactory manufacturing conditions or changes in the market situation, or it is carried out in connection with a planned manufacturing rationalization. It is useful to proceed according to the following steps [1–5]:

- Definition of the objectives
- Determination of the field and period of the investigation
- Definition of the data to be recorded and availability check
- Test recording and data accuracy check
- Recording of the manufacturing flow
- Evaluation and documentation
- Carrying out and monitoring of corrective measures.

The entire analysis and evaluation is based on recording and evaluating the throughput elements defined in Chapt. 3. Figure 5.2 gives a short summary of the above explanations, and makes it clear that (with the exception of the first operation) a single feedback event per operation is sufficient for this analysis, which reports both the end of the throughput element concerned and the beginning of the next one [3]. Because of the small operation time percentage of the total lead time, the scheduled operation time can still be equated with the actual operation time, despite existing deviations. From this it follows that the interoperation time (*TIO*) is the difference between the lead time (*TL*) and the operation time (*TOP*). If, for certain applications, the distinction between operation time and interoperation time does not suffice, it is recommended to carry out a separate investigation of the individual field in question, for example with the help of work measurement studies [6, 7].

Figure 5.2 shows a sample calculation of the lead time of Operation 2. The order lead time is then either the sum of the single operation lead times, or the difference between the release date and the order completion date. If desired, and if the data are available, an order entry date and an order exit date can be defined. From this one can determine how long an order stays in the manufacturing control system.

Three types of evaluations can be carried out. The throughput elements of a manufacturing order permit an order-related evaluation, whereas the throughput elements of a single work place, a group of work places, a cost center, or a job shop allow a work center-related evaluation. In case all throughput elements are evaluated independently of the order or work center to which they belong, the evaluation will be operation-related.

Figure 5.3 [3] shows the general data flow chart of such an evaluation. It is based on a program that is part of the DUBAF (Durchlaufzeit- und Bestandsanalyse in

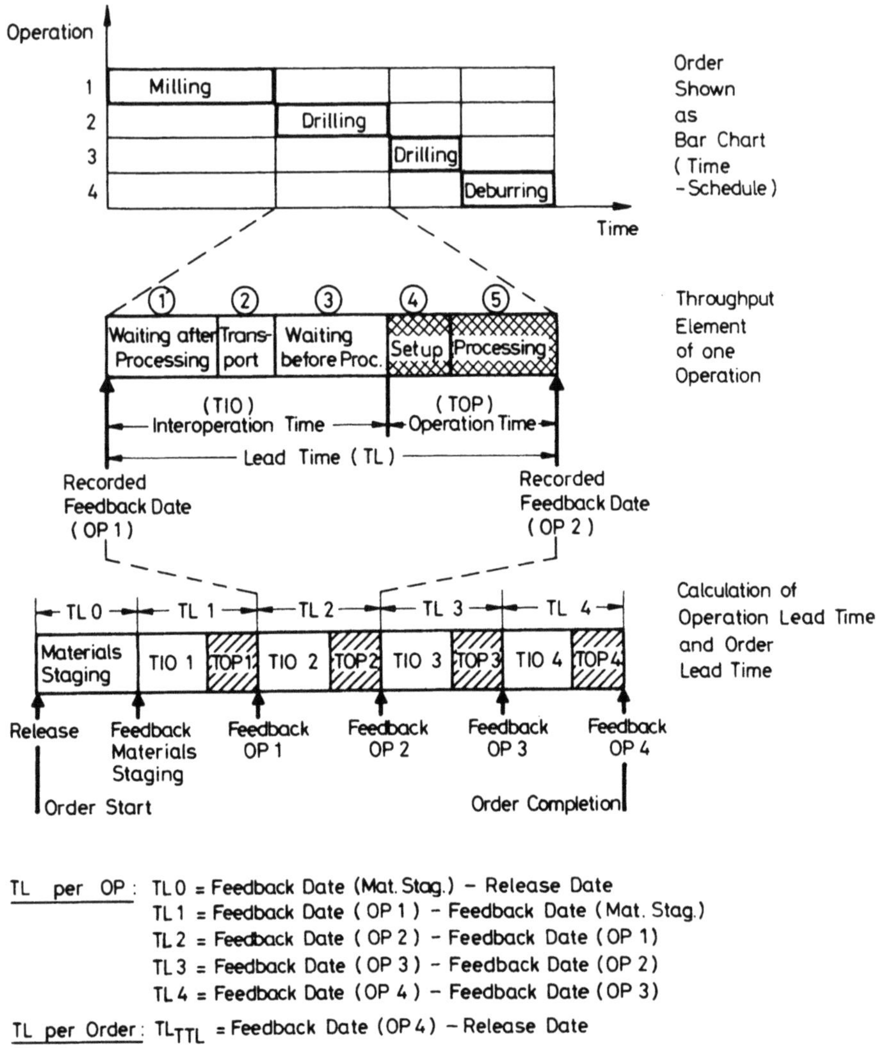

Fig. 5.2. Definition and calculation model for operation and order lead time (After Bechte [4] and Dombrowski [3])

der Fertigung = FlowTime and Inventory Analysis in Manufacturing) method, and which was developed under Kettner at the Institut für Fabrikanlagen, University of Hannover, in collaboration with the Deutsche Gesellschaft für Betriebswirtschaft (German Business Management Society) and renowned industrial companies [1, 2]. It is constantly being used and improved [3–5].

One can distinguish two program blocks. In the 'Recording' block, which is carried out in the plant in question, the feedback data from a defined field of investigation are collected over a continuous period of time, typically half a year, and stored on a suitable medium, generally a magnetic tape. In addition, information on the shop calendar, certain work center information – such as work center number, capacity, and unit cost, if necessary – and other manufacturing information must be

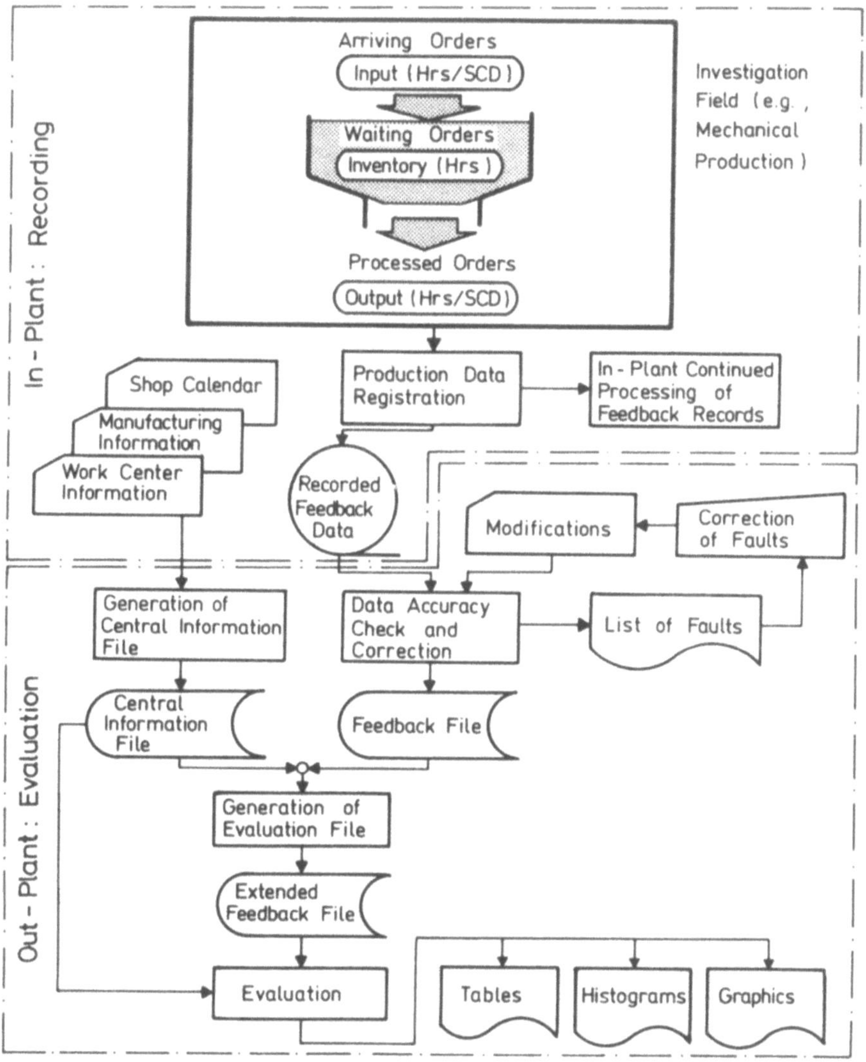

Fig. 5.3. General data flow chart of a flow time and inventory analysis. SCD: Shop Calendar Day

collected for the investigation period [1]. The 'Evaluation' program block is performed outside the plant, i.e. either in a qualified corporate department aided by the computer center or in a suitable external consultancy bureau.

First, the raw data are subjected to a plausibility check. Any faults must be corrected either by the evaluating authority, or in the factory on the basis of a list of faults. The corrected operation-related and order-related feedback data form the feedback file.

The other data, such as the shop calendar, capacity data, etc., are stored in the central information file. From these two files the evaluation file, which is an extended feedback file, is produced. Evaluations in the form of lists, frequency distribu-

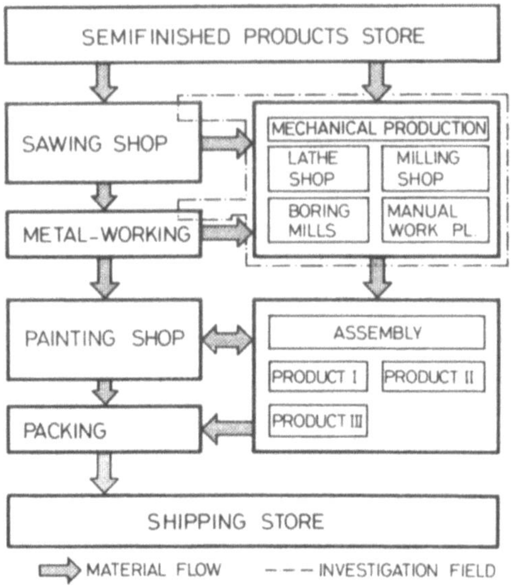

Fig. 5.4. Selecting the investigation field

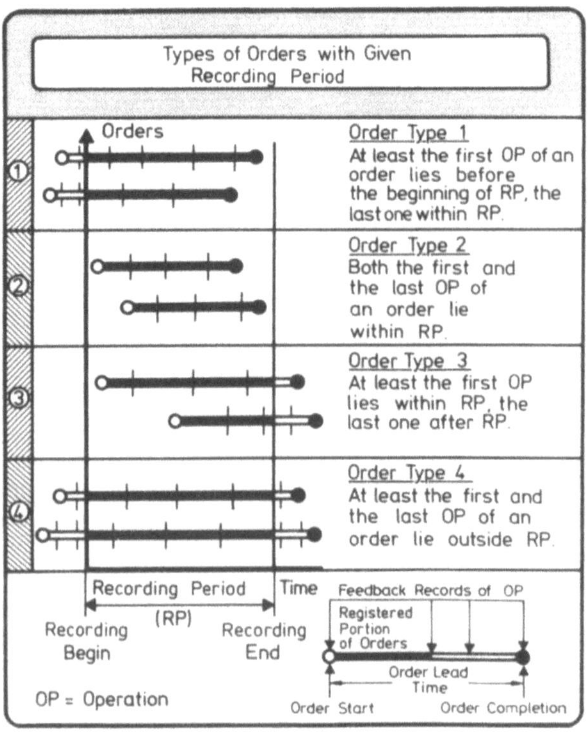

Fig. 5.5. Defining the data collection period

tion histograms, and graphics (throughput diagrams) are then generated from this file. A number of important aspects are now examined in detail.

When defining the objectives, a broad overview should be given preference over single aspects. This is the only way in which to discover weak points in the manufacturing flow or recognize capacity bottlenecks with their causes and effects and to avoid overly high data recording costs. Single aspects can always be dealt with later – at a lower cost and limited to individual weak points.

The field of investigation should be organizationally and spatially closed, and should include all work centers passed by the orders in question. This means that, to some extent, operations from outside the closed field must also be considered. Figure 5.4 shows such a case [3]. Besides the mechanical manufacturing, which was the area under investigation, work centers in the sawing and the metal-working shops also had to be considered, because certain parts (e.g. turned and sheet-metal parts) have to undergo certain operations there.

In principle, such an analysis is not limited to manufacturing processes, but can also be extended to processes before and after manufacturing. The main field of investigation has, however, always been job shop manufacturing. References to other process analyses are given in Hackstein [6] and elsewhere.

The determination of the data collection period should be given particular consideration. Since a representative number of both operations and orders must be evaluated, the considerably longer time of order throughput should be relevant here. There are basically four different types of orders (Fig. 5.5) [3].

Only in Case 2 are complete orders recorded. In all other cases the orders are 'cut' either at the beginning (Case 1), at the end (Case 3), or at both sides (Case 4). The fully recorded operations of the order types 1, 3, and 4 can, however, be used for operation- and work-center-related evaluations.

The approximate value for the length of the data collection period is double the mean order flow time. It is estimated by using empirical values, by checking samples of routing sheets, or with a rough estimate from the order stock and the capacity, because the equation developed in Chapt. 4 – 'mean lead time (range) equals mean inventory divided by mean performance' – is also valid for an entire shop.

Since there are always sequence transpositions, and because of the great variances in lead times, one must expect that within limited data collection periods there is always quite a large portion of incomplete orders, a fact that may lead to an unknown error when calculating the order lead times. Figure 5.6 shows the authentic distribution of the four order types from Fig. 5.5 over the investigation period from an extensive factory investigation [5].

At the beginning of the data collection period, the portion of complete orders (Order Type 2) is very small; it then increases with time, and then drops again at the end of the data collection period. Between shop calendar days 1010 and 1070 their portion is relatively constant, so this period was chosen as the reference period for the evaluation. As was expected, the number of unfinished orders (Order Type 3) – i.e. the orders started but not completed within the data collection period – increases continuously, while the number of leftover orders (Order Type 1), which were started before the data collection period and completed within it, decreases continuously. The portion of through orders (Order Type 4) is, in this case, more than 20% of the fully recorded orders.

To obtain a reliable value for the mean order flow time, the portions of the four order types are plotted over the performance curve, i.e. the output curve of all orders processed in this period [5]. This produces the so called *dynamic order-type curve* (Fig. 5.7). One can see clearly that the mean flow time for the complete orders is

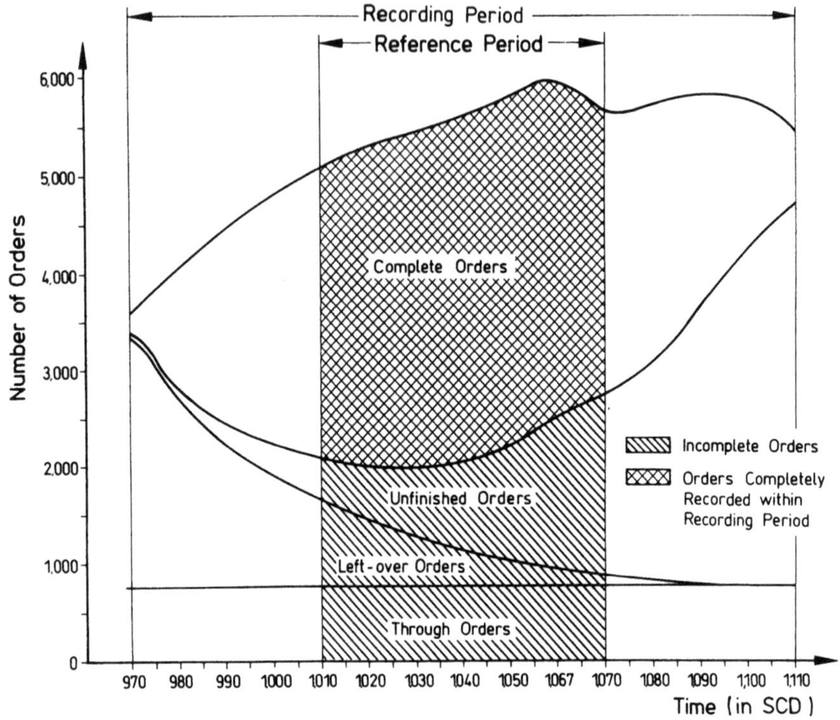

Fig. 5.6. Development of the order type proportions in the course of a lead time and inventory analysis (After Nyhuis [5])

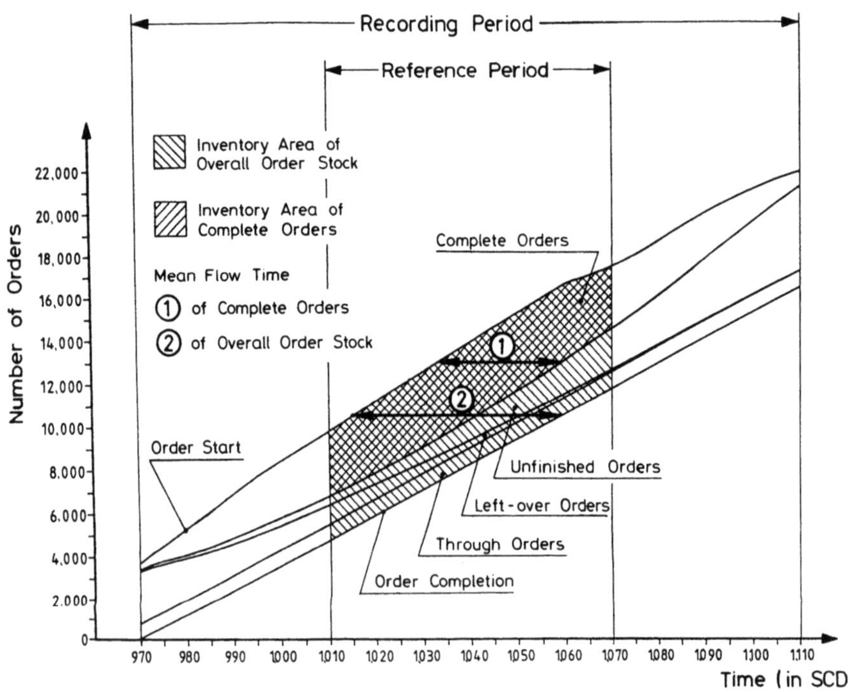

Fig. 5.7. Performance-related dynamic order-type curve of a lead time and inventory analysis (After Nyhuis [5])

Identification	1. Order No. 2. Operation No. 3. Work Center No. (Actual)
Dates	4. Order Staging Date for Gateway Operation 5. Completion Date of All the Operations (Actual)
Time	6. Process Time per Unit (Planned) 7. Setup Time (Planned)
Quantity	8. Quantity (Actual), Yield 9. Scrap Volume
Value	10. Materials Value per Unit before Processing *

* To Be Registered in Case of Capital Tie-Up Only

Fig. 5.8. Minimum data record for DUBAF analysis (After Dombrowski [3])

considerably shorter than the lead time of the overall order stock. In this case, the flow time value for the complete orders was 27 workdays, as against 44 workdays for all orders in the inventory. When the analysis is based on only one single data recording, such a check on the recorded data is to be recommended, so as to avoid too favourable, and thus unrealistic, order flow time values.

After having determined the investigation period, one must decide on the data to be recorded. For investigation in unsplit manufacturing lots, which is most frequently the case, the data given in Fig. 5.8 are sufficient [1, 23]. The figure shows the identification of the data record (order number, operation number according to the operation sheet, work center number), the dates data, the time data, and the quantity data for determining the throughput element, as well as the materials value per processing unit, if capital tie-up is also to be calculated. The accuracy of the dates should be in reasonable relation to the flow time. In most cases, it is sufficient to give the day of the feedback event (cf. Sect. 3.8).

As stated above, it is recommended to carry out a test recording after having determined these dates. Although the companies concerned often believe that the necessary data are all available, or at least can be provided by the DP department without many problems, this is generally not the case when it comes to collecting data. In some cases one must even create special recording forms. Data storage problems become easier with the use of production data collection systems, but the data are largely not provided in the necessary combination, and a special arrangement for regular storage must be made for the data collection period. The actual data collection period which, as experience shows, is six to twelve months long in mechanical shops, should begin only when a test recording and evaluation over four to twelve weeks has shown that the data are reliable and available in the necessary form.

The next important step is a quality check on the recorded data. Obviously, this can only be carried out on criteria which can be checked for plausibility. Such checking criteria from an authentic investigation are shown in Fig. 5.9 [3].

1 All variables have to be available; values must be greater than zero.

2 All dates have to be located within investigation period.

3 Differences in volume between operations of the same order ?

4 Operations sequence transposition ?

5 Multiple feedback records of the same operation ?

6 All of the work centers have to be located within investigation field.

7 All operations of an order must have the same actual order release date.

8 Order completeness ?

Fig. 5.9. Check criteria for collected feedback data (After Dombrowski [3])

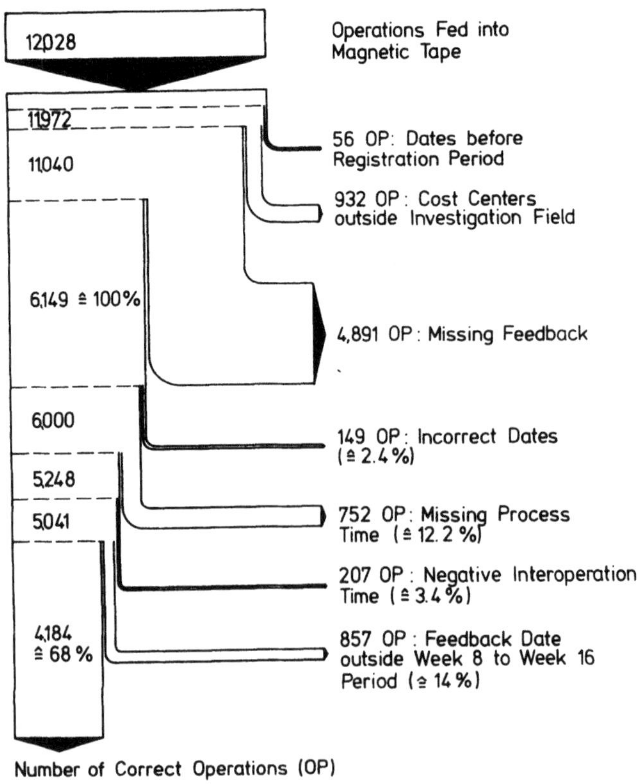

Fig. 5.10. Example results of a feedback data accuracy check (After Dombrowski [3])

The points on this checklist can above all reveal weak spots in feedback discipline, but feedback accuracy can not be investigated without comparing the data with other data recorded especially for this purpose.

Figure 5.10 shows the results of an authentic quality check on approximately 12000 operations, carried out in a metalworking shop [3]. Only about half of the available feedback data could be evaluated, 2.4% of them had incorrect dates, 12.2% did not give a process time, and 3.4% had negative interoperation times.

All investigation costs are frequently justified by just one such evaluation, because many companies, after having introduced a production data collection system, live in the false belief that the recorded data reflect the reality. Often, the cause of the data inaccuracies is the fact that the same data are used for wage calculation and for feedback into the shop floor control system. Therefore, it is vital to uncouple these two feedback functions.

5.3.2 Forms of Evaluation and Representation

The recorded throughput elements should be edited for maximum clarity, for which the evaluation criteria, key data, and forms of representation, as shown in Table 5.1, can serve as guidelines.

As criteria for the evaluation, three possibilities have proved useful. Order-related evaluation provides information on the order flow times, the number of operations per order, and the order schedule performance (if the planned due dates were also recorded). Operation-related evaluation informs above all about weighted and unweighted flow times, their frequency distribution, and the operation time percentage of flow time. Work center-related evaluation illustrates the throughput behavior of the work centers in the levels from the single work place to the entire investigation field.

The evaluated key data are primarily the inventory, flow time, utilization, and lateness as they were developed from the throughput diagram in Chapt. 4.

Depending on the purpose of the investigation, the results can be represented in the form of more or less detailed tables, histograms, and statistical evaluations of single key values. The dynamic processes at the work centers can, in most cases, be described by giving key data for the single periods, and with the help of the throughput diagram.

An example of hierarchically sorted tabulation of important work center key data is shown in Table 5.2. The table gives the quantities processed, setup times,

Table 5.1. Typical kinds of evaluation and representation for collected process data

Evaluation criteria	Key data	Representation forms
• Order	• Order Time	• Hierarchically Structured Tables
• Operation	• Lead Time	• Statistical Evaluations of Characteristic Data
• Work System Plant Job Shop Cost Center Work Center Single Work Place	• Inventory • Performance • Capacity • Schedule Deviation • Capital Flow	• Frequency Diagrams • Throughput Diagrams • Transport Matrices • Period-Related Key Data

Table 5.2. Example production data evaluation with the SORTAB program

PERFORMANCE TABLE STANDARD LIST - INVESTIGATION FIELD 203 - INVESTIGATION PERIOD PER 07.DEC. .. TO 05.MAY ..

JSNO (-)	COCNO (-)		QUANTITY (ITEMS)	TS OP (HRS)	TOP OP (HRS)	TIO OP (SCD)	TL OP (SCD)	TOPC OP (%)	TL CUM (SCD)	NOOP CUM (-)	ADVAL OP (DM)	CAPTU OP (DMxJEAR)
36	3611	TOTAL	17669.0	292.3	1706.1	448.0	586.0	6094.4	1289.0	430.0	92025.5	3189.6
		MEAN VALUE	96.0	1.6	9.3	2.4	3.2	33.1	7.0	2.3	500.1	17.3
		STANDARD DEV.	145.9	1.1	10.8	1.0	1.3	22.7	5.8	1.2	578.9	47.6
		NUMBER OF VALUES	184	184	184	184	184	184	184	184	184	184
	3613	TOTAL	275098.0	583.0	5762.2	2814.0	3161.0	12857.2	7349.0	2082.0	285369.1	20119.1
		MEAN VALUE	382.6	0.8	8.0	3.9	4.4	17.9	10.2	2.9	396.9	28.0
		STANDARD DEV.	688.9	0.8	10.5	2.7	2.8	16.6	6.3	1.2	548.7	66.4
		NUMBER OF VALUES	719	719	719	719	719	719	719	719	719	719
	3614	TOTAL	86623.0	468.6	5427.9	946.0	1428.0	9720.3	2495.0	641.0	314935.3	9511.3
		MEAN VALUE	312.7	1.7	19.6	3.4	5.2	35.1	9.0	2.3	1137.0	34.3
		STANDARD DEV.	423.7	1.5	23.1	2.6	3.3	28.7	6.6	1.4	1267.4	88.6
		NUMBER OF VALUES	277	277	277	277	277	277	277	277	277	277
	3621	TOTAL	186829.0	1538.8	7338.5	1672.0	2266.0	13912.1	3630.0	857.0	476270.6	11666.8
		MEAN VALUE	439.6	3.6	17.3	3.9	5.3	32.7	8.5	2.0	1120.6	27.5
		STANDARD DEV.	753.5	2.4	18.1	3.0	3.5	19.8	5.3	1.0	1137.7	49.4
		NUMBER OF VALUES	425	425	425	425	425	425	425	425	425	425
	3622	TOTAL	102091.0	855.0	3894.2	1390.0	1716.0	13380.3	2534.0	731.0	171732.8	3382.2
		MEAN VALUE	203.8	1.7	7.8	2.8	3.4	26.7	5.1	1.5	342.8	6.8
		STANDARD DEV.	348.9	1.8	8.9	1.5	1.7	23.6	5.6	1.0	392.7	23.5
		NUMBER OF VALUES	501	501	501	501	501	501	501	501	501	501
	3631	TOTAL	.10E+07	4182.3	14293.1	4493.0	5451.0	31038.0	5451.0	1225.0	657484.0	3732.4
		MEAN VALUE	821.2	3.4	11.7	3.7	4.4	25.3	4.4	1.0	536.7	3.0
		STANDARD DEV.	1426.8	1.0	14.9	1.8	2.3	17.1	2.3	0.0	687.1	16.9
		NUMBER OF VALUES	1225	1225	1225	1225	1225	1225	1225	1225	1225	1225
	3641	TOTAL	.16E+07	2894.0	18272.4	7695.0	9692.0	48280.0	26213.0	7888.0	691226.0	50899.8
		MEAN VALUE	570.2	0.7	6.4	2.7	3.4	16.3	9.2	2.8	241.6	17.8
		STANDARD DEV.	1078.8	0.8	10.0	6.5	2.3	18.8	6.0	1.3	372.2	55.9
		NUMBER OF VALUES	2861	2861	2861	2861	2861	2861	2861	2861	2861	2861
36		TOTAL	.33E+07	10014.8	56694.4	19458.0	24300.0	135283.1	48961.0	13854.0	.27E+07	102501.2
		MEAN VALUE	533.9	1.5	9.2	3.1	3.9	21.9	7.9	2.2	434.3	16.6
		STANDARD DEV.	1043.8	1.6	13.2	4.7	2.6	20.3	5.8	1.3	674.7	52.2
		NUMBER OF VALUES	6192	6192	6192	6192	6192	6192	6192	6192	6192	6192

A B B R E V I A T I O N S :

JSNO	JOB SHOP NUMBER	TOPC OP	OPERATION TIME COMPONENT
COCNO	COST CENTER NUMBER	TL CUM	LEAD TIME CUMULATED
TS OP	SET UP TIME PER OPERATION	NOOP CUM	NUMBER OF OPERATIONS CUMULATED
TOP OP	OPERATION TIME PER OPERATION	ADVAL OP	ADDED VALUE PER OPERATION
TIO OP	INTEROPERATION TIME PER OPERATION	CAPTU OP	CAPTAL TIE-UP
TL OP	LEAD TIME PER OPERATION		

operation times, interoperation times, actual lead times, the number of processed operations, the added value, and the capital tie-up, each with its cumulative value, mean value (related to the number of operations), and standard deviation, for Work Centers 3611, 3613, 3614, 3621, 3622, 3631, and 3641 in Job Shop 36 for the investigation period 7 December to 5 May.

A statistics program allows a closer investigation in any single key value. The 6192 operations recorded in Table 5.2, for example, were evaluated as to their *TL*/OP (lead time per operation, i.e. the simple lead time). Table 5.3A shows the frequency table and the corresponding frequency diagram, and Table 5.3B shows the characteristic data of this distribution. For the latter, generally only the mean value, the median value, and the standard deviation are of interest. They are calculated using the equations developed in Chapt. 3.

Table 5.4 shows the development of fundamental key data over the investigation period. Besides the single values for the entire investigation period of the whole Job Shop 36, which are, in part, known from Table 5.2, it gives the single values for eight *periods*. Each period consists of 14 days (i.e. two weeks), of which 10 are workdays. All operation times and lead times were calculated in SCDs. The performance and load values correspond to the output and input for each period, respectively. This table can be set up for each level from the single work place to the entire plant, and represents one of the most important results of such an investigation.

Table 5.3. (A) Example frequency table and histogram of key data (Example: actual lead time per operation); (B) Example characteristic data of a key data evaluation (Example: actual lead time per operation)

A) Frequency Table and Histogram

** X = 24 TL OP (SCD) ACTUAL LEAD TIME PER OPERATION **

FREQUENCY TABLE						HISTOGRAM OF RELATIVE VALUES
GROUP	GROUP CLASSIFICATION FROM	UP TO	FREQUENCIES ABSOLUTE	RELATIVE	CUMUL.	0 5 10 15 20 25 30 %
	LESS	0	0	0.0	0.0	| | | | | | |
1	0	1	0	0.0	0.0	
2	1	2	0	0.0	0.0	
3	2	3	1913	30.9	30.9	*********************************
4	3	4	1813	29.3	60.2	*******************************
5	4	5	837	13.5	73.7	****************
6	5	6	571	9.2	82.9	*********
7	6	7	332	5.4	88.3	*****
8	7	8	236	3.8	92.1	****
9	8	9	159	2.6	94.7	***
10	9	10	82	1.3	96.0	*
11	10	11	65	1.0	97.0	*
12	11	12	64	1.0	98.1	*
13	12	13	29	0.5	98.5	
14	13	14	22	0.4	98.9	
15	14	15	22	0.4	99.2	
16	15	16	12	0.2	99.4	
17	16	17	6	0.1	99.5	
18	17	18	6	0.1	99.6	
19	18	19	0	0.0	99.6	
20	19	20	8	0.1	99.8	
21	20	21	1	0.0	99.8	
22	21	22	4	0.1	99.8	
23	22	23	2	0.0	99.9	
24	23	24	2	0.0	99.9	
25	24	25	1	0.0	99.9	
26	25	26	3	0.0	100.0	
27	26	27	0	0.0	100.0	
28	27	28	0	0.0	100.0	
29	28	29	0	0.0	100.0	
30	29	30	1	0.0	100.0	
31	30	31	1	0.0	100.0	
32	31	32	0	0.0	100.0	
33	32	33	0	0.0	100.0	
34	33	34	0	0.0	100.0	
35	34	35	0	0.0	100.0	
36	35	36	0	0.0	100.0	
37	36	37	0	0.0	100.0	
38	37	38	0	0.0	100.0	
39	38	39	0	0.0	100.0	
40	39	40	0	0.0	100.0	
GREATER OR EQUAL		40	0	0.0	0.0	| | | | | | |
		TOTAL ->	6192	100.0	100.0	0 5 10 15 20 25 30 %

B) Characteristic Data

CHARACTERISTIC DATA		
1.	NUMBER OF GROUPS	40
2.	GROUP WIDTH	1
3.	LOWER LIMIT	0
4.	UPPER LIMIT	40
5.	NUMBER OF VALUES	6192
6.	NUMBER OF VALUES LESS THEN LOWER LIMIT	0
7.	NUMBER OF VALUES GREATER OR EQUAL UPPER LIMIT	0
8.	NUMBER OF VALUES WITHIN GROUP LIMITS	6192
9.	MINIMUM VALUE	2
10.	MAXIMUM VALUE	30
11.	RANGE OF VALUES	28
12.	TOTAL	.243E+05
13.	SUM OF SQUARES	.136E+06
14..	MEAN VALUE	3.924
15.	VARIANCE	.651E+01
16.	STANDARD DEVIATION	2.55
17.	COEFFICIENT OF VARIATION	65.01
18.	MEDIAN	3.65
19.	LOPSIDENESS	3.47
20.	EXCESS	16.86

Table 5.4. Sample evaluation of feedback data over several periods, using the BETA program

MANUFACTURING ANALYSIS - PLANT 203 - TOTAL VALUES SHOP 36 PAGE 1

PERIOD FROM 205	TOTAL	1 - 218	2 - 232	3 - 246	4 - 260	5 - 274	6 - 288	7 - 302	8 - 316	ABBREVIATIONS:
...CAPACITY...										
TTL (HRS)	66045	8650	8570	7425	8400	8450	8270	8140	8140	TOTAL CAPACITY
...OUTPUT...										
TTL (HRS)	56694	8190	6360	6190	8063	8018	6916	7365	5593	TOTAL OUTPUT
MEAN (%)	85.8	94.7	74.2	83.4	96.0	94.9	83.6	90.5	68.7	MEAN UTILIZATION
SD (%)	113.8	96.3	108.2	112.4	114.5	120.0	126.0	131.7	93.1	STANDARD DEVIATION UTILIZATION
HRS/SWC	667	96	75	73	95	94	81	87	66	OUTPUT PER WORK CENTER
...STANDARD TIMES...										
TO (HRS)	9.16	7.18	8.83	9.90	10.28	9.53	9.27	10.09	9.24	OPERATION TIME MEAN VALUE
SD (HRS)	13.19	8.34	12.26	14.22	13.24	13.03	13.78	17.76	13.2	OPERATION TIME STANDARD DEVIATION
TS (MIN)	97.04	90.73	78.72	103.13	110.46	100.87	92.24	104.38	98.96	SETUP TIME MEAN VALUE
SD (MIN)	98.57	91.5	83.9	101.44	102.14	102.67	107.41	99.3	98.03	SETUP TIME STANDARD DEVIATION
TP (MIN)	0.85	0.85	0.80	0.91	0.87	0.79	0.87	0.93	0.78	PROCESS TIME PER UNIT MEAN VALUE
SD (MIN)	1.70	1.57	1.82	1.80	2.04	1.28	1.53	2.11	1.22	PROCESS TIME PER UNIT STANDARD DEVIATION
TO (SCD)	2.82	1.65	2.60	3.01	2.71	2.68	3.16	4.12	2.78	OPERATION TIME MEAN VALUE
SD (SCD)	3.42	1.53	3.14	3.12	2.66	2.77	3.61	5.53	3.27	OPERATION TIME STANDARD DEVIATION
OP TTL	6192	1141	720	625	784	841	746	730	605	TOTAL NUMBER OF OPERATIONS
OP NO ST	43	4	4	4	4	9	12	1	5	NUMBER OF OPERATIONS WITHOUT STANDARD TIME
OP/SWC	73	13	8	7	9	10	9	9	7	NUMBER OF OPERATIONS PER SINGLE WORK CENTER
...LOT SIZE...										
MEAN (-)	533.9	398.3	562.9	542.4	580.4	598.2	535.4	540.8	586.1	MEAN VALUE
SD (-)	1043.8	662.9	1008.0	1007.6	1209.5	1296.9	1001.9	974.9	1187.6	STANDARD DEVIATION
...LOAD (DIRECT)...										
TTL (HRS)	54440	6952	5628	7668	7517	7293	7235	6431	5716	TOTAL
MEAN (%)	82.4	80.4	65.7	103.3	89.5	86.3	87.5	79.0	70.2	MEAN VALUE RELATED TO CAPACITY
SD (%)	137.5	111.7	122.9	148.6	158.1	150.0	143.5	125.1	131.9	STANDARD DEVIATION RELATED TO CAPACITY
...INVENTORY (DIRECT)...										
INI (HRS)	5506	5506	4268	3636	5014	4469	3744	4063	3128	INITIAL VALUE
MEAN (SCD)	5.23	6.05	4.65	4.86	5.73	5.41	5.2	5.22	4.68	MEAN VALUE RELATED TO CAPACITY
SD (SCD)	3.76	3.51	4.61	3.58	3.92	3.75	3.70	3.42	3.09	STANDARD DEVIATION RELATED TO CAPACITY
...LEAD TIMES (DIRECT)...										
MUN (SCD)	3.92	4.08	4.58	4.00	4.12	3.81	3.47	3.60	3.63	MEAN VALUE UNWEIGHTED
SD (SCD)	2.55	2.04	3.54	3.55	2.45	2.16	2.09	2.29	1.96	STANDARD DEVIATION UNWEIGHTED
MW (SCD)	5.95	5.26	6.43	6.40	5.89	5.58	5.70	6.71	5.78	MEAN VALUE WEIGHTED
SD (SCD)	4.06	2.32	4.36	5.18	3.38	3.14	3.62	5.78	3.86	STANDARD DEVIATION WEIGHTED
TA (SCD)	6.08	6.07	6.91	5.64	5.87	5.73	6.00	6.30	6.22	ADVANCE TIME MEAN VALUE
SD (SCD)	0.92	1.47	0.71	0.59	0.57	0.51	0.67	0.93	0.67	ADVANCE TIME STANDARD DEVIATION
RAN (SCD)	5.93	6.40	5.95	5.69	5.97	5.71	5.67	5.77	6.30	RANGE MEAN VALUE
SD (SCD)	8.95	9.05	8.21	6.84	9.41	9.16	11.21	7.66	9.10	RANGE STANDARD DEVIATION
ITC (SCD)	0.15	0.33	0.96	0.05	0.10	0.03	0.34	0.53	0.08	INVENTORY TREND COMPONENT
SC (SCD)	0.13	0.80	0.48	0.76	0.02	0.15	0.30	0.41	0.44	SEQUENCE COMPONENT
...POSITION...										
MEAN (-)	2.1	2.0	2.5	2.1	2.2	2.1	2.0	1.9	2.1	MEAN VALUE
SD (-)	1.4	1.2	1.4	1.4	1.5	1.5	1.3	1.3	1.3	STANDARD DEVIATION
...CAPITAL FLOW...										
INP (TDM)	5512.3	839.2	669.0	691.7	772.4	711.1	671.3	650.9	506.9	INPUT
OUT (TDM)	5624.4	783.9	839.6	531.7	809.0	767.7	714.9	643.9	533.7	OUTPUT
INI (TDM)	367.1	367.1	422.4	251.8	411.8	375.2	318.6	275.0	282.0	INITIAL VALUE
MEAN (TDM)	329.3	414.9	314.2	329.1	395.3	332.0	300.8	280.5	267.5	MEAN VALUE
MAT (TDM)	549.1	88.1	39.0	70.5	92.0	73.1	74.5	72.6	39.1	VALUE OF MATERIAL (MATERIALS COST)
PEV (TDM)	2689.0	387.1	298.1	292.3	384.3	383.5	327.2	350.3	266.2	ADDED VALUE (WORK COSTS)

The throughput diagram with the inventory trend of Work Center 3641 is shown in Fig. 5.11, and was plotted automatically from the recorded data. In addition to the input and output curves it contains the release curve, thus also allowing us to determine, besides the direct inventory at the work center concerned, the indirect inventory at the upstream work centers. In the release curve, one notices a periodical occurrence of high-volume orders, which is less distinct in the input curve, when the orders reach the work center. This indicates lot splittings after release, or after one of the upstream processing steps. The horizontal lines in the curves mark the non-productive days, since the time axis is calendar time, not shop calendar time.

Finally, from the evaluation file one can deduce the material flow between the

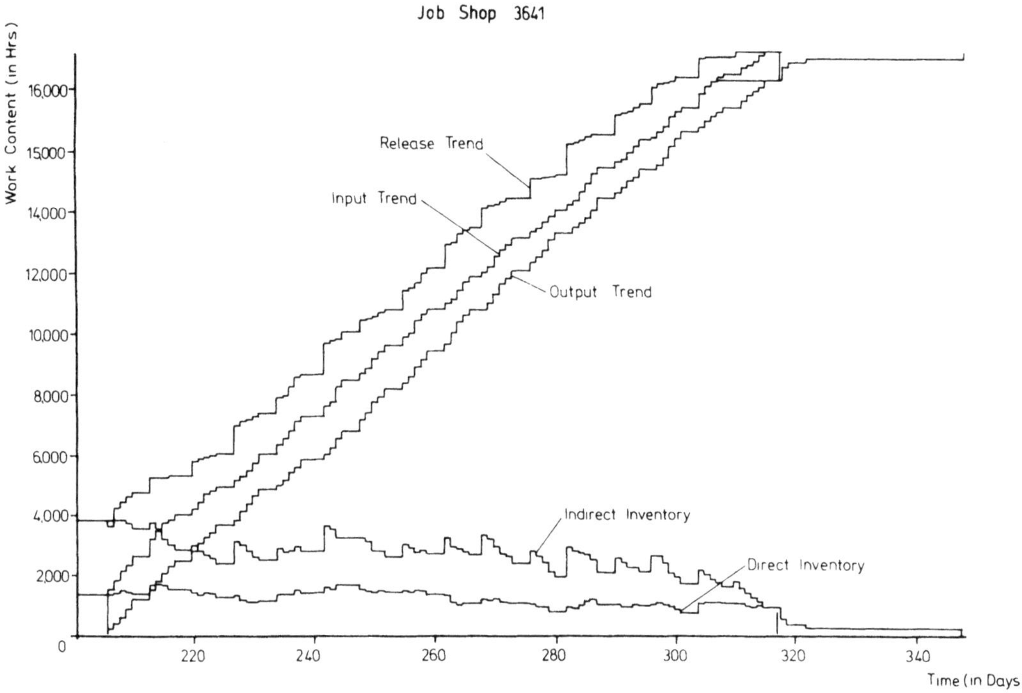

Fig. 5.11. Drawing up a throughput diagram using the AIZPLT program

Table 5.5. Evaluation of the material flow relations between different cost centers using the TRAMAT program. STPT: starting point; ENPT: end point; UNPT: unknown point

RELATION MATRIX WITH SUCCESSOR VALUES - PLANT 203 - MATERIAL FLOW ON COST CENTER LEVEL

FACTOR 1.000 UNIT OF MEASURE [-] FREQUENCY VALUES GIVEN IN NUMBERS OF MOVED LOTS

	STPT	UNPT	3512	3520	3631	3622	3621	3614	3611	3613	3641	3705	3711	3710	3720	3730	3552	3530	3540	3560	3551	3105	ENPT	SUM
STPT			16		1559	572	157	139	52	76	172	136		14	13	31	1	2		11		1		2952
UNPT																								0
3512			15			1		2		5	9			1										33
3520				6							1						2			6				15
3631						26	187	2	9	68	1188			1			74	2		1			1	1559
3622				6		54	67	36	62	88	290			6		1	118	2		1		1	2	734
3621						14	37	38	7	86	184			2			183							551
3614						4	27	22	11	51	172						53						1	341
3611						8	4	38	12	60	56			1			47	3	1			1	1	232
3613						20	25	13	28	22	404	2				1	325	1		4		1	15	861
3641						9	28	27	18	311	738			10	1	3	2110		2	5	3	1	38	3304
3705						2	2	1	1	15	1	28	8	61	5	29	15	15						183
3711										1		6		1										8
3710						5	10	4	9	35	16	7		43	2	15	44	1		19		2	20	232
3720						2		4	7	2	1			4	9					1			1	31
3730						12	3	1	5	19	27	3		1		10	14		1	9			7	112
3552			2	3		2	2		2	5	13			73		17		106	10	1095	69	142	1526	3067
3530						2	1	13	7	11	6	1		9	1	5	2			26	18	3	34	139
3540									1	1										3	2		9	16
3560						1	1	1	1	2	25			4			2	3	2	368	50	34	1086	1580
3551										3				1						1			139	144
3105											1						77	4		30	2	64	72	250
ENPT																								0
SUM	0	0	33	15	1559	734	551	341	232	861	3304	183	8	232	31	112	3067	139	16	1580	144	250	2952	16344

STPT = Starting point ENPT = End Point UNPT = Unknown Point

work centers. When the data records are arranged by work center numbers and the respective downstream work center numbers, we can see which material quantities flow from the work center concerned to the downstream work center, and from which work centers material flows to the work center concerned. The result can be represented in a matrix such as that shown in Table 5.5. In this case, the frequency values correspond to the number of lots moved during the investigation period, i.e. the number of orders that flowed from each work center to the other work centers, or vice versa. Instead of the number of lots, the number of parts, hours, or dollars can also be used.

The material flow matrix thus becomes a valuable tool for discovering weak points in the material flow, and for initiating improvements in work center or storage area layout, and in transport organization [8, 9]. For this purpose, the matrix should be transformed, with the aid of a computer program, into a graphic representation of the material flow. Figure 5.12 shows the material flow representation of all values from the matrix in Table 5.5, whereas, in Fig. 5.13, for the sake of clarity, all material flows with less than 100 lots in the investigation period have been suppressed.

The material flow representations show a major flow trend, which can be made

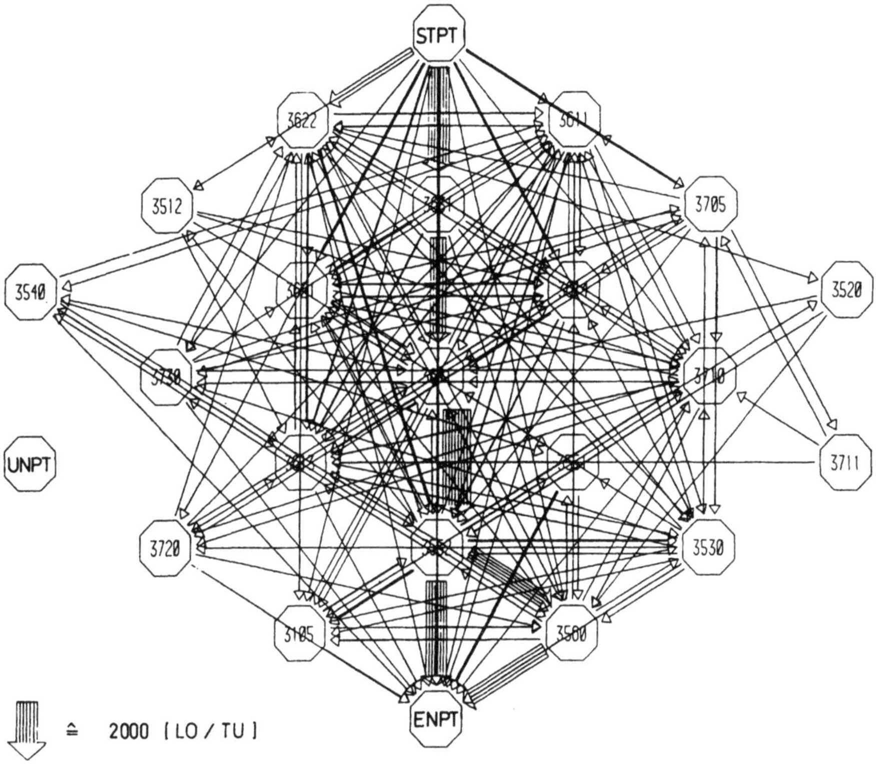

Fig. 5.12. Movement relations graph (all data taken from Table 5.5). LO/TU: lots per time unit; STPT. starting point; ENPT: end point; UNPT: unknown point

visible by arranging the points of material flow in a certain order, as is the case in the Sankey diagram widely used in power plant engineering. If each work center in this material flow is now represented as a funnel, we obtain Fig. 5.14, which was first published by Kettner and Bechte as the funnel model [10].

It represents an analogy model of the manufacturing flow, and has become the symbol of load-oriented or flow-oriented manufacturing control. Since a throughput diagram can be attributed to each funnel as well as to the entire system, the funnel model is not only a symbolic representation but can also be exactly calculated using an arithmetical model. This allows us to monitor and control the manufacturing flow on the basis of the mathematical relationships between the key values developed from the throughput diagram in Chapt. 4. This is dealt with in some detail below.

5.3.3 Representation of the Results

The results of the manufacturing analysis in a job shop can best be ordered as shown in Table 5.6. According to the objectives of the investigation, the data base with the values given here has to be specified first. The feedback errors found in this process offer the first indications of possible improvements in production data collection.

The section order data characterizes the order quantity processed in the inves-

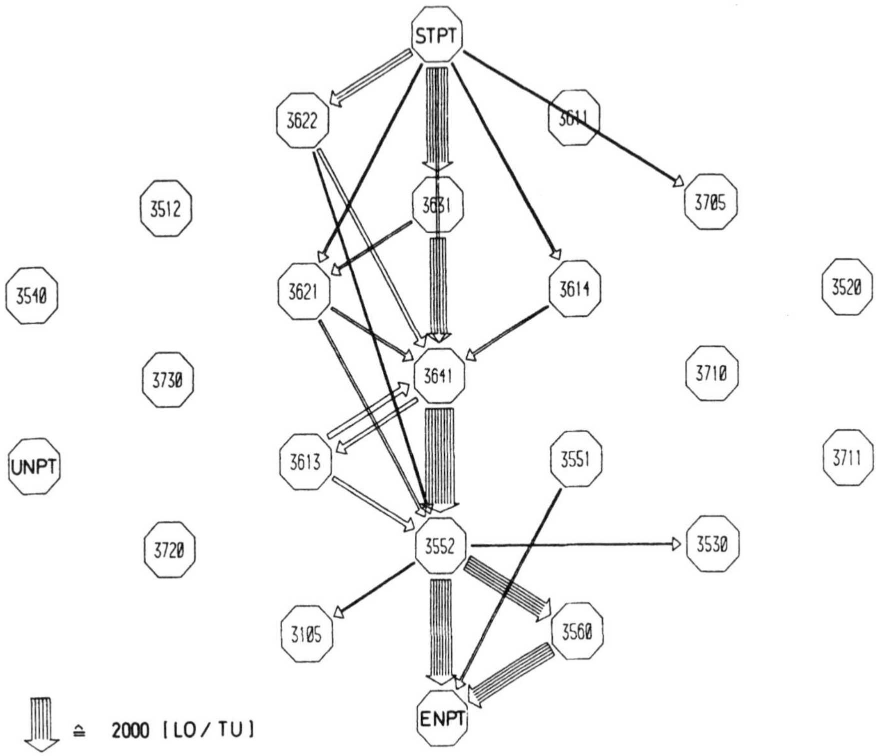

Fig. 5.13. Evaluation of material flow relations with 100 or more movement units (corresponding to 82% of all the movements shown in Table 5.5)

tigation period, using the specified criteria. As forms of representation, frequency distribution diagrams and Pareto curves should be used.

Figure 5.15A shows an example from the examination of order time per operation. The distribution graph of the 7306 evaluated operations is lopsided to the left, which is typical of these evaluations, the mean value of the unweighted order time here being 6.1 hours. If one plots the number of working hours contained in each group, the resulting distribution is considerably flatter, with a rise in work content in the high groups and a mean value of 55.5 hours. This is the value for the weighted mean order time.

The Pareto curve for this frequency distribution is produced by plotting cumulatively the relative frequency values of the numbers of operations and the numbers of working hours respectively, as shown in Fig. 5.15B. The first group value, containing all order time values up to 2 hours, will serve as our example. We can see from Fig. 5.15A that about 52% of all operations, but only about 8% of the work content, belong in this group. These two values give us the first point of the Pareto curve: its value is 2 hours. The other points are determined by cumulating the respective individual values.

Here again, the so called 80–20 rule known from materials management – which says that 80% of an item distribution in a warehouse constitutes only 20% of the value of these items – holds true. In our case, 80% of the number of operations corresponds to 25% of the work content of all operations. The next 15% of the

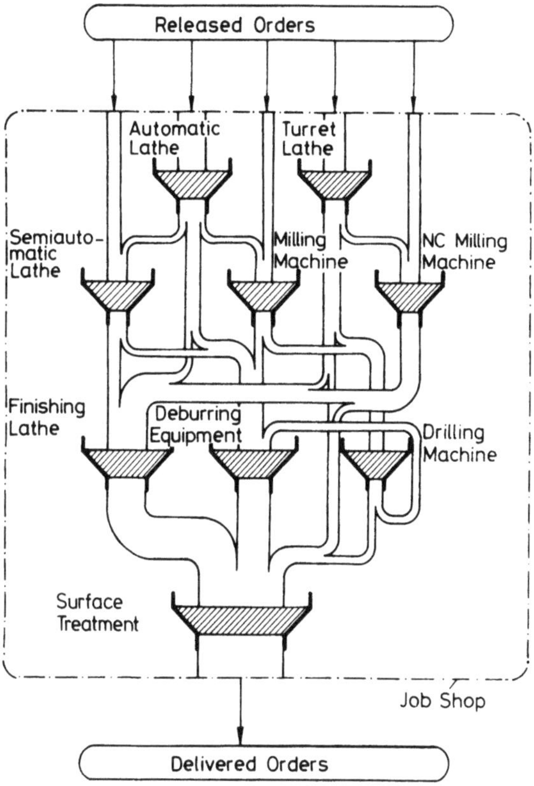

Fig. 5.14. Funnel model of job shop production (After Kettner and Bechte [10])

operations contains a further 30% of the work content, and the last 5% of the operations corresponds to 45% of the work content of all operations. Lot size values show a very similar distribution. These distributions offer the first indications of how to differentiate between operations when establishing the operation sheet, when determining the standard time, and when monitoring schedules.

In the representation of results, emphasis should be placed on the flow times of the orders, operations and different work system levels, which were treated at some length in Chapt. 3. The comparison of the values of single cost centers and work centers often leads to interesting results. As an example, Fig. 5.16 shows extracts from the hierarchy of some work-center-related key data from a manufacturing analysis [11].

Besides performance and inventory, the weighted and unweighted mean actual lead times were also evaluated. The long weighted mean lead time of 14.3 workdays in Job Shop 6 is chiefly due to the long lead time of Cost Center 614 (18.8 workdays). If one looks at the work centers contained in these cost centers, one can see that, once again, a few single work centers are responsible for the high average.

Since, in this case, the work centers are exclusively NC machines, the question arises as to whether the observed lead times were meant to be so long when the machines were purchased. An explanation could be found in the wide scattering of the lead time distribution for the entire cost center 614, shown in Fig. 5.17.

If one classifies the lead time groups according to the four priority classes ex-

Table 5.6. Essential evaluation results from a manufacturing analysis. –: useless; ×: common practice; +: for particular work centers only

Evaluation result	Evaluation level					
			Work system			
	Order	Operation	Plant	Job shop	Cost center	Work center
Data base						
– Investigation period						
– Orders, operations	×	×	–	–	–	–
– Investigation field	–	–	×	×	×	
– Feedback errors	×	×	×	×	×	×
Order data						
– Quantities (Items, in., oz., etc.)	×	×	–	–	–	–
– Time (Cost per item, setup time)	×	×	–	–	–	–
– Lot size	×	×	–	–	–	–
– Number of operations	×	–	–	–	–	–
Lead times						
– Weighted/unweighted	×	×	×	×	×	×
– Operation time	×	×	×	×	×	×
– Interoperation time	–	×	×	×	×	×
– Operation time percentage	×	×	×	×	×	×
Manufacturing process						
– Period-related key data		–	×	×	×	×
– Throughput diagrams	×	–	×	×	+	+
Dates						
– Deviations: planned/actual	×	–	–	–	–	–
Material flow						
– Relation matrix	–	–	×	+	+	–
– Material flow graph	–	–	×	+	+	–
– Funnel model	–	–	×	+	+	–

plained in Fig. 2.17, one can see that about 38% of the work content stayed at this cost center for longer than 30 days, and that only 13% of the work content was processed in a time of up to 10 days. From the distribution of the unweighted lead time, which is also included, one can see how important it is to evaluate the weighted lead time. The number of orders processed within 10 days amounted to about 39%. This gives the impression that the 'important' orders, namely the express and normal orders, are processed in a tolerable time. Only a comparison with the weighted values shows how mistaken the assumption was. To improve the situation one should try to find out why C and D orders have to wait so long, and whether perhaps organizational shortcomings in the NC programming and tooling supply are the cause.

After the analysis of flow time and its components according to their frequency distribution, the next important step is the representation in periods of the manufacturing flow in the investigation period. This has been illustrated with the list of key data shown in Table 5.4, and with the throughput diagrams containing inventory trends, throughput elements, and lateness detailed in Chapt. 4.

Figure 5.18 shows a throughput diagram with the throughput elements of Work Center 614 460, which belongs to Cost Center 614 given in Fig. 5.16 and which

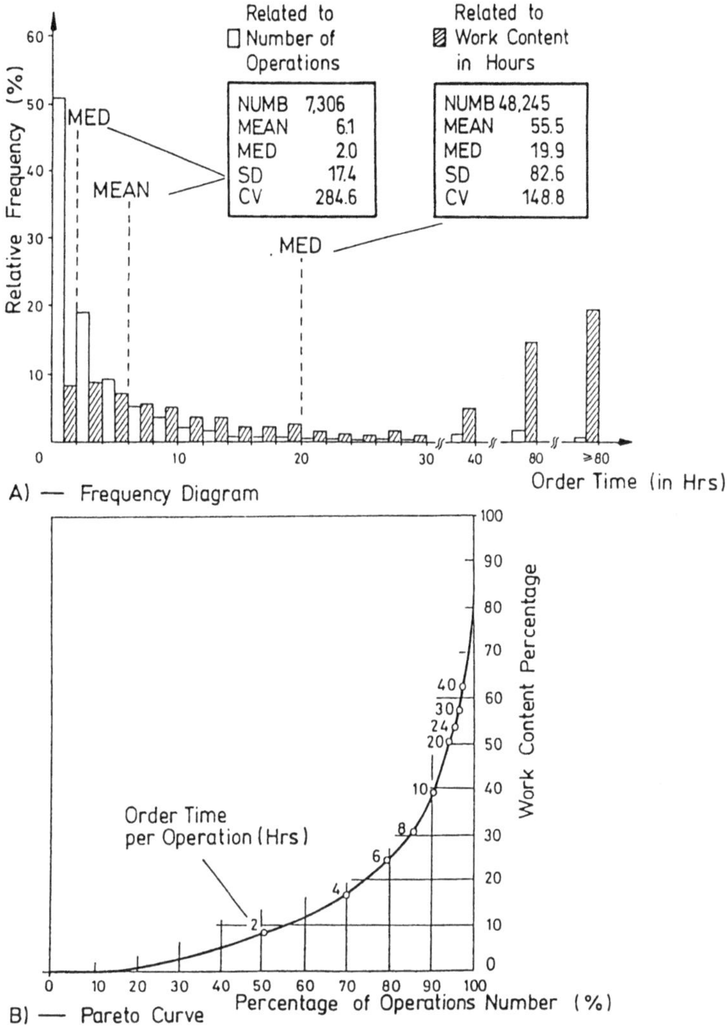

Fig. 5.15. Evaluation of order time distribution per operation for an aggregate plant

has the very high weighted mean lead time value of nearly 23 workdays. We can see wide fluctuations in the input curve, caused by large variations in the work contents of the individual orders, and by the irregular input dates of these orders. The output curve shows similar fluctuations. The situation is aggravated by strong order sequence transpositions, which lead to wide variances in lead times. From this, some recommendations for improvement can immediately be deduced which, above all, aim at better balanced input and output trends. This is dealt with in Sect. 5.5.

After having analyzed the manufacturing flow, one should check the lateness of the orders. For this, a representation as shown in Fig. 1.5, which was taken from such an evaluation, can be used.

Representation of the material flow for the aggregate plant and for the job shops and cost centers in question (also described above) is the final step in the representation of results.

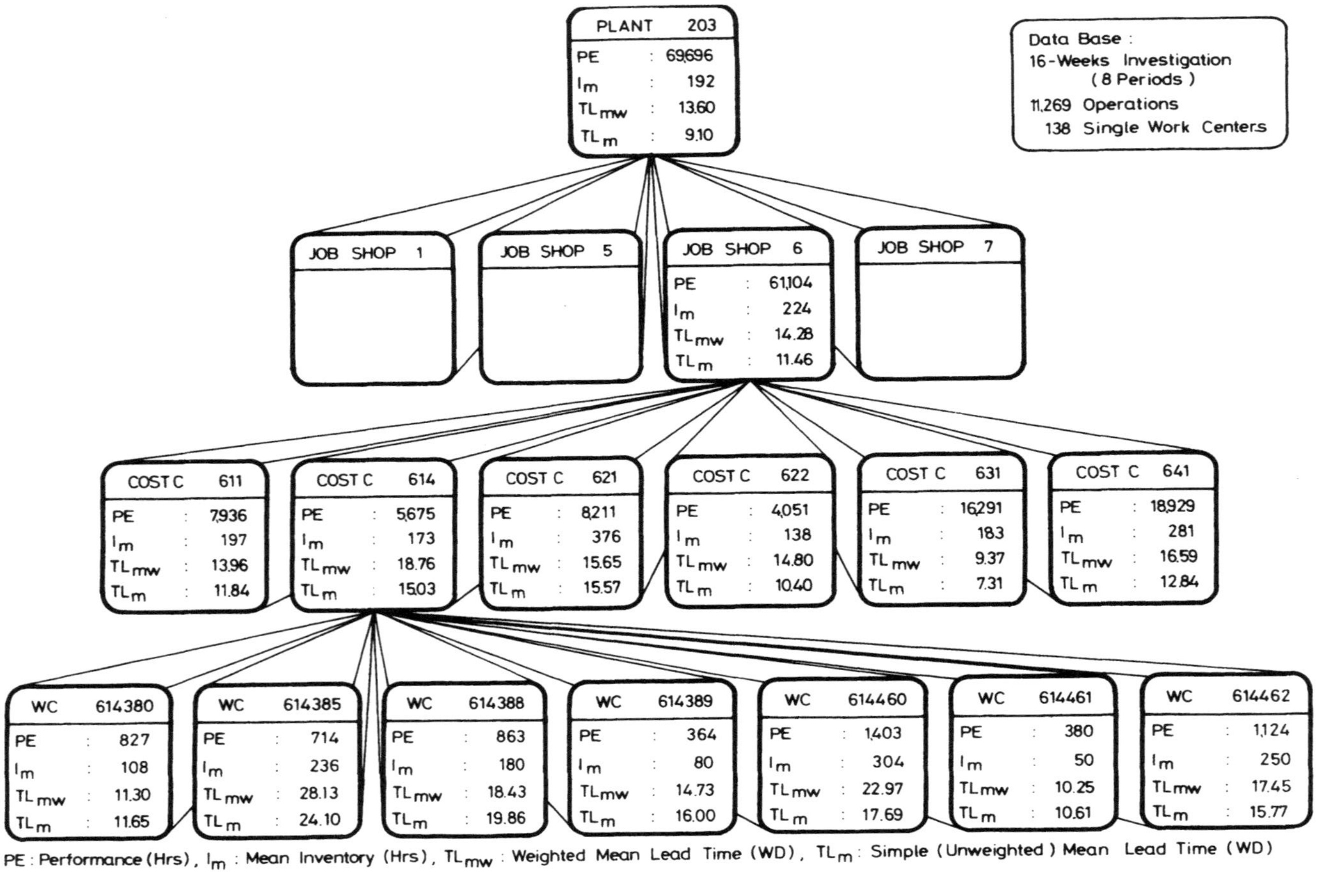

Fig. 5.16. Hierarchic structure of manufacturing key data: utilization, mean inventory, simple and weighted mean flow time (data taken from a metalworking shop) (After Lorenz [11])

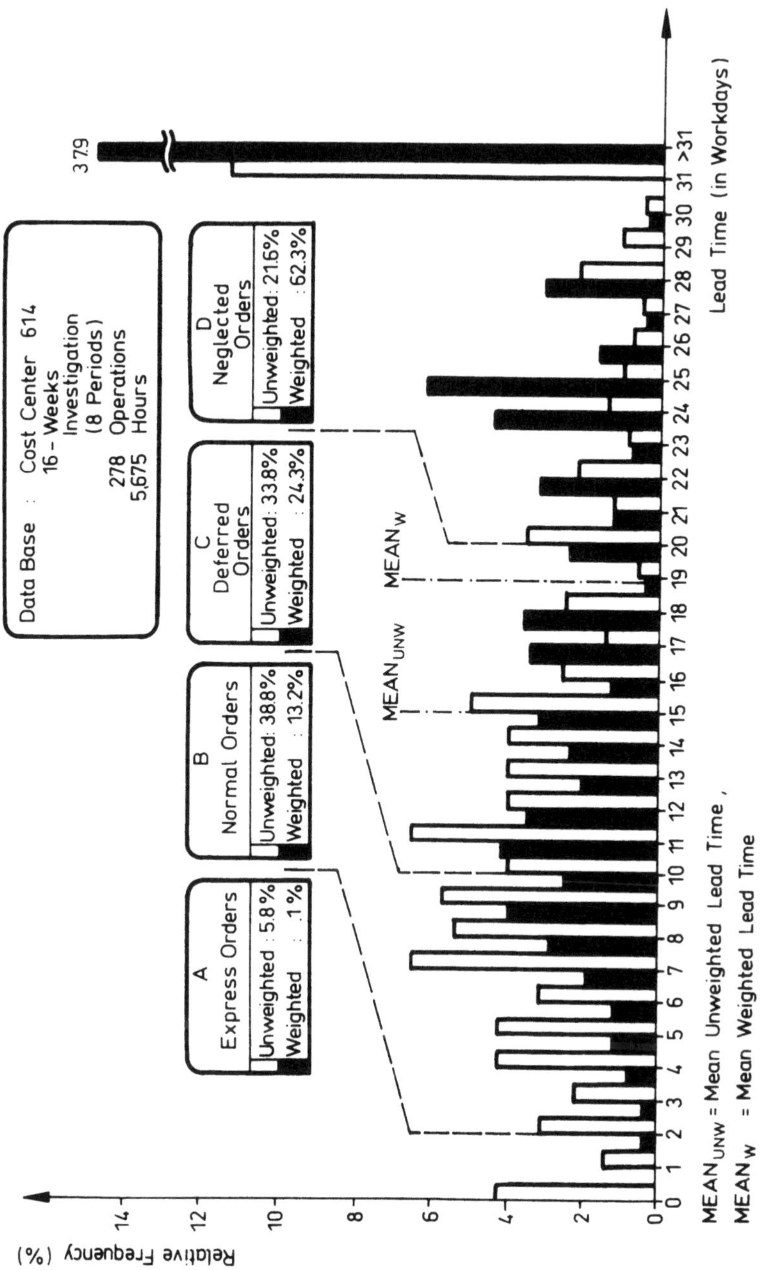

Fig. 5.17. Flow time distribution of an NC cost center (After Lorenz [11])

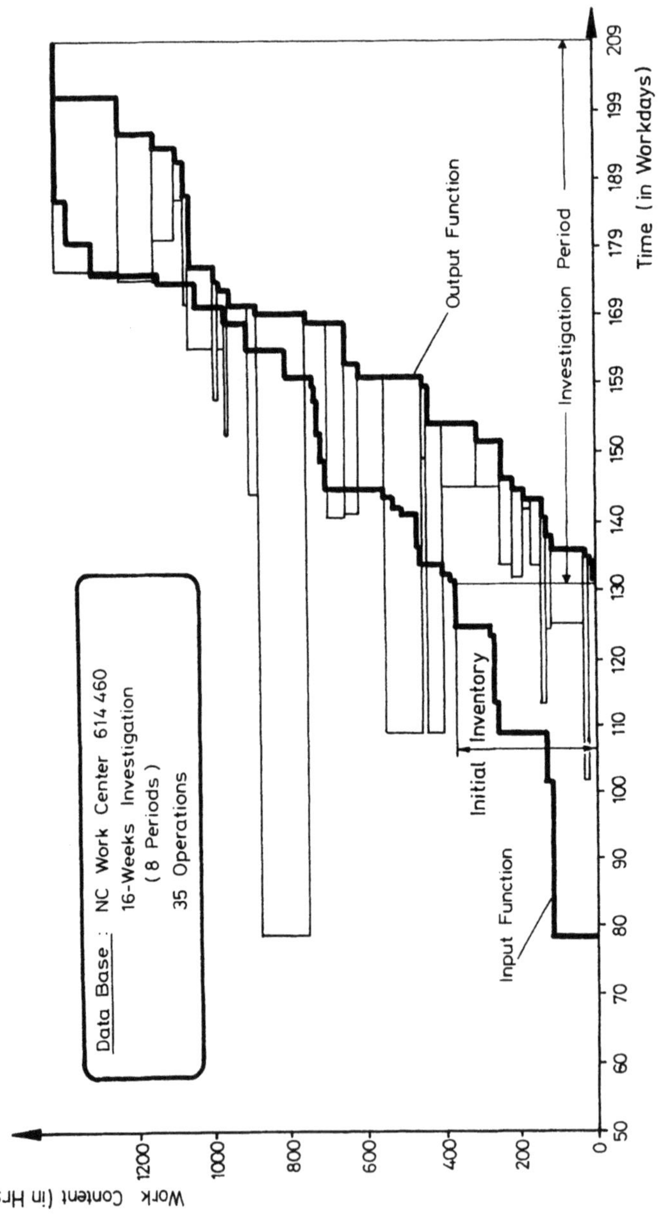

Fig. 5.18. Sample throughput diagram for an NC work center (After Lorenz [11])

Those responsible for the shops, for manufacturing control and for materials management are often amazed at the results, and there are often discussions about whether it is appropriate to choose weighted lead times as monitored variables, or whether the investigation period or field of investigation were chosen properly.

The most important measure taken as a consequence of such investigations is undoubtedly permanent monitoring of the flow time, inventory, utilization, and lateness objectives with an autonomous monitoring system. Before we introduce a concept for and an example of such a system, we first make some general propositions for improving the manufacturing flow, which can be deduced from the above statements.

5.3.4 General Rules and Possibilities for Improving Manufacturing Flow

Improvement of the manufacturing flow can generally be looked upon as a matter of avoiding waiting times. Table 5.7 gives general rules and the ensuing approaches towards improvement [12].

Starting with the product design, one should ask the question whether the product can be manufactured in fewer structural stages; one pre-assembly and one final assembly stage would be ideal. This would mean greater market flexibility for the company, since storage would perhaps only have to take place on the level of modules and, due to fewer intermediate stages, in-plant flow times would decrease, thus semi-finished products inventory tie-up would be reduced. This question should be dealt with in close cooperation between the design and process planning departments when designing or redesigning a product.

In operations planning, one must make sure that as few changes in routing as possible are necessary in the throughput of parts and products. This leads to numerous suggestions for investment and layout planning. When purchasing new machine tools, for example, one should ensure that as many parts as possible can be processed completely in one or two setups by a machine, even if this means that individual operations are less cost-effective. For example, this might mean the use of powered drilling and milling tools in the turret head of a lathe. One should also consider installing fixtures or cheaper machines directly beside an existing main machine tool for necessary auxiliary operations such as drilling, deburring, cleaning, etc., thus

Table 5.7. Guidelines for manufacturing process improvement

Avoid waiting times by the following means

- Keep number of product levels as small as possible
  - → Keep to only one pre-assembly as well as one final assembly
- Reduce changes in work centers in manufacturing process
  - → Provide each work center to process various operations
  - → Arrange production and pre-assembly to be product oriented
- Keep work contents small and well balanced
  - → Cut setup times at bottlenecks
  - → Review lot size formulas
- Shorten queue lengths at work centers
  - → Monitor inventory
  - → Control input

eliminating interoperation times and achieving better utilization of manpower during machine run time. After the material flow analysis has shown the interlinkages between single work places, one should consider creating process-oriented parts-manufacturing cells and subassembly cells. Such sub-shops are, to a large extent, self controlling in their organization, and greatly relieve the shop floor control system. They can reduce lead times by 30–50% and more, and make the factory flexible [13].

Lot sizing is another important factor. Our discussion of throughput diagrams and weighted lead times made it clear that a great variance in work center order times has a negative effect on schedule deviation. Since only a small number of lots deviate strongly in work content from the mean work content at a work center, these should be analyzed more closely. If the work center is a bottleneck, and its capacity cannot be increased further, one should look for technical or organizational measures for setup time reduction [28]. Often, an exact analysis of the setup procedures reveals surprising possibilities for their improvement [14]. Another approach is determining the optimum lot size. A close examination often shows that a reduction in lot sizes by 20–30% causes a rise in unit cost by only 3–5% or even less. Thus, if work center capacity is sufficient one can reduce lot sizes, even if this means longer setup times [29, 30].

Finally, one can see from the throughput diagram that inventories should be as low as possible at the work centers, since this results in short queues and lead times. With a given capacity, inventories are determined only by order release. Inventories must consequently be controlled and permanently monitored. This can be carried out by technical facilities such as interlinked manufacturing equipment. Inventory-controlling manufacturing control methods can also perform this task. (This is dealt with in detail in Chapt. 6.)

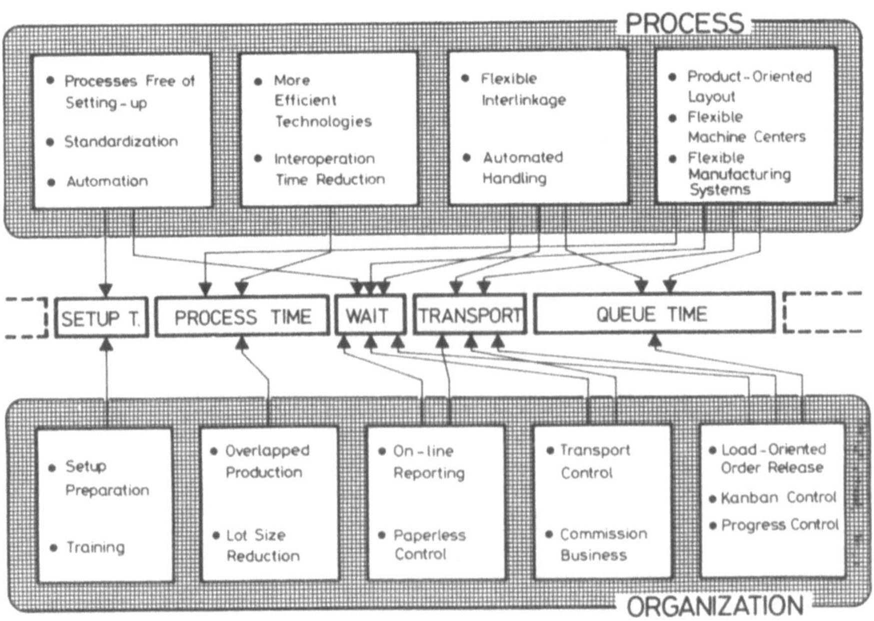

Fig. 5.19. Technical and organizational approaches to manufacturing flow time reduction

Figure 5.19 assigns these and other proposals to the individual components of lead time in the throughput element, distinguishing between process-oriented and organization-oriented approaches [12]. The sequence of corrective measures is determined by their potential for improvement: waiting time reduction has the highest priority; operation time and transit time come second. The choice between a technical or an organizational measure must be based on economic considerations, i.e. the best profit expense ratio. Finally, besides the mean value, the variance of the individual components of lead time must be constantly monitored and improved.

It becomes clear from the above statements on the improvement of manufacturing flow that, without detailed knowledge about the actual flow processes, corrective measures cannot be successful. Therefore, we now want to introduce a continuous monitoring system, which was first developed for use in manufacturing, but can, in principle, also be used for other production activities.

5.4 A Continuous Monitoring System for Manufacturing Flow

5.4.1 Objectives and Concept

An overall analysis of the manufacturing flow as discussed above costs a great deal of time and money. Figure 5.20 shows the schedule of a manufacturing analysis performed in a shop with about 90 work centers. The processes at the machines were

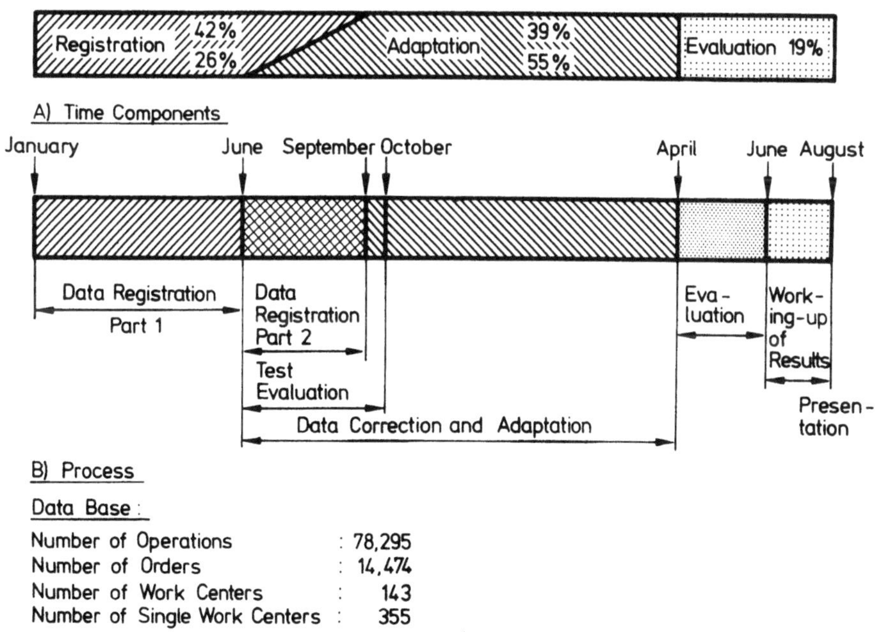

Fig. 5.20. Schedule of a single detailed manufacturing analysis (After Holzkämper [15])

recorded over a period of eight months in order to obtain reliable order flow time values [15]. In this particular case, an exceptionally large amount of data correction and adaptation were necessary, but even in more favorable cases results cannot usually be expected before one or one-and-a-half years after the beginning of data recording.

This cost is justified by the fundamental knowledge gained and the possible corrective measures which ensue; but a single analysis does not, of course, permit any modifications to be made to the on-going manufacturing flow.

The logical consequence is to use such an analysis – on a smaller scale – as a permanent monitoring system. The fundamental idea is to calculate the key data of the job shop which are relevant for the manufacturing flow – mainly inventory, utilization, flow time, and lateness – periodically from the feedback data of the plant, and to visualize them in a kind of process window. First concepts for this were published by the Institut für Fabrikanlagen [15–20], and the first practical application was realized by Bechte on the basis of his experiences with DUBAF [21, 22]. Such a monitoring system is a new function in MRD, which must be incorporated into an existing shop floor control system, as shown in Fig. 5.21 [15].

The prerequisite for this is an operational production data collection (PDC) system. The monitoring system must first check the data supplied by this system for plausibility and completeness using check criteria like those shown as an example in Fig. 5.9. This data check not only ensures correct data for the monitoring system, but also supplies objective criteria for judging the quality of the feedback system. Another important prerequisite for the implementation of a monitoring system is the issue of planned key data, since the calculation of key data merely from the actual status does not reveal deviations from a desired status. The planned data are taken from the planned process status as defined by the resources planning and manufacturing control departments. This planned process status must therefore be reproduced in the monitoring system.

The monitoring system itself has several subfunctions. First, the actual process status must be permanently stored and periodically updated with data from the PDC system. From these data, key data – essentially flow time, inventory, utilization, and lateness – must be calculated and documented. Finally, actual and planned data must be compared and the reason for deviations found. If possible, proposals for improvement should be developed. The last-mentioned function could be called a kind of diagnosis; for this reason, the whole system is called the *monitoring and diagnostic system.*

As Fig. 5.22 indicates, the monitoring system can be used in two respects [15]. In work center-related monitoring, the object of monitoring is the manufacturing flow at individual work centers, cost centers, job shops, or even the entire monitored field. The emphasis is placed on the behavior of the system as a whole, which is described by input and output data and by the key data inventory, output, flow time, and lateness and their interrelationships. The throughput diagram serves as the model for representing the order throughput.

Order-related monitoring, on the other hand, deals with the sequence of work centers through which one individual order has passed, or is going to pass. Here, the emphasis lies on the assessment of past and future schedule progress, therefore order lead time, lateness, scrap volume, etc., are preferred as key data. In contrast to conventional progress monitoring, a statement is made on the probability that the unfinished operations meet a certain date. This can be done on the basis of the expected throughput behavior of the downstream work centers.

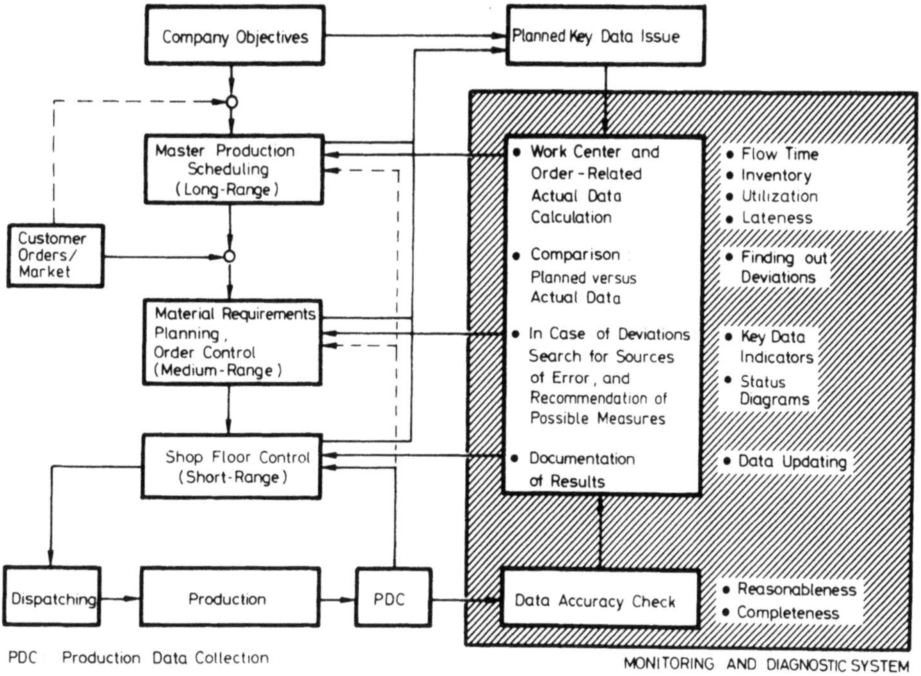

Fig. 5.21. Integration of the monitoring and diagnosis system into the manufacturing control process (After Holzkämper [15])

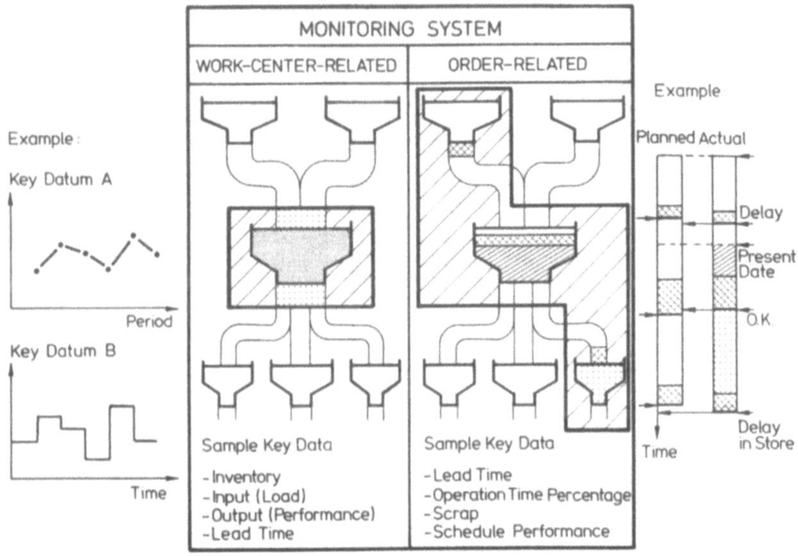

Fig. 5.22. Evaluation levels of the monitoring and diagnostic system (After Holzkämper [15])

The monitoring and diagnostic system described here refers only to shop orders. It could conceivably be used, however, to also include purchase orders, thus allowing customer orders to be monitored at all points.

In the following we describe the above-mentioned monitoring system as realized by Bechte, with its focus on representing the planned and actual processes in a plant and its job shops.

It does not yet include any diagnostic functions.

5.4.2 Example of a Continuous Monitoring System

The system's realization starts with a representation of the monitored field by describing input and output with their planned and actual dates. From this, the necessary key data (inventory, output, flow time, and lateness) are calculated. To reduce the amount of data to be held, the essential data are stored period by period in a monitoring table, which is constantly updated according to the changes in the current period, and stored in a powerful PC. It is only in this computer that reports and diagrams are produced, as required, in the form chosen by the user.

In the following, the basic points are demonstrated using the sample work center from Chapts. 3 and 4. The user sees three components of a continuous monitoring system (Fig. 5.23): the monitoring table contains the essential key data, in a concise and descriptive form; a simplified monitoring diagram represents order throughput; in a key data graph, selected key data are plotted in a bar chart. Figure 5.23 illustrates the three components with the values from our sample work center (see Chapt. 4).

First, the structure of the monitoring table in Fig. 5.23A is explained. The investigation period known from Table 4.1 (Calendar Day 205 to 232) has been divided into four periods with 7 calendar days (1 week) each (Line 1), or 5 shop

	KEY DATA (ACTUAL)		PERIOD 1	PERIOD 2	PERIOD 3	PERIOD 4	PERIOD 1-4
			1	2	3	4	5
1	Calendar Days	(Day) (Days)	205-211 7	212-218 7	219-225 7	226-232 7	205-232 28
2	Workdays	(SCD) (SCD)	30-34 5	35-39 5	40-44 5	45-49 5	30-49 20
3	Input per Period	(Hrs)	34.9	24.4	96.3	29.3	184.9
4	Input Cumulated	(Hrs)	130.7	155.1	251.4	280.7	280.7
5	Output per Period	(Hrs)	37.9	31.4	30.1	46.6	146.0
6	Output Cumulated	(Hrs)	37.9	69.3	99.4	146.0	146.0
7	Change in Inventory	(Hrs)	-3.0	-7.0	+66.2	-17.3	+38.9
8	Mean Inventory	(Hrs)	104.1	83.7	86.3	147.0	105.3
9	Mean Range	(SCD)	13.7	13.3	14.3	15.8	14.4
10	Weighted Mean Lead Time	(SCD)	16.5	8.1	15.5	8.2	11.8
11	Weighted Mean Lateness	(SCD)	-9.5	+1.2	-15.4	-1.3	-5.8
12	Mean Loading	(%)	95	79	75	117	91

A) Monitoring Table

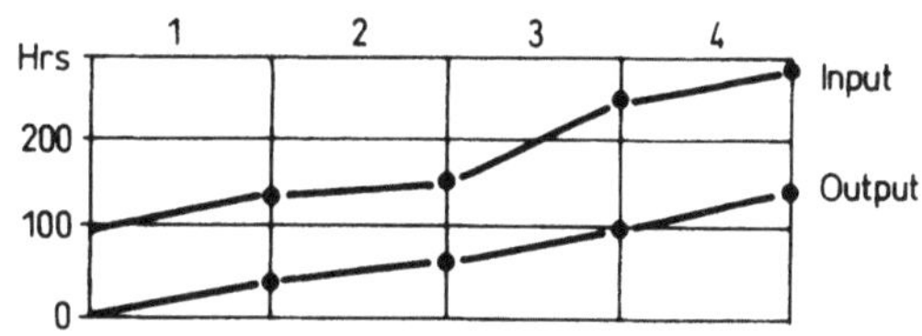

B) Monitoring Diagram

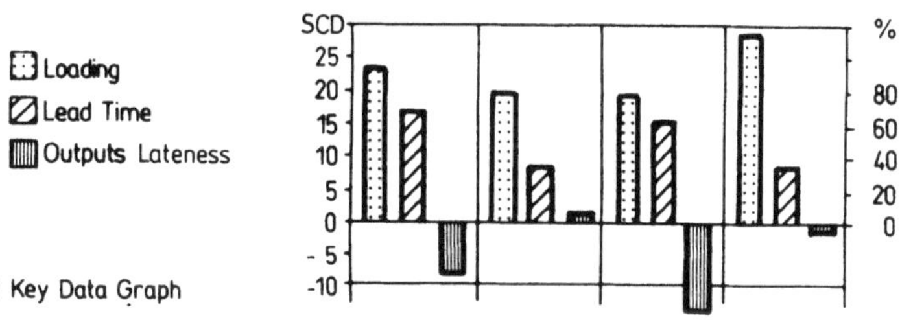

C) Key Data Graph

Fig. 5.23. Components of a permanent work center monitoring system

calendar days each (Line 2), respectively. A detailed calculation of the input, output, inventory, and range values can be followed in Fig. 5.24. Columns 2 and 3 contain input and output per shop calendar day (the values are taken from Table A2).

Beginning with the initial inventory, the inventories at the end of each day (Column 5) can be calculated from the difference between the inputs (Column 2) and outputs (Column 3) on that day. The inventory area (Column 4) for one workday is calculated from the initial inventory of that day (= inventory at the end of the previous day) multiplied by one workday.

If one divides the inventory areas of one period (total of Column 4) by the respective number of workdays, one obtains the mean inventory. When the inventory area is divided by the output in that period, this gives us the mean range in shop calendar days. Finally, both key values were calculated for the entire reference period (Periods 1–4): 105.3 hours (mean inventory) and 14.4 SCD (mean range).

PERIOD 1

DAY (SCD)	INPUT (Hrs)	OUTPUT (Hrs)	AREA (Hrs×SCD)	INVENTORY AT THE END OF DAY (Hrs)
1	2	3	4	5
		INITIAL INVENTORY 95.8		
30	9.6	.0	95.8	105.4
31	3.8	.5	105.4	108.7
32	13.6	11.4	108.7	110.9
33	7.9	19.2	110.9	99.6
34	.0	6.8	99.6	92.8
TOTAL 5	34.9	37.9	520.4	92.8

MEAN INVENTORY $I_{m1} = \frac{520.4}{5} = 104.1$ Hrs

MEAN RANGE $R_{m1} = \frac{520.4}{37.9} = 13.7$ SCD

PERIOD 2

TOTAL	5	24.4	31.4	418.4	85.8

MEAN INVENTORY $I_{m2} = \frac{418.4}{5} = 83.7$ Hrs

MEAN RANGE $R_{m2} = \frac{418.4}{31.4} = 13.3$ SCD

PERIOD 3

TOTAL	5	96.3	30.1	431.7	152.0

MEAN INVENTORY $I_{m3} = \frac{431.7}{5} = 86.3$ Hrs

MEAN RANGE $R_{m3} = \frac{431.7}{30.1} = 14.3$ SCD

PERIOD 4

TOTAL	5	29.3	46.6	734.9	134.7

MEAN INVENTORY $I_{m4} = \frac{734.9}{5} = 147.0$ Hrs

MEAN RANGE $R_{m4} = \frac{734.9}{46.6} = 15.8$ SCD

MEAN INVENTORY

$$I_{m1\text{-}4} = \frac{520.4 + 418.4 + 431.7 + 734.9}{4 \times 5} \times \frac{\text{Hrs} \times \text{SCD}}{\text{SCD}}$$

$$I_{m1\text{-}4} = \frac{2105.4}{20} = 105.3 \text{ Hrs}$$

MEAN RANGE

$$R_{m1\text{-}4} = \frac{520.4 + 418.4 + 431.7 + 734.9}{37.9 + 31.4 + 30.1 + 46.6} \times \frac{\text{Hrs} \times \text{SCD}}{\text{Hrs}}$$

$$R_{m1\text{-}4} = \frac{2105.4}{146} = 14.4 \text{ SCD}$$

Fig. 5.24. Calculation of mean inventory and mean range per period at a work center (individual values taken from Table 4.2)

Furthermore, the monitoring table contains the change in inventory (Line 7). It characterizes the inventory change from period to period, and was calculated as the difference between the inventory values at the ends of the periods. The initial inventory in Period 1 (= final inventory of previous period), for example, was 95.8 hours; the inventory at the end of Period 1 was 92.8 hours. The inventory was reduced by 3 hours, thus Line 7, Column 1 of Fig. 5.23A gives minus 3.0.

	PERIOD 1			
	ORDER NO.	ORDER TIME TO (Hrs)	LEAD TIME TL (SCD)	AREA (HrsxSCD)
	1	2	3	4
	115	.5	17	8.5
	119	11.4	9	102.6
	110	15.4	28	431.2
	125	3.8	2	7.6
	120	6.8	11	74.8
TOTAL	5	37.9		624.7

WEIGHTED MEAN LEAD TIME $TLM_1 = \frac{624.7}{37.9} = 16.5$ SCD

Period 2:

TOTAL	4	31.4		253.1

WEIGHTED MEAN LEAD TIME $TLM_2 = \frac{253.1}{31.4} = 8.1$ SCD

Period 3:

TOTAL	5	30.1		467.7

WEIGHTED MEAN LEAD TIME $TLM_3 = \frac{467.7}{30.1} = 15.5$ SCD

Period 4:

TOTAL	6	46.6		382.6

WEIGHTED MEAN LEAD TIME $TLM_4 = \frac{382.6}{46.6} = 8.2$ SCD

WEIGHTED MEAN LEAD TIME $TLM_{1-4} = \frac{624.7 + 253.1 + 467.7 + 382.6}{146} \times \frac{Hrs \times SCD}{Hrs}$

$$TLM_{1-4} = \frac{1728.1}{146} = 11.8 \text{ SCD}$$

Fig. 5.25. Calculation of weighted mean actual lead time per period at a work center (individual values taken from Table 3.4)

Another important monitoring variable is the weighted mean actual lead time (*TLM*). It is calculated in Fig. 5.25 from the lead time area (Column 4: order time multiplied by lead time).

As a comparison, the values calculated in Appendix A, Tables A2 and A4, which are given in shop calendar days, can be used.

Like Fig. 5.24, Fig. 5.25 is set up in periods. In each period, the lead time areas are added up and divided by the output of the respective period, which produces the weighted mean lead time.

Utilization (Line 12, Fig. 5.23A, here called mean loading) can be calculated with a simple equation: utilization equals output rate divided by capacity. Here, capacity was 8 hours per SCD. In Period 4, utilization has the curious value of 117%. Such values can occur when outputs are reported for which some processing has already been carried out in previous periods.

Line 11 in Fig. 5.23A shows another important value of the monitoring table: weighted mean lateness of outputs. This value tells us whether completed orders are 'in schedule'. Its calculation can be found in Fig 5.26. The value of 5.8 SCDs means that an added capacity of about 6 days would have been necessary to make up for the delay.

The other flow time dependent values such as advance time, inventory trend

PERIOD 1						
ORDER NO.	PROCESS END (ACTUAL) $t_{PE\,act}$ (SCD)	PROCESS END (PLANNED) $t_{PE\,plan}$ (SCD)	ORDER TIME OF OUTPUTS TO (Hrs)	SCHEDULE DEVIATION SD (SCD)	SCHEDULE DEVIATION AREA (POS) TO × SD (Hrs×SCD)	SCHEDULE DEVIATION AREA (NEG) TO × SD (Hrs×SCD)
1	2	3	4	5	6	7
115	31	20	.5	- 11	-	5.5
119	32	29	11.4	- 3	-	34.2
110	33	13	15.4	- 20	-	303.0
125	33	39	3.8	+ 6	22.8	-
120	34	29	6.8	- 5	-	34.0
5	TOTAL		37.9	33	22.8	381.7

$OPSD_{mw} = \frac{22.8}{37.9} = .6$ SCD | $ONSD_{mw} = \frac{381.7}{37.9} = 10.1$ SCD | $OSD_{mw_1} = OPSD_{mw} - ONSD_{mw} = -9.1$ SCD

PERIOD 2						
4	TOTAL		31.4	+ 5	44.7	7.7

$OPSD_{mw} = \frac{44.7}{31.4} = 1.4$ SCD | $ONSD_{mw} = \frac{7.7}{31.4} = 0.2$ SCD | $OSD_{mw_2} = OPSD_{mw} - ONSD_{mw} = +1.2$ SCD

PERIOD 3						
5	TOTAL		30.1	- 43	6.3	470.0

$OPSD_{mw} = \frac{6.3}{30.1} = 0.2$ SCD | $ONSD_{mw} = \frac{470.0}{30.1} = 15.6$ SCD | $OSD_{mw_3} = OPSD_{mw} - ONSD_{mw} = -15.6$ SCD

PERIOD 4						
6	TOTAL		46.6	+ 6	75.0	136.2

$OPSD_{mw} = \frac{75.0}{46.6} = 1.6$ SCD | $ONSD_{mw} = \frac{136.2}{46.6} = 2.9$ SCD | $OSD_{mw_4} = OPSD_{mw} - ONSD_{mw} = -1.3$ SCD

$$OSD_{mw\,1-4} = \frac{\sum_{i=1}^{4} OSD_i \times OUT_i}{\sum_{i=4}^{4} OUT_i} = \frac{-9.5 \times 37.9 + 1.2 \times 31.4 + (-15.4) \times 30.1 + (-1.3) \times 46.6}{37.9 + 31.4 + 30.1 + 46.6} \times \frac{SCD \times Hrs}{Hrs}$$

$$OSD_{mw\,1-4} = \frac{-846.5}{146.0} = -5.8 \text{ SCD}$$

$OPSD_{mw}$ = Positive Mean Weighted Schedule Deviation of Outputs

$ONSD_{mw}$ = Negative Mean Weighted Schedule Deviation of Outputs

OSD_{mw} = Mean Weighted Schedule Deviation of Outputs

OUT = Output in Period P

Fig. 5.26. Calculation of the most important schedule performance key data per period at a work center

component and sequence component are not mentioned at this point; these would result in too much key data for practical application of a monitoring system. These values are more useful for diagnosis purposes in deviation analyses (see Sect. 5.5).

From Lines 4 and 6 of Fig. 5.23A, we can now plot the monitoring diagram (Fig. 5.23B). However, the input and output trends are not given in detail with each single work content, but as a straight line between two points. As defined, the output curve starts at zero and follows the values of cumulated output (Line 6, Fig. 5.23A). The input curve starts at the initial inventory value of 95.8 hours, and follows the values of cumulated input.

After having calculated the key data, one can represent them as columns in a key data graph (Fig. 5.23C). For more clarity, only the values of utilization, weighted lead time, and weighted lateness of the outputs are given in our example.

The monitoring diagram and the key data graph give quite a good insight into what has happened at this work center in the four periods in question. The heavy input in Period 3 has increased the inventory in Period 3 by nearly half the value of the previous periods, which has corresponding effects on the range in Period 4. The absolute values of the lead time and its wide variations are unsatisfactory and should be better controlled.

The throughput diagram can now be extended to include events before and after the work center concerned, as described in Chapt. 4, and as Fig. 5.27 shows in an extended monitoring diagram. Since it is now the exit throughput curve that starts at zero (it characterizes the dates of completion of the orders which were processed by the work center in question), the other curves are displaced according to the individual inventories between the respective curves.

These processes from the extended throughput diagram are, in Fig. 5.28, represented by means of the funnel model [21]. Between the scheduled entry date and the release date, the orders make up the dispatch inventory with a corresponding dispatch lead time. The released orders do not generally flow directly to the work center in question, which means that, until the input date, they are an indirect inventory for this work center and have an indirect lead time. The difference between the input and output curves is the direct inventory with the respective direct lead time. To find how much work for the work center is still in the 'pipeline'

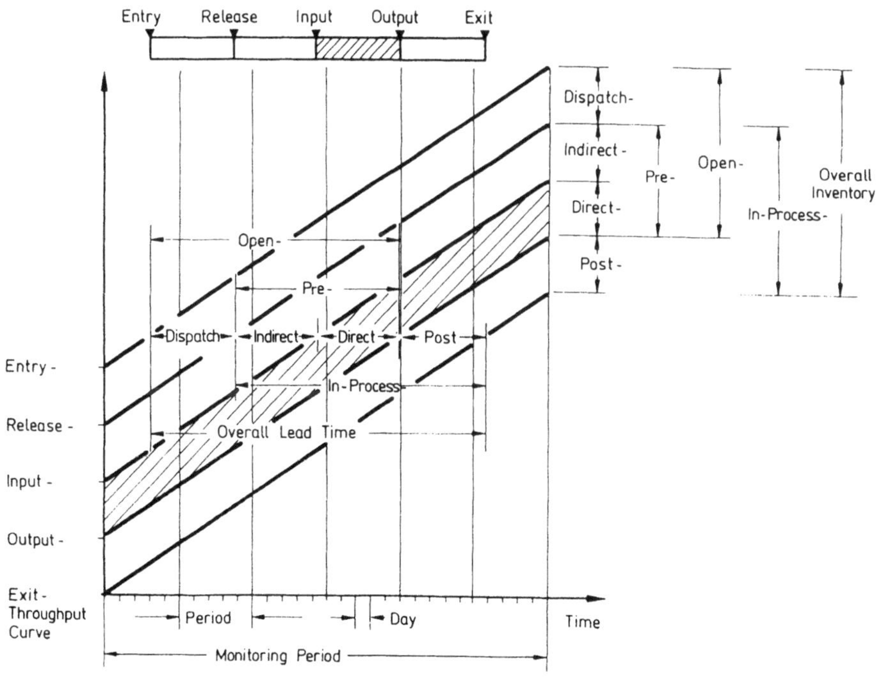

Fig. 5.27. Definition of types of inventory and lead time in the monitoring throughput diagram of a work center (After Bechte [17])

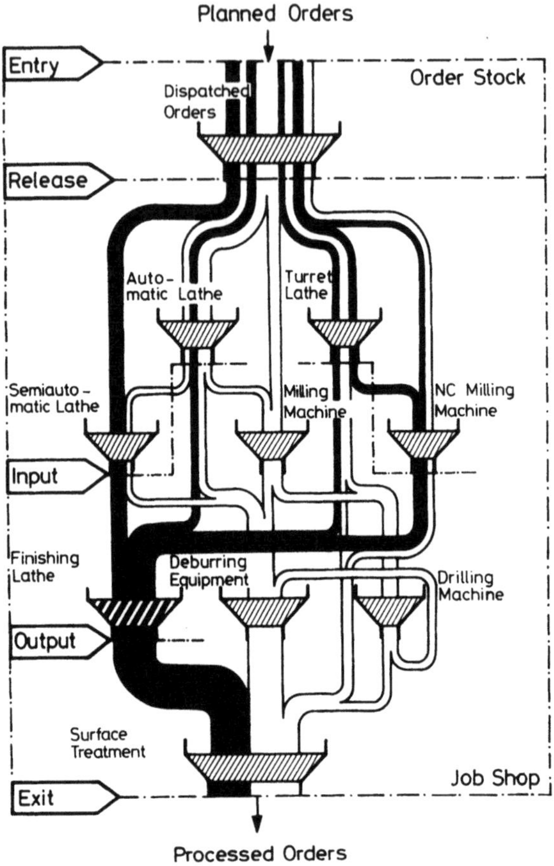

Fig. 5.28. Order flow times at a work center, shown in a funnel model (After Bechte [17])

between the entry and the input, the values open inventory and open lead time, pre-inventory and pre-lead time were defined in Fig. 5.27. Between output and exit, the orders usually pass other work centers, and 'exit' is generally defined as the date of input into intermediate store or assembly store. This is the post inventory with the post lead time. The difference between release and exit is characterized as the in-process inventory and the in-process lead time, and the difference between entry and exit of the orders passing the work center in question as the overall inventory and the overall lead time, respectively.

The monitoring table and the monitoring diagram can, however, not only be set up for actual values, but also for the values planned for the future.

Thus the monitoring table is expanded, as outlined in Fig. 5.29, and this permits a comparison between the planned and the actual procedure at this work center. The value 'lateness planned' can be interpreted as a maximum permitted deviation value.

In practice, the monitoring tables are compiled on the basis of the reported transaction data. Table 5.8 shows the data planned for the monitoring system KOSYF (German: 'Kontroll-System Fertigung' – Monitoring System for Manufacturing),

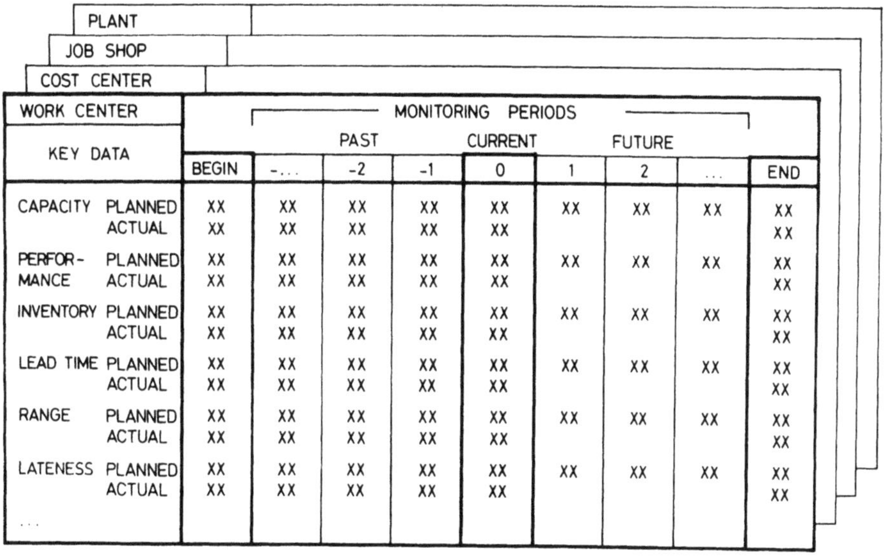

Fig. 5.29. Structural principle of a monitoring diagram

which are divided into Part A with order-related and Part B with operation-related characteristics [21].

The data for the monitoring table derived from this list are shown in Table 5.9. The order data describe the work content to be processed, or which has already been processed, with the order time, lot size and the mean position of the work center in the flow of the processed orders. The weighted mean lateness of inputs characterizes the schedule performance of the upstream work centers. The lateness of outputs is also given. The output lateness of the final work centers is the lateness of the entire order.

The movement data describe the five throughput curves. Besides the actual data of the orders in the inventory with information about the number of operations, the quantity of parts, and the total standard time, the planned data are also given, but only as standard hours. With the actual data, one can judge not only the required capacity, but also the dispatching costs (number of operations) and the transport and inspection costs (number of operations and quantity). The comparison between planned and actual times reveals deviations in and weak points of the manufacturing flow. The monitoring tables can also be condensed according to the hierarchy of the monitored load centers, and can thus be used for evaluations on different levels of decision.

Finally, the key data describe the actual status of inventory, range, and weighted actual lead time for the eight defined inventory and lead time types per period and for the entire monitoring period. The planned values, however, are only calculated for the in-process and direct inventories and lead times, respectively, since these values can best be planned as objectives at the order entry.

If the rated capacity data are also available they can easily be integrated into the system. By comparing them with the output data, period-related utilization rates can be calculated.

Table 5.8. Structure of the movement data record used for the generation of a monitoring table (After Bechte [17])

No.	Description		Remarks
Part A	Order-related data		
1	Order no.		
2	Quantity entry		Planned quantity on entry
3	Quantity release		Actual quantity on release
4	Date entry		
5	Date release	(actual)	
6	Date exit	(actual)	
7	Date release	(planned)	
8	Date exit	(planned)	
9	Lead time dispatch		Entry – release
10	Lead time in-process		Release – exit
11	Lead time overall		Entry – exit
12	Lead time dispatch (planned)		
13	Lead time in-process (planned)		
14	Lead time overall (planned)		
Part B	Operation-related data		
15	Operation no.		
16	Operation numerator		
17	Work center no.	(planned)	
18	Setup time	(planned)	
19	Time per unit	(planned)	
20	Order time	(planned)	
21	Quantity input	(actual)	When required only
22	Quantity output	(actual)	When required only
23	Scrap	(actual)	When required only
24	Work center no.	(actual)	When required only
25	Setup time	(actual)	When required only
26	Time per unit	(actual)	When required only
27	Order time	(actual)	When required only
28	Date input	(actual)	
29	Date output	(actual)	
30	Date input	(planned)	
31	Date output	(planned)	
32	Delay input		Planned input – actual input
33	Delay output		Planned output – actual output
34	Indirect lead time	(actual)	Release – input
35	Direct lead time	(actual)	Input – output
36	Post-lead time	(actual)	Output – exit
37	Pre-lead time	(actual)	Release – output
38	Open lead time	(actual)	Entry – output
39	Indirect lead time	(planned)	
40	Direct lead time	(planned)	
41	Post-lead time	(planned)	
42	Pre-lead time	(planned)	
43	Open lead time	(planned)	

Table 5.9. Contents of KOSYF monitoring table (After Bechte [17])

Order data
- Work center no.
- Number of workdays per period
- Order time
- Operation time
- Mean lot size
- Mean position
- Weighted mean lateness of inputs
- Weighted mean lateness of outputs

Movement data per period, and cumulative
- Work (actual)
  - in number of operations
  - in quantity (items, yards, lb., ...)
  - in standard time (hrs)
- Work (planned)
  - in standard time (hrs)
    of entry, release, input, output, and exit curve

Key data per period, and for the entire monitoring period
- Inventory at the end of the period (actual)
  - in items
  - in quantity
  - in standard hours
- Mean inventory (actual)
- Mean range (actual)
- Mean lead time (actual)

for dispatch-, indirect-, direct-, post-, open-, pre-, in-process and overall inventory, and lead time
- Inventory at the end of the period (planned)
  - in standard hours
- Mean inventory (planned)
- Mean range (planned)
- Mean lead time (planned)

for direct and in-process inventory

5.4.3 Results and Use of a Continuous Monitoring System

For the user, the complete monitoring tables, with their more than 100 key data per period, are much too extensive. On the other hand, a final decision on the choice of key data at the time of system installation is neither useful nor possible. One should therefore compile a report on the key data that are relevant for the respective user. This report should not exceed one page per work center, and should contain an appropriate summary of key data.

Table 5.10 shows, as an example, the 'Transactions and Inventories' report for one work center over 12 periods with 4 or 5 workdays each. Weeks 40–47 represent past periods, Week 48 (NOW) is the current week, and Weeks 49–51 represent the next three periods. The values in Column END refer to the eight weeks before Week NOW.

The key data basically describe the five throughput curves shown in Fig. 5.27 and a choice of key data between these curves. For this, the whole report has been divided into five key data blocks.

The first block describes the course of the entry curve with the load value per

Table 5.10. Example of a KOSYF monitoring table for movement and inventory data (After Bechte [17])

D161116	84 47 04			MACHINE	161116							TRANSACTIONS	&	INVENTORIES		
WEEK			BEG	40	41	42	43	44	45	46	47	NOW	49	50	51	END
WORK DAYS	PAST	*SCD		5	5	5	5	4	5	5	4	5	5	5	5	58
LOAD	ENT	*HRS		53	18	33	49	6	56	31	94	17				340
LOAD -CUM	ENT	*HRS	248	301	319	352	401	407	463	494	588	605	605	605	605	588
LOAD-P-CUM	ENT	*HRS	248	301	319	352	401	407	463	494	588	605	605	605	605	588
OPERAT	STOCK		35	37	21	19	15	4	18	28	44					44
QUANTI	STOCK	*TPC	73	98	52	29	45	22	72	59	148					148
LOAD	STOCK	*HRS	73	91	53	30	50	27	77	71	156					156
A-L TIME	STOCK	*SCD		5.1	4.4	6.6	3.7	4.4	3.0	-1.7	5.7					4.0
LOAD	REL	*HRS		35	56	56	29	29	6	37	9					257
LOAD -CUM	REL	*HRS	175	210	266	322	351	380	386	423	432	432	432	432	432	432
LOAD-P-CUM	REL	*HRS	92	152	194	226	248	295	346	382	409	447	470	470	470	409
OPERAT	TRANS		17	27	32	34	37	41	33	25	16					16
QUANTI	TRANS	*TPC	44	57	88	73	93	112	94	88	68					68
LOAD	TRANS	*HRS	41	53	75	71	87	106	88	78	63					63
A-L TIME	TRANS	*SCD		6.3	8.0	6.4	9.3	8.0	14.0	16.7	16.9					10.8
LOAD	INP	*HRS		23	34	60	13	10	24	47	24	2				235
LOAD -CUM	INP	*HRS	134	157	191	251	264	274	298	345	369	371	371	371	371	369
LOAD-P-CUM	INP	*HRS	74	96	148	198	226	245	317	346	382	409	449	470	470	382
OPERAT	ONHAND		32	22	27	40	36	27	8	7	6					6
QUANTI	ONHAND	*TPC	50	40	50	79	70	72	36	32	20					20
LOAD	ONHAND	*HRS	54	41	50	72	63	65	25	33	16					16
LOAD-P	ONHAND	*HRS	6	12	13	4	7	1	28	7	4		5	1		4
A-L TIME	ONHAND	*SCD		8.8	9.2	7.0	7.3	9.0	8.2	11.7	2.7					7.8
A-LT -P	ONHAND	*SCD		2.4	2.2	2.2	2.2	2.5	2.0	2.6	2.0	2.1	2.0	2.2	3.0	2.2
LOAD	OUT	*HRS		36	25	38	22	8	64	39	41	5				273
LOAD -CUM	OUT	*HRS	80	116	141	179	201	209	273	312	353	358	358	358	358	353
LOAD-P-CUM	OUT	*HRS	68	84	135	194	219	244	289	339	378	409	444	469	470	378
OPERAT	DOWNST		53	40	42	50	50	46	52	52	62					62
QUANTI	DOWNST	*TPC	85	67	71	95	80	65	95	130	143					143
LOAD	DOWNST	*HRS	80	71	76	96	80	66	104	123	138					138
A-L TIME	DOWNST	*SCD		6.3	10.6	7.7	11.0	10.7	13.4	8.5	7.4					9.3
LOAD	EXT	*HRS		45	20	18	38	22	26	20	26	5				215
LOAD -CUM	EXT	*HRS		45	65	83	121	143	169	189	215	220	220	220	220	215
LOAD-P-CUM	EXT	*HRS	9	69	86	129	183	208	239	285	322	368	405	447	470	322

period (LOAD ENT), its cumulative actual trend (LOAD-CUM ENT), and the planned trend (LOAD-P-CUM ENT). All data are given in standard hours. The actual and planned entry curves are the same, because the orders are created at the time of their entry. The other four blocks are in principle always the same as each other, except for some additional planned values.

The second block describes what happens between entry and release of the orders. First, the last three key data of this block characterize the trend of the release curve with the same values as were used for the entry curve in the first block. These are LOAD REL, LOAD-CUM REL, and LOAD-P-CUM REL. Then, the inventory between the entry and release curves is calculated as the number of operations (OPERAT STOCK), as the quantity (QUANTI STOCK) contained in these operations (in thousand pieces, TPC), and in standard hours (LOAD STOCK). Finally, the weighted average (= mean) lead time between entry and release is given in shop calendar days (A-L TIME STOCK).

The third and fourth blocks contain the same key data as the second block, but they give the processes between the input and release curves and the output and input curves, respectively. Thus they provide data on the indirect and direct inventories, respectively, and the direct and indirect lead times, respectively. Since the entire key data sheet is mainly meant for work center control, the fourth block also contains the planned values for the direct inventory (LOAD-P ONHAND) and the direct weighted average (= mean) lead time (A-LT-P ONHAND).

The fifth block of the monitoring report shows what happens between the output and the exit of all orders processed by this work center, and uses the same key data as the second and third blocks.

This monitoring report aims at characterizing the work center together with upstream and downstream processes. For a closer examination of the direct processes, Table 5.11 gives a more detailed description of the direct inventory in four key data blocks.

The first block describes the input in number of operations per period (OPERAT INP), and cumulated (OPERAT-CUM INP), in thousand items per period (QUANTI INP), and cumulated (QUANTI-CUM INP), and the work (in hours) coming in per period (LOAD INP). Besides the actual work input per period, the planned input per period (LOAD-P INP) and the cumulative values of these two magnitudes (LOAD-CUM INP and LOAD-P-CUM INP) are given. The last value in this block describes the lateness of the input curve, here called the average schedule delay (A-DELAY INP).

Whereas the second block describes what happens between the input and output curves, the third block characterizes the output curve using the same key data as for the input curve. In the second block, in addition to the above-described data on the inventory, namely the number of operations (OPERAT ONHAND), the quantity (QUANTI ONHAND), and the work in actual (LOAD ONHAND) and planned values (LOAD-P ONHAND), also the average inventory is given with actual (A-LOAD ONHAND) and planned values (A-LOAD-P ONHAND). Further, there are the actual (A-TIME S ONHAND) and planned average range (A-TS-P ONHAND), as well as the above-mentioned actual (A-L TIME ONHAND) and planned weighted average lead time (A-LT-P ONHAND).

The fourth block contains information on the in-process orders, namely the average lot size in pieces (A-LOT SIZE), the average order time in hours – here called the average run time (A-RUN TIME) – as well as the operation time in shifts, here also called A-RUN TIME.

Table 5.11. Example of a KOSYF monitoring table for the direct inventory of a work center (After Bechte [17])

D161116	84 47 04			MACHINE 161116								ONHAND INVENTORY INPUT OUTPUT				
WEEK			BEG	40	41	42	43	44	45	46	47	NOW	49	50	51	END
WORK DAYS PAST		*SCD		5	5	5	5	4	5	5	4	5	5	5	5	58
OPERAT	INP			10	19	32	17	8	9	12	15	2				122
OPERAT-CUM	INP		85	95	114	146	163	171	180	192	207	209	209	209	209	207
QUANTI	INP	*TPC		23	34	70	12	9	24	49	30	2				251
QUANTI-CUM	INP	*TPC	135	158	192	262	274	283	307	356	386	388	388	388	388	386
LOAD	INP	*HRS		23	34	60	13	10	24	47	24	2				235
LOAD-P	INP	*HRS		22	52	50	28	19	72	29	36	27	40	21		308
LOAD -CUM	INP	*HRS	134	157	191	251	264	274	298	345	369	371	371	371	371	369
LOAD-P-CUM	INP	*HRS	74	96	148	198	226	245	317	346	382	409	449	470	470	382
A-DELAY	INP	*SCD		8.2	5.1	9.8	1.1	3.3	-1.2	1.0	0.2					4.4
LOAD -CUM	REL	*HRS	175	210	266	322	351	380	386	423	432	432	432	432	432	432
OPERAT	ONHAND		32	22	27	40	36	27	8	7	6					6
QUANTI	ONHAND	*TPC	50	40	50	79	70	72	36	32	20					20
LOAD	ONHAND	*HRS	54	41	50	72	63	65	25	33	16					16
LOAD-P	ONHAND	*HRS	6	12	13	4	7	1	28	7	4		5	1		4
A-LOAD	ONHAND	*HRS		49	44	62	67	66	41	36	24					49
A-LOAD-P	ONHAND	*HRS		9	23	24	12	8	24	21	13	12	15	10		14
A-TIME S	ONHAND	*SCD		6.8	8.8	8.3	14.9	34.7	3.2	4.7	2.4					6.8
A-TS -P	ONHAND	*SCD		2.7	2.3	2.1	2.3	1.3	2.6	2.1	1.3	2.0	2.1	2.0	2.0	2.1
A-L TIME	ONHAND	*SCD		8.8	9.2	7.0	7.3	9.0	8.2	11.7	2.7					7.8
A-LT -P	ONHAND	*SCD		2.4	2.2	2.2	2.2	2.5	2.0	2.6	2.0	2.1	2.0	2.2	3.0	2.2
OPERAT	ONHAND		32	22	27	40	36	27	8	7	6					6
OPERAT	OUT			20	14	19	21	17	28	13	16	4				148
OPERAT-CUM	OUT		53	73	87	106	127	144	172	185	201	205	205	205	205	201
QUANTI	OUT	*TPC		33	24	41	21	7	60	53	42	7				281
QUANTI-CUM	OUT	*TPC	85	118	142	183	204	211	271	324	366	373	373	373	373	366
LOAD	OUT	*HRS		36	25	38	22	8	64	39	41	5				273
LOAD-P	OUT	*HRS		16	51	59	25	25	45	50	39	31	35	25	1	310
LOAD -CUM	OUT	*HRS	80	116	141	179	201	209	273	312	353	358	358	358	358	353
LOAD-P-CUM	OUT	*HRS	68	84	135	194	219	244	289	339	378	409	444	469	470	378
A-DELAY	OUT	*SCD		0.1	-0.7	1.3	-0.3	-2.5	-3.0	-1.0	2.7					-0.4
OPERAT	DOWNST		53	40	42	50	50	46	52	52	62					62
A-LOT SIZE		*PCS		1673	1713	2154	988	406	2130	4078	2679					1902
A-RUN TIME		*HRS		1.8	1.8	2.0	1.1	0.4	2.3	3.0	2.5					1.8
A-RUN TIME		*SHI		0.3	0.3	0.4	0.3	0.1	0.5	0.7	0.6					0.5
A-INV RATIO				2.2	3.0	2.5	2.4	2.6	3.4	2.9	3.9					2.7
A-POSITION				3.0	3.0	3.2	3.4	2.8	3.0	2.4	3.3					3.0
LOAD-P-CUM	EXT	*HRS	9	69	86	129	183	208	239	285	322	368	405	447	470	322

Finally, the fourth block also gives the key data which characterize the position of the work center in the flow of the processed orders. The first is the ratio of the sum of direct and indirect inventories to the direct inventory (called the average inventory ratio, A-INV-RATIO); the second is the average position (A-POSITION) of this work center in the routing of the processed orders.

The two described monitoring reports seem very extensive at first but, after a certain period of familiarization, they offer valuable help for planning, manufacturing control, and dispatching of work, because of their logical structure and because they are always the same from a single work place up to the entire plant.

For a quick detection of irregularities, however, graphics are much more suitable. Therefore, it is practical to visualize the process using the throughput diagram in a simplified form.

For the same periods as in the previous figure, Figs. 5.30A and B show the five throughput curves for the machine monitored in Tables 5.10 and 5.11 (Machine 16116). These are monitoring diagrams of (5.30A) the actual and (5.30B) the planned processes, like the ones produced by the graphics software of the PC [21]. In both graphs, the entry curve is by definition the same, so that the difference from the other curves is evident.

Although, in the actual process, orders were released much earlier than planned, the actual performance was only almost half the planned performance until Week 47, and the mean direct lead time (7.8 SCDs) is more than three times as long as the planned value (2.2 days) (see Table 5.10, 4th key data block, last column). The reason for the poor performance is probably that the capacity in the first four weeks was too small, since there never was an underload at the work center. The long actual lead time, on the other hand, is doubtlessly due to the unbalanced release and to the lack of coordination of input with output.

Consequently, the range (called TS = Time Supply) increases particularly drastically in Weeks 43 and 44, as can be seen in the monitoring graph of lead times in the direct inventory (Fig. 5.31).

As a whole, the work center shows a persistently uncontrolled behavior compared with the planned throughput.

The tables and graphs disprove the common argument that such evaluations are only a reprocessing of past events which can no longer be changed anyway. Our last example proves that the opposite is true: the planned status is unrealistic to start with, and the actual status reveals that the controlling is carried out unsystematically. The use of a monitoring system thus makes it possible to perform specific modifications to the manufacturing flow with the help of the existing planning and control system.

As has already been mentioned, a monitoring system of this type must be closely interlinked with the existing data processing systems. Figure 5.32 outlines the software and hardware components of a possible solution. The existing feedback records must be processed, together with the order data, in a program to produce transaction data. Each data record describes one throughput element containing the data shown as an example in Table 5.8. This program also checks the data for plausibility. The subsequent program produces the monitoring tables from the transaction data, with monitoring parameters controlling the evaluation periods and the reduction levels. With the help of a file transfer program, the data are transferred from the central computer system, on which the programs run more economically because of the large amounts of data and computing operations, to a common PC with integrated hard disk and diskette stations. A standard lists

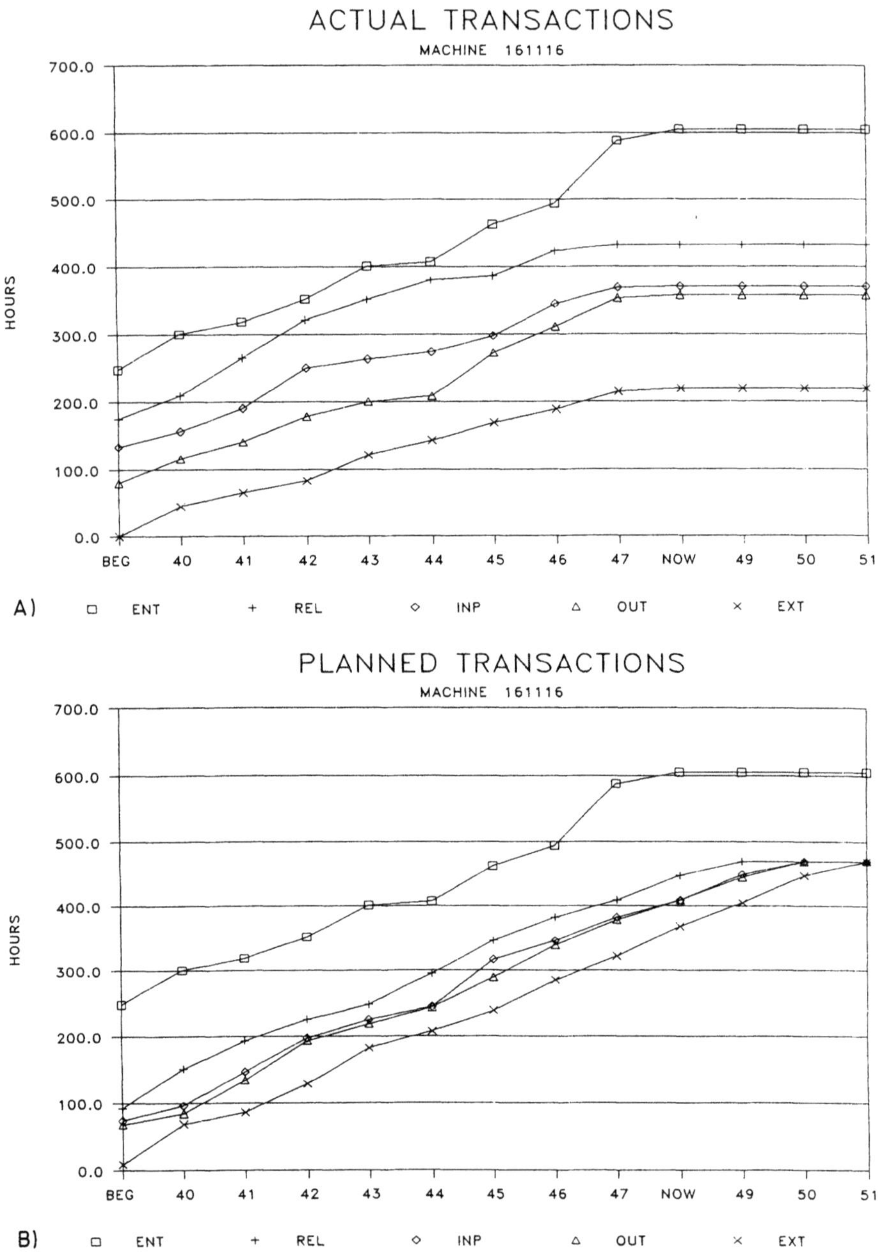

Fig. 5.30. KOSYF monitoring diagram for the movements and inventories at a work center. (A) Actual versus (B) Planned Status (After Bechte [17])

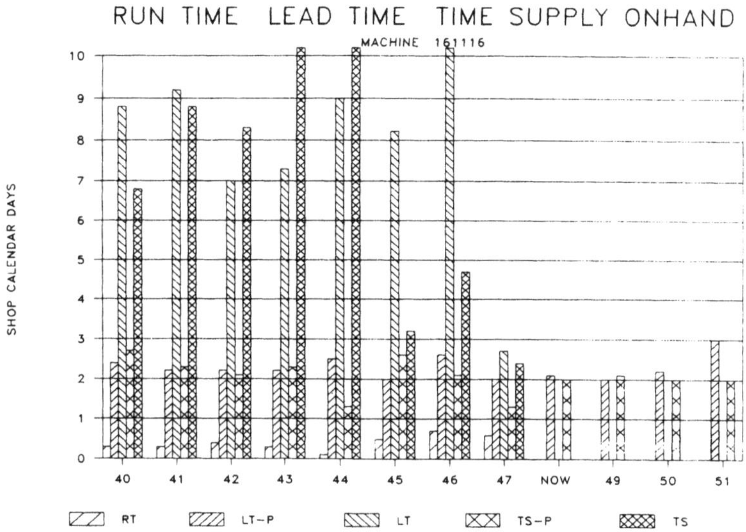

Fig. 5.31. KOSYF monitoring graph for selected lead time data in the direct inventory. RT: run time (= operation time); LT-P: planned lead time; LT: actual lead time; TS-P: planned time supply (= planned range); TS: actual time supply (= range)

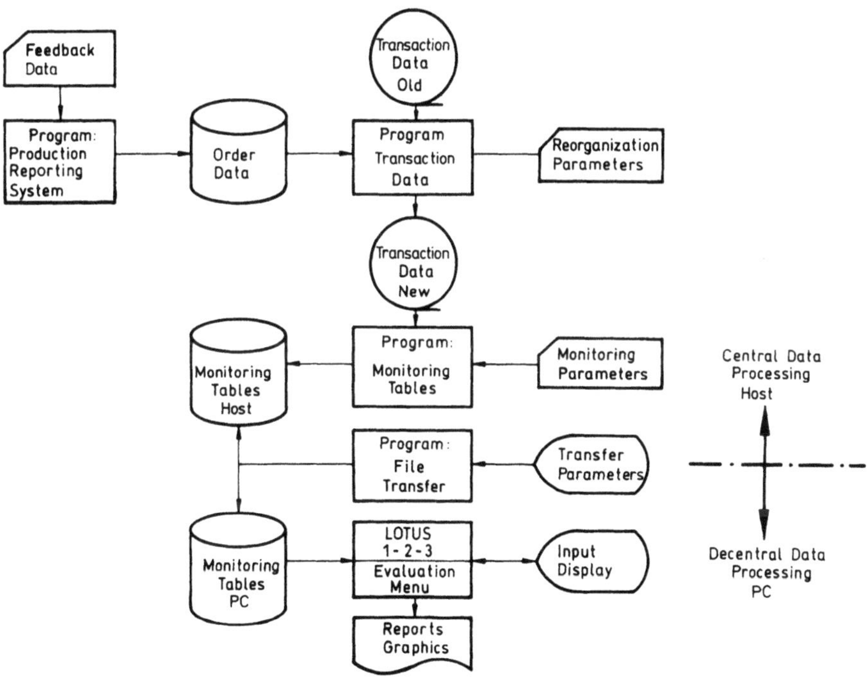

Fig. 5.32. Information flow in the KOSYF monitoring system (After Bechte [17])

Key Data		User: Management	Sales Department	Order Planning	Production Scheduling	Production Control	Dispatcher (Foreman)
Order-Related		X	X	X	X	X	X
Work-Center-Related	Plant	X			(X)	(X)	
	Job Shop	X	(X)	X	X	(X)	
	Cost Center	(X)		X	X	X	
	Work Center				X	X	X

(X) When Required

Fig. 5.33. Monitoring system fields of application

and graphics software installed on this computer produces user reports and graphs which meet previously defined criteria, and which can be represented on a color monitor or printed out on a printer or plotter. Today, such PC systems are common in most job shops, thus permitting a decentralized use and processing of the monitoring reports.

The different ranges of application of a monitoring system are outlined in Fig. 5.33. Whereas the management is interested in the overall order-related and plant-related key data, the sales and planning departments will favor order throughput data, so that they may plan the exact delivery dates. Finally, the monitoring reports and graphs are a useful aid for the production activity control department.

A monitoring system of the type explained above is a valuable component for flow-oriented manufacturing control. The results from the system must, however, be interpreted before one can take the corresponding measures. This step, which could be compared with a diagnosis, requires a great deal of expertise and knowledge, and at present can only be carried out by staff in the shops, or in the planning or manufacturing control departments. The expert systems being used increasingly in other fields can also provide help here. In the last section of this chapter, however, we introduce some considerations aimed at a further differentiation of the mean inventory and the weighted mean lead time so as to be able to trace them back to their respective causes.

5.5 Diagnosis of the Manufacturing Flow in the Throughput Diagram

5.5.1 Breaking down Inventories

The in-process inventory at a work center is determined – as shown above – by the input and output curves. This, however, does not permit any statements to be made on the minimum inventory required by a work center, and on the influential factors for the size of this minimum inventory. To answer these questions, the inventory can be further broken down into several components, following a proposal made by Lorenz [19].

Figure 5.34 shows a throughput diagram with fictitious input and output curves. The inventory area (*AI*) between input and output is now divided into four types of areas.

The basis area (*AB*) corresponds to the rectangle 'initial basis inventory multiplied by reference period'. The basis inventory is defined as the initial inventory minus the order time of the order that came in on the last day of the previous period. This is based on the idea that the lines which limit the basis inventory represent the ideal process. In the ideal process, it is possible that the order in question is already being processed, and therefore it no longer belongs to the basis inventory.

If, as a next step, the input and output curves are straightened to form an ideal line, this produces the flow area (*AF*), when the work input is more or less than the output in the reference period. The flow area can therefore be positive or negative, and thus it stands for the increase or decrease of the inventory in the reference period.

Since both the input and output curves neither start nor end with the ideal input and output lines, but earlier or later in time, the control area (*AC*) is produced. In our example, all input foot points are too early and all output head points are too late. This produces an inventory-adding control area in the input (here positive by definition) and in the output (here negative by definition). Both areas can, however, have positive or negative values.

Finally, the figure shows an area type called the lot area (*AL*). It is produced by the finite and varying work contents of the input or output orders. Since the work content of the lots is always greater than zero, both the input and the output lot areas are inventory-adding and per definition always positive.

The current inventory $I(T)$ at any time T can thus always be broken down into the above-defined inventory types. At the end of the reference period, due to the flow inventory built up over the period, a new basis inventory, here called 'final basis inventory', is formed. It represents the 'initial basis inventory' for the next reference period. (How the final inventory is determined is explained later.)

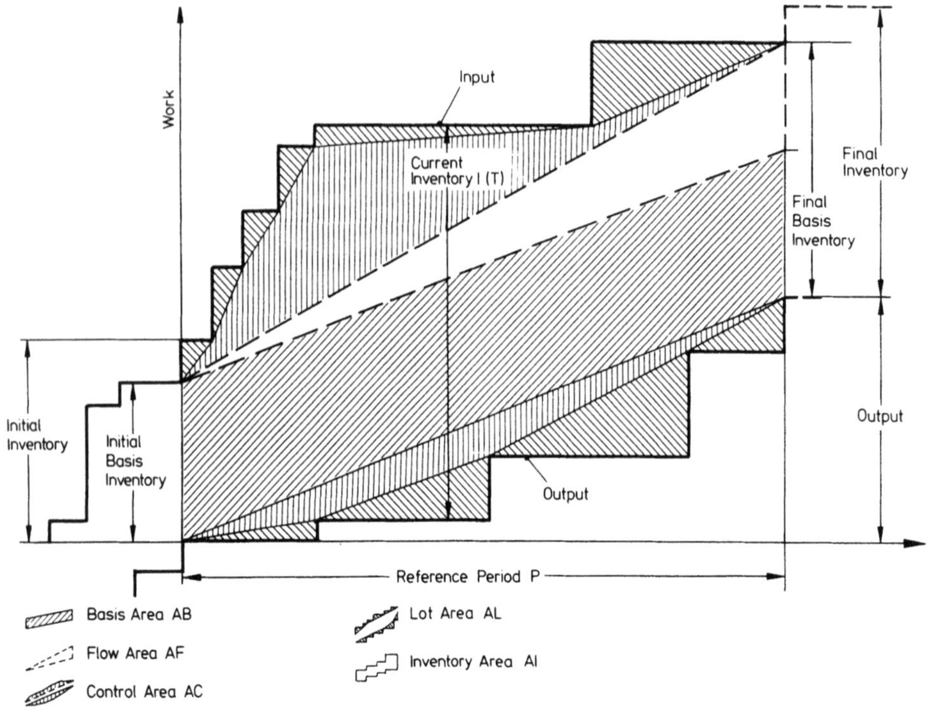

Fig. 5.34. Division of the throughput diagram into its components (After Lorenz [19])

From these inventory area types we can calculate the respective inventory components. It is advisable to carry out separate calculations for the input and output processes (Fig. 5.35).

This gives the following equation for the mean inventory (I_m):

$$\begin{aligned} I_m = IBI &+ (IFL_{minp} - IFL_{mout}) \\ &+ (IC_{minp} - IC_{mout}) \\ &+ (IL_{minp} - IL_{mout}) \end{aligned} \tag{5.1}$$

where:

Initial Basis Inventory (IBI) = AB/P
Mean Flow Inventory (IFL_m) = $(AF_{inp} - AF_{out})/P$
Mean Control Inventory (IC_m) = $(AC_{inp} - AC_{out})/P$
Mean Lot Inventory (IL_m) = $(AL_{inp} - AL_{out})/P$
P = Period.

It is not unusual for two or more input or output events to happen at the same

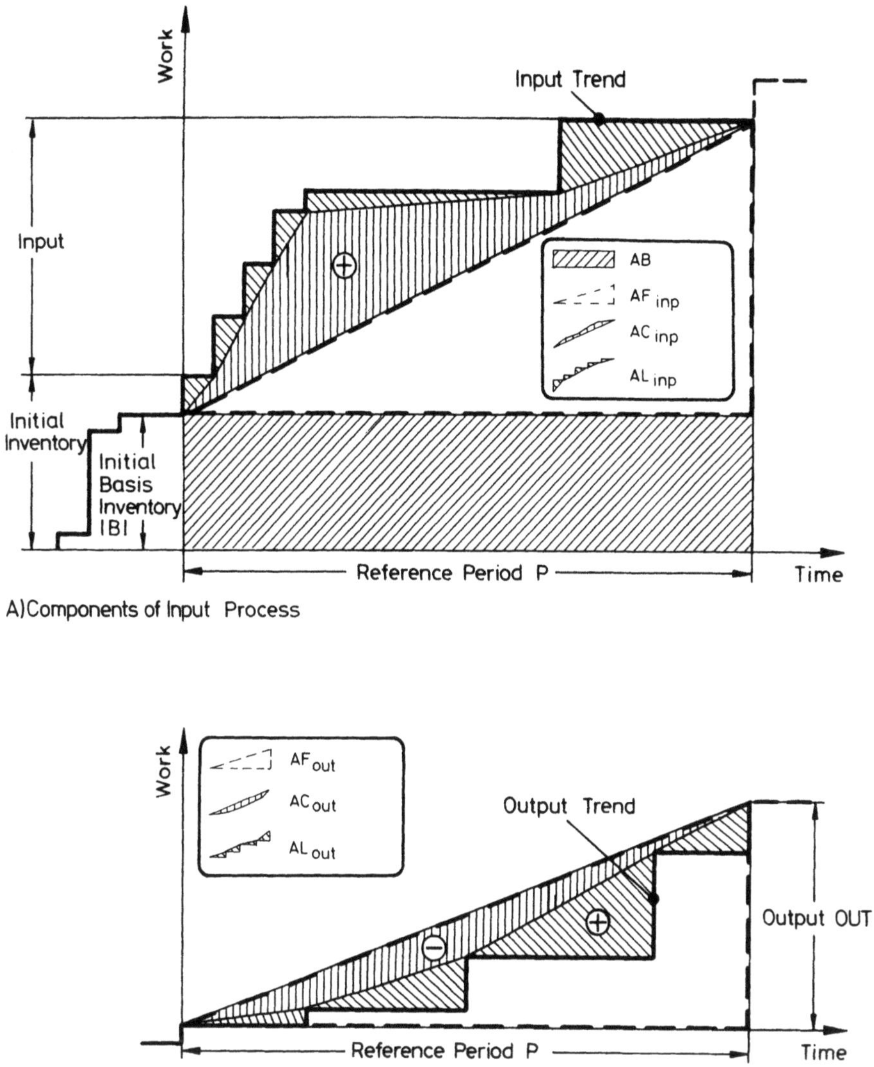

B) Components of Output Process

Initial Basis Inventory	IBI	
Mean Flow Inventory	IFL_m	$= AF / P$
Mean Control Inventory	IC_m	$= AC / P$
Mean Lot Inventory	IL_m	$= AL / P$

$$\text{Mean Inventory } I_m = IBI + IFL_{m\,inp} + IC_{m\,inp} + IL_{m\,inp} - IFL_{m\,out} - IC_{m\,out} - IL_{m\,out}$$

Fig. 5.35. Division of the input and output processes into their components (After Lorenz [19])

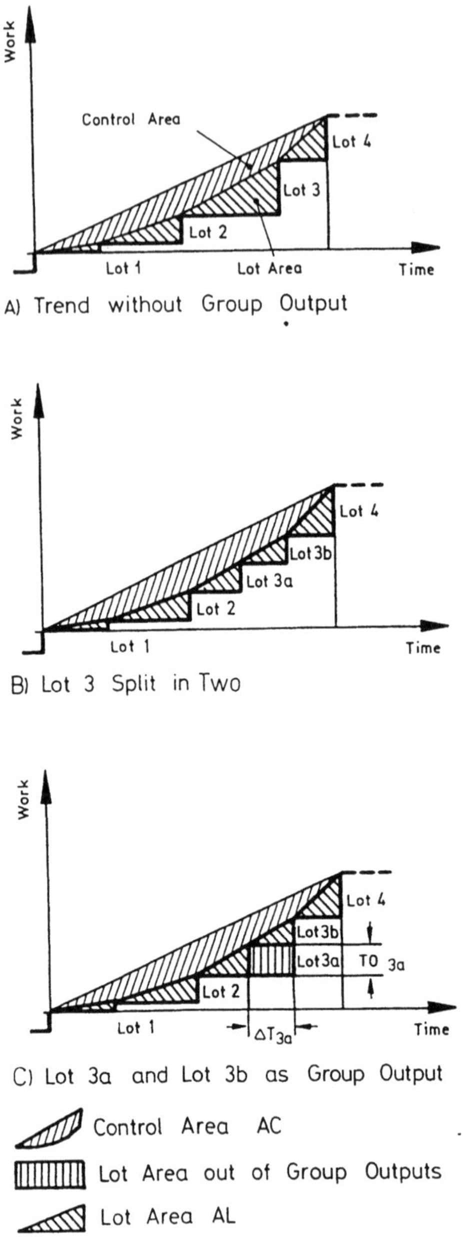

Fig. 5.36. Control and lot components in case of group outputs

time at a work center. Figure 5.36 demonstrates this case with the sample output curve from Fig. 5.34.

In Case (A), the output curve is the same as in Fig. 5.34. In Case (B), Lot 3 is divided up into two sublots whose outputs are arranged to be reported at such times that the control area is not changed. However, the lot area decreases because the early output of Lot 3a reduces the inventory. Case (C) assumes that the output of

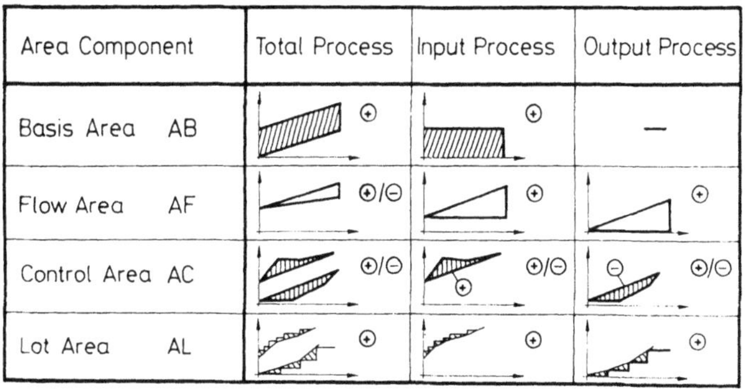

⊕ Positive Sign Possible Only
⊖ Negative Sign Possible Only
⊕/⊖ Positive or Negative Sign Possible

Fig. 5.37. Sign rules for the area components in the throughput diagram

Sublot 3a is reported at the same time as that of Sublot 3b: the lot area is increased by the rectangular areas resulting from the group outputs, although the (unnecessary) inventory tie-up is not caused by the work content of the respective order, but only by its (unnecessarily) late output report date. The same is true for the input process.

If this effect is not accounted for, the resulting error is in the range of 1–3% of the entire inventory area, as a check with our practical sample data has shown. If the error is related to the lot area, it can be 5–40%.

In accordance with Lorenz, who did not separately calculate this area component and included the lot area and this component in his 'lot and group inventory area', in the following we neglect this error. A rough estimate of this component should, nevertheless, be carried out to show all possibilities for inventory reduction.

When calculating the area components using Eqn. (5.1), special attention has to be paid to the signs (Fig. 5.37). The basis area is always positive. The flow area of the entire process is the difference between the input and output flow areas, which are by definition positive. It can therefore have either a positive or a negative sign. The control area is positive if it lies above the ideal input or output line, and negative if it lies below.

Finally, the lot area is always positive, since the work content of the orders is always greater than zero, and because the lot areas always have an inventory-adding effect in the input as well as in the output processes.

Besides the signs, the start and end points of the various curves must also be exactly defined when dividing up the inventory area. For the case in which there were both input and output events on the last day of the previous period as well as on the last day of the period in question, the area division as shown in Fig. 5.34 is unambiguous. The case in which there were no events on both last days is shown in Fig. 5.38.

In this case, the initial basis inventory is determined by plotting the input and output lot areas beyond the period limits. The initial basis inventory (IBI) is then the distance between the two lines limiting the lot area at the time T_0. The same applies for the end of the period. It is, however, possible that the work contents and dates of

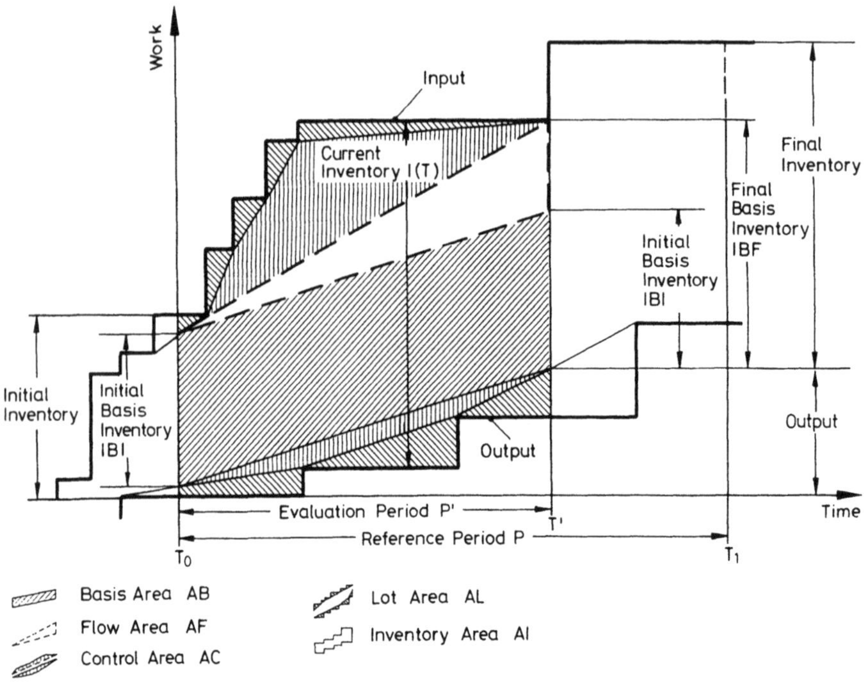

Fig. 5.38. Division of the inventory area into its components when there are no inputs and outputs on the last day of the reference period

the next input and output events for the time T_1 are not yet known. There are two alternative solutions: either one waits for the next input and/or output event or the period end T' is moved back to a suitable point in time. Figure 5.38 shows the second possibility (the evaluation period P' is shorter than the reference period P), whereas our practical example in Appendix B uses the first method.

In practice, one normally evaluates several periods at a time. For former periods, it is always possible to evaluate the entire reference period, since the first input and output events of the next period are known. Only for the last period is it sometimes necessary to shorten the reference period to obtain a suitable evaluation period, as was shown in Fig. 5.38.

5.5.2 Breaking down Lead Times

For a continuous diagnosis of the manufacturing flow at a work center, it seems appropriate to convert the inventory components per period into range components by dividing them by the respective output of the period.

For the mean range (R_m)

$$R_m = RB_m + RF_m + RC_m + RL_m$$

$$R_m = \frac{1}{OUT} \times [AB + (AF_{inp} - AF_{out}) + (AC_{inp} - AC_{out}) + (AL_{inp} + AL_{out})] \quad (5.2)$$

where:

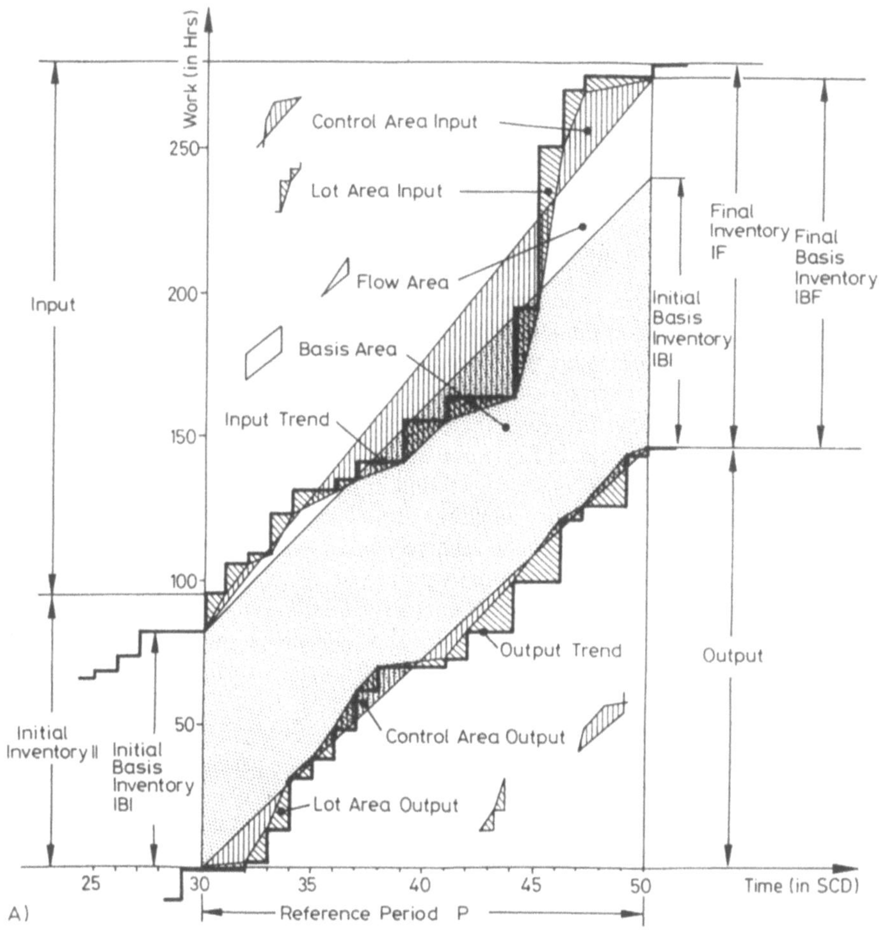

	Basis Area AB	(Hrs × SCD)	= 1,636.0	=	77.7 %
+	Flow Area AF (Input – Output)	(Hrs × SCD)	1,936.0 – 1,458.3 = 478.0	=	22.7 %
+	Control Area AC (Input – Output)	(Hrs × SCD)	– 248.4 + 11.8 = –236.6	=	–11.2 %
+	Lot Area AL	(Hrs × SCD)	123.0 + 104.8 = 227.8	=	10.8 %
=	Inventory Area AI	(Hrs × SCD)	= 2,105.4	=	100.0 %

B)

Fig. 5.39. Graph of the inventory components at a work center

RB_m = mean basis range
RF_m = mean flow range
RC_m = mean control range
RL_m = mean lot range
OUT = output in reference period
AB = basis area
AF = flow area of input or output function
AC = control area of input or output function
AL = lot area of input or output function.

Figure 5.39A represents the area components of the in-process inventory for our sample throughput diagram. One can distinguish the positive flow area, due to the fact that there is more input than output. In the input process, widely varying control and lot areas overlap each other, whereas the output process is comparatively steady, because, in this case, there are hardly any idle or down times in the order processing. The values in Fig. 5.39B are taken from Table 5.12. (The tables and figures mentioned in Table 5.12 can be found in Appendix B.)

If no major disruptions occur and if the output events are reported punctually, the output control area must always be relatively small, whereas the output lot area is determined by the mean value and the variance of the processed orders. The input control area is caused by the irregular interarrival dates and by the sequence transpositions of the orders. ('Interarrival date' and 'interprocessing date' are concepts from queueing theory, and stand for the dates between two inputs and outputs, respectively.)

Figure 5.40 details the development of the flow, control, and lot inventories (input and output), in single periods and for the entire reference period, of the work center represented in Fig. 5.39. We can see that the inventory components suffer wide variations at times. Variations in the flow inventory (Fig. 5.40A) are due to poor coordination of input with output. The strong deviations from the ideal control component (Fig. 5.40B), which would be zero, are caused by uneven input and

Table 5.12. Calculation of inventory components, ranges, and essential key data of a work center (per period or for the entire reference period)

KEY DATA	FORMULA SYMBOL	UNIT	CALCULATION	PERIOD 1	PERIOD 2	PERIOD 3	PERIOD 4	PERIOD 1 - 4
1 Initial Inventory	II	(Hrs)	Table B1	95.8	92.8	85.8	152.0	95.8
2 Input	INP	(Hrs)	Table B1	34.9	24.4	96.3	29.3	184.9
3 Output	OUT	(Hrs)	Table B1	37.9	31.4	30.1	46.6	146.0
4 Final Inventory	IF	(Hrs)	Table B1	92.8	85.8	152.0	134.7	134.7
5 Length of Period	P	(SCD)		5.0	5.0	5.0	5.0	20.0
6 Initial Basis Inventory	IBI	(Hrs)	Figure B1	81.8	88.9	77.2	84.3	81.8
7 Final Basis Inventory	IBF	(Hrs)	Figure B1	88.9	77.2	84.3	129.6	129.6
8 Basis Area	AB	(HrsxSCD)	PxIBI	409.0	444.5	386.0	421.5	1,636.0
9 Mean Basis Range	RBm	(SCD)	AB/OUT	10.8	14.2	12.8	9.0	11.2
10 Flow Area	AF	(HrsxSCD)	(IBF-IBI)xP/2	17.8	-29.3	17.7	113.3	478.0
11 Mean Flow Range	RFm	(SCD)	AF/OUT	0.5	-0.9	0.6	2.4	3.3
12 Mean Flow Inventory	IFLm	(Hrs)	AF/P	3.6	-5.9	3.5	22.7	23.9
13 Control Area	AC	(HrsxSCD)	Fig.B1 and Fig 5-44	48.0	-35.1	-31.7	116.5	-236.6
14 Mean Control Range	RCm	(SCD)	AL/OUT	1.3	-1.1	-1.1	2.5	-1.6
15 Mean Control Inventory	ICm	(Hrs)	AL/P	9.6	-7.0	-6.3	23.3	-11.8
16 Lot Area	AL	(HrsxSCD)	Fig.B1 and Fig 5-44	45.8	38.4	59.9	83.7	227.8
17 Mean Lot Range	RLm	(SCD)	AL/OUT	1.2	1.2	2.0	1.8	1.6
18 Mean Lot Area	ILm	(Hrs)	AL/P	9.2	7.7	12.0	16.7	11.4
19 Inventory Area	AI	(HrsxSCD)	AB+AF+AC+AL	520.6	418.6	432.0	735.0	2,105.2
20 Mean Range	Rm	(SCD)	AI/OUT	13.7	13.3	14.4	15.8	14.4
21 Mean Inventory	Im	(Hrs)	AI/P	104.1	83.7	86.4	147.0	105.3
22 Inventory Trend Component	ITC	(SCD)	Tab. B2 and Tab. A4	5.2	-0.5	0.0	-2.5	0.4
23 Mean Advance Time	TAm	(SCD)	Rm+ITC	18.9	12.8	14.4	13.3	14.8
24 Sequence Component	SC	(SCD)	Tab. B2 and Tab. A5	-2.5	-4.7	1.2	-5.0	-3.0
25 Weighted Mean Lead Time	TLm	(SCD)	TAm+SC	16.4	8.1	15.6	8.3	11.8

Note: Deviations behind Decimal Point Are Due to Rounding Errors

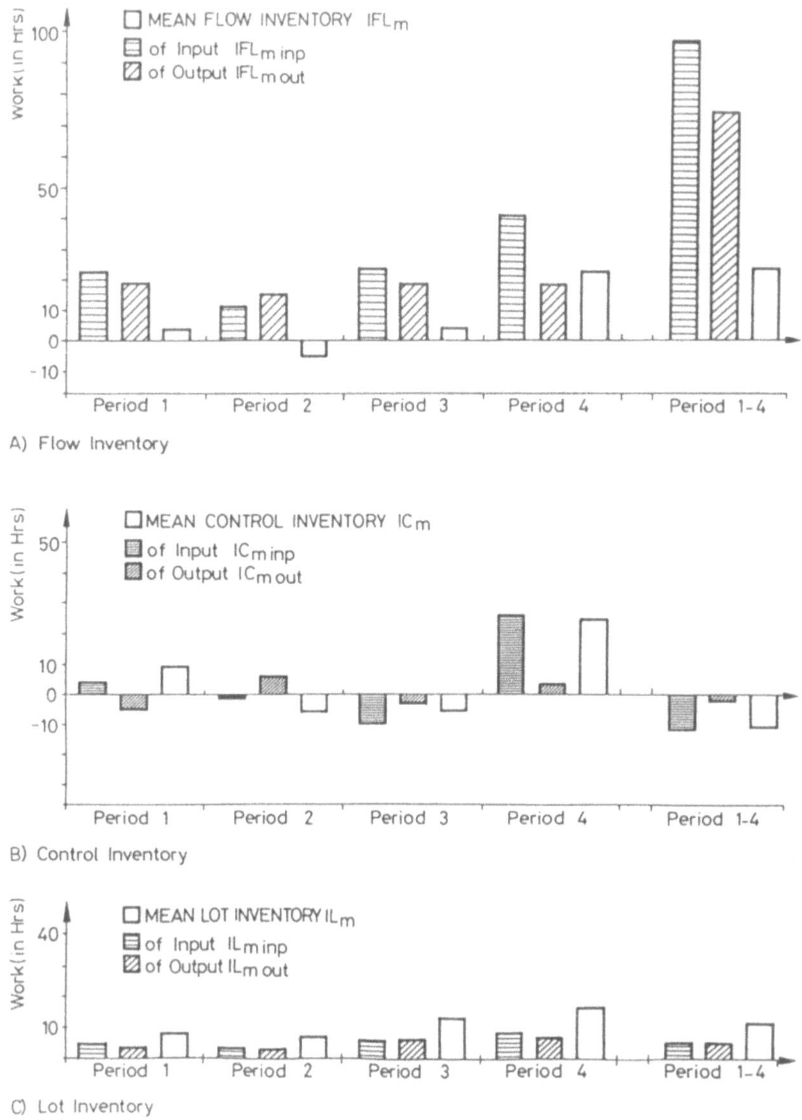

Fig. 5.40. Graph of inventory components at a work center (input and output)

output trends. The most spectacular rise in input control inventory occurred from Period 3 to Period 4. As we can see from Fig. 5.39, this is due to the extremely high input at the end of Period 3. The lot inventory (Fig. 40C) has no such extreme variations, but nevertheless the values scatter quite considerably. This is caused by variations in the work content of the single orders.

The resulting flow, control, and lot components in the input and output processes are summarized in Fig. 5.41. The basis inventory was added, resulting in the mean inventory.

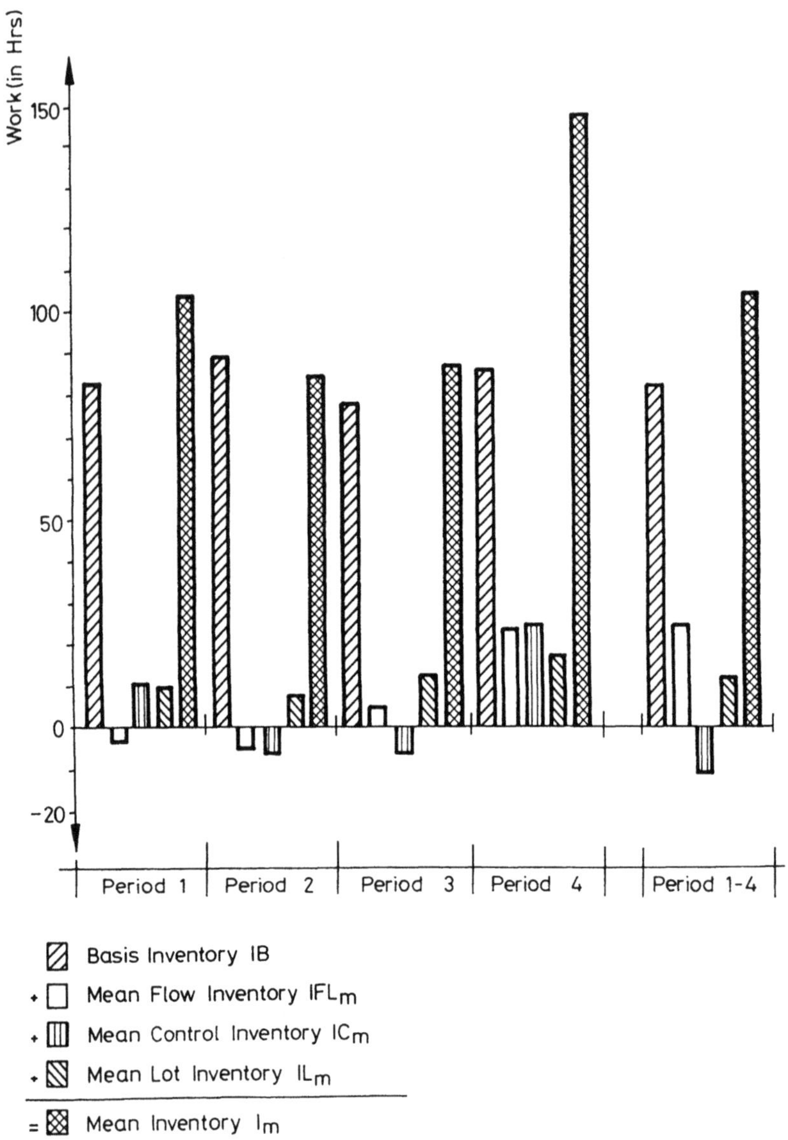

Fig. 5.41. Graph of mean inventory and inventory components (resulting from the input and output trends)

The single values for Figs. 5.40 and 5.41 are taken from Table 5.12. The table gives the symbol and the formula for all key data, or a reference where the formula can be found. We shall therefore not deal with the quite substantial calculations on which Table 5.12 is based. More detailed information and a graphic representation in *periods* of the breaking up of inventories in the throughput diagram can be found in Appendix B.

Note that the key data summarized for Periods 1 to 4 (i.e. for the entire reference period) do not coincide with the values calculated in Chapt. 4. Although the calculations were carried out on the basis of the same data, the evaluation in Chapt. 4 was based on calendar days (i.e. including the non-productive days) and in this chapter on SCDs (i.e. exclusively in workdays). For a direct comparison see Tables A1 to A7 in Appendix A.

The values developed above for the diagnosis of inventories and lead times can be entered into a diagnosis table, as in Table 5.13 (for the throughput diagram from Fig. 5.39). It is designed in accordance with the monitoring table in Fig. 5.23 and, besides the inventory, the range and their components, it also contains the sequence component and the inventory trend component, so as to be able to include the weighted mean actual lead time.

Table 5.13. Diagnosis table for work center lead time

DIAGNOSIS KEY DATA		PERIOD 1	PERIOD 2	PERIOD 3	PERIOD 4	PERIOD 1 - 4
1 Calendar Days	(Day)	205-211	212-218	219-225	226-232	205-232
2 Workdays	(SCD)	30-34 (5)	35-39 (5)	40-44 (5)	45-49 (5)	30-49 (20)
3 Mean Inventory	(Hrs)	104.1	83.7	86.4	147.0	105.3
4 Basis Range	(SCD)	10.8	14.2	12.8	9.0	11.2
5 Flow Range	(SCD)	0.5	-0.9	0.6	2.4	3.3
6 Control Range	(SCD)	1.3	-1.1	-1.1	2.5	-1.6
7 Lot Range	(SCD)	1.2	1.2	2.0	1.8	1.6
8 Mean Range	(SCD)	13.7	13.3	14.4	15.8	14.4
9 Sequence Component	(SCD)	-2.5	-4.7	1.2	-5.0	-3.0
10 Inventory Trend Component	(SCD)	5.2	-0.5	0.0	-2.5	0.4
11 Weighted Mean Lead Time	(SCD)	16.4	8.1	15.6	8.3	11.8

Note: Deviations behind Decimal Point Are Due to Rounding Errors

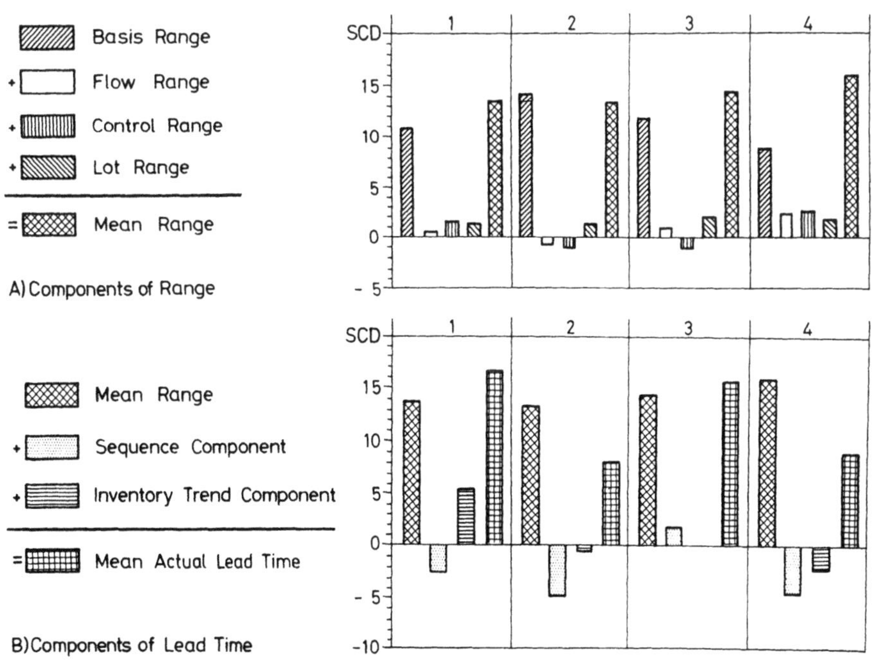

Fig. 5.42. Diagnosis graph for range and lead time key data

These values can also be plotted in bar charts, as shown in Fig. 5.42A for range and in Fig. 5.42B for lead time.

It has by now become clear that there are a number of clearly definable and calculable factors which influence the inventory and the actual lead time at a work center. These findings must now be used to take the respective measures.

5.5.3 Measures to be Deduced from Monitoring and Diagnosis Results

From analysis, monitoring, and diagnosis of the manufacturing flow, several measures for a lasting inventory and flow time reduction can be deduced. They have already been mentioned in Table 5.7; Fig. 5.43 summarizes the main points. The distance between the ideal input and output curves – the basis inventory – depends only on the frequency distribution of the interoperation times. The interoperation time is made up of waiting times caused by inspection and transport events. Inspection and transit times must therefore be reduced and harmonized with the help of organizational measures. Then, inventory control must make sure that the basis inventory level found to be realistic is adhered to. This inventory control must be integrated into the manufacturing control process.

When input and output in a period do not coincide, a positive or negative flow inventory is produced. Basically, there are two possible ways of avoiding flow inven-

LEAD TIME COMPONENT OUT OF	LIMIT CRITERION	MEASURE	
Basis Inventory	Transit and Inspection Time	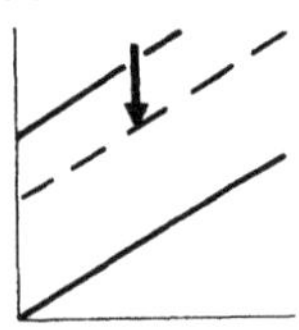	• Inventory Control • Transport Organization • Inspection Organization
Flow Inventory	Difference between Input and Output	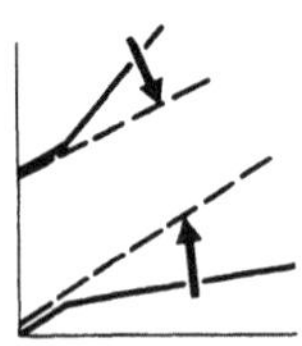	• Release Control • Capacity Leveling
Control Inventory	Interarrival Date		• Finite Control Dispatching • FIFO
Lot Inventory	Order Time Variance	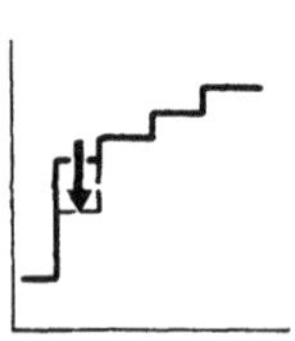	• Smoothing of Lot Work Contents • Setup Time Reduction • Lot Splitting

Fig. 5.43. Methods of lasting flow time and inventory reduction

tories. Either one must make sure that, with a given capacity, the input of work flow (load) in the next period is the same as the expected output of work flow, or capacity must be adjusted in such a way that the expected output corresponds to the desired input. The first measure can be combined with the inventory control process. Capacity control, however, requires a separate process, since capacity changes can, firstly, cause flow time deviations from the allowance values, due to the changes in operation times, and secondly, result in changes in output dates and thus in the scheduling of the orders in question.

The control inventory is mainly determined by the input curve, since in the output curve the throughput elements have practically no room to move. This is because, in practice also, the operation elements are usually processed without interruptions. With such a continuous processing, the control area of the output curve follows directly from the distribution of the operation times and thus of the order times. This does not apply for the input process. It is determined by events which happen to occur at various other upstream work centers, and by the sequence in which the incoming orders are processed. Obviously, a sequence transposition increases the control inventory, and for this reason the FIFO priority rule is recommended in principle. The input dates can not be controlled at the work center in question, but must be determined at the upstream work centers. At the work center considered, this aim can, however, be achieved by finite control, e.g. when an inquiry system reminds the transport control system that the respective orders are due.

The one component of inventory which manufacturing control can not influence directly is the lot inventory. It is produced by the order time variance of the input and output lots respectively. An infinite number of lots with infinitely small, equal work contents, i.e. a continuous flow of work, would be the ideal case. Since this is not possible in job-shop production, one should try to come as near this ideal as possible under the given circumstances. The order contents of the lots are determined during job and material planning from the point of view of cost, which is known as economic lot sizing. In this process, one tries to minimize the total costs for each lot. It is not yet common also to include the lot inventory, that is, the inventory carrying costs caused by the work in process. It is generally advisable first to reduce the variance of work contents. Either the lots must be divided in such a way that a certain order time of, for example, 16 hours, corresponding to an operation time of two days with single-shift work, is not exceeded, or the setup costs must be examined more closely to reduce them with technical or organizational measures, and thus be able to process smaller lot sizes. Recently, a formula for harmonizing the lot sizes was developed, with promising results [29, 30]. Note, however, that by reducing lot sizes more orders are produced and, when the lots are merely divided, more frequent setups require additional capacity. As a second measure, besides reducing the order time variance, the mean work content of the orders must be reduced. This can generally not be achieved with order planning or manufacturing control measures, but with investment in flexible machine centers and systems, in which different single parts can be processed in whatever sequence one requires without setup (or, to be more exact, the setup is carried out parallel to the machine running time). The influence of the lot size upon the order throughput is examined in more detail in Sect. 8.2.1.

As was explained in Sect. 4.4, besides these four components of inventory, two other components – the sequence component and the inventory trend component – also have an effect upon the mean flow time in a period. The sequence component disappears when there are no more sequence transpositions. All the measures which reduce the control inventory also reduce the sequence component of flow time. The inventory trend component is influenced by several factors. Since, to achieve an inventory trend component of zero, the initial and final inventory areas must be equal, this means that the flow inventory as well as the control and lot inventories of the input and output curves must also be equal.

5.6 Use of Graphics for the Representation of Throughput Diagrams and Key Data

The graphic representations of the throughput diagram and the key data derived from it have clearly shown how descriptive the visualization of the complex processes in job-shop production can be. The cost for producing such graphics on color monitors and color plotters is constantly decreasing; their use is therefore not limited to scientific applications or to single presentation events, but they are increasingly useful also for everyday factory practice. The following representations are meant to give a general idea of such graphics and invite the reader to make designs of their own.

Two series of graphs are given as examples, and these could be used in manufacturing analyses or in a monitoring system. How these graphs were produced is of lesser importance; therefore, only references will be given to the respective places where the calculations are dealt with. The graphics will be considered more from the viewpoint of an imagined user who intends to use them for short-, medium- and long-term measures to improve the manufacturing flow. The user should know the fundamentals of the funnel model and of the key data in the throughput diagram, but does not need to know in detail about how these diagrams are calculated. This approach, comparable with the Financial Controlling Method, is increasingly employed in larger companies.

5.6.1 Representation of the Results of Manufacturing Analyses

The plates outlined below represent results of manufacturing analyses. They were produced as demonstration models with the help of standard graphics software on a prototype version. The data were taken from a manufacturing analysis and evaluated using the software DUBAF (see Sect. 5.3).

Plate 1 shows the throughput diagram of a work center. The throughput elements were also included into the graph.

Plate 2 evaluates the flow time distribution of the orders which were processed by the work center from Plate 1 in the period covered by the throughput diagram. Additional statistical information is given on the work content of the orders and on the maximum and minimum values, the mean value, median, and the standard deviation of the weighted flow time per operation. Plate 3 shows the respective values for the weighted order time.

The importance of the Pareto curve for a clear description of the weighted and unweighted lead time and order time distributions was already explained above.

Plate 4 shows the Pareto distribution of the order times of all orders completed by the work center from Plate 1 in the period given there.

The last two plates in the manufacturing analysis section are meant to illustrate the lateness of orders. Plate 5 shows a conventional representation of the due date deviation of 8761 orders, with the respective statistical values.

Plate 6 illustrates the order lateness in a throughput diagram containing the actual and planned output dates of the orders. The stochastic nature of the magnitude 'lateness' becomes evident. Since each order is 'weighted' with its work content (i.e. since it is shown as a rectangle with the sides lateness and order time), one obtains a representation of the weighted schedule deviation as developed in Sect. 4.4.7. In addition the backlog measured in hours is shown.

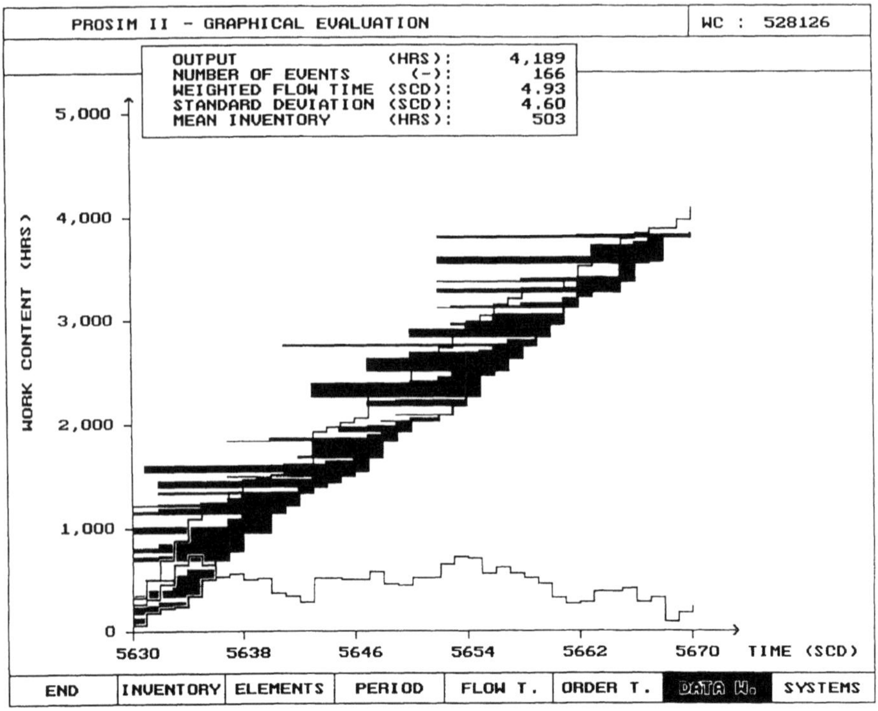

Plate 1. Inventory trend in the throughput diagram of a work center

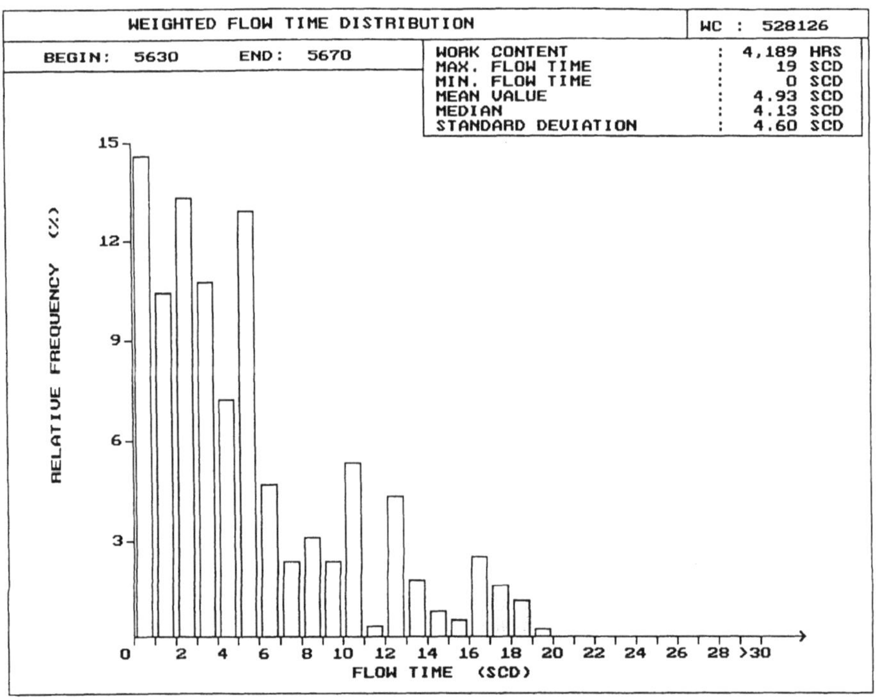

Plate 2. Distribution of work center flow time

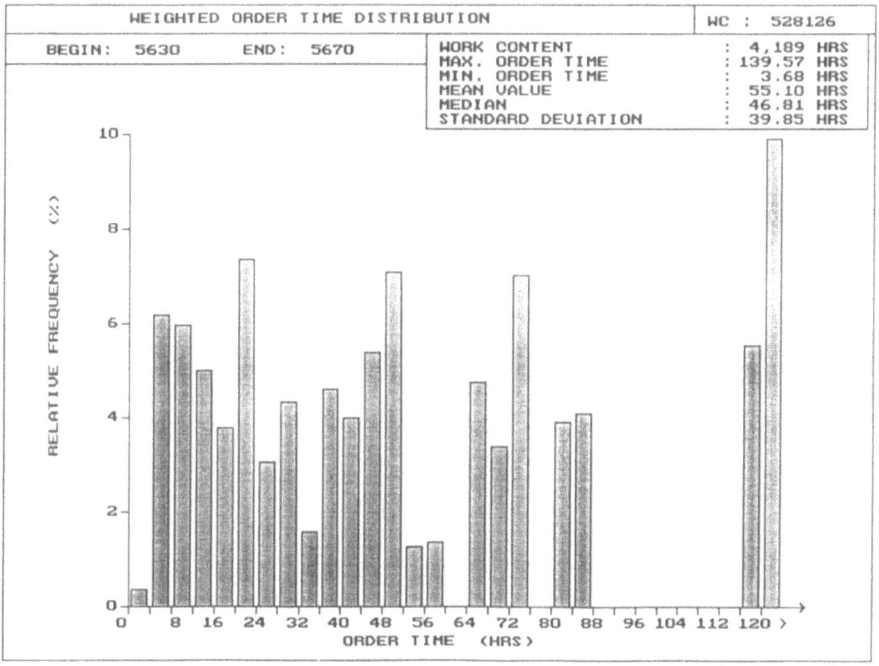

Plate 3. Pareto curve of operation time at a work center

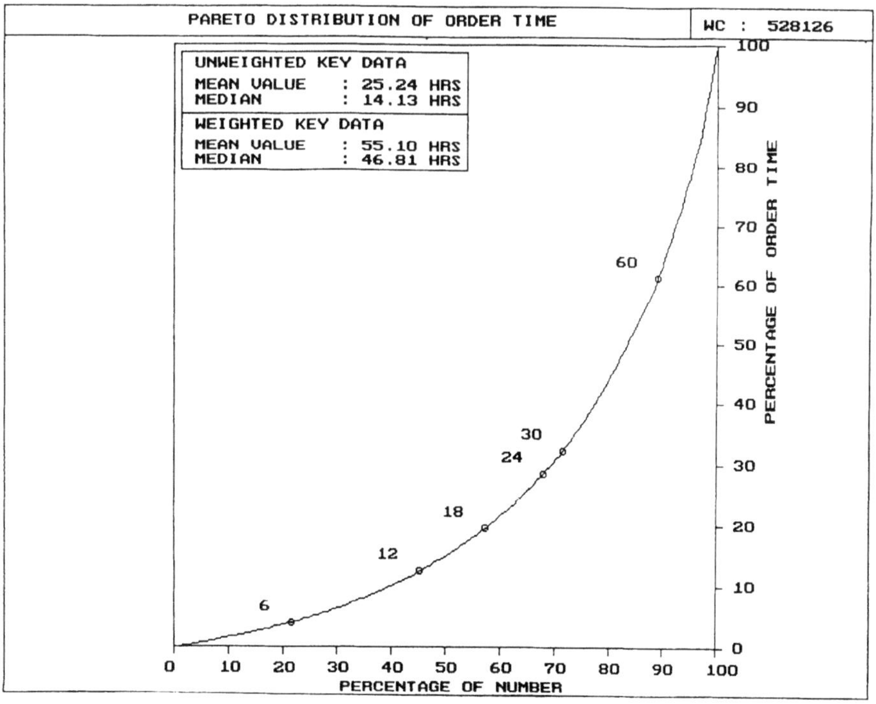

Plate 4. Deviation diagram of some work center key data (MEAN: mean value of 8 periods)

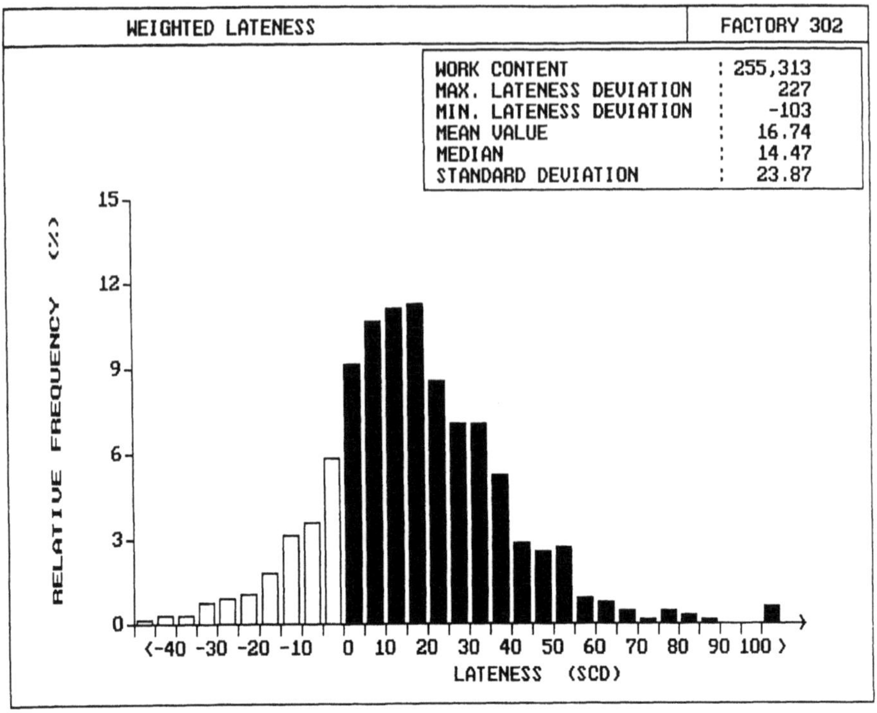

Plate 5. Distribution of lateness for a group of orders

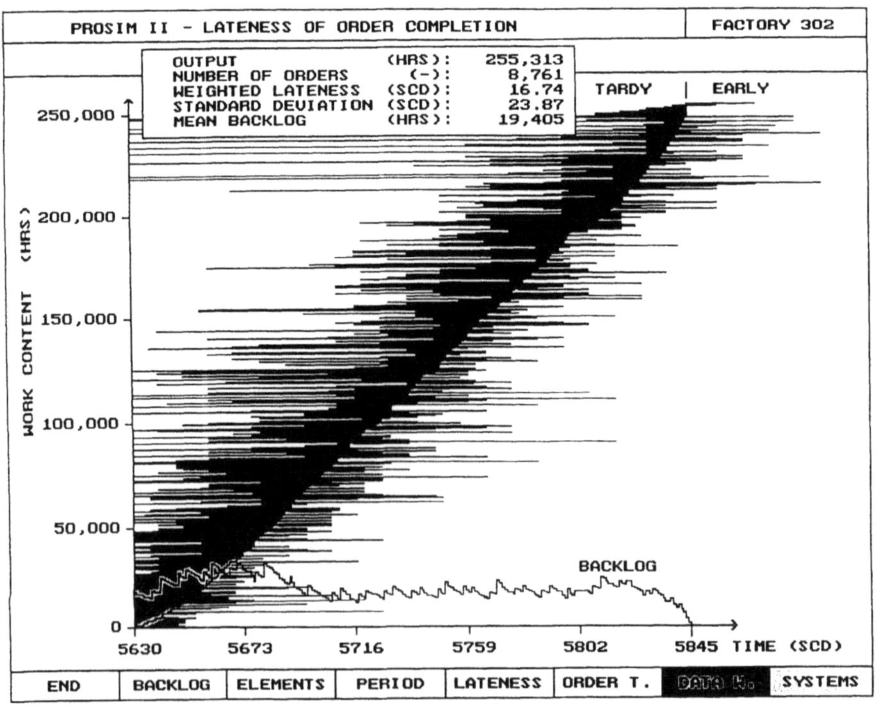

Plate 6. Diagram of unweighted lateness for a group of orders

5.6.2 Graphics for Medium-Term Manufacturing Flow Monitoring

The plates outlined below were produced using the monitoring system KOSYF presented in Sect. 5.4, with an evaluation frequency of 1 week and an evaluation period of 12 weeks [21]. The data were taken from a shop producing plastic foils – generally in five operations, with a large range of variants, and in different lot sizes [22].

First, Plates 7 and 8 show the five throughput curves of the entire shop from Calendar Week 44 (marked BEG = beginning) to Week 04 of the subsequent year. The graphs show the situation until the end of Week 51 (marked NOW = actual week).

Whereas Plate 7 represents the planned status based on the scheduled lead time values, Plate 8 shows the actual status which, of course, does not permit any statements on the time after the 52nd week, which is why the curves are horizontal after this point. The entry curves (ENT) are the same for both diagrams, since they represent the starting point of scheduling.

The planned status in Plate 7 allows a critical evaluation of the schedule. It is obvious that scheduling was unrealistic here: first, the release reacts only as late as Week 52 to the increased order entry from Week 46 on, which is too late. Second, unrealistically small indirect and direct inventories, and thus too short lead times, are planned. The actual status in Plate 8 shows that, on the one hand, orders were released earlier than planned but, on the other hand, the actual delivery (represented by the EXT curve) lies behind schedule by nearly 9000 hours. This can also be seen from the fact that the starting point of the planned exit curve is not at zero but at about 9000 hours.

Plates 9 and 10 show the situation in the same shop 16 months later: the NOW week is Calendar Week 16. The scheduling in Plate 9 shows more realistic values than two years ago, but there are still delivery delays of about 7000 hours, which can be seen from the difference between the planned (about 42000 hours) and actual (about 35000 hours, Plate 10) values of the exit curves. The release, too, is not yet properly harmonized with the order entry.

The actual lead times are reflected in the actual inventories, whose components are shown in Plate 11 (for the same period as in Plate 10); release fluctuations manifest themselves above all in fluctuations in the indirect inventory (called TRANS).

What is interesting is the development of the actual lead time and the delays as a function of the inventory and release behavior. Plate 12 is a week-by-week representation of the planned lead time values for the in-process inventory between order release and order exit (LT-P) and the respective actual lead times (LT) as well as the release delay (DEL REL) and the exit delay (DEL EXT). From Week 9 on, the planned lead time was gradually reduced from 21 to 15 days, but the actual values did not follow this standard in the long-term, since the in-process inventory as the sum of the direct, indirect, and post-inventories was not reduced accordingly (cf. Plate 11). This resulted in exit delays (DEL EXT) of 8 to 11 days which were only partly due to the release delay (DEL REL).

To gain better control over the schedule performance, the company monitored the planned and actual release trends, and the planned and actual exit trends in a separate throughput diagram (Plate 13). Between Weeks 9 and 13, release was almost on schedule, so Plate 12 shows only relatively small delays for these weeks.

Whereas the diagrams introduced above give a general survey and serve to support order planning and release processes, the monitoring of the direct inventory tells us something about the situation in the shop itself. An example is shown in Plate 14, which is similar to Plate 12, but monitors the lead times and lateness of the direct inventory. While in Weeks 5 to 11 a comparison of the planned and actual lead times shows a satisfactory lead time performance, the mean output delay (DEL OUT) in the same period (6 to 9 days) is about double the mean direct lead time, which is obviously due to the input delay (DEL INP), which, in turn, results from the release delay.

These graphs make it especially clear that flow time and lateness are two objectives which must be monitored separately, and controlled using different measures. While the flow time is actually a function of the mean inventory and capacity, the lateness is subject to two other factors: the deviation between the planned and actual lead times, and the deviation between the planned and actual release dates.

Plates 7–14 see pages 196–199

5.7 Implications for Manufacturing Control

Even if the basic structure of the throughput diagrams and their different forms seem simple in the beginning, it becomes clear from practical examples that it is not always easy to draw the correct conclusions from the deviations of planned and actual values, and to find the right measures for improvement.

It was, however, shown in Sect. 5.5, where inventories and flow times were broken down into their components, that the flow time at a work center is influenced by a small number of clearly describable factors, and these can be deduced from the relationship between the input and output curves.

The most important task of a manufacturing control system is to control the basis and flow inventories, i.e. to balance input against output. Measures for finite control of the control inventory should not be taken until this has been implemented. As a parallel development, measures for capacity planning and lot size planning must be worked out.

The throughput diagram is especially well suited as a model for such control systems, since the effects which measures taken to change one objective (e.g. the flow time) have on other objectives (e.g. utilization and lateness) can be described and, above all, checked with the help of the monitoring and diagnostic system.

In the following chapter, another component for manufacturing control in job-shop production and batch production – which has become known under the name of 'load-oriented order release' – will be introduced. It controls above all the mean inventory, and thus directly influences the mean flow times of the operations and orders.

A first version of the system was 'discovered' by Jendralski during his simulation studies [24], and developed into an easily usable system by Bechte [25]. Buchmann integrated it into a PPC system [26], and Erdlenbruch improved some of its essential characteristics [27]. The sections below are based on these four publications.

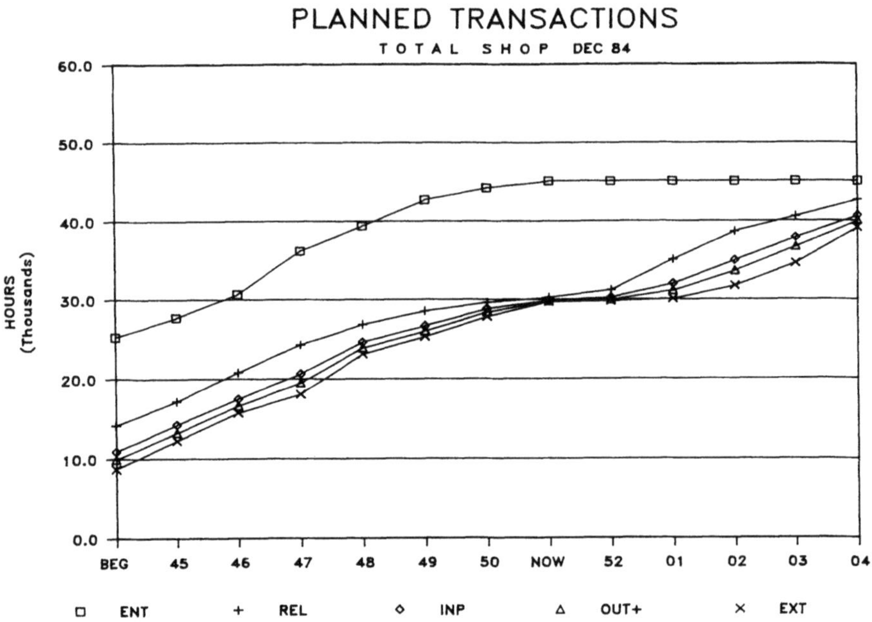

Plate 7. Throughput diagram of total shop (planned status) from week 44 to week 04. ENT: Entry; REL: Release; INP: Input; OUT: Output; EXT: Exit (After Bechte [21])

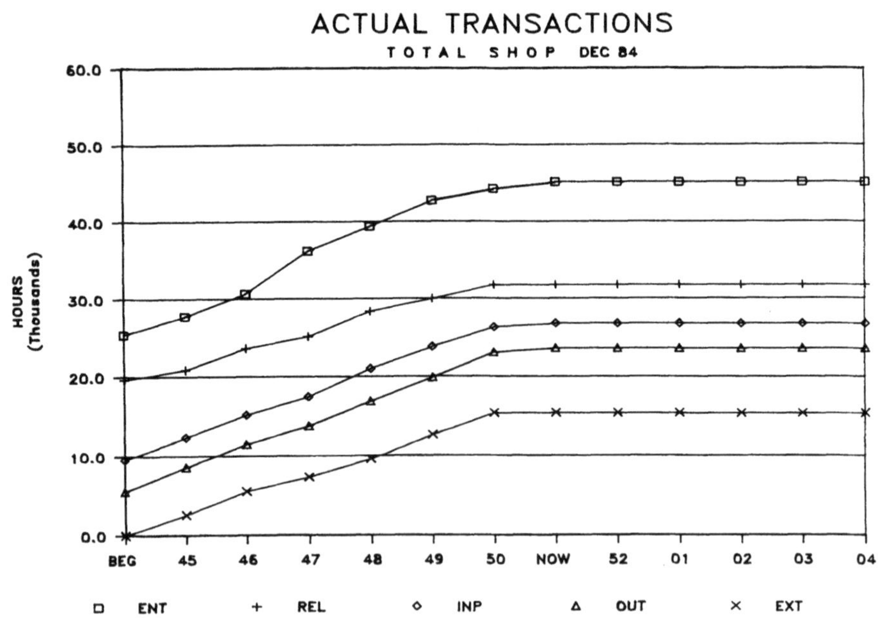

Plate 8. Throughput diagram of total shop (actual status) from week 44 to week 04 (After Bechte [21])

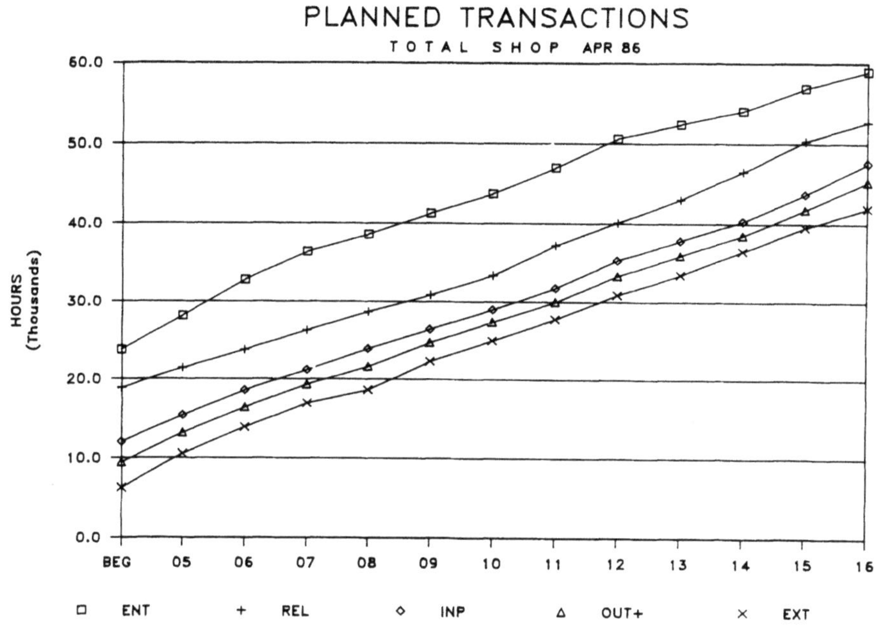

Plate 9. Throughput diagram of total shop (planned status) from week 04 to week 16 (After Bechte [21])

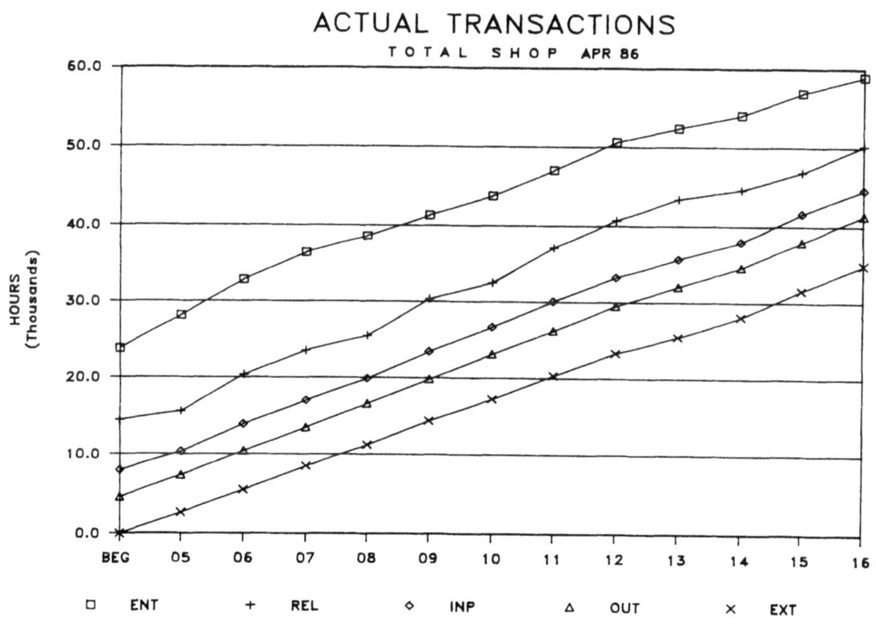

Plate 10. Throughput diagram of total shop (actual status) from week 04 to week 16 (After Bechte [21])

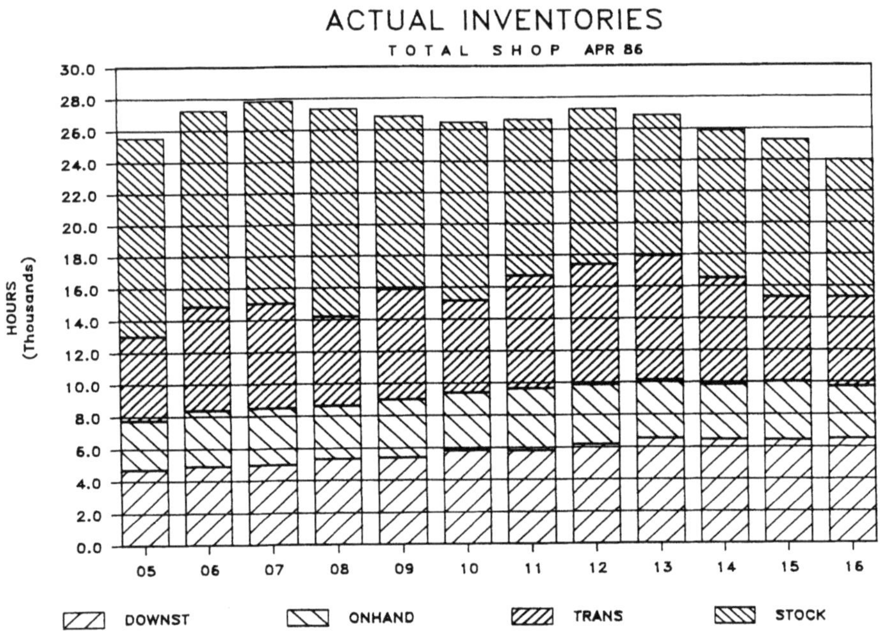

Plate 11. Inventory trend of total shop (actual status) from week 05 to week 16. STOCK: on-stock inventory; TRANS: indirect inventory; ONHAND: direct inventory; DOWNST: down stream inventory (After Bechte [21])

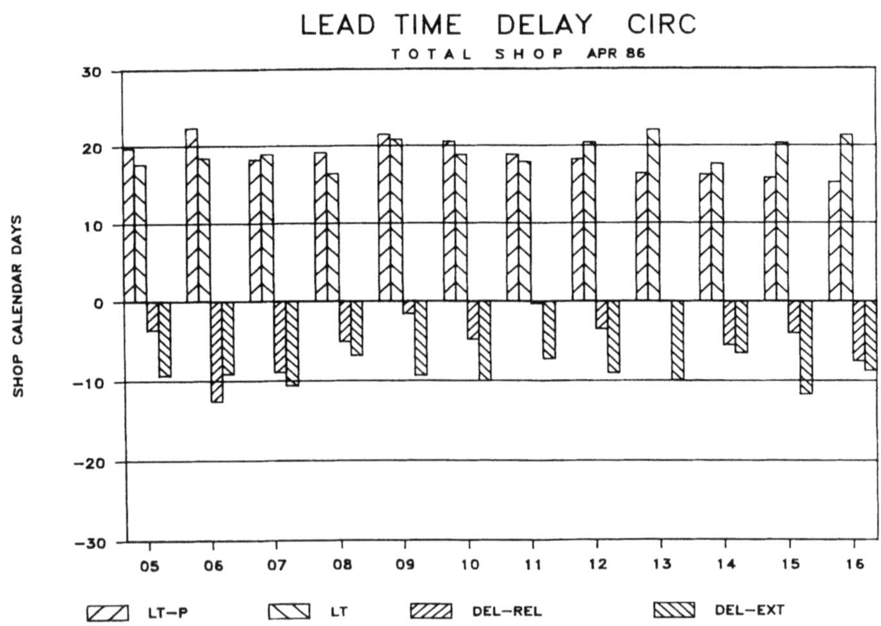

Plate 12. Trends of planned and actual lead time and of the release and exit delays of the in-process inventory of total shop from week 05 to week 16. LT-PLAN: planned lead time; LT: actual lead time; DEL-REL: release delay; DEL-EXT: exit delay (After Bechte [21])

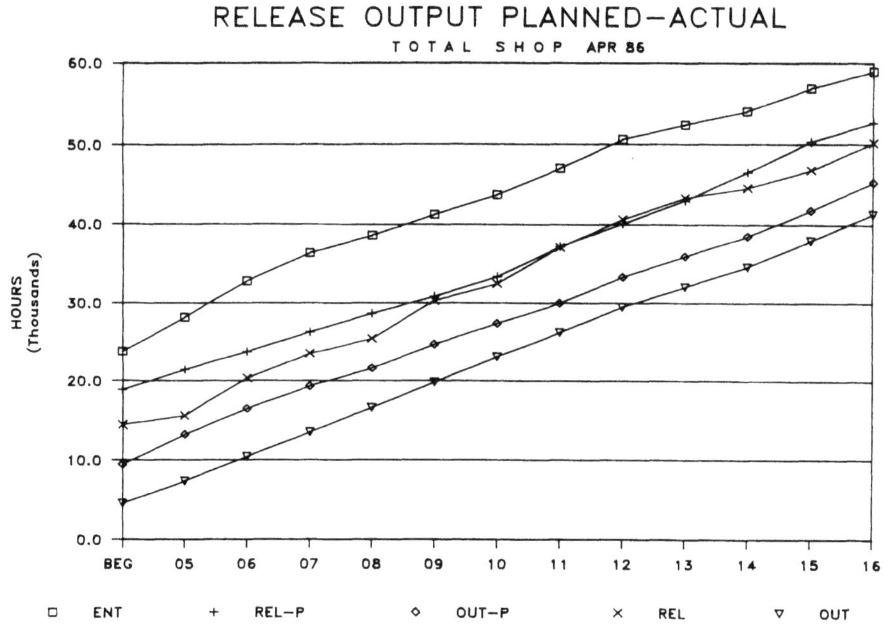

Plate 13. Planned versus actual release trends and planned versus actual ouput trends for total shop from week 04 to week 16 (After Bechte [21])

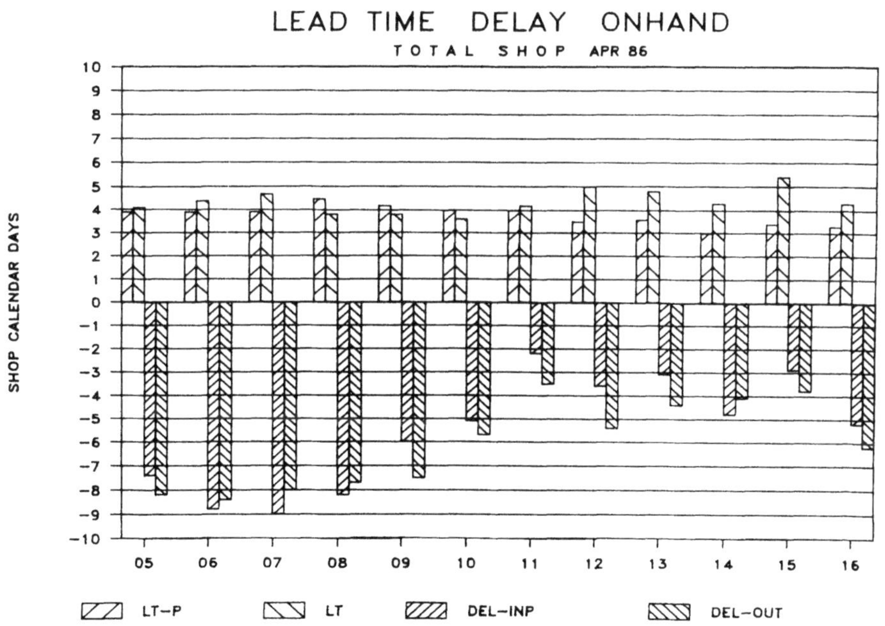

Plate 14. Planned versus actual lead time trends and input versus output delays for total shop from week 05 to week 16 (After Bechte [21])

5.8 References

[1] Kettner, H. (ed.). Neue Wege der Bestandsanalyse im Fertigungsbereich (Recent Inventory Analysis Methods in Manufacturing). Fachbericht des Arbeitsausschusses Fertigungswirtschaft (AFW) der Deutschen Gesellschaft für Betriebswirtschaft (DGfB) (report). IFA (Institut für Fabrikanlagen der Technischen Universität Hannover), Hannover, 1976.

[2] Kettner, H, Kreutzfeldt, H-F. DUBAF – eine Methodenbeschreibung zur Durchlaufzeit-und Bestandsanalyse im Fertigungsbereich (DUBAF – A Method of Lead Time and Inventory Analysis in Manufacturing). *Werkstattstechnik*, **68** (3): 1978, 157–162.

[3] Dombrowski, U. Durchlaufzeit- und Bestandsanalyse im Fertigungsbereich (DUBAF) (Lead Time and Inventory Analysis in Manufacturing (DUBAF)). In: Dokumentation zum Fachseminar 'Statistisch orientierte Fertigungssteuerung' des Instituts für Fabrikanlagen der Universität Hannover am 14./15. 1984 in Hannover (seminar paper). IFA, Hannover, 1984, pp 190–214.

[4] Bechte, W. Durchlaufzeit- und Bestandsanalysen als bewährter Einstieg in die Rationalisierung von Fertigungsabläufen (Lead Time and Inventory Analyses as a Tried and Tested Lead-in to Manufacturing Flow Rationalization). In: Dokumentation zum Fachseminar "Statistisch orientierte Fertigungssteuerung" des Instituts für Fabrikanlagen der Universität Hannover am 14./15. 1984 in Hannover (seminar paper). IFA, Hannover, 1984, pp 215–225.

[5] Nyhuis, F. Fertigungsablaufanalyse benötigt für gute Ergebnisse systematische Planung (Systematic Planning as a Prerequisite for Manufacturing Flow Analysis). *Maschinenmarkt*, **90** (20): 1984, 442–445.

[6] Hackstein, R. *Produktionsplanung und -steuerung (PPS) – Ein Handbuch für die Betriebspraxis (Production Planning and Control (PPC) – A Manual for Manufacturing)*. VDI-Verlag, Düsseldorf, 1984. Verlag, Düsseldorf, 1984.

[7] Bobenhausen, F. Analyse der arbeitsvorgangsbezogenen Durchlaufzeitstruktur in Betrieben der Einzel- und Kleinserienfertigung (Analysis of Operation-Related Lead Time Components in Job-Shop and Small-Batches Production). Dissertation Universität Dortmund, 1985 (doctoral thesis).

[8] Buß, P. Integrierte Planung der Lagerstruktur bei Neuplanung von Betrieben mit Werkstattfertigung (Integrated Warehouse Planning when Planning New Job-Shop Production Plants). Dissertation Universität Hannover, 1985 (veröffentlicht in: Fortschritt-Berichte der VDI-Zeitschriften, Reihe 2, Nr. 26, Düsseldorf, 1985) (doctoral thesis).

[9] Greim, H-R. Reorganisationsplanung der Zwischenlagerstruktur in Betrieben mit Werkstattfertigung (Reorganisation of Intermediate Stores in Job-Shop Production). Dissertation Universität Hannover, 1985 (veröffentlicht in: Fortschritt Berichte der VDI-Zeitschriften, Reihe 2, Nr. 110, Düsseldorf, 1986) (doctoral thesis).

[10] Kettner, H, Bechte, W. Neue Wege der Fertigungssteuerung durch belastungsorientierte Auftragsfreigabe (Recent Manufacturing Control Methods by Use of Load-oriented Order Release). *Zeitschrift des Vereins Deutscher Ingenieure*, **123** (11): 1981, 459–466.

[11] Lorenz, W. Organisatorische Maßnahmen zur Steuerung und Kontrolle des Fertigungsablaufs bei NC-Maschinen in einer Werkstattfertigung (Organizational Measures for Manufacturing Flow Control and Monitoring of NC Machines in Job Shop Production). Vortrag zum Seminar 'NC-Technologie-Forum NC/CNC/DNC-Einsatz in der betrieblichen Praxis' der Gesellschaft für Management und Technologie (gfmt) am 25./26. 06. 1984 in München (seminar paper).

[12] Wiendahl, H-P. Erprobte Methoden zur Reduzierung von Durchlaufzeiten in der Produktion (Tried and Tested Methods for Lead Time Reduction in Manufacturing). *Industrielle Organisation*, **53** (9): 1984, 391–395.

[13] Autorenkollektiv (several authors). Methodik und Praxis der Durchlaufzeitverkürzung in der Einzel- und Kleinserienfertigung (Lead Time Reduction in Job-Shop and Small-Batches Production – Methods and Applications). In: Tagungsunterlage zur VDI/ADB-Fachtagung am 7./8. 03. 1985 in Fellbach bei Stuttgart (proceedings). VDI-Verlag, Düsseldorf, 1985.

[14] Nyhuis, F. Rüstzeitanalyse – Voraussetzung für eine systematische Rüstzeitreduzierung (Setup Time Analysis – A Prerequisite for Systematic Setup Time Reduction). Vortrag zum Fachseminar 'Statistisch orientierte Fertigungssteuerung' des Instituts für Fabrikanlagen der Universität Hannover (seminar paper). IFA, Hannover, 1984, pp 242–261.

[15] Holzkämper, R. Konzeption eines Kontroll- und Diagnosesystems zur Überwachung des Fertigungsablaufs (A Model of a Manufacturing and Diagnostic System for Manufacturing Flow Monitoring). *Zeitschrift für Wirtschaftliche Fertigung*, **79** (9): 1984, 451–455.

[16] Holzkämper, R. Voraussetzungen für die Realisierung eines Kontroll- und Diagnosesystems zur organisierten Fertigungsablaufüberwachung (Prerequisites for Acquirement of a Monitoring and

Diagnostic System for Manufacturing Flow Monitoring). *Zeitschrift für Wirtschaftliche Fertigung*, **80** (6): 1985, 238–243.

[17] Bechte, W. Arbeitsinhalt-Zeit-Funktionen – ein Kontrollinstrument für die Fertigungssteuerung (Relationships between Work Content and Time – A Means for Monitoring in Manufacturing Control). *Fertigung und Betrieb/Industrial Engineering*, **32** (2): 1983, 7–14.

[18] Wiendahl, H-P. Beeinflußbarkeit von Durchlaufzeiten, Beständen, Leistung und Termintreue mit Hilfe von PPS-Systemen (How to Influence Lead Times, Inventories, Performance, and Schedule Performance with the Aid of PPC Systems). *VDI Bericht*, 490; 1983, 85–91.

[19] Lorenz, W. Differenzierte Bestandsanalyse im Fertigungsbereich – eine neue Methode zur Bestandskontrolle (Sophisticated Inventory Analysis in Manufacturing – A Recent Inventory Monitoring Method). *Arbeitsvorbereitung*, **20** (4): 1983, 104–107.

[20] Lorenz, W. Differenzierte Durchlaufzeitanalyse im Fertigungsbereich – Ein Verfahren zur Durchlaufzeitkontrolle (Sophisticated Lead Time Analysis in Manufacturing – A Lead Time Monitoring Technique). *Arbeitsvorbereitung*, **20** (5): 1983, 144–149.

[21] Bechte, W. Rechnergestütztes Durchlaufzeit- und Bestandskontrollsystem (KOSYF) als Basis einer flußorientierten Fertigungssteuerung (Computer-Aided Lead Time and Inventory Monitoring System as a Basis of a Flow-Oriented Manufacturing Control). Krautzig und Bechte, Unternehmensberater, Hannover, 1985.

[22] Schumacher, E. Darstellung und Überwachung von Kennzahlen auf PC's – Eine realisierte Lösung (Key Data Representation and Monitoring on PCs). Kongreß PPS '85 des Ausschuß für Wirtschaftliche Fertigung Bad Soden am 6.-8. 11. 1985 in Böblingen (conference paper).

[23] Schumacher, E, Bechte, W. Design and Implementation of Floworiented Manufacturing Control System in a Plastic Foils Factory. *First World Congress of Production and Inventory Control*, Vienna, May 27–29 1985.

[24] Jendralski, H. Kapazitätsterminierung zur Bestandsregelung in der Werkstattfertigung (Capacity Scheduling for Inventory Control in Job-Shop Production). Dissertation Technische Universität Hannover, 1978 (doctoral thesis).

[25] Bechte, W. Steuerung der Durchlaufzeit durch belastungsorientierte Auftragsfreigabe bei Werkstattfertigung (Lead Time Control in Job-Shop Production by Use of Load-oriented Order Release). Dissertation Universität Hannover, 1980 (veröffentlicht in: Fortschritt-Berichte der VDI-Zeitschriften, Reihe 2, Nr. 70, Düsseldorf, 1984) (doctoral thesis).

[26] Buchmann, W. Zeitlicher Abgleich von Belastungsschwankungen bei der belastungsorientierten Fertigungssteuerung (Temporal Leveling of Fluctuations in Load with Load-Oriented Manufacturing Control). Dissertation Universität Hannover, 1983 (veröffentlicht in: Fortschritt-Bericht der VDI-Zeitschriften, Reihe 2, Nr. 63, Düsseldorf, 1983) (doctoral thesis).

[27] Erdlenbruch, B. Grundlagen neuer Auftragssteuerungsverfahren für die Werkstattfertigung (Principles of Recent Order Control Techniques in Job-Shop Production). Dissertation Universität Hannover, 1984 (veröffentlicht in: Fortschritt-Berichte der VDI-Zeitschriften, Reihe 2, Nr. 71, Düsseldorf, 1984) (doctoral thesis)..

[28] Wiendahl, H-P, Frühwald, C, Fu, Z. Computer Aided Analysis, Diagnosis, and Planning of Setup Process. *2nd German Japanese Symposium Shortening of Setup Time*. Bad Soden, Germany, September 10 1990.

[29] Wiendahl, H-P. Throughput-Oriented Lot Sizing. *Annals of the CIRP*, International Institution for Production Engineering, **39** (1) 1990.

[30] Wiendahl, H-P. A New Logistic Based Lot Sizing Technique. Fundamentals and Case Studies. *34th International Conference Proceedings*. October 20–25 1991, Seattle, WA. American Production and Inventory Control Society, Falls Church, VA.

Chapter 6

Load-Oriented Order Release

6.1 Abstract

The term 'Load-oriented order release' stands for a method that controls flow times in a shop floor by controlling the actual input of work versus the planned output. The problem of managing not only the first work centers but also the downstream work centers will be solved using a mathematical approach which is easy to understand and to program, and which will be described in detail. The effect of the release method on the throughput of orders is then studied carefully by means of extensive computer simulations.

6.2 Fundamental Relationships

It was shown in Sect. 4.4.6, with the help of an area balance in the throughput diagram, that given certain preconditions, the weighted mean lead time of a work center is in direct ratio to its mean inventory and in inverse ratio to its mean performance. Figure 6.1 shows an authentic throughput diagram overlayed with the ideal input and output lines.

Assuming that the ideal input and output lines are parallel (steady-state process), we can draw the hatched triangle for which the following geometrical equation applies:

$$\tan\alpha = \frac{I_m}{TL_m}$$

$$\tan\alpha = PE_m$$

$$TL_m = \frac{I_m}{PE_m} \tag{6.1}$$

where:

TL_m = weighted mean lead time
I_m = mean inventory
PE_m = mean performance.

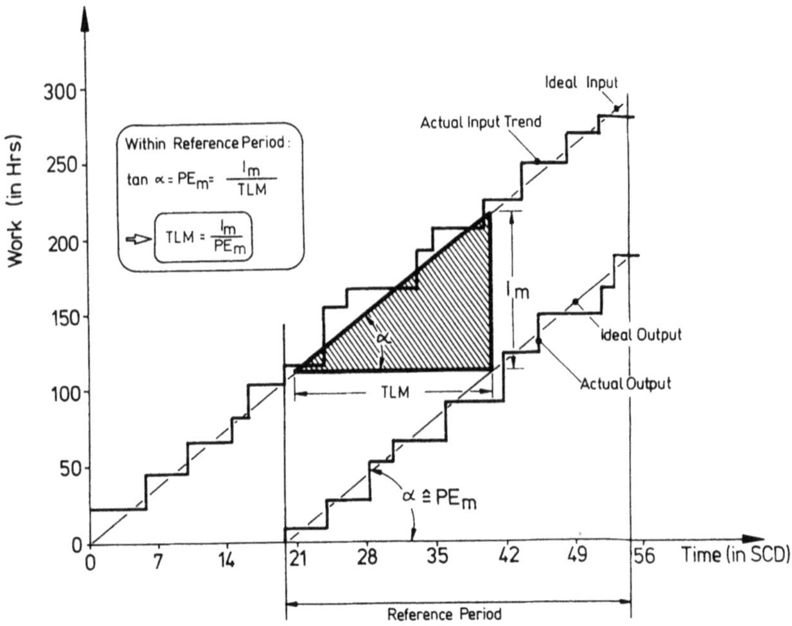

Fig. 6.1. Analytical relationships between lead time, performance, and inventory in production. SCD: Shop Calendar Day; PE_m: Mean Performance; I_m: Mean Inventory; TLM: Weighted Mean Lead Time

Lead time is understood here as a generic term for both actual lead time (= flow time) and planned lead time (= allowance). Performance, sometimes also called throughput, is defined as the output rate measured in standard hours per time unit (usually hours per workday).

This equation, also called the *funnel formula*, can not only be used for analytical purposes but also for control purposes. If one wants to ensure a specific flow time for a work center, one must always ensure that a certain mean inventory is maintained. This can, in turn, be achieved by inputting only so much work during a reference period as is expected to leave the work center in the same period. A change in mean flow time can be achieved either by a change in input (i.e. the load), or by a change in output (i.e. the performance). The performance is determined by the master production schedule, and is usually fixed for a longer period (e.g. several weeks). This means that mean inventory is a fundamental control variable for mean flow time.

If the inventory at a work center, and thus the vertical distance between the input and output curves, is varied within wide limits, a corresponding variation of the mean flow time value will result. From this, characteristic production curves can be developed. Figure 6.2 shows the general shape of such curves. It clearly reveals the change in mean performance and mean weighted lead time as a function of mean inventory. Above a certain inventory value – here called the 'critical value' – there is no major increase in performance because there is always enough work to avoid breaks in processing. Below this value, idle time will occur with increasing frequency.

When the inventory is increased beyond the 'critical value', the lead time will increase in direct ratio to the inventory, since the performance remains constant; the

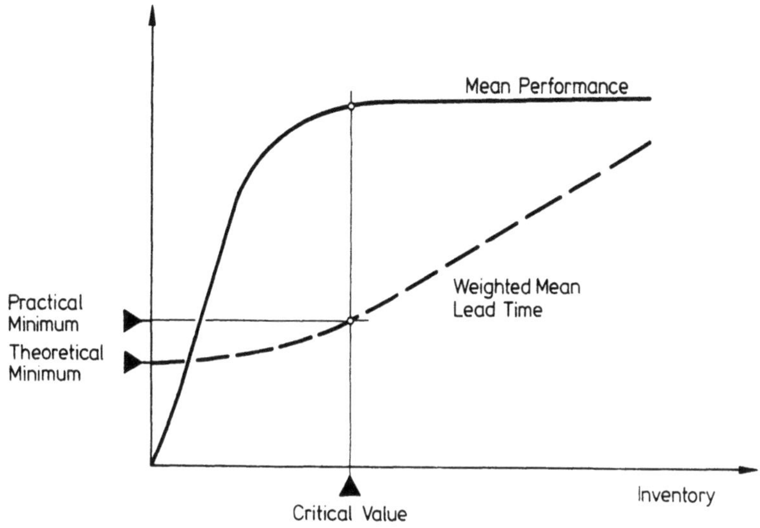

Fig. 6.2. Lead time and performance as functions of the inventory in production (After Bechte [2])

work center then obeys the 'funnel formula'. Below this value, the lead time decreases with the inventory until it reaches the value of the weighted mean operation time plus the mean transit time.

This leads us to a simple conclusion: the inventory of each work center must be controlled in such a way that, on the one hand, idle time is avoided, but on the other hand, a short lead time is achieved. As will be shown in Sect. 6.6, the realistic minimum lead time depends only on the mean value and range of variation of the operation times and on that of the interoperation times.

To put this principle into practice, one must take into account a number of constraints. The first is that the input and output curves in an authentic shop are not ideal. As was shown in Sect. 5.5, the influence of the real production flow manifests itself in three factors:

- Usually, the input and output amounts of work coming from released orders are not equal in every period, so any shop floor control system has to look for a technique to balance input against output continuously, i.e. to establish a control loop.
- The input and output curves of the work centers within a period generally do not follow a straight line, thereby producing an unsteady inventory level. This phenomenon has to be controlled on the dispatcher level.
- The work content of input and output orders vary quite strongly at times, mainly because of a lot-sizing procedure which does not consider logistic objectives. This effect cannot be handled by the shop floor control system, but by the materials management system.

The second important constraint is the flexible interlinking of the work centers in a shop, due to the fact that the operations routings of the processed orders often change. As was shown in Chapt. 2, it becomes all the more difficult to pre-determine the input and output dates of a certain lot at a certain work center, as the lot passes a large number of work centers between the present work center and the work center in question, and as this event lies in the future.

Third, some orders of a manufacturing process are scheduled in such a way that they must be processed together at a certain date – for example the upper and lower parts of a gearbox – but have different routings before and after that.

The most important of the three constraints is the flexible interlinking of the work centers. This means that a control method must be developed which controls the inventories, and thus the flow times, at all work centers of the shop in such a way that the allowances presumed in scheduling can really be achieved for every order. For this, the decisive factor in the throughput of the orders is obviously the date at which the orders are released. Based on this knowledge, Kettner and Bechte [1] developed the principle of 'load-oriented order release', which is described in detail in the following sections.

6.3 Procedure

First, we will again look at a single work center. Figure 6.3 shows the situation at the end of a schedule period [2, 3]. On the left of the figure we can see sections of the input and output curves from the recent past, and the future ideal throughput diagram for the next period. Further, a medium-term scheduled performance (PE_m) was assumed, to which a scheduled output (OUT) corresponds. Finally, an assumed scheduled mean lead time (TLM) must be achieved. Since, in the ideal throughput diagram, the input and output curves are parallel, the scheduled mean inventory (I_m) is constant over the entire schedule period (P).

The real initial inventory, however (called the 'leftover inventory (ILO)' in the figure), deviates from the planned mean inventory: in the figure it is shown to be larger than the latter. Thus, the work to be released is not the planned input INP but rather Load Limit LL minus Leftover Inventory ILO.

The sum of the planned mean inventory and the planned output is called the load limit (LL); the difference between the load limit and the leftover inventory is called the release (REL). The method developed from this is called *load-oriented order release*. Unlike conventional capacity scheduling methods, this method does not try to schedule single orders along the scheduled output curve with an accuracy to the day or the hour, but performs a period-by-period balance on the basis of the expected inputs and outputs.

This approach is known from commercial accounting and is applied in bookkeeping. On the right, Fig. 6.3 symbolically shows the loading account of the work center at the 'present date'. The cross-hatched area represents the leftover inventory, and the blank area corresponds to the release. The inputs planned for the subsequent period are booked (released) onto the account up to the load limit. During the schedule period, orders enter and leave so that at the end of the period we get a new balance with a new leftover inventory.

One important characteristic of this method is obvious: only one account is kept for each work center, and this is updated in every period. Thus, the conventional method of keeping several accounts for several future periods is no longer necessary. The account corresponds exactly to the funnel, and observing the processes on this account allows us to draw a precise throughput diagram for this work center.

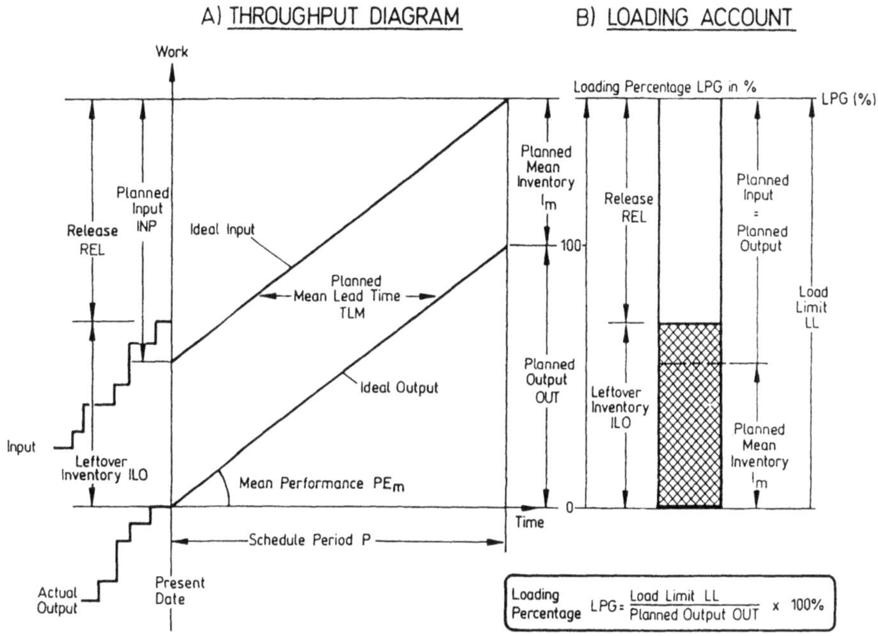

Fig. 6.3. Throughput model of load-oriented order release for one work center (After Bechte [2] and Erdlenbruch [3])

The following relationships are valid for a work center (cf. Fig. 6.3):

$$INP + I_m = OUT + I_m \tag{6.2}$$

$$REL + ILO = OUT + I_m \tag{6.3}$$

$$LL = OUT + I_m \tag{6.4}$$

$$REL = LL - ILO \tag{6.5}$$

where:

REL = work released for one schedule period (in hours)
LL = load limit (in hours)
OUT = scheduled output in the schedule period (in hours)
I_m = scheduled mean inventory (in hours)
ILO = leftover inventory at the beginning of the schedule period (in hours)
INP = input within the schedule period (in hours).

To avoid having to redetermine the load limit for every change in the scheduled performance, it should be related to the scheduled output. The value calculated in this way is called the loading percentage (*LPG*), and is defined as:

$$LPG = \frac{LL}{OUT} \times 100\% = \frac{I_m + OUT}{OUT} \times 100\%$$

$$LPG = \left(1 + \frac{I_m}{OUT}\right) \times 100\% \tag{6.6}$$

This value is, however, not only related to the inventory values OUT and I_m, but also to the time values TLM and P, as can be seen from Fig. 6.3:

$$\text{From} \quad \frac{I_m}{TLM} = \frac{OUT}{P}$$

$$\text{follows} \quad \frac{I_m}{OUT} = \frac{TLM}{P} \tag{6.7}$$

so that the following is valid for the loading percentage:

$$LPG = \left(1 + \frac{TLM}{P}\right) \times 100\% \tag{6.8}$$

where:

TLM = scheduled weighted mean lead time (in workdays)
P = length of schedule period (in workdays).

This relationship can also be shown in a diagram (Fig. 6.4). We can see the curves resulting from arbitrary values of TLM and P, and the curves produced when the

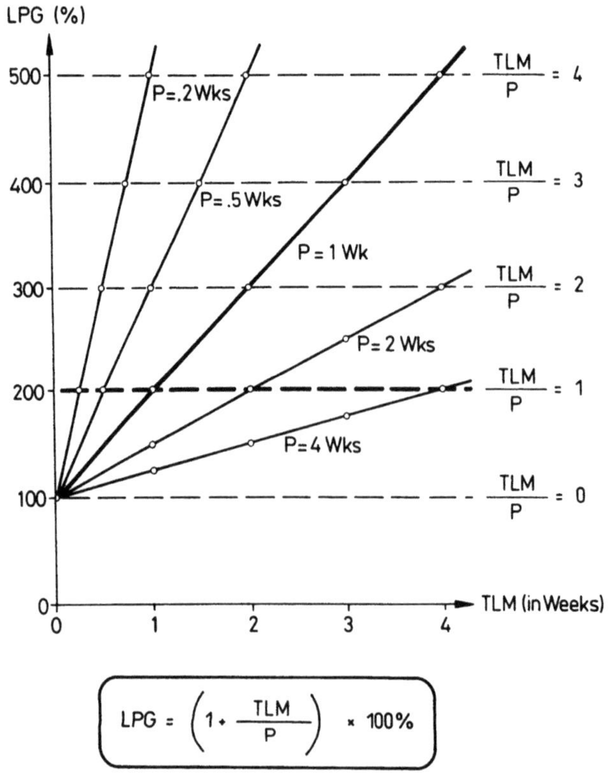

Fig. 6.4. Loading percentage as a function of scheduled weighted mean lead time and schedule period P

quotient from *TLM* and *P* is an independent variable. The diagram is independent of the time dimension (here the values are given in weeks), but the dimension must be the same for *TLM* and *P*.

In mechanical engineering shops, the length of one schedule period is usually 1 week; thus, with the usual mean lead times of 1 to 2 weeks per operation, loading percentages of 200 to 300 result. In other cases they can be considerably lower. With a schedule period of 10 days and a scheduled lead time of 4 days, for example, *LPG* would be 140.

Figure 6.3 raises the question of how to schedule the input of orders which, assuming a certain lead time, are not expected until the next period or the period after next, or even later. How can we ensure that there is available capacity for these orders, although the period in question is not yet planned?

To answer these questions, the lead time scheduling of the orders and the recalculation of the load of future periods to account for the period in question must be included in the release. The steps to be taken are shown in Fig. 6.5 [2].

The starting point for the release is given by the type, quantity and due dates of the shop floor orders determined during order planning. Further, the routing for the orders, specifying the quantity, operation sequence, work center number, as well as the setup time and time per unit, must be known. The first step in the release process is the backward scheduling of all issued shop orders which have not yet been released. The lead time value (allowance) is assumed to be the planned weighted mean lead time predetermined per work center. These values are basically determined according to the principles outlined in Sect. 5.5.3 and 6.2. A suggestion for a programmable calculation method as a component of load-oriented order release is presented in Sect. 6.6.

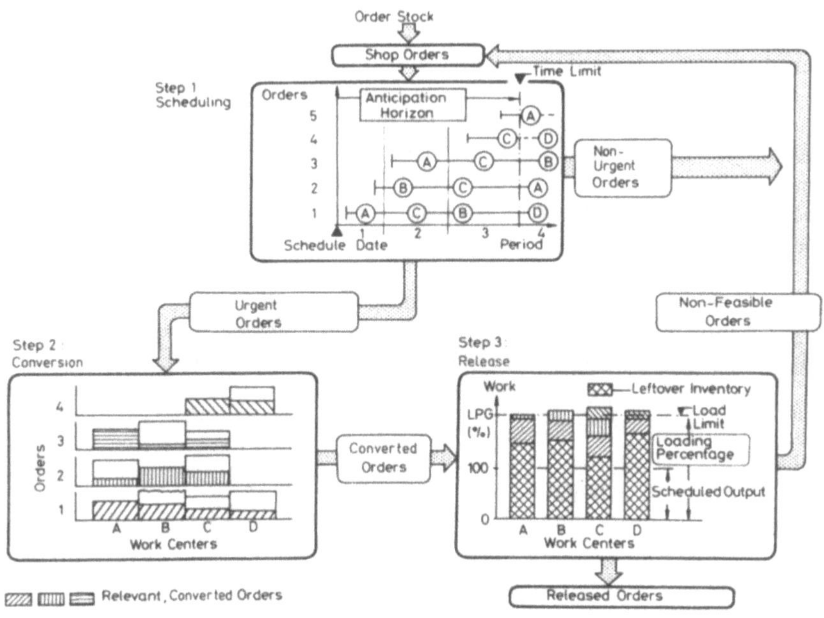

Fig. 6.5. Steps of load-oriented order release (After Bechte [2])

Backward scheduling results in a list of orders before release, arranged in the sequence of scheduled start dates. The start dates of the first orders can even lie in the past. In this list, those orders which lie below a time limit are classified as urgent orders. The time difference between this limit and the time when planning is carried out is called the *anticipation horizon*, and should be measured in numbers of periods.

When determining the time limit, the following considerations are decisive. In principle, only those orders whose start dates lie in the subsequent schedule period are due. Since it is unlikely, however, that these orders will fill the accounts of the load centers up to the load limit, a certain anticipation of orders in the subsequent schedule periods is permitted. Finite loading, which is the next step, will ensure that only those orders beyond the present schedule period that use no more than the remaining capacities up to the load limit will be released. Two to three schedule periods have been found to be a practical value for the anticipation horizon, but this has not been mathematically determined. It can, however, generally be said that the anticipation horizon should be greater the more the capacity requirement of the current order stock deviates from the long-term mean values by which the capacities are determined.

The first step in load-oriented order release produces, first, a list of the non-urgent orders. These are deferred and left in the inventory of unreleased orders until, in the next planning cycle, they are again subjected to the release process, together with the other issued but not yet released orders (including new ones). Second, one is given a list of the urgent orders, with their planned start dates, the operations sequences with standard times and work centers, and the planned exit dates, arranged according to the planned starting dates. Unlike many conventional planning systems, the list does not contain any other priority data. The only criterion for priority is the planned start date. This date may also lie in the past.

The next step in load-oriented order release is to evaluate whether the urgent orders will actually meet the scheduled inventory conditions at the individual work centers after their release. This means that a check is made as to whether the load limit of any work center passed by the released order is expected to be exceeded as a result of the release. In Fig. 6.5, this is the third step marked 'Release'.

For this, the first operation of Order 1 in the urgent list is loaded onto the load account of the corresponding Work Center A, with its standard time *TO*. Usually, this will not cause the balance on the load account to reach the load limit, so that the second operation of Order 1 can be loaded, this time onto the account of Work Center C.

In our example, this second operation of Order 1 will probably not enter Work Center C before the second schedule period. The load account is, however, only kept for the very next schedule period. On the other hand, the possibility that the operation reaches Work Center C in the first period cannot be ruled out. This would be the case if, for some reason, the order had been processed as a rush order at A, which is not an infrequent case, as has been shown in the practical lead time investigations dealt with above. To account for such a possibility, one should know how probable this case is. If this probability is known, the load value of the second operation can be multiplied by this probability factor, and the resulting load value can be loaded onto the load account in the next schedule period. This calculation is called *conversion of order times*. If this conversion is carried out for each operation, one can be sure that the load account is, on average, loaded correctly.

This conversion of the operation times of future orders, in order to obtain their probability value in the next schedule period, is the second step, marked 'Conversion' in Fig. 6.5. It is probably the most important idea behind the whole concept of load-oriented order release. It was developed by Bechte during his research work and described, although only briefly, in his doctoral thesis [2]. Since this conversion process is hard to understand at first, it is described in detail below, once we have dealt with the third step of Fig. 6.5.

After the second operation, the third operation is loaded onto the load account of Work Center B with its converted order time, and a check is made as to whether the load limit is exceeded. The same is done with the fourth and last operation. If this test loading reveals that the load limit of Work Center D is not exceeded, either, this order is released, which means that it is entered into the list of released orders, and the Load Accounts A, B, C, and D of the four work centers concerned are loaded with the converted order times of the corresponding operations.

Now the second order is tested in the same manner, followed by the third, the fourth, etc. As soon as the load limit of an account is exceeded for the first time, this account is blocked. The next operation arriving at this blocked account is refused together with all the other operations of the same order, and the entire order is entered into the list of non-feasible orders. Together with the non-urgent orders and the rescheduled orders it is re-entered into lead time scheduling in the next planning cycle, and probably released, since the work center in question has had some output in the meantime and can therefore take new input. Each order is only released once.

As a result of load-oriented release, lists of non-urgent, urgent, non-feasible, and released orders are produced. In some justified cases, however, the method permits orders which have not been released to be started, for example, if special arrangements for extra capacity at the bottlenecks are made.

6.4 Conversion of the Orders to be Loaded

The aim of conversion is to answer the question of how probable it is that an order will reach a certain work center in the next period if it first has to pass other work centers. For this, we shall first look at one work center WC_p (Fig. 6.6). In period P, the orders which are in the leftover inventory (ILO) or under release (REL) are ready for processing. However, orders can only be processed up to a total order time equal to the outputs (OUT) in the same period.

Then, the probability $POUT_p$ for any order from $ILO + R$ to be processed in the next period is obviously:

$$POUT_p = \frac{OUT}{ILO + REL}$$

where:

OUT = output in one period
ILO = leftover inventory at the beginning of the period
REL = released work.

However, Eqns. (6.3) and (6.6) say that $ILO + REL = I_m + OUT$ and

$$LPG = \frac{I_m + OUT}{OUT} \times 100$$

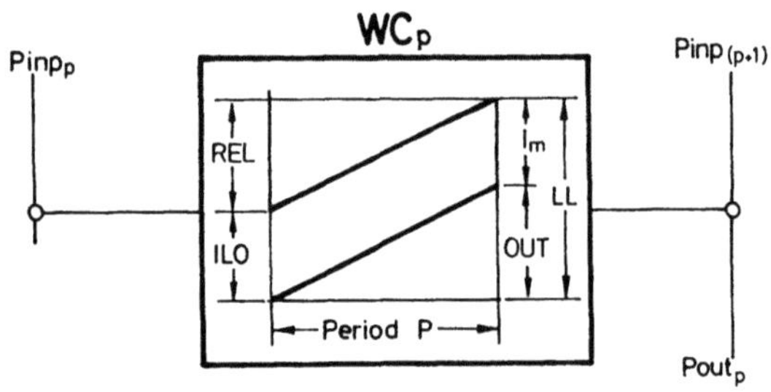

Output Probability at Work Center WC_p :

$$Pout_p = Pinp_p \times \frac{OUT}{REL + ILO}$$

$$Pout_p = Pinp_p \times \frac{OUT}{LL}$$

$$\boxed{Pinp_{(p+1)} = Pout_p = Pinp_p \times \frac{100}{LPG}}$$

WC_p = Work Center p
Pinp = Input Probability
Pout = Output Probability

REL = Release
ILO = Leftover Inventory
I_m = Mean Inventory
OUT = Planned Output
LL = Load Limit

Fig. 6.6. Output probability for an order at a work center

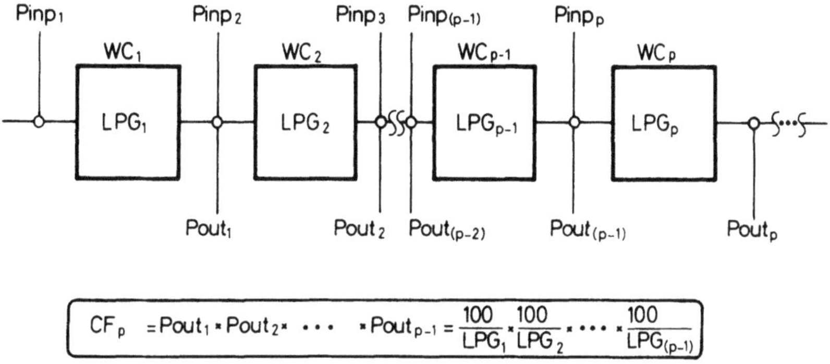

LPG = Loading Percentage
Pinp = Input Probability
Pout = Output Probability
CF = Conversion Factor for Order Content

Fig. 6.7. Conversion factor and input probability of an order at a work center

where:

I_m = scheduled mean inventory
LPG_p = loading percentage of work center p

so that the following is valid for the output probability for an order which has already reached the work center p:

$$POUT_p = \frac{100}{LPG_p} \tag{6.9}$$

Let us now look at a series of work centers, WC_1 to WC_p. What is the probability that an order waiting before WC_1 will reach WC_p in the subsequent period (Fig. 6.7)? To get there it first has to pass all upstream work centers, WC_1 to WC_{p-1}. The probability for this is:

$$PINP_p = POUT_1 \times POUT_2 \times \cdots \times POUT_{(p-1)}$$

which is called the conversion factor CFp. It now follows from Eqn. (6.9) that

$$CF_p = \frac{100}{LPG_1} \times \frac{100}{LPG_2} \times \cdots \times \frac{100}{LPG_{(p-1)}} \tag{6.10}$$

If all work centers have the same loading percentage, the equation can be simplified as follows:

$$CF_p = \left(\frac{100}{LPG}\right)^{p-1} \tag{6.11}$$

If the order content (TO) of the corresponding operation is now multiplied by this conversion factor, the resulting converted order content is a measure for the probable load (L_p) that this order will represent for this work center (WC_p):

$$L_p = CF_p \times TO \tag{6.12}$$

We now show how conversion works, using an order O1 as an example (Fig. 6.8)

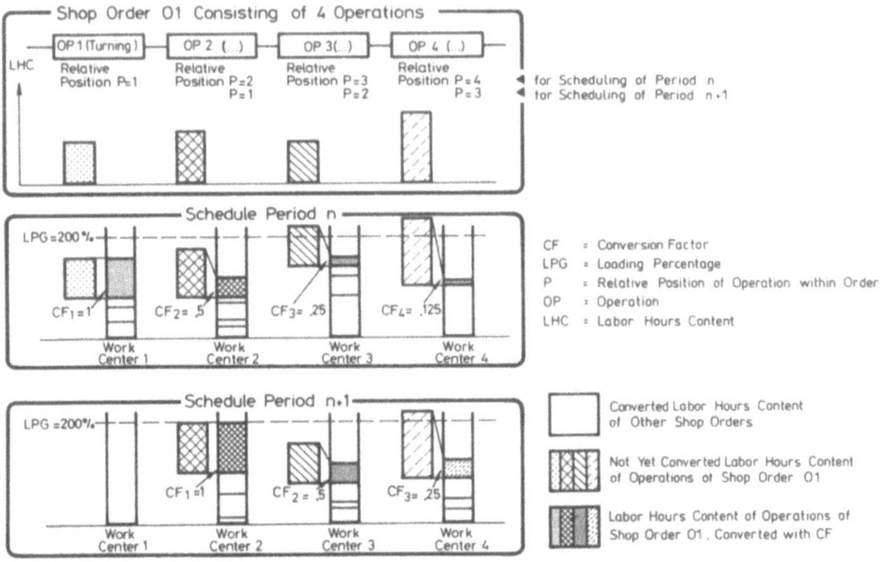

Fig. 6.8. Conversion of the work content of operations during load-oriented order release (After Lorenz [4])

[4]. The order must pass four work centers. Let us assume that Order O1 with Operations OP1 to OP4 lies before Work Center 1 in Schedule Period *n*. We now want to determine the conversion factor at all four work centers.

To simplify the calculation, let the loading percentage of all four work centers be 200. Then CF_1 obviously equals 1, because work can be started at any time, since the order is available with a probability of 100%. Operation 2 cannot, however, be started before the next period, after Operation 1 has been completed. Since *LPG* is 200, this has a probability of only 0.5, because the work center can only complete half the available work. The conversion factor is thus $100/200 = 0.5$. At the third work center, operations 1 and 2 must be completed before Operation 3 can be started in the next period (*n*). CF_3 is therefore $0.5 \times 0.5 = 0.25$. Accordingly, CF_4 equals $0.5 \times 0.5 \times 0.5 = 0.125$.

We now look at the period $n + 1$. Let us assume that Operation 1 is completed and the order lies before Work Center 2. The conversion factor for Operation 2 is now 1. It is now called CF_1, because the index *p* describes the relative position of the operation in the production progress. Accordingly, CF_2 is now 0.5 and CF_3 is 0.25. In a way, work content 'increases', the nearer it gets to the work center concerned.

6.5 Sample Demonstration of the Release Process

The following, more detailed example serves to demonstrate the release process. It is based on a list of 12 orders, whose start dates lie below the time limit (which is not considered here), and which are therefore urgent orders.

Table 6.1. List of urgent orders before release (Period 1)

Order Number	Planned Start Date	Op No 10		Op No 20		Op No 30		Op No 40		Op No 50		Loading Sequence
		TO	WC	TO	WC	TO	WC	TO	WC	TO	WC	
(-)	(SCD)	(Hrs)	(-)	(Hrs)	(-)	(Hrs)	(-)	(Hrs)	(-)	(Hrs)	(-)	
3001	501	10	B	15	A	20	E	80	D	80	C	1
3002	510	20	D	30	C	30	A	40	E			8
3003	505	20	B	10	C	40	D	20	E			5
3004	520	40	A	40	D	60	C	40	E			12
3005	503	20	A	30	B	60	C	80	D	40	E	3
3006	504	10	C	10	D	30	A	80	B	80	E	4
3007	502	5	D	10	C	40	B	40	A			2
3008	515	40	E	60	D	40	C	60	A	80	B	10
3009	507	15	B	20	C							7
3010	513	5	A	20	D	30	E					9
3011	506	20	A	20	B	20	C	80	D			6
3012	519	20	B	40	C	40	A	40	E	40	D	11

Op No = Operation Count Number
TO = Order Time in Standard Hours
WC = Work Center Number

Table 6.2. List of work centers before order release (Period 1)

Work Center Number >	A	B	C	D	E
Leftover Inventory before Period 1 in Hours	30	40	35	25	30
Capacity for Period 1 in Hours per Week	40	50	40	30	20
Capacity for Period 2 in Hours per Week	40	40	45	50	30
Load Limit in Period 1 (LPG = 200%)	80	100	80	60	40
Load Limit in Period 2 (LPG = 200%)	80	80	90	100	60

Table 6.1, listing these orders, gives their order numbers, their planned start dates given as shop calendar days (SCD), and the respective operation sequences with operation numbers, order time *TO* (in hours) and references to the corresponding work centers (A to E).

Table 6.2 gives the data required for order release, namely the capacity for the next two periods (1 and 2) and the initial inventory on the respective loading account, here called the 'Leftover Inventory before Period 1'.

We must now load as many urgent orders as possible for the next two periods, and mark the released and deferred orders. For the sake of simplification, we will continue to assume that the loading percentage is 200 at all work centers, that the dispatching rule followed is FIFO, and that the utilization is 100%. Further, we assume that the orders representing the leftover load will not have to be processed at other work centers after their completion.

In a first step, the orders are listed in the sequence of their planned start date (listing by priority). This gives the loading sequence shown in Table 6.1, last column.

The next step is the conversion of the order times of the individual operations (Table 6.3). All first operations are not converted, since they are immediately available. All order times of the second operations are multiplied by $100/LPG = 0.5$; all those of the third operations by $0.5 \times 0.5 = 0.25$; all those of the fourth oper-

Table 6.3. List of urgent orders with converted load before order release (Period 1)

Serial Number = Sequence	Order Number (-)	Op No 10 TO (Hrs)	Op No 10 WC (-)	Op No 20 TO (Hrs)	Op No 20 WC (-)	Op No 30 TO (Hrs)	Op No 30 WC (-)	Op No 40 TO (Hrs)	Op No 40 WC (-)	Op No 50 TO (Hrs)	Op No 50 WC (-)	Release Code R: Released X: Deferred
1	3001	10	B	7.5	A	5	E	10	D	5	C	R
2	3007	5	D	5	C	10	B	5	A			R
3	3005	20	A	15	B	15	C	10	D	2.5	E	R
4	3006	10	C	5	D	7.5	A	10	B	5	E	R
5	3003	20	B	5	C	10	D	2.5	E			X
6	3011	20	A	10	B	5	C	10	D			R
7	3009	15	B	10	C							R
8	3002	20	D	15	C	7.5	A	5	E			X
9	3010	5	A	10	D	7.5	E					X
10	3008	40	E	30	D	10	C	7.5	A	5	B	X
11	3012	20	B	20	C	10	A	5	E	2.5	D	X
12	3004	40	A	20	D	15	C	5	E			X

Op No = Operation Count Number
TO = Converted Order Time in Standard Hours
WC = Work Center Number

ations by 0.5 × 0.5 × 0.5 = 0.125 and finally all those of the fifth operations by 0.5 × 0.5 × 0.5 × 0.5 = 0.0625.

In Fig. 6.9 we can see the accounts of the Work Centers A to E which, to begin with, contain the leftover inventories from Table 6.2. Next, the converted order times of the urgent orders from Table 6.3 are loaded onto the accounts. The first operation of Order 3001, for example, loads 10 hours onto Account B. It is now physically present as a first operation at the work center. Account A is loaded with the (converted) load of 7.5 hours for the second operation, Account E with 5 hours for the third operation, Account D with 10 hours for the fourth and, finally, Account C is loaded with 5 hours for the fifth operation. Since none of the operations for the order concerned has exceeded the respective load limit, Order 3001 is given a release code (R) in Table 6.3. From the second operation on, the order time for all orders was converted, so the orders are so far not physically available at the work centers. The respective load elements in Fig. 6.9 are those not marked by dotted backgrounds.

The next orders on the urgent list are now loaded. Orders 3007 and 3005 can also be released. With Order 3006, however, the load limit of Work Center E is exceeded for the first time; the order is still released, but the loading account is blocked from now on. The fourth operation of the next order (3003) meets the blocked Account E and the entire order is rejected, although three of the four operations could have been processed. The order is deferred and given the code X. Orders 3011 and 3009 still meet open loading accounts, but Order 3002 and the following orders are refused by all work centers. In this release cycle, six out of 12 orders have been released and the other six deferred.

Now the job shop begins its work according to the FIFO dispatching rule. Every work center works the capacity hours given in Table 6.2 within Period 1. Therefore, Work Center A can, besides the leftover load, also process half of Order 3005, since this order is physically available, as the loading account (Fig. 6.9) shows, although Orders 3001 and 3007 have higher schedule priorities. Work Center B can, besides the leftover load, also process all of Order 3001; Work Center C processes a part of

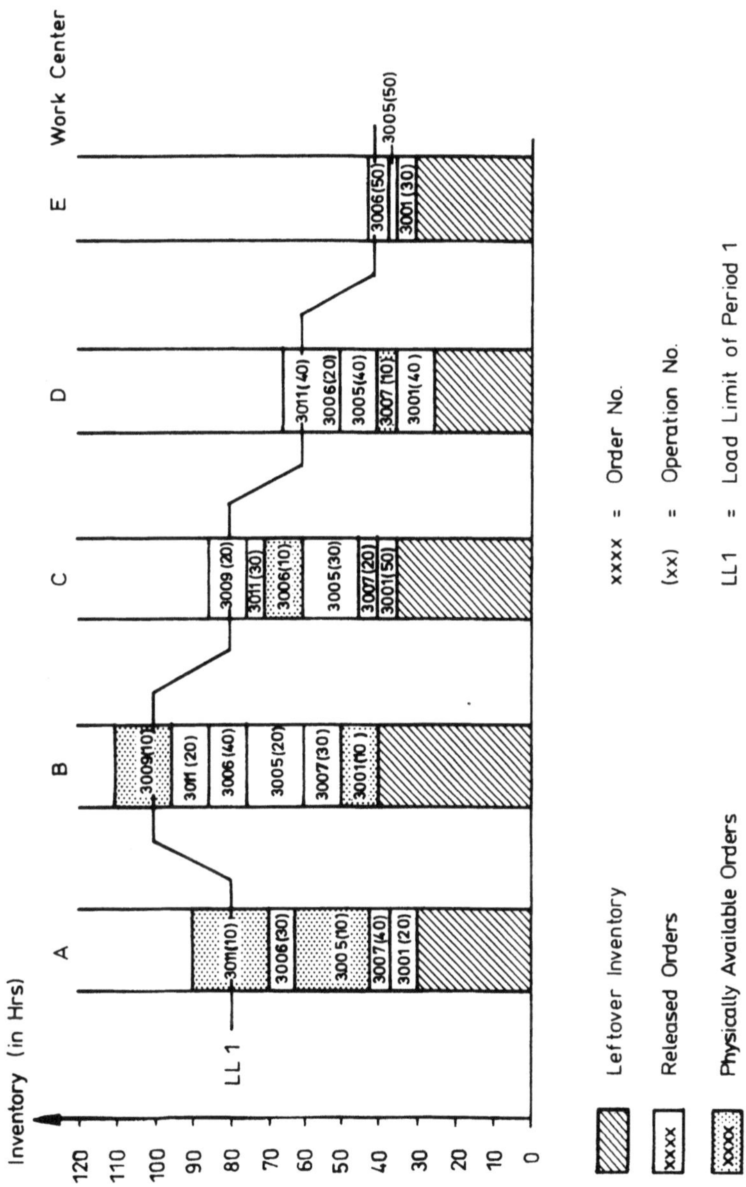

Fig. 6.9. Balance of loading accounts before release in period 1 after loading of converted orders

Table 6.4. List of urgent orders with converted load before order release (Period 2)

Serial Number = Sequence	Order Number (-)	Op No 10 TO (Hrs)	Op No 10 WC (-)	Op No 20 TO (Hrs)	Op No 20 WC (-)	Op No 30 TO (Hrs)	Op No 30 WC (-)	Op No 40 TO (Hrs)	Op No 40 WC (-)	Op No 50 TO (Hrs)	Op No 50 WC (-)	Release Code O: Open R: Released X: Deferred
1	3001	*	B	15	A	10	E	20	D	10	C	O
2	3007	*	D	10	C	20	B	10	A			O
3	3005	20	A	15	B	15	C	10	D	2.5	E	O
4	3006	10	C	5	D	7.5	A	10	B	5	E	O
5	3003	20	B	5	C	10	D	2.5	E			R
6	3011	20	A	10	B	5	C	10	D			O
7	3009	15	B	10	C							O
8	3002	20	D	15	C	7.5	A	5	E			R
9	3010	5	A	10	D	7.5	E					R
10	3008	40	E	30	D	10	C	7.5	A	5	B	X
11	3012	20	B	20	C	10	A	5	E	2.5	D	X
12	3004	40	A	20	D	15	C	5	E			R

Op No = Operation Count Number
TO = Converted Order Time in Standard Hours
WC = Work Center Number
* = Completed Operation

Order 3006, and Work Center D all of Order 3007. Work Center E cannot fully process its leftover load; part of it must be carried over to the subsequent period.

For release planning it is important that the completion of orders is reported before the planning is carried out. With larger orders, it is advisable to report parts of the order, i.e. the hours processed up to the end of the schedule period, so as not to block the work center. This possibility has also been made use of in our example.

This is the status of our release planning at the beginning of the next period. A new release cycle begins. First, the list of urgent orders is updated (Table 6.4). The first operation of Order 3001 could be processed. This, however, increases the load values of the remaining operations, since their relative positions have changed. Compared with the preceding period, the values are doubled because of the 200% load limit, as a comparison with Table 6.3 shows. The first operation of Order 3007 was also completed at Work Center D; the load values of the other operations are doubled. Other operations were not completed, and therefore the load values of the other orders do not change compared with the previous period.

The next step is the loading process. Let us assume that no new orders are issued. The capacity values are, however, changed, as can be seen in Table 6.2. If the loading percentage stays the same, this means new load limits for the subsequent period. In Fig. 6.10, they are marked *LL*2. Now the urgent orders are again loaded on top of the leftover inventories, until the loading accounts are blocked. The calculation can be followed using Table 6.4. Four more orders can be released; two remain blocked.

Another important result of the release process, besides a list of the orders to be released, is a list of the orders blocked as 'non-feasible'. It shows which are the work centers that block the release. Table 6.5 shows this list after the first release cycle in our example. Besides the order number and the number of unsuccessful release attempts, it contains the work center indicators in the same sequence in which they were reported as being blocked during the test loading. This gives the dispatcher very concrete indications as to which measures could be taken to achieve the order's release.

When applying this process in practice, the question arises as to which value should be chosen as the load limit. This is the subject of the following section.

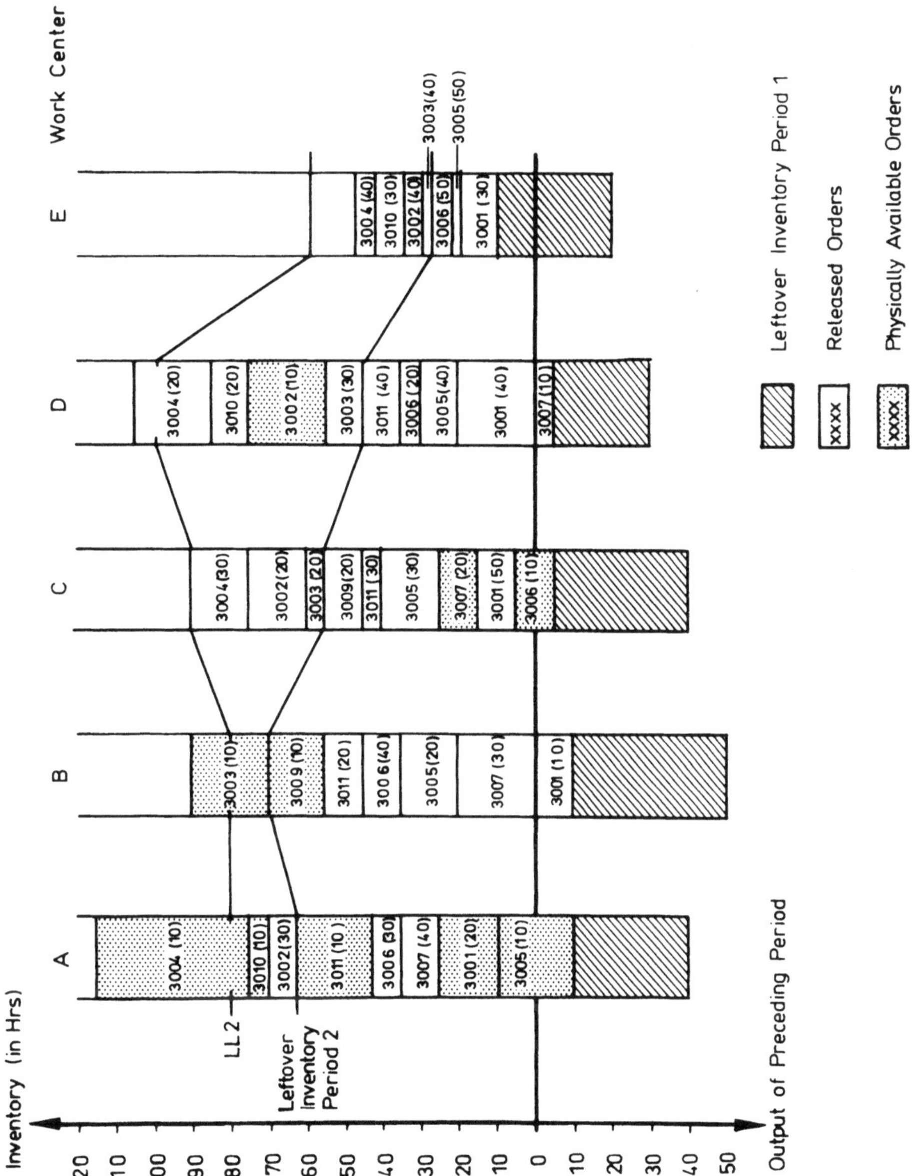

Fig. 6.10. Balance of loading accounts before release in period 2 after loading of converted orders

Table 6.5. List of deferred orders after period 1, and sequence of deferring work centers

Order Number	Attempts to release	Deferring Work Centers A	B	C	D	E
3003	1					1
3002	1	3		2	1	4
3010	1	1			2	3
3008	1	4	5	3	2	1
3012	1	3	1	2	5	4
3004	1	1		3	2	4

6.6 How to Choose the Load Limit and Loading Percentage Values

The load limits and the loading percentages can be either the same or different for all work centers. When developing the load-oriented order release method, Jendralski [6] and Bechte [2] assumed uniform values for the aggregate shop. Erdlenbruch [3] showed that, with very widely varying operation times, it can be useful to choose individual load limit values. It must, however, be ensured that these values are harmonized with each other in lead time scheduling, and do not render the process too complicated.

Let us again first look at a single work center to find the 'correct' load limit value and the 'correct' loading percentage, and to see how these values can be determined. In Sect. 6.7, the single work center lead time values are then linked together with the lead time scheduling for the entire orders.

The question concerning the 'correct' load limit of a work center can best be discussed using the throughput model shown in Fig. 6.3. For a given capacity, the load limit obviously only depends on the mean inventory, from which, using the 'funnel formula' (mean weighted lead time equals mean inventory divided by mean performance), a mean weighted lead time can be determined. This makes our question that of the 'correct' inventory at a work center. In the throughput diagram, the answer to this question seems to be very easy: the input and output trends must be such that the work center is very probably always working.

With the help of Fig. 6.11, we now try to find a method for determining a realistic scheduled lead time. Figure 6.11A shows fictitious throughput elements of a work center which are positioned in such a way that their operation elements succeed each other without intervals, which is always the case when a process is not disrupted.

The maximum lead time (TL_{max}) of this work center is then only determined by the maximum value of the sum of the operation time (TOP) and the interoperation time (TIO). Per definition, the interoperation time includes the transit and inspection time (TT). Given a process without disruptions and without sequence transpositions, only the maximum transit and inspection time (TT_{max}) must be considered, so that we can construct an ideal process consisting of throughput elements with only the components TOP and TT_{max}.

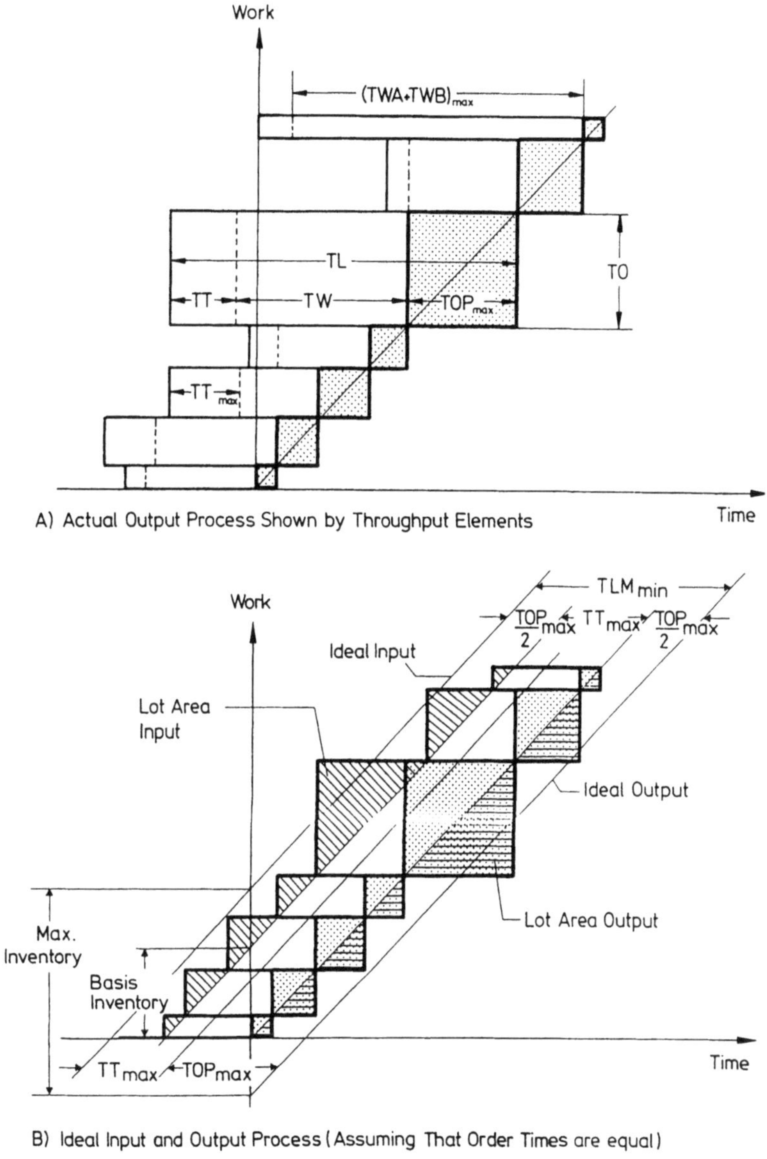

Fig. 6.11. Ideal order throughput at a work center in case of erratic order times

Figure 6.11B shows this idealized process, which has the following characteristics (see also Sect. 5.5.1):

- The maximum lead time of all throughput elements being considered is the sum of the maximum operation time of these elements and the maximum transit and inspection time.
- The input flow area is equal to the output flow area, since the amounts of work coming in and going out are always the same. The resulting flow area is therefore zero.

- The output control area is zero, since, given uniform capacity, the operation elements succeed each other on a straight line without gaps.
- The input control area is also zero, since the inputs arrive at the same intervals as those at which the outputs leave, which is why the input and output curves are congruent.
- Over longer stretches of time, the input and output lot areas are of the same value and correspond to the operation time area. The resulting lot area from input and output does not, however, correspond to this value when the process is looked at period by period, because of the offset in time between the orders in the input and output.
- With a given capacity, the basis inventory area results only from the maximum transit and inspection time.

The maximum inventory of a work center in a period, given a fixed capacity and a trouble-free flow without sequence transpositions, only depends on the maximum transit and inspection time, and on the maximum operation time in this period.

Because of the arbitrary and sometimes widely varying order times, which directly determine the variance of the operation times, it does not make sense to determine the minimum inventory and lead time values per period according to the maximum order time value. Instead, it is advisable to carry out a longer-range evaluation of the order times, for example, over several periods using the Pareto curve explained above. In Table 6.6, this was carried out for the sample work center from Chapt. 4, which processed the orders given in Columns 1 and 2. (The values are taken from Table 4.1, and are arranged in increasing order time sequence.)

Figure 6.12 shows the Pareto distribution of these values. Assuming that the maximum operation time value should be that which includes 95% of the orders, we find a value of about 14 hours of order time, which corresponds to an operation time of 14/8 = 1.8 workdays (given a capacity of eight hours per day at this work center). Assuming further a maximum permissible transit and inspection time of two days,

Table 6.6. Calculation of Pareto curve key data of order time at a work center

Order Number	Order Time TO
(-)	(Hrs)
1	2
1	.5
2	2.1
3	2.1
4	2.6
5	3.8
6	4.2
7	5.1
8	5.1
9	5.3
10	6.8
11	6.9
12	7.4
13	7.7
14	8.8
15	9.6
16	11.4
17	13.6
18	13.8
19	13.8
20	15.4
20	146.0

Group of same Order Time	Values per Group			Work per Group		
	Individual Value		Cumulative	Individual Value		Cumulative
(Hrs)	(-)	(%)	(%)	(Hrs)	(%)	(%)
3	4	5	6	7	8	9
1	1	5.0	5.0	0.5	0.3	0.3
2	0	0.0	5.0	0.0	0.0	0.3
3	3	15.0	20.0	6.8	4.7	5.0
4	1	5.0	25.0	3.8	2.6	7.6
5	1	5.0	30.0	4.2	2.9	10.5
6	3	15.0	45.0	15.5	10.6	21.1
7	2	10.0	55.0	13.7	9.4	30.5
8	2	10.0	65.0	15.1	10.3	40.8
9	1	5.0	70.0	8.8	6.0	46.8
10	1	5.0	75.0	9.6	6.6	53.4
11	0	0.0	75.0	0.0	0.0	53.4
12	1	5.0	80.0	11.4	7.8	61.2
13	0	0.0	80.0	0.0	0.0	61.2
14	3	15.0	95.0	41.2	28.2	89.5
15	0	0.0	95.0	0.0	0.0	89.5
16	1	5.0	100.0	15.4	10.5	100.0
Total >	20	100.0	100.0	146.0	100.0	100.0

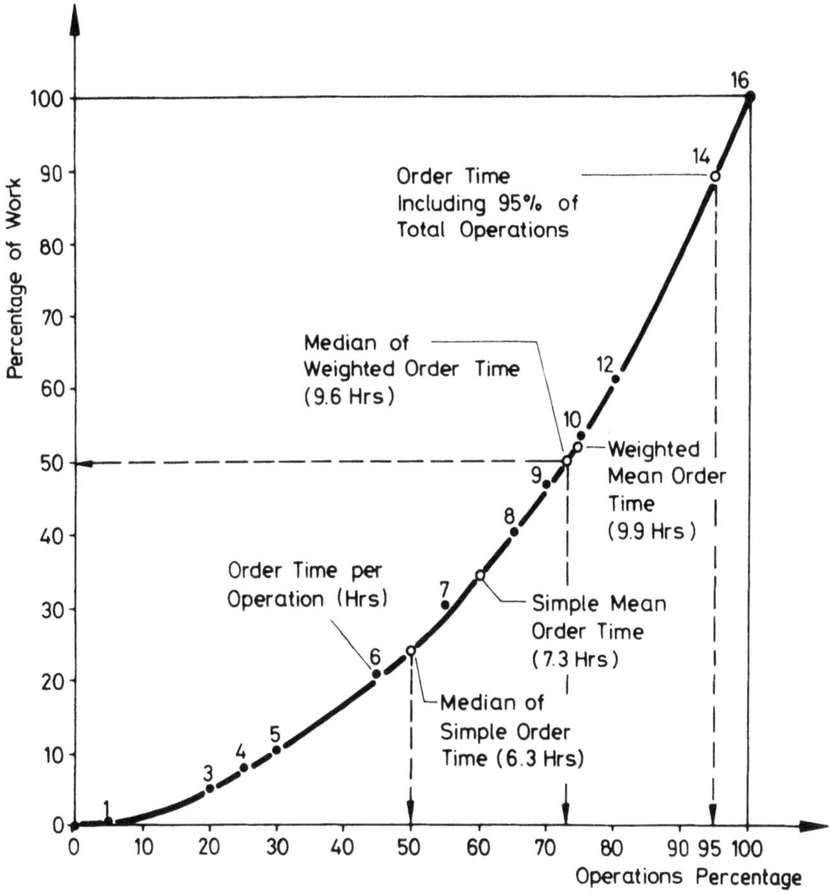

Fig. 6.12. Pareto curve of order time for processed lots at a work center

which could be the typical value for this individual shop, the ideal mean lead time of this work center is 3.8, or about four workdays.

Because there will always be offsets in the manufacturing flow due to sequence transpositions for rush orders, breaks in production, delayed deliveries, etc., an added minimum buffer time of one day seems adequate, so that the realistic scheduled value of the weighted mean lead time (*TLM*) for this work center is five days.

In Fig. 6.13, an ideal process was constructed on the basis of the output curve measured in Chapt. 4, using a mean lead time value of five days. We can see the considerable reserves compared with the real input curve and thus with the actual lead time value of 11.1 workdays.

From this we can conclude that using the mean value and the variances of the transit and operation times is sufficient for determining the scheduled load limit. From these two values, a weighted mean scheduled lead time (*TLM*) can be calculated.

The load limit can then be determined as:

Given: mean scheduled lead time (*TLM*), mean scheduled performance (PE_m), and length of schedule period (*P*):

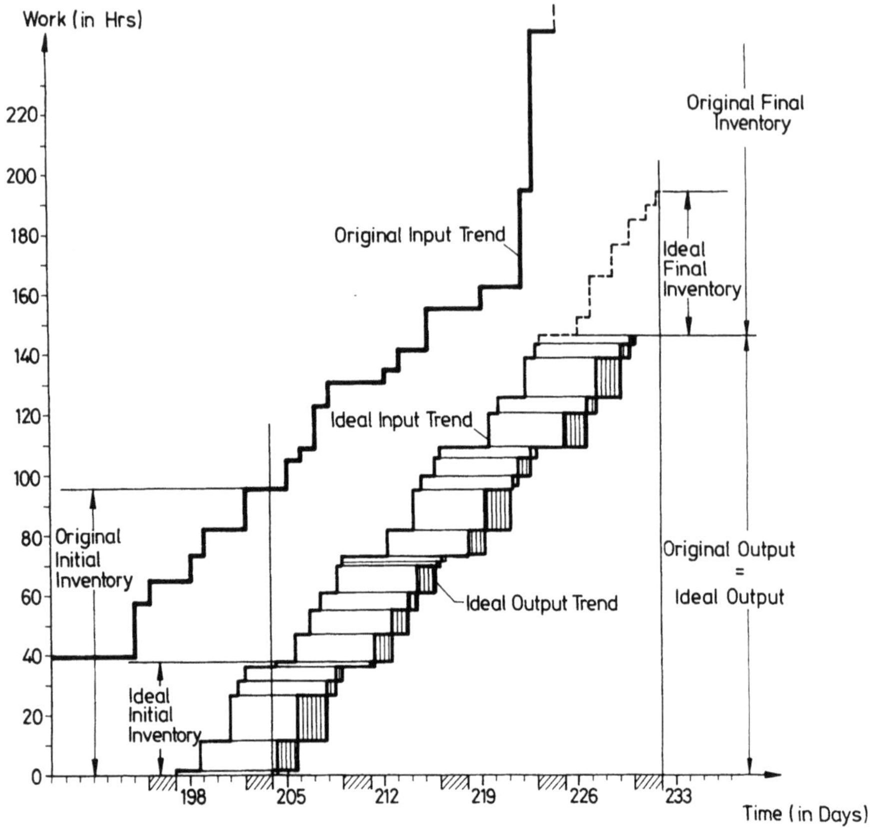

Fig. 6.13. Construction of the scheduled throughput diagram for a work center (practical example) (After Bechte [2])

Find: load limit (LL)

Eqn. (6.1): From $TLM = \frac{I_m}{PE_m}$

it follows that $I_m = TLM \times PE_m$

Eqn. (6.4): From $LL = OUT + I_m$

and $OUT = PE_m \times P$

it follows that

$$LL = PE_m \times P + I_m$$
$$= PE_m \times P + TLM \times PE_m$$

Result: $LL = PE_m(TLM + P)$ (6.13)

The loading percentage can be calculated from Eqn. (6.8):

$$LPG = \left(1 + \frac{TLM}{P}\right) \times 100\%$$

In our example, given a PE_m of 8 hrs/SCD, a TLM of 5 SCD, and a P of 5 SCD, the load limit would be:

$$LL = 8\ \text{hrs/SCD} \times (5\ \text{SCD} + 5\ \text{SCD}) = 80\ \text{hrs}$$

The loading percentage is then:

$$LPG = \left(1 + \frac{5\ \text{SCD}}{5\ \text{SCD}}\right) \times 100\% = 200\%$$

For practical application, this means that the interoperation time and operation time distributions must be monitored using the monitoring system described in Sect. 5.4. This helps to keep track of long-term changes – e.g. due to improvements in the manufacturing processes or in sequencing and transport organization – and allows for adjustments to the loading percentages.

Figure 6.14 gives a list of some methods for determining the load limit. An estimative method which is not based on the interoperation and operation time distributions, but only on the mean value of the weighted operation time, is given as the first technique in the figure, and was proposed by Erdlenbruch [3, 5]. This method uses three components of the throughput element, namely the transit, buffer, and operation time percentages of the total lead time. The five components described in Fig. 3.5 have thus been merged into three components. These three components are now understood as weighted mean values, which change only over a longer period of time [3].

Figure 6.15A shows the original mean operation lead time value of an aggregate mechanical manufacturing shop (13.5 workdays). The weighted mean lead time and

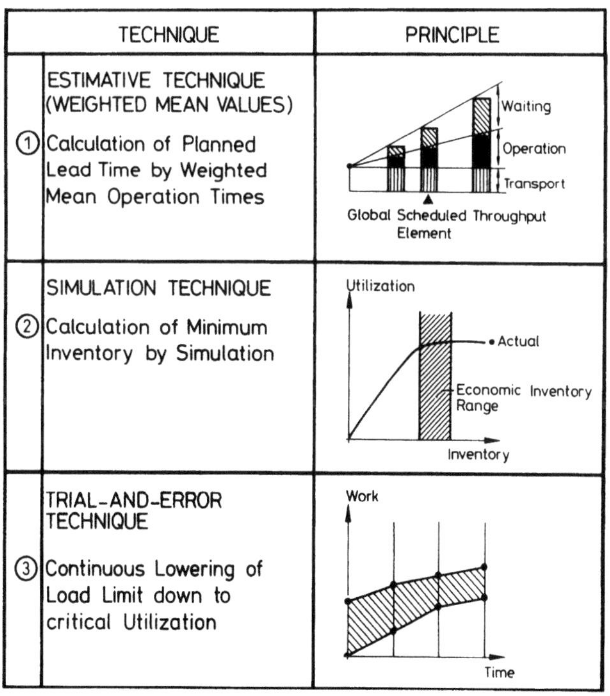

Fig. 6.14. Techniques of determining the load limitation within load-oriented order release

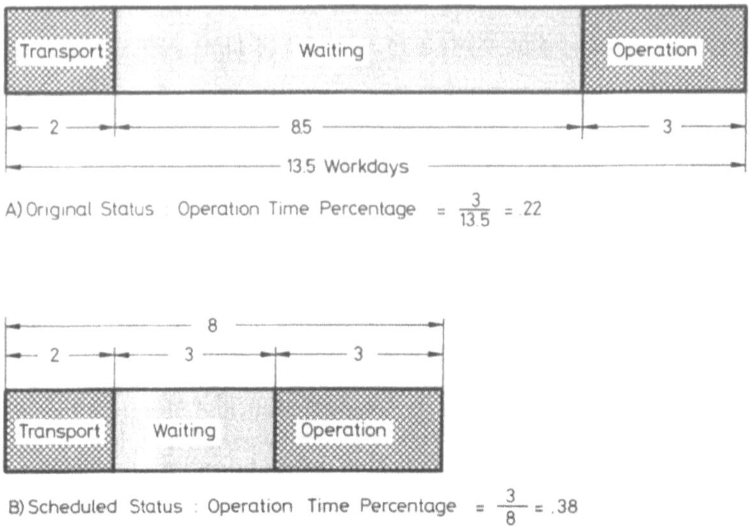

Fig. 6.15. Mean values of throughput element components (data taken from a mechanical manufacture) (After Erdlenbruch [3])

the weighted mean operation time are recorded values; the transit time percentage was estimated. To determine the scheduled status, the following must be considered. The transit time is independent of changes in inventory. Further, simulation and experience have shown that a buffer time of about the same value as the weighted mean operation time is normally sufficient to compensate for the above-discussed effects of extreme order times, rush orders, delays in transport, machine down times, etc.

Thus, a buffer factor of 2 (i.e. buffer time equal to weighted mean operation time) results in a mean scheduled throughput element of $3 \times 2 + 2 = 8$ workdays, consisting of the mean transit time, the weighted mean operation time, and the mean waiting time (Fig. 6.15B). This value is now considered to be a global scheduled throughput element (see also Fig. 6.16).

At some work centers there are wide deviations from the mean weighted operation time value. Therefore, the individual scheduled values for the single work centers are fixed by multiplying the individual weighted mean operation time by the buffer factor explained above (in this case 2). The individual buffer time value determined in this way, when added to the individual mean operation time value and the transit time value, results in the individual scheduled lead time value for every work center. Figure 6.16 shows the principle for this calculation in the form of a graph. We can see how the scheduled lead time (TLM) changes with the work center-dependent weighted mean operation time (TOP).

The effects of individual load limits on lead time, as compared with uniform load limits, can be vizualized with a simple example [5]. Let us assume that an order has three operations with work contents of 52, 39, and 22 hours, respectively. Table 6.7 gives the loading percentages. The standard loading percentage is 250; the individual ones are 480, 420, and 240. Using Eqns. (6.9) and (6.10), the conversion factors are calculated: 0.4 with the standard loading percentage, and 0.21, 0.23, 0.42 with the

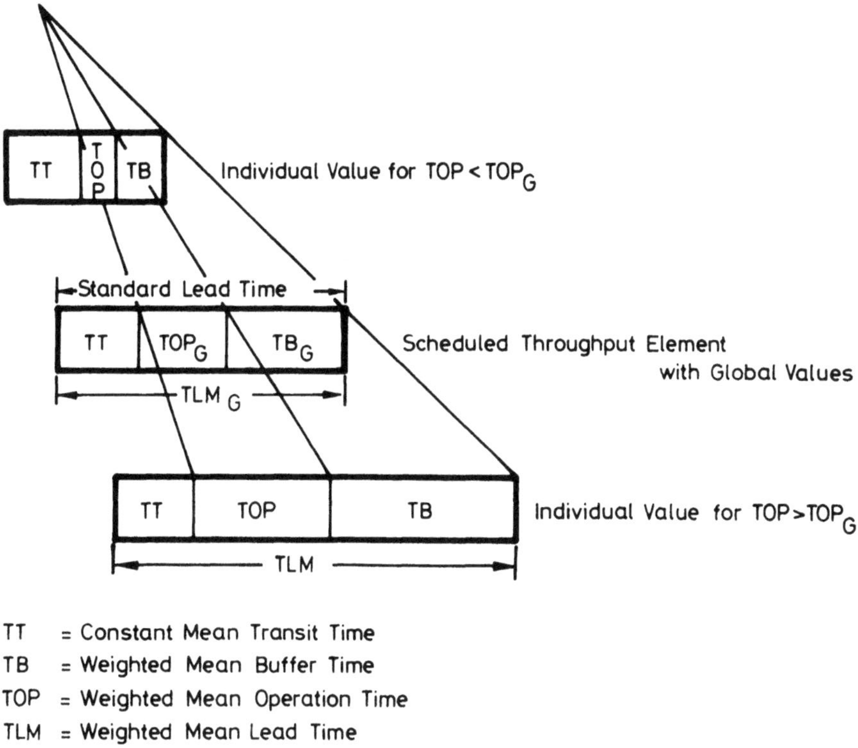

Fig. 6.16. Derivation of work-center-specific scheduled lead time (After Erdlenbruch [3])

Table 6.7. Lead time calculation for an order with extremely erratic operation order times – standard versus individual loading percentage

Key Data		A: Standard Load Limit			B: Individual Load Limit		
		Op 1	Op 2	Op 3	Op 1	Op 2	Op 3
1		2	3	4	5	6	7
1 Loading Percentage	(%)	250	250	250	480	420	240
2 100/LPG	(-)	0.4	0.4	0.4	0.21	0.24	0.42
3 Conversion Factor	(-)	1.0	0.4	0.16	1.0	0.2	0.1
4 Order Time	(Hrs)	52.0	39.0	22.0	52.0	39.0	22.0
5 Order Time Converted	(Hrs)	52.0	15.6	3.5	52.0	8.2	1.1
6 Planned Mean Lead Time	(SCD)	7.5	7.5	7.5	19.0	16.0	7.0
7 Operation Time	(SCD)	6.5	4.9	2.8	6.5	4.9	2.8
8 Transit Time	(SCD)	2.0	2.0	1.0	2.0	2.0	1.0
9 Buffer Time	(SCD)	-1.0	0.6	3.7	10.5	9.1	3.2

Base: Period (P) = 5 SCD, Daily Capacity (CDAY) = 8 Hrs/SCD

individual loading percentages. From this, the converted order times (TOD) can be determined (Line 5 in Table 6.7).

The scheduled lead times (TLM) were calculated by transforming Eqn. (6.8) as follows:

$$LPG = \left(1 + \frac{TLM}{P}\right) 100\%$$

It follows that:

$$TLM = \left(\frac{LPG}{100} - 1\right) \times P \tag{6.14}$$

where:

LPG = loading percentage
P = length of period.

As expected, the scheduled lead times are the same in case A (standard load limit), and in case B (individual load limits) they vary widely.

It is interesting to note how TLM is made up. As was defined in Fig. 6.13, it consists of the operation time (TOP) (calculated in Table 6.7, Line 7), the transit time (TT) (assumed to be 2 workdays for OP1 and OP2 and 1 workday for OP3), and the buffer time (TB). Table 6.7, Line 9, shows the resulting buffer time values for both cases.

Figure 6.17 compares the lead times and their components for Cases A and B. In Case A, Operation 1 demonstrates the main problem of the Standard Loading Method, namely that it can lead to a negative buffer time which, of course, cannot

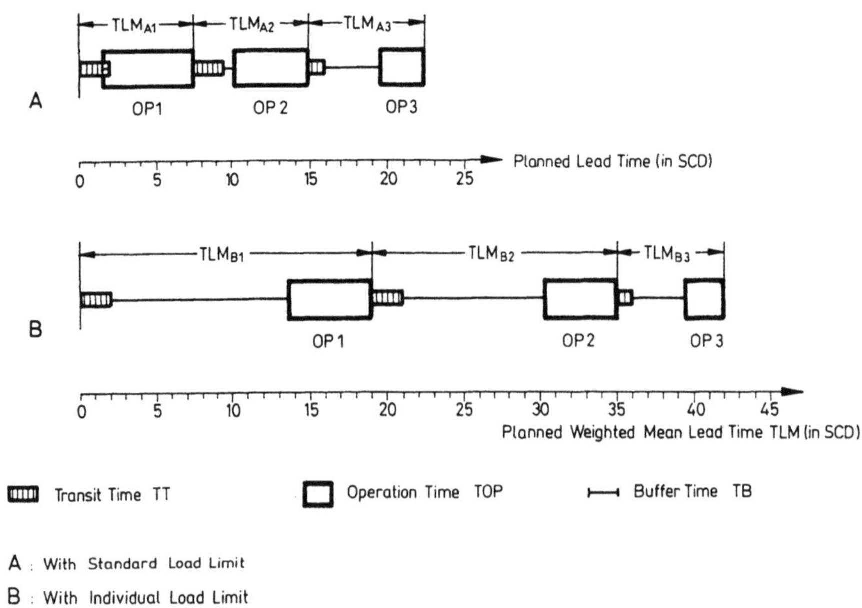

Fig. 6.17. Components of the lead time of an order (After Erdlenbruch [3])

be achieved. Operation 2 has a relatively short buffer time, and Operation 3 a buffer time of about the same length as the operation time.

With the Standard Loading Method, the order throughput is poorly scheduled altogether, and will certainly lead to problems when put into practice. In Case B, on the other hand, the work centers processing large work contents are given correspondingly large buffer times, while those of Work Center 3 are shorter but still sufficient.

Thus, the use of individual load limits is always advisable when the evaluation of the operation times shows that there are considerable differences between the work centers to be controlled.

A second method for determining load limits is simulation techniques, as shown in Fig. 6.14. These can, for example, be based on event-oriented, discrete models, such as those used by Jendralski [6], Bechte [2], Buchmann [7], and Erdlenbruch [3] for their investigations into load-oriented order release. This is, however, still very costly and – for this purpose – not an economical method.

In contrast to this, a queueing model as developed by Lorenz [9], which is based on the throughput diagram and its inventory components, seems to be very promising. Its forecast accuracy is satisfactory at a relatively low cost. More detail on the model itself will be given in Sect. 9.6. Figure 6.18 shows the characteristic curves of a single work place, which have been calculated using this queueing model [9]. We can see the bend in the performance curve suggesting an optimum inventory of about 280 hours, which is not far below the originally observed mean inventory

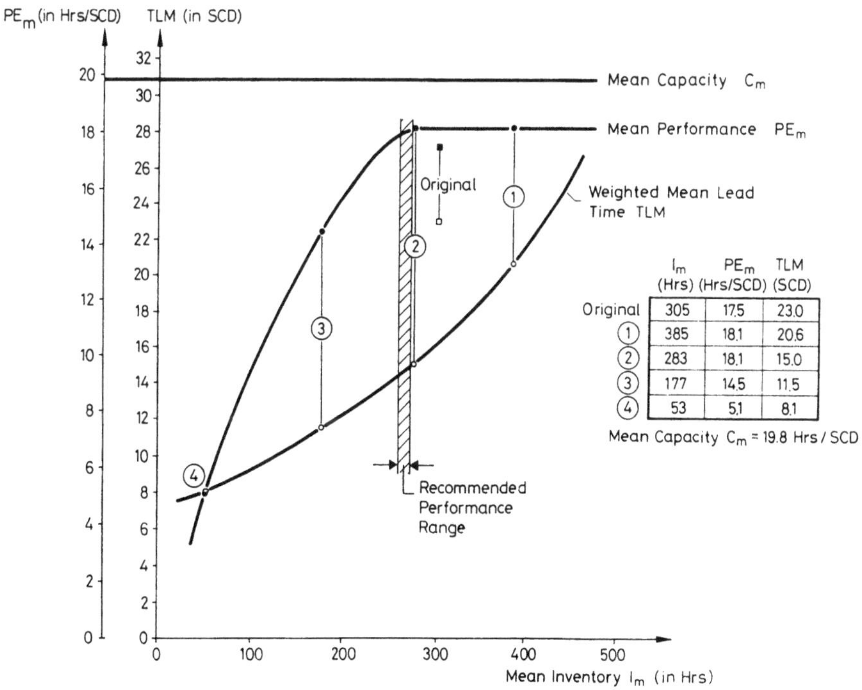

	I_m (Hrs)	PE_m (Hrs/SCD)	TLM (SCD)
Original	305	17.5	23.0
(1)	385	18.1	20.6
(2)	283	18.1	15.0
(3)	177	14.5	11.5
(4)	53	5.1	8.1

Mean Capacity C_m = 19.8 Hrs / SCD

Fig. 6.18. Simulated characteristic production curves for an NC work center (After Lorenz [9])

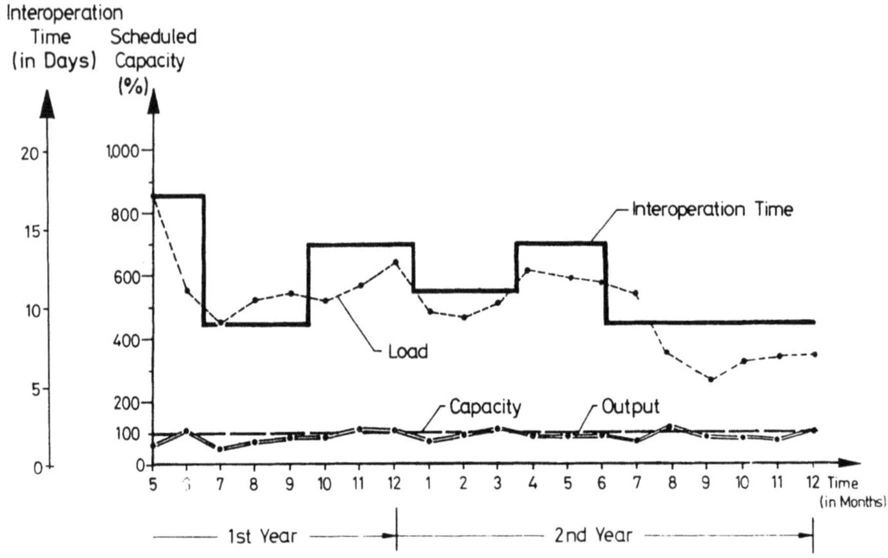

Fig. 6.19. Interoperation time, load, and output trends at an NC lathe work center (After Ritter [10])

of 305 hours. Practical experience, however, has not yet been gathered using this method.

Finally, a method especially suitable for practical application is that which Fig. 6.14 calls the *trial-and-error* technique. With the help of the monitoring system, one slowly approaches the bend in the performance curve by gradually lowering the load limit. In the scheduled throughput diagram, one can estimate how long it will take for a change of the load limit to become noticeable, namely at least approximately the mean previous flow time. A constant and irregular change of the load limit is therefore not recommended.

Figure 6.19 shows authentic measurements from an NC lathe center. One can see how the interoperation time and the utilization changed with the loading percentage [10]. Obviously, there was a too drastic lowering of the load after Month 6 of the first year, which caused an underload in Month 7, but which nearly halved the interoperation time. When the load was increased again, the utilization became better, but the interoperation time also rose considerably. After two more attempts, the work center's interoperation time balanced out at a mean value of 9 days.

Of the three methods described for determining the load limit, the trial-and-error method best suits practical requirements, and is the only one employed by users of the load limit technique so far. The estimative methods are useful when lead time limits need to be calculated, and measures to determine their approximate values are required. Simulations, however, will not be possible until the advent of a new generation of PPC systems. The so called 'simulation features' offered in PPC systems so far are in most cases not genuine simulations, in the sense that feedback information is produced. One should rather speak of test loadings. Whichever method is chosen, it is important to constantly compare scheduled and actual values, and exercise caution when changing load limits.

The mean scheduled values for work center lead times – however they may have been determined – must now be used in lead time scheduling, to link order control with manufacturing control.

6.7 Interlinking Order Control with Manufacturing Control

If, during lead time scheduling, the allowance values for every work center are used for the corresponding operations of an order, a correspondence between the operation-related and the work-center-related lead times as mean values is guaranteed.

As was explained in Sect. 6.3, the lead time scheduling for the shop orders is carried out from the due date backwards, operation per operation, as specified in the operation sheet. All operations are scheduled with the scheduled allowances of the corresponding work centers and added up to calculate the order lead time, together with three other time values (Fig. 6.20) [3]. These are the staging time between the beginning of the first operation and the release of the order, needed for staging the work papers and the material, the anticipation time between the release and the material planning, needed for checking the requirements and resources of materials, NC programs, tooling, etc., and a safety time between the planned end date and the required end date, needed, for example, for delivery to the required destination (store, assembly, shipping department, etc.). The total of all these times is the replacement time, which is a value maintained in the materials planning files for every manufactured part.

Figure 6.20 also indicates how, as discussed above, every operation time can be calculated from its three components. Instead of the overall lead time, one takes

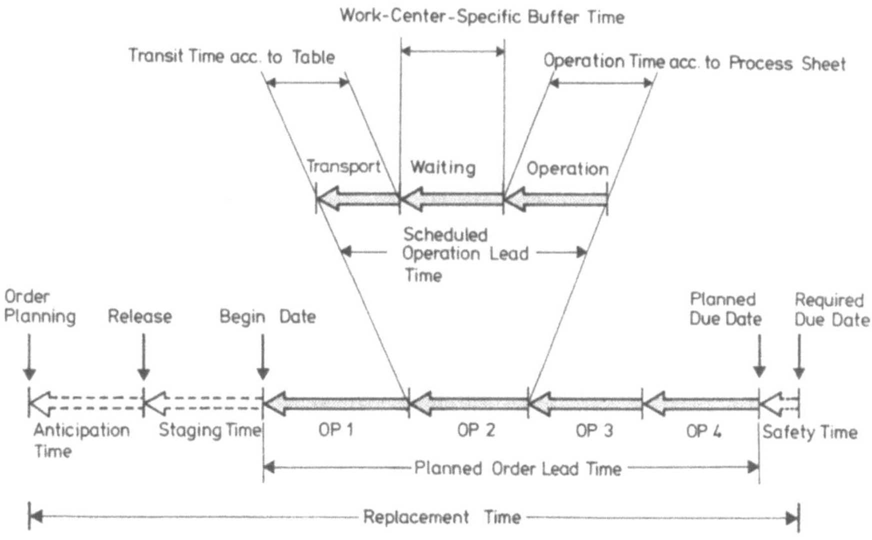

Fig. 6.20. Determination of scheduled order lead time (After Erdlenbruch [3])

only the work-center specific buffer time from the work center table, adds the operation time (calculated from the order time in the routing sheet) and the transit time, which, in many shops, is kept in a transit time table.

The release dates and the planned due dates enter into the manufacturing control process. The load limit guarantees that whenever the order enters the individual work centers it meets an inventory situation which corresponds to the planned lead time. Now the flow time control loop is closed.

6.8 Effects of the Load Limit and Time Limit Parameters in Simulation and in Practice

6.8.1 Simulation of Manufacturing Processes as an Aid in the Testing of Control Algorithms

In contrast to technical control processes, an extensive testing of manufacturing control methods and of the effects of their parameters is not possible in authentic shops, since, first, it is not possible to reproduce manufacturing situations several times and, second, the testing of extreme limit values cannot be permitted for economic reasons. This is where simulation can be of assistance.

Studies of known simulation methods carried out by the Institut für Fabrikanlagen have shown that they did not create sufficiently realistic situations. Thus, the institute developed several simulation packages [6, 2, 7], of which the system PROSIM, designed and programmed by Erdlenbruch, represents the state of the art at present [3, 11]. The system has been further developed and implemented on a 32-bit PC.

The structure of this system, as used for examining the load-oriented manufacturing control method, is shown in Fig. 6.21 [3]. The main program calls up four subprogram modules.

The *Simulation Begin* module opens a central information file containing general data on the study to be carried out, control parameters for the planned experiments, descriptions of files and records, a personnel attendance report, descriptions of the work centers, the order records, the evaluated key data of the work centers, and the data for generating throughput diagrams.

The *Data Provision* module runs the file of manufacturing orders. These data on orders processed in an authentic shop were gathered using feedback information and the corresponding work papers.

The *Event Control* module controls the single phases and steps of the manufacturing flow and manages the corresponding files. This module periodically starts the *Scheduling* module which, in turn, simulates the steps of the method examined, in this case the steps of load-oriented manufacturing control. It could also simulate another control method used in practical application.

After scheduling is finished, the *Event Control* module starts another module, called *Shop Simulation*. This module processes the order quantity provided by the Scheduling module, using the simulated manufacturing structure, and produces a feedback record for each single event. These records have the same structure as those registered in the authentic plant.

The *Simulation End* module closes the files after all orders have been processed,

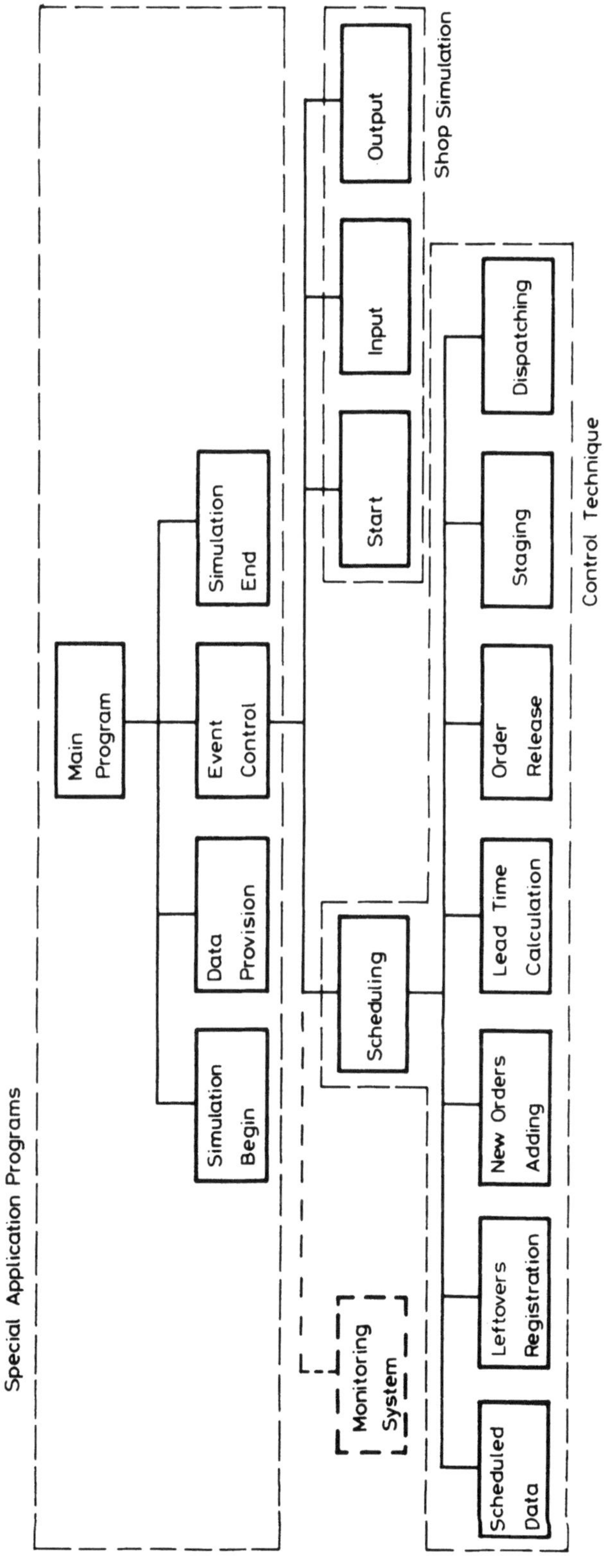

Fig. 6.21. Functional tree of a simulation model

stores the data needed for evaluation in the central information file, and finishes the simulation.

The simulation is event-oriented; the simulation clock is advanced for each subsequent date. Therefore, there is no connection between the simulated time and the simulation computation time. The model does not account for disruptions, since these are short in proportion to the lead times. The loss of capacity due to disruptions, however, is given as a percentage in the capacity data of the work center files.

For the interpretation of the simulation results, it is important to understand how the shop processes are simulated. This is therefore explained using Fig. 6.22.

In this figure we can see the single steps an order goes through during the processing of its operations. After the order has been released (here called 'Start'), the

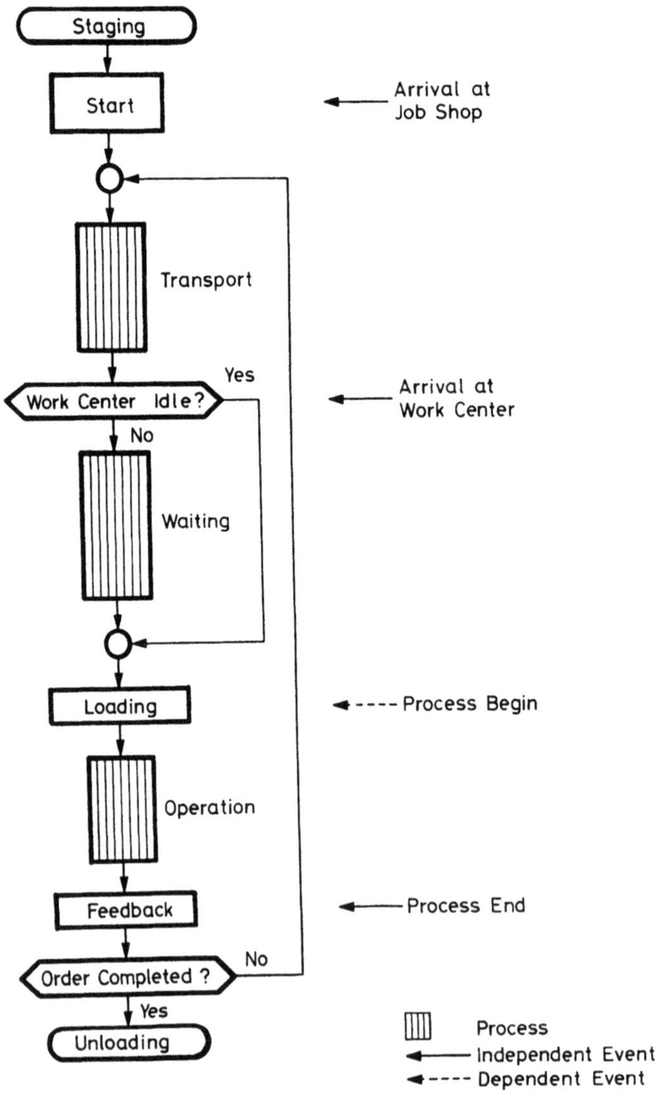

Fig. 6.22. Process simulation within the simulation model (After Erdlenbruch [12])

first transport activity to the first work center is carried out as specified in the operation sheet. If the respective work center is idle, the order is loaded into it; if not, the order enters the queue according to its priority. The sole priority criterion is the due date scheduled for the operation at this work center. After leaving each work center, the order is asked whether this was its last operation. If the answer is 'yes', the order is unloaded, otherwise it is transported to the downstream work center. How long an order waits in the queue depends solely on the total quantity of orders waiting for processing and on its relative priority. Each event in each operation and order is entered in a data record file which remains in the system.

A common test for the quality of the model is to simulate the manufacturing flow using the data recorded in the investigated shop, without using any particular order control. In this process, the capacity is determined according to the actual performance to obtain a reference status. This means that the real capacities may be higher. The actual dispatching procedure can not be simulated completely, since many influential factors which are impossible to reconstruct, such as rush orders or personnel absenteeism, may have overlayed the dispatching rule, if any. For model testing purposes, one should use the so-called slack-time rule, which arranges the orders at a work center according to their delay. This helps to keep the due dates.

The data collected during the model test are now compared with the data actually recorded. Figure 6.23 compares the actual and simulated output and inventory trends of the factory described in Chapt. 3, Table 3.5, and of a typical work center [3]. The total performance data correspond very well, and the inventory deviations are well below 10%, too, a fact which speaks for the accuracy of the model.

It is especially important to check on the distribution of the order and operation lead times. As an example, Fig. 6.24 compares the actual values with the data produced in the model test [3]. The model does not represent rush orders of extremely high priority and orders with extremely long lead times. Therefore, the middle groups are stronger in the model than in reality, which results in a lower coefficient of variation. On the whole, one can say that this model is well adapted to reality.

The original model allows the simulation of shops of up to 200 work centers, and was installed on a large-scale computer. The number of orders and the length of the manufacturing period which the model can simulate are virtually only limited by computer power.

To make the simulation process clearer, it was programmed on a PC in such a way that one can follow the individual simulation phases [12]. After the program has been started (Fig. 6.25), it first displays the three parameters of order release, namely the period length (here 200 hours), the time limit (here with an anticipation horizon of three period lengths), and the load limit (here 1.5 = a loading percentage of 150). Following the menu, the user can now either start the release planning (Option 0), change the three mentioned parameters (Option 1), change the capacities (Option 2), call up the list of orders (Option 3), or end the simulation (Option 4).

For the following demonstration we have used data from eight work centers from an authentic shop. The program simulated 40 periods with 80 hours each (= 3200 working hours), so that the system could reach a certain rhythm. At the end of each period, all accounts were corrected, the data were saved, and the release planning was started automatically. Figure 6.26 shows a reduced order list after the lead time scheduling has been carried out.

The table has one line per work center (WC_1 to WC_8) and one 'Total' line for the total system. The order stock in each line is divided into three groups. The first group (Columns 2 and 3) gives the direct inventory of each work center for the

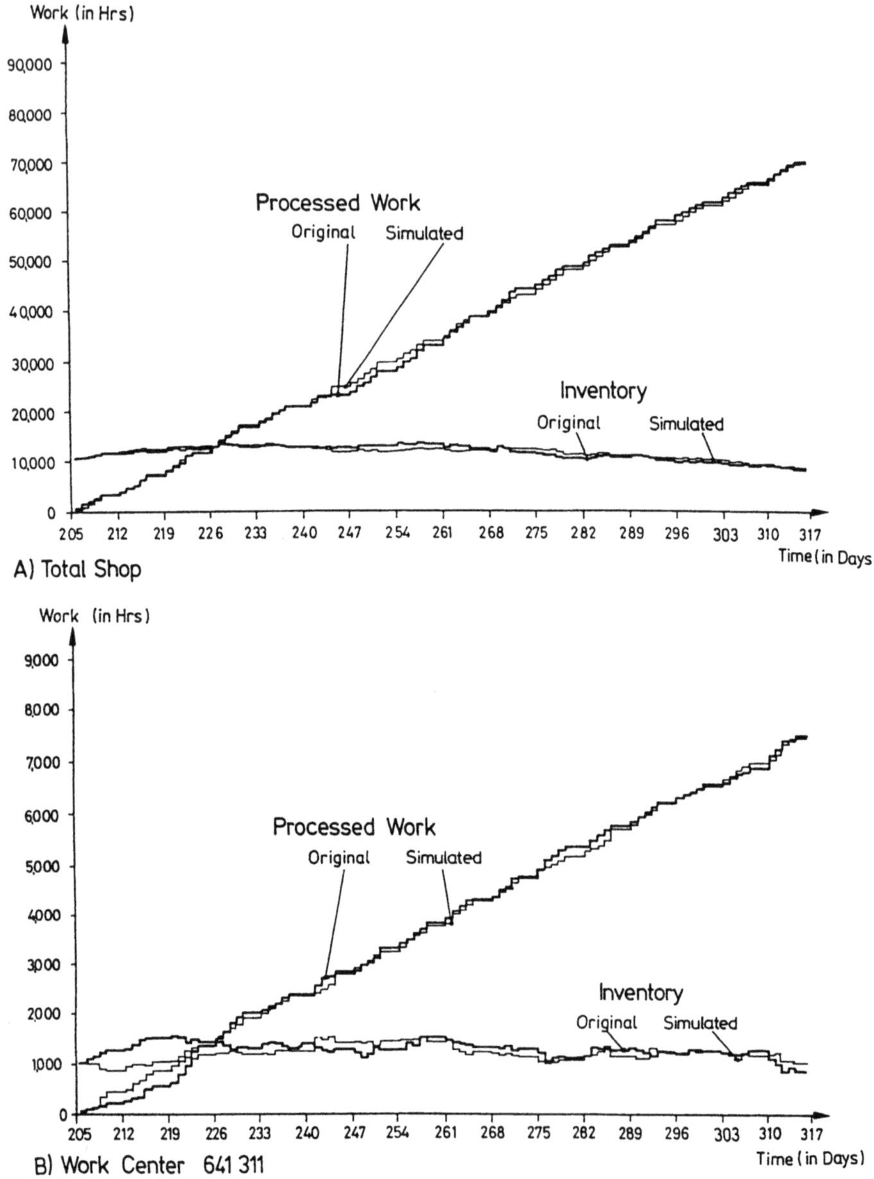

Fig. 6.23. Actual versus simulated output and inventory trends for a work center and for the total shop (After Erdlenbruch [12])

following period, which ends with Hour 1325. With a period length of 200 hours, this means that the period starts at Hour 1125. (From this we can see that the model has worked for five periods already, starting not at zero but at Hour 125.)

The work load is described by two values. The first, *OP* (Column 2), tells us how many operations (lots) are waiting before the work center concerned, and the second, *TIME* (Column 3), states the order times contained in the lots. In the 'Total' line, the *OP* value does not mean the total of the lots of all work centers (which would be 10 + 10 + 11 + 11 + 12 + 7 + 13 + 5 = 79 in the next period), but the

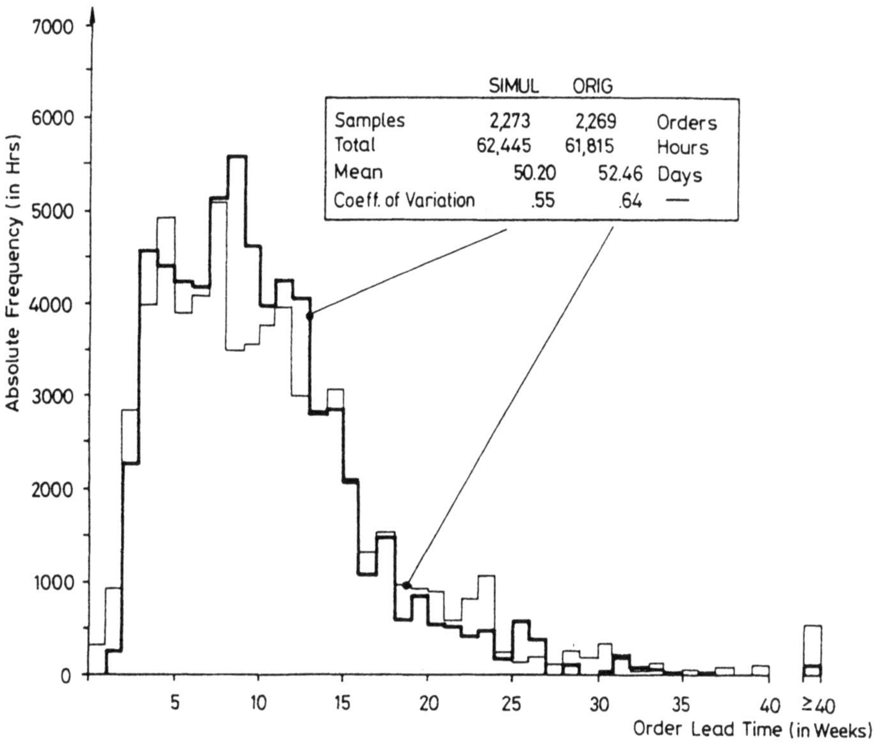

A) Orders

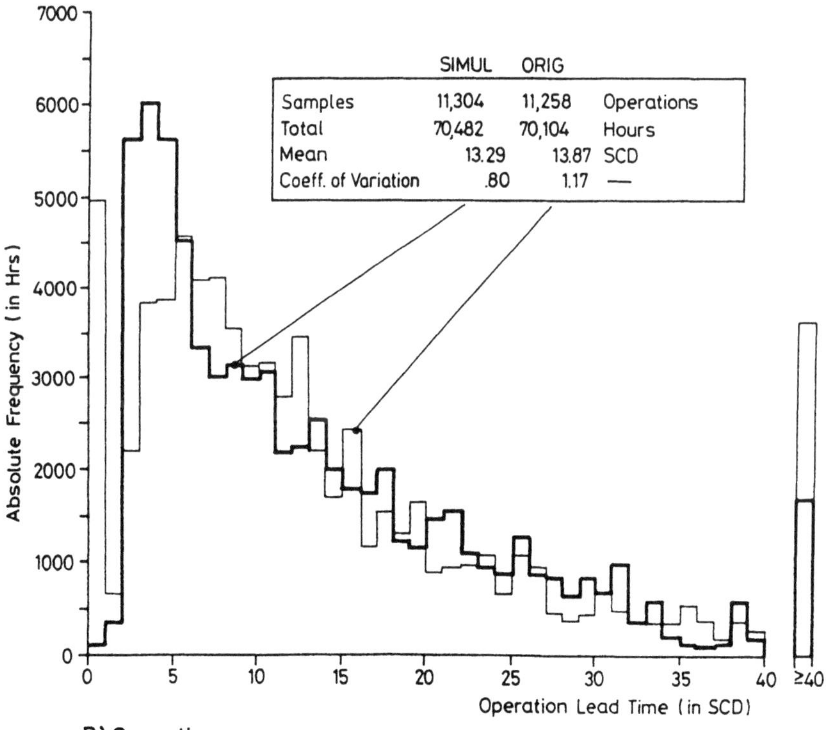

B) Operations

Fig. 6.24. Actual versus simulated lead time distribution for the total shop (After Erdlenbruch [12])

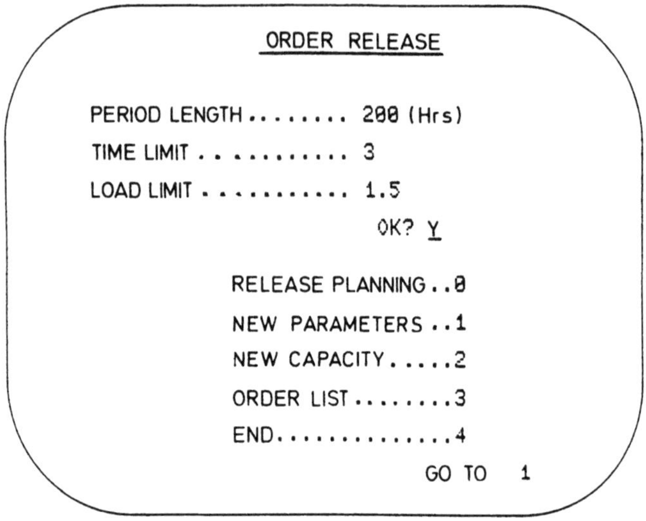

Fig. 6.25. Order release parameters and possible corrective actions on the simulation process on a personal computer (After Erdlenbruch [12])

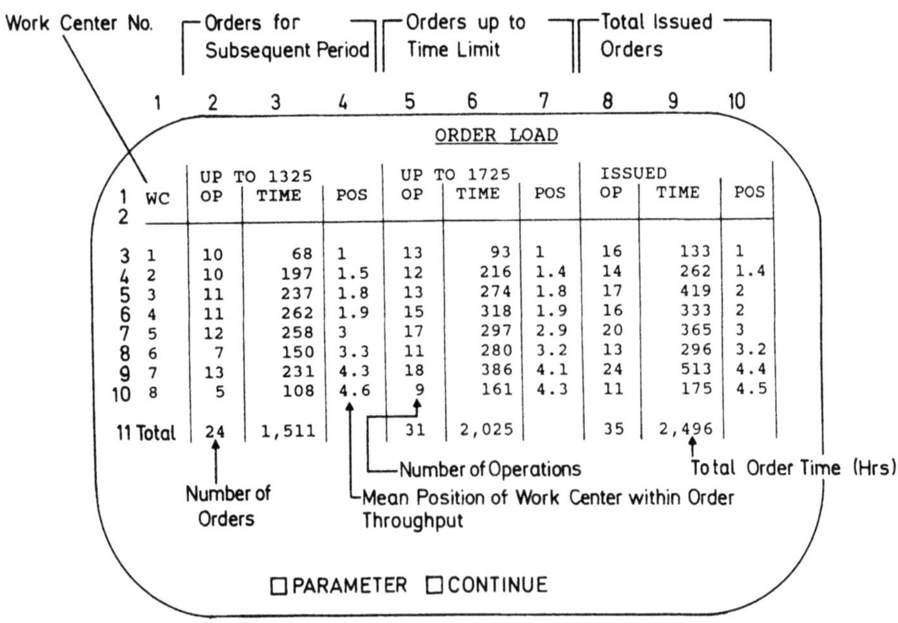

		UP TO 1325			UP TO 1725			ISSUED		
1	WC	OP	TIME	POS	OP	TIME	POS	OP	TIME	POS
2										
3	1	10	68	1	13	93	1	16	133	1
4	2	10	197	1.5	12	216	1.4	14	262	1.4
5	3	11	237	1.8	13	274	1.8	17	419	2
6	4	11	262	1.9	15	318	1.9	16	333	2
7	5	12	258	3	17	297	2.9	20	365	3
8	6	7	150	3.3	11	280	3.2	13	296	3.2
9	7	13	231	4.3	18	386	4.1	24	513	4.4
10	8	5	108	4.6	9	161	4.3	11	175	4.5
11	Total	24	1,511		31	2,025		35	2,496	

Fig. 6.26. Pre-release order load, classed into groups of queueing orders (up to period end 1325), orders within the time limitations (up to period end 1725), and total issued orders

number of shop orders, here 24. Column 4 (*POS*) gives the mean position of the work center concerned in the flow of the 24 orders. The work centers in Lines 2 to 10 are arranged according to this value.

The second group of the order stock (Columns 5 to 7) contains all orders which lie below the time limit, which also includes the orders from the first group. The time limit ends at Hour 1725. (The period begins at Hour 1125 and the anticipation horizon is three times the period length, so the time limit is at Hour $1125 + 3 \times 200 = 1725$.)

Finally, the third order group contains all orders issued up to the present date (in this case Hour 1125).

Now the order release component loads the orders – arranged according to start

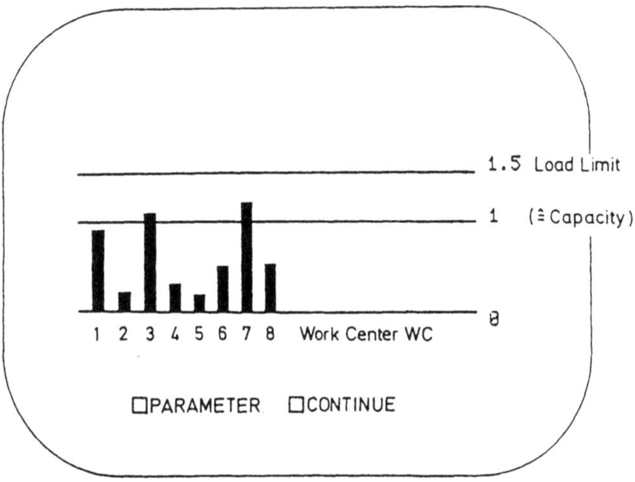

A) Leftover Load before Release

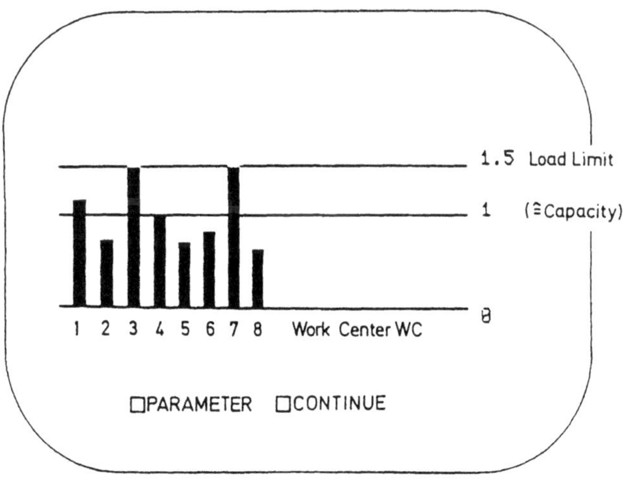

B) Load after Release

Fig. 6.27. Load at the eight work centers of a job shop pre-release versus post-release dates – into the loading accounts, operation by operation. Figure 6.27A shows the balances of the eight loading accounts before the release.

The two horizontal lines symbolize the capacity (1) and the load limit (1.5). Work Centers 1, 3, and 7 have work for about one period; the other work centers have only very little work on-hand. After the release, the loading situation is as shown in Fig. 6.27B.

Work Centers 3 and 7 are loaded up to their load limits, whereas the other work centers could not be supplied with enough orders that were below the time limit. To achieve a better utilization, the user could interfere with the program at this point by changing either the time limit or the load limit. In principle, a change in capacities is also possible (cf. Fig. 6.25, Option 2), but this is not demonstrated here since they are not parameters of load-oriented order release.

If the result of the order release is accepted, the actual simulation of the order throughput begins. Funnel symbols, one for each work center, appear on the screen (Fig. 6.28).

In our example there are eight funnels. The column to the right of each funnel symbolizes the direct inventory (in hours) which is currently in the funnel. The number of orders corresponding to this inventory is represented by circles lying in the funnels. The black circles are the waiting orders, while the white ones are the orders in process. The column to the left of each funnel represents the work (in standard hours) which is contained in the released order stock for this work center, but which can possibly still be lying at the upstream work center (indirect inventory plus direct inventory). Work Center 5, for example, is processing only one single order (the right column is low), but there is still a great deal of work coming (the left column is high).

Now the simulation program runs 40 periods with 80 hours each. The place and time of every event are recorded. Besides the evaluation of the actual process, five simulation runs with different load limits were carried out. The files produced after each simulation were evaluated to calculate the characteristic values. As an example

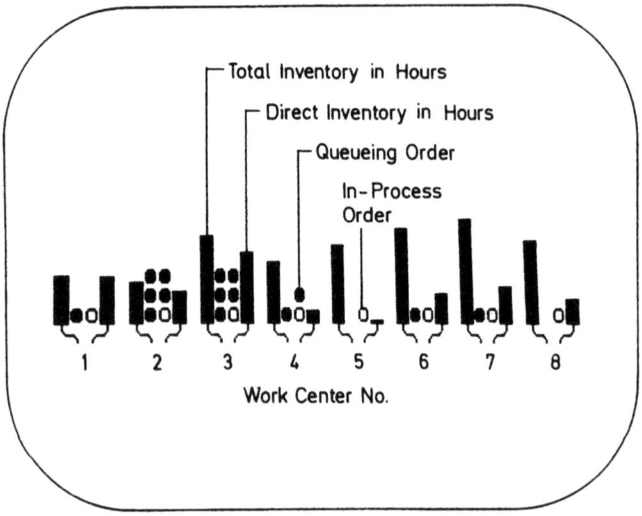

Fig. 6.28. Orders, total and direct inventory, shown in the funnel model of a job shop

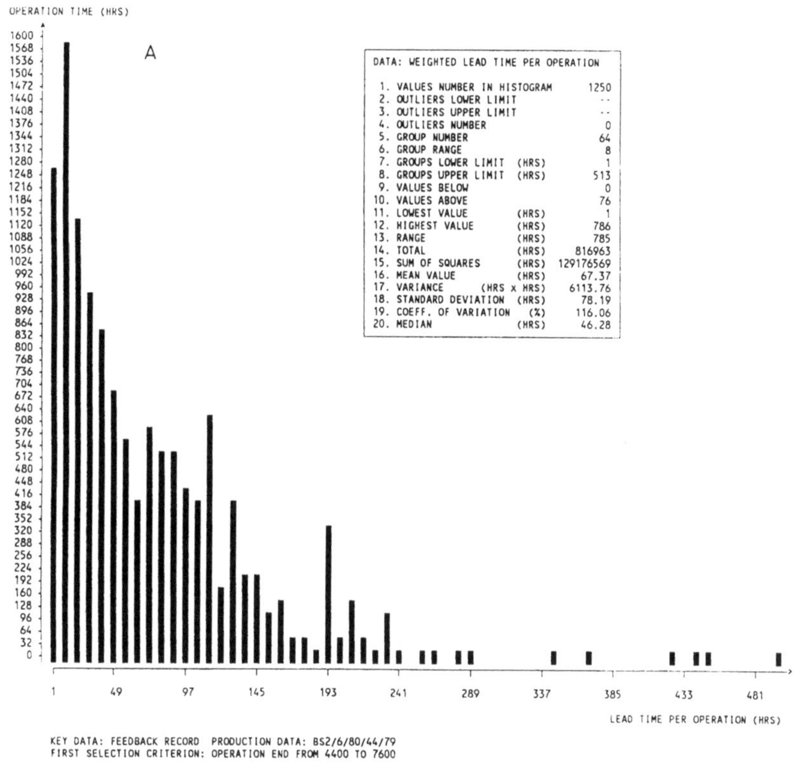

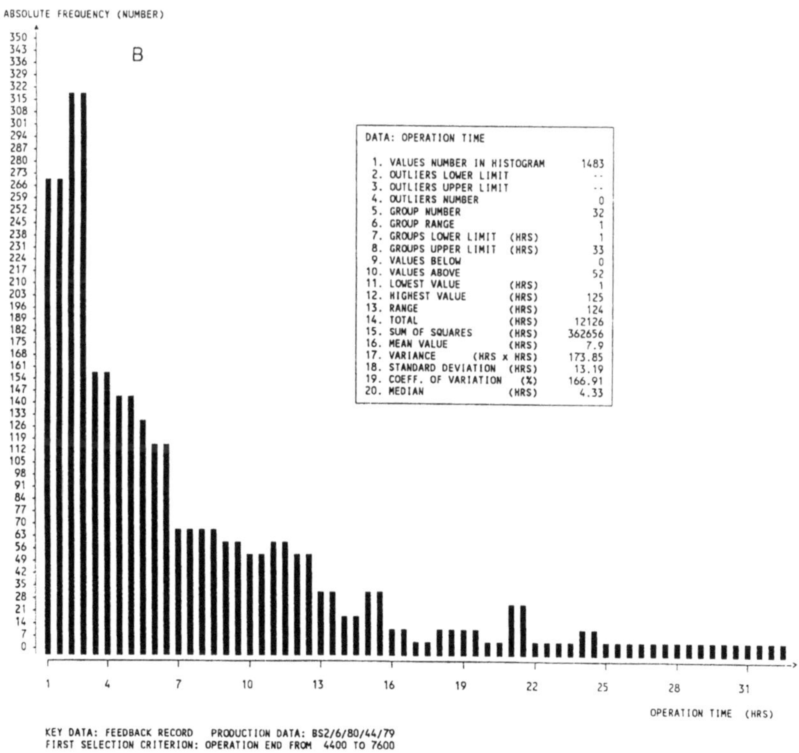

Fig. 6.29. (A) Distribution of weighted lead time per operation (simulation 2); (B) distribution of operation time per operation (simulation 2) (After Erdlenbruch [12])

of the evaluation of a simulation (here simulation no. 2), Fig. 6.29A shows the distribution of the weighted lead times, with the key data explained in Chapt. 3, in a bar chart of the group values. In this example, the weighted mean lead time per operation is 67.4 hours which, with a capacity of eight hours per day, corresponds to about 8.4 workdays.

The second important distribution to be evaluated is that of the operation times. Figure 6.29B shows the frequency distribution and the key data for simulation no. 2. The mean value is 7.9 hours, while the standard deviation is 13.2 hours. The number of values corresponds to the number of operations (here 1483): therefore, the total value corresponds to the number of standard hours contained in these operations (here 12 126). Since the work is finished within the simulation period, the total value is equal to the total output within the simulation period.

The most important data from the two frequency distribution diagrams are then entered into a table of results (Table 6.8). We will pick out the values from Simulation 2. The output (12 126 hours) is taken from the table in Fig. 6.29B, value no. 14, as are the mean operation time (TOP_m) of 7.9 hours (value no. 16) and the standard deviation of the operation time ($TOPS$) of 13.2 hours (value no. 18). The weighted mean operation time (TOP_{mw}) can be calculated as was shown in Chapt. 3, Eqn. (3.22).

One can also use an equation derived by Erdlenbruch [3]:

$$TOP_{mw} = TOP_m + \frac{TOPS_2}{TOP_m} \tag{6.15}$$

where:

TOP_m = simple mean operation time
$TOPS_2$ = variance of operation time.

Thus, for Simulation 2 (the variance value was taken from the table in Fig. 6.29B, no. 17), it follows that

Table 6.8. Simulation results for eight work centers with different load limitations. (Base: 8 work centers, investigation period 3200 hours, ca. 12 000 hrs of order time, ca. 1550 operations.)

Value		Original	Results of Simulation Run 1	2	3	4	5
OUT	(Hrs)	12038	9739	12126	12833	13677	13899
TOPm	(Hrs)	7.6	7.2	7.9	8.1	8.5	8.5
TOPs	(Hrs)	11.4	12.4	13.2	14.0	14.7	14.7
TOPmw	(Hrs)	24.6	28.5	29.9	32.4	33.8	34.0
TLM	(Hrs)	132.5	35.1	67.4	100.3	130.4	174.1
TLMs	(Hrs)	225.0	12.7	78.2	135.0	179.4	316.2
Im	(Hrs)	498.1	106.7	255.4	402.2	557.3	756.2
Um	(%)	47	38	47	49	51	51
LPG	(%)	266	144	185	225	263	318

Base: 8 Work Centers, Investigation Period 3200 Hrs
ca. 12000 Hrs of Order Time, ca 1550 Operations

$$TOP_{mw} = 7.9 \text{ hrs} + \frac{173.9 \text{ hrs} \times \text{hrs}}{7.9 \text{ hrs}} = 29.9 \text{ hrs}$$

In this special case, it was not necessary to distinguish between order time (TO) and operation time (TOP), since the time grid was given in hours instead of in calendar days or workdays. This avoids having to calculate TOP from TO; both key values have the same numerical value. This approach is often used in simulations, since it eliminates daily capacity, which is not of interest here, as an influential factor.

The next values in the table of results (Table 6.8) are the weighted mean lead time (67.4 hrs), taken from Fig. 6.29A, and the corresponding standard deviation (78.2 hrs).

The mean inventory, which is another important value, can be derived from Fig. 6.29A as follows. Normally, the exact calculation is carried out as shown in Sect. 4.4.2 and exemplified in Table 4.2, namely by using the area between the input and output curves from the beginning to the end of the investigation period.

It was, however, also shown in Sect. 4.4.6 that the following applies for longer periods of time:

$$TLM = TA_m = R_m = \frac{I_m}{PE_m} \tag{6.16}$$

where:

TLM = weighted mean lead time
TA_m = mean advance time
R_m = mean range
I_m = mean inventory
PE_m = mean performance.

It follows that:

$$I_m = TLM \times PE_m = TLM \times \frac{OUT}{P \times n}$$

where:

OUT = output in reference period
P = length of period
n = number of periods in reference period.

With the values from Simulation 2, it follows that:

$$I_m = 67.4 \text{ hrs} \times \frac{12\,126 \text{ hrs}}{80 \times 40 \text{ hrs}} = 255.4 \text{ hrs}$$

The utilization (U_m) can be calculated from the output and the capacity available (8 hrs/hrs, since there are 8 work centers). The capacity in the reference period is therefore 8 hrs/hrs × 3200 hrs = 25 600, so that, in Simulation 2, the utilization is 12126 hrs (OUT)/25600 hrs = 0.47, or 47%.

The actual mean loading percentage follows from Eqn. (6.8):

$$LPG = \left(1 + \frac{TLM}{P}\right) \times 100\%$$

The period length is 80 hours, so that for Simulation 2 it follows that

$$LPG = \left(1 + \frac{67.8 \text{ hrs}}{80.0 \text{ hrs}}\right) \times 100\% = 185\%$$

The loading percentage defined during the order release was 200. The difference between the defined and the calculated values can be explained as follows: the calculated loading percentage is the mean value of the loading percentages from eight work centers. With the given mixture of orders, it was obviously not possible to load all work centers up to their load limits, so that at some work centers the mean inventory was lower than might be expected with the defined load limits.

It is also possible that the calculated loading percentage value is higher than the defined value. This might be caused by in-system leftover loads: either there are still orders at the work center which were released earlier, or orders with very large work contents were released. In both cases, the total load can exceed the load limit. As was explained above, a loading account is only blocked after the limit has been exceeded for the first time. Otherwise, high-volume orders would always be rejected.

Figure 6.2 showed that, from repeated simulations, characteristic output and weighted mean lead time curves can be developed. Figure 6.30 shows these curves for our simulation example with eight work centers.

The output (OUT) decreases, as expected, with the mean inventory (I_m), first

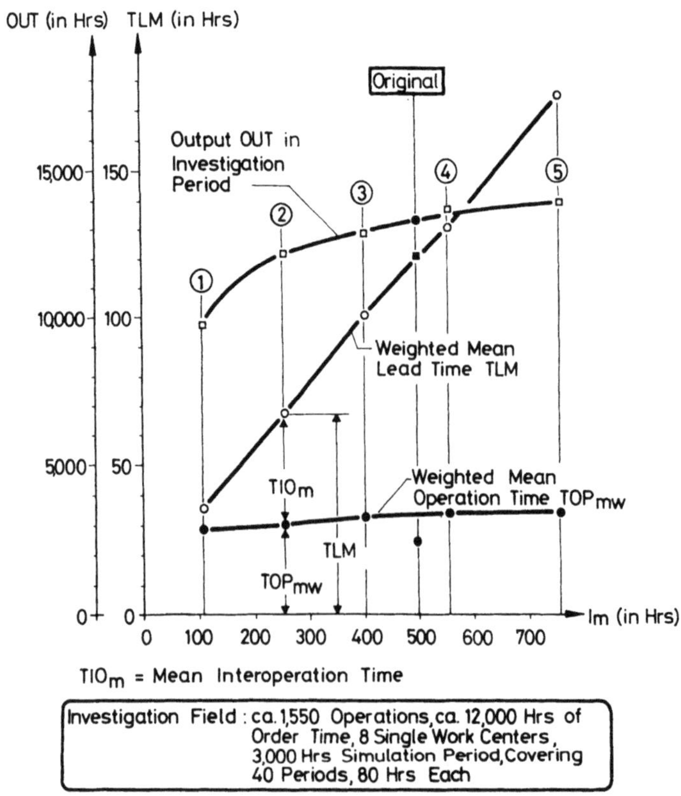

Fig. 6.30. Characteristic curves for processed orders (OUT) and weighted mean lead time per operation (TL) of a work center with eight machines (After Erdlenbruch [12])

slowly, then more rapidly. Theoretically, it ends at the zero point of the coordinate system. The weighted mean lead time (*TLM*) consists of the weighted mean operation time (TOP_{mw}) and the mean interoperation time (TIO_m). The transit time was, for the sake of simplicity, assumed to be zero, so that with decreasing inventory and thus with decreasing waiting time, the lead time approaches the weighted operation time value. The mean interoperation time (TIO_m) is thus purely a waiting time due to the queues at the individual work centers.

Our example is a special case, since the output curve approaches a value which corresponds to a utilization of only about 50%. Besides the above-discussed difference between the defined loading percentages and those actually obtained, this is further evidence of the fact that this is an extremely unevenly loaded group of work centers, of which a few are bottlenecks. From Fig. 6.27B we can assume that Work Centers 3 and 7 are the ones in question. Simulations with eight work centers are, however, not representative for the processes in an authentic shop; they are more suitable for demonstration, testing, and training purposes. For testing the influence of the load limit on the throughput of orders, we therefore give an example of a more extensive simulation.

6.8.2 Effects of the Load Limit

To examine the effects of the load limit, the shop introduced in Chapt. 3, Table 3.5, will be used as the test object. The feedback data recorded over a period of about six months were first evaluated and then used for running the simulation program. Before starting the simulations, a model test as explained in the previous section was carried out. This was followed by the actual simulation experiments. The order throughput was started again, but this time under controlled conditions. In our example, about 3000 shop orders with a total of about 16 000 operations and a work content of 69 700 hours were processed on 50 work centers (90 single work places). The parameter to be examined was the load limit, which was successively fixed at six different values, namely 100%, 150%, 200%, 250%, 300%, and 400%. The period length was five workdays.

Figure 6.31 summarizes the results of the simulations [3]. The graph also includes the original output, lead time, and operation time values as measured during the investigation. The representation is not the one used by Erdlenbruch, but is given as characteristic curves like those used in Fig. 6.2. We can see the output curve, plotted by connecting the single results. It can be transformed into the characteristic line for the mean output rate by dividing the values on the work axis (vertical axis) by the simulated period of 16 weeks. This calculation shows, for example, that on average $62\,000/16 = 3875$ hours were processed per week, which corresponds to a mean daily output of $3875/(5 \times 90) = 8.6$ hours per workday and per work place. This means that the work centers mainly worked in single shifts with eight hours per day.

The horizontal axis represents the mean inventory (I_m) of the investigation period, which followed from the individual loading percentage values. We can see that the output curve, starting at $LPG = 400$ drops slowly at first, but falls more rapidly from about 250% on. This can be explained by the fact that, as the inventory decreases, the funnels 'run empty' (which means that the work centers are not working), first only at times and at some work centers, then increasingly often and at more work centers. The potential output according to the capacity reported by the work centers within the investigation period is also given in the diagram. The vertical distance

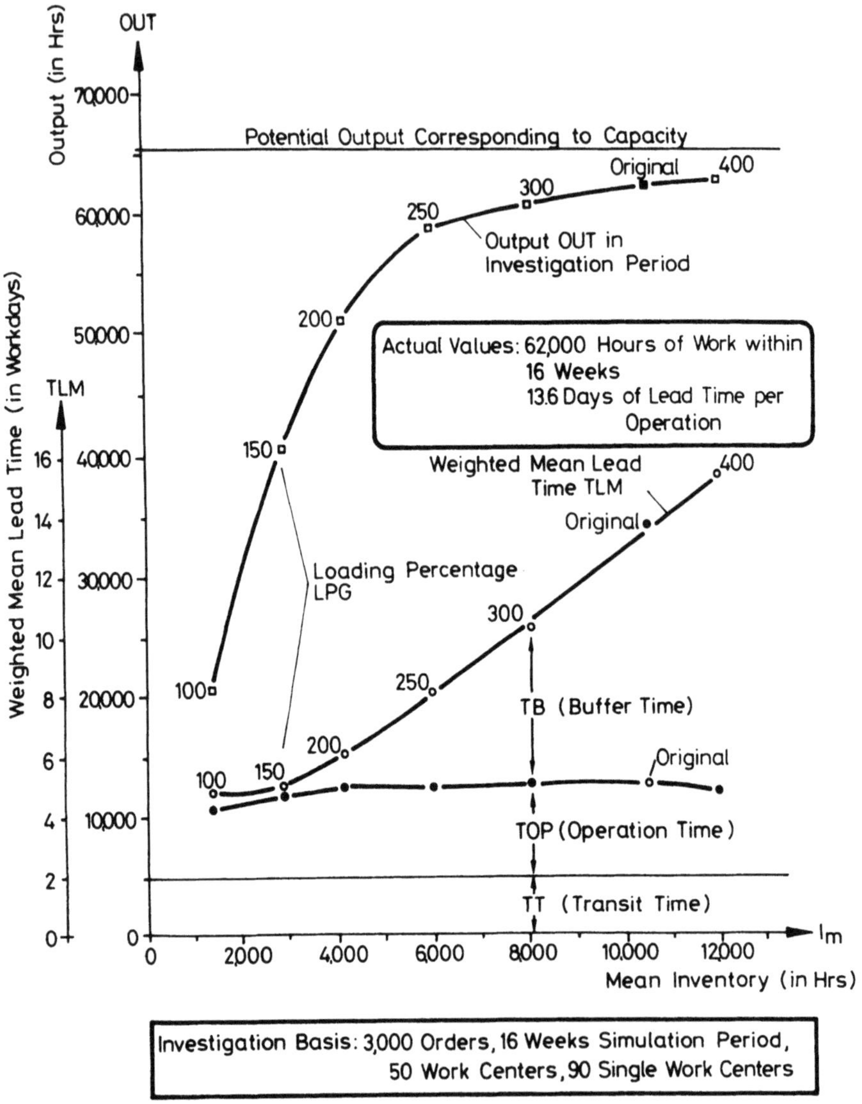

Fig. 6.31. Characteristic curves for processed orders, weighted mean lead time per operation, and their components, at different load limitations (After Erdlenbruch [12])

between the potential output and the actual (simulated) output curve thus represents the theoretical loss in capacity during the individual simulations.

The second curve is the characteristic of the weighted mean lead time per operation (*TLM*). We can see the three components: the transit time (*TT*) was assumed to be two days in all simulations, so as not to simulate too favorable conditions compared with the real situation. The (weighted) operation time (*TOP*) was calculated from the work contents of the individual operations. It therefore varies slightly from simulation to simulation, since, due to the finite simulated period, it is not possible to process exactly the same orders each time. The buffer time (*TB*) is the difference between *TLM* and the sum of *TOP* and *TT*.

The lead time curve is not linear, either. Starting at about 15.3 days at $LPG = 400$, it drops almost linearly at first, but then gradually approaches the TOP plus TT value from about $LPG = 250$ on. The reason is that, when inventories decrease while the work centers are still loaded to capacity, the lead time obeys the 'funnel formula' (weighted mean lead time equals mean inventory divided by mean performance). At the point $LPG = 400$, for example, the values are as follows: $PE_m = 63\,000$ hours/16 × 5 workdays = 787.5 hrs per workday. The mean lead time is then 12 000 hrs/787.5 hrs/day = 15.2 workdays, which corresponds very closely with the value produced by simulation (15.3 workdays).

If, however, the inventory falls below the utilization limit, the components of lead time which do not depend on the inventories (i.e. on the lengths of the queues) become increasingly important. These components are the operation time and the transit time, which depend solely on the capacity and on the transport facilities, respectively.

From both curves, it follows that the most reasonable loading percentage lies between 250 and 300. With $LPG = 300$, for example, an inventory and lead time reduction of 25% compared with the actual process can be obtained, whereas the utilization would decrease by 4% only. If this was a practical case, the bottleneck cost centers would probably be operated at a loading percentage of around 300, the others at 250.

From this example we can see most clearly that the dilemma which is said to exist between a good utilization and short lead times is not really a matter of either–or, but a matter of finding the economic optimum on the shop-specific operation curves.

An examination of the lead time distribution and the dispatching procedure will further illustrate the importance of the load limit as a parameter of load-oriented manufacturing control.

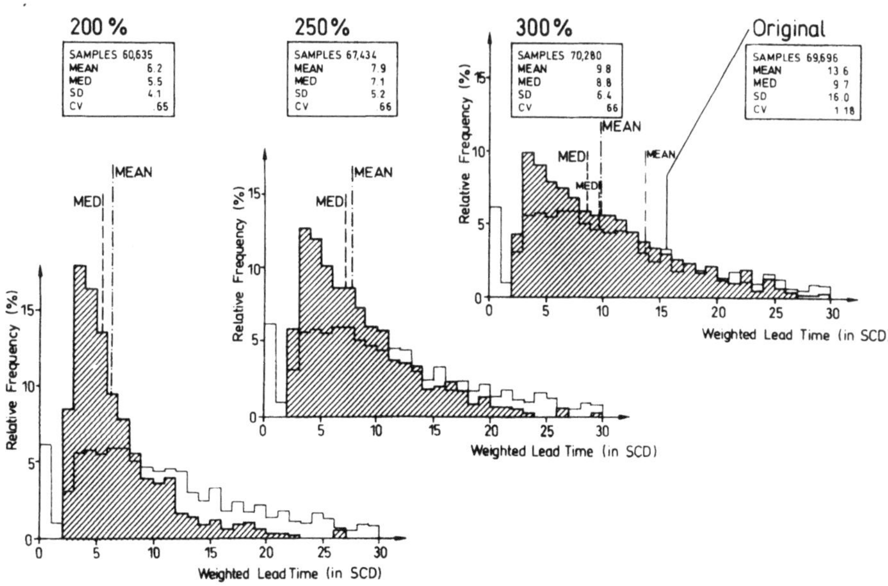

Fig. 6.32. Distribution of simulated operation lead times in the total shop for different loading percentages (After Bechte [2])

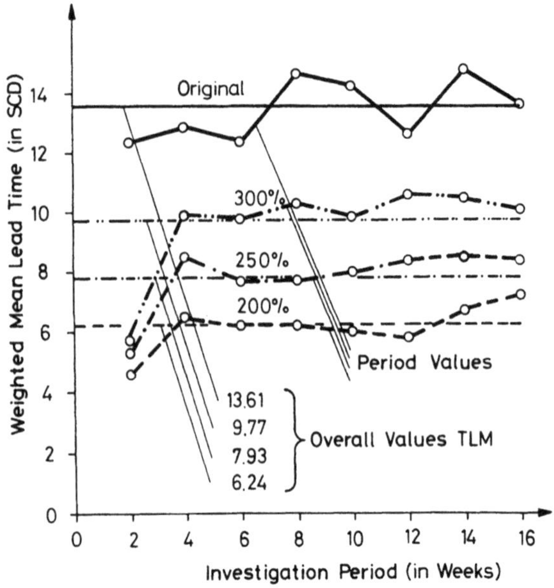

Fig. 6.33. Trends of simulated weighted mean lead time in the total shop for different loading percentages (After Bechte [2])

Figure 6.32 compares operation lead time distributions at three different loading percentages with the actual process [2]. This simulation has been carried out with values from the same shop as in Fig. 6.31, but not with the same orders. Therefore, the reference value for the weighted mean lead time, 14.1 workdays per operation, is slightly higher than the 13.6 workdays in Fig. 6.31. We can see that the distributions become markedly steeper and more symmetrical with decreasing load limit, which also results in lower standard deviation values. This is due to the fact that with decreasing inventories there are fewer possibilities for sequence transpositions. The FIFO dispatching rule is therefore enforced, thus causing less variance.

To evaluate mean key data over longer periods of time (here 16 weeks = 8 periods), one should look at the period-related mean values. Figure 6.33 shows the simulated weighted mean lead time curves for the whole shop described in Table 3.5. The evaluation was taken from the simulation on which Fig. 6.32 is based [2].

We can see that the actual lead times varied by over two days from a long-term mean of 13.6 days. In the simulations with different loading percentages, the values were first much lower, as the simulation model first had to find a normal rhythm. The lead time curves of the simulated processes are much more even, which is due to a better dispatching discipline than in the actual process, because the fixed dispatching rule was strictly followed.

The range of lead time variations was reduced to about one day, which, in turn, improves the schedule performance of the orders. Figure 6.34 (left) [2] compares the *work contents* of the outputs at different load limits with the actual process. On the right, the figure shows the lateness distribution. Here, the mean order lead time is about 10 weeks. With a loading percentage of 300, the orders are, on average, completed 0.75 weeks earlier than in reality, with a standard deviation of 1.47 weeks. At $LPG = 250$, the mean lateness is 0.9 weeks and the standard deviation 1.68

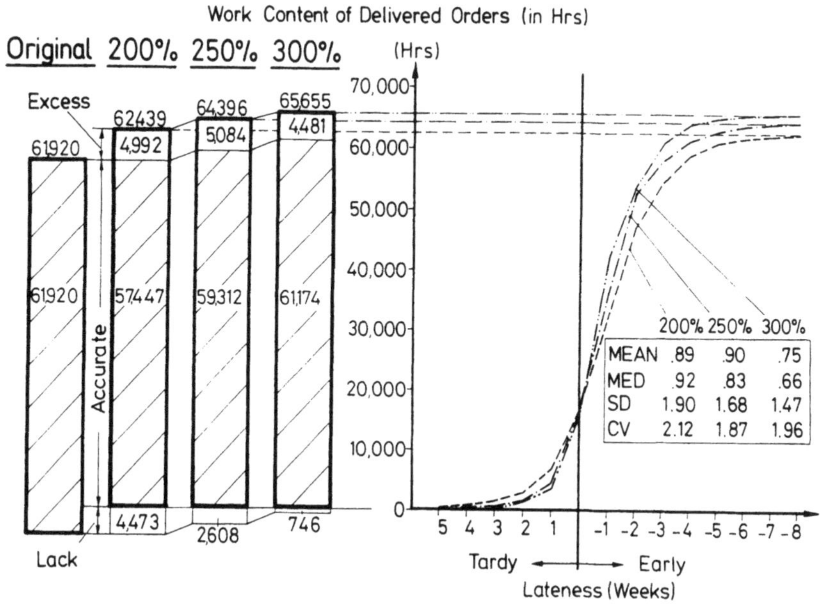

Fig. 6.34. Delivered orders and schedule deviations for different loading percentages (After Bechte [2])

weeks: the orders are, on average, completed earlier than at $LPG = 300$. Note, however, that the orders processed were not exactly the same. This is why the values 0.89 and 1.90 at $LPG = 200$ can only be used with certain reservations, since the proportion of orders that correspond with the actual process has further decreased compared with $LPG = 250$ or 300. It can, however, generally be said that the order lead times and the schedule performance improve with decreasing load limits.

This phenomenon can also be observed in the throughput diagrams of the individual work centers. As an example, Fig. 6.35A–C shows the throughput diagram of Work Center 622 341 which, according to Table 3.5, consists of two chucking lathes [13].

The actual process (Fig. 6.35A) shows an irregular input pattern and widely varying lead times. With a simulated loading percentage of 300 (Fig. 6.35B), the distance between the input and output curves has decreased markedly. The weighted mean lead time has decreased by half (from about 20 days to about 10 days), and there are no longer any extremely long individual lead times, which is due to the dispatching discipline maintained in the simulation. With a further reduction of LPG to 200 (Fig. 6.35C), lead time and inventory values are further improved, but now the utilization is poorer. The work center is not always provided with work, for example around the Days 220, 240, 280, and 300. This means that to avoid underload the optimum loading percentage lies between 200 and 300 for this work center. It must, however, be taken into account that, in practice, a work center will not be allowed to run idle.

Up to now, our considerations have been based on a uniform load limit for all work centers. As was shown in our discussion about the 'correct' load limit, it makes more sense to set different load limits since, in some shops, the mean operation times of individual cost centers and work centers can vary widely. The effects of this are shown in Fig. 6.36, which compares the output and mean operation lead time curves simulated by Erdlenbruch for standard and individual load limits [2].

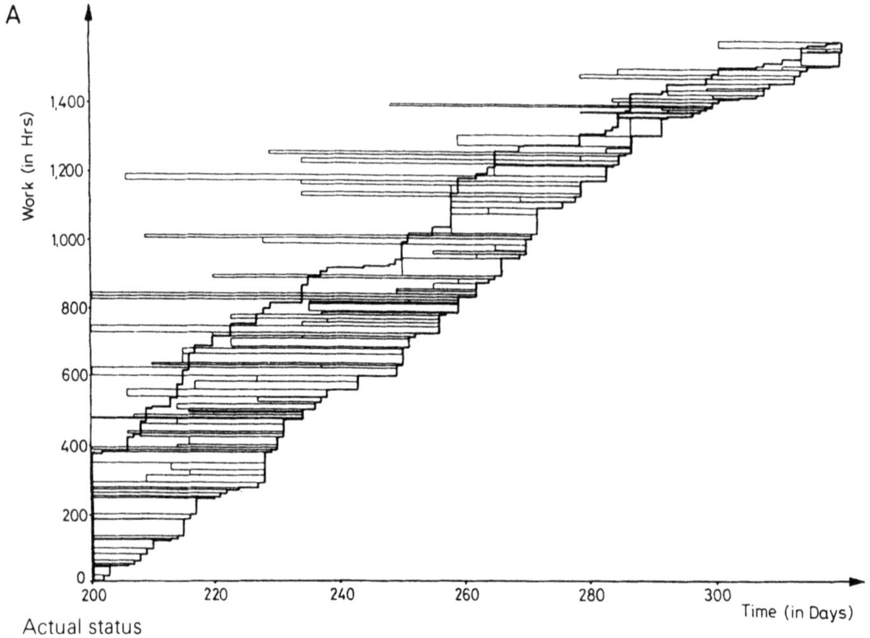

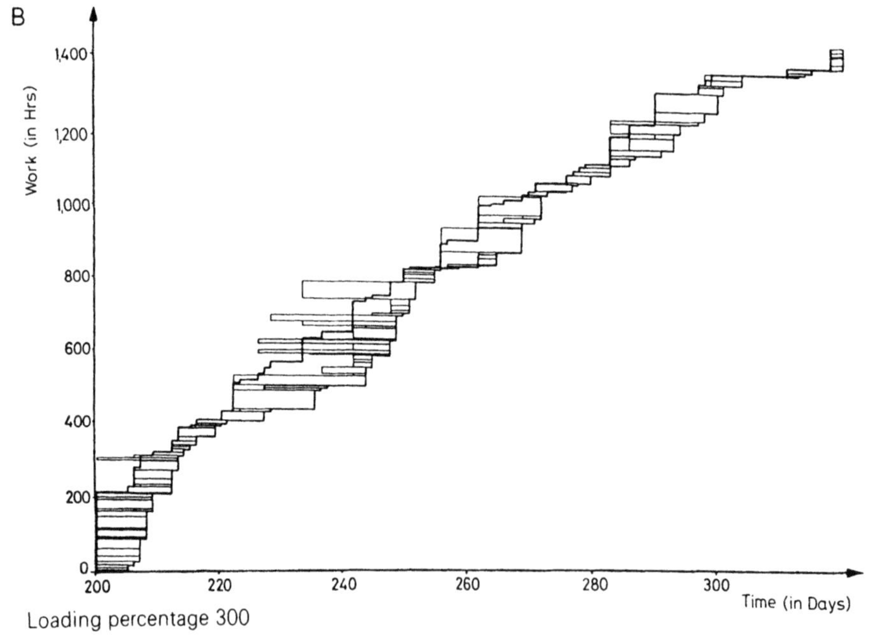

Fig. 6.35. Work center arrival and output trends in throughput diagrams (After Krautzig [13])

continued

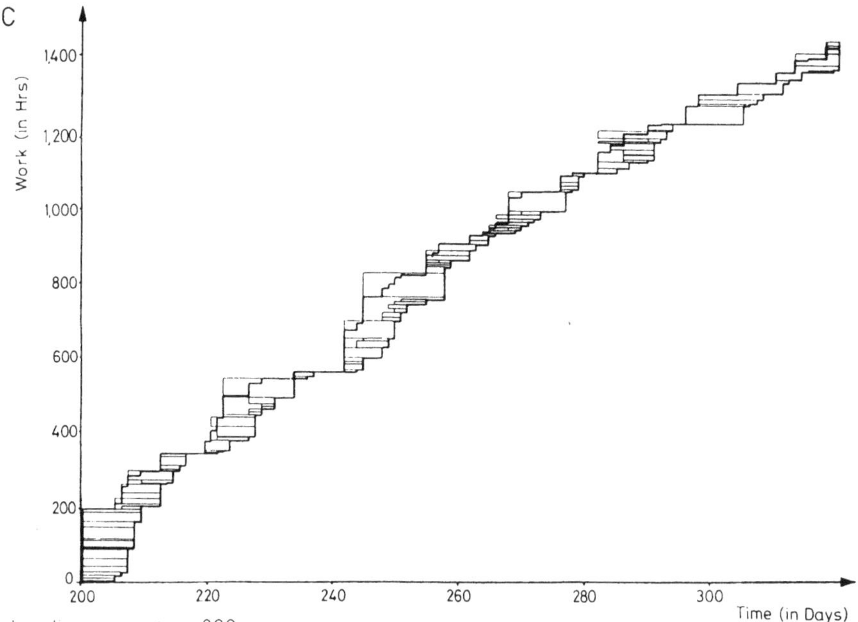

Loading percentage 200

Fig. 6.35 (*continued*)

The output curve ('processed orders *OUT*') shows that, above a mean load limit of *LPG* = 200, more work can be processed if individual load limits are set than if the load limit is the same for all work centers. The operation time of the processed operations is about 10% longer, which indicates that with a standard load limit, larger-scale orders are given lower priority. In this case, capacities can be used better with individual load limits, but the mean lead times are only slightly improved. The buffer times, however, are shorter, due to the longer mean operation times.

Generally it can be said that the mathematical relationship between lead time, inventory and utilization, which was found using analytical methods, is proved by the realistic simulation, and that the load limit, or the loading percentage, is the crucial parameter of load-oriented manufacturing control.

A good deal of practical experience in the use of this method has meanwhile been gathered in electrical engineering, electronics, and precision machinery plants. In one printed circuit board (PCB) production shop, for example, the use of load-oriented order release meant that the mean flow time could be improved from 48 days to 22 days with a weekly release, and to 16 days when the release frequency was changed to twice a week. Additional measures were taken to improve feedback accuracy and dispatching discipline. In other shops of the same plant, which produce PCBs, filter chokes and coils, similar improvements were attained [14].

In another plant, where high-grade fastening elements such as dowels, pins, nails, etc., are produced, the mean item flow times were reduced from 33 days to 23 days within about six months; over the same period, the work-in-process, i.e. the job-shop inventory, decreased by 25%, and the orders with delay were reduced from 1100 to 500, i.e. by about 50% [15]. Since all load-oriented order release systems also monitor the lead time and inventory trends as a 'side result', a lasting improvement is guaranteed. More results can be found elsewhere [16].

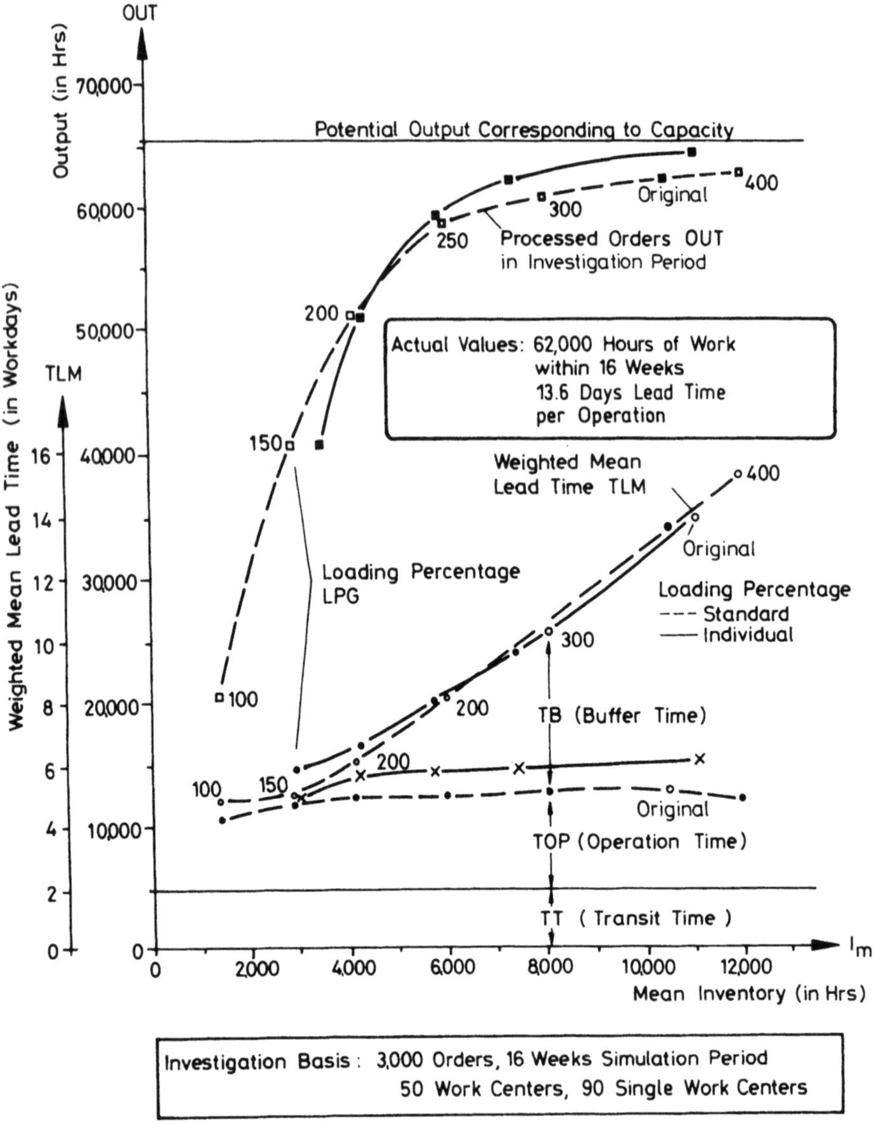

Fig. 6.36. Characteristic production curves – standard versus individual load limitations (After Erdlenbruch [12])

6.8.3 Effects of the Time Limit

The time limit, or the anticipation horizon, was introduced as the second important parameter of load-oriented order release. The anticipation horizon is the period of *time in* which the start dates of the urgent orders lie. The beginning of this time is the scheduled date 'today'; its end is called the *time limit*. Lorenz examined the effect of the anticipation horizon on the mean performance and mean lead time. To do this, he employed a simulation model whose reliability had been proved with a model test using actual data and event-oriented simulation (see Sect. 9.6). He took four different anticipation horizons (0, 2, 4, and 6 weeks) and different loading percentages [8]. The results are shown in Fig. 6.37.

The simulation also used data from the model shop introduced in Chapt. 3, so that the results are comparable with Erdlenbruch's characteristic curves in Fig. 6.31. We can see that the influence of the anticipation horizon on the performance and lead time is greater the higher the loading percentage and thus the mean inventory are. Both Erdlenbruch's and Buchmann's simulations [7] proved that an anticipation horizon of three periods is the optimum, becoming less influential as inventories decrease.

Since the influence of the anticipation horizon is obviously dependent on how well the capacity required by the current mix of orders is coordinated with the

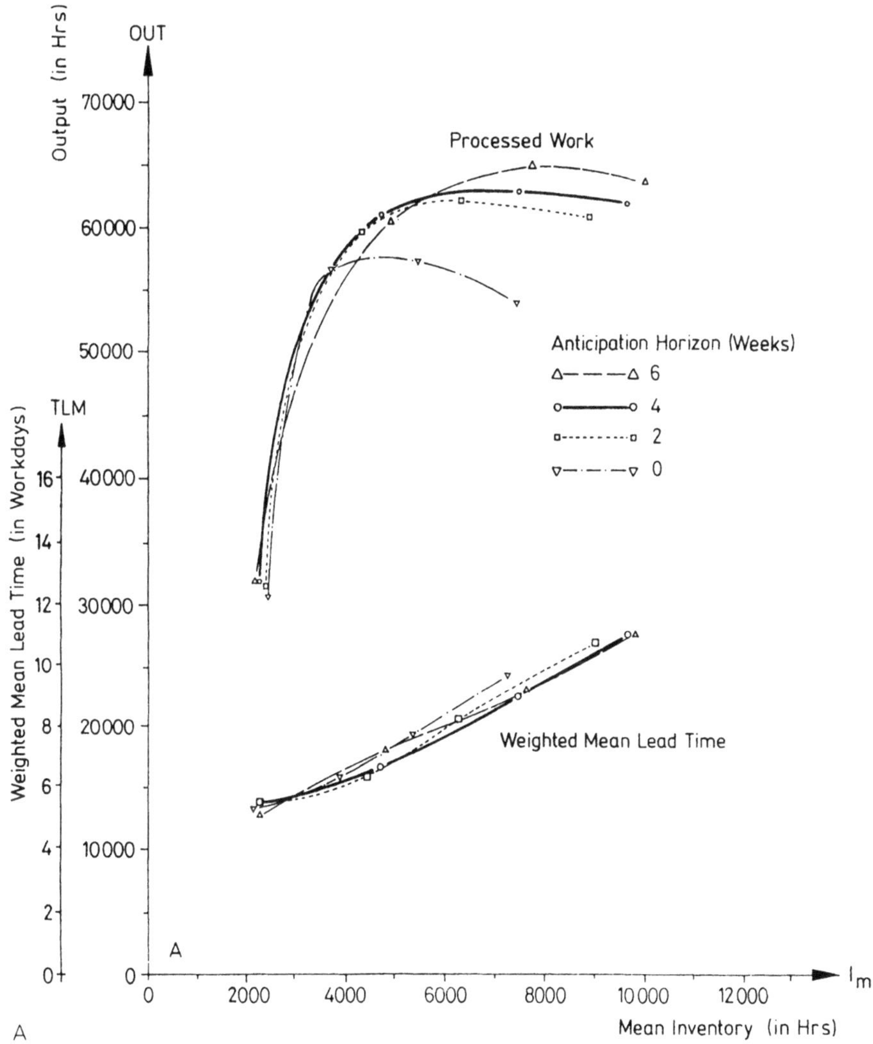

Fig. 6.37. (A) Effects of the anticipation horizon on the processed work (OUT) and on weighted mean lead time with different loading percentages; (B) effects of the anticipation horizon (key data) (After Lorenz [9])

continued

LPG	AH	Mean Inventory (Hrs)	Mean Lead Time (SCD)	Mean Output (Hrs)
LPG = 400%	△ AH = 6 Wks	9850	11.0	63042
	○ AH = 4 Wks	9650	11.0	62845
	□ AH = 2 Wks	9000	10.7	61660
	▽ AH = 0 Wks	7250	9.7	54431
LPG = 300%	△ AH = 6 Wks	7650	9.2	64306
	○ AH = 4 Wks	7500	9.0	63990
	□ AH = 2 Wks	6250	8.1	62094
	▽ AH = 0 Wks	5350	7.8	57868
LPG = 200%	△ AH = 6 Wks	4800	6.7	59685
	○ AH = 4 Wks	4700	6.6	60040
	□ AH = 2 Wks	4450	6.3	59369
	▽ AH = 0 Wks	3950	6.2	56406
LPG = 100%	△ AH = 6 Wks	2250	5.1	30810
	○ AH = 4 Wks	2250	5.5	30613
	□ AH = 2 Wks	2250	5.5	30850
	▽ AH = 0 Wks	2150	5.3	30613

LPG = Loading Percentage
AH = Anticipation Horizon

B

Fig. 6.37 *(continued)*

capacity available in the subsequent period, the 'correct' value cannot be found using a simple estimative method, as has been shown above for the load limit. In this case, the trial-and-error method is more convenient. Simulations and experience have shown that three periods are a reasonable starting value [7, 14].

6.9 Controller Analogy of Load-Oriented Order Release

For a better illustration of the 'load limit' and 'time limit' control parameters, there exists an analogy model, derived from the funnel model, consisting of three interconnected funnels (Fig. 6.38).

The bottom funnel contains the orders released for a work center which, in turn, include orders physically present at the work center, orders which are still at the upstream work center, and orders whose output was reported at the upstream work

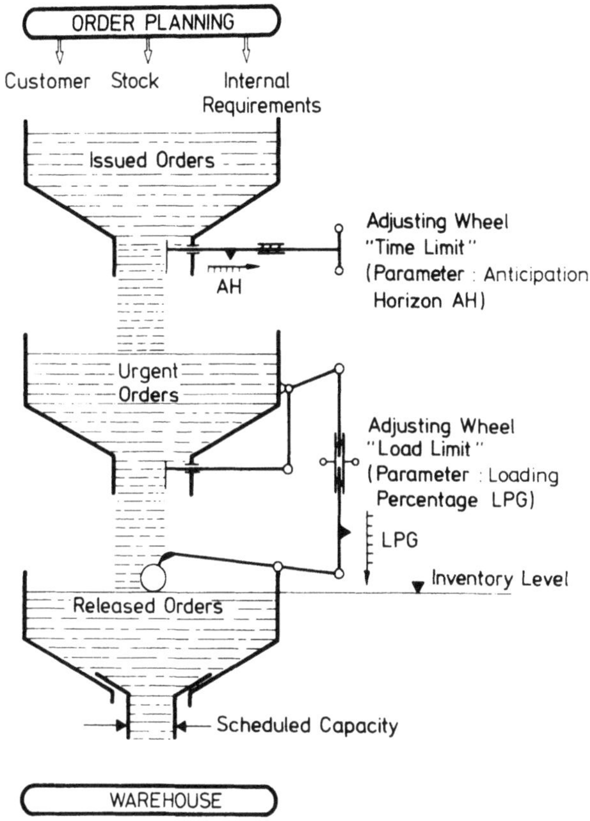

Fig. 6.38. Control-loop analogy of load-oriented order release

center but which are still involved in inspection or transport processes. The capacity of the funnel is fixed.

An inventory level sensor transmits the current inventory value to the slide valve of the middle funnel, which contains the urgent and feasible orders. The outflow of work in the bottom funnel controls the inflow of orders from the middle funnel. The inventory level in the bottom funnel – and thus the mean lead time – is determined by the adjusting 'load limit' wheel, which can change the distance between the upper and lower rod linkages. This keeps the inventory on the required level, independently of capacity fluctuations.

The upper funnel contains the overall issued order stock, the issued orders. The adjusting 'time limit' wheel admits only those orders which lie within the fixed anticipation horizon into the inventory of urgent orders.

In reality, the orders are neither infinitesimal nor continuous but flow stochastically, with finite order times, through the system. However, the model correctly reflects the idea of load-oriented order release; the method 'thinks' integrally, i.e. it does not follow single elements deterministically along their throughput, but enforces the required scheduled lead time by controlling the inventory level. With this method, individual orders can flow faster or more slowly through the system, to the advantage or disadvantage of the other orders. This is usually controlled by priority rules, which determine the final sequencing in the queues.

6.10 Priority Rules and Order Sequencing in Load-Oriented Order Release

Great importance has been attached to order sequencing in common manufacturing control systems. Stimulated by the development of operations research methods, various dispatching rules were created and their effect on the schedule performance

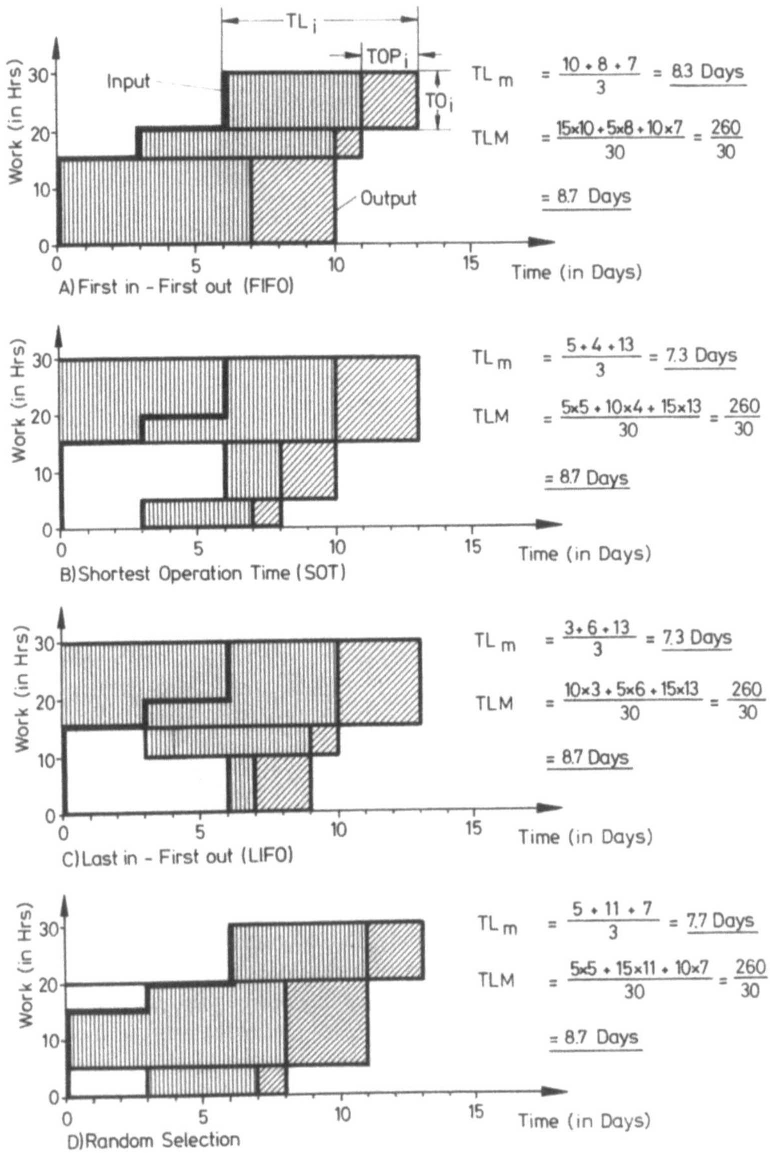

Fig. 6.39. Effects of different dispatching rules on weighted and unweighted mean lead time

has been investigated in numerous research projects. A critical review in a plant, however, revealed that "... the effects of the scheduling rules available to the user (priority rules, order release rule, downtime rule, pull-ahead rule) are not clear, and this does not permit their efficient use" [18].

If one examines the influence of dispatching rules in the throughput diagram, one discovers that they have no effect on the weighted mean lead time. This will be demonstrated with a simple example in Fig. 6.39.

Figure 6.39A shows the throughput diagram of a work center which has to process three orders with different work contents. The orders are processed according to the FIFO rule, i.e. in the sequence of their arrival. The simple mean lead time (*TLM*) is 8.3 days, and the weighted mean lead time (*TLM*) is 8.7 days. In Case B, the orders are processed according to the SOT (shortest operation time) rule, i.e. the order with the smallest work content is processed first, then the order with the second smallest, then the largest order. The unweighted mean lead time drops to 7.3 days, a seeming improvement. The weighted mean lead time, however, has the same value as with FIFO. The LIFO rule (last-in – first-out) (Fig. 6.39C) also brings only a seeming improvement, while the random selection rule seems to cause a poorer mean work center lead time value.

The use of the 'correct' (i.e. the weighted mean lead time) value shows that the priority rule has absolutely no effect on the mean work center lead time, since this is related to the capacity, and as, according to the funnel formula, the mean work center lead time only depends on the mean inventory and the mean performance.

There is, however, an indirect influence of the sequencing rules on lead time. It can be seen, for example, in the throughput diagrams in Figs. 6.35A–C. It is obvious that in the actual process (Fig. 6.35A) there is a large variation in the lengths of the throughput elements at the work center, while in the two simulations (Figs. 6.35B, C), the variation is markedly lower. Note, however, that the slack-time rule was always applied in the simulations, which arranges the orders in a queue in the sequence of their priority. The following can be concluded from this: priority rules do influence the variance of weighted lead times, but do not influence the mean weighted value.

Bechte [2] conducted interesting investigations on this subject. He compared the variance of the weighted mean lead times in the actual process with the variance in simulations using FIFO and with those in simulations using the slack-time rule, each with several mean inventory levels [2].

We shall first look back at Fig. 6.32. If we compare the actual weighted mean lead time (*TLM*) and its standard deviation (SD) with the simulated values at the different load limits, we see that, in the simulation, SD changes proportionally with the mean value (MEAN). Figure 6.40 compares the values of the three simulated processes (*LPG*s of 200, 250, and 300) [2].

The coefficient of variation (CV = SD/MEAN) shows this relationship clearly. The standard deviation and coefficient of variation in the actual process are strikingly different from the simulated values. The standard deviation is disproportionately higher than in the simulation, which is due to the application of several different dispatching rules. Further investigations have been conducted by Haupt [17].

It can generally be said that the lead time variance depends primarily on the weighted mean lead time.

This will be proved with an example (Fig. 6.41) [14]. Case (A) shows three work centers at which there are long queues. Each deviation from the FIFO rule means a much shorter lead time for the lot which is given priority, but also causes numerous other transpositions and thus great lead time variances. With short queues (Case B),

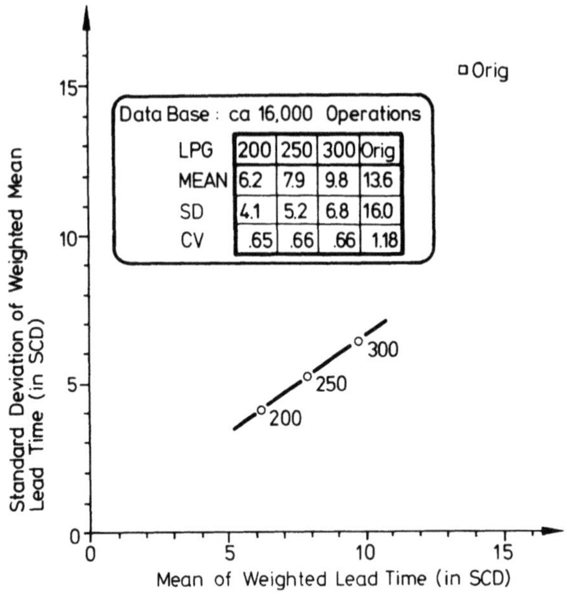

Fig. 6.40. Different standard deviations at different mean values of weighted mean lead time (After Bechte [2])

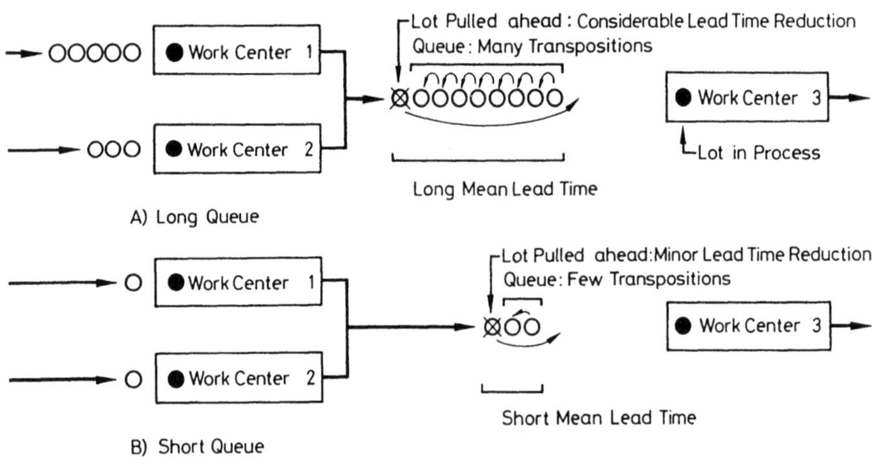

Fig. 6.41. Effects of queue length on lead time and sequence transpositions (After Ulfers [14])

pull-aheads have much less positive (possible lead time reduction for one lot) as well as negative effects (transpositions of the other lots).

From this we can draw two more conclusions. First, the lower the mean inventory and the shorter the queue, the less effective a priority rule becomes. Second, as inventories decrease, the FIFO rule is enforced more and more. It is thus the 'natural' rule and causes the least lead time variances of all priority rules.

In practice, the FIFO rule can not always be followed, since unavoidable external events (e.g. rush orders) or internal events (e.g. breakdowns) inevitably cause se-

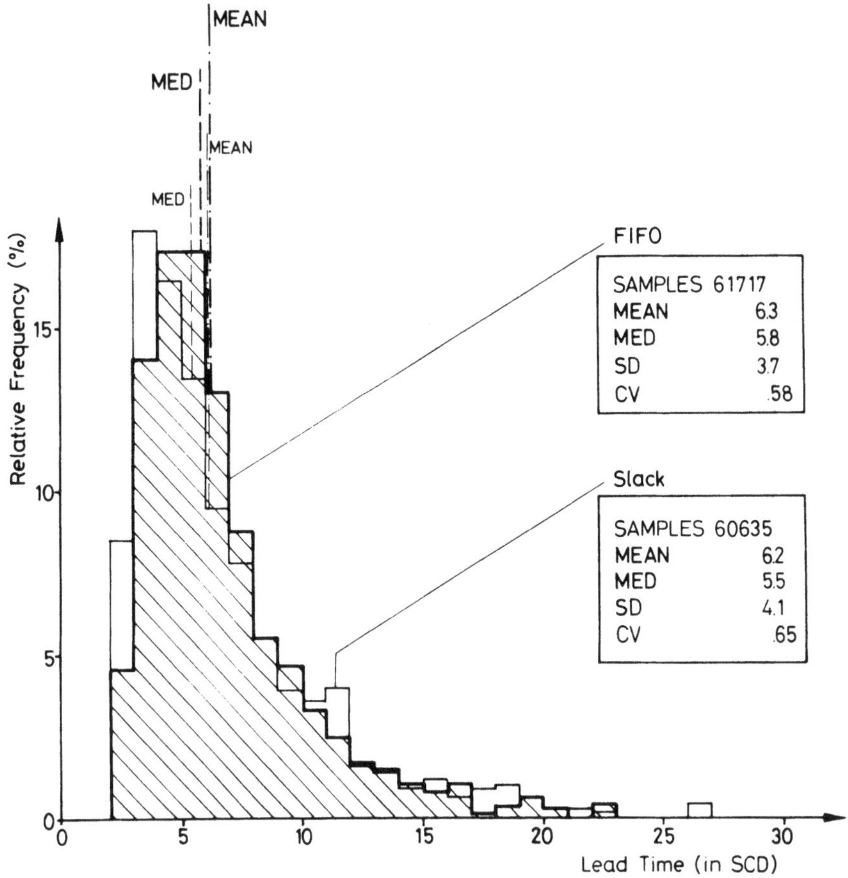

Fig. 6.42. Distribution of simulated lead time per operation for the total shop–slack time versus FIFO rule (loading percentage 200) (After Bechte [2])

quence transpositions. In such cases, a due-date-oriented rule such as slack time or the Modified Due Date (MDD) rule [20, 21] should be followed.

Bechte [2] has compared the effects of the FIFO rule and those of the slack-time rule on the operation lead times and on the order schedule performance at a loading percentage of 200. Figure 6.42 compares the lead time frequency distributions under the two rules at the 50 work centers of our sample shop [2]. While the mean values are practically the same, the standard deviation with FIFO (3.7 workdays) is, as expected, lower than with slack-time dispatching (4.1 workdays). In the actual process, the standard deviation was 16.0 workdays (cf. Fig. 6.32).

Figure 6.43 shows the effects of the two rules on the schedule performance of about 3000 processed orders [2]. The actual dates were taken as reference values. The processed orders were not the same as in the actual process, but they had practically the same work content. With FIFO dispatching, there is a positive lateness of 0.95 weeks, and the standard deviation of the lateness is 2.24 weeks. Approximately 18 000 of about 60 000 hours (i.e. about 25% of the order hours) were delivered behind schedule.

With the slack-time rule, the mean lateness is only slightly lower – 0.89 weeks; but

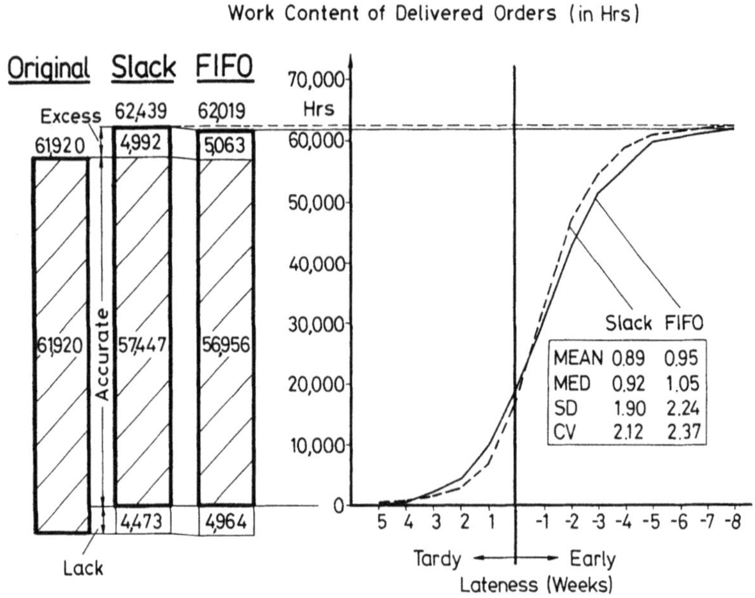

Fig. 6.43. Delivered orders and schedule deviation – slack time versus FIFO rule (loading percentage 200) (After Bechte [2])

the standard deviation drops from 2.24 weeks to 1.90 weeks, i.e. by about 18%. Thus, the 'slack' curve (broken line) is somewhat steeper than the FIFO curve (unbroken line). The number of order hours completed behind schedule has also decreased. From this we can clearly see that a greater variance of lead times must be accepted for the sake of a better schedule performance, a fact which, in practice, requires additional control and scheduling time.

To sum up, the following points can be made concerning priority rules in load-oriented order release:

- Keeping the inventories at the work centers at the lowest reasonable level favors the 'natural' FIFO dispatching sequence and makes priority rule selection moot.
- Priority rules which deviate from FIFO do not change the mean weighted work center lead times, but increase their variance.
- In principle, the FIFO rule should be applied whenever possible, since it causes the least lead time variance and needs no finite control measures.
- In exceptional cases, deviations from FIFO are permitted. In these cases, a due-date-oriented rule should be applied. The additional control time and cost should be carefully weighed against the expected gains in time and cost.
- Other priority rules are not recommended. Calculations concerning alleged improvements are based only on the unweighted lead time, and do not take into account the inventory situation.

Note that priority rules have been discussed here only from the point of view of a quick and on-time order throughput. With complex manufacturing units, such as flexible manufacturing systems, there may be other factors which influence the dispatching sequence. Such factors, for example the availability of tooling and fixtures (see Sect. 8.6), may have priority over an 'optimal' order throughput because of their importance for the utilization of expensive equipment.

6.11 References

[1] Kettner, H, Bechte, W. Neue Wege der Fertigungssteuerung durch belastungsorientierte Auftragsfreigabe (Recent Production Control Methods by Use of Load-Oriented Order Release). *Zeitschrift des Vereins Deutscher Ingenieure*, **123** (11): 1981, 459–466.

[2] Bechte, W. Steuerung der Durchlaufzeit durch belastungsorientierte Auftragsfreigabe bei Werkstattfertigung (Lead Time Control in Job-Shop Production by Use of Load-Oriented Order Release). Dissertation Universität Hannover, 1980 (veröffentlicht in: Fortschritt-Berichte der VDI-Zeitschriften, Reihe 2, Nr. 70, Düsseldorf, 1984) (doctoral thesis, published in the above VDI report).

[3] Erdlenbruch, B. Grundlagen neuer Auftragssteuerungsverfahren für die Werkstattfertigung (Principles of Recent Order Control Techniques in Job-Shop Production). Dissertation Universität Hannover, 1984 (veröffentlicht in: Fortschritt-Berichte der VDI Zeitschriften, Reihe 2, Nr. 71, Düsseldorf, 1984) (doctoral thesis, published in the above VDI report).

[4] Lorenz, W. Begrenzung des Auftragsbestandes – Angemessene Fertigungsbestände und -durchlaufzeiten durch belastungsorientierte Fertigungssteuerung (Limiting the Order Stock – Appropriate In-Process Inventories and Lead Times by Use of Load-Oriented Manufacturing Control). *Schweizer Maschinenmarkt*, **84** (23): 1984, 28–32.

[5] Erdlenbruch, B. Belastungsorientierte Auftragsfreigabe – Grundlagen, Verfahren, Weiterentwicklung, Voraussetzungen für den praktischen Einsatz (Load-Oriented Order Release – Principles, Methods, Expansion, Prerequisites for Practical Application). In: Dokumentation zum Fachseminar "Statistisch orientierte Fertigungssteuerung" des Instituts für Fabrikanlagen der Universität Hannover (IFA), 1984 (seminar proceedings). IFA, Hannover, 1984, pp 20–45.

[6] Jendralski, H. Kapazitätsterminierung zur Bestandsregelung in der Werkstattfertigung (Capacity Scheduling for Inventory Control in Job-Shop Production). Dissertation Technische Universität Hannover, 1978 (doctoral thesis).

[7] Buchmann, W. Zeitlicher Abgleich von Belastungsschwankungen bei der belastungsorientierten Fertigungssteuerung (Temporal Leveling of Fluctuations in Load with Load-Oriented Manufacturing Control). Dissertation Universität Hannover, 1984 (veröffentlicht in: Fortschritt-Berichte der VDI-Zeitschriften, Reihe 2, Nr. 63, Düsseldorf, 1983) (doctoral thesis, published in the above VDI report).

[8] Lorenz, W. Entwicklung eines arbeitsstundenorientierten Warteschlangenmodells zur Prozeßabbildung der Werkstattfertigung (Working-Hour-Related Queueing Theory as a Representation of the Manufacturing Process in Job-Shop Production). Dissertation Universität Hannover, 1984 (veröffentlicht in: Fortschritt-Berichte der VDI-Zeitschriften, Reihe 2, Nr. 72, Düsseldorf, 1984) (doctoral thesis, published in the above VDI report).

[9] Lorenz, W. Organisatorische Maßnahmen zur Steuerung und Kontrolle des Fertigungsablaufs bei NC-Maschinen in einer Werkstattfertigung (Organizational Measures for Control and Monitoring of NC Machines in Job-Shop Production). Vortrag vom Seminar der gfmt am 25–26 Juni 1984 in München (seminar paper).

[10] Ritter, K.-H. Belastungsorientierte Auftragsfreigabe (Load-Oriented Order Release). Institut für Angewandte Arbeitswissenschaft, 96: 1983, 34ff.

[11] Lüssenhop, Th. Ein modulares Simulationsmodel – Instrument bei der Entwicklung neuer Methoden der Produktionssteuerung (A Modular Simulation Model as an Instrument for the Development of New Production Control Techniques). In: Dokumentation zum Fachseminar "Statistisch orientierte Fertigungssteuerung" des Instituts für Fabrikanlagen der Universität Hannover, 1984 (seminar proceedings). IFA, Hannover, 1984, pp. 128–142.

[12] Erdlenbruch, B. Auswirkungen von Fertigungssteuerungsmaßnahmen – Simulation auf der Basis realer Betriebsdaten (The Effects of Manufacturing Control Measures – A Simulation Based on Authentic Data). Vortrag Nr. 8 zum Seminar 'Neue Wege der Fertigungssteuerung' des Instituts für Fabrikanlagen der Universität Hannover am 16–17 März 1982 in Hannover (seminar proceedings). IFA, Hannover, 1982.

[13] Krautzig, J. Planung der Kapazitätsreservierung unter Berücksichtigung der Personalflexibilität bei Werkstattfertigung (Capacity Allocation in Job-Shop Production with Regard to Personnel Flexibility). Dissertation, Universität Hannover, 1981 (doctoral thesis).

[14] Ulfers, H-A. Erfahrungen mit der belastungsorientierten Auftragssteuerung ABS bei Siemens (The Load-Oriented Order Control System ABS in Use at Siemens). In: *Fachbericht zum Seminar 'Praxis der belastungsorientierten Fertigungssteuerung'* des Instituts für Fabrikanlagen der Universität Hannover (seminar proceedings). IFA, Hannover, 1986, pp 149–169.

[15] Knecht, R. Zwei Jahre belastungsorientierte Auftragsfreigabe (Two Years of Load-Oriented Order Release). In: *Fachbericht zum Seminar 'Praxis der belastungsorientierten Fertigungssteuerung'* des Instituts für Fabrikanlagen der Universität Hannover (seminar proceedings). IFA, Hannover, 1986, pp 119–148.

[16] Wiendahl, H-P. (ed.). *Praxis der belastungsorientierten Fertigungssteuerung* (*Practical Application of Load-Oriented Manufacturing Control*). Hanser, München, 1992.

[17] Haupt, R. A Survey of Priority Rule-Based Scheduling. *OR Spektrum* (11): 1989, 3–16.

[18] Pabst, H-J. Analyse der betriebswirtschaftlichen Effizienz einer computergestützten Fertigungssteuerung mit CAPOSS-E in einem Unternehmen der Einzel- und Kleinserienfertigung (Economic Efficiency Analysis of a Computer-Aided Manufacturing Control System with CAPOSS-E in Job-Shop and Small-Batch Production). Dissertation Universität Erlangen, 1985 (veröffentlicht in: Europäische Hochschulschriften, R.V., Vol. 645, Frankfurt/Bern/New York, 1985) (doctoral thesis, published in the above journal).

[19] Wiendahl, H-P, Lüssenhop, Th. Wirkung von Prioritätsregeln – eine kritische Betrachtung (Effects of Priority Rules – A Critical View). *Zeitschrift des Vereins Deutscher Ingenieure*, **131** (1): 1989, 36–41.

[20] Baker, KR, Bertrand, JWM. An Investigation of Due-Date Assignment Rules with Constrained Tightness. *Journal of Operations Management*, **1** (3): 1981, 109–120.

[21] Baker, KR, Kanet, JJ. Job Shop Scheduling with Modified Due Dates. *Journal of Operations Management*, **4** (1): 1983, 11–22.

Chapter 7
Schedule-Oriented Capacity Planning and Control

7.1 Abstract

Load-oriented order release assumes that the capacities of the relevant work centers are given and balanced in advance, otherwise the algorithm will reject the orders which may overload bottleneck capacities. Using the throughput diagram, this chapter describes how the capacity adjustment can be achieved. It shows the necessary degree of flexibility of personnel operating the work centers.

7.2 The Problem and the Method

In load-oriented order release it is assumed that the capacities are fixed. The reason for this is that during long-range planning, which is carried out long before the order release, the available capacity is adapted to the required capacity. This works surprisingly well over longer periods and at high levels, i.e. on the level of job shops or for the whole plant. On the level of individual load or cost centers, however, more or less large deviations will occur in the course of time, and they will either cause schedule deviations or lead to short-term capacity adjustments.

Second, the lateness is not monitored as an independent variable in the order release methods described so far. On the contrary, one assumes that shorter lead times and a more even distribution of load will result in a smaller variance in the flow times and thus a smaller schedule deviation. This assumption is correct in principle, and could be proved in simulations and in practice [1–3]. However, a considerable proportion of orders can still not be finished on schedule, despite careful planning and sufficient capacity. This is due to events which are unforeseen when schedule and capacity planning are carried out, and which occur after the orders have been started. These events are either short-term alterations in orders (quantity changes, due date changes, or rush orders), or disruptions in capacity (personnel absenteeism, machine breakdown, lack of materials or facilities).

To cope with these problems, the exact capacity requirements at each point in time beyond the next period should be known in advance, so that the individual work center capacities can be adapted wherever possible.

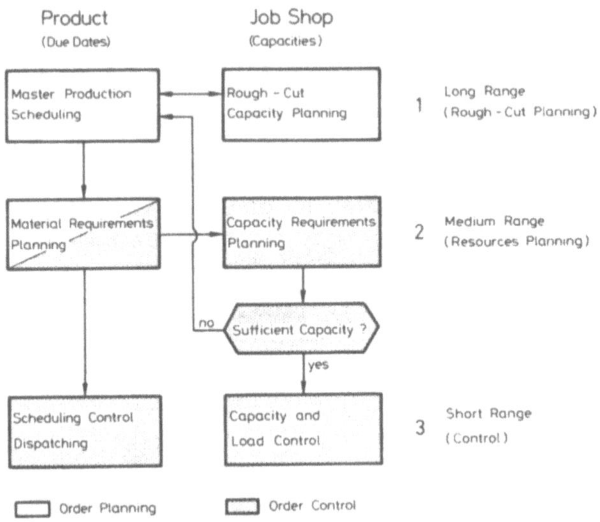

Fig. 7.1. Functions of manufacturing control (After Plossl [4])

When the order release is carried out it is too late for such capacity planning, since larger-scale capacity modifications – such as two-shift work, the re-allocation of personnel, or subcontracting – often require several weeks' notice for planning, and must be maintained over a longer period of time, e.g. at least one or two weeks.

The order release can only show deviations between the required and available capacity, and can then either reject orders (with the help of the load limit) or pull orders ahead (within the time limit). Only minor capacity corrections are possible at this stage – for example by introducing overtime on a small scale, or by employing cross-trained workers.

This clearly shows that, especially in manufacturing companies with rapidly changing mix of orders, capacity planning must be carried out immediately after order planning. The capacities must be within the scope given by long-range planning, and they must also take into account the shop's possibilities for capacity adjustment. This is all the more important, the shorter the intended flow times become. Short flow times cause very small buffer times. As a result, uneven manufacturing flows caused by order time variances and changes in schedule cannot be leveled out as smoothly as with longer flow times.

Figure 7.1 shows Plossl's [4] approach, which stresses the parallel organization of order and capacity planning. Erdlenbruch used this model as a basis for the development of new control methods [1]. One can see two functional chains: the scheduling of the products, and the capacity planning for the shop. Both chains contain long-, medium-, and short-range planning stages. Whereas load-oriented order release, in the short-term control phase, coordinates the given capacities with the given order due dates, a comparable method does not exist for resources planning and rough-cut planning.

To develop such a method, it is necessary to consider the throughput of shop orders not only from release to delivery, but also as part of the entire order network (Fig. 7.2) [5]. In demand-controlled order planning (as opposed to planning by time

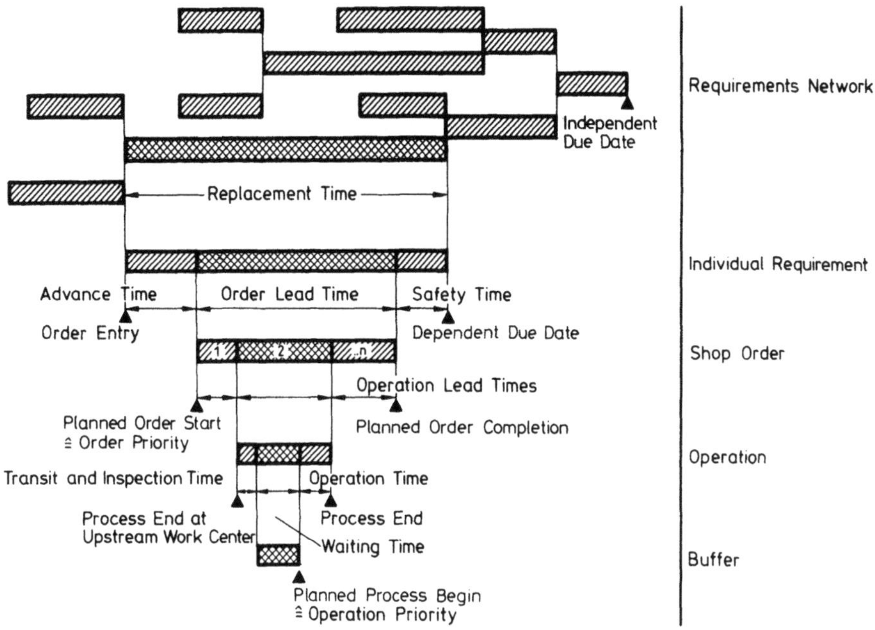

Fig. 7.2. Components of throughput planning (After Erdlenbruch [5])

phased order point), one proceeds from the known due date of the product (independent requirements due date) and, for each manufacturing level of the product, one calculates the dependent requirements (quantitatively) using the bill-of-material tree and (in view of the schedule) using the replacement times. This is shown in the order network in Fig. 7.2.

The replacement times for the individual requirements are composed of the known order lead time (allowance) and two more time components (Fig. 7.2, centre): the advance time allows orders to be started earlier – within the time limit – than would otherwise be necessary for meeting the due date. If this time were not contained in the replacement time, the load-oriented order release could not pull orders ahead, because material and facilities would not be ready. The safety time, on the other hand, has the function of compensating for unavoidable variances in the order lead time. This safety time, however, must not be regarded as a waiting time buffer for the individual operations; the order due date must be planned at the requirement date minus the safety time.

The single operations, in turn, contain the components operation time, waiting time, and transit time, as discussed above. Besides transportation, the transit time can also cover quality inspection and minor post-operations such as deburring or cleaning. The resulting planned operation start date also serves as the sorting criterion in the work center queue, and thus determines the sequence of the operations (operation priority).

By integrating individual operations, manufacturing orders, and replacement times into a network, it is possible to determine realistic lead times.

The lead times can also be determined the other way round: from the long-term production schedule, one first determines the weighted mean operation times for the individual load centers. The necessary mean waiting time and mean transit time are then added to the operation time. The waiting time is calculated from the mean weighted operation time and a proportionality factor, which lies around 1.0, depending on the variances in operation time (for this factor cf. Sect. 6.6). The mean transit time is a load-center specific empirical value. In job-shop production, it lies between 0.5 and 1 day. With the weighted mean lead time per load center determined in this way and with a given period length, the load limits, the respective loading percentages, and thus the mean inventories, are also fixed. When the values are compared with the lead times actually recorded, one can see at an early date whether a change in the order mix will affect lead times at the work centers and thus change the order lead times. After the actual due dates for the end products have been fixed, a lead time scheduling using the calculated operation lead times will ensure that the values can, on average, be met.

These considerations do not, however, tell us anything about the capacity requirements. Therefore, the start dates determined by lead time scheduling must be combined with the standard times known from the routing sheets, so that a capacity load may be determined. For this, load centers must first be defined.

7.3 Defining the Load Centers

Single work places are not normally controlled individually. Machines of the same kind are grouped into load centers. In load-oriented manufacturing control, each load center corresponds to a funnel in the funnel model.

An order controlled within the shop concerned can either pass through this load center or not. Only in special cases does a load center consist of one single work place, for example, a processing center which is unique in the shop. The sample plant in Table 3.6 also contains a series of such single work centers.

A second example of how work centers in a manufacturing shop can be structured into load centers is given in Fig. 7.3 [6]. More than 300 single work places in a PCB shop are organized into only 50 load centers. Depending on their type, the boards, which have been prepared elsewhere, now pass different insertion and inspection centers, but only one soldering and washing line; some of the boards are washed several times [6].

The material flow matrix presented in Chapt. 5 (Table 5.5 and Fig. 5.12) can help in creating such funnel models with their load centers. This matrix is produced by a representative evaluation of feedback data from the manufacturing process. For newly planned manufacturing processes, the matrix is developed from a comparable manufacturing process and its corresponding route sheets.

A distinction is made between different types of load centers [6–8]. Most load centers are of the normal load center type. Their capacity is primarily determined by the personnel attendance time or by the availability of machine capacity, for example of an automatic machine. Their machines either process work of the same kind (e.g. turning or grinding), or they combine operations which necessarily belong together, such as soldering and cleaning or milling and deburring. For these centers, only one load account is kept.

In some load centers, the capacity depends on both machinery and manpower at the same time. This is the case if, for example, the number of machines is greater than the number of operators. Figure 7.4 outlines this case [8]. Six machine centers of a

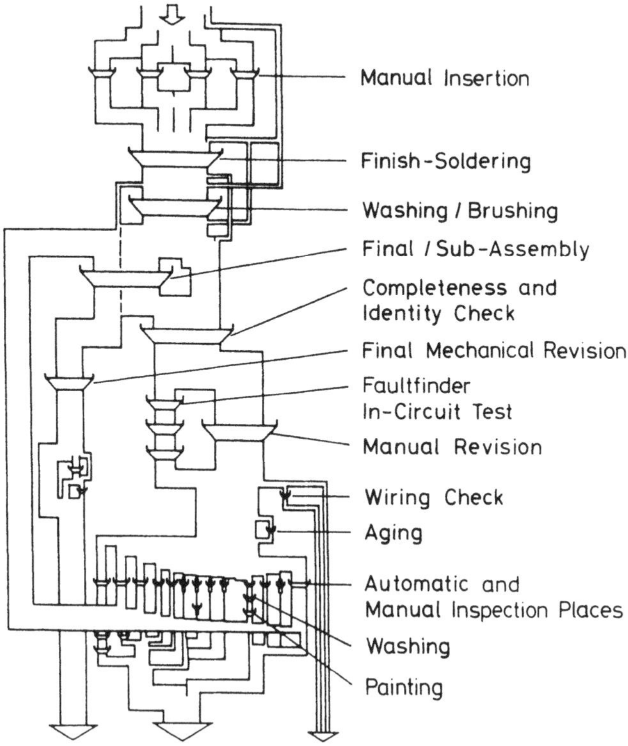

Fig. 7.3. Funnel model of a PCB production (After Ulfers [6])

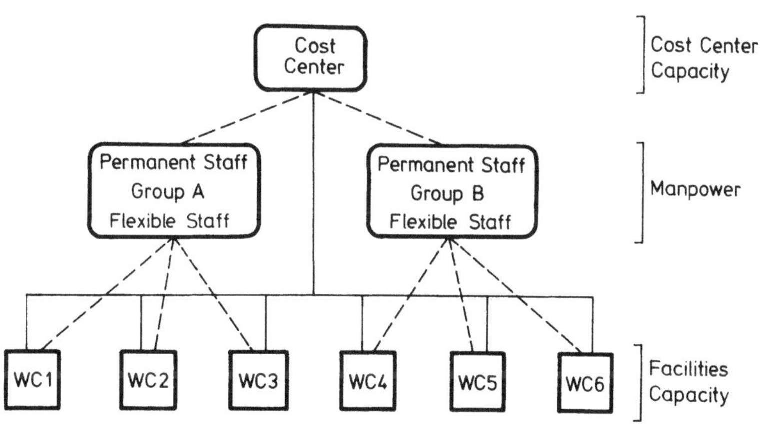

Fig. 7.4. Interlinkage of manpower and facilities capacity in a cost center with multiple-machine operation (After Ritter [8])

cost center are assigned to two staff groups which, in turn, consist of permanent staff and a few flexible workers. The latter are persons who can move and work at any one of the three centers, depending on the loading situation. In such a case, besides the capacity accounts for the technical facilities, a collective personnel capacity account is kept, which is called an *overall load center* [6]. The order release then checks whether the (converted) work content of the order does not exceed the load

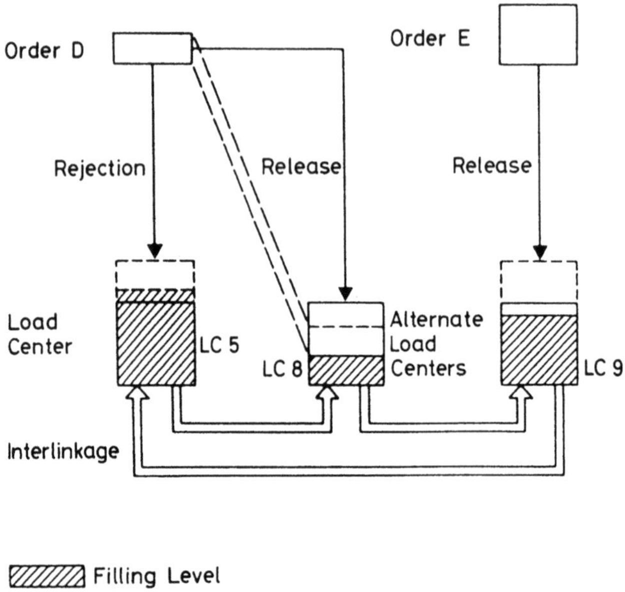

Fig. 7.5. Effects of alternate load centers (After Ulfers [6])

limit of the respective machine capacity or the load limit of the collective personnel capacity.

An important capacity balancing measure is a changeover of personnel from an underloaded to an overloaded load center [9]. The load center taking on the load is then called the *alternate load center*, and the possibility for such a changeover is marked in the master data of both load centers concerned. Figure 7.5 outlines an example of three such load centers [6]. Order D is first rejected at LC 5. LC 8 is identified as the alternate load center, so that the order can be released and booked onto the load account of LC 8. Other changeover possibilities are indicated from LC 8 to LC 9, and from LC 9 to LC 5.

There are some other work centers that need not be checked for free capacities during the release, since only a small part of their capacity is ever used. These are called *non-critical load centers* [6]. Nevertheless, load limits and planned lead times are still defined for these centers so that the corresponding orders can be planned correctly. The planned lead times are, in turn, used to calculate the corresponding load limits so as to convert and load the work contents of the orders concerned.

Finally, there are load centers whose lead times are virtually independent of the work contents of the processed orders, since they consist of process-dependent waiting times. Examples are heat treatment processes for paint curing or for materials aging. These are process-dependent load centers; their planned lead times are constant [6].

The grouping of work centers into load centers must be carried out very carefully. On the one hand, too many and too small centers cause an unnecessarily large number of feedback events and allocation lists. On the other hand, grouping too many work centers together leads to large 'funnel contents', and the orders might be processed according to local optimization criteria instead of according to their due date priority.

7.4 Schedule-Oriented Capacity Planning Procedure

From our discussion in Chapt. 2 it became clear why conventional capacity scheduling works so poorly in practical application. In Figures 7.6A and B, we can see the usual process of lead time scheduling, after which the single work centers were loaded in an uninterrupted sequence by shifting the operation elements within the

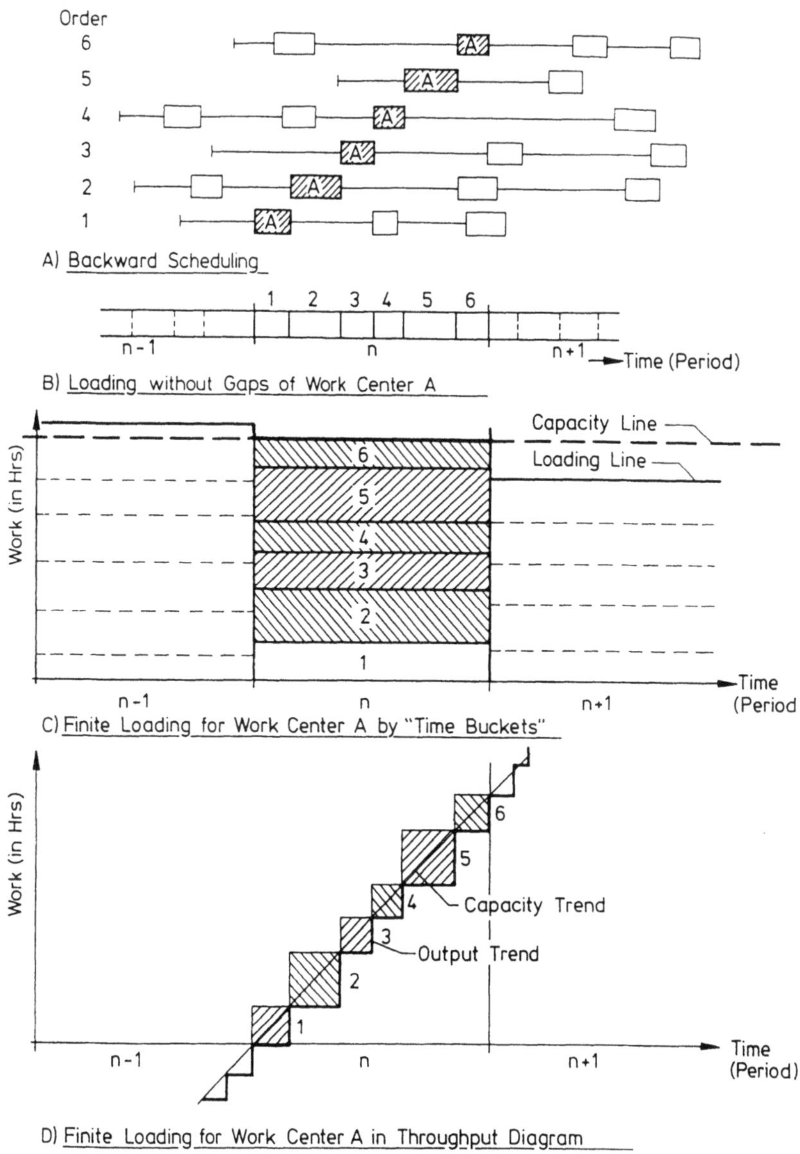

Fig. 7.6. Conventional finite loading in time buckets and in the throughput diagram

buffer times. Figure 7.6C shows the finite loading calculation resulting from this, which in turn gives us the loading level (represented by the loading line). A comparison with the capacity level (represented by the capacity line) shows apparent underloads or overloads in the individual periods.

If this is plotted in a throughput diagram, we get Fig. 7.6D. We can see the output curve and the capacity curve. If the orders are processed without disruptions, the operation elements join each other at the corners so the capacity curve is congruent with a line formed by joining the diagonal lines of each element.

On the one hand, conventional capacity planning is one-dimensional, since it considers the output curve but not the input curve. On the other hand, it is also deterministic, since it expects the orders to be processed in exactly the planned schedule and sequence. As we know, however, this is only the case for a very short time, due to the inevitable variance in the lead times, and because of the other stochastic influences on the manufacturing flow.

Figure 7.6D suggests a solution for the problem. The output curve can be seen as a cumulated capacity requirement, and the input curve as a cumulated capacity supply. The advantage of this cumulative approach is that sequence transpositions have practically no effect on the output curve, since it is irrelevant for the capacity requirement when an order in the queue is processed. The sequence transpositions only have an effect on the schedule deviation.

This cumulative capacity planning approach is not new; it was used by Brankamp as a decision criterion in his capacity alignment model [10]. Figure 7.7 shows how Ritter [8] used this approach. In the company concerned, the overall order stock is subjected to lead time scheduling period by period. The load is cumulated using the planned end dates and plotted automatically for each load center. The planned capacity is plotted in the same diagram.

On April 22nd of that year, the overall known order stock for this lathe center until the end of March of the subsequent year was about 73 500 hours. The planned

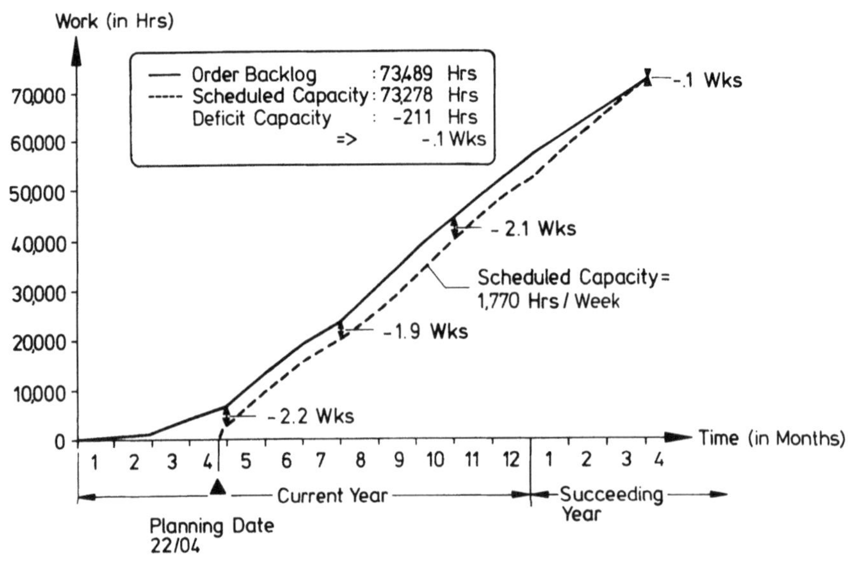

Fig. 7.7. Cumulative curves of order stock and scheduled capacity of a cost center (chucking lathe) (After Ritter [8])

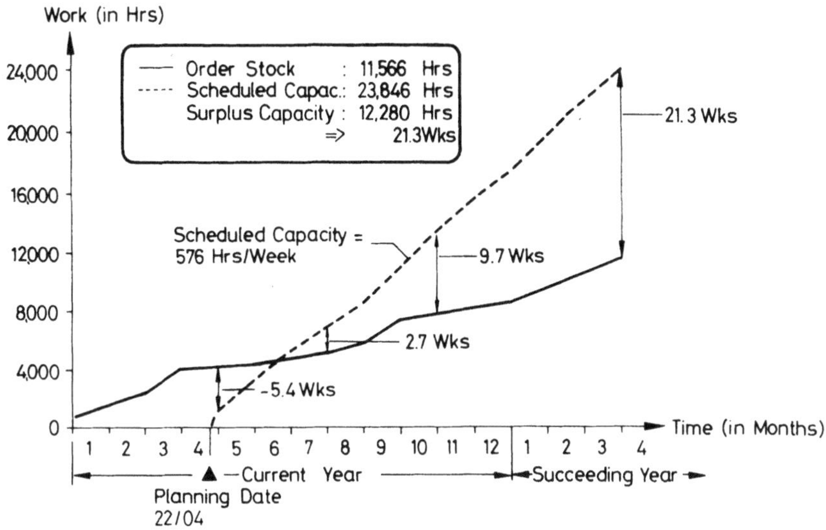

Fig. 7.8. Cumulative curves of order stock and scheduled capacity of a cost center (assembly department) (After Ritter [8])

capacity for the same period was about 73 300 hours. Thus there was a deficit of 200 hours at the end of the planning period. If, at any time, one divides the difference between the cumulated order stock and the cumulated performance by the planned output, this calculates the range of this work center at that time (for the definition of range, cf. Sect. 4.4). For example, the range was −2.2 weeks at the end of April of the current year, which means that the orders are expected to have a mean delay of 2.2 weeks, since capacity lags behind the capacity requirement. With the order stock as known on the planning date, there will presumably be a delay until the end of the current year; then it will gradually decrease as the order stock is reduced.

Figure 7.8 shows an entirely different situation. The cost center shown here is an assembly capacity. The schedule delay of the orders is much higher for the same planning date as in Fig. 7.7. It will presumably be reduced within two months because the requirement is relatively small compared to the capacity. After that time, it is advisable to reduce the capacity.

The solution that was outlined for determining the schedule-oriented capacity requirement is, however, still not ideal: it is normally not possible to bring the capacity curve exactly in line with the planned output curve.

This is where the method suggested by Erdlenbruch can be applied [1]: the mean work center throughput element, which was already shown in Figs. 6.16 and 6.20, contains the three components 'Operation', 'Transport', and 'Waiting', and these are shown again in Fig. 7.9A.

Now we can imagine the following model for the processing of an authentic order [5]:

- The operation must be finished no later than on the date defined in the lead time scheduling (Fig. 7.9A).
- The lot must reach the work center concerned no later than on the date 'latest transport end', but can also arrive there earlier.

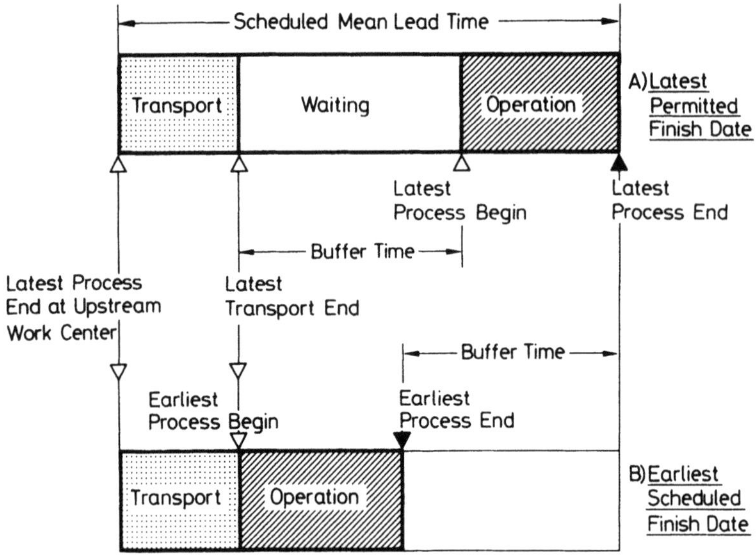

Fig. 7.9. Model throughput element for lead time scheduling (After Erdlenbruch [5])

- An operation can then begin earlier than planned, i.e. as soon as it is available, and on condition that it is the top-priority order in the queueing order stock (Fig. 7.9B).
- Thus, every operation has a time buffer corresponding to the planned lead time minus the transit time and the operation time.

If one now sorts the orders according to increasing planned output dates, one can use their work contents and the dates 'earliest process end' and 'latest process end' to plot a capacity planning diagram consisting of two output curves. These curves show most clearly and comprehensibly the time and inventory buffer within which the operations can be moved at this work center. A kind of capacity requirement corridor is created. It is the aim of capacity planning to locate the future capacity curve within this corridor, wherever feasible.

Figure 7.10 shows a capacity planning diagram developed according to these considerations for an individual work center during a simulation [1]. In the actual process, the capacity remained constant. In the simulation, the same amount of work was scheduled in the investigation horizon of 112 days (= 16 weeks), but the maximum capacity was at times 23 hours per day at every single work place. We can see that the large work hour requirement around Day 260 can probably not be processed on schedule, despite the high capacity.

Figure 7.11 is the result of the simulation using the capacity curve defined in Fig. 7.10. The load limit was fixed at a loading percentage of 250 [1]. Then the orders were released according to the load-oriented order release algorithm. The flow of orders was then simulated as described in Sect. 6.8, and the throughput diagrams of each work center were plotted. Besides the capacity curve, the figure also shows the release and input curves as well as the output curve, which lags somewhat behind the planned output. This is because – as we saw in Fig. 7.10 – the maximum capacity as provided was not sufficient around Day 260.

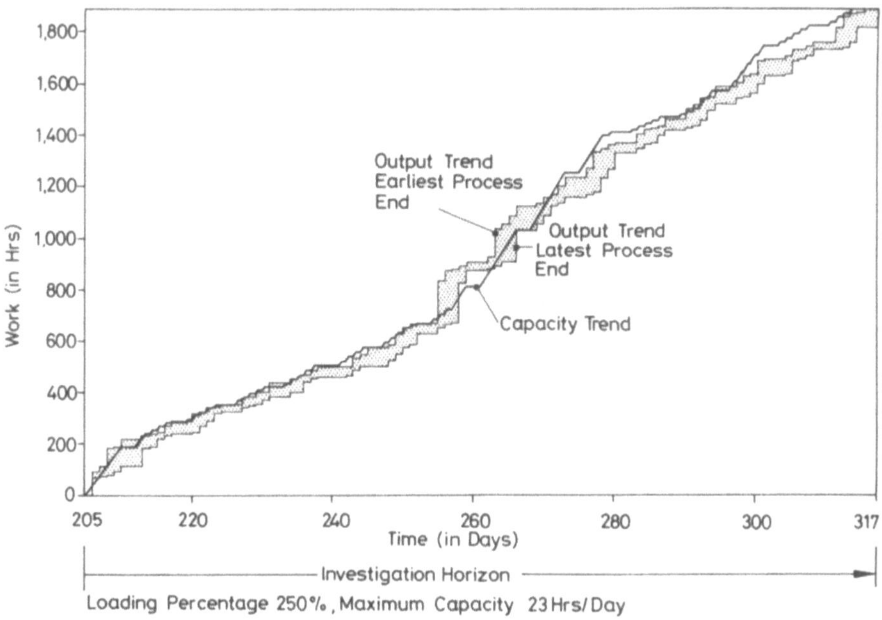

Fig. 7.10. Capacity planning diagram for a milling work center (After Erdlenbruch [5])

The effects of a flexible capacity on the order schedule deviation are interesting. Figure 7.12 shows the lateness of the orders in the shop whose characteristic production lines were already presented in Fig. 6.36 [1]. The recorded values are taken as scheduled delivery dates. The first simulation used load-oriented order release with individually adjusted load limits.

Now a comparative simulation was carried out which, in addition, used schedule-oriented capacity planning as explained above. Where the first simulation with individual load limits had already brought about a mean improvement of 7.6 days in comparison to a standard load limit, flexible capacity further increased this value by about 0.4 days, making 8 days altogether. What is even more important is that the standard deviation of the lateness nearly halved from about 13.2 days to about 6.6 days. This shows the positive effects of this process module on schedule performance. The improvement can also be expressed very clearly in a schedule deviation diagram of the orders. Figure 7.13 shows the diagram for the simulation with individual load limits, but with fixed capacities [1]. Unlike in the schedule deviation diagram of the work centers – as shown in Fig. 4.19 – it is not the actual output dates, but the planned output dates that are chosen as the reference curve from which the schedule deviation areas are plotted. The actual output curve was also plotted into the diagram, so that the mean earliness of 7.6 calendar days, known from Fig. 7.12, can roughly be estimated. What shows up very clearly now is the strong variance in the schedule deviation.

By way of comparison, Fig. 7.14 shows the simulated order throughput using schedule-oriented capacity planning [1]. Besides the somewhat earlier processing, we can see a great reduction in the variance of the schedule deviation. There are still

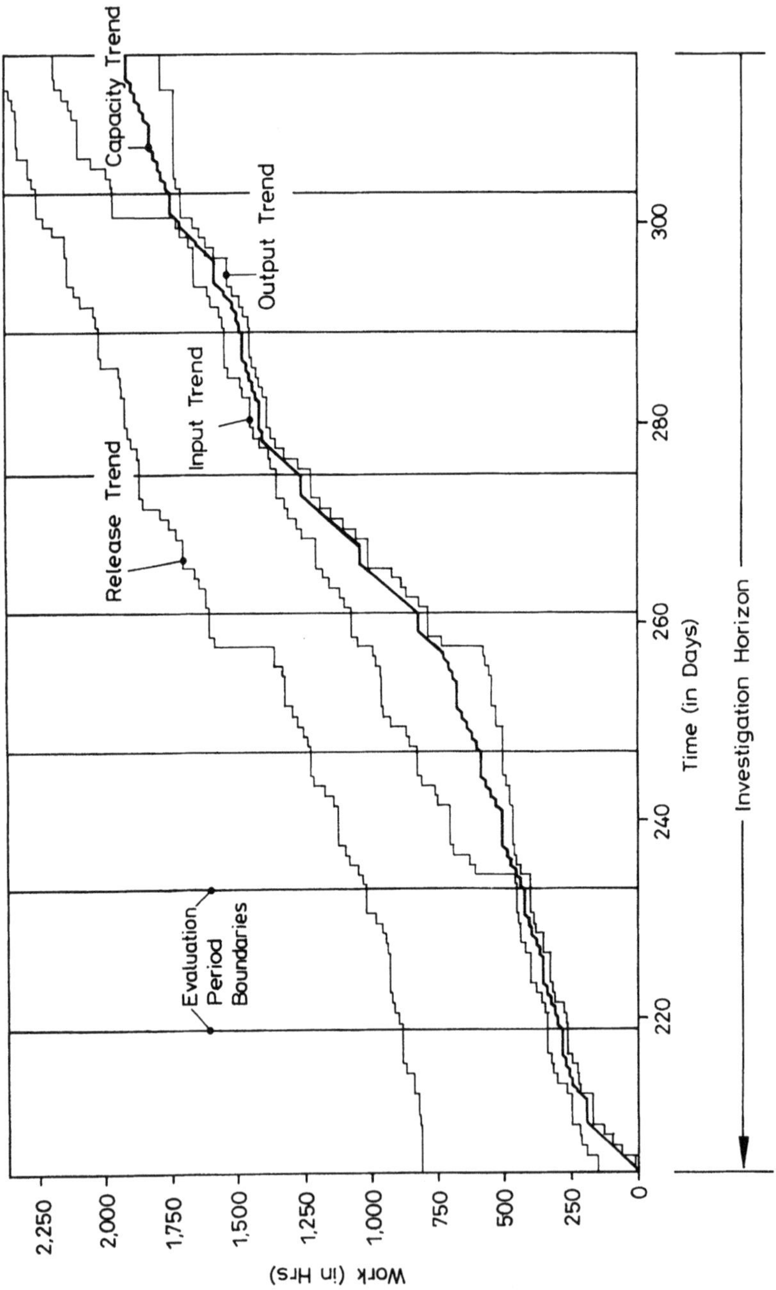

Fig. 7.11. Capacity monitoring diagram for a milling work center (After Erdlenbruch [5])

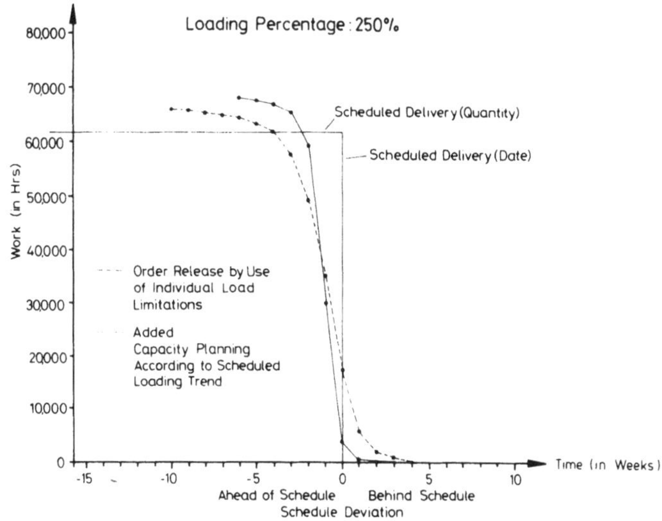

Results:		Capacity		
		Constant	Variable	
Mean Lateness	:	-7.64	-7.98	Days
Standard Deviation	:	13.15	6.64	Days

Fig. 7.12. Distribution of order lateness – constant versus variable capacity (After Erdlenbruch [5])

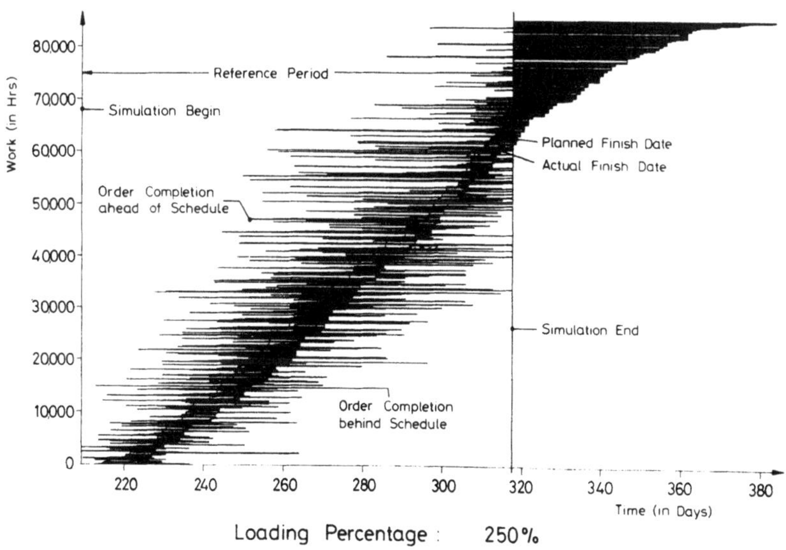

Fig. 7.13. Lateness diagram with constant capacity (After Erdlenbruch [5])

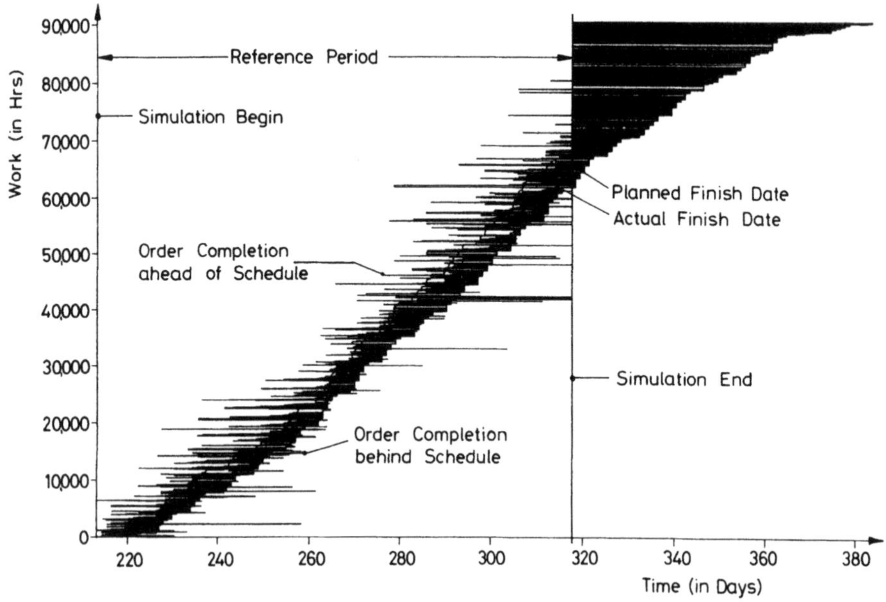

Fig. 7.14. Lateness diagram with variable capacity (After Erdlenbruch [5])

great schedule deviations for single orders, but these have become so few compared to the total number that it is now possible to take corrective measures.

To sum up, the following can be said about schedule-oriented capacity planning. The cumulative approach using the 'capacity corridor' avoids over-exact load planning. First, it makes clear when and where bottlenecks are likely to occur, and which capacity adjustment will be necessary. Second, it shows whether the adjustment lies within the scope of what is possible in the plant concerned. Finally, if it is not possible for the capacity curve to run completely within the 'corridor', the effects on the schedules of single orders can also be assessed [5].

Even this planning cannot, of course, completely balance out extreme changes in the order mix, or extensive changes in the due dates. The effects of such disturbances, however, become clear much earlier, so that it is possible to take measures much less intuitively than was the case before. In addition, the effectiveness of load-oriented order release is improved. Schedule-oriented capacity planning interlinks the 'utilization' and 'schedule performance' objectives within the framework of the 'lead time' objective as controlled by the mean inventory. The better the flexibility of capacity control over time, the smaller the schedule deviation will be.

Capacity planning itself is effective on all three levels of manufacturing control. In long-range planning, the capacity requirements are determined roughly on the basis of the production program. The trend of required capacity and available capacity over time is not, as a rule, taken into account. In medium-range planning, the cumulative capacity requirement can be defined as shown above, on the basis of the known required quantities and dates for the shop orders. These capacity data are, in turn, the standards for the order release on the short-range control level. The deviations occurring on this level manifest themselves as orders which are either rejected or pulled ahead, and can be balanced out with short-term capacity adjustments, provided that these are possible.

This was the last component in a concept of a load-oriented manufacturing control loop, which logically combines the four objectives utilization, inventory, lead time, and lateness. The monitoring system based on the throughput diagram adds to this concept a regular comparison of planned and actual values, and permits the user to identify the causes of deviations and take appropriate corrective action.

7.5 References

[1] Erdlenbruch, B. Grundlagen neuer Auftragssteuerungsverfahren für die Werkstattfertigung (Principles of Recent Order Release Techniques in Job-Shop Production). Dissertation Universität Hannover, 1984 (veröffentlicht in: Fortschritt-Berichte der VDI-Zeitschriften, Reihe 2, Nr. 71, Düsseldorf, 1984) (doctoral thesis, published in the above VDI journal).

[2] Knecht, R. Zwei Jahre belastungsorientierte Auftragsfreigabe bei Firma Hilti (Two Years of Load-oriented Order Release at HILTI). In: Wiendahl, H-P. (ed.). *Praxis der belastungsorientierten Fertigungssteuerung* (*Practical Load-Oriented Manufacturing Control*). Institut für Fabrikanlagen der Universität Hannover (IFA), Hannover, 1986, pp 119–148.

[3] Möller, G. Reduzierung der Durchlaufzeit durch Einsatz eines PPS-Systems (Lead Time Reduction by Use of a PPC System). *Zeitschrift für Wirtschaftliche Fertigung*, **81** (3): 1986, 140–143.

[4] Plossl, GW. *Manufacturing Control – The Last Frontier for Profits*. Reston, USA, 1973.

[5] Erdlenbruch, B. Aufbau eines Fertigungssteuerungsystems zur Kapazitäts-, Durchlaufzeit- und Bestandsplanung – Terminorientierte Kapazitätsplanung – Belastungsorientierte Auftragsfreigabe (A Production Control System for Capacity, Lead-Time, and Inventory Planning; Schedule-Oriented Capacity Planning; Load-Oriented Order Release). In: Wiendahl, H-P. (ed.). *Praxis der belastungsorientierten Fertigungssteuerung* (*Practical Load-Oriented Manufacturing Control*). IFA, Hannover, 1986, pp 181–195.

[6] Ulfers, H-A. Belastungsorientierte Auftragssteuerung – Erfahrungen mit dem ABS-Verfahren in einer Flachbaugruppenfertigung (Load-Oriented Order Control – Practical Application of the ABS System in a PCB Production). *Zeitschrift des Vereins Deutscher Ingenieure*, **126** (4): 1984, 71–77.

[7] Buchmann, W. Zeitlicher Abgleich von Belastungsschwankungen bei der belastungsorientierten Fertigungssteuerung (Temporal Leveling of Load Fluctuations with Load-Oriented Manufacturing Control). Dissertation Universität Hannover, 1983 (veröffentlicht in: Fortschritt-Berichte der VDI-Zeitschriften, Reihe 2, Nr. 63, Düsseldorf, 1983) (doctoral thesis, published in the above VDI journal).

[8] Ritter, K-H. Voraussetzungen zum Einsatz der belastungsorientierten Auftragsfreigabe (Prerequisites for Load-Oriented Order Release). In: Wiendahl, H-P. (ed.). *Praxis der belastungsorientierten Fertigungssteuerung* (*Practical Load-Oriented Manufacturing Control*). IFA, Hannover, 1986, pp 53–71.

[9] Krautzig, J. Planung der Kapazitätsreservierung unter Berücksichtigung der Personalflexibilität bei Werkstattfertigung (Capacity Allocation in Job-Shop Production with Regard to Personnel Flexibility). Dissertation, Universität Hannover, 1981 (doctoral thesis).

[10] Brankamp, K. *Ein Terminplanungssystem für Unternehmen der Einzel-und Serienfertigung* (*A Scheduling System for Job-Shop and Batch Production*), 2nd edn. de Gruyter, Würzburg/Wien, 1973.

Chapter 8

Implementation of Load-Oriented Manufacturing Control

8.1 Abstract

To implement load-oriented manufacturing control in a shop, some prerequisites have to be fulfilled, same as for any shop floor control system. After discussing these prerequisites, this chapter concentrates on the data flow diagrams and interfaces to MRP, thus enabling the readers to program such a system on their own. Some advice is given on how to introduce this concept, and what economic results can be expected.

8.2 Prerequisites

Load-oriented manufacturing control replaces the conventional process steps of 'backward scheduling', 'finite loading', 'capacity alignment', and 'sequencing' introduced in Chapt. 2. It controls the lead time, inventory, utilization, and schedule deviation variables in an overall process model, represented by the throughput diagram or the funnel model, and monitors the variables with a monitoring system based on the same model. Order release is carried out on the level of the short-term job-shop control (Fig. 8.1) [1]. Before this, there is the order planning level, as discussed above, which has the aim of generating 'feasible' job-shop and purchase orders, and manufacturing planning, which has the task of adjusting the capacity requirements to the available capacity for a longer term.

The job-shop control on the lowest level will work better, the better the previous planning steps were. There are, however, a few other prerequisites for the use of load-oriented manufacturing control, and these are dealt with in the following subsections [1–4].

8.2.1 Influence of the Lot Size on Inventories and Mean Lead Time

The lot size is normally defined by cumulating the known requirements at the order planning stage. Repetitive order costs and the costs of the stocks of finished products

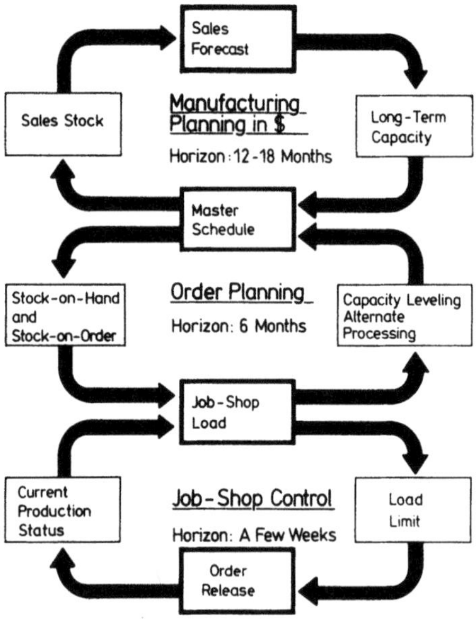

Fig. 8.1. Three levels of manufacturing control (After Ritter [1])

must be minimized in each case. This makes the lot size a variable that cannot be influenced by manufacturing control. Only in special cases (i.e. when there are delays) should it be changed by so-called *lot splitting.*

This method, known as economic lot sizing, has undergone increasing criticism, since it underestimates all the hidden costs of inventory. Therefore, a 'lot size one' is often demanded. 'Lot size one', however, also means that no repetitive order costs are allowed to occur, which, with manufactured parts, would demand a setup time of zero.

The question is whether, from the viewpoint of load-oriented manufacturing control, 'lot size one' always makes sense, particularly since the following must be taken into account:

- Compared with lotwise processing, there are many more orders to control.
- When there are bottleneck machines at which a setup is not possible during running time, there will be a loss in capacity which reduces the performance of the total shop.
- Frequent changes in types of parts cause higher scrap rates.

Let us recall our considerations on determining the load limit at individual load centers. In Fig. 6.11, we saw that the mean lead time only depends on the variation of transit times and the variation of weighted operation times. Since the lot size directly influences the order time, and thus the operation time, it also has a direct effect on the weighted mean lead time per operation. This effect can be quantified with the lot inventory (see Sect. 5.5). Figure 8.2 shows the disruptive influences of large-sized lots using the data from an authentic work center [2]. Large lots lead to deviations from the ideal input and output curves (straight lines of dashes and dots), and cause fluctuations in inventories and lead times.

From this we can conclude that from the point of view of manufacturing control,

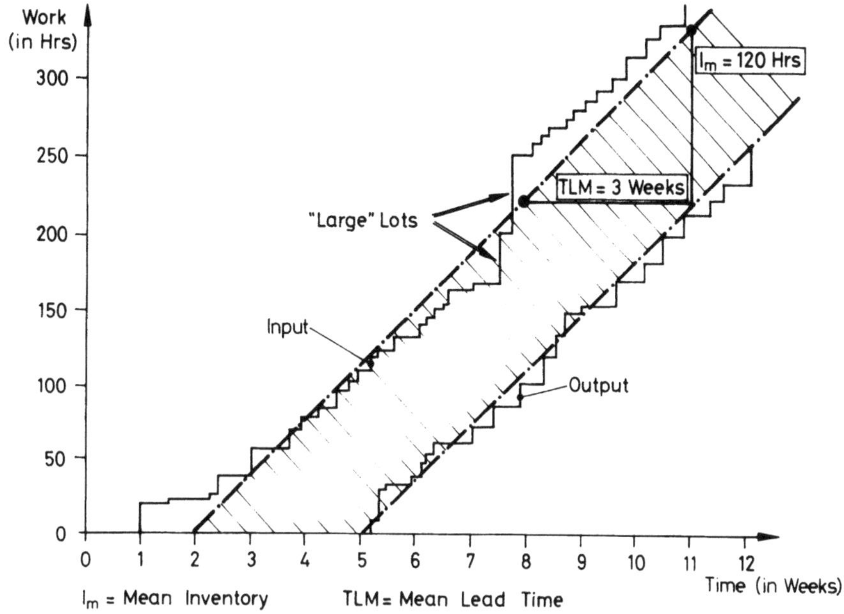

Fig. 8.2. Effects of large lots on lead time and inventory (After Kettner and Bechte [2])

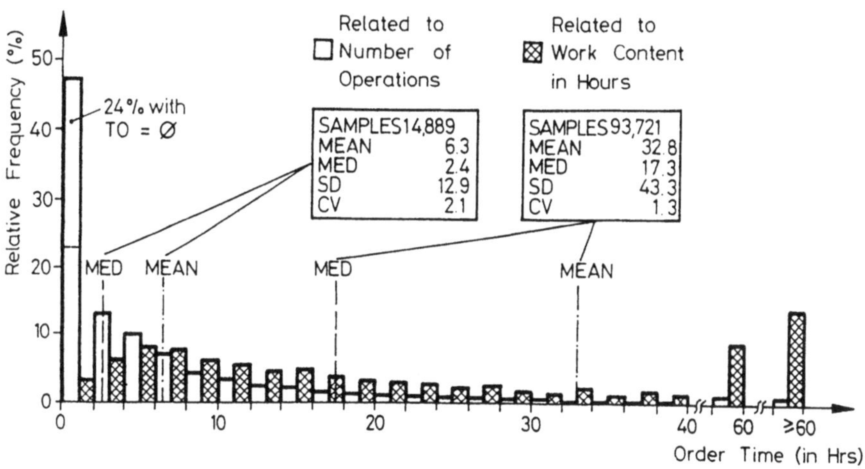

Fig. 8.3. Distribution of order time per operation by number of operations and by work content (After Bechte [6]). □: Related to number of operations; ▩: related to work content in hours

it is important to apply a limit to the mean value and the variation of the weighted mean lead time.

To do this, let us examine the order time distribution in our sample factory for all evaluated operations. Figure 8.3 shows the frequency distribution, and Fig. 8.4 the corresponding Pareto distribution [6]. (The calculation of a Pareto curve is described in Sect. 6.6, Table 6.6.)

We can see that 80% of the operations have order times of up to about 9 hours,

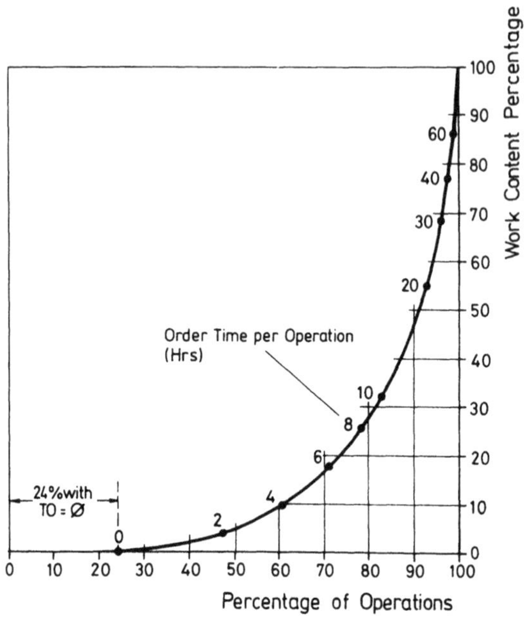

Fig. 8.4. Pareto curve of order time per operation for the sample factory (After Bechte [6])

and only represent about 28% of the work content, whereas 10% of the high-volume operations need more than approx. 17 hours of order time, and constitute about 53% of the total work content. This means that a very small portion of the lots determines the variation in the operation time. From this it follows that dividing the few large lots into several small lots could have a considerable effect on the variation in operation time, and thus on its mean value. Thus, instead of following the 'lot size one' concept, one should try to reach a 'lot size harmonization'. In our example, a lot limitation to a work content of 20 hours would only affect 7% of the number of lots and a limitation to 40 hours would only affect 3%.

To estimate the effects of such a lot limitation on the lead time, inventory, utilization, and schedule deviation variables, several simulations were carried out with the data from our sample factory (Table 3.5). The work content was limited to 40 hours per lot [6, 7].

Figure 8.5 shows the weighted mean lead time curves and the output curves with and without lot splitting [7]. We can see from the output curve that the output is consistently higher with lot splitting than without it. This is in part due to an increased number of setups, and to an increase in real output because of better work center utilization. In the simulated period of 16 weeks, additional setup time accounted for 1600 hours, which corresponded to a setup time increase of 14% and an increase in total output of 2.3% [6].

For a *LPG* less than 300, the distance between the two curves increases. This is because, due to the more uniform order times, there are fewer idle times at downstream work centers. The increase in output soon outweighs the capacity lost due to the additional setup times. At the loading percentage (*LPG*) of 200 with lotsize limitation, for example, the output was nearly as high as at *LPG* = 250 without lotsize limitation, but the inventory was reduced from some 6500 hours to some

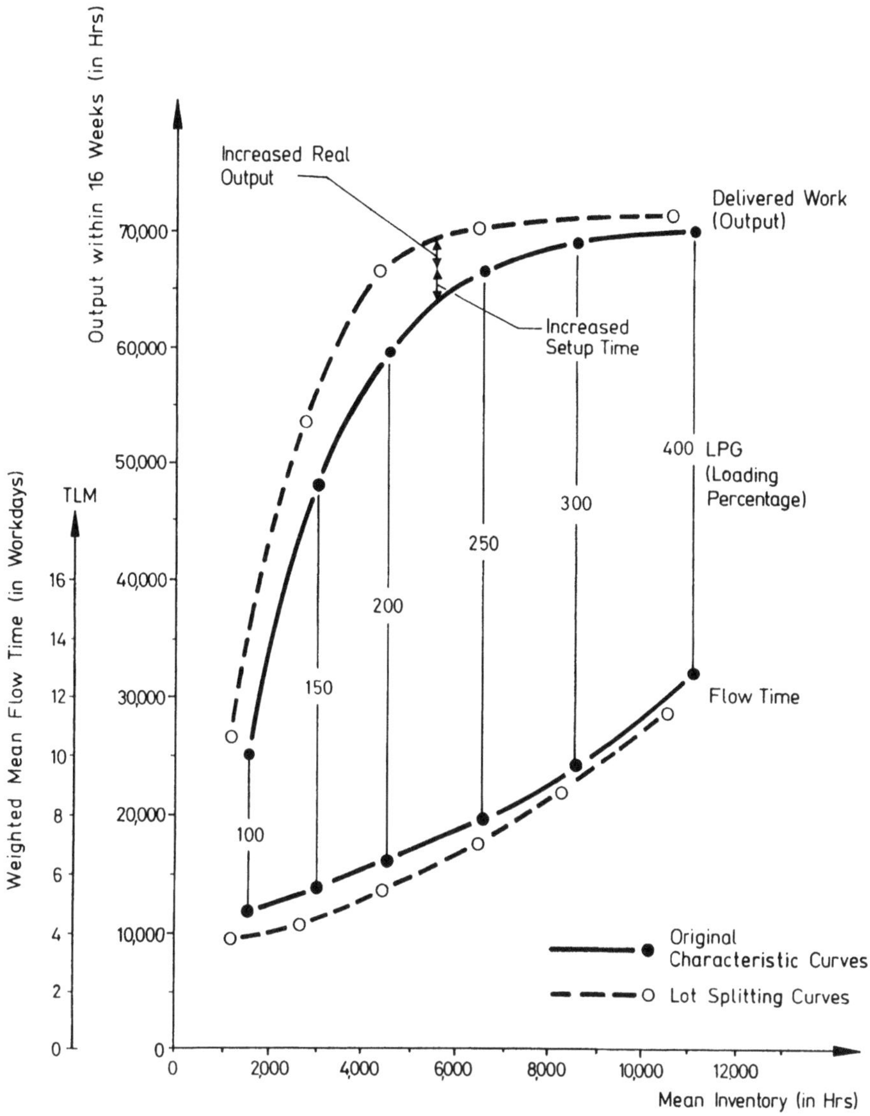

Fig. 8.5. Characteristic production lines in case of lot splitting (After Bechte [6] and Erdlenbruch [7])

4300 hours (by about 30%), and the weighted mean lead time at the work centers was reduced from about 8 days to about 5.5 days (also by about 30%). This is a remarkable potential for improvement.

Figure 8.6 compares the distributions of the simulated operation lead times at $LPG = 200$ with and without lotsize limitation [6]. The weighted mean has decreased from 6.2 to 5.5 workdays per operation, and the standard deviation has dropped from 4.1 to 2.9 workdays, which is remarkable. We can see from the distribution that this is due to the considerably smaller number, or even the lack of values in the lead time groups above 10 days. (The fact that the groups of up to two days are not filled is due to a minimum interoperation time of two days assumed in the simulation.)

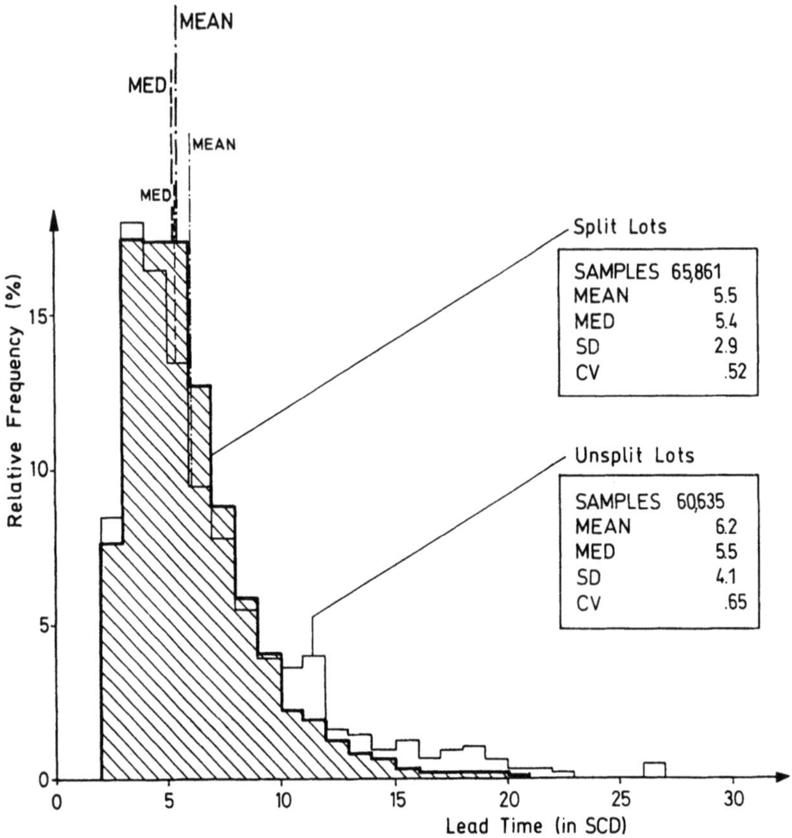

Fig. 8.6. Distribution of operation lead time in the sample factory – split versus unsplit lots – for a loading percentage of 200 (After Bechte [6])

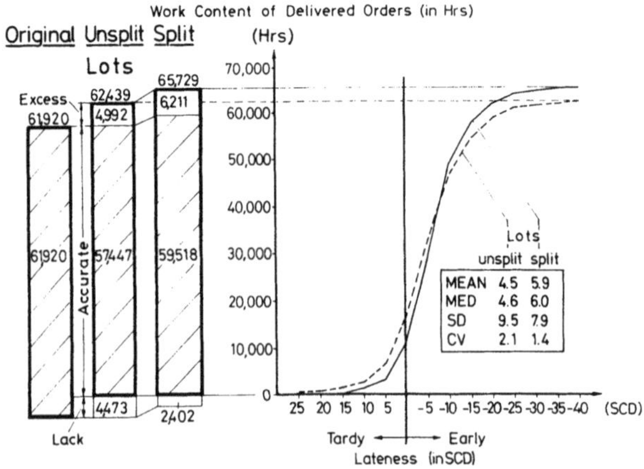

Fig. 8.7. Delivered orders and lateness – split versus unsplit lots – for a loading percentage of 200 (After Bechte [6])

The reduction in the weighted mean lead time per operation and in its variance also has a positive effect on the delivery performance (Fig. 8.7) [6].

At *LPG* = 200, the mean earliness improves from 4.5 workdays with unsplit lots to 5.9 workdays, i.e. by about 25%. The standard deviation decreases from 9.5 to 7.9 workdays, i.e. by about 17%. The proportion of delayed order hours drops from about 15000 to about 10000, i.e. by about 30%.

The approach was further developed by Bertrand, who proposed to consider not only the capital tie-up caused by the storage of finished parts in the lot sizing process, but also the capital tie-up due to the processing, waiting and storage during manufacturing [50]. Nyhuis developed a new lot-sizing formula, called *throughput-oriented lot-sizing*, that implements this idea, and which is based on the funnel theory and the operating curves described in Sect. 6.8 [51]. References to English articles on Nyhuis' formula are given in Chapt. 5 ([29] and [30]).

To conclude, we can say that the harmonization of the order times has a strong positive effect on the lead time and the schedule performance. Therefore, one should consider a limitation of work contents when defining the lot sizes during job planning.

Similar recommendations are given by manufacturing control methods according to the Kanban principle. There, the work content of one lot should not exceed half a workday.

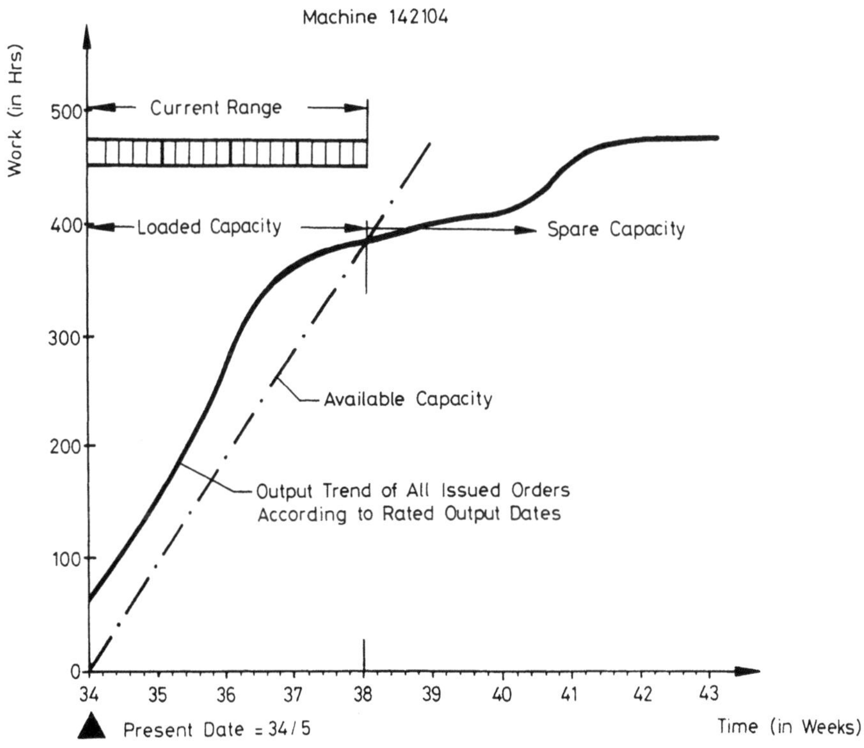

Fig. 8.8. Determination of the current range for one machine by use of the capacity and planned output curves (After Bechte [6])

8.2.2 Orders Must Have a Due Date

For determining the planned due dates of the individual operations, the due date of the order must be known. If materials planning is based on unrealistic replacement times, it is inevitable that at the release stage, orders which should be released are instead deferred. Being ironic, one could call this an 'order release prevention algorithm'. Permanent lead time monitoring with the help of the above-described monitoring system, however, reveals the reasons for frequent order rejections.

However, there will still be more or less large deviations between the lead times planned on the basis of the load limits, and the actual lead times caused by the current inventory situations at the work centers. To check whether a certain due date of an order is still realistic in a given situation, one can, in addition to lead time scheduling with planned values, define the actual ranges of the work centers concerned [8]. To do this, the current planned output curve and current rated capacity curve at the work centers concerned are compared as shown in Fig. 7.7. Figure 8.8 shows the graphic determination of the so called current range at a sample work center with a capacity of about 20 hours per workday. (The range determined in this way is not identical with the mean range defined in Sect. 4.4.3.1.) At the end of Week 38, the planned output curve and the rated capacity curve intersect; from this date on, spare capacity is available.

Now the throughput of the order in question is simulated, as shown in Fig. 8.9 [8]. First, the current ranges of the work centers needed for this order are plotted as

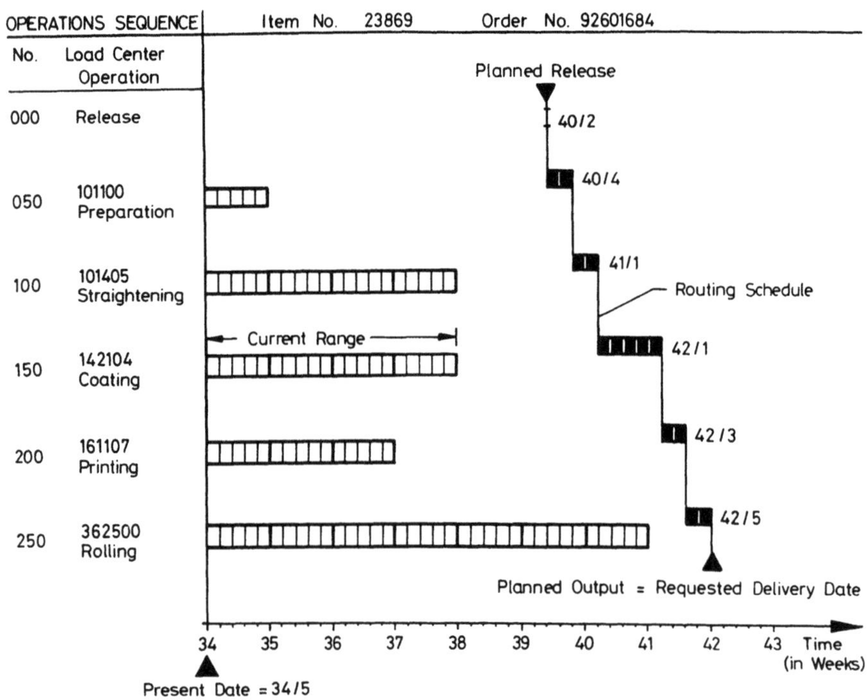

Fig. 8.9. Determination of order bottlenecks in case of fixed delivery date, by use of the current range (example without bottlenecks) (After Bechte [6])

bars on the time axis, starting from the present date. This shows us the earliest date at which capacity is available at each work center, and how much capacity there is. As a second step, a production run schedule with the planned lead times of the work centers concerned is set up. This schedule is positioned on the time axis in such a way that it does not fall below the longest range of all the work centers, thus giving a realistic delivery date (here the end of Week 41). At the critical Work Center 362500, for Operation 250, there is still a buffer, so that delivery at the requested date (end of Week 42) can be promised.

Several data can be gathered from such a graph. If the realistic delivery date lies before the requested date (= planned output date), a definite date can be fixed between them. This also fixes the release date. If, however, the realistic delivery date lies after the requested date, this must be caused by one or more work centers. Figure 8.10 shows such a situation.

In this case, the load situation of the load centers was the same as in the previous figure, but the requested delivery date now lies at the end of Week 39. We can see that bottlenecks are to be expected at Operations 100 and 250. (Operation 150 is not critical, because Load Center 142104 is idle from Day 38/5 on, and the lead time of five days for Operation 150 contains only one day of operation time.) Thus, either the delivery date must be shifted or the necessary additional capacity must be created. Only in exceptional cases should other orders in the queues be deferred. On the whole, great importance should be attached to realistic planned lead time values and thus realistic planned due date values.

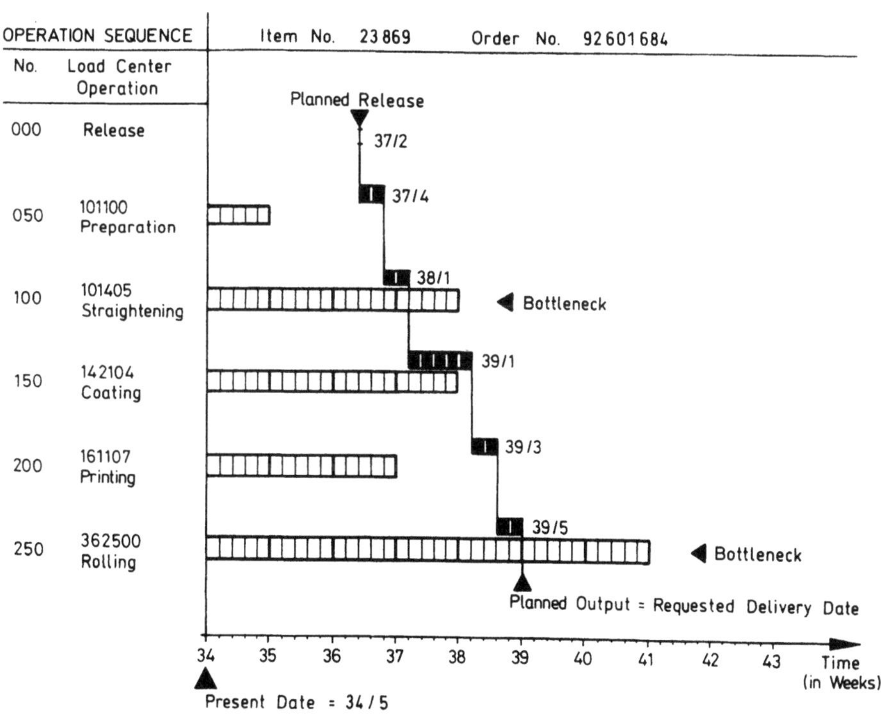

Fig. 8.10. Determination of order bottlenecks in case of given delivery date, by use of the current range (example with two bottlenecks) (After Bechte [6])

8.2.3 An Operation Sheet with Standard Times Must Exist

It is obvious that load-oriented manufacturing control cannot work without standard times, since its very principle is based on the examination of input and output data of work at the work centers. It is, however, not always useful or possible to calculate standard times for all operations, for example when the effort involved in generating, handling, and processing the data is in no proportion to the possible benefit. In such cases, fixed planned lead times are defined, but the work centers concerned are not included in the release process so that they cannot cause order rejection.

8.2.4 Material, Tooling, Fixtures and NC Programs Must Be Available

A suitable availability check within the materials planning system (before shop floor control) must ensure that material, tools, fixtures and NC programs are ready at the time of order release, or at the individual dates.

One interesting way to carry out simultaneous availability planning is to include all elements that are necessary for an order in the master route sheet [9]. Figure 8.11A shows a simple example of the manufacturing routing of a part X. Materials A and B are joined in Operations 1 and 2, Material C and Material E, which was processed in a separate operation, are added in Operation 3 and, finally, Material D comes in during Operation 4. The finished product X is the end of the chain. The routing sheets X and F are developed from this (Fig. 8.11B). The intermediate product F consists of Material E, and is processed with Tool 11 in one operation.

For each component of the operation sheet, namely the work centers, the personnel center, the tooling, and the material, an inventory account with inputs and outputs is kept (Fig. 8.12). At the scheduling stage, all accounts are simultaneously subjected to an availability check. The material account, for example, contains the present stock-on-hand and the current purchase orders on the inventory side (inputs), and all known allocations (loadings) on the requirement side (outputs). In a schedule-oriented congruence calculation, all future inputs and outputs are checked, and one can see the inventories that are available at any date. If one of the components is not available, all components are postponed until all of them are available [9]. (These inputs and outputs are not identical with the respective concepts in the throughput diagram, but correspond to the planned output curve or capacity curve in the throughput diagram.)

8.2.5 Machine Availability and Personnel Capacity Must Be Known

This point is especially important, but it is one which is neglected in many firms. Loading control cannot be reliable if only the monthly averages of the available capacities are used. Only if their daily values are known can a schedule-oriented capacity planning help to make use of the current flexibility in the factory (see Chapt. 7).

The question is often asked as to how disruptions in the manufacturing flow due to lack of personnel, material, or production facilities, or to machine breakdowns, can be taken into account within production planning and control.

First, it is useful to find out about the type, the frequency and the duration of

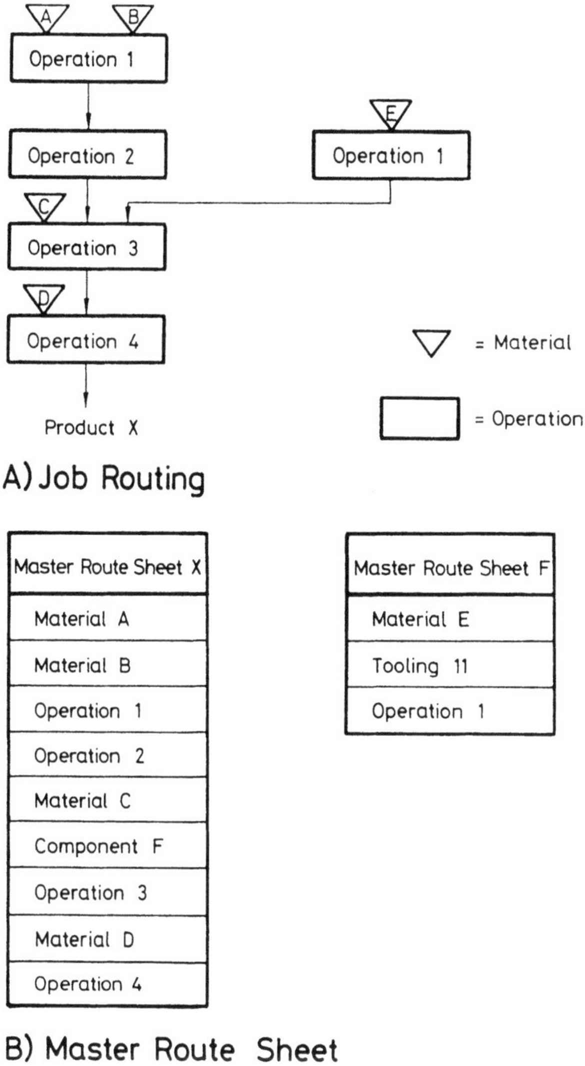

Fig. 8.11. Master route sheet including materials and tooling requirements (After Kazmaier [9])

disruptions. Figure 8.13 shows the results of a four month evaluation in a mechanical job-shop production consisting of 390 single work places. On average, there were about 11 trouble events a day, 60% of which were of an organizational and 40% of a technical origin.

A more differentiated evaluation at the cost centers (Table 8.1) shows that only 19% of all disruptions occurred directly at the machines (3% machine breakdown, 4% failure of tooling, 12% failure of fixtures). Most of the disruptions were concentrated in the automatic turning shop, in the large-sized parts production (processing of complicated gear boxes with automatic change of drilling heads), and in the milling shop. The average interval between trouble events for all shops was about 1.5 hours for two-shift work.

First, such a disruption causes a loss in capacity. Experience shows that, in a

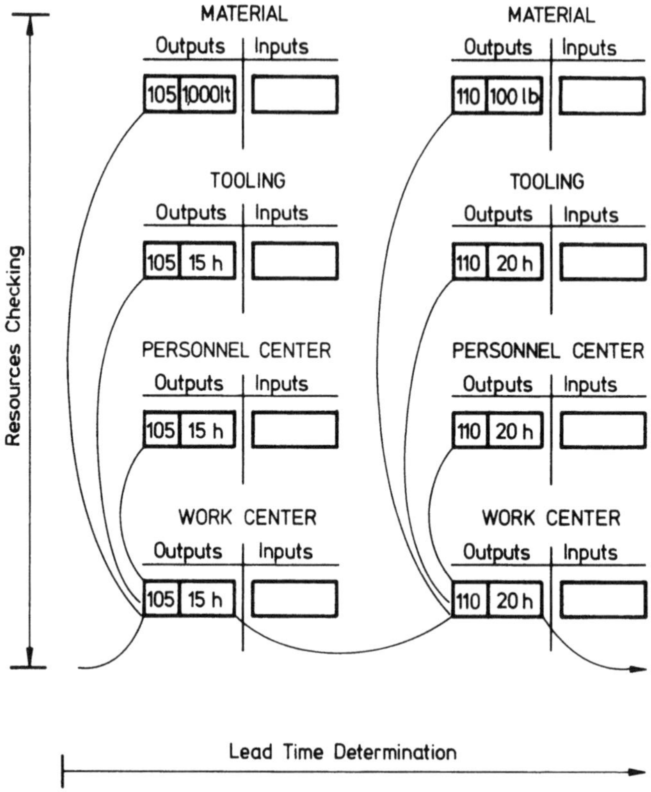

Fig. 8.12. Simultaneous resources checking within manufacturing control (After Kazmaier [9])

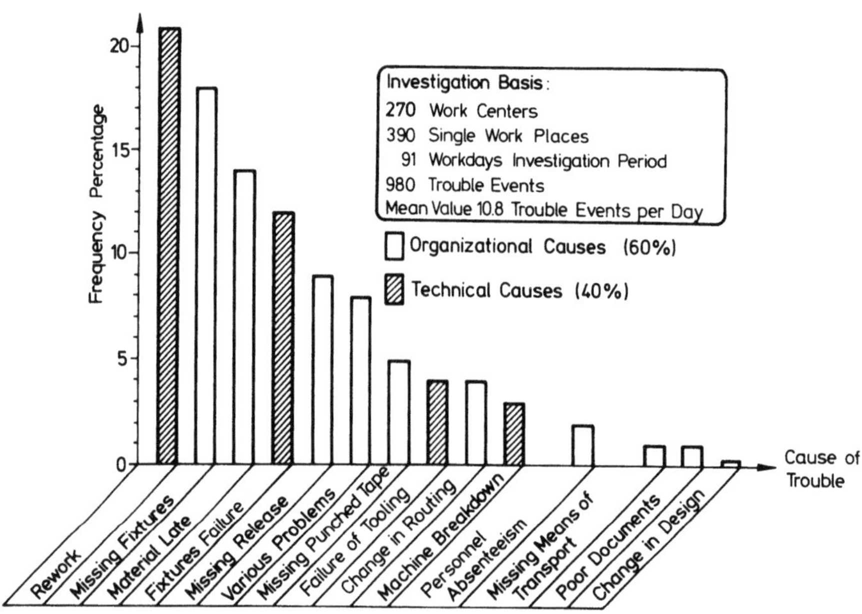

Fig. 8.13. Trouble-spot frequency in a mechanical manufacturing shop

Table 8.1. Causes and frequency of trouble spots in a mechanical manufacturing shop

Overall: 91 Days Investigation Period, 270 Work Centers, 390 Single Work Centers

	Total Shop		COST CENTER												
			Dressing	Hardening	Automatic Lathe	Turret Lathe	Turning	Large-Sized Parts Production	Gear Wheel Working	Drilling Shop	Milling Shop	Grinding Shop	Gear Case Production	Parts Metal-working Shop	Inspection
Cause of Trouble	abs	%	1102	1113	1122	1123	1124	1125	1126	1127	1128	1129	1144	1149	1362
1. Personnel Absenteeism	19	2					3		3	1		8			4
2. Machine Breakdown	25	3	13		1		3		1			6	1		
3. Failure of Tooling	39	4				10	1	4		7	12	5			
4. Fixtures Failure	116	12	1		9	1	15	29	2		47	9	3		
5. Missing Fixture	177	18			17			105		28	27				
6. Missing Punched Tape	49	5					8	40		1					
7. Missing Means of Transport	5	1										1			4
8. Poor Documents	5	1	2		2										1
9. Missing Release	93	9				1	12	24		1	8	3			44
10. Change in Design	2	.5						1							1
11. Change in Routing	38	4				2	1	2			3	1	29		
12. Various Problems	74	8			5	1	4	2		3	1		3		57
13. Material Late	136	14	22	1	66	1	17	8		3		6			12
14. Rework	202	21		1	1		7	5		2	2	4	1		179
Total Quantity	980	100	38	2	101	16	69	220	6	46	100	43	37	0	302
Mean Trouble Spots Frequency per Day	10.8	-	.4	.02	1.1	.2	.8	2.4	.06	.5	1.1	.5	.4	0	3.2

well-managed shop, this must be calculated as being 1.5%–2% of the available capacity. Second, the order throughput of the lot is interrupted, or not even started. The machines usually stand idle for about 10 to 20 minutes; larger-scale defects, which lead to several days of idle time, are not so frequent. The lot itself can be delayed for several days, for example when a fixture or a tool is not available. In the statistical mean flow time, however, this does not have noticeable effects, so that the buffer time included in the allowance can be considered to be sufficient.

However, with increasing automation and interlinkage of machines into automatic manufacturing systems, such as flexible manufacturing or assembly systems, more attention should be paid to such disruptive factors. At the current state of the art, mean intervals of 15–30 minutes between trouble events must be expected for flexible manufacturing systems, for example [10, 11]. For automatic assembly systems running at cycle times of between 4 and 20 seconds, intervals of four to six minutes have been experienced [12]. Since most disruptions are short, an appropriate trouble monitoring and maintenance organization must make sure that the planned capacity is actually available.

8.2.6 Operation Feedback Must Be Complete and Sufficiently Accurate

Punctual and accurate reporting of finished orders are also of great importance. They become more important the shorter the flow times and the lower the inventories are. It is not a question of feedback accuracy to the minute for the outputs, but of complete information on the work status at the time when work is released and allocated. With the common flow times of five days per operation, a feedback accuracy to the day is quite sufficient. For realistic capacity monitoring and planning, it is also important that rush orders, which bypassed the release process, are reported as outputs, since values of past periods are used for determining the planned values of subsequent periods.

To conclude, the prerequisites for load-oriented manufacturing control are the same as those necessary for every other manufacturing control system if it is to produce useful information.

8.3 Program Modules of Load-Oriented Manufacturing Control

8.3.1 Overview

Figure 8.14, supplementing a model suggested by Buchmann [3,4], integrates the components discussed above into a data flow chart of load-oriented manufacturing control, and also explains the preceding functions of material management. The functions explosion of bill of materials, requirements calculation, order calculation, and the generation of the shop orders during master production scheduling are shown in the top part of the figure.

The current shop orders are entered periodically, e.g. weekly, first into the 'Capacity Planning' module, where the schedule-oriented capacity requirements are defined. Then, in the 'Release Planning' module, the orders are classified as either feasible or deferred, on the basis of the current capacity data and the respective loading percentages and time limits. The feasible orders enter the 'Sequencing' module, which defines the short-term sequence in which orders should be processed at the load centers, e.g. in daily intervals. Finally, from the current feedback data, the last module, 'Monitored Data Calculation', must calculate the control tables explained above (see Sect. 5.4), which, in turn, are the basis for the monitoring graphs and diagrams.

8.3.2 Capacity Planning

Capacity planning includes several modules (Fig. 8.15): first, periodical checks must be made to determine whether the weighted mean lead times, which influence the loading percentages, are still correct on the basis of the current order mix. To do this, the weighted mean order time for the individual load center is evaluated, then divided by the planned capacity to find the mean operation time; then, mean scheduled buffer time and transit time are added to calculate the weighted mean scheduled lead time per load center.

At this stage, one is already made aware of the high-volume orders, for which, if there is an order time limit (for example, 20 hours), lot splitting is recommended. A comparison of these calculated lead time values with the current planned values on the one hand, and with the current actual values supplied by the monitoring system

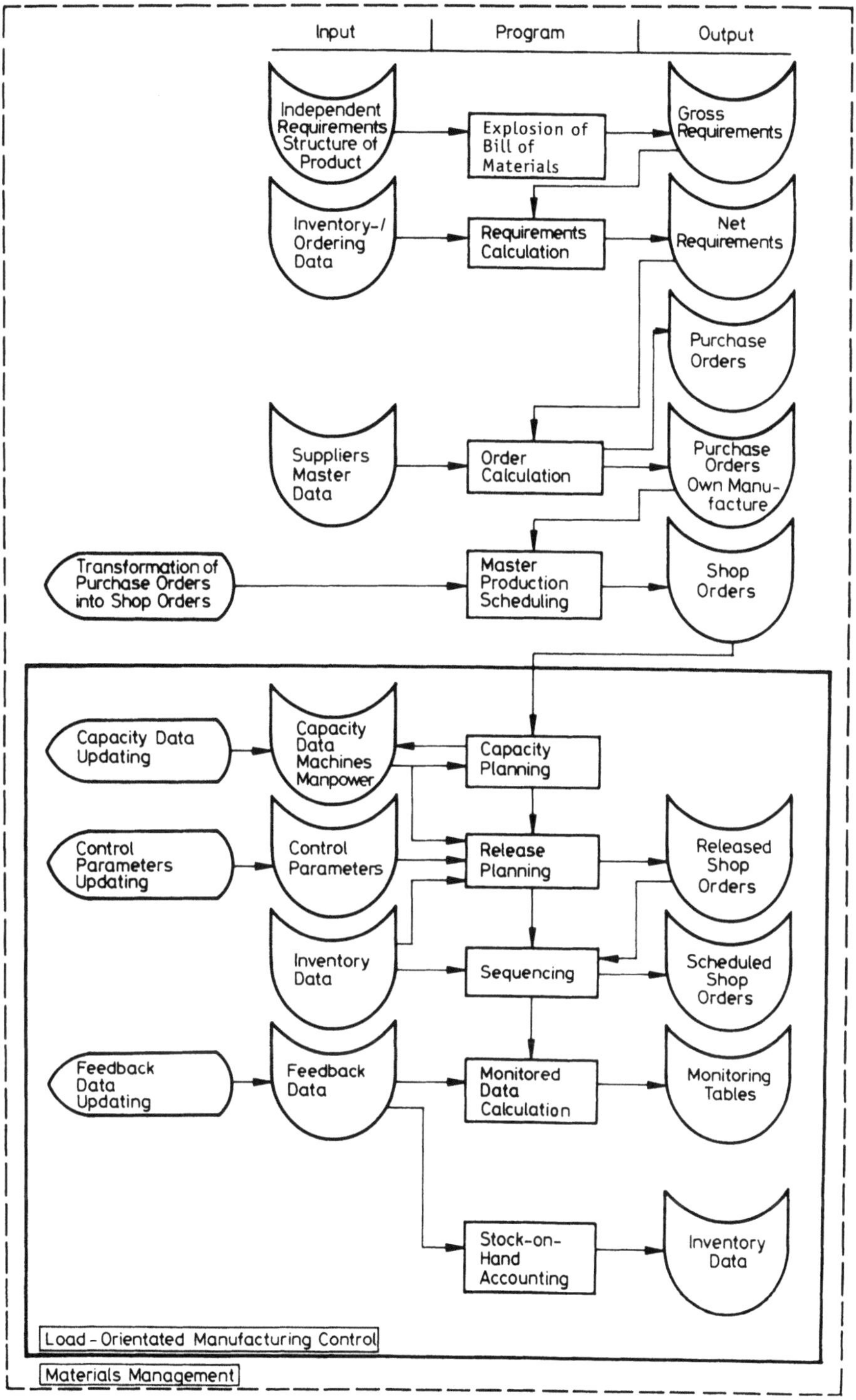

Fig. 8.14. Data flow chart of load-oriented manufacturing control

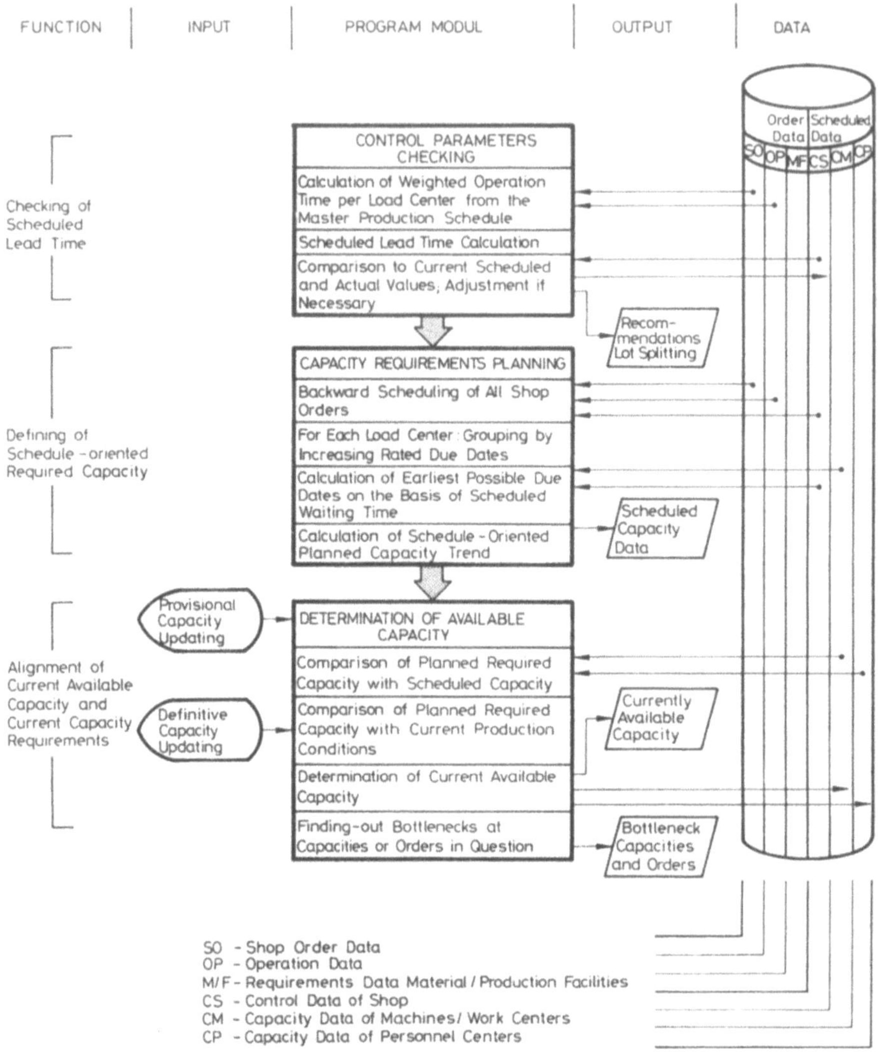

Fig. 8.15. Data flow in capacity planning

on the other hand, can help to determine whether it is useful to change the planned values. The values should, however, be changed at intervals not shorter than approximately the mean order flow time, so that their effects at the individual work centers can become noticeable, and that the resulting fluctuations in inventory, flow time and utilization can still be monitored. Therefore, this module does not have to be used in each planning period, but only every four periods or less.

The actual capacity planning itself begins in the next module. As described above (see Sect. 7.4), two output curves per load center are generated, and the resulting schedule-oriented capacity trend defined. Note that in allowance scheduling, the schedule-independent lead time values must be converted into calendar-day values, so that non-productive days such as weekends, holidays, vacations, etc. can be taken into account.

In the next step, the schedule-oriented capacity requirements must be compared, capacity group by capacity group, with the capacities defined in the annual master route sheet, and with the current possibilities to change capacity, if any.

The compromise which is then found between manufacturing control, order control, and the shop enters the capacity file of the next program module as the relevant available capacity. Bottlenecks and underload at individual capacity centers, and orders which may suffer a delay, must be highlighted. Since this calculation includes the overall issued order stock, corrective action can be taken, if necessary, for the next planning cycle. The calculation also shows the order inventory which is behind schedule for each load center, and thus illustrates the schedule deviation which has been calculated by the monitoring system. This 'Capacity Planning' program module supplies reliable, up-to-date scheduled capacity data to the subsequent release module.

8.3.3 Release Planning

The release planning module has three functions performed by three sub-modules (Fig. 8.16) [4]. First, the start dates and up-to-date lead time values of all issued orders are calculated; from the result, the high-priority order stock within the anticipation horizon is determined. The orders in this stock are arranged in the sequence of increasing start dates.

The second sub-module, 'Order Release Planning', uses the conversion algorithm (see Chapt. 6) to check whether the load limits of the individual load accounts are exceeded by the release of these urgent orders. For the orders which are still feasible, the system checks whether the necessary production facilities are available at the dates requested. A list of released orders and a list of deferred orders are set up, arranged according to order priority, and made available to the next sub-module, 'Release Recommendations Listing'.

This sub-module groups the released, i.e. feasible, orders according to their load centers, and generates the very important list of rejected orders, in which the causes of rejection – either capacities or production facilities – are given.

Figure 8.17 shows an example of such a list of rejected orders. One can immediately see which load center has rejected how many orders. This list is therefore a very important aid to short-term capacity planning. Since there is still at least one period between the release and the order start, and since the lead time includes buffer times, it is possible to take appropriate measures to ensure due-date performance for important orders.

The lists of released and rejected orders must be seen as recommendations only. The final decision on release or withholding should be left to the manufacturing controller with his personal experience and knowledge of the plant and of individual orders. This fulfils the demand for a flexible and adequate computer-aided manufacturing control, in which the manufacturing flow is not controlled by a supposedly 'optimal' but obscure computer program, but by a competent and qualified production manager.

The order release, which works period-by-period, is followed by the 'Sequencing' module (as outlined in Fig. 8.18), which usually works on a shift-by-shift or day-by-day basis.

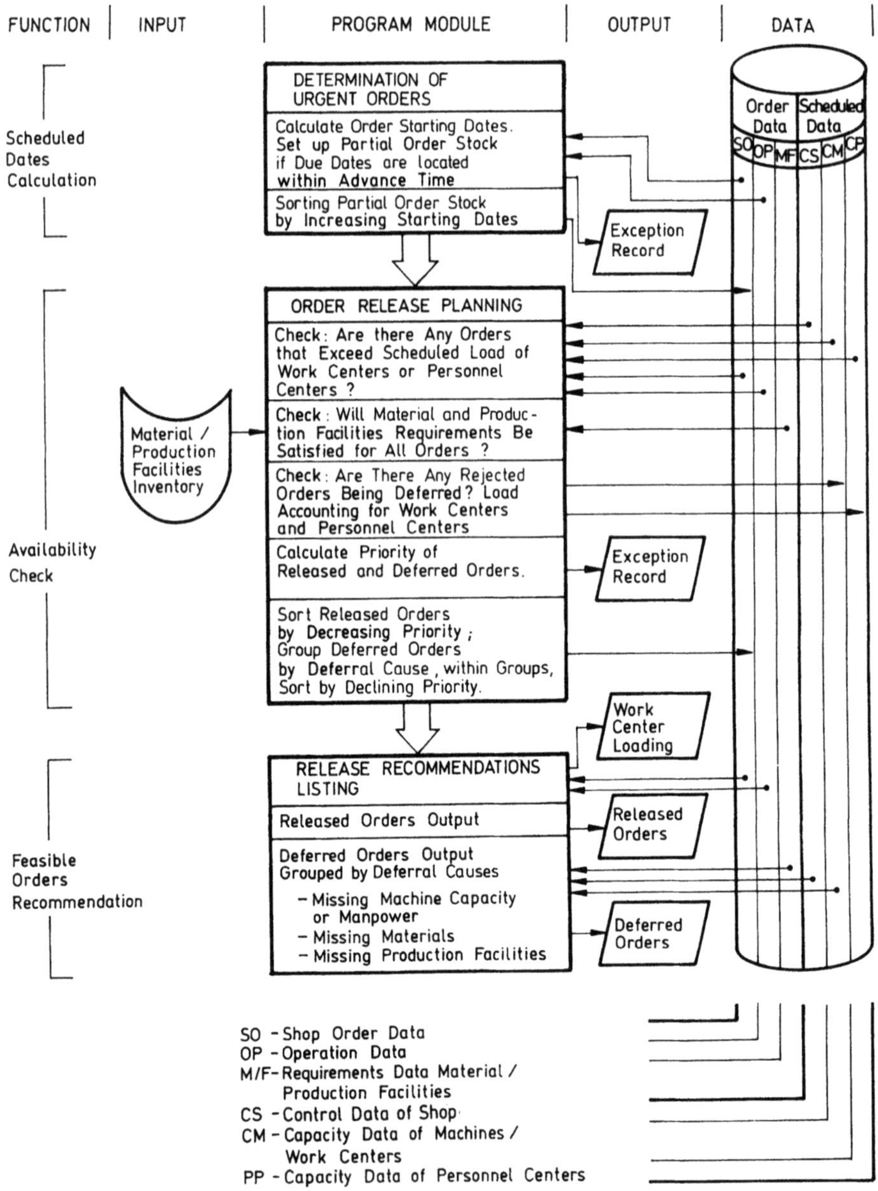

Fig. 8.16. Data flow in release planning (After Buchmann [3])

8.3.4 Sequencing

In the 'Sequencing' module, the manufacturing control allots packages of work – also called *daily* or *weekly slices* – to the individual work places, since the release usually refers to entire work centers only (Fig. 8.18) [4]. In this module, too, the calculations should be simple and clear, and the results should only be regarded as recommendations to be followed within the given planning margin, wherever possible.

```
UNRELEASED   ORDERS                                      WEEK / MONTH / YEAR

                                              REJECTING  LOAD  CENTER
                                                         W W E H R A A A
                                                         0 0 0 2 0 0 0 0
                                                         7 8 4 8 8 3 6 3

SUBJ NO        LOT NO   QTY   COST   STAD   DUED

S 348 - B 47   67       100   55     205    285          . . X . .
S 348 - B 49   87       200   55     ...    ...          . . X . .
. ...                                                    . . X . .
. ...                                                    . X . . .
. ...                                                    X . . . .
. ...                                                    . X . . .
. ...                                                    . . X . .
. ...          . ..     ..    ...    ..     ...   ...    . . . . .

                                                           1   1
NUMB LOTS   43   QTY   2540                      NUMBER  4 5 6 4 6 4 0 0

COST = COST CENTER, STAD = STARTING DATE, DUED = DUE DATE
```

Fig. 8.17. Fundamental structure of the list of rejected orders, with rejecting load centers (According to Siemens AG)

The module also has several sub-modules. The first sub-module, after having been started by the order planning specialist, in turn starts the 'Shop Papers Request', i.e. drawings, NC programs, routing sheets, and material slips. If feedback records are still used due to the lack of on-line recording, they can also be called up for the individual operations at this stage. Since the papers are requested as late as possible, alterations in routing sheets, dates, and quantities can be carried out until immediately before order release, if necessary. The next sub-module, 'Sequencing', sorts the orders in the queues for each load center in the sequence of their priorities on the basis of the current feedback data on the actual production progress. With this sorted order stock, the planning part of load-oriented order release is completed.

The next sub-module, 'Dispatching', is used to allocate the orders to the individual work places, for example by the foreman of the respective cost center. Within the released order stock there is a certain room for maneuver, which is equivalent to the range in days (discussed in Chapt. 4). If, for example, the mean order inventory at a work center with five machines working 16 hours a day is $I_m = 240$ hours, then the range is 3 days. The range is meant to allow a compromise to be made between an exact processing according to schedule priority, and necessary room for maneuver when the machines are loaded, which takes into account setup families, special skills of individual workers, and quality aspects of individual workpieces. This module can be complemented by color graphics as shown in Sect. 5.6.

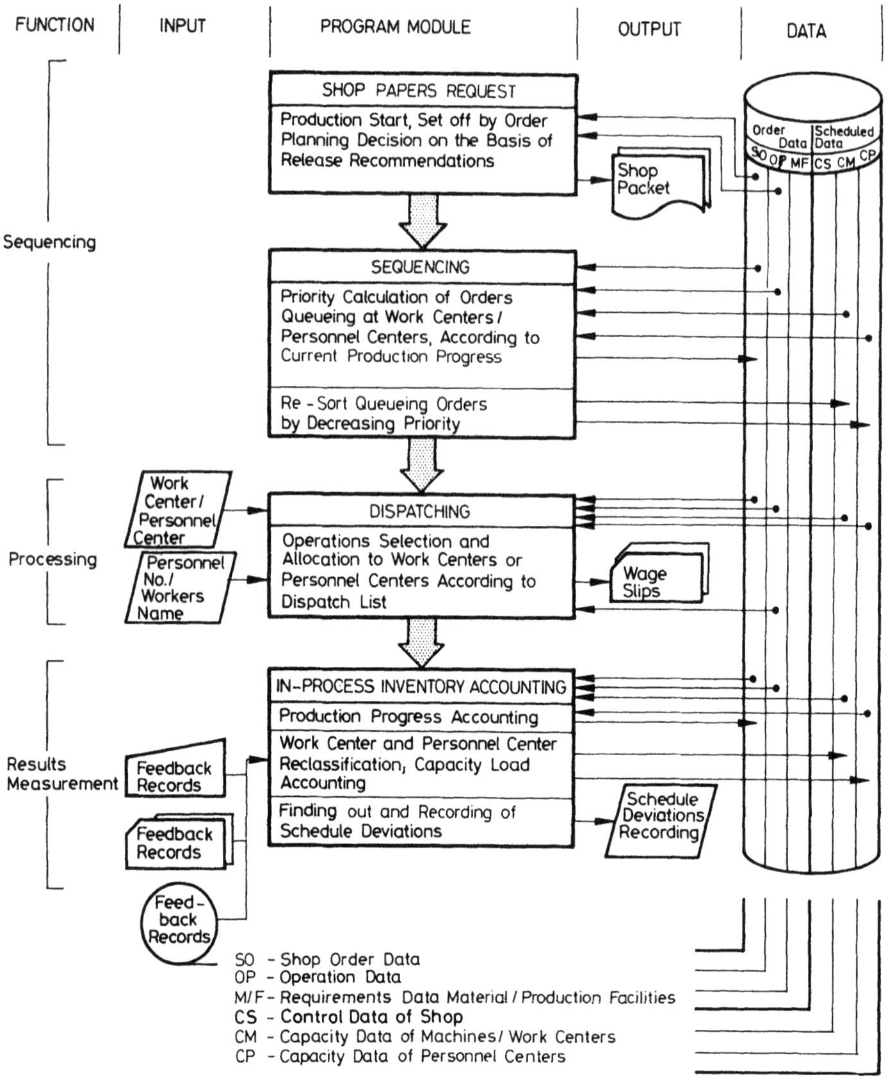

Fig. 8.18. Data flow in sequencing (After Buchmann [3])

When the order has been assigned to the worker and the machine, it is 'in the system', and, if necessary, the wage slip can be printed. If the personnel-based wages data and the process-based feedback data are treated separately in the computer, one can fulfil the frequent demand for both up-to-date order progress data and a completely separate wage calculation system.

The final sub-module, 'In-process Inventory Accounting', measures the results of the current process and updates the data. The movements of the lots from work center to work center can be recorded from the feedback data, and the order times of the subsequent operation can be booked as input, or the order times of the operation just finished can be booked as output onto the respective load accounts of the upstream or downstream work centers. Deviations from the planned work center location, alterations in quantities, and non-plausible feedback data must also be recorded.

8.3.5 Calculation of Monitored Data

The 'Monitored Data Calculation' module is based on the feedback data, and has several functions. First, it is a kind of quality check of the feedback data, checking them for reasonableness. The criteria for this were shown in Fig. 5.9. Second, deviations between planned and actual values must be discovered as early as possible, so that the throughput of the released orders can be influenced, if necessary. Finally, the module allows the control parameters of load-oriented order release to be checked and thus ensures realistic due dates for future orders. However, the pre-

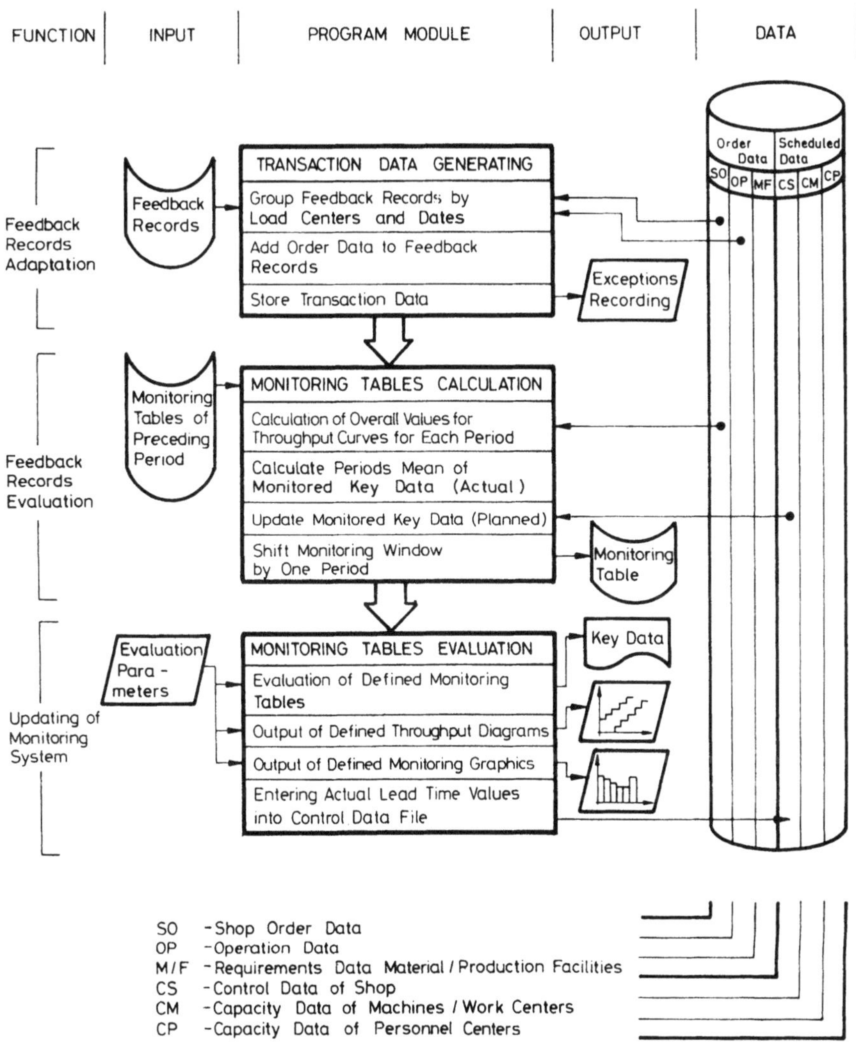

Fig. 8.19. Data flow in monitored data calculation

requisites for efficient monitoring are up-to-date order data, from which the monitored data can be calculated. Figure 8.19 shows the sub-modules needed for this calculation. First, the so-called transaction data must be generated by supplementing the feedback data on the finished orders with the corresponding work content data.

The 'Monitoring Tables Calculation' sub-module sums up the work hours of the dispatched, released, input and output orders reported per load center in the previous period. Then the monitored key values, namely inventory, flow time, utilization, and lateness, can be calculated using the formula developed in the Chapters 4 and 5, and compared to the respective planned values. Only a limited number of periods is represented in the 'monitoring window', i.e. the bottommost throughput curve is shifted towards the zero point of the diagram. Each time this is done, the oldest period and its values disappear from the relevant monitoring table, but they are stored on tapes or the host memory disc for long-term evaluation.

The next sub-module, 'Monitoring Tables Evaluation', either generates a small number of monitoring tables, throughput diagrams and monitoring graphs on a regular basis, or all data are available on a PC for on-demand examination (cf. Sect. 5.4). It is useful to update the flow time values periodically to pass them on to the 'capacity planning' and 'release planning' modules. Very widely varying flow time values (i.e. due to a large number of rush orders) should be smoothed exponentially. If a complete monitoring system is not planned, at least the actual weighted mean lead time per load center should be calculated.

Finally, Fig. 8.20 explains the planning horizon and the planning intervals of the individual modules. In this example, one square represents one week. This means

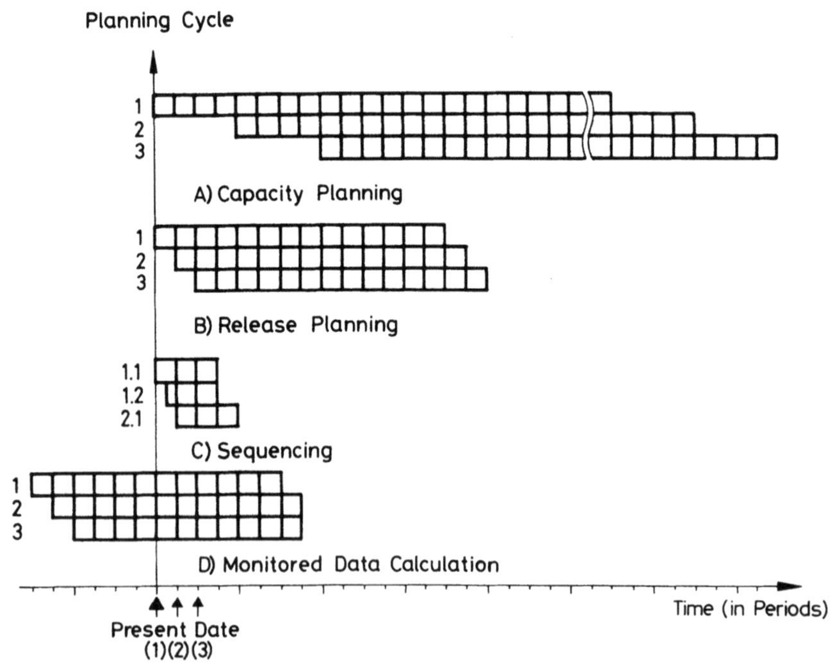

Fig. 8.20. Planning intervals and planning horizon of the program modules in load-oriented manufacturing control

that capacity planning is carried out monthly, and includes the existing as well as the planned orders, if necessary. The release planning only deals with current orders, but does this weekly, and the sequencing module only includes the released orders, and this twice a week. The monitoring data module carries out a weekly evaluation of the planned and actual data from a defined stretch of time in the past, and adds the planned data for a defined future stretch of time.

8.4 System-to-User Interfaces and Hardware Configuration

The description of load-oriented manufacturing control and the respective software products showed that most of the software modules are sequential processes, and are not influenced by the user.

The interfaces between the user and the system essentially involve the following [4, 13].

- Updating of control parameters, capacity data, and feedback data.
- Starting individual modules and evaluating the results.
- Influencing orders and load centers in case of changes or disruptions in the order flow or in capacities, but only when this is unavoidable.

The calculations themselves are simple, but the programs can nevertheless be complex, since every plant has a few special requirements which must be fulfilled [13]:

- Not only machine load centers, but also personnel centers and load centers for production facilities and their interlinkage must be specified.
- Alternate routing and subcontracting must be possible.
- Shop calendars, variations in personnel attendance times, and factors which influence capacity must be taken into consideration.
- A material availability check must be allowed for.

The programs themselves are either centralized or decentralized, depending on the distribution of the available hardware and production sites, and on the number of load centers and shops to be controlled.

Figure 8.21 is an overview of a standard solution of the centralized type offered by one software company. It is a modular, integrated planning, control and monitoring system based on a common data base, and designed for production companies [14]. It consists of the main functions of Sales, Design, Materials Management, Scheduling, Finance and Accounting, Personnel Management, and a few horizontal functions. Above these there is a superior Top Management function.

Each main function is divided into numerous sub-functions, as illustrated in Fig. 8.22 for the 'Scheduling' subsystem. The Manufacturing Control module corresponds to the load-oriented order release with its steps as explained above, complemented by Shop Order Administration, which is an indispensable service function within this module.

This software system illustrates how load-oriented order release is closely interlinked with the other subfunctions of manufacturing control. Capacity planning, as described in Chapt. 7, is to be found in the 'Supply Planning' module. The moni-

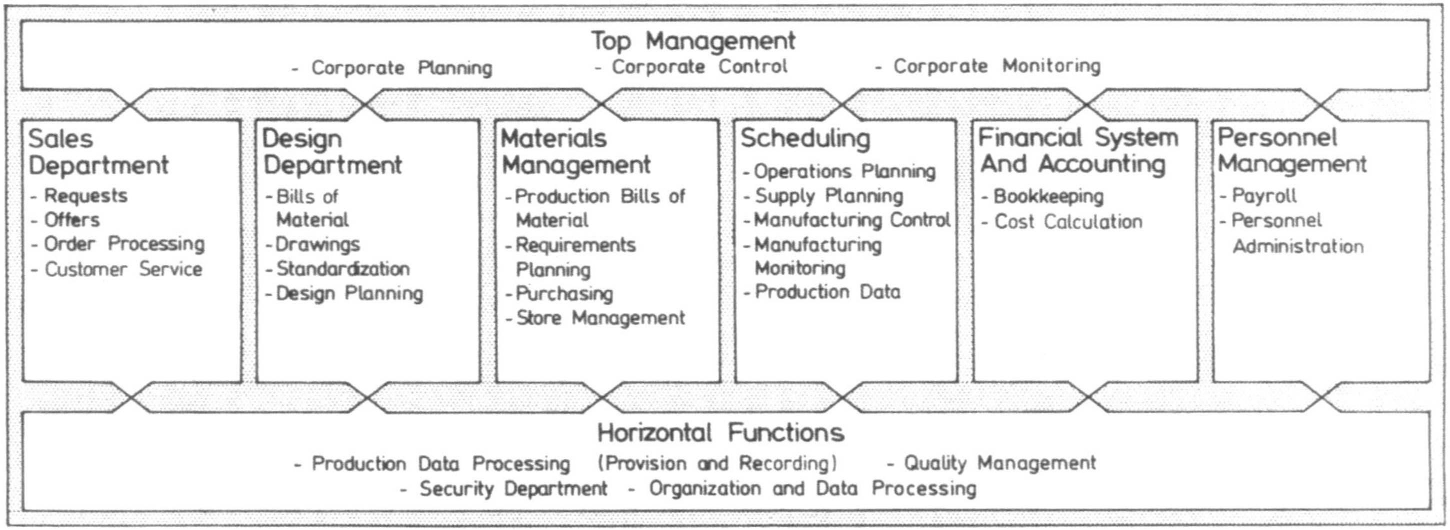

Fig. 8.21. Modules of the PSK 2000 system (Courtesy of Strässle, Stuttgart)

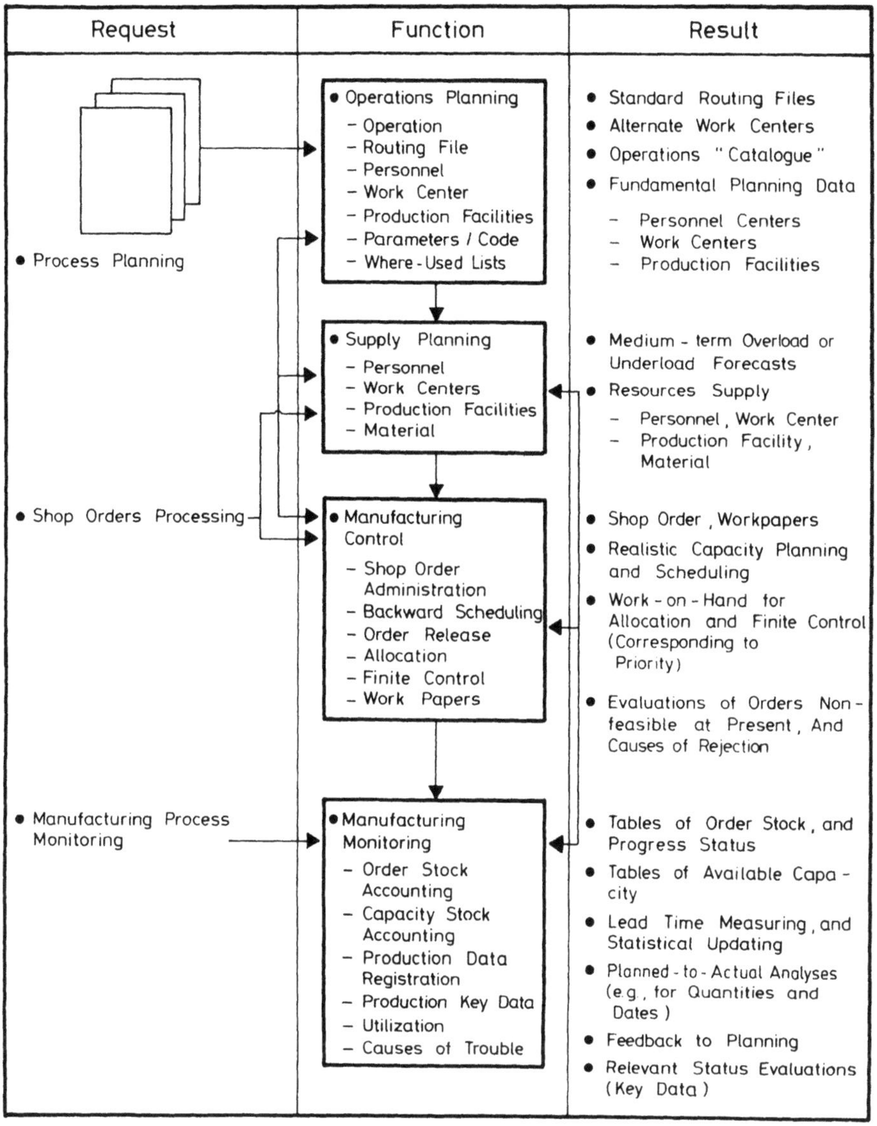

Fig. 8.22. Scheduling subsystem of PSK2000 (Courtesy of Strässle, Stuttgart)

toring system is part of the 'Manufacturing Monitoring' module, which in this program is only represented in the form of a few special statistical key data. Due to the modular design of the system, however, it is also possible to implement the graphs and tables described above.

Manufacturing control systems which run on PCs or workstations are becoming more and more important. Via a data interface, the central mainframe computer periodically sends the PC the latest data on the order stock and on the order's progress status.

Figure 8.23 shows an example of such a PC system [15]. The order data are transmitted from the MRP system to the PC. There, capacity requirements planning

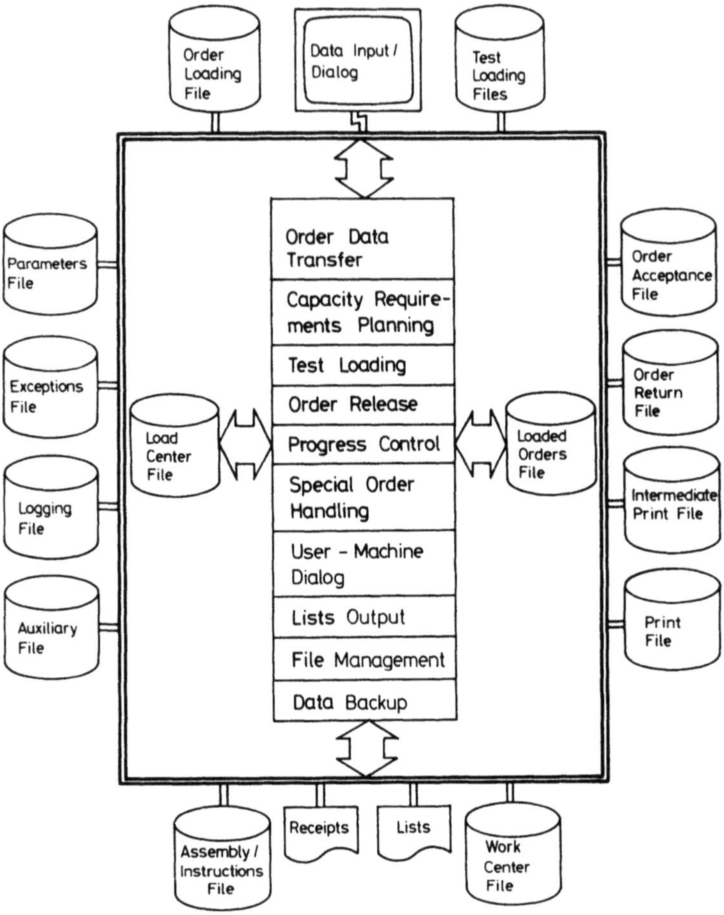

Fig. 8.23. Programs and files of the ABS-PC software (Courtesy Siemens AG)

and the order release calculation are carried out. The order progress is monitored using integrated production data collection terminals. The other functions serve to influence individual orders (scrapping, freezing due to disruptions, defreezing, splitting), and for dialog purposes (enquiries, changing the control data, and starting program modules and statistical evaluations), as described above.

Other commercial implementations of the method have been programmed by Bechte [8] and Erdlenbruch [13]. At the end of 1990, ten commercial versions were available. The automobile manufacturer BMW has developed its own software component for in-house use [16].

Since the current versions of load-oriented manufacturing control need much less computing time than the conventional finite scheduling methods, all orders are normally rescheduled, which means that they are completely recalculated. Only changes in single orders due to special conditions are carried out as net change calculations, since their relatively small influence on the total time and load situation does not justify a total rescheduling.

8.5 Implementation Strategies

The implementation of load-oriented manufacturing control requires that everybody, from company and shop management down to the manufacturing controller and the foreman, give up those traditional concepts that were described in Chapt. 1 as the myths of manufacturing control. Also, the prerequisites for the effective use of load-oriented manufacturing control, as given in Sect. 8.2, must be fulfilled.

A readiness to rethink the present methods is not always achieved by theoretical discussion or by giving examples from other companies, since one often thinks that the situation in one's own company is better than it really is. Figure 8.24, which is

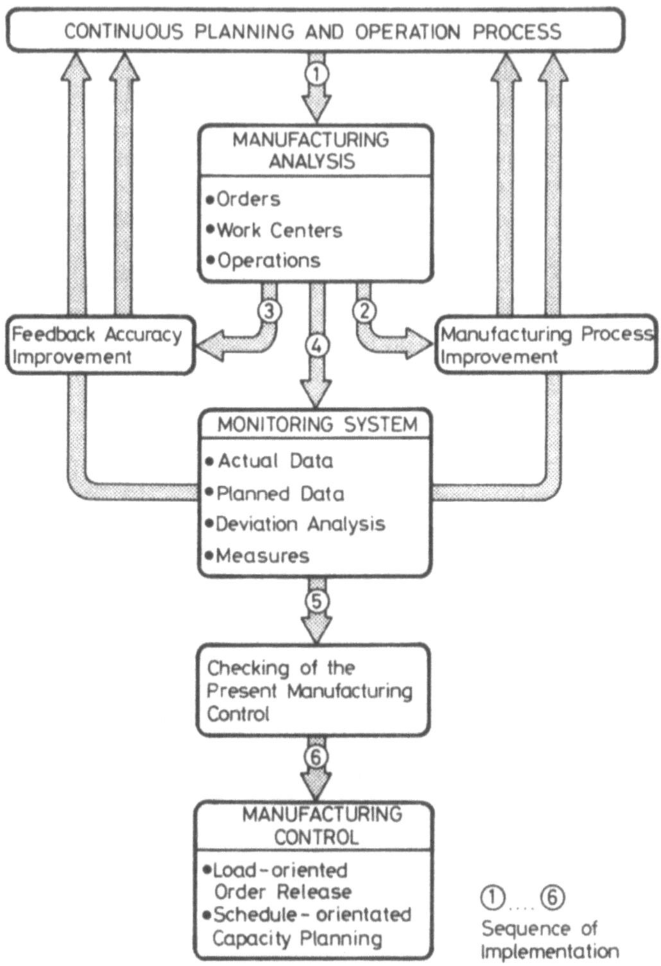

Fig. 8.24. Steps in the implementation of load-oriented manufacturing control

based on Fig. 5.1, shows one way of implementing the method step-by-step, each step giving directly usable intermediate results.

The first step, the fundamental manufacturing analysis (as described in Sect. 5.3), aims at obtaining reliable data on flow times and lateness, and explaining the relationships between inventories, lead times, and utilization. The concept of weighted lead time is, at first, especially difficult to handle, but throughput diagrams with data from one's own plant can help in overcoming these difficulties. Moreover, weak points in the feedback system can be discovered and indications can be given on how to improve feedback, and how to divide up the capacities into load centers.

Just as often, the manufacturing analysis leads to technical and organizational improvements which shorten lead times and reduce inventories (Step 2 in Fig. 8.24). They include product designs with as few assembly levels as possible, and processes which cause as few changes in routing as possible [17] (cf. also Sect. 5.3.4). Also, when facilities and investments are being planned, all possibilities of lead time reduction should be fully utilized [18].

Figure 8.25 shows some ways of improving the production flow, which are based on the general throughput element at a work center [19]. When facilities are planned, it is no longer sufficient merely to optimize the material flow and minimize transport; the non-productive waiting times must also be minimized, and this is done by the integration of setup, processing, waiting, and transport. Besides a product-oriented layout in accordance with the requirements of group-technology (so called *machining cells*), the interlinking of work centers by automated guided vehicles (AGV) or flexible conveyor belts also improves the order throughput. Not only does it cause shorter flow times, but it also makes an uncontrolled accumulation of inventories and constant sequence transpositions less likely. Push-through racks at the work centers have also proved successful, since, besides demanding a monitoring of inventories, they enforce processing in FIFO sequence.

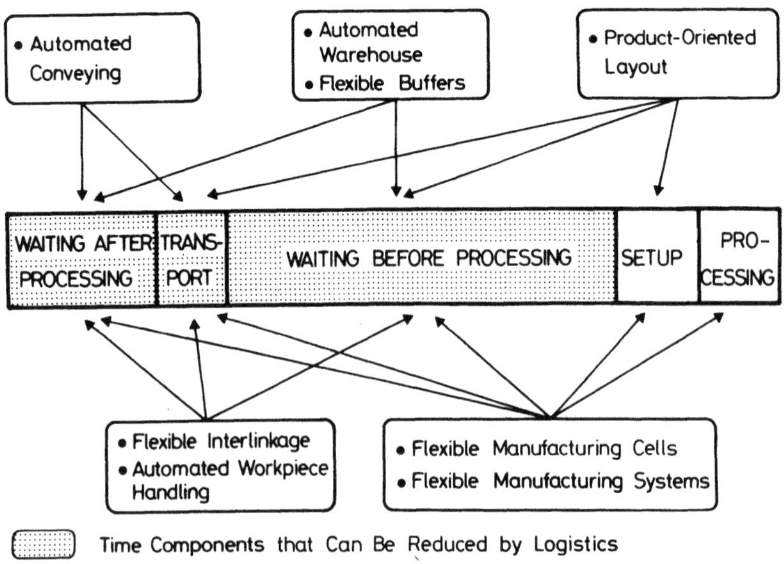

Fig. 8.25. Approaches to improvement of the manufacturing process within facilities and investment planning

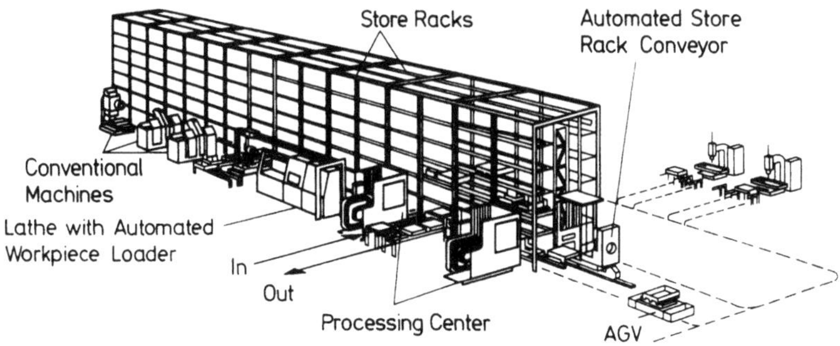

Fig. 8.26. Job shop with integrated automated storage and conveying system (Bygg-Transport, Sweden)

Besides automating transport, if one also wants to automate the storage of orders between the work centers, there is often the problem of different interfaces between machines with different levels of automation. Figure 8.26 shows an interesting solution for this problem. An automated storage and conveying system is integrated into the shop, so that the whole setup could be called an automated job-shop [19].

The store rack has the function of limiting inventories; its defined storage capacity represents, so to speak, a physical load limit. The individual machines are coupled to the storage system by means of roller conveyors. Due to the restricted amount of space available on the conveyors (e.g. 2–4 pallets), the amount of work held in store is limited, and must be processed in the planned FIFO sequence. The parts can be taken from the pallet either manually or with the help of an automated pick-and-place device, such as a gantry robot. A change of pallets with the parts fixed on the pallet is also possible. This is, in principle, the setup of a flexible manufacturing system [20]. However, it permits the use of conventional machines without the automated change of parts, tools, or NC programs, and allows them to be replaced, step-by-step, by more highly automated equipment, eventually including the staging of raw materials and tools [21].

Besides these possibilities of improving the manufacturing flow by technical measures, attention must be paid to the feedback data (Step 3 in Fig. 8.24).

Section 3.8 and Fig. 5.9 give some details on this subject. Only when the quality of the feedback data is satisfactory can the monitoring system be implemented (Step 4). The system permits a permanent monitoring of the relevant key data and their improvement with the help of the existing manufacturing control and production activity control system (see Sect. 5.4).

Eventually, one will discover that the existing manufacturing control system as a whole is in need of improvement (Step 5). Only at this point should the implementation of load-oriented manufacturing control be carefully considered (Step 6). Note that it is by no means necessary to control every single work place using this method. Flexible manufacturing systems, manufacturing cells, or production lines, for example, can be taken as a single funnel, which is then controlled using another method, e.g. the Japanese Kanban system.

The above procedure for implementation is, of course, not equally suitable for all companies, since an implementation time of at least one year must be reckoned for each step. If preparatory steps have already been taken in one field or another, a different sequence of steps can perhaps be chosen. A minimum of two years must, however, be taken into account for the implementation of all the components of load-oriented manufacturing control, before measurable and permanent improvement can be expected [13].

8.6 Load-Oriented Manufacturing Control in Automated Production Sites

8.6.1 Controlling Flexible Manufacturing Systems

As the automation of production processes increases, a reliable design and control of capital-intensive production equipment becomes more and more important. This is especially true for flexible manufacturing systems (FMS), whose number is growing, particularly in the US, Japan, and Germany [22–24].

This leads us to the question of whether the control approach based on the funnel model can also be used for these systems. At the Technische Universität of Magdeburg (Germany), Gottschalk [24, 25], together with industrial partners, for the first time carried out practical studies to examine whether the control of FMS can be improved using the load-oriented approach. The criterion for comparison between the old and the new methods was that the required output should be performed with minimum order lead times [25].

The system examined is an FMS for prismatic parts, such as gear boxes, flanges, and manifolds for diesel engines. Its layout is shown in Fig. 8.27. For economic reasons, the range of parts to be processed was divided into small prismatic parts (up to 480 × 400 × 350 mm) and medium prismatic parts (up to 980 × 950 × 900 mm), and two corresponding groups of machines were defined. Both groups share the same store and transport system [25].

From the total range of 499 small- and 99 medium-sized parts processed by the system, representative parts were chosen for examination. Some examples and general data are shown in Table 8.2.

Then, simulations were carried out with three variants of order release:

- Order release according to external priorities (as usual)
- Load-oriented order release (ten-day release cycle)
- Load-oriented order release (daily release).

The investigation period was one month.

The simulation results showed that when the release method was implemented in the program 'work center loading' module, a reduction in weighted mean lead time by about 15% and in mean inventory by about 9% could be expected. The output of the system would drop by less than 1%.

Further, it became clear that two special features of FMS have to be considered in the release process. First, the availability of tooling must be checked. Figure 8.28 shows how a so-called 'tooling limit' can be inserted into the control model. With the tooling limit, the program checks whether the set of tools needed for the part is available in the tool magazine, before the part is released. This avoids down times at the processing center due to tooling changes. The tooling limit is especially effective for additional rush-orders, which in this case consisted of parts missing in the assembly department. In the simulations, only 2.2% of the additional orders were unable to pass the tooling limit.

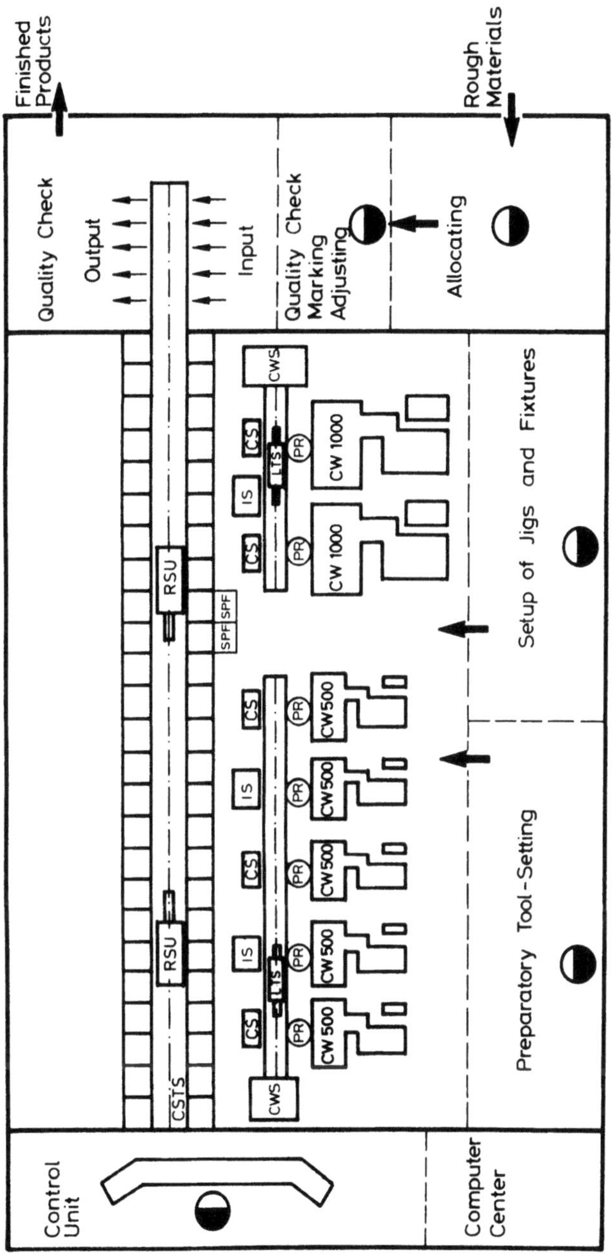

CSTS	Combined Store and Transport System	IS	Inspection Station
LTS	Linear Transport System	SPF	Staging Area for Pallets and Fixtures
RSU	Rack Serving Unit	PR	Pallets Replacing
CWS	Cooling and Washing Station	CW	Machining Center
CS	Clamping Station	◐	Employee

Fig. 8.27. Layout of a flexible manufacturing system (After Gottschalk [25])

Table 8.2. Production activity specification for a flexible manufacturing system. (A) General data; (B) representative parts for simulation (excerpt) (After Gottschalk [25])

A)

	Size (mm)	Types of Parts	Items per Year
Component 1	: 480 × 400 × 350	499	127,553
Component 2	: 980 × 950 × 900	99	6,039
Lot Sizes	: 5–250 Items per Lot		
Number of Lots	: ca. 5,900 Lots per Year		

B)

No.	Designation	Lot Size	Lots per Year	Number of Operations	Processing Time (hr/Lot)
1	Gear Case	25	1	1	4.5
2	Flow Indicator	50	1	1	13.9
3	Filter Base	25	1	1	6.1
4	Slide Case	10	2	1	4.6
5	Gear Case	50	1	1	12.1
6	Gear Case	50	1	1	8.1
.					
.					
.					
.					
.					
49	Pawl	50	1	1	6.3
.					
.					
.					

Second, when load-oriented manufacturing release is used in an FMS, the process time must be split into system-internal and system-external components (Fig. 8.29).

In conventional manufacturing, the setup time and the process times per lot have to be loaded onto only one account. In an FMS, however, these times must be divided between system-internal accounts (machine loading accounts) and system-external accounts (tool-setting accounts, fixture setup accounts, marking/adjusting accounts), each having a load limit.

This controls the flow of parts between the processing centers and the other centers. If, during marking and adjusting, the parts gained too much advance over the processing centers, this unnecessarily increases the flow time; too great an advance at the other centers increases the tie-up of tooling and fixtures.

The studies have proved that the method is, in principle, also suitable for automated manufacturing.

8.6.2 Integration into CIM Concepts

Since about 1984, CIM (Computer Integrated Manufacturing) has been widely discussed in Europe. In these discussions, the integration of production planning and control into the flow of information within the company plays an important role. There are two different approaches to the subject: one aims mainly at the data processing problems, with a focus on the integration of different hardware and software,

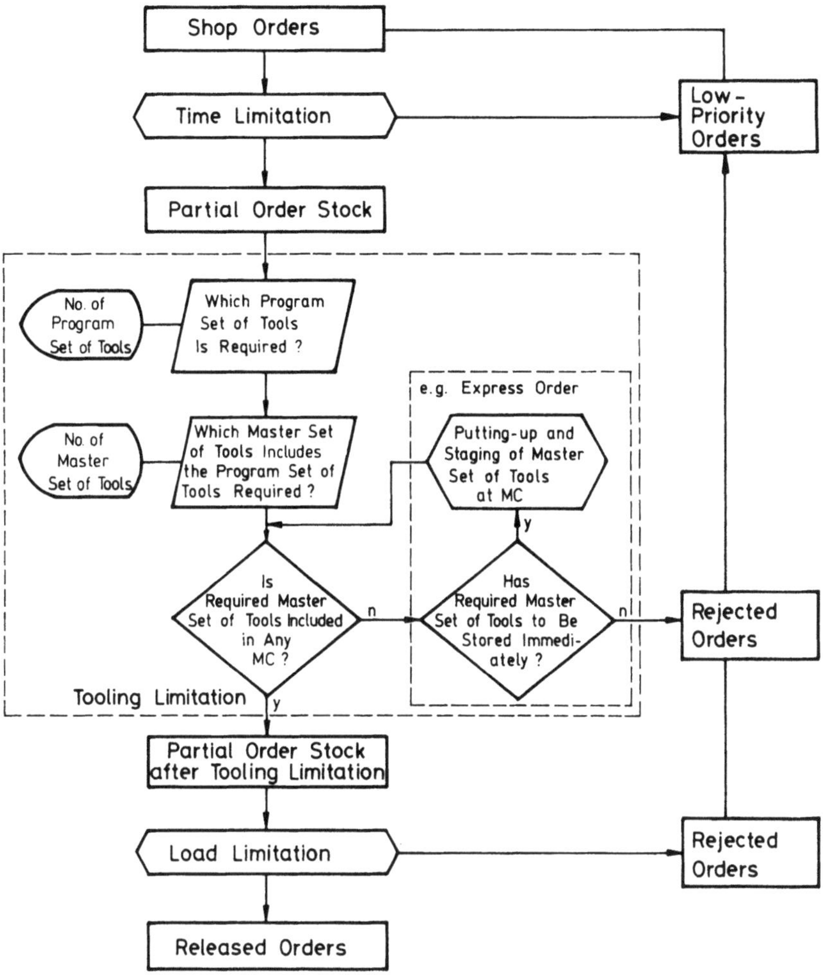

Fig. 8.28. Load-oriented control model of a flexible manufacturing system (with time, tooling, and load limitations) (After Gottschalk [25])

while the other concentrates on a strategic objective to interlink the organization of the many functions of a plant.

Figure 8.30 shows a frequently cited representation of a CIM concept introduced by Waller, which emphasizes the automation aspect [26]. The figure has been complemented by a few points referring mainly to production planning and control.

In accordance with market demands, the corporate management defines the requirements to be met by production planning and control (PPC), and in the development and the design of new products (CAD = Computer Aided Design). Operations planning (CAP = Computer Aided Planning), observing the requirements of quality assurance (CAQ = Computer Aided Quality Assurance), transforms the design papers, such as drawings and bills of material, into operation sheets and test plans, which, together with the quality data, flow into the production process (CAM = Computer Aided Manufacturing) as master data, process data, and control data. The task of PPC is to define scheduled shop and purchase orders from the continu-

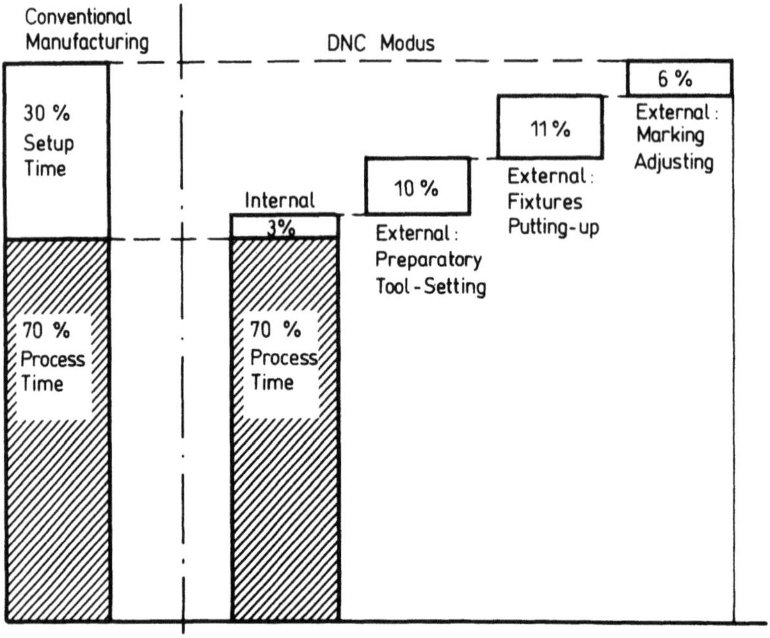

Fig. 8.29. Classification of the operation time components at a flexible manufacturing system into internal and external (After Gottschalk [25])

ous order flow. Manufacturing control is not assigned to PPC here, but to the manufacturing process itself, together with process control. This is to be seen as a monitoring of the product quality, as well as a control of the economic objectives inventory, flow time, utilization, and lateness.

Another frequently cited representation of the CIM concept is that by Scheer (Fig. 8.31) [27, 28]. It points out the interrelation between PPC and the product design (CAD and CAP) and manufacturing (CAM) systems. Based on this model, fundamental studies for the development of a CIM prototype system were carried out to prove that it is possible to put CIM concepts into practice using existing standard system modules. CIM was not seen as a monolithic product, but as a sum of technical tools and organizational measures for the achievement of a more economical development, design, and manufacturing of new products [29]. This permits the gradual implementation of CIM in companies with varying automation levels, and with an uneven distribution of computer facilities.

In his models, Scheer examines three types of data [28]. First, there are the order-independent master data or fundamental data stored in bill-of-material files, operation sheet files, and production facilities files. Second, there are data relating to the customer orders, giving technical and planning information. Third, data on the manufacturing orders are used to control locations, times, and quantities during the manufacturing of parts. The handling of these data flows, which sometimes have widely differing requirements, will be a crucial point in the further development of CIM.

It is remarkable that, although many contributions have been made on the subjects of computer-assisted models and methods of product design [30] and the design of technical processes and plants [31, 32], there have, as yet, been none on the

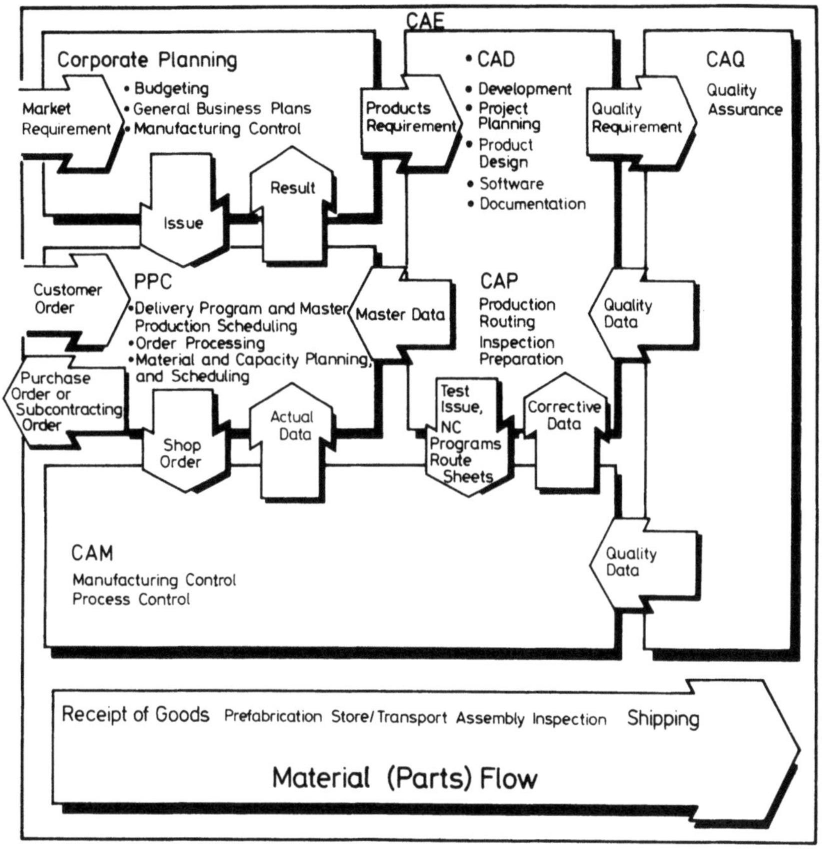

Fig. 8.30. Functions of computer integrated manufacturing (CIM) (After Waller [26])

modeling, planning and control of the production flow in the aggregate plant from the point of view of PPC [33, 34].

The funnel model, or the throughput diagram, is such a PPC model which is suitable for CIM. For a better understanding it is first necessary to define the fundamental functions of manufacturing control from the point of view of an integrated information flow (Fig. 8.32). One can distinguish five fundamental levels [48].

On the level of execution, energy and information are used to transform material into parts, modules, and products in the real production process. The only important point as far as manufacturing control is concerned is the completion of defined operations at defined dates and with defined quantities at the individual work centers; the process is modeled by the throughput diagram.

Recording the manufacturing flow, from the point of view of manufacturing control, primarily means determining the current physical and temporal progress of the manufacturing process. In many cases today, the completion dates are still reported manually at production data collection (PDC) terminals. With increasing automation, the data are collected by the control units of the automated warehouse, transport, handling, processing and assembly systems. When it is a matter of transportation of single parts or lots on pallets, or of fixtures in boxes or magazines, these changes in location and quantity can be recorded on transaction messages giving

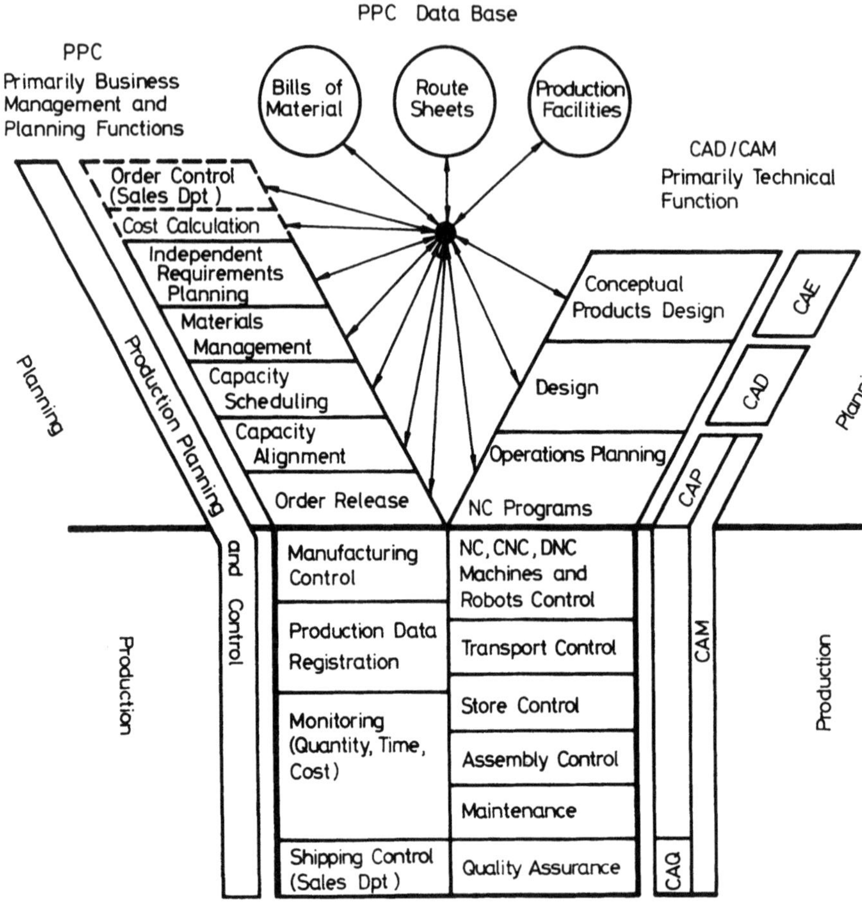

Fig. 8.31. Information systems in production within the framework of CIM (After Scheer [27])

places and times. For manufacturing and assembly processes, it is important also to measure the numbers of good and scrapped parts or of parts that have to be reworked. These are identified and reported by in-process measuring [35].

On the level of monitoring and production activity control, these transaction data, or movement data, are used for visualizing the process and calculating the process-dependent and order-dependent key data. For this function, a so called *Electronic Leitstand* is used increasingly [49]. The corresponding monitoring system for the single work centers was described in Sect. 5.4. It is likely that in the future there will be an order monitoring component which can monitor the throughput not only of single-level manufacturing processes, but also of complex orders with several structural levels – including all manufactured and purchased parts. The diagnostic system, which was outlined in Sect. 5.5, examines deviations between the planned and actual processes. It is conceivable to have an expert system at this point which carries out the diagnosis of the manufacturing flow and its throughput diagrams in order to, first, find the causes of the detected deviations and, second, suggest short-term measures for improvement [36].

The actual control functions are to be found on the next level. The quantities to

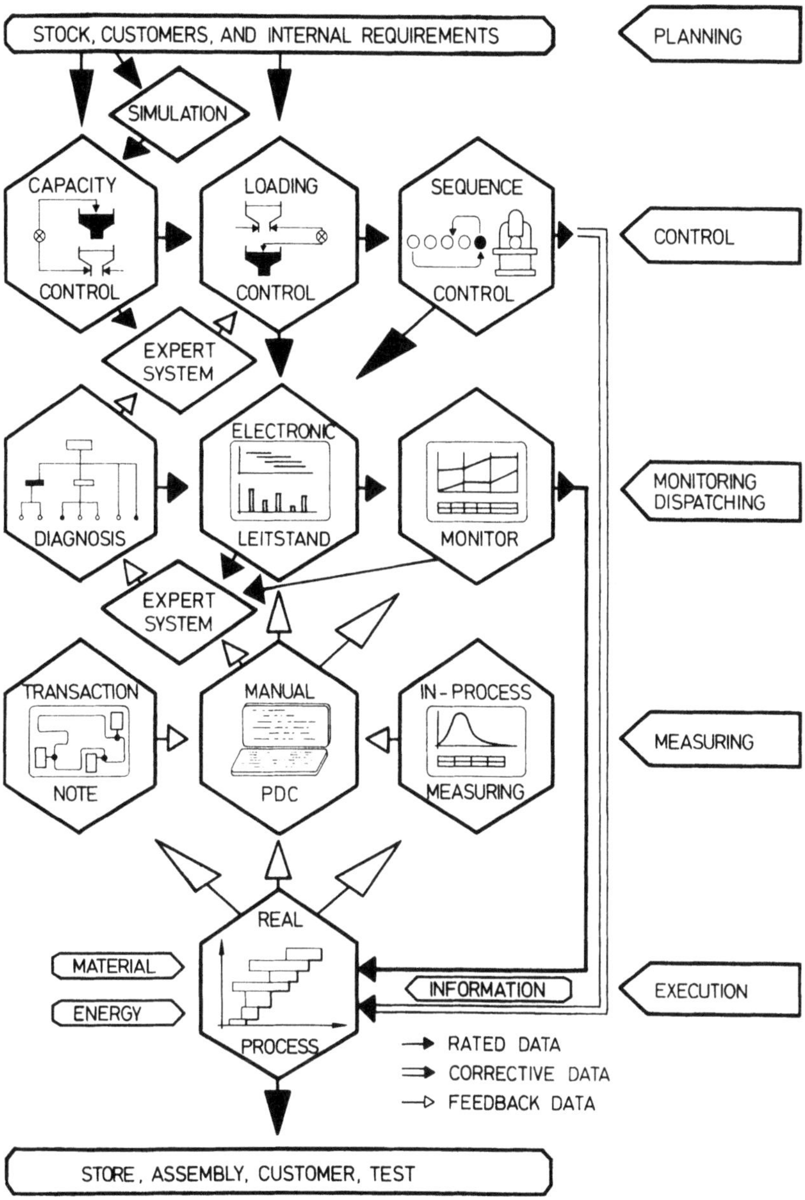

Fig. 8.32. Manufacturing control functions in a computer-integrated plant

be controlled are the capacity, the load (and thus the flow time), and the processing sequence. As was explained in detail above, other control methods apart from load-oriented manufacturing control, such as the Kanban system for example, can be used in some cases (see also Chapt. 9). On this level, too, an expert system might be used to suggest changes of parameters in the control algorithms, on the basis of data from the diagnostic module on the previous level.

On the planning level, the types, quantities, and due dates of purchase orders and manufacturing orders are defined. For this purpose, simulation methods or queueing models can be useful. Whereas simulation is becoming an increasingly important aid in planning, designing, and controlling production facilities (see [31, 32, 37]), it is not yet being implemented as an instrument of manufacturing control for entire shops, although it seems very promising.

On the whole, the manufacturing of single products must be controlled according to the same principles as those applied to continuous processes. Short flow times and in-time schedule performance must be regarded as market-related quality features, while the inventory in the shops and the utilization of capacities must be regarded as company-related quality features. Whereas the funnel model has so far only been used for the manufacturing flow, it is also suitable for all other departments of the company where there is an input and output of work, and where orders compete for capacities in interlinked structures. Examples are warehouse and transport systems or design departments.

As such complex systems are very expensive, the question of cost effectiveness is one which arises very frequently. Whereas the results of the first evaluations of new technologies are available [38], studies on the cost effectiveness of PPC systems are still very difficult to carry out. We therefore present a few ways of evaluating the financial and organizational effects of the implementation of a new manufacturing control method such as load-oriented manufacturing control.

8.7 Effects of Load-Oriented Manufacturing Control

8.7.1 Cost Effectiveness

When determining the cost effectiveness of manufacturing control systems, the installation, maintenance, and operation costs are compared with the expected savings. In this calculation, one must take into account not only the manufacturing and assembly processes, but the overall order throughput, such as the stages between entry and release, and those between completion and shipping of the orders. A close observation of the capital tied up during the throughput of an order can serve as a means of determining possible savings to be made by the introduction of an improved manufacturing control system (Fig. 8.33).

From these observations, a series of cost-saving measures can be deduced. These mainly consist of changes in inventories (e.g. stock-on-hand, space required, scrapping volumes) and changes in lead times (e.g. volume of alterations, frequency of disruptions, forecast reliability, quality). What is especially important, and at the same time hard to quantify, is the effect of reductions in delivery time on market share, price structuring, penalties, etc.

Figures are available to prove the great potential for rationalization offered by an improved manufacturing control. One mechanical engineering company, for example, has calculated that, out of its annual turnover of DM340 ($212.5) million, there is a cost potential of DM32.7 ($20.5) million which can be influenced by manufacturing control (Table 8.3).

An inventory and flow time analysis, carried out essentially as explained in Sect. 5.3, showed that, with a better manufacturing control system, the manufacturing

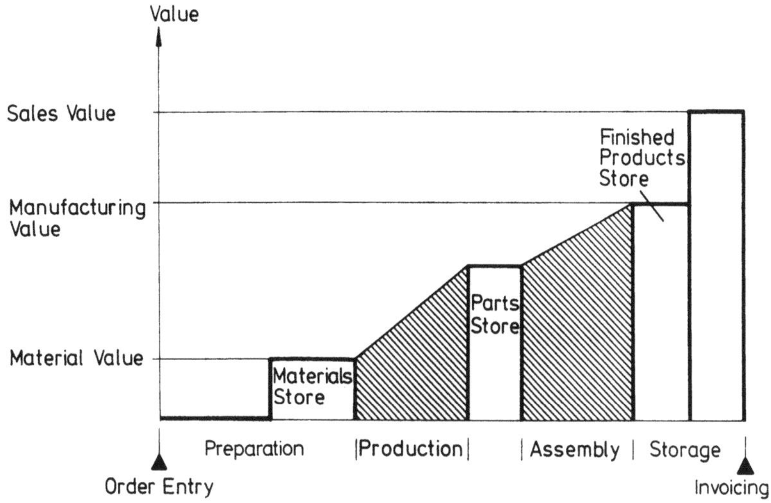

Fig. 8.33. Capital tie-up during order lead time (According to Siemens AG)

Table 8.3. Potential for savings due to a PPC system in a mechanical engineering company (After Pabst [39])

Type of Cost	Potential (Million $)	Feasible Savings (Million $)
Interest Costs	9.3	1.7
Penalties	0.8	0.4
Extended Production Capacity	7.5	2.5
Overtime Costs	2.9	0.4
Total	20.5	5.0

Annual Turnover: 212.5 Million $

and assembly flow times can be reduced by about 30% without loss in performance. This results in interest savings of about DM2.7 ($1.7) million per year, which are due to a lower in-process capital, as well as to reductions in backorders – the latter leading, in turn, to an expected 50% decrease in penalties. The extended production line costs are of special importance here. It could be proved that, because of supposed capacity bottlenecks, a great deal of work was subcontracted, although most of it could have been manufactured in-house. Finally, an improved manufacturing control system could also cut overtime costs, although not to the same extent as inventory-related costs [39].

Whereas interest costs and penalties are due to inadequate scheduling control, extended production line costs and overtime costs indicate poor capacity planning.

The case described above is typical of an order-dependent job-shop production, whereas in batch production with numerous variants, the central problems are unsatisfactory delivery times and inventories (work-in-process). From current experience with load-oriented manufacturing control, a 30% reduction in lead time and inventory and a 50% reduction in backorders seems realistic [8, 40–42]. What is more, computing times are considerably shorter for load-oriented order release than for conventional capacity scheduling methods. Although one cannot make general statements because there are so many different hardware and software environments, one can expect, for shop floor control, a reduction in computing time by 30%–50% compared to conventional methods [13, 43].

Furthermore, once manufacturing control is in operation, disrupted orders or rush orders no longer have to be followed individually through the factory, since the manufacturing controllers can now concentrate on analysing the deviations detected by the monitoring system and on taking respective corrective action.

8.7.2 Qualification and Motivation

Since the early 1970s, 'deterministic' manufacturing control has come under increasing criticism especially in Europe (see also Chapt. 2). This is the consequence of the 'Taylorization' of the production process, which means that it was subdivided into ever simpler, almost meaningless single steps. As the processes become more complex, their control also becomes more complex and less flexible. Above all, as workers lose insight into the processes with which they are involved, they also lose their motivation to change the conditions [44]. It was soon realized that this development was leading to a dead end [45].

In this connection, the conventional manufacturing control methods are also being questioned from ergonomical and sociological points of view [46]. Besides the technical problems of centralized planning discussed above, such as order time variance, engineering changes, and disruptions, emphasis is also being placed on the necessity to let the program carry out only a planning outline, and leave the finite scheduling in the shops to the personnel concerned.

Today, one can observe a general trend towards small and autonomous manufacturing and assembly units which are provided with a production program for a certain period in advance by production planning and control. The details, however, including all technical, dispatching, and quality-assurance functions, are the responsibility of the individual shops [47].

Load-oriented manufacturing control supports this development, since:

- it visualizes the production process and its essential key objectives
- it illustrates the interdependencies between these objectives
- it reveals opportunities for corrective measures
- it has only a small number of control parameters, and these are clearly understandable
- it shows the effects of changes in parameters and 'infringements of the rules', and
- it helps the experienced operator instead of ruling him.

This concludes the description of load-oriented manufacturing control and its application. Regardless of the degree of automation and interlinkage, the funnel model on which the method is based is valid for every production process in which individual orders with different work contents must be processed at several different work centers, and where these orders have to compete with other orders for capacities.

The program modules outlined above are, however, especially profitable for job-shop production of single products or small batches. Therefore, the following chapter will briefly describe the other manufacturing control methods, and contrast them with the load-oriented approach.

8.8 References

[1] Ritter, K-H. Voraussetzungen zum Einsatz der belastungsorientierten Auftragsfreigabe (Prerequisites for the Implementation of Load-Oriented Order Release). In: Wiendahl H-P. (ed.). *Praxis der belastungsorientierten Fertigungssteuerung* (*Practical Load-Oriented Manufacturing Control*). Institut für Fabrikanlagen der Universität Hannover (IFA), Hannover, 1986, pp 53–73.

[2] Kettner, H, Bechte, W. Neue Wege der Fertigungssteuerung durch belastungsorientierte Auftragsfreigabe (Recent Production Control Methods by Use of Load-Oriented Order Release). *Zeitschrift des Vereins Deutscher Ingenieure*, **123** (11): 1981, 459–466.

[3] Buchmann, W. Konzeption eines Programmsystems zur belastungsorientierten Fertigungssteuerung (A Concept for a Computer System for Load-Oriented Production Control). *Zeitschrift für Wirtschaftliche Fertigung*, **77** (12): 1982, 390–394.

[4] Buchmann, W. Realisierung von Programmsystemen zur belastungsorientierten Fertigungssteuerung (Implementation of Computer Systems for Load-Oriented Production Control). *Zeitschrift für Wirtschaftliche Fertigung*, **77** (11): 1982, 374–377.

[5] Wildemann, H. *Just-In-Time in Deutschland* (*Just-In-Time Implementation in Germany*). Universität Passau, Germany, 1985.

[6] Bechte, W. Steuerung der Durchlaufzeit durch belastungsorientierte Auftragsfreigabe bei Werkstattfertigung (Lead Time Control in Job-Shop Production by Load-Oriented Order Release). Dissertation, Universität Hannover, 1980 (veröffentlicht in: Fortschritt-Berichte der VDI-Zeitschriften, Reihe 2, Nr. 70, Düsseldorf, 1984) (doctoral thesis, published in the above VDI journal).

[7] Erdlenbruch, B. Auswirkungen von Fertigungssteuerungsmaßnahmen – Simulation auf der Basis realer Betriebsdaten (The Effects of Production Control Measures – Simulations Based on Authentic Production Data). Vortrag Nr. 8 zum Seminar 'Neue Wege der Fertigungssteuerung' des Instituts für Fabrikanlagen der Universität Hannover am 16–17 März 1982 in Hannover (seminar proceedings). Hannover, 1982.

[8] Bechte, W. Kontroll- und Planungssystem zur belastungsorientierten Fertigungssteuerung im Dialog-Konzept und Realisierung des Systems KPS-F (A Dialog-Driven Monitoring and Planning System for Load-Oriented Production Control. Implementation of the System KPS-F). In: Wiendahl, H-P. (ed.), *Praxis der belastungsorientierten Fertigungssteuerung* (*Practical Load-Oriented Manufacturing Control*). IFA, Hannover, 1984, pp 89–118.

[9] Kazmaier, E. Berücksichtigung der Belastungssituation im Rahmen eines neuen PPS-Systems auf der Basis einer dialogorientierten Ablaufplanung (A Recent PPC System which Accounts for the Existing Loading Status, Based on Dialog-Oriented Sequencing). In: Tagungsunterlagen zum Fachseminar 'Statistisch orientierte Fertigungssteuerung' des Instituts für Fabrikanlagen der Universität Hannover, 1984, pp 81–97 (conference proceedings).

[10] Wiendahl, H-P, Springer, G. Untersuchung des Betriebsverhaltens flexibler Fertigungssysteme (Analysis of FMS in Operation). *Zeitschrift für Wirtschaftliche Fertigung*, **81** (2): 1986, 95–100.

[11] Streifinger, E. Zuverlässigkeit und Verfügbarkeit komplexer Fertigungsanlagen (Reliability and Availability of Complex Production Facilities). *Zeitschrift des Vereins Deutscher Ingenieure*, **127** (21): 1985, 865–870.

[12] Wiendahl, H-P, Ziersch, W-D. Untersuchung des Störverhaltens automatischer, verketteter Montageanlagen (Analysis of Disruptions in Automated, Interlinked Assembly Facilities). *Werkstattstechnik*, **72** (5): 1982, 275–279.

[13] Erdlenbruch, B. Aufbau eines Fertigungssteuerungssystems zur Kapazitäts-, Durchlaufzeit- und Bestandsplanung – Terminorientierte Kapazitätsplanung – Belastungsorientierte Auftragsfreigabe (A Production Control System for Capacity, Lead Time, and Inventory Planning. Schedule-Oriented Capacity Planning. Load-Oriented Order Release). In: Wiendahl, H-P (ed). *Praxis der belastungsorientierten Fertigungssteuerung* (*Practical Load-Oriented Manufacturing Control*). IFA, Hannover, 1986, pp 181–195.

[14] Kölle, J. Gesamtsystem zum Planen – Steuern – Kontrollieren von Fertigungsunternehmen PSK 2000 – Ein Überblick (PSK 2000 – An Overall System for Planning, Controlling, and Monitoring of Manufacturing Companies). In: Seminarunterlage zum AWF-Workshop 'Belastungsorientierte Fertigungssteuerung', 8–9 Mai 1985, Eschborn, Taunus (seminar proceedings).

[15] Herrmann, N. Das System 'Auftragsfreigabe mit Belastungsschranke' (ABS-X) auf PC-Basis (The System ABS-X – Order Release by Use of Load Limits – for PCs). In: Seminarunterlage zum AWF-Seminar 'Belastungsorientierte Fertigungssteuerung', 24–25 Februar 1986, Bad Soden, Taunus (seminar proceedings).

[16] Storfinger, R. Die belastungsorientierte Fertigungssteuerung für mechanische Fertigungen in der

BMW AG (Projekt BORA) (Load-Oriented Production Control in Mechanical Manufacturing – The Project BORA at BMW). In: Wiendahl, H-P. (ed.), *Praxis der belastungsorientierten Fertigungssteuerung* (*Practical Load-Oriented Manufacturing Control*). IFA, Hannover, 1986, pp 197–218.

[17] Wiendahl, H-P. Erprobte Methoden zur Reduzierung von Durchlaufzeiten in der Produktion (Tried and Tested Methods for Lead Time Reduction in Manufacturing). *Industrielle Organisation*, **53** (9): 1984, 391–395.

[18] Autorenkollektiv (several authors): Methodik und Praxis der Durchlaufzeitverkürzung in der Einzel- und Kleinserienfertigung (Methods and Applications of Lead Time Reduction in Job-Shop and Small-Batches Production). In: Tagungsunterlage zur VDI/ADB-Fachtagung, 7–8 März 1985, Fellbach (conference proceedings).

[19] Wiendahl, H-P, Enghardt, W. Logistikgerechte Fabrik, rechnergestützt geplant (Logistics in a Factory by Computer Aided Planning). *Zeitschrift für Wirtschaftliche Fertigung*, **80** (3): 1985, 131–136.

[20] Selmer, W, Thümmler, K, Mehrens, K. Durchlaufzeiten in der variantenreichen Kleinserienfertigung – Erfahrungen mit einer Transport- und Regalanlage als Kernstück eines flexiblen Fertigungssystems – Zielsetzung und Konzept – Einbindung in den betrieblichen Ablauf – Erfahrungen und Empfehlungen (Lead Times in Small-Batches Production – Application of a Store Rack and Transport System as the Crucial Element of a Flexible Manufacturing System). In: Tagungsunterlage zur VDI/ADB-Fachtagung, 7–8 März 1985, Fellbach (conference proceedings).

[21] Wiendahl, H-P, Enghardt, W. Rechnergestützte Layoutplanung unter Einbeziehung praxisgerechter Randbedingungen (Computer-Aided Layout Planning which Accounts for Practical Requirements). In: Rechnerunterstützte Fabrikplanung für die Gestaltung der Produktion von morgen. VDI/ADB-Fachtagung, 24–25 October 1985, Fellbach/b. Stuttgart, Düsseldorf, (proceedings).

[22] Mertins, K. Entwicklungsstand flexibler Fertigungssysteme – Linien-, Netz- und Zellenstrukturen (State of the Art of Flexible Manufacturing Systems – Line, Network, and Cell Structures). *Zeitschrift für Wirtschaftliche Fertigung*, **80** (6): 1985, 249–265.

[23] Steinhilper, R. Flexible Fertigungssysteme in der Praxis – Erwartungen und Einsatzerfahrungen (Applications of Flexible Manufacturing Systems – Expectations and Experiences). *Technische Zeitschrift für Metallbearbeitung* **79** (8): 1985, 37–45.

[24] Gottschalk, E, Wirth, S. Neue flexible Teilefertigungssysteme der DDR im internationalen Vergleich (Recent Flexible Systems for Parts Production in the GDR). *Fertigungstechnik und Betrieb*, **36** (4): 1986, 213–221.

[25] Gottschalk, E. Untersuchungen zum Einsatz der belastungsorientierten Fertigungssteuerung in flexiblen Fertigungssystemen (Studies for the Implementation of Load-Oriented Production Control in Flexible Manufacturing Systems). In: Wiendahl, H-P. (ed.), *Praxis der belastungsorientierten Fertigungssteuerung* (*Practical Load-Oriented Manufacturing Control*). IFA, Hannover, 1986, pp 235–251.

[26] Waller, S. Die automatisierte Fabrik (The Automated Factory). *Zeitschrift des Vereins Deutscher Ingenieure*, **125** (20): 1983, 838–842.

[27] Scheer, A-W. Vorgehensweise und Möglichkeiten zur Realisierung eines CIM-Konzeptes unter Berücksichtigung vorhandener EDV-Instrumente (Implementation of a CIM Concept in View of Existing Data Processing Systems). In: Kongreß PPS 85 des Ausschusses für wirtschaftliche Fertigung (AWF), 6–8 November 1985, Böblingen (proceedings).

[28] Scheer, A-W. Anforderungen an Datenverwaltungssysteme in CIM-Konzepten (Requirements on Data Management Systems in CIM Concepts). In: Tagungsbericht der 18. IPA-Arbeitstagung 'Produktionsplanung und -steuerung in der CIM-Realisierung', 22–23 April 1986, Stuttgart (proceedings). Springer-Verlag, Berlin/Heidelberg, 1986, pp 121–141.

[29] Bauernfeind, U. Realisierung von CIM-Konzepten mit Standard-Komponenten (Implementation of CIM Concepts Using Standard Modules). *Zeitschrift für Wirtschaftliche Fertigung*, **80** (9): 1985, 397–403.

[30] Spur, G, Krause, F-L. *CAD-Technik* (*Computer Aided Design* (*CAD*)). Hanser, München, 1984.

[31] Spur, G, Hirn, W, Seliger, G, Viehweger, B. Simulation zur Auslegungsplanung und Optimierung von Produktionssystemen (Simulation in Layout Planning and Optimization of Manufacturing Systems). *Zeitschrift für Wirtschaftliche Fertigung*, **77** (9): 1982, 446–452.

[32] Warnecke, H-J, Zipse, Th, Zeh, K-P. Rechnergestützte Planung und digitale Ablaufsimulation flexibler Fertigungssysteme (Computer-Aided Planning and Digital Simulation for Flexible Manufacturing Systems). In: Tagungsbericht der 17. IPA-Arbeitstagung 'Flexible Fertigungssysteme' 1984, Stuttgart (proceedings). Springer-Verlag, Berlin/Heidelberg, 1984, pp 197–207.

[33] Warnecke, H-J. (ed.), Produktionsplanung, Produktionssteuerung in der CIM-Realisierung (Production Planning and Control in CIM Systems). Tagungsbericht der 18. IPA-Arbeitstagung 'Produktionsplanung und -steuerung in der CIM-Realisierung', 22–23 April 1986, Stuttgart (proceedings). Springer-Verlag, Berlin/Heidelberg, 1986.

[34] Geitner, U. Rechnerunterstützte Produktion – Ein Überblick über den Stand von Praxis und Forschung (Computer-Aided Manufacturing – State of the Art in Research and Practice). *Zeitschrift für Wirtschaftliche Fertigung*, **81** (1): 1986, 9–14.

[35] Brankamp, K, Bongartz, B. *Der moderne Stanzbetrieb – Vom Sensormonitoring zur Geisterschicht (The Progressive Punching Shop – From Sensor Monitoring to the Man-less Shift)*. VDI-Verlag, Düsseldorf, 1986.

[36] Wiendahl, H-P, Ludwig, E. A Knowledge Based Diagnostic System for Production Processes – Basics and Concepts. In: Advances in Production Management Systems. Proceedings 4th IFIP TC5/W65.7 International Conference on Advances in Production Management Systems, Espoo, Finland, 20–22 August 1990. Elsevier (North-Holland), Amsterdam/New York, 1991.

[37] Deutsche Gesellschaft für Logistik (ed.). Simulationstechnik und Logistik (Simulation Systems and Logistics). Fachtagung der Deutschen Gesellschaft für Logistik, 3–4 June 1986, Dortmund (proceedings).

[38] Wildemann, H et al. Strategische Investitionsplanung für neue Technologien in der Produktion (Investment Planning for New Manufacturing Technologies). Tagungsbericht zum 2. Fertigungswissenschaftlichen Kolloquium an der Universität Passau, 5–7 März 1986 (proceedings).

[39] Pabst, H-J. Analyse der betriebswirtschaftlichen Effizienz einer computergestützten Fertigungssteuerung mit CAPOSS-E in einem Unternehmen der Einzel- und Kleinserienfertigung (Economic Efficiency of Computer-Aided Production Control with CAPOSS-E in Job-Shop and Small-Batches Production). Dissertation Universität Erlangen, 1985 (veröffentlicht in: Europäische Hochschulschriften, R.V., Vol. 645, Frankfurt/Bern/New York, 1985) (doctoral thesis, published in the above periodical).

[40] Ulfers, H-A. Erfahrungen mit der belastungsorientierten Auftragssteuerung ABS bei Siemens (The Load-Oriented Order Control System ABS in Use at Siemens). In: Fachbericht zum Seminar 'Praxis der belastungsorientierten Fertigungssteuerung' des Instituts für Fabrikanlagen der Universität Hannover, Hannover, 1986, pp 149–169 (seminar proceedings).

[41] Knecht, R. Zwei Jahre belastungsorientierte Auftragsfreigabe bei Firma Hilti (Two Years of Load-Oriented Order Release at HILTI). In: Fachbericht zum Seminar 'Praxis der belastungsorientierten Fertigungssteuerung' des Instituts für Fabrikanlagen der Universität Hannover, Hannover, 1986, pp 119–148 (seminar proceedings).

[42] Büttner, R. Kontroll- und Planungssystem zur belastungsorientierten Fertigungssteuerung im Dialog – Anwendererfahrungen (A Monitoring and Planning System for Load-Oriented Production Control – User Experiences). In: Wiendahl, H-P. (ed.), *Praxis der belastungsorientierten Fertigungssteuerung (Practical Load-Oriented Manufacturing Control)*. IFA, Hannover, 1986, pp 75–88.

[43] Zwanzger-Brehm, B. Vergleich eines deterministischen und statistischen Kapazitätsabstimmungsverfahrens mit Hilfe der Simulation und realer Betriebsdaten (Deterministic vs. Statistical Load Leveling Technique – A Comparison by Use of Simulation and Authentic Production Data). Dissertation Technische Universität Braunschweig, 1984 (Doctoral thesis).

[44] Warnecke, H-J. Taylor und die Fertigungstechnik von morgen (Taylor and Tomorrow's Production Engineering). In: FTK '85, schriftliche Fassung der Vorträge zum Fertigungstechnischen Kolloquium, 10–11 Oktober 1985, Stuttgart. Springer-Verlag, Berlin/Heidelberg, 1985, pp 1–12 (proceedings).

[45] Brödner, P. Fabrik 2000. *Alternative Entwicklungspfade in die Zukunft der Fabrik (Alternatives for Development towards the Future Factory)*. Edition Sigma, Berlin, 1985.

[46] Manske, F, Wobbe-Ohlenburg, W. Fertigungssteuerung im Maschinenbau aus der Sicht von Unternehmensleitung und Werkstattpersonal (Production Control in Mechanical Engineering from the Point of View of the Management and the Job-Shop Personnel), Teil 1: Totalplanung oder Rahmenplanung ? *Zeitschrift des Vereins Deutscher Ingenieure*, **127** (11): 1985, 395–402; Teil 2: Konsequenzen für Arbeiter, Meister und Steuerungspersonal. *Zeitschrift des Vereins Deutscher Ingenieure*, **127** (12): 1985, 457–462; Teil 3: Organisatorische und personalwirtschaftliche Gestaltungshinweise für Werkstattsteuerungssysteme. *Zeitschrift des Vereins Deutscher Ingenieure*, **127** (13): 1985, 489–494.

[47] Tress, DW. Kleine Einheiten in der Produktion – Wer wachsen will, muß kleiner werden (Small Production Units – Shrinking for Growth ?). Zeitschrift für Organisation **55** (3): 1986, 181–186.

[48] Wiendahl, H-P. Integration von PPS-Systemen in CIM-Konzepte (Integration of PPC Systems into CIM Concepts). *VDI-Berichte*, 611(1986) 207–231.

[49] Kanet, JJ, Sridharan, V. The Electronic Leitstand: A New Tool for Shop Scheduling. *Manufacturing Review*, **3** (3): September 1990, 161–170.

[50] Bertrand, JWM. Multiproduct Optimal Batch Sizes with In-Process Inventories and Multi Work Centers. *IEE Transactions*, Juni 1985, 157–163.

[51] Nyhuis, P. Durchlauforientierte Losgrößenbestimmung (Throughput Oriented Lot Size Determination). Dissertation University of Hannover 1991. Published as *Fortschritt-Berichte Zeitschrift des Vereins Deutscher Ingenieure*, **2** (225): Düsseldorf, 1991.

Chapter 9

Comparison of Load-Oriented Manufacturing Control with Other Methods

9.1 Abstract

This chapter describes other well known methods of manufacturing control such as Kanban, Job-Progress Number, Queueing Models, and the OPT system, explains their fields of use as compared with Load-Oriented Manufacturing, and outlines their differences from the latter. It will be shown that all methods can be explained by means of the throughput diagram. The Electronic Leitstand, which is a kind of Gantt chart-based scheduling tool on a computer screen, is also briefly explained as a relatively new device for finite control of jobs once they have been released by shop floor control.

9.2 Overview

The fundamental tasks of a manufacturing control system developed on the basis of the funnel model are valid for any process. The aim is to control the order throughput in such a way that the orders pass the shop as quickly as possible and on time (short flow times, good schedule performance). This requires short queues at the work centers (low inventories). At the same time, however, both unnecessary waiting times at the bottleneck capacities and an idling of work centers must be avoided (good and even utilization).

As was proved above, one essential parameter for controlling these key values is the mean inventory at the individual work centers. Controlling the 'mean work center inventory' parameter is quite simple, depending on how the manufacturing flow is organized, in which way the work centers are interlinked, how the lead time is distributed, and how many different workpieces (and thus operations) have to be considered.

Two extreme production situations will serve as examples. Figure 9.1 shows a fictitious *time-phased* production where the orders are processed in one day by four work centers working eight hours a day. The transit time (TT) is assumed to be two

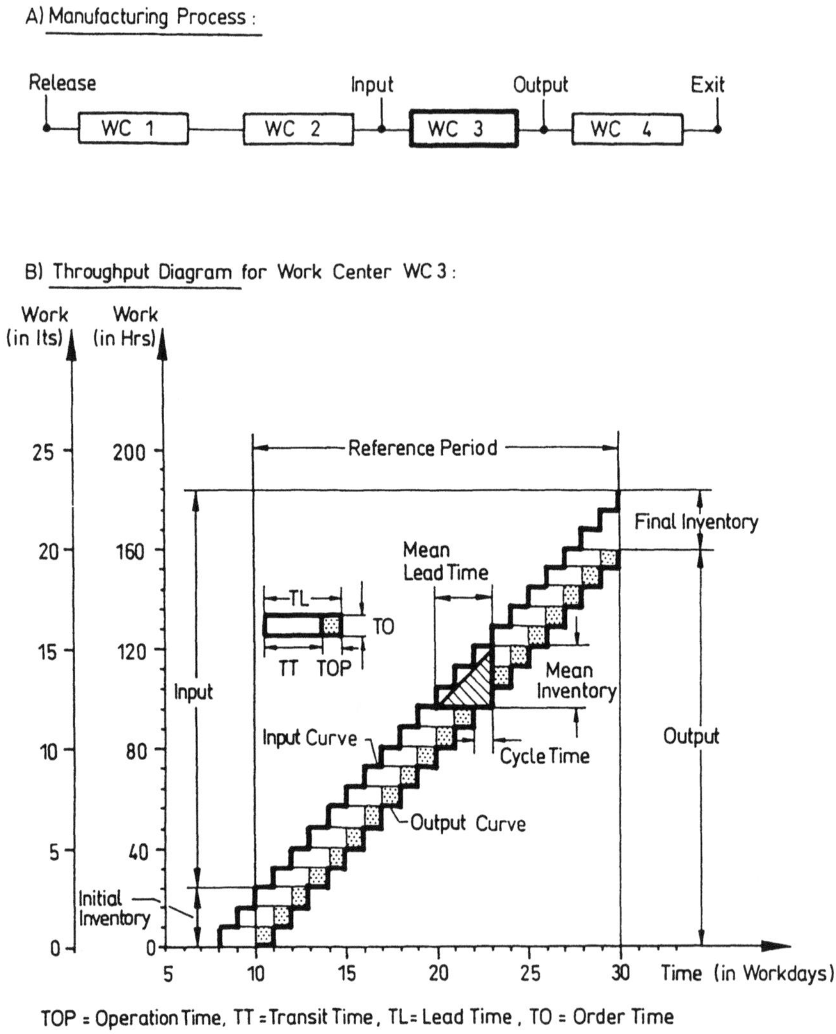

C) Process Characteristics:

① Equal Work Hours Content
⟹ No Operation Time Variance

② No Waiting, No Breakdowns
⟹ No Buffer Stock

③ Regular Input and Output
⟹ Regular Loading to Capacity

④ Resulting from ①, ② and ③
⟹ No Lead Time Variance

⑤ No Sequence Transposition
⟹ together with ③ ⟹ No Schedule Deviation

Fig. 9.1. Manufacturing process with regular input and output trends (time-phased process) days. Figure 9.1B shows the work center throughput diagram of Work Center 3. Each order is processed immediately on arrival. This leads to an ideal process without buffer times which allows a uniform, 100% utilization of the capacity, and in which, due to the constant operation and lead times the orders do not suffer any schedule delay. In this case, the work amount can also be given in items instead of hours.

Such a process does not exist in practice. Over a longer period, there will at least be a few disruptions in a real process. Besides, there are few products where only one single model is manufactured; thus, setup processes and changes in cycle times are required. There are, however, some examples of time-phased production lines, e.g. in the manufacture of engine housings, gear boxes, etc., which are, in principle, very similar to our ficticious process. They usually have intermediate buffers which compensate for minor disturbances. It is obviously not necessary to control such a process with the help of the funnel model: there are simpler solutions for this.

The other extreme of a single-unit production process is shown in Fig. 9.2. This could be a shop with machines for very large parts with long and varying operation times, such as machine tool bases, turbine housings and rotors, etc. (Fig. 9.2A). The problems this poses for manufacturing control can be seen in a throughput diagram of such a heavy-duty work center (Fig. 9.2B). The work contents vary widely, and often reach values which require operation times of one week and longer. The workpieces are manufactured with 'lot size one', and produced only once in the same design. The inputs and outputs at the machines concerned vary widely, and for scheduling reasons, sequence transpositions are often necessary. Such processes are characterized by long waiting times, uneven utilization of capacities, and great variations in lateness (Fig. 9.2C).

In this case, mean inventory is not useful as a control parameter, since there can be extreme variations from period to period. Also, the release of orders with such a large work content – orders which are nearly always the A-parts of the corresponding product and as such are those relevant for the due date – will not be deferred because a load limit is apparently exceeded. Generally, one can say that load control can no longer be used in such a case, because the deviation from a 'mean' process is too large.

All other types of manufacturing process fall between these extremes.

For this reason, several methods have emerged for the control of batch production processes. Figure 9.3 shows an overview of the different approaches; in this illustration, load-oriented manufacturing control replaces the conventional method.

The methods are classified qualitatively using two parameters. The first is the production principle, which is determined by the quantity of parts manufactured per year. The values range between the quantity one, which means custom production, and several million, e.g. in large-batch production. The second parameter is the number of different variants (models), which is associated with the variance of operation times.

In principle, two types of methods can be distinguished. In a simplified representation, the first group includes the control methods for job-shop and repetitive production, which are mostly organized according to the principle of intermittent production and controlled using conventional manufacturing control as described in Chapt. 2. The other group contains methods for continuous production and high-volume production, where the productive equipment is mostly arranged in

A) Manufacturing Process:

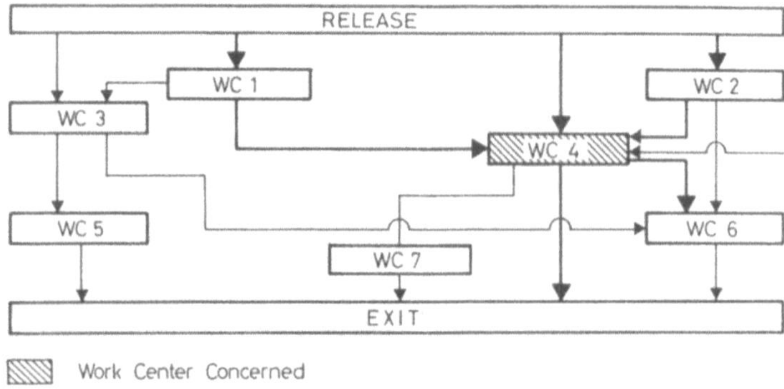

B) Throughput Diagram for Work Center WC 4:

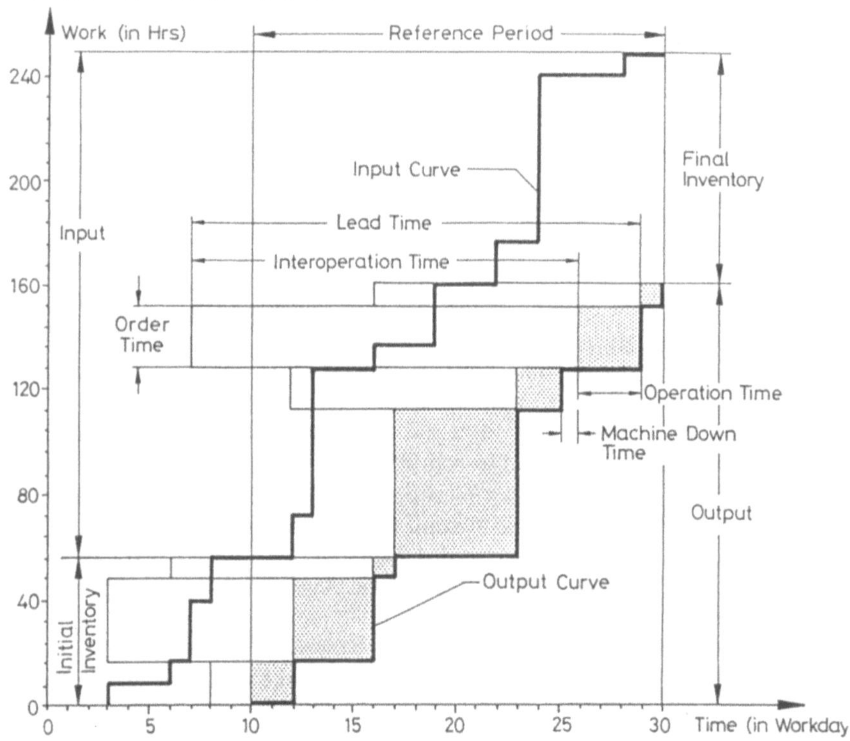

C) Process Characteristics

1. Unequal Work Hours Content
   ⟹ Wide Operation Time Variance
2. Long Waiting Times, Caused by Large Buffer Stock, Breakdowns, and Frequent Sequence Transpositions
   ⟹ together with (1) ⟹ Long and Widely Varying Lead Times, and ⟹ Large Buffer Stock
3. Irregular Input
   ⟹ Erratic Inventory and Range
   ⟹ Irregular Capacity Utilization
4. Considerable Sequence Transpositions
   ⟹ Large and Varying Schedule Deviations
5. Irregular Output
   ⟹ Risk of Waiting Times at Downstream Work Centers

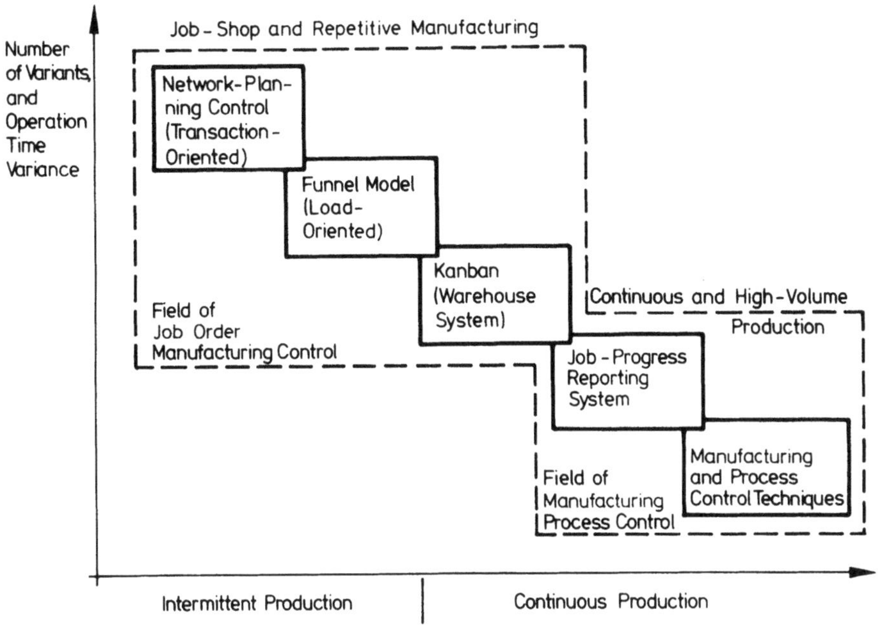

Fig. 9.3. Manufacturing control techniques (survey)

a product-dependent, operation-oriented continuous production sequence and is more or less automatically interlinked, and where control is being increasingly carried out directly by computer. Between the two groups, manufacturing is product-oriented and has a limited number of variants. There, automated flexible manufacturing systems or flexible manufacturing cells are becoming more and more common [1, 2].

In this rough classification, load-oriented manufacturing control is located between transaction-oriented network-planning control and the Japanese Kanban principle. As the production situation becomes less complex, the job-progress reporting system and the technological process control can be applied. Special methods, such as the queueing models, are not included in the figure, since they have not yet been successful in practice. The reasons for this will be explained after the methods shown in Fig. 9.3 have been discussed in detail (see Sect. 9.6).

It has not yet been explicitly pointed out that load-oriented manufacturing control is designed for so called *single-level* manufacturing. The time and quantity requirements placed on the parent assembly level of the products are deliberately not included in the control data; such interlinkages are guaranteed exclusively by the due dates. As it offers a high statistical probability for on-time due date performance, load-oriented manufacturing control effectively supports the schedule performance of multiple-level products. In the next section we examine the other methods.

◁ **Fig. 9.2.** Manufacturing process with extremely irregular input and output trends

9.3 The Kanban Principle

The Japanese Kanban system, which was developed by the Toyota company, is a method which has been the basis of intense discussion, and has been used with success in all industrial countries since the late 1970s. Its aims are low in-process inventories, short flow times and good schedule performance, and these aims are achieved with an astonishingly low amount of control [3]. In the Federal Republic of Germany, the concept was promoted above all by H. Wildemann, in his presentations at congresses, in lectures and papers, and in working groups [4].

The basic idea behind the Kanban system is to structure the production of a group of standard products or parts with defined variants into self-controlling cycles according to the warehouse principle. The emphasis is placed on a 'continuous flow' of the orders. Figure 9.4 compares the Kanban control principle with conventional central manufacturing control [4].

Each control center (e.g. rough machining, finishing, pre-assembly, and final assembly) is preceded by an inventory buffer containing an exactly defined quantity of the parts or materials needed for manufacturing the product concerned or its components. If the inventory in these buffers falls below an exactly defined minimum, the 'customer' orders a defined quantity of the necessary components from the 'supplier', to be delivered at a defined date. For this, he uses an order card called the order 'Kanban' (Japanese: kanban = card, billboard, sign). After the arrival of the order card, the supplier starts manufacturing the quantity of components defined in the Kanban, and delivers them in the desired quality in a standardized container on the date defined by the orderer. Generally, the quantity defined on the Kanban corresponds to the contents of one container (lot). Thus, the Kanban system uses the

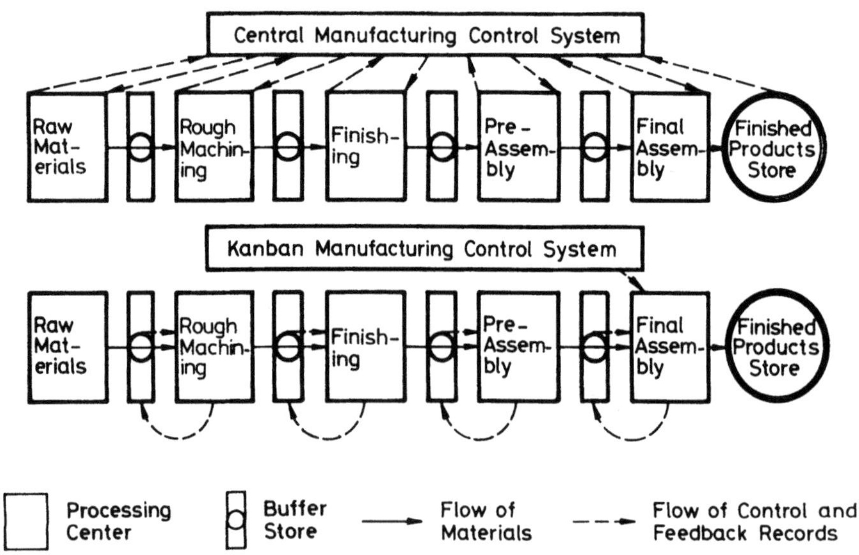

Fig. 9.4. Information and material flow – central versus Kanban manufacturing control system (After Wildemann [4])

pull principle (the customer gets the required parts from the supplier) instead of the conventional push principle (the supplier delivers his work to the customer).

The essential elements of the system are:

- production is based on historical demand, not future requirements
- interlinked, self-controlling cycle
- pull principle for the subsequent step
- flexible working of staff and equipment
- short-term control carried out by the operator
- Kanban card as a special source of information.

The essential organizational rules are:

- the user may neither order more material than he needs nor order too early
- the supplier may not deliver more parts than needed before ordered to do so, may not deliver more than the ordered quantity, and may not deliver faulty products
- the controller must assign evenly balanced loads to the shops and issue an appropriate number of Kanban cards (as few as possible) into the control cycles.

According to various publications, the following prerequisites must be met before Kanban control can be used [4]:

- harmonization of the master production schedule to obtain evenly distributed, small work contents per lot
- the productive equipment must be arranged in a process-oriented sequence and, if possible, the work rhythm must be uniform in the entire job shop
- production facilities must be easily available and setup times must be short
- scrap-rates must be low and quality control must be done by self-checking
- staff must be highly motivated and qualified.

Like the funnel model, the Kanban principle evidently uses the inventory as a control parameter, by circulating only a limited number of Kanbans per control cycle.

The following equation is used for calculating the number of Kanbans in circulation (NK) [5]:

$$NK = \frac{DD \times TL \times (1 + \alpha)}{CC} \tag{9.1}$$

where:

DD = daily demand in items per day
TL = scheduled lead time in days
$1 + \alpha$ = part-related safety factor
CC = container content in items.

The inventory (I) within a control cycle is, by definition, constant so that

$$I = NK \times CC \tag{9.2}$$

Furthermore, the work performed per day (PE_m) in a control cycle must correspond to the daily demand (DD) so that

$$PE_m = DD \tag{9.3}$$

Also, with constant order time, the unweighted (TL_m) and weighted mean lead times (TLM) are the same (see Sect. 3.6.2). If Eqn. (9.2) is now substituted for (9.1) and (9.3), we obtain:

$$I = PE_m \times TL\,(1 + \alpha) \tag{9.4}$$

The funnel formula was:

$$I_m = PE_m \times TLM$$

Except for the expression $(1 + \alpha)$, the two equations are identical. The difference can be explained as follows. By definition, the funnel formula only accounts for the inventory between the output from the upstream work center and the work center concerned. In a Kanban control cycle, the Kanban inventory (I) also contains the inventory from the returning empty containers, which is accounted for by the factor α in the Kanban formula.

This means that Kanban control is an inventory-oriented manufacturing control method with special constraints, a fact which allows a simple control process. The sum of the Kanbans in circulation is control parameter no. 1 in Kanban control. It controls the mean inventory and thus the lead time (in the funnel model this is done by the load limit). The second important control parameter is the order content per Kanban, which is kept constant by defining the number of parts per container. The order content of half a day is given as the recommended value.

Figure 9.5 shows the effects of the two Kanban control parameters in a fictitious throughput diagram. Since the dispatching rule used is FIFO, there are no sequence transpositions at the work centers, so that all the prerequisites for low inventories, short lead times, and good schedule performance are fulfilled.

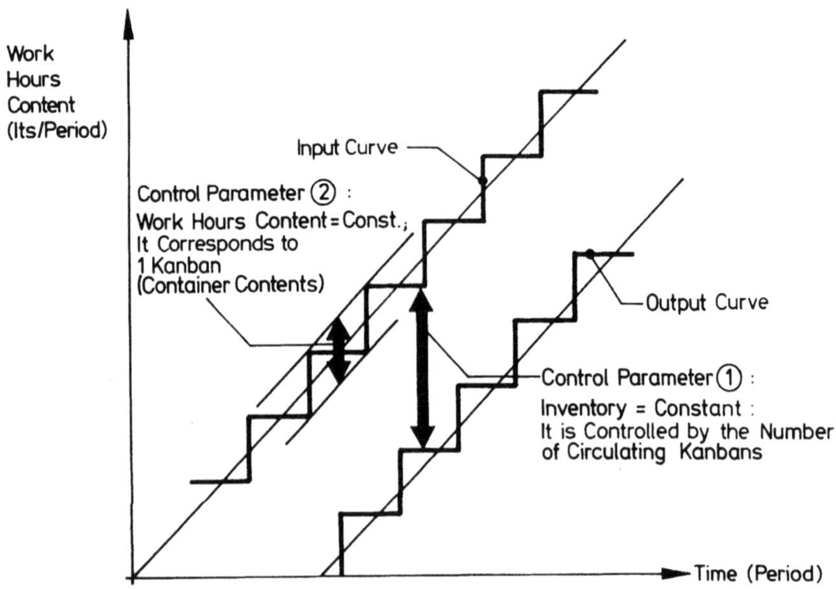

Fig. 9.5. Effects of the Kanban system

However, the Kanban philosophy includes more than simple control cycles: besides the manufacturing flow, it also controls the assembly and outside suppliers [4]. It is seen as part of the present just-in-time philosophy of production control. As a result of its clear organization of the manufacturing flow, its simple control principle, and the inclusion of planning and quality-assurance events, it ensures high product quality, reduces all non-productive waiting times, and allows staff to identify with 'their' product. If the prerequisites for using this system can be fulfilled, it is highly recommended.

9.4 The Job-Progress Number System

The job-progress number system, originally used in the automotive industry, has proved successful in large-batch production for decades [7, 8]. The basic idea behind it can be explained with a simple example (Fig. 9.6) [7].

Let us assume that our sample manufacturing process consists of the steps parts production, intermediate storage, and assembly. In each of these steps, called 'checking blocks', the output, i.e. the number of parts (or subassemblies, or products) over a certain period of time, is counted with the help of so called *progress numbers*. These are used to plot an output curve for each checking block. Note that the output is counted in items, not in hours. This is possible because the diagram is only valid

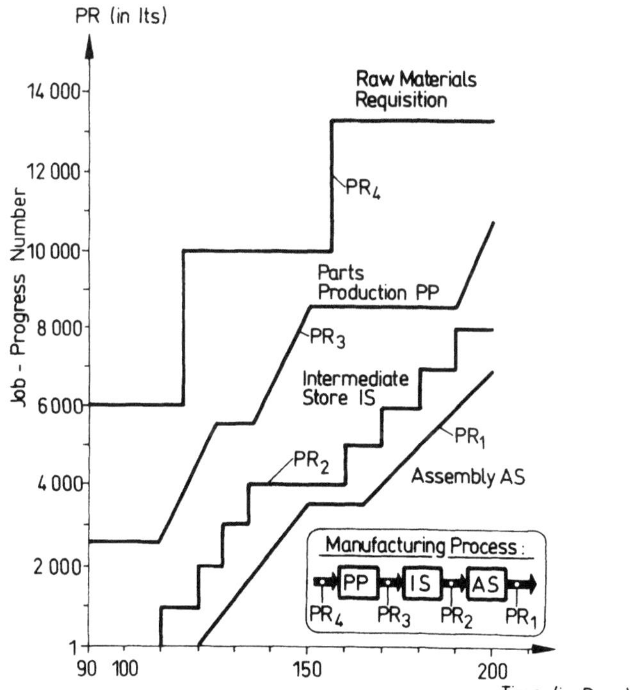

Fig. 9.6. Manufacturing control by job-progress records (principle) (After Gottwald [7])

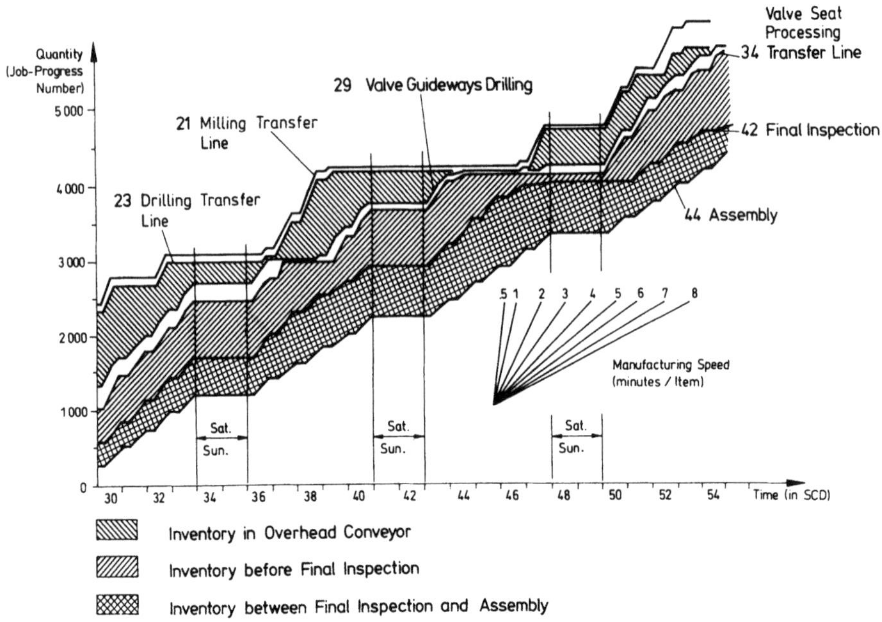

Fig. 9.7. Mechanical manufacturing of grey cast-iron parts shown in a job-progress diagram (After Heinemeyer [8])

for one single part, and because counting the parts is very easy, especially when they are mounted on pallets and moved with the help of an automated transport system. The parts have a mean flow time in each block, the block throughput time. Since the output from one block coincides with the input into the next block, the entire process can be represented by a set of progress curves. The capacity must be calculated in such a way that the block throughput times and performances are realistic.

The example shown in Fig. 9.7 is the large-batch production of a cast-iron housing, controlled with progress records [9]. As we can see from the hatched areas, inventories build up periodically in the transport and storage system (an overhead conveyor) and before and after the final inspection. The buffer function of the conveyor is necessary because the outputs of Work Centers 21 (milling) and 23 (drilling) are higher than that of Work Center 29 (drilling). This is why the output curve for Work Centers 21 and 23 shows horizontal sections; during these times, another housing was processed. On the whole, the inventories in the system develop irregularly.

The advantage of the job-progress number system is, first, that inputs and outputs of the individual checking blocks are examined cumulatively and, second, that the planned and the actual progress records are compared at the checking points, thus simplifying control.

A prerequisite is that the productive equipment must be arranged in accordance with the product flow, so that the manufactured parts must be practically the same over a longer period of time (several weeks or months). In practice, however, this is only possible in very few industrial situations. Therefore, the system is not an alternative for load-oriented manufacturing control.

9.5 Finite Control with a Graphic Control Unit (Electronic Leitstand)

As explained in Figs. 9.1 and 9.2, load-oriented manufacturing control gives unreliable information on the flow time of a lot or part if the inputs or outputs at the individual work centers are irregular. This can be remedied by multiplying the period length as compared to the other shops, but doing the planning in the same rhythm as in the other shops, with an overlapping of the periods.

If the number of parts and load centers is relatively small and the parts are expensive, one can still use Gantt charts or network plans, which have been used successfully for decades. Due to the large amount of manual calculation and graphic representation, these methods have, however, been used only at larger intervals – e.g. monthly – or only for special cases – e.g. for planning a large part which was the most relevant for schedule performance.

In addition, for utilization planning and activity control of the orders, electronic control units (so-called Leitstands) have been developed to act as the connecting link between planning and processing [10]. An interlinkage between planning data and feedback data, as indicated in Fig. 8.32, using an interactive control unit with an interactive-mode graphic computer workstation, can increase the transparency of the order throughput, and permit a quick calculation of alternatives and rapid detection of bottlenecks.

The idea was first successfully put into practice by the Fraunhofer-Institut für

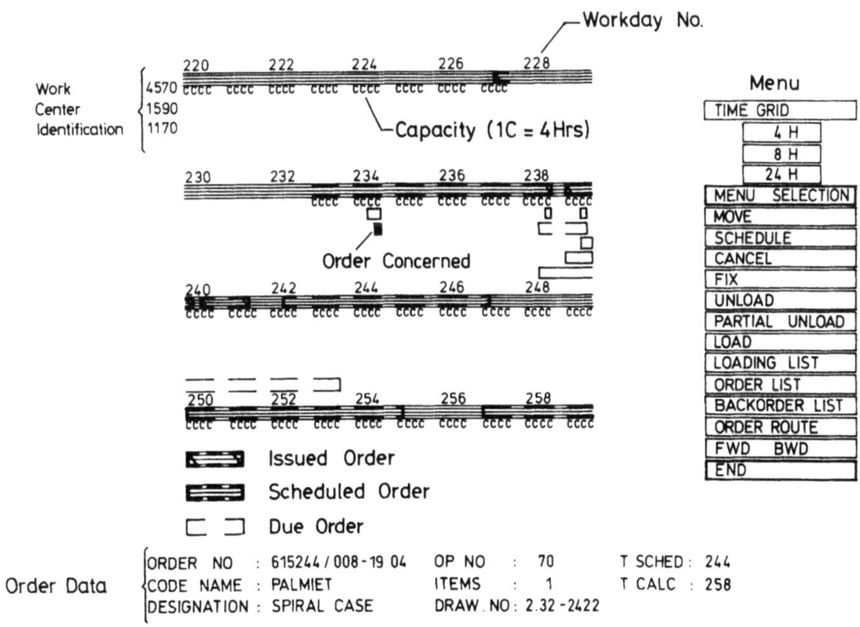

Fig. 9.8. Order scheduling at a graphic control unit (IPA Stuttgart/Voith Heidenheim)

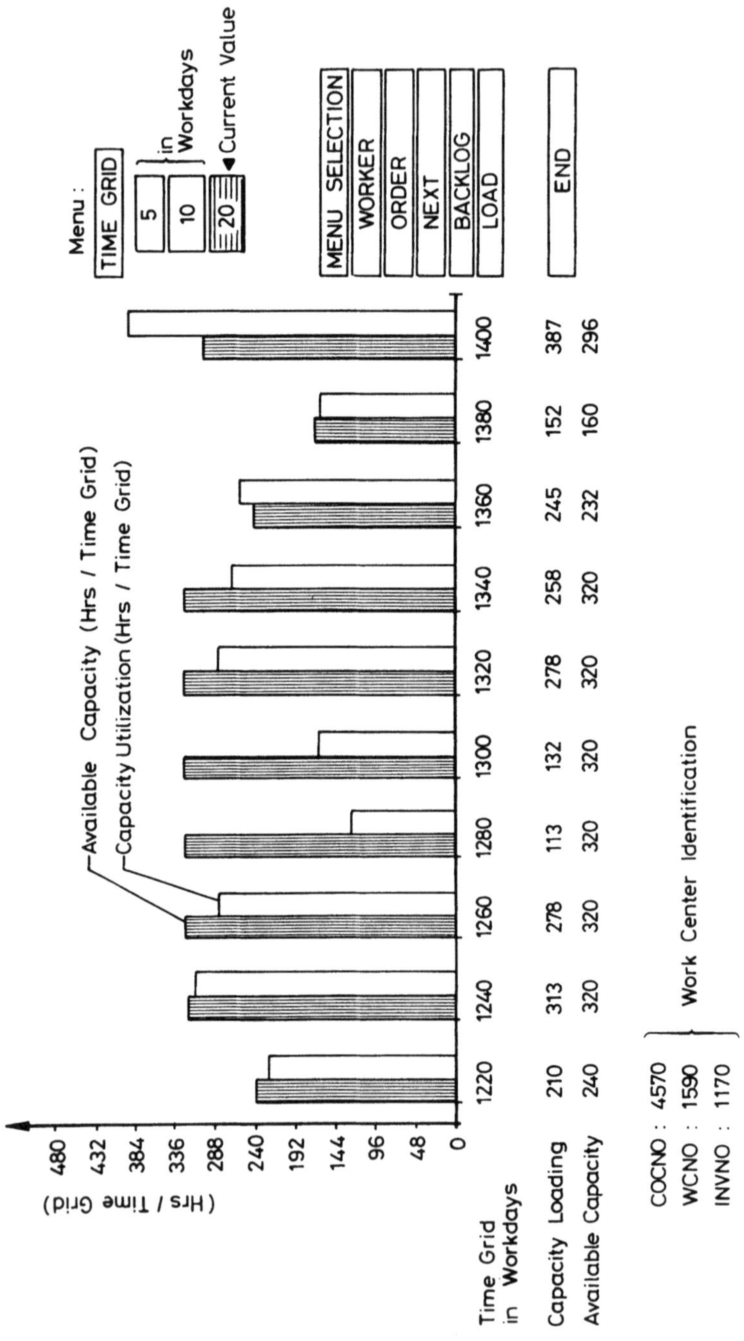

Fig. 9.9. Capacity planning at a graphic control unit (IPA Stuttgart/Voith Heidenheim)

Produktionstechnik und Automatisierung (IPA) at the University of Stuttgart, using a high-resolution color-graphics screen [11, 12].

As a starting point, the system is given the data on the manufacturing orders with the sequences of operations and the corresponding standard times and work centers. The orders are now subjected to lead time planning, using daily capacity data. Figure 9.8 shows the screen display for the order planning at one work center. The time axis is divided into four lines with 10 workdays each. It describes the planned capacity (here marked with Cs; $1C = 4$ hours) and the orders planned for this work center (visualized with different bars). Besides the physically existent and the planned orders, one can also see the orders which are still to be planned (here called 'due orders').

The order being considered is in competition with planned orders and 'due orders'. Using the menu, the planner now moves the order concerned and the competing orders, until a compromise between the scheduled date (T SCHED = Workday 224) and the date calculated from the loading (T CALC) is found. After the new schedule has been entered, the computer automatically changes the schedules of all orders affected by the move, since the order interlinkage is also stored in its memory.

Immediately afterwards, the current level of utilization can be vizualized (Fig. 9.9) by comparing the available capacity with the required capacity (here called capacity utilization) for each workday.

Such electronic control units, or 'Leitstands', are likely to be an important means of complementing the previous planning aids at the point of intersection between finite control and activity control. They are the next level of detail after the throughput diagram. Whereas load control aims at controlling and monitoring the throughput of the orders and the capacities as a whole, the graphic control unit supports the deterministic finite control of single orders and single work places. (For an English reference see [49] in Chapt. 8.) Load-oriented manufacturing control, the monitoring system, and the graphic control unit can therefore be usefully combined (cf. Fig. 8.32).

9.6 Queueing Models

One method to account for the stochastic nature of manufacturing flows is to use the queueing theory. Despite the considerable amount of literature published on the subject of these models and their successful use. (e.g. in telephone and computer networks), they are not yet used very widely in manufacturing control for custom production and small-batch production.

As extensive studies of 25 queueing models have shown, this is primarily due to the fact that important prerequisites which are fundamental for the use of existing queueing models cannot be met in practice, especially in job shop production [13]. These prerequisites are:

- the manufacturing flow must have leveled off at a certain status
- input and output events must flow independently of each other and at random
- no priorities for the processing of manufacturing orders in the queue.

In the studies performed by Lorenz [13], the lead times that were calculated using

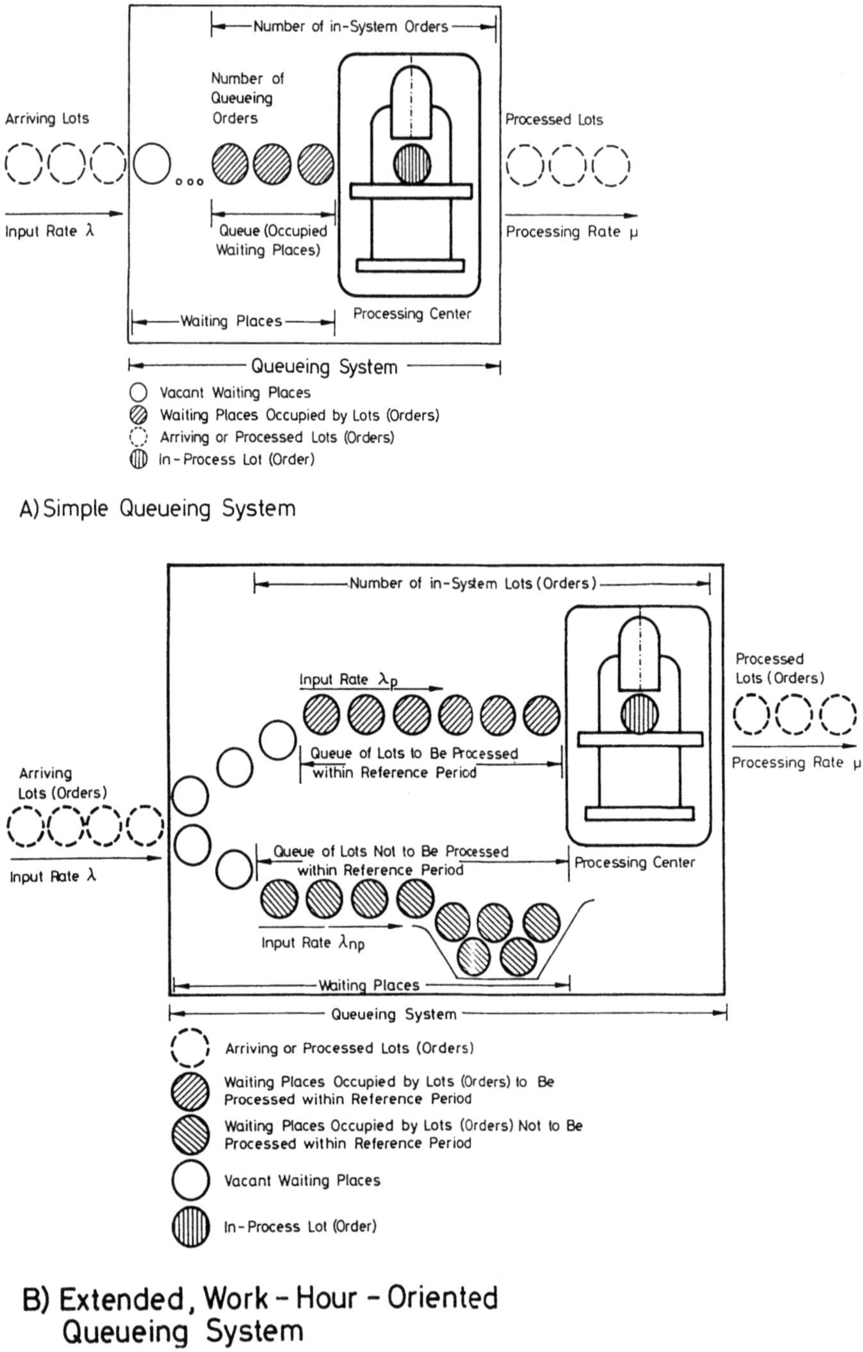

Fig. 9.10. Queueing model for job shop manufacturing (After Lorenz [14])

the queueing models being examined (in queueing theory, one speaks of waiting times) were, on average, about half those measured in a real process. With some models, the lead times were even negative [13].

Based on the components of inventory and lead time (basis, flow, control, and lot components, see Sect. 5.5), Lorenz developed a combined queueing model. Its fundamentals are shown in Fig. 9.10, where they are contrasted with the previous model [14].

The essential difference is that one distinguishes between a queue to be processed (output) and a queue not to be processed (leftover inventory), and that the inter-arrival and interprocessing times are weighted with the work content of the corresponding orders. Therefore, besides the flow time, this model can also calculate the inventory and the output trend. The computing times for determining lead time, inventory, and performance amounted to 3%–5% of the time needed when using the simulation model described in Sect. 6.8.1.

The new queueing model has been tested with actual and simulated data from a sample factory, for several mean inventories and using several priority rules. On the whole, a good adaptation to the actual or simulated processes was achieved. As a data base, the representative shop with 50 work centers, known from previous examples, was used. The investigation period consisted of eight two-week periods, which corresponds to a total of 400 individual values for each key datum in the actual process and in each simulated process.

On average, the 25 tested models overestimated only 6% of the actual lead time values but underestimated about 90%. Of the calculated waiting time values, only 4% were within the relative deviation range of ±20% acceptable for technical and organizational purposes. There were significant differences between the individual models, but even in the most realistic queueing model, only 5% of the values were within the ±20% range. On the other hand, the work-hour-oriented queueing model overestimated or underestimated the actual lead times in only 4.5% and 0.4% of the cases, respectively, whereas over 95% of the calculated values were within the range of ±20% of relative deviation. The mean relative deviation was 5%.

A good correspondence between the values calculated with the new queueing model and the real or simulated values could also be proved for mean inventory and mean output, so that the model seems to be a reasonable substitute for simulation techniques. Figure 9.11 compares the characteristic curves that were calculated for our sample shop using the queueing model ('analytical model') and those recorded or simulated. Note, however, that the queueing model produces 'only' mean lead time, inventory, and output values, so that throughput diagrams cannot be constructed.

One application of the queueing model could be the quick calculation of characteristic operation curves to determine the 'bend' of the perfomance curve. The output and lead time curves in Figs. 6.18 and 6.37, for example, were calculated using this model.

9.7 The OPT System

Since 1982, major companies in the US have been cited to use a production control system named OPT (Optimized Production Technology), a software developed in

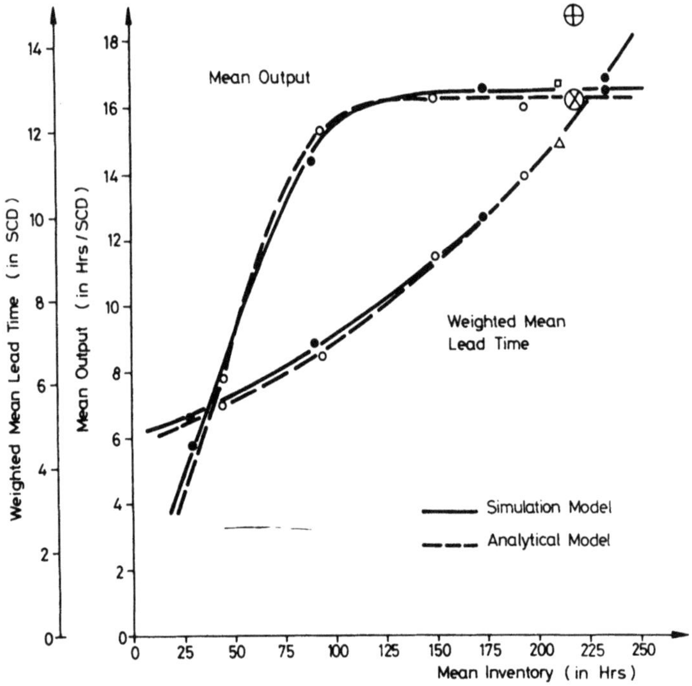

Simulation Model

	Mean Inventory (Hrs)	Mean Lead Time (SCD)	Mean Output (Hrs/SCD)
Actual Status	219	⊕ 14.87	⊗ 16.23
● RSI = 3 Wks	234	12.97	16.81
● RSI = 2 Wks	174	10.06	16.57
● RSI = 1 Wks	90	6.99	14.33
● RSI = 0 Wks	29	5.23	5.76

Analytical Model

	Mean Inventory (Hrs)	Mean Lead Time (SCD)	Mean Output (Hrs/SCD)
Experiment	212	△ 11.75	□ 16.68
○ RSI = 3 Wks	193	10.95	15.91
○ RSI = 2 Wks	150	9.03	16.20
○ RSI = 1 Wks	94	6.63	15.20
○ RSI = 0 Wks	45	5.47	7.75

Fig. 9.11. Characteristic production lines – work-hour-oriented queueing model versus event-oriented simulation (After Lorenz [14]). RSI: Range of Scheduled Inventory

Table 9.1. Manufacturing control – conventional versus OPT rules (After Goldratt [20])

Conventional	OPT
1. Balance capacity, then try to maintain flow.	1. Balance flow, not capacity.
2. The level of utilization of any worker is determined by his own potential.	2. The level of utilization of a non-bottleneck is not determined by its capacity, but by another bottleneck in the system.
3. Activation and utilization of workers are the same.	3. Activation and utilization of a resource are not synonymous.
4. An hour lost at a bottleneck is just an hour lost at that resource.	4. An hour lost at a bottleneck is an hour lost for the total system.
5. An hour saved at a non-bottleneck is an hour saved at that resource.	5. An hour saved at a non-bottleneck is just a miracle.
6. Bottlenecks temporarily limit throughput but have little impact on inventories.	6. Bottlenecks govern both throughput and inventories.
7. Splitting and overlapping of batches should be discouraged.	7. The transfer batch may not, and many times should not, be equal to the process batch.
8. The process batch should be constant both in time and along its route.	8. The process batch should be variable, not fixed.
9. Schedules should be determined sequentially by: * Pre-determined batch size. * Calculating lead time. * Assigning priorities, setting schedules according to lead time. * Adjusting the schedules according to apparent capacity constraints by repeating the above 3 steps.	9. Schedules should be established by looking at all of the constraints simultaneously. Lead times are the result of a schedule and cannot be pre-determined.
Motto The only way to reach a global optimum is by ensuring local optima.	Motto The sum of the local optima is not equal to the global optimum.

Israel by E.M. Goldratt and others [16, 17]. It was first presented in the Federal Republic of Germany in 1984 [15]. The system is designed to guide and control the production flow, concentrating not on single events but on the performance of the entire shop [18]. The suppliers, Creative Output, consider the OPT approach to be a considerable improvement on the American MRP (Material Requirement Planning) system and the Japanese Kanban system [19, 20].

Table 9.1 illustrates the contrasts between the conventional planning rules and the rules of OPT [18].

The first OPT rule points to the importance of the manufacturing flow, as against the conventional emphasis on capacity. Rules 2 to 6 emphasize the importance of adjusting the manufacturing flow to bottleneck capacities.

To achieve this, the order network is divided into two subnetworks (Fig. 9.12), one containing the critical capacities and the other containing those that are not critical. Quality control is also to concentrate on the bottleneck capacities, in contrast to the alleged principle of indiscriminate distribution in Japanese quality philosophy. Further, setup times are to be avoided at bottleneck capacities, whereas setup time reductions at non-bottleneck capacities are said to cause idle times.

To speed up the throughput, one distinguishes between a processing lot and a transport lot (Rules 7 and 8). The processing lot is a whole-number multiple of the

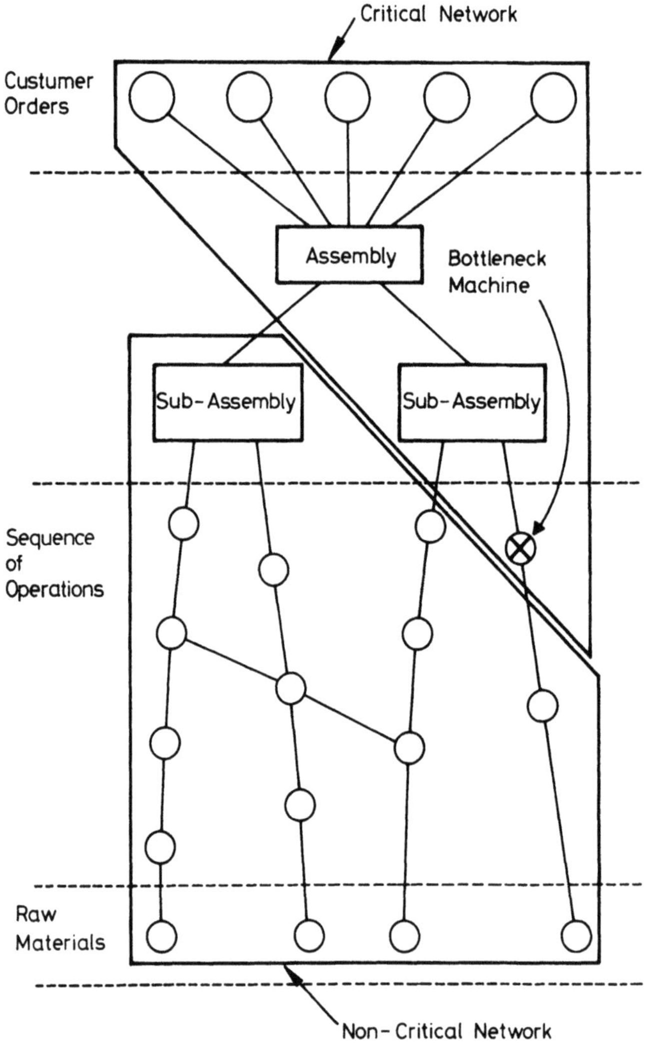

Fig. 9.12. OPT network model (After Goldratt [20])

transport lot, and must be variable during the throughput. Finally, the flow time of an order is the result of planning calculation and not definable in advance (Rule 9).

The system is available as a commercial software product, and is supposed to enable the user to implement the fundamentals of Just-in-Time production very quickly. It claims to take into account the interrelationships between inventories, lot sizes, flow times, and capacity, and to enable the user to control inventories, cash flow and, ultimately, profits. This is done by a 'copyrighted' algorithm, which has not yet been explained [18].

For the present, only the following can be observed.

- The emphasis on the bottleneck capacities is reasonable; however, these capacities change continually, a fact which requires a network recalculation each time.

- The distinction between processing lots and transport lots does, in principle, allow a shorter order throughput and lower inventories, since standstills at downstream machines due to large processing lots are less probable. However, there are additional costs for controlling and re-joining the sublots.
- The essential inventory, flow time, utilization, and schedule deviation objectives are not represented and monitored constantly – at least not in a way that is visible for the user.
- The user cannot use simple control parameters to give priority to a certain objective
- The system produces an 'optimal' production plan which is not transparent to the user. It must be doubted that there really is only one cost-minimized plan, since the same minimum total production costs can be achieved with different combinations of utilization and inventory.

A more detailed analysis of the OPT method is not possible with the information currently available.

The comparison of load-oriented manufacturing control with other new or known manufacturing control methods discussed or used today has shown that the funnel model and the throughput diagram derived from it describe virtually all production environments. The monitoring system can also be used for all manufacturing processes. Only the schedule and capacity planning can, when the equipment is arranged according to a strictly defined production schedule, be carried out more easily using the Kanban method or using the job-progress reporting system. When, on the other hand, flow times are extremely long and vary widely, a control using Gantt charts or network plans is to be recommended, so that data on interlinked orders can be carried along through the process.

On the whole, it can be observed that all modern control methods take the relationships between flow time, inventory, and utilization into account and that inventories are considered to be the 'root of all evil'. In principle, the bottom limits for inventory and lead time can generally not be pre-determined, as became clear in the discussion of these two objectives and their components in the throughput diagram.

9.8 References

[1] Spur, G. Entwicklungstendenzen rechnerintegrierter Fabrikstrukturen in Europa (The Trends of Computer Integrated Structures in Europe's Factories). In: Tagungsband zur EMO, Hannover, 1985.

[2] Warnecke, H-J. Taylor und die Fertigungstechnik von morgen (Taylor and the Manufacturing Technology of Tomorrow). In: FTK '85, schriftliche Fassung der Vorträge zum Fertigungstechnischen Kolloquium, 10–11 Oktober 1985, Stuttgart, pp 1–12 (proceedings). Springer-Verlag, Berlin/Heidelberg, 1985.

[3] Shingo, S. *Study of Toyota Production System.* Japan Management Association, Tokyo, 1981.

[4] Wildemann, H et al. *Flexible Werkstattsteuerung durch Integration von KANBAN-Prinzipien (Flexible Shop-Floor Control by Integrated Kanban Principles).* CW-Publikationen, München, 1984.

[5] Wantuck, KA. The ABCs of Japanese Productivity Production and Inventory Management. *Review and APICS News*, **9**: 1981, 22–28.

[6] Shah, K. Anwendungsmöglichkeiten einer materialflußorientierten Fertigungssteuerung in deutschen Unternehmen nach japanischen Kanban-Prinzipien (Possible Uses of Material-Flow-Oriented Manufacturing Control in German Factories according to Japanese Kanban Principles). In: Seminar 'Flexible Werkstattsteuerung und Betriebsdatenerfassung in der Einzel- und Kleinserienfertigung'

der gfmt (Gesellschaft für Management und Technologie), März 1982, München (seminar proceedings).

[7] Gottwald, MK. Produktionssteuerung mit Fortschrittszahlen (Manufacturing Control with Progress Records). In: Seminar 'Neue PPS-Lösungen' der gfmt (Gesellschaft für Management und Technologie), 12–13 Oktober 1982, München (seminar proceedings).

[8] Heinemeyer, W. Fortschrittszahlen – ein Ansatz zur Steuerung in der Serienfertigung (Job-Progress Records – A Control Model for Batch-Production). In: Tagungsunterlage zum Fachseminar 'Statistisch orientierte Fertigungssteuerung' (seminar proceedings). Institut für Fabrikanlagen (IFA) der Universität Hannover. Hannover, 1984, pp 98–127.

[9] Kettner, H. (ed.). Neue Wege der Bestandsanalyse im Fertigungsbereich (Recent Inventory Analysis Methods in Manufacturing). Fachbericht des Arbeitsausschusses Fertigungswirtschaft (AFW) der Deutschen Gesellschaft für Betriebswirtschaft (DGfB) (report). Institut für Fabrikanlagen der Technischen Universität, Hannover, 1976.

[10] Simon, R. Der Fertigungsleitstand in der Produktion – Organisatorische Voraussetzungen und Einführungsprobleme (The Manufacturing Control Unit in Production – Organizational Prerequisites and Problems with its Implementation). Teil 1: Die theoretischen Anforderungen. *Arbeitsvorbereitung*, **17** (1): 1980, 10–13; Teil 2: Berichte über Einsätze in der Praxis. *Arbeitsvorbereitung*, **17** (2): 1980, 53–56; Teil 3: Einführung und Auswirkung. *Arbeitsvorbereitung*, **17** (3): 1980, 85–89.

[11] Warnecke, H-J, Aldinger, L. Werkstattsteuerung – ein Einsatzgebiet für interaktive Farbmonitor-Systeme (Application of Interactive Color Display Systems in Shop Floor Control). In: Dokumentation zur Fachtagung CAMP 1983 in Berlin (proceedings). VDI-Verlag, Düsseldorf, 1983.

[12] Seeber, H. Möglichkeiten zur Durchlaufzeitverkürzung in der kundenbezogenen Einzelfertigung von Großmaschinen (How to Reduce Lead Times in Custom Production of Heavy-Duty Machines). In: Verein Deutscher Ingenieure-Fachtagung 'Methodik und Praxis der Durchlaufzeitverkürzung in der Einzel- und Kleinserienfertigung', Fellbach, 1985, pp 133–146 (proceedings).

[13] Wiendahl, H-P, Lorenz, W. Analyse von Warteschlangenmodellen mit realen Betriebsdaten einer Werkstattfertigung (An Analysis of Queueing Models Using Authentic Job-Shop Manufacturing Data). *Werkstattstechnik*, **74** (10): 1984, 619–623.

[14] Lorenz, W. Entwicklung eines arbeitsstundenorientierten Warteschlangenmodells zur Prozeßabbildung der Werkstattfertigung (Working-hour-related Queueing Theory as a Representation of the Manufacturing Process in Job-Shop Production). Dissertation Universität Hannover, 1984 (veröffentlicht in: Fortschritt-Berichte der VDI-Zeitschriften, Reihe 2, Nr. 72, Düsseldorf, 1984) (doctoral thesis, published in the above VDI journal).

[15] Autorenkollektiv (several authors). OPT ... eine Antwort Amerikas auf Kanban ? (OPT – America's Answer to Kanban ?) Symposium der Gesellschaft für Fertigungssteuerung und Materialwirtschaft e.V., 19 September 1984, München (proceedings).

[16] Lowndes, JC. Production Management Concept Cuts Delays, Budget Overruns. *Aviation Week & Space Technology*, May 13 1985.

[17] Powel, B. Boosting Shop-Floor Productivity by Breaking All the Rules. *Business Week*, November 26 1984.

[18] Meshar, A. OPT – Auf dem Weg zu neuen Lösungen (OPT – On the Way Towards New Solutions). In: Produktionsmanagement – heute realisiert. Jahrestagung '85 der Gesellschaft für Fertigungssteuerung und Materialwirtschaft e.V., 25 Januar 1985, Heidelberg (proceedings).

[19] Fox, B. MRP, Kanban or OPT. What's best? Inventories & Production, **2** (4) July–August: 1982; (Part I of: OPT, An Answer for America). OPT, An Answer for America; Part II: **2** (8) November–December: 1982; Part III: **3** (1) January–February: 1983; Part IV: **3** (2) March–April: 1983.

[20] Goldratt, EM, Cox, J. *The Goal – Excellence in Manufacturing*. North River Press, Inc., 1985.

[21] Baumgarten, H et al. RKW-Handbuch Logistik. Integrierter Material- und Warenfluß in Beschaffung, Produktion und Absatz. Ergänzbares Handbuch für Planung, Einrichtung und Anwendung logistischer Systeme in der Unternehmenspraxis (Integrated Material and Product Flow in Purchase, Production, and Sales. A Handbook for Planning, Implementation, and Use of Logistics Systems). Erich Schmidt Verlag, Berlin.

[22] Wildemann, H. (ed.). Just-In-Time-Produktion in Deutschland (Just-In-Time Production in Germany). Bericht zur Tagung, 26–27 September 1985, Böblingen (proceedings). Universität Passau, 1985.

Chapter 10

Summary

In view of the international competition among the industrialized countries, companies are becoming more and more interested in production planning and control (PPC). Short lead times, on-time schedule performance, and low inventories are the most important objectives, whereas the importance of utilization, on which much emphasis was placed in the past, has accordingly decreased.

Although electronic data processing is now widely used for PPC purposes, specialists feel that the systems and methods used to-date have their weak points. Especially in manufacturing control, expensively produced allocation lists become obsolete too soon, and very often additional informal control systems are developed by the foremen, so that orders can be processed on time despite numerous changes and disruptions. In addition, studies have shown that, in practice, mean flow times are much longer and schedule deviations much greater than was previously assumed. The fact that interest in the Japanese Kanban system is very great shows that manufacturing companies feel the need for simpler and more effective manufacturing control systems.

Starting from this situation, the aim of this book has been to find a new approach to manufacturing control, and to develop system components which together fit into an overall concept of load-oriented manufacturing control.

First, the current predominant method of manufacturing control was analysed; it was found that it does not take into account the statistical nature of the manufacturing flow, but merely imitates the conventional manual file-cards method. The weak points of the current method lead us to a new method of manufacturing control, characterized as being 'model-oriented', 'more transparent', 'simpler to handle', and as one which 'makes more use of human experience'.

One crucial concept for describing the manufacturing flow is lead time. As an introduction to the new manufacturing control approach, lead time was therefore analysed and defined in detail, and its statistical features explained with numerous data from practical studies.

The findings from these studies then lead to a universal model of the production process, which was developed step-by-step. The model, which is based on the idea that every work center or system can be seen as a funnel, is mathematically represented in what is known as the throughput diagram. This throughput diagram graphically and numerically represents the work input and output at the work center over time. The construction of such a diagram was shown, using an actual example. Further, it was shown how the four key values of inventory, lead time, utilization, and schedule deviation – as well as other key values – can be calculated from the diagram and represented graphically.

The first application of the throughput diagram is in analysis, monitoring, and diagnosis of the manufacturing process. The necessary calculations and procedures were shown in a comprehensible manner, and examples given. Practical use of this diagram is made in a permanent monitoring system, the structure and results of which were described, and in a diagnostic method which elucidates the reasons for inventory tie-up in manufacturing, and thus supplies the crucial parameters for controlling the manufacturing flow. These parameters are the mean inventory, the mean value and variance of the order time, the dispatching rules, and the capacity trend over time. The results of such analysis, monitoring, and diagnostic systems can also be represented as color graphs on computer screens, and plotted automatically. Examples of this were given.

The subsequent chapters described the practical application of the control parameters in several modules of a manufacturing control method. The task of load-oriented order release, besides providing a realistic lead time scheduling, is to ensure that the work centers are always provided with a defined work inventory. The method developed to achieve this is based on a cumulative calculation of the inputs and outputs in the subsequent planning period, and makes use of the statistically calculable relationships between inventory, flow time, and output rate at a work center. The method was described mathematically in great detail, and exhaustive information was given concerning the choice of the control parameters load limit and time limit. The effects of the method were illustrated with the help of simulations and practical examples.

Until now, the importance of dispatching rules has often not been fully recognized. Chapter 6 showed that, in a well-controlled manufacturing flow, dispatching according to the FIFO rule leads to the smallest variations in mean lead times, so that this rule should always be used, except for those few cases where due-date-oriented rules can be applied.

The problem of capacity control has not yet been solved satisfactorily. Based on the curve of planned outputs from the throughput diagram, a capacity planning and control module is developed which, besides improving utilization, also improves the schedule performance of the orders.

For practical application, the question is how to implement these modules in a load-oriented manufacturing control system. This problem was therefore dealt with in a separate chapter (Chapt. 8). First, this chapter made clear the prerequisites for using the method, and dealt especially with the problem of lot sizing with respect to flow times and small flow time variances. Then the load-oriented manufacturing control system itself was described, with its components capacity planning, order release, sequencing, and calculation of monitoring data. Since, besides a laboratory version, there are also an increasing number of commercially used systems, the system is only a model. The description was therefore supplemented by examples of typical systems, some running on PCs with interfaces to a larger computer. The chapter also gave information on the implementation of such systems, and showed how the basic idea behind the funnel model fits into a framework of measures for reducing lead times and inventories.

The modules presented serve above all for the manufacturing control of single-stage production processes in intermittent job-shop and batch production. However, since the process model on which the method is based is of a universal nature, its inherent planning, control, and monitoring possibilities go beyond these applications. Taking into account the increasing tendencies towards production automation, Chapt. 8 first discussed the possible uses of load-oriented manufacturing control in the loading planning of flexible manufacturing systems. Then the existing modules and foreseeable future developments were integrated within a CIM concept using a function-oriented model of manufacturing control. Among the foreseeable but not yet existent modules are integrated expert systems for process diagnosis and parameter control, and simulation modules for decision support in the case of alternative production strategies.

Load-oriented manufacturing control replaces the conventional lead time and capacity scheduling methods. The potential user is therefore especially interested in a comparison with other methods. The final chapter therefore compared the load-oriented method with Kanban control, with the job-progress records method, with finite control at a graphic control unit, with queueing models, and with the OPT system (discussed in the US since 1982). Differences and similarities were revealed.

The load-oriented manufacturing control method thus contributes to a flexible, simple and error-tolerant manufacturing control. It is based on a universal and simple model of the manufacturing process and, due to its statistical approach, it ensures a quick and even order throughput. The method supports the personnel involved in manufacturing control and the management, instead of patronizing them; their responsibility and experience can be put to good use.

This is a manufacturing control method which, using few variables, allows a company to react to changing situations by putting more or less emphasis on individual objectives without neglecting overall economic efficiency.

Appendix A

Table A1. (corresponds to Table 4.1.) Calculation of the input and output curves of a work center, based on the feedback records of the reference period (shop calendar days 30–49) in the order of the input dates. (A) Data base; (B) Inventory input and output trend

A) Data Base | B) Inventory Input and Output Trend

Line Number	Order Number (-)	Input Date tPEU (SCD)	Output Date tPE (SCD)	Order Time TO (Hrs)	Inventory on SCD 30 (Hrs)	Input between SCD 30 and SCD 49 (Hrs)	Cumulative Overall Input (Hrs)	Output between SCD 30 and SCD 49 (Hrs)	Inventory on SCD 49 (Hrs)
	1	2	3	4	5	6	7	8***)	9
1	103	947 #	59	3.3	3.3		-208.4		3.3
2	104	955 #	125	13.1	16.4		-195.3		16.4
3	102	996 #	25	26.4			-168.9		
4	101	997 #	997 #	18.8			-150.1		
5	105	998 #	14	17.8			-132.3		
6	106	998 #	0	2.1			-130.2		
7	107	998 #	18	34.9			- 95.3		
8	109	3	13	70.0			- 25.3		
9	110	5	33	15.4	31.8		- 9.9	15.4	
10	113	11	25	30.3			20.4*)		
11	115	14	31	.5	32.3		20.9	.5	
12	114	17	28	11.4			32.3		
13	116	18	41	7.4	39.7		39.7	7.4	
14	119	23	32	11.4	51.1		51.1	11.4	
15	120	23	34	6.8	57.9		57.9	6.8	
16	118	24	36	7.7	65.6		65.6	7.7	
17	112	25	52	3.4	69.0		69.0		19.8
18	117	25	43	4.2	73.2		73.2	4.2	
19	121	26	37	8.8	82.0		82.0	8.8	
20	108	29	43	13.8	95.8		95.8**)	13.8	
21	124	▲ 30	35	9.6		9.6	105.4	9.6	
22	125	31	33	3.8		13.4	109.2	3.8	
23	123	32	48	13.6		27.0	122.8	13.6	
24	126	33	40	2.6		29.6	125.4	2.6	
25	127	33	36	5.3		34.9	130.7	5.3	
26	122	35	50	3.8		38.7	134.5		23.6
27	131	36	41	2.1		40.8	136.6	2.1	
28	128	36	69	4.7		45.5	141.3		28.3
29	132	38	45	13.8		59.3	155.1	13.8	
30	135	40	45	6.9		66.2	162.0	6.9	
31	140	43	46	5.1		71.3	167.1	5.1	
32	136	43	51	26.9		98.2	194.0		55.2
33	141	44	60	57.4		155.6	251.4		112.6
34	142	45	49	2.1		157.7	253.5	2.1	
35	129	45	63	14.1		171.8	267.6		126.7
36	143	45	53	2.9		174.7	270.5		129.6
37	145	46	48	5.1		179.8	275.6	5.1	
38	134	▼ 49	96	5.1		184.9	280.7		134.7
				TOTAL >	95.8	184.9	280.7	146.0	134.7

▲ ▼ Reference Period

\# Refers to Shop Calendar of previous Period with 1000 SCD

*) Zero Point of Input Curve

**) Beginning of Input Curve in Reference Period

***) In order to plot the Output Curve, TO Values (Column 4) have to be rearranged according to Output Dates (Column 3)

Table A2. (corresponds to Table 4.2.) Calculation of inventory trend, mean inventory, and mean range at a work center (in hours or shop calendar days)

Day (SCD)	Input INP (Hrs)	Output OUT (Hrs)	Inventory Area AI (Hrs x SCD)	Inventory at the End of Day I (Hrs)
1	2	3	4	5
			Initial Inventory	95.8
30	9.6	.0	95.8	105.4
31	3.8	.5	105.4	108.7
32	13.6	11.4	108.7	110.9
33	7.9	19.2	110.9	99.6
34	.0	6.8	99.6	92.8
35	3.8	9.6	92.8	87.0
36	6.8	13.0	87.0	80.8
37	.0	8.8	80.8	72.0
38	13.8	.0	72.0	85.8
39	.0	.0	85.8	85.8
40	6.9	2.6	85.8	90.1
41	.0	9.5	90.1	80.6
42	.0	.0	80.6	80.6
43	32.0	18.0	80.6	94.6
44	57.4	.0	94.6	152.0
45	19.1	20.7	152.0	150.4
46	5.1	5.1	150.4	150.4
47	.0	.0	150.4	150.4
48	.0	18.7	150.4	131.7
49	5.1	2.1	131.7	134.7
			Final Inventory	134.7
20	184.9	146.0	2,105.4	< TOTAL

Mean Inventory Im = AI/P = (2,105.4 Hrs x SCD)/20 SCD = 105.3 SCD

Mean Range Rm = AI/OUT = (2,105.4 Hrs x SCD)/146 Hrs = 14.4 SCD

Table A3. (corresponds to Table 4.3.) Calculation of mean advance time at a work center, based on the outputs in the reference period and the inputs up to the output level in the reference period (in shop calendar days). (A) In order of input days (Reference period SCD 30–49); (B) in order of output days

A) In Order of Input Dates (Reference Period SCD 30 - 49)

Order Number (-)	Input Date tPEU (SCD)	Order Time TO (SCD)	End of Reference Period Minus Input Date (SCD)	Advance Time Area A1 (Hrs x SCD)
1	2	3	4	5
113	11	20.4*)	38	775.2
115	14	.5	35	17.5
114	17	11.4	32	364.8
116	18	7.4	31	229.4
119	23	11.4	26	296.4
120	23	6.8	26	176.8
118	24	7.7	25	192.5
112	25	3.4	24	81.6
117	25	4.2	24	100.8
121	26	8.8	23	202.4
108	29	13.8	20	276.0
124	30	9.6	19	182.4
125	31	3.8	18	68.4
123	32	13.6	17	231.2
126	33	2.6	16	41.6
127	33	5.3	16	84.8
122	35	3.8	14	53.2
131	36	2.1	13	27.3
128	36	4.7	13	61.1
132	38	4.7**)	11	51.7
	TOTAL >	146.0		3,515.1

B) In Order of Output Dates

Order Number (-)	Output Date tPE (SCD)	Order Time TO (Hrs)	End of Reference Period Minus Output Date (SCD)	Advance Time Area A2 (Hrs x SCD)
6	7	8	9	10
115	31	.5	18	9.0
119	32	11.4	17	193.8
110	33	15.4	16	246.4
125	33	3.8	16	60.8
120	34	6.8	15	102.0
124	35	9.6	14	134.4
118	36	7.7	13	100.1
127	36	5.3	13	68.9
121	37	8.8	12	105.6
126	40	2.6	9	23.4
131	41	2.1	8	16.8
116	41	7.4	8	59.2
108	43	13.8	6	82.8
117	43	4.2	6	25.2
135	45	6.9	4	27.6
132	45	13.8	4	55.2
140	46	5.1	3	15.3
123	48	13.6	1	13.6
145	48	5.1	1	5.1
142	49	2.1	0	0.0
	TOTAL >	146.0		2,050.4

*) Order 113 : Only 20.4 of Total 30.3 Hours are relevant for Reference Period SCD 30 - 49

**) Order 132 : Only 4.7 of Total 13.8 Hours are relevant for Reference Period SCD 30 - 49

Mean Advance Time TAm = ATM/OUT = (A1 - A2)/OUT
= (3,515.1 - 1,345.2)/146
= (2,169.9 Hrs x SCD)/146 Hrs
= 14.9 SCD

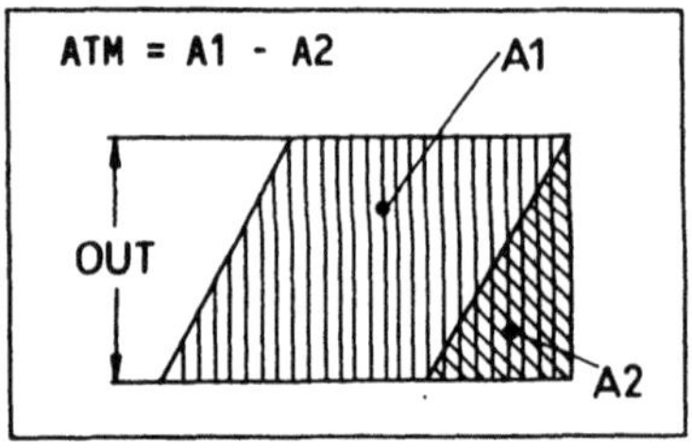

Sketch for Calculation of ATM

Table A4. (corresponds to Table 4.4.) Calculation of the weighted mean lead time of a work center, based on the feedback records in the order of the output dates (in shop calendar days)

Order Number (-)	Input Date tPEU (SCD)	Output Date tPE (SCD)	Order Time TO (Hrs)	Lead Time TL (SCD)	Lead Time Area ALT (Hrs x SCD)
1	2	3	4	5	6
115	14	31	.5	17	8.5
119	23	32	11.4	9	102.6
110	5	33	15.4	28	431.2
125	31	33	3.8	2	7.6
120	23	34	6.8	11	74.8
124	30	35	9.6	5	48.0
118	24	36	7.7	12	92.4
127	33	36	5.3	3	15.9
121	26	37	8.8	11	96.8
126	33	40	2.6	7	18.2
131	36	41	2.1	5	10.5
116	18	41	7.4	23	170.2
108	29	43	13.8	14	193.2
117	25	43	4.2	18	75.6
135	40	45	6.9	5	34.5
132	38	45	13.8	7	96.6
140	43	46	5.1	3	15.3
123	32	48	13.6	16	217.6
145	46	48	5.1	2	10.2
142	45	49	2.1	4	8.4
20	< TOTAL >		146.0	284	1,728.1

Weighted Mean Lead Time TLM = ALT/OUT
= (1,728.1 Hrs x Days)/146 Hrs = 11.8 Days

Table A5. (corresponds to Table 4.5.) Calculation of the inventory trend component (in shop calendar days)

A) Initial Inventory Area AII (Historical Input Curve from Zero to Initial Inventory)

Order Number (-)	Input Date tPEU (SCD)	Order Time TO (Hrs)	Time up to SCD 29 (SCD)	Initial Inventory Area AII (Hrs x SCD)
1	2	3	4	5
113	11	20.4*)	18	367.2
115	14	.5	15	7.5
114	17	11.4	12	136.8
116	18	7.4	11	81.4
119	23	11.4	6	68.4
120	23	6.8	6	40.8
118	24	7.7	5	38.5
112	25	3.4	4	13.6
117	25	4.2	4	16.8
121	26	8.8	3	26.4
108	29	13.8	0	0.0
11	< TOTAL >	95.8		797.4

*) 20.4 Hrs out of 30.3 Hrs

B) Final Inventory Area AFI (Historical Input Curve from Output (SCD 49) up to Final Inventory)

Order Number (-)	Input Date tPEU (SCD)	Order Time TO (Hrs)	Time up to SCD 49 (SCD)	Final Inventory Area AFI (Hrs x SCD)
1	2	3	4	5
132	38	9.1*)	11	100.1
135	40	6.9	9	62.1
140	43	5.1	6	30.6
136	43	26.9	6	161.4
141	44	57.4	5	287.0
142	45	2.1	4	8.4
129	45	14.1	4	56.4
143	45	2.9	4	11.6
145	46	5.1	3	15.3
134	49	5.1	0	0.0
10	< TOTAL >	134.7		732.9

*) 9.1 Hrs out of 13.8 Hrs

Range of Initial Inventory RII = AII/OUT = (797.4 Hrs x SCD)/146 Hrs = 5.5 SCD

Range of Final Inventory RFI = AFI/OUT = (732.9 Hrs x SCD)/146 Hrs = 5.0 SCD

Inventory Trend Component ITC = (AII - AFI)/OUT = (797.4 - 732.9 Hrs x Days)/146 Hrs = .4 SCD

Table A6. (corresponds to Table 4.6.) Calculation of the sequence component (in shop calendar days)

A) Initial Inventory Additional Area

A1) Lead Time Area of Initial Inventory

Order Number	Input Date tPEU	Order Time TO	Cumul. Input up to SCD 29	Time up to SCD 29	Lead Time Area of Initial Inventory ALTI
(-)	(SCD)	(Hrs)	(Hrs)	(SCD)	(HrsxSCD)
1	2	3	4	5	6
103	947 #	3.3	3.3	82	270.6
104	955 #	13.1	16.4	74	969.4
110	5	15.4	31.8	24	369.6
115	14	0.5	32.3	15	7.5
116	18	7.4	39.7	11	81.4
119	23	11.4	51.1	6	68.4
120	23	6.8	57.9	6	40.8
118	24	7.7	65.6	5	38.5
112	25	3.4	69.0	4	13.6
117	25	4.2	73.2	4	16.8
121	26	8.8	82.0	3	26.4
108	29	13.8	95.8	0	0.0
12	<TOTAL>	95.8	95.8	TOTAL >	1,903.0

A2) Initial Inventory Area

Order Number	Input Date tPEU	Order Time TO	Time up to SCD 29	Initial Inventory Area AII
(-)	(SCD)	(Hrs)	(SCD)	(HrsxSCD)
1	2	3	4	5
113	11	20.4*)	18	367.2
115	14	.5	15	7.5
114	17	11.4	12	136.8
116	18	7.4	11	81.4
119	23	11.4	6	68.4
120	23	6.8	6	40.8
118	24	7.7	5	38.5
112	25	3.4	4	13.6
117	25	4.2	4	16.8
121	26	8.8	3	26.4
108	29	13.8	0	0.0
11	<TOTAL>	95.8	TOTAL >	797.4

Refers to Shop Calendar of previous Period with 1000 SCD

*) 20.4 Hrs out of 30.3 Hrs

Initial Inventory Additional Area AIIA = ALTI - AII
= 1,903.0 - 797.4 Hrs x SCD = 1,105.6 Hrs x SCD

B) Final Inventory Additional Area

B1) Lead Time Area of Final Inventory

Order Number	Input Date tPEU	Order Time TO	Cumul. Input up to SCD 49	Time up to SCD 49	Lead Time Area of Final Inventory ALTF
(-)	(SCD)	(Hrs)	(Hrs)	(SCD)	(HrsxSCD)
1	2	3	4	5	6
103	947 #	3.3	149.3	102	336.6
104	955 #	13.1	162.4	94	1231.4
112	25	3.4	165.8	24	81.6
122	35	3.8	169.6	14	53.2
128	36	4.7	174.3	13	61.1
136	43	26.9	201.2	6	161.4
141	44	57.4	258.6	5	287.0
129	45	14.1	272.7	4	56.4
143	45	2.9	275.6	4	11.6
134	49	5.1	280.7	0	0.0
10	<TOTAL>	134.7	280.7	TOTAL >	2,280.3

B2) Final Inventory Area

Order Number	Input Date tPEU	Order Time TO	Time up to SCD 49	Final Inventory Area AFI
(-)	(SCD)	(Hrs)	(SCD)	(Hrs x SCD
1	2	3	4	5
132	38	9.1*)	11	100.1
135	40	6.9	9	62.1
140	43	5.1	6	30.6
136	43	26.9	6	161.4
141	44	57.4	5	287.0
142	45	2.1	4	8.4
129	45	14.1	4	56.4
143	45	2.9	4	11.6
145	46	5.1	3	15.3
134	49	5.1	0	0.0
10	<TOTAL>	134.7	TOTAL >	732.9

Refers to Shop Calendar of previous Period with 1000 SCD

*) 9.1 Hrs out of 13.8 Hrs

Final Inventory Additional Area AFIA = ALTF - AFI
= 2,280.3 - 732.9 = 1,547.4 Hrs x SCD

Sequence Component SC = (AIIA - AFIA)/OUT
= (1,105.6 - 1,547.4)/146 = (- 441.8 Hrs x SCD)/146 Hrs = - 3.0 SCD

Table A7. (corresponds to Table 4.7.) Calculation of the schedule deviation key data of the work center output function (in shop calendar days)

Order Number (-)	Actual Completion Date tPEa (SCD)	Planned Completion Date tPEr (SCD)	Order Time TO (Hrs)	Order Lateness Output OLO (SCD)	Weighted Lateness Output OLO x TO (Hrs x SCD)	Simple Quadratic Deviation (OLOm-OLO)² (SCD²)	Weighted Quadratic Deviation (OLOmw-OLO)² x TO (SCD² x Hrs)	Rated Output: In Order of Rated Due Dates (SCD)	Rated Output: Cumulative Order Time (Hrs)
1	2	3	4	5	6	7	8	9	10
115	31	20	0.5	11	5.5	59.3	13.5	13	15.4
119	32	29	11.4	3	34.2	0.1	89.4	18	29.2
110	33	13	15.4	20	308.0	278.9	3,105.3	20	29.7
125	33	39	3.8	-6	-22.8	86.5	529.1	29	37.1
120	34	29	6.8	5	34.0	2.9	4.4	29	43.9
124	35	38	9.6	-3	-28.8	39.7	743.4	29	55.3
118	36	35	7.7	1	7.7	5.3	177.4	35	63.0
127	36	39	5.3	-3	-15.9	39.7	410.4	35	67.2
121	37	37	8.8	0	0.0	10.9	296.0	37	76.0
126	40	39	2.6	1	2.6	5.3	59.9	38	85.6
131	41	44	2.1	-3	-6.3	39.7	162.6	39	89.4
116	41	29	7.4	12	88.8	75.7	284.5	39	94.7
108	43	18	13.8	25	345.0	470.9	5,087.2	39	97.3
117	43	35	4.2	8	33.6	22.1	20.3	39	110.9
135	45	48	6.9	-3	-20.7	39.7	534.3	44	113.0
132	45	44	13.8	1	13.8	5.3	318.0	44	126.8
140	46	49	5.1	-3	-15.3	39.7	394.9	48	133.7
123	48	39	13.6	9	122.4	32.5	139.3	49	138.8
145	48	54	5.1	-6	-30.6	86.5	710.1	53	140.9
142	49	53	2.1	-4	-8.4	53.3	201.7	54	146.0
20	< TOTAL >		146.0	65	846.8	1,393.8	13,281.8	TOTAL >	146.0

Total Negative Weighted Lateness (Earliness) = 148.8 SCD x Hrs

Total Positive Weighted Lateness (Tardiness) = 995.6 SCD x Hrs

Simple Mean Output Lateness OLOm = 65 SCD/20 = 3.3 SCD

Weighted Mean Output Lateness OLOmw = (846.8 SCD x Hrs)/146 Hrs = 5.8 SCD

Negative Weighted Mean Output Lateness OLONmw = (148.8 SCD x Hrs)/146 Hrs = 1,0 SCD

Positive Weighted Mean Output Lateness OLOPmw = (995.6 SCD x Hrs)/146 Hrs = 6.8 SCD

Simple Standard Deviation of Output Lateness OLOS = $\sqrt{1393.8 \text{ SCD}^2/20}$ = 8.3 SCD

Weighted Standard Deviation of Output Latneness OLOSw = $\sqrt{13{,}281.8 \text{ (SCD}^2 \text{ x Hrs)}/146 \text{ Hrs}}$ = 9.5 SCD

Appendix B

Table B1. (A) Calculation of the input and output curves, based on the feedback records of each period in the order of the input dates (period 1). Table B1 corresponds to Table 4.1, but the calculations have been carried out period-for-period. (For the calculation see Chapt. 4.); (B) calculation of the input and output curves, based on the feedback records of each period in the order of the input dates (periods 2, 3 and 4)

DATA BASE | PERIOD 1

	Line Number	Order Number (-)	Input Date tPEU (SCD)	Output Date tPE (SCD)	Order Time TO (Hrs)	Inventory at the End of SCD 29 (Hrs)	Cumulative Input from SCD 30 to 34 (Hrs)	Output from SCD 30 to 34 (Hrs)	Inventory at the End of SCD 34 (Hrs)	Total Input regarding Period 1 (Hrs)
		1	2	3	4	5	6	7	8	9
Before Period 1	1	103	947 #	59	3.3	3.3			3.3	
	2	104	955 #	125	13.1	16.4			16.4	
	3	102	996 #	25	26.4					
	4	101	997 #	997 #	18.8					
	5	105	998 #	14	17.8					
	6	106	998 #	0	2.1					
	7	107	998 #	18	34.9					
	8	109	3	13	70.0					
	9	110	5	33	15.4	31.8		15.4		-9.9
	10	113	11	25	30.3					20.4
	11	115	14	31	0.5	32.3		0.5		20.9
	12	114	17	28	11.4					32.3
	13	116	18	41	7.4	39.7			23.8	39.7
	14	119	23	32	11.4	51.1		11.4		51.1
	15	120	23	34	6.8	57.9		6.8		57.9
	16	118	24	36	7.7	65.6			31.5	65.6
	17	112	25	52	3.4	69.0			34.9	69.0
	18	117	25	43	4.2	73.2			39.1	73.2
	19	121	26	37	8.8	82.0			47.9	82.0
	20	108	29	43	13.8	95.8			61.7	95.8
Period 1	21	124	30	35	9.6		9.6		71.3	105.4
	22	125	31	33	3.8		13.4	3.8		109.2
	23	123	32	48	13.6		27.0		84.9	122.8
	24	126	33	40	2.6		29.6		87.5	125.4
	25	127	33	36	5.3		34.9		92.8	130.7
Period 2	26	122	35	50	3.8					
	27	131	36	41	2.1					
	28	128	36	69	4.7					
	29	132	38	45	13.8					
Period 3	30	135	40	45	6.9					
	31	140	43	46	5.1					
	32	136	43	51	26.9					
	33	141	44	60	57.4					
Period 4	34	142	45	49	2.1					
	35	129	45	63	14.1					
	36	143	45	53	2.9					
	37	145	46	48	5.1					
	38	134	49	96	5.1					
Total						95.8	34.9	37.9	92.8	130.7

\# Refers to Shop Calendar of previous Period with 1000 SCD

PERIOD 2				PERIOD 3				PERIOD 4			
Cumulative Input from SCD 35 to 39 (Hrs)	Output from SCD 35 to 39 (Hrs)	Inventory at the End of SCD 39 (Hrs)	Total Input regarding Period 2 (Hrs)	Cumulative Input from SCD 40 to 44 (Hrs)	Output from SCD 40 to 44 (Hrs)	Inventory at the End of SCD 44 (Hrs)	Total Input regarding Period 3 (Hrs)	Cumulative Input from SCD 45 to 49 (Hrs)	Output from SCD 45 to 49 (Hrs)	Inventory at the End of SCD 49 (Hrs)	Total Input regarding Period 4 (Hrs)
10	11	12	13	14	15	16	17	18	19	20	21
		3.3				3.3				3.3	
		16.4				16.4				16.4	
			-5.6								
		23.8	1.8		7.4						
			13.2								
			20.0								
	7.7		27.7								
		27.2	31.1			19.8	-0.3			19.8	
		31.4	35.3		4.2		3.9				
	8.8		44.1				12.7				
		45.2	57.9		13.8		26.5				-3.6
	9.6		67.5				36.1				6.0
			71.3				39.9				9.8
		58.8	84.9			33.4	53.5		13.6		23.4
		61.4	87.5		2.6		56.1				26.0
	5.3		92.8				61.4				31.1
3.8		65.2	96.6			37.2	65.2			23.6	35.1
5.9		67.3	98.7		2.1		67.3				37.2
10.6		72.0	103.4			41.9	72.0			28.3	41.9
24.4		85.8	117.2			55.7	85.8		13.8		55.7
				6.9		62.6	92.7		6.9		62.6
				12.0		67.7	97.8		5.1		67.7
				38.9		94.6	124.7			55.2	94.6
				96.3		152.0	182.1			112.6	152.0
								2.1	2.1		154.1
								16.2		126.7	168.2
								19.1		129.6	171.1
								24.2	5.1		176.2
								29.3		134.7	181.3
24.4	31.4	85.8	117.2	96.3	30.1	152.0	182.1	29.3	46.6	134.7	181.3

Table B2. (A) Calculation of the inventory trend component and the sequence component for period 1. Tables B2A–D closely resemble Table 4.6B. In Chapt. 4, the calculation is explained in detail. For calculation of the inventory trend component and the sequence component, the lead time area of the orders in the initial and the final inventories as well as the initial and final inventory areas must be known from the historical input trend. When calculating in periods, it suffices to calculate the final values, since the final values of a period are the same as the initial values of the subsequent period. The initial values have been taken from Table A5 in Appendix A, since it gives the necessary values in shop calendar days, as opposed to Table 4.6A; (B) calculation of the inventory trend component and the sequence component for period 2; (C) calculation of the inventory trend component and the sequence component for period 3; (D) calculation of the inventory trend component and the sequence component for period 4

A) Period 1: SCD 30 to 34

Order Number (-)	Input Date tPEU (SCD)	Order Time TO (Hrs)	Time up to SCD 34 (SCD)	Area (HrsxSCD)
1	2	3	5	6
103	947 #	3.3	87	287.1
104	955 #	13.1	79	1034.9
116	18	7.4	16	118.4
118	24	7.7	10	77.0
112	25	3.4	9	30.6
117	25	4.2	9	37.8
121	26	8.8	8	70.4
108	29	13.8	5	69.0
124	30	9.6	4	38.4
123	32	13.6	2	27.2
126	33	2.6	1	2.6
127	33	5.3	1	5.3
12	< TOTAL >	92.8	TOTAL >	1798.7

Order Number (-)	Input Date tPEU (SCD)	Order Time TO (Hrs)	Time up to SCD 34 (SCD)	Area (HrsxSCD)
1	2	3	4	5
116	18	1.8 *)	16	28.8
119	23	11.4	11	125.4
120	23	6.8	11	74.8
118	24	7.7	10	77.0
112	25	3.4	9	30.6
117	25	4.2	9	37.8
121	26	8.8	8	70.4
108	29	13.8	5	69.0
124	30	9.6	4	38.4
125	31	3.8	3	11.4
123	32	13.6	2	27.2
126	33	2.6	1	2.6
127	33	5.3	1	5.3
13	< TOTAL >	92.8	TOTAL >	598.7

Lead Time Area of Final Inventory AAFI1 = 1,798.7 - 598.7 = 1,200.0 Hrs x SCD

Lead Time Area of Initial Inventory AAII1 = AFIA0 **) = 1,105,6 Hrs x SCD ***)

Sequence Component SC1 = (AAII1 - AAFI1)/OUT1 = (1,105.6 - 1,200.0)/37.5 = -2.5 SCD

Inventory Trend Component ITC1 = (AII1 - AFI1)/OUT1 = (AFI0 - AFI1)/OUT1
= (797.4 - 598.7)/37.9 = 5.24 SCD

\# Refers to Shop Calendar of previous Period with 1000 SCD

*) 1.8 from 7.4 Hrs

**) Lead Time Area and Final Inventory Area correspond to the Value of the Final Inventory of the previous Period

***) Value is taken from Table A 5

Table B2 (*continued*)

Period 2 : SCD 35 to 39

12	< TOTAL >	85.8	TOTAL >	1964.0	12	< TOTAL >	85.8	TOTAL >	615.8

Lead Time Area of Final Inventory AAFI2 = 1,964.0 - 615.8 = 1,348.2 Hrs x SCD

Lead Time Area of Initial Inventory AAII2 = AFIA1 = 1,200.0 Hrs x SCD

Sequence Component SC2 = (AAII2 - AAFI2)/OUT2 = (1,200.0 - 1,348.2)/31.4 = -4.7 SCD

Inventory Trend Component ITC2 = (AII2 - AFI2)/OUT2 = (AFI2 - AFI2)/OUT2
= (598.7- 615.8)/31.4 = -0.5 SCD

Table B 2 B

Period 3 : SCD 40 to 44

11	< TOTAL >	152.0	TOTAL >	1928.0	13	< TOTAL >	152.0	TOTAL >	614.5

Lead Time Area of Final Inventory AAFI3 = 1,928.0 - 614.5 = 1,313,5 Hrs x SCD

Lead Time Area of Initial Inventory AAII3 = AFIA2 = 1,348.2 Hrs x SCD

Sequence Component SC3 = (AAII3 - AAFI3)/OUT3 = (1,348.2- 1,313.5)/30.1 = 1.2 SCD

Inventory Trend Component ITC3 = (AII2 - AFI3)/OUT3 = (AFI3 - AFI3)/OUT3
= (615.8 - 614.5)/30.1 = -0.0 SCD

Table B 2 C

Period 4 : SCD 45 - 49

10	< TOTAL >	134.7	TOTAL >	2280.3	10	< TOTAL >	134.7	TOTAL >	732.9

Lead Time Area of Final Inventory AAFI4 = 2,280.3 - 732.9 = 1,,547 Hrs x SCD

Lead Time Area of Initial Inventory AAII4 = AFIA3 = 1,313.5 Hrs x SCD

Sequence Component SC4 = (AAII4 - AAFI4)/OUT4 = (1,313.5 - 1,547.2)/46.6 = - 5.0 SCD

Inventory Trend Component ITC4 = (AII4 - AFI4)/OUT4 = (AFI3 - AFI4)/OUT4
= (614.5 - 723.9)/46.6 = -2.5 SCD

Table B 2 D

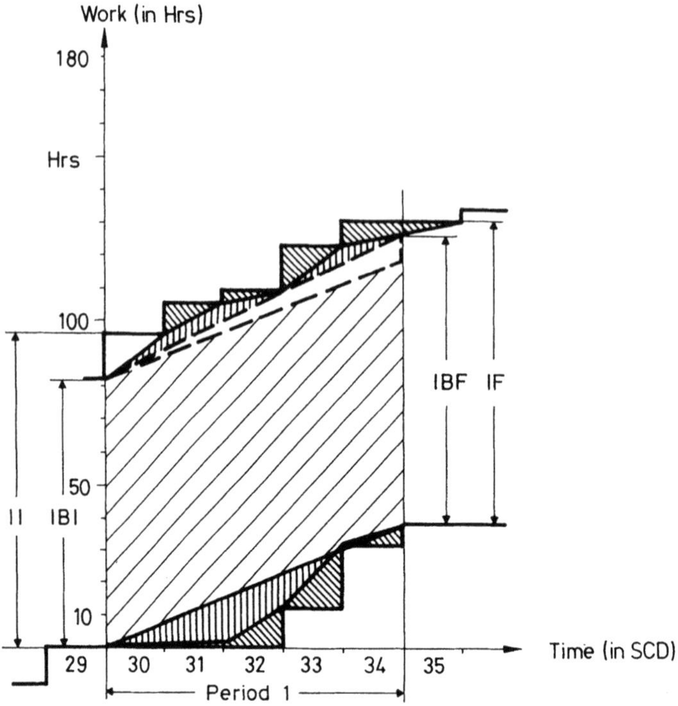

Initial Inventory	II	(Hrs)	95.8
Input	IN	(Hrs)	34.9
Output	OUT	(Hrs)	37.9
Final Inventory	IF	(Hrs)	92.8
Initial Basis Inventory	IBI	(Hrs)	81.8
Final Basis Inventory	IBF	(Hrs)	88.9

	Basis Area		AB	(Hrs x SCD)		409.0
+	Flow Area	(Input – Output)	AF	(Hrs x SCD)	111.9 – 94.3 =	17.6
+	Control Area	(Input – Output)	AC	(Hrs x SCD)	16.6 –(–31.3)=	48.0
+	Lot Area	(Input – Output)	AL	(Hrs x SCD)	26.3 + 19.5 =	45.8
=	Inventory Area		AI	(Hrs x SCD)		520.4

(D)

Fig. B1 (*continued*)

continued

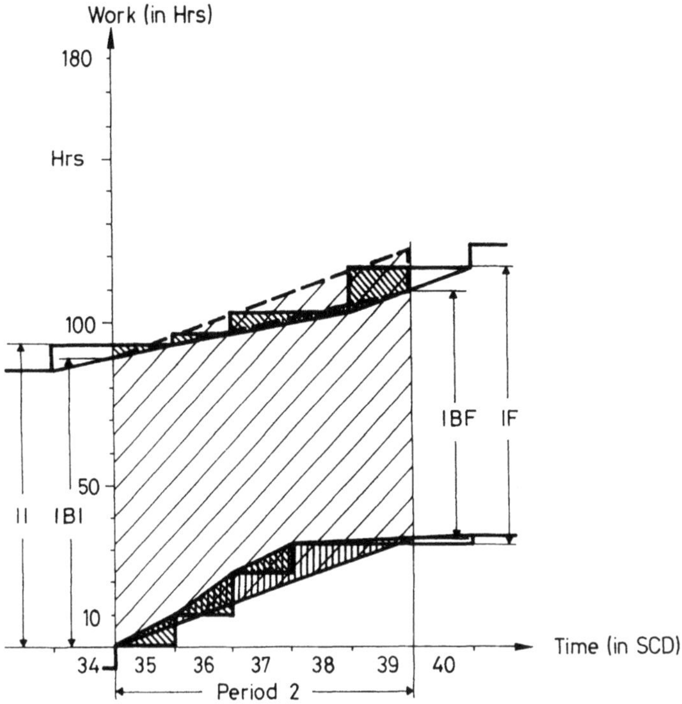

Initial Inventory	II	(Hrs)	92.8
Input	IN	(Hrs)	24.4
Output	OUT	(Hrs)	31.4
Final Inventory	IF	(Hrs)	85.8
Initial Basis Inventory	IBI	(Hrs)	88.9
Final Basis Inventory	IBF	(Hrs)	77.2

	Basis Area		AB	(Hrs x SCD)	444.5
+	Flow Area	(Input - Output)	AF	(Hrs x SCD)	53.6 - 82.8 = -29.2
+	Control Area	(Input - Output)	AC	(Hrs x SCD)	-5.5 - 29.6 = -35.1
+	Lot Area	(Input - Output)	AL	(Hrs x SCD)	21.0 + 17.4 = 38.4
=	Inventory Area		AI	(Hrs x SCD)	418.6

(B)

Fig. B1 (*continued*)

continued

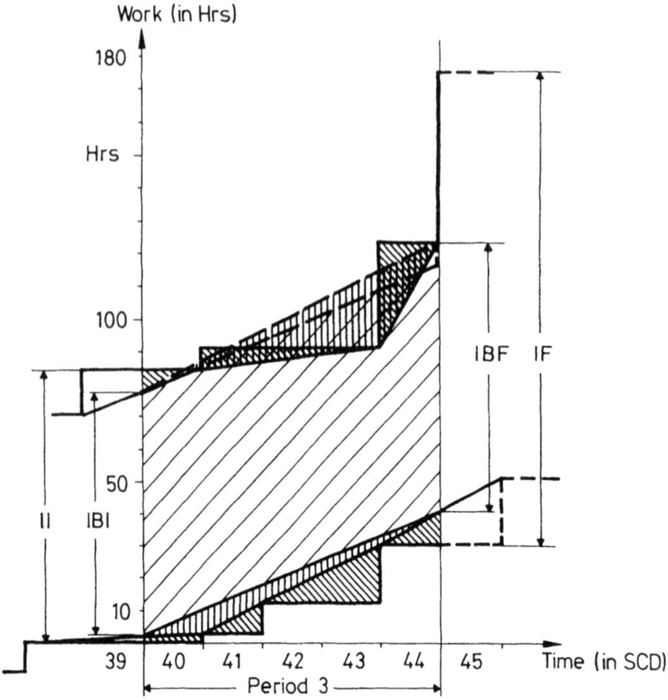

Initial Inventory	II	(Hrs)	85.8
Input	IN	(Hrs)	96.3
Output	OUT	(Hrs)	30.1
Final Inventory	IF	(Hrs)	152.0
Initial Basis Inventory	IBI	(Hrs)	77.2
Final Basis Inventory	IBF	(Hrs)	84.3

Basis Area		AB	(Hrs x SCD)	386.0
+ Flow Area	(Input - Output)	AF	(Hrs x SCD)	114.5 - 96.8 = 17.7
+ Control Area	(Input - Output)	AC	(Hrs x SCD)	-50.2 -(-18.5) = -31.7
+ Lot Area	(Input - Output)	AL	(Hrs x SCD)	29.8 + 30.1 = 59.9
= Inventory Area		AI	(Hrs x SCD)	431.9

(C)

Fig. B1 (*continued*)

continued

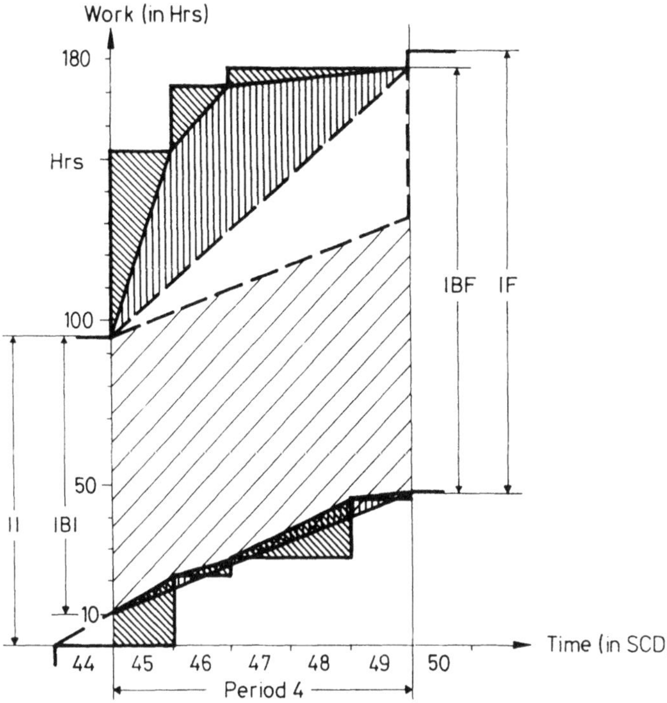

Initial Inventory	II	(Hrs)	152.0
Input	IN	(Hrs)	29.3
Output	OUT	(Hrs)	46.6
Final Inventory	IF	(Hrs)	134.7
Initial Basis Inventory	IBI	(Hrs)	84.3
Final Basis Inventory	IBF	(Hrs)	129.6

	Basis Area		AB	(Hrs x SCD)	421.5
+	Flow Area	(Input - Output)	AF	(Hrs x SCD)	204.0 - 90.6 = 113.4
+	Control Area	(Input - Output)	AC	(Hrs x SCD)	128.8 - 12.3 = 116.5
+	Lot Area	(Input - Output)	AL	(Hrs x SCD)	45.9 - 37.8 = 83.7
=	Inventory Area		AI	(Hrs x SCD)	735.1

(A)

Fig. B1. Division of the inventory area into its components (periods 1, 2, 3, and 4). Corresponding to the area components defined in Sect. 5.5.1, Figs. B1A–D show the division of the inventory into its components basis, flow, control, and lot inventory, for four periods. At the period limits, the division is carried out as shown in Fig. 5.39. The calculation of the inventory and area values is not shown in detail, but it can easily be performed using the period throughput diagrams and with simple geometrical considerations. (A) Inventory Components for Period 1; (B) Inventory Components for Period 2; (C) Inventory Components for Period 3; (D) Inventory Components for Period 4

continued

Subject Index

A

actual lead time,
  frequency distribution 57, 58, 72
actual mean lead time per operation 82
advance time, mean 104
allowance = planned lead time 41
anticipation horizon 209
arrivals and completions, cumulative plot 91
authentic lead time measurements 75

B

backward scheduling 17, 19f.

C

capacity 115f.
capacity adjustment 25
capacity alignment 26, 28
capacity monitoring diagram 274
capacity planning 292ff.
capacity planning diagram 273
capacity scheduling 18, 25
characteristic production curves 229, 252
characteristic production lines
  with lot splitting 283
check criteria for feedback data 138
closed-loop scheduling cycle 15
computer integrated manufacturing
  (CIM) 313
continuous monitoring system 158
control loop of load-oriented
  order release 255
control-loop model of production
  scheduling and control 14
conversion factor (of an order) 213
conversion of order times 211
conversion of work content 214
converted orders 210
cumulative curves of order stock and
  scheduled capacity 270f.
cumulative plot of
  arrivals and completions 91
current range, determination 285

D

daily capacity 20
data collection period 134f.
data record for DUBAF analysis 137
dispatching rules, effects on lead time 256
distribution of flow time 30, 151
distribution of order time, evaluation 149
distribution of simple and weighted mean
  actual lead time per operation 81
distribution of simple and weighted
  order lead time 85
distribution of simulated lead time 259
DUBAF analysis, data record 137

E

Electronic Leitstand 333
essential key data of work center,
  calculation 182
extended work center throughput diagram 123

F

feedback data, accuracy check criteria 138
--, evaluation 142
finite loading 17, 25
flow time = actual lead time 39
flow time, components 38, 41
--, distribution 30, 151
flow time and inventory analysis 133
flow time reduction, approaches 154
flow times,
  frequency distribution 6, 30, 40, 151

frequency distribution (absolute)
of actual lead time 57
frequency distribution (relative)
of actual lead time 58, 72
frequency distribution of
flow times 6, 30, 40, 151
frequency distribution of simple and
weighted order time 83
frequency distribution of simple vs.
weighted lead time 55, 57, 58
funnel formula 117, 204
funnel model 47, 147, 164, 267

G

graphic control unit 333

I

improvement of the
manufacturing process 306
--, guidelines 153
infinite loading 17
initial inventory 206
input process, components 177
inspection time 20f.
interoperation time TIO 20, 21
inventory, types 163
inventory area, components 180
inventory components 181ff.
inventory trend (in throughput diagram) 190
inventory trend component ITC 109

J

job-progress diagram 332
job-progress number system 331

K

Kanban manufacturing control 328
Kanban system, effects 330
key data, histogram 141
key data graph 159
key data of work center, calculation 182

L

large lots, effects 279, 281
lateness, calculation 118
--, weighted mean 117f.
lateness diagram 192, 275
lateness distribution 6, 7
lateness of orders 6, 7, 189, 192, 275
lateness of orders as a function of
capacity planning 275
lead time, simple vs. weighted mean 47ff., 60
--, types 163
--, weighted mean 48, 60, 106
lead time (actual),
calculation of simple and weighted 71
lead time components 38, 41
--, definition 122
lead time measurements, authentic 75
lead time scheduling 16, 19, 272
load and output curves 93
load centers 266
load limit LL 206, 220
--, effects 245
--, standard vs. individual 249ff.
--, techniques of determination 225
load limitations, different 242
load-oriented manufacturing control,
data flow chart 293
--, implementation steps 305
load-oriented order release 206, 210
loading percentage LPG 208, 220
long-range planning 13
lot size, influence on
inventories and lead time 279
lot waiting time 43

M

manufacturing analysis, schedule 155
manufacturing control, functions 264, 315
--, levels 280
manufacturing control techniques, survey 327
manufacturing flow analysis 131
manufacturing flow monitoring 129
manufacturing key data, structure 150
manufacturing process improvement 306
--, guidelines 153
manufacturing process monitoring 130
master route sheet 289
material flow matrix 144
material flow relations
between cost centers 144ff.
mean advance time 104
--, calculation 105
mean flow time per operation 7, 8
mean inventory, calculation 102, 160
mean lead time,
simple vs. weighted 47ff., 60
--, weighted 48, 60, 106
mean operation time,
simple vs. weighted 59
--, weighted 61
mean range 103f.
--, calculation 102, 160
medium-term planning 13
model test 235
monitored data, calculation 299
monitoring and diagnosis system 157

monitoring diagram 159, 172
--, structural principle 165
monitoring graph 173
monitoring system, components 159
--, information flow 173
monitoring table 159, 168, 170
movement data record 166
movement relations 145

N

non-feasible orders 210
non-urgent orders 210

O

one-dimensional throughput element 42f., 48
operation lead time 20, 78
--, distribution 284
--, model for definition and calculation 132
operation overlapping 23
operation splitting 24
operation time TOP 20, 38, 59
operation time, simple mean 59
--, simple vs. weighted mean 59
--, weighted mean 61
operation time percentage TOPP 62
OPT network model 340
OPT system 337ff.
order bottlenecks, determination 286f.
order flow time distribution 6
order lead time (scheduled), determination 231
order lead time 41, 53
--, components 53
--, model for definition and calculation 132
order release parameters 238
order stock and scheduled capacity,
cumulative curves 270f.
order throughput diagram 121
--, derivation 125
order time TO 20
order time, evaluation of distribution 149
order time per operation, distribution 281
order times, conversion 211
output curve 93
output process, components 177
overlapping of operations 23

P

Pareto curve, calculation 222
Pareto curve
of order time 84, 189, 191, 223, 282
performance PE 115
planning horizon 300
planning intervals 300
position of a work center, weighted mean 125
process time TP 20
processed orders, characteristic curve 244
production data evaluation 140
production planning and control model 15f.

Q

quality-time-cost cycle 34
queue length, effects on lead time 258
queuing models 335ff.
queuing time 20f.

R

range, mean 103f.
relative frequency distribution
of actual lead time 58, 72
release planning 295f.
released orders 210

S

schedule period 18
scheduled lead time, determination 231
scheduled throughput diagram,
construction 224
scheduling cycle, closed-loop 15
sequence component SC of
weighted lead time 111
sequence component, calculation 113
sequencing 296ff.
setup time TS 20, 41
setup time reduction 24
shop calendar 69ff.
shop orders 210
shop simulation 232
short-term planning 14
simple and weighted actual lead times,
calculation 71
simple and weighted mean actual lead
time per operation, distribution 81
simple and weighted order lead time,
distribution 85
simple and weighted order time,
frequency distribution 83
simple mean operation time 59
simple vs. weighted lead time,
frequency distribution 55, 57, 58
simple vs. weighted mean lead time 47ff., 60
simple vs. weighted mean operation time 59
simulated lead time, distribution 259
simulated operation lead times,
distribution 247
simulated weighted mean lead time,
curves 248
simulation model 233
simulation results with different

load limitations 242
simultaneous resources check 290
single-unit production 325
splitting of operations 24

T

throughput curves 189, 196
throughput diagram 94ff.
throughput diagram, components 175
--, construction 97
--, example 152, 250
throughput diagram (scheduled), construction 224
throughput element, accuracy in calculation 73
--, mean values of its components 226
--, one-dimensional 42f., 48
throughput model 207
throughput planning, components 265
time per unit TU 20
time-phased processes 324
transit time TT 20f., 38
two-dimensional throughput element 43, 48

U

urgent orders 210
utilization L 115

V

vicious cycle of mistakes in manufacturing control 7, 9

W

waiting after processing 20f., 41
waiting before processing 20f., 41
waiting time 38
weighted lead time, distribution 241
weighted mean actual lead time (TLM), calculation 161
weighted mean lateness 117f.
weighted mean lead time 48, 60, 106
weighted mean lead time of work center, calculation 107, 161
weighted mean operation time 61
weighted mean position of a work center 125
weighted operation time 61
work center flow time, distribution 190
work center lead time 21, 75ff.
--, components 44
--, measurement 45
work center throughput diagram, extended 123
work content, conversion 214
work-center-specific scheduled lead time 227

Z

Z-diagram 91f.

Printing: Mercedesdruck, Berlin
Binding: Buchbinderei Lüderitz & Bauer, Berlin

MIX
Papier aus verantwortungsvollen Quellen
Paper from responsible sources
FSC® C105338

If you have any concerns about our products,
you can contact us on
ProductSafety@springernature.com

In case Publisher is established outside the EU,
the EU authorized representative is:
Springer Nature Customer Service Center GmbH
Europaplatz 3, 69115 Heidelberg, Germany

Printed by Libri Plureos GmbH
in Hamburg, Germany

Foreword

Engineering materials are mainly used for structures. Therefore high-strength, stiffness and sufficient toughness are of prime importance.

For a long time engineers thought first in terms of metals. Material scientists developed alloys tailored to the needs of industry.

Ceramics are known to be brittle and therefore not suitable in the first place for structural application under stress.

Polymers with their low modulus became attractive when reinforced with high-strength fibres. Composites processed by polymer, metal or ceramic matrices and high-strength reinforcements have been introduced into many sectors of industry.

Engineering materials for structural applications fulfil a function: they withstand high stresses, temperatures, fatigue, creep etc. But usually we do not call them functional materials.

Functional material serve applications apart from classical engineering fields. Electricity conducting materials, semi conductors, memory alloys and others are called functional materials. Both categories of materials - structural and functional - cover the aspects of the DLR-Seminar 1991 on

"Advanced structural and functional materials".

The seminar is focused on advanced materials in research and development or at the edge of being introduced into the market. Preference was given to materials with a dual character: structural and functional, some of them more structural, some more functional in their potential.

It was necessary to select typical and representative families of during the years by R + D activities somewhat matured materials. It was not intended to cover all possible innovations in a two half-days seminar.

The main purpose of the seminar is to bring modern ideas and techniques of engineering science to the attention of DLR scientists and engineers as well as colleagues from industry and academia of many different disciplines in order to stimulate progress in the aerospace market and beyond. The seminar should be an opportunity for learning and understanding the big potential of advanced structural and functional materials.

W.G.J. Bunk

Contents

Nanocrystalline Materials
Herbert Gleiter, Universität Saarbrücken 1

Ceramics for Structural Purposes
Richard Brook, Max-Planck-Institut für Metall-forschung, Stuttgart 39

Thermostructural Ceramic Matrix Composites: An Overview
Roger Naslain, Laboratoire des Composites Thermo-structuraux, F-Pessac 51

Intermetallics - Fundamentals and Prospects
Hans-Jürgen Engell; Armin von Keitz; Gerhard Sauthoff, Max-Planck-Institut für Eisenforschung, Düsseldorf 91

Shape Memory Alloys
Erhard Hornbogen, Ruhr-Universität, Bochum 133

Self-Reinforcing Polymer Composites
Gerhard Wegner, Max-Planck-Institut für Polymerfor-schung, Mainz 165

Functional Molecular Materials derived from Organic Conducting Polymers
Francis Garnier, Laboratoire des Matériaux Moléculaires, C.N.R.S., F-Thiais 175

Piezoelectric Composites
Helmut Thomann, Siemens AG, München 199

Superconducting Materials
J. Georg Bednorz, IBM-Forschungslabor, CH-Rüschlikon
(not available at the time of printing)

Nanocrystalline Materials

H. Gleiter
Universität des Saarlandes, Gebäude 43, Institut für Neue Materialien, 6600 Saarbrücken, Germany

ABSTRACT

Nanocrystalline solids are polycrystals the crystal size of which is a few (typically 1 to 10) nanometers so that 50% or more of the solid consists of incoherent interfaces between crystals of different crystallographic orientations. Materials consisting primarily of internal interfaces represent a separate state of solid matter because the atomic arrangements formed in the cores of interfaces are known to be arrangements of minimum energy in the potentials field of the adjacent crystal lattices. The boundary conditions imposed on the atoms in the interfacial cores by the adjacent crystal lattices, result in atomic structures in the interfacial cores which cannot be formed elsewhere (e.g. in glasses or perfect crystals). Nanocrystalline materials seem to be of interest for the following four reasons:

(1) Nanocrystalline materials exhibit atomic structures which differ from the two known solid state structures: the crystalline and the glassy state.

(2) The properties of nanocrystalline materials differ (in some cases by several orders of magnitude) from those of glasses and/or crystals with the same chemical composition.

(3) Nanocrystalline materials seem to permit the alloying of conventionally insoluble components.

(4) If small (1 to 10 nm diameter) glassy droplets are consolidated (instead of small crystals), a new type of glasses, called nanoglasses, is obtained. Such glasses seem to differ structurally from glasses generated by rapid solidification.

INTRODUCTION

Research on "Nanocrystalline Materials" has been initiated less than a decade ago. Thus, the term nanocrystalline materials, may not be familiar to all readers. Let us therefore start by addressing the questions:

(i) What are nanocrystalline materials and(ii) why are they of interest?

Many properties of solids are known to depend on their atomic structure. Hence, if we succeed in generating a solid with a novel atomic structure, this solid is likely to exhibit novel properties as well. As nanocrystalline materials seem to have an atomic structure deviating from the presently available solid state structures (1), they may open the way for generating solids with new properties. The basic reasons suggesting nanocrystalline materials to have a new type of atomic structure are summarized in Figs. 1 and 2. The two figures indicate the atomic arrangements in the vicinity of an edge dislocation in a cubic crystal (Fig. 1) and an incoherent grain boundary between two cubic crystals tilted relative to one another by 36.9° about a common [100] axis (Fig. 2a). The atomic structure of the boundary was deduced from a high resolution electron micrograph (Fig. 2b, section A) of a 36.9° [100] tilt boundary between NiO crystals (Merkle et al 1987). NiO has a NaCl type structure. The cores of both lattice defects exhibit the following two characteristic features. A reduced atomic density and a modified atomic structure in comparison to the perfect lattice. In fact, the core of the edge dislocation may be described in terms of a channel comprising seven atoms in every cubic plane (Fig. 1).

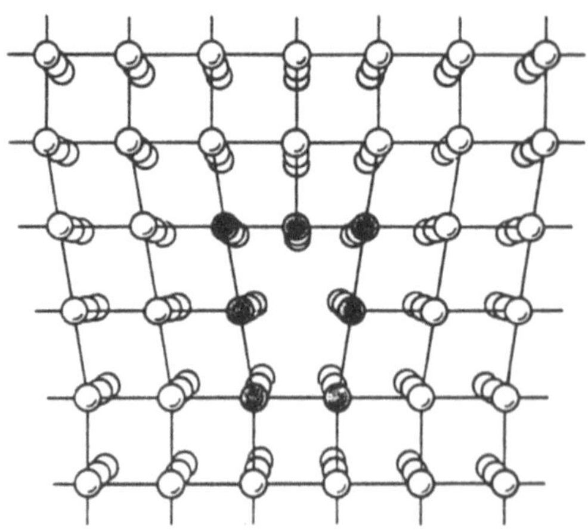

FIG. 1 Edge dislocation in a simple cubic crystal. In the dislocation core, the atomic density and coordination is changed in comparison to the perfect cubic lattice. In fact, the core of the dislocation may be interpreted as a channel formed by seven atoms as indicated in Fig. 1, whereas the crystal has a cubic structure.

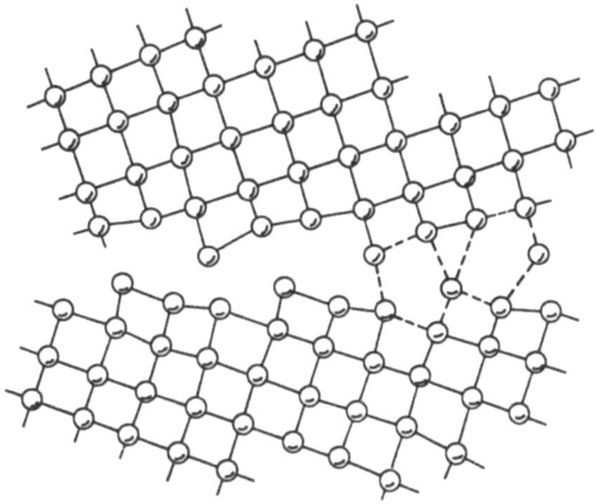

FIG. 2a Atomic structure in the core of grain boundary between two crystals tilted relative to one another by 36.9° about a common [100] direction. The structure was deduced (2) from the high resolution electron micrograph shown in Fig. 2b (section A). The boundary core structure may be described as a two-dimensional periodic array of two different pentagonal polyhedra (indicated on the right side of the boundary).

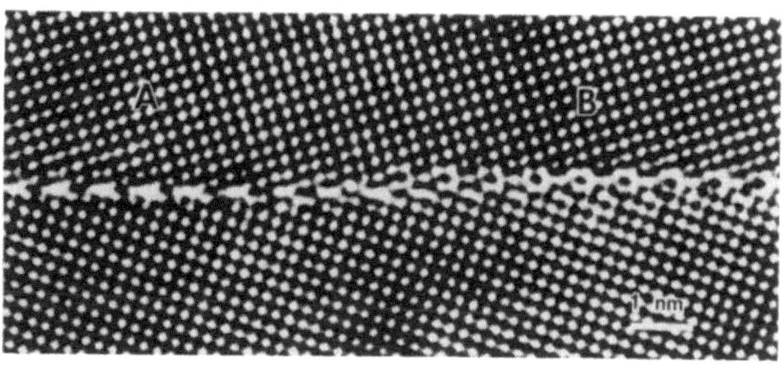

FIG. 2b Electron micrograph of a symmetric 36.9° [100] tilt grain boundary in NiO (2). The black regions represent the atomic positions. In the boundary core the atomic density and coordination is changed in comparison to the perfect lattice. The boundary structures in the facets A and B are different due to the vertical displacement of the boundary plane relative to the two crystals.

The structure of the grain boundary core shown in Fig. 2a represents a periodic sequence of two different pentagonal polyhedral units tilted relative to one another whereas the two crystal lattices have a cubic structure. The newly formed atomic arrangements in the boundary core result in interatomic spacings

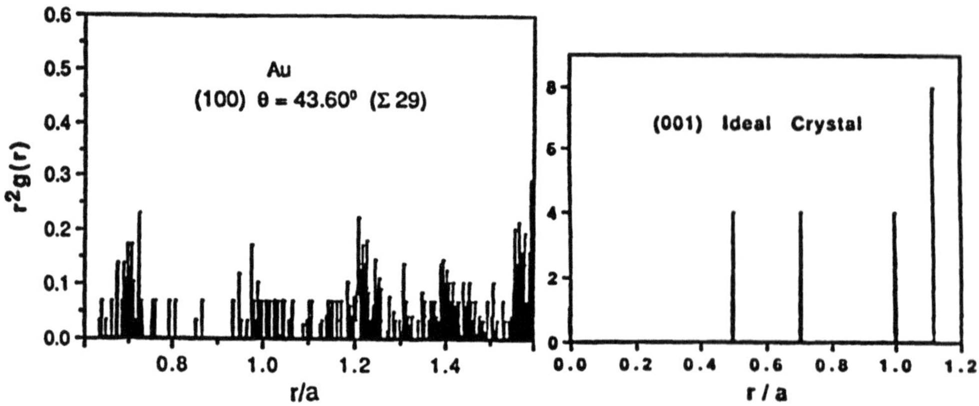

FIG. 2c Radial distribution function, $r^2g(r)$. for the atoms in the core of a 43.6° (100) (Σ29) boundary as described by an embedded atom potential for gold (left side). The figure on the right, displays the radial distribution function of an ideal crystal on (001)(80).

that cannot exist in the perfect lattice. For example Fig. 2c displays the distribution of interatomic spacings in the core of a 36.9 [001] boundary in gold in comparison to the spacings in the perfect lattice of the same material. Obviously, the boundary core is characterized by a broad distribution of interatomic spacings and atomic arrangements which cannot be formed in the perfect lattice. The physical reason for the reduced atomic density and the modified atomic arrangements in the boundary cores may be seen by considering, for example, the structure of the grain boundary core shown in Fig. 2a. The boundary core is the region where two crystals with different crystallographic orientations are joined together. Due to the different orientations, the two adjacent crystal lattices match poorly. The poor matching results in an atomic density in the core region which is lower than in the perfect lattice. The density reduction depends on the crystallographic parameters of the boundary such as the orientation relationship between both crystals, the inclination of the boundary plane relative to both crystal lattices as well as on the interatomic forces i.e. the type of chemical bonding. Typically, the boundary core density is found to be reduced by about 15 to 40% relative to the density of the perfect lattice. In fact, the boundary core density is far lower than in the liquid or glassy state and becomes comparable to the densities of highly compressed gases. The physical reasons for the reduced density and the modified coordination are the constraints excerted by the adjacent crystal lattices on the atoms in the defect cores. Without these constraints the core structures would collaps. In liquids or glasses no constraints of this sort exist and, hence, their structures differ from the one in the cores of defects. There is also an other way to see the difference between the structure of the nanocrystalline and the glassy state. In the glassy state, the disorder induced by thermal energy is frozen in. The disorder in nanocrystalline materials results from the imcompatibility introduced into the lattice in the form of defects and does not require any thermal energy. Between two given crystals, a large variety of different boundary core structures may be formed because the boundary core structure of a boundary depends on the orientation relationship between both crystals and the boundary inclination. In other words, each time the orientation relationship between both crystals and/or the inclination of the boundary plane is varied, a new boundary core structure with different interatomic spacings (Fig. 2c) is generated.[1] If this is so, then the question arises: Why are these numerous boundary core structures and the reduced density not noticed if we study a polycrystalline solid? If the average crystal size is 10 μm or more, a straightforward estimate indicates that a volume fraction of grain boundary cores in such a polycrystal is less than 10^{-4}, and hence, the atomic arrangements in boundary cores nor-

[1] An example may be seen in Fig. 2b. The boundary plane in section B is displaced relativ to A. This displacement results in different boundary core structures in A and B.

mally escape our attention. Nevertheless, precise measurements have revealed the expected effects (3), (4), (5), (77). Thus, if we wish to study and/or utilize the grain boundary core structures, we have to enhance the volume fraction of boundary cores in a polycrystal. This goal may be achieved by reducing the crystal size to a few lattice constants (i.e. a few nanometers) so that the volume fraction of boundary cores approaches 50% or more. If this is done, we obtain solids which consist to a large extent of boundary cores, and hence the atomic structure and the properties of such solids are controlled primarily by the structure and properties of the grain boundary core regions. Solids of this type have been proposed about ten years ago and were called nanocrystalline solids (1), (6) because they are obtained by reducing the crystal size of a polycrystal to a few nanometers. Other terms used in the literature are: ultrafine grained materials, nanophase or nanometer-sized solids. Naturally, they could have equally well be termed "defect core solids". A two-dimensional model of a nanocrystalline material is displayed in Fig. 3. For the sake of clarity, atoms in the various

FIG. 3 Atomic structure of a two-dimensional nanocrystalline material. The structure was computed by following the procedure given in ref. 74. The interaction between the atoms was represented by a Morse potential fitted to gold. The atoms in the central portions of the "crystals" are indicated in black. The ones in the boundary core regions are represented by open circles.

crystals are indicated as full circles whereas the atoms in the boundary cores are represented by the open symbols although both types of atoms are assumed to be chemically identical. Structurally, the material consists of the following two components: the crystals with different crystallographic orientations but identical structure and a network of boundaries, the structure of which differs from boundary to boundary and is characterized by the following two features:

(i) The average atomic density in the boundary regions is about 70% of the crystal density. The boundary density cannot be enhanced by inserting additional atoms because the local free volumes of the vacant sites are too small so that interatomic penetration would result if an atom is inserted.

(ii) The boundaries exhibit a broad distribution of nearest neighbour configurations and of interatomic spacings ranging from closely packed atoms to widely spaced ones.

In nanocrystalline materials the situation is structurally complimentary to the one in single crystal or a coarse grained polycrystals. In the single crystal or the coarse grained polycrytals, the crystal lattice represents the basic structural component and the lattice defect cores (e.g. dislocation cores, grain boundarie cores) are perturbations limited to a small volume fraction of the material. In nanocrystalline materials, the priorities are inverted: the defect cores represent the essential structural structural component and the residual crystalline regions are required for generating the boundary cores.

In this article we focus attention on the discussion of the basic concepts and features of nanocrystalline materials. The experimental and technical details have been documented in several reviews (7)-(11), (49), (42) . As the field finds itself in a state of rapid expansion, many of our present views may have to be modified in the years to come.

HISTORICAL DEVELOPMENT

Although the generation of a new type of solid state structure by utilizing the atomic arrangements in defect cores is less than a decade old, precursor developments can be traced back in history of Materials Science to the turn of this century, and in nature even to the very early state of the universe. The significance of submicroscopic microstructures for mechanical magnetic and electronic properties of solids has apparently been recognized first by Sorby, Tschernoff, Osmond, Howe, Sauveur and others before the turn of the century. Their studies between 1870 and about to 1900 led to the conclusion that the fine-scale microstructure retained after the allotropic transformation of iron alloys gives martensite its hardness. The discovery of precipitation hardening by Wilm in 1906 was the first observation suggesting that the correlation between microstructure and properties (originally proposed for ferrous alloys only) applies to non-ferrous materials as well. Wilm quenched an Al-Cu-Mg-Mn alloy and noticed (after a long weekend) a substantial hardening relative to the as-quenched state. However, it was not until 1919 when Merica, Waltenberg and Scott proposed that the hardening resulted from precipitation of a new phase on a submicroscopic level. Numerous observations in the subsequent years substantiated and generalized this view and led to the classification of the properties of solids with different types of chemical bonding into structure-sensitive and non-sensitive ones.

In the following years, this classification played an important role in promoting the idea of lattice defects and their significance for crystalline solids. In fact, the physical understanding of the mechanism by which nanometer-sized microstructures affect the properties of solids received a remarkable boost after World War II, from the advent of the theory of lattice defects - in particular dislocation theory - and from the availability of new high resolution research techniques such as electron and field ion microscopy. Both developments eludicated the physical basis for understanding the correlation between the structure-sensitive properties and the microstructure of solids.

As a matter of fact, the development of most high strength and high temperature materials available today is based on the results of those studies. When it was recognized that dislocations play a similar role for the mechanical behaviour of materials as do domain walls or flux lines for ferromagnetic or superconducting properties, respectively, it became apparent that ferromagnetic and superconducting properties can be manipulated, too, by suitably varying the microstructure (12). In fact, the pinning of ferromagnetic domain walls or of flux lines in type II superconductors by finely dispersed precipitates leads to magnetic materials with high coercive forces and superconductors with high critical current densities (12). Instead of reducing the motion of dislocations by precipitates from supersaturated solid solutions, small second phase particles introduced by means of powder metallurgy or by extrusion of two phase coarse-grained structures (13), (14) may be used as well. The enhanced defect density in irradiated, in highly cold worked (15) as well as in fine-grained materials (16) results in similar effects because the defects inhibit dislocation motion. A closely related approach in the field of ceramics generated by the sol-gel method was the proposal, to use heterogeneity on a nanometer scale (17)-(22). This approach was based on the hypothesis that diphasic or multiphasic ("nanocomposit") xerogels which are heterogeneous on a nanometer scale store more energy than a single phase gel and thus exhibit new properties.

About 1970, a second period of developments in the area of nanometer-sized microstructures started when it was recognized that specifically tailored nanometer-sized microstructures permit the generation of solids with new atomic and/or electronic structures. These developments seem to bring Feynman's dream of nanotechnology (23) closer to reality. In the area of semiconducting materials extensive research activities were initiated on a new two-dimensional class of materials with a nanometer-scale microstructure by the proposal of an "engineered" semiconductor superlattice (24), (25) consisting of alternate thin, fully coherent layers with different dopings or different compositions leading to new electronic structures. These superlattices represent an extension of double- or multibarrier structures where quantum effects prevail because the layers have a thickness of a few lattice constants. Several recent reviews of the present state of understanding of the physics and application of quantum well structures and superlattices have been compiled in ref. (26). The nanocrystalline materials discussed in this paper represent the second example of a tailored nanometer-sized microstructures. Although the charateristic dimension (at least in one direction) are comparable in the case of semiconductor superlattices and nanocrystalline materials, both groups of materials differ in the underleing physical concepts. Semiconductor superlattices are based on quantum size effects due to the small thickness of the layers that are joined together by fully c o h e r e n t interfaces. Incoherent interfaces would be detrimental. In nanocrystalline materials the situation inverted. It is the peculiar atomic structures formed in the cores of i n c o h e r e n t interfaces (or other defect cores) that are utilized. The small dimensions of the crystals are required to obtain a large volume fraction of incoherent interfaces. However, quantum size effects play no role.

Long before, materials with nanometer-sized microstructures have been synthesized by man, they have been formed in several natural composites. For example, human teeth are an exquisitely tailored composite at the 1-2 nm level of fibrils of hydroxyapatite topotactically grown onto collagen. A similar structure is found in natural corals and in the shells of abalone. Semicrystalline polymers are composed of plate shaped crystalline regions with thicknesses of typically 1-3 nm. The crystalline regions are separated by "amorphous" layers consisting of tie molecules between adjacent crystallites as well as macromolecules folded in a regular or irregular fashion. Probably the oldest solids with a nanometer-sized microstructure are primitive meteorites in which aggregates of carbon atoms or silicon carbide of 5nm or less are found. These aggregates are believed to be formed by condensations in the early age of the solar system of in interstellar space (27), (28). Asbestos, opals and calcedon represent minerals with a nanometer-sized microstructure formed on the surface of the earth many millions years before any man-made substance of this type was available.

The earliest applications of materials with nanometer-sized microstructures habe been revealed in ancient Egypt and China where the size of the pigments of certain colours has been reduced into the submicron regime by grinding or milling in order to enhance the intensity of the colours.

PREPARATION

The experimental arrangement used most frequently to prepare nanocrystalline solids is a modified gas condensation method (Fig. 4), where the material (e.g. Fe) is evaporated into an inert gas atmosphere (e.g. He, pressure about 0.1...1 kPa).

Through interatomic collisions with the He atoms, the evaporated Fe atoms lose kinetic energy and condense in the form of small crystals a few nanometers across which accumulate on a vertical cold finger as a loose powder. After restoring high vacuum ($<10^{-6}$ Pa) the powder is stripped off and funneled into a piston and anvil device where it is compacted (pressure up to 5 GPa) into a pellet. Sputtering, electron gun or laser evaporation may be employed instead of thermal evaporation. Other methods which have been utilized to obtain fine grained powders suitable for generating nanocrystalline solids are ball milling (e.g. applied to Ru), vapour phase synthesis, sliding wear, decomposition of suitable compounds and precipitation from solutions or gels. In the latter cases, special care has to be taken to remove surface contaminants. A systematic study of the impurity content of nanocrystalline solids prepared by inert gas condensation revealed that gaseous and metallic impurities were less than 5% and He less than 50 ppm. The density of metallic nanocrystalline materials after consolidation varied between 75% and more than 90% of the crystalline density (depending on the material), approaching 100% after grain growth. The distribution of crystal sizes was determined by transmission electron microscopy and small angle X-ray or neutron scattering. A typical distribution (for nanocrystalline Pd) is presented in Fig. 5. For further details we refer to ref 40.

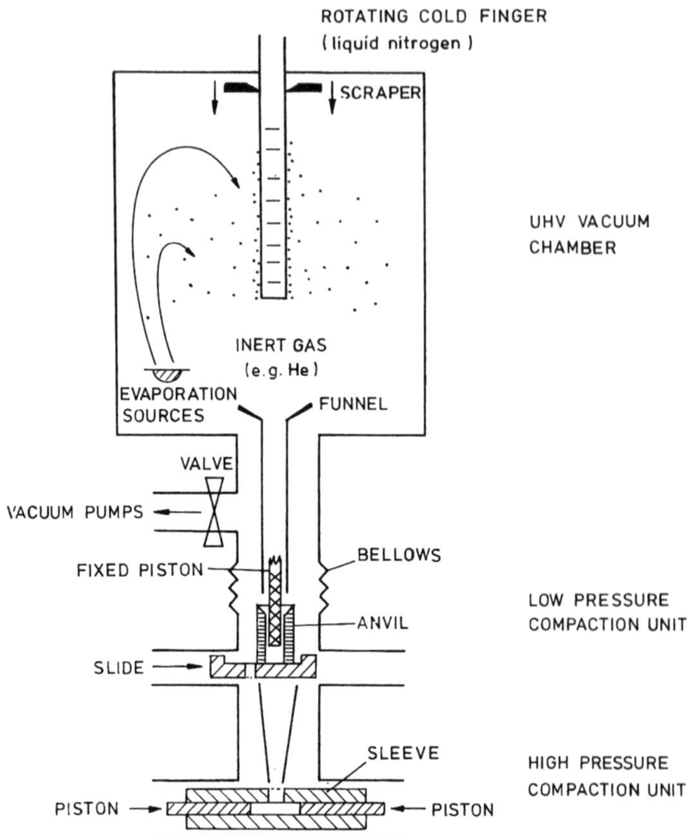

FIG. 4 Schematic drawing of a gas-condensation chamber for the synthesis of nanocrystalline materials. The material evaporated condenses in the inert gas in the form of small crystallites which are subsequently transported via convection to the liquid nitrogen filled cold finger. The powder of crystallites is finally scraped from the cold finger, collected via the funnel and consolidated first in the low-pressure compaction device and then in the high-pressure compaction unit. Both units are kept under UHV conditions. Instead of an evaporation device, a sputtering source has been utilized successfully.

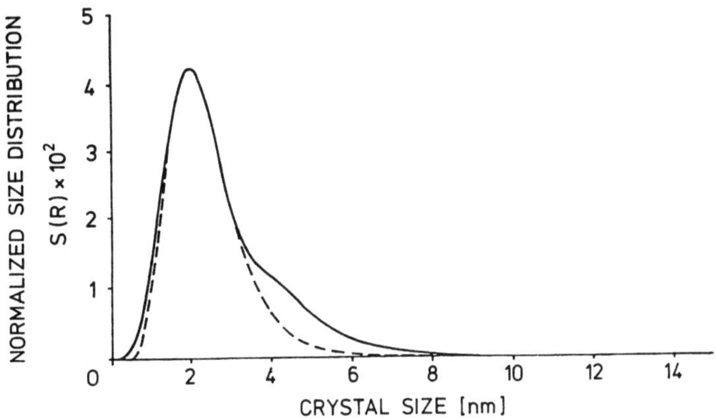

FIG. 5 Normalized size distribution of nanocrystalline Pd obtained from the small angle scattering measurements. The broken line represents the log-normal distribution.

STRUCTURE

In the previous paragraph, the two characteristic features of a solid consisting of a large volume fraction of grain boundary cores were pointed out to be (i) a reduced atomic density in the boundary regions and (ii) a broad spectrum of interatomic spacings coupled with a large variety of the nearest neighbour configurations in comparison to the perfect crystal.

Both features seem to be born out by the experimental observations. High resolution microscopy (2), and X-ray diffraction on Au-bicrystals (5), (30) indicate a reduction of the atomic density in the core region between 15 and 40%. Similar numbers are suggested by recent theoretical studies on boundaries in bicrystals (78). Small angle scattering measurements on nanocrystalline Pd (29), (31), and TiO_2 (32) indicated the average density in the boundary cores (width between 0.5 and 1 nm) to be reduced by about 20 to 40% relativ to the crystal density. Information about the atomic arrangement in the boundary cores of nanocrystalline materials was obtained by static and dynamic studies. EXAFS measurements (33) indicate a reduction of the number of nearest, next nearest etc. neighbours averaged over all atoms of a nanocrystalline material relative to a single crystal (Figs. 6a and 6b).

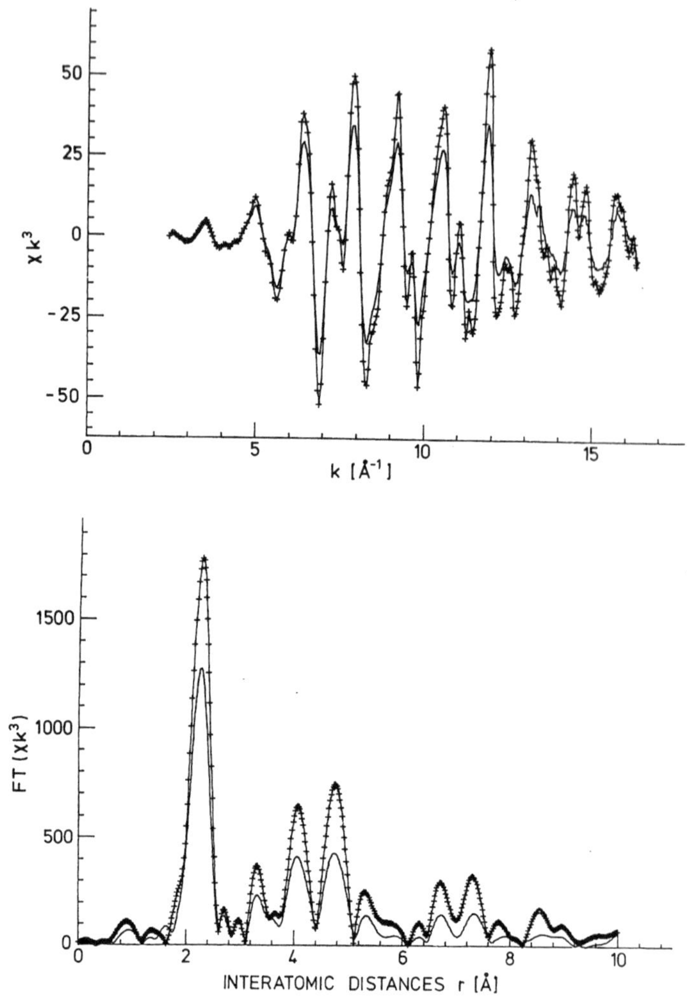

FIG. 6a,b The weighted EXAFS oscillations χk^3 (Fig. 6a) and the corresponding Fourier transform FT (χk^3), (Fig. 6b, phase shift is not included) of a nanocrystalline Cu sample (—) (crystal diameter: 10 nm) in comparison to polycrystalline Cu (+++). In the nanocrystalline sample the amplitude of the EXAFS oscillations and the FT (χk^3) are reduced relative to in the polycrystalline Cu (33).

This result may be unterstood, if the structure in the boundary cores differ from the structure in the lattice. In fact, the experimentally observed reduction of the EXAFS amplitudes and the corresponding Fourier Transform (Figs 6a, b) agrees with the idea (Fig 3) of a broad distribution of different nearest neighbour configurations in the boundary cores. A similar conclusion is suggested by Mössbauer spectroscopy measurements on nanocrystalline FeF_2 (34). The quadrupole splitting distribution of nanocrystalline FeF_2 is broadened (Fig. 7) relative to the one of an FeF_2 single crystal indicating a variety of atomic configurations with non-lattice symmetry to exist in the boundary cores.

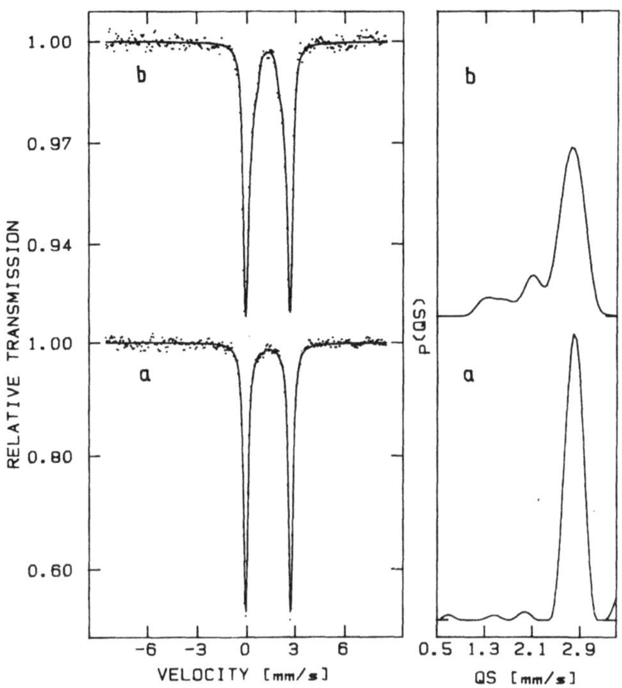

FIG.7 The Mössbauer spectra and quadrupole splitting (Q.S.) distributions for coarse grained (a) FeF_2 powder and (b) nanocrystalline FeF_2 (77).

X-ray diffraction studies (35) on nanocrystalline Fe have been interpreted in terms of a boundary core structure in which the atoms are displaced from the ideal lattice sites of both crystals (Fig. 8).

These displacements result in a diffuse x-ray diffraction background. However, the absence of this background in the diffraction curves of nanocrystalline Pd or Cu suggests that the atomic

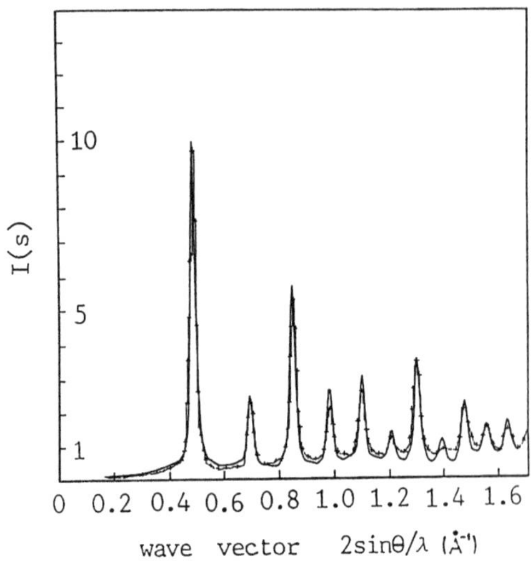

FIG. 8 Comparison of the measured (-- + --) and computed (—) interference functions of nanocrystalline Fe. The model system selected for computing the interference function is a mixture of 6-nm (75 vol.%) an 4-nm (25 vol.%) Fe crystals in which the boundary atoms are displaced from the ideal lattice sites in random directions. The displacement distances were chosen as follows. The atoms in the boundary core were displaced by 50% and in the two subsequent layers by 25% of the nearest neighbour distance relative to the lattice sites they occupied originally. The layers of displaced atoms were used to model the ensemble average of the structure of all boundary cores. The atomic density in the core region was adjusted to the measured boundary density (29), (35).

structure in the boundary cores of nanocrystalline materials may depend on the chemical composition. This conclusion agrees with the well established different atomic relaxations at free surfaces and in the dislocations cores of chemically different materials. In other words, in a nanocrystalline material the boundary core structures (and hence the properties of this material) may be manipulated by varying the chemical composition in the boundary cores e.g. by incorporating solute atoms into the boundary cores. The modification of the boundary properties due to boundary segregation of solute atoms is well known for numerous alloys. For example, the embrittlement of Cu by Bi segregation to the boundaries or the enhanced ductility of Ni_3Al if B is present in the boundaries. The structural deviation between the crystals and the boundary cores are also born out by methods probing the dynamic displacement of the atoms in the boundary cores. Measurements of the Debye temperature by Mössbauer spectroscopy indicate a reduction by about 100K in nanocrystalline Fe (76) and about 210K in nanocrystalline FeF_2 (36). The Debye temperatures of Fe and FeF_2 are 467K and 298K, respectively. These results seem consistent with inelastic neutron scattering measurements suggesting additional low frequency phonons in nanocrystalline Ni in comparison to a coarse grained polycrystal of the same material. The low frequency phonon modes may be the reason for the enhanced specific heat, c_p, (38), (39) of nanocrystalline materials. For example, in the case of 6nm Pd, c_p is enhanced relative to a coarse grained polycrystal by about 40% at 300K. Measurements of the elastic constants by bending, sound wave propagation or Brillouin scattering indicate a reduction of the Youngs and the shear modulus by 15% or more (37). The reduction cannot be caused by residual porosity because the results of the last two methods depend litte or not at all on porosity. However all of the observations reported may be understodd by assuming (i) the average atomic density in the boundary cores to be reduced relativ to the perfect crystal and (ii) this reduction to result in atoms that are more weakly coupled to their neighbours than the ones in the perfect lattice.

Although the above discussion has been limited on manipulating the atomic structure of solids by means of incorporating a high density of grain boundaries, basically the same arguments hold for other lattice defects as well, such as dislocations, interphase boundaries etc.. In recent years, the research activities on the production, the of atomic structure and the properties of nanocrystalline materials increased remarkably. The progress made has been documented in several review articles (7) - (11), (40), (41).

PROPERTIES

If the atomic structure of nanocrystalline solids differs from crystals and glasses, the structure-dependent properties of nanocrystalline solids are expected to deviate from the properties of the chemically identical substances in the glassy and/or crystalline state. Table I summarizes some properties of nanocrystalline solids in comparison to the ones of crystals and glasses with the same chemical composition.

Table I: Comparison between some properties of nanocrystalline glassy and crystalline solids with the same chemical composition.

Properties	Chemical composition	Nanocryst. solids	Glassy solids	Crystal. solids	
Diffusion (300K)	$^{67}Cu/Cu$ Ag/Cu	10^{-19} $2\cdot 19^{-19}$	$\approx 10^{-36}$	10^{-39} 10^{-39}	m^2s^{-1}
Solid solubility	Bi/Cu H/Cu	4 10^{-1}		$<10^{-4}$ 10^{-3}	%
Plastic deformation	TiO_2 CaF_2	>100		<0.1	%
Saturation magnetization (4K)	Fe	130	215	220	emu/g
Electronic specific heat (0.1..1K)	Cu	1.6	≈1	1.1	$10^{-5}J/gK^2$
Charge carrier density	Cu	0.6	≈1.2	1.24	electrons/ atom

In the case of the properties listed in Table I, the crystalline and glassy state deviate by 5% or less. However, the corresponding properties of the nanocrystalline solids differ between 50% and twenty orders of magnitude. Such large deviations are hard to understand unless the atomic structure of nanocrystalline solids is assumed to differ from the structures of crystals and/or glasses.

Figs. 9 and 10 illustrate the remarkable enhancement of the diffusion in a nanocrystalline metal and the plasticity of a nanocrystalline ceramic material. Both effects may be of interest for technological applications. In the last two lines of Table I properties are listed that are controlled by the electronic structure of the solids. The remarkable change of the electronic specific heat and of the observed reduction of the charge carrier density (by means of Hall measurements) suggest that nanocrystalline solids not only differ structurally but also electronically from the crystalline and the glassy state with the same chemical composition.

More than one hundred papers have been published so far dealing with the various aspects of the structure and the properties of materials containing a high density of defects, preferably grain boundaries. Hence a comprehensive and critical discussion of all

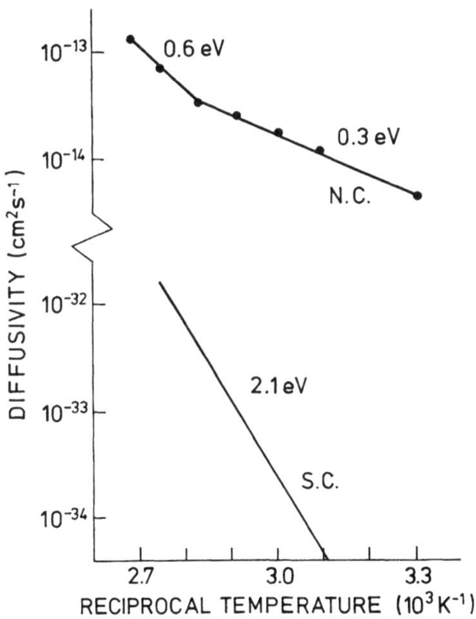

FIG.9 Diffusivity of Ag in a Cu single crystal (S. C.) and in nanocrystalline Cu (N.C. 10 nm average crystal size). The numbers indicated in the figure represent the activation enthalpies for both diffusion processes (42).

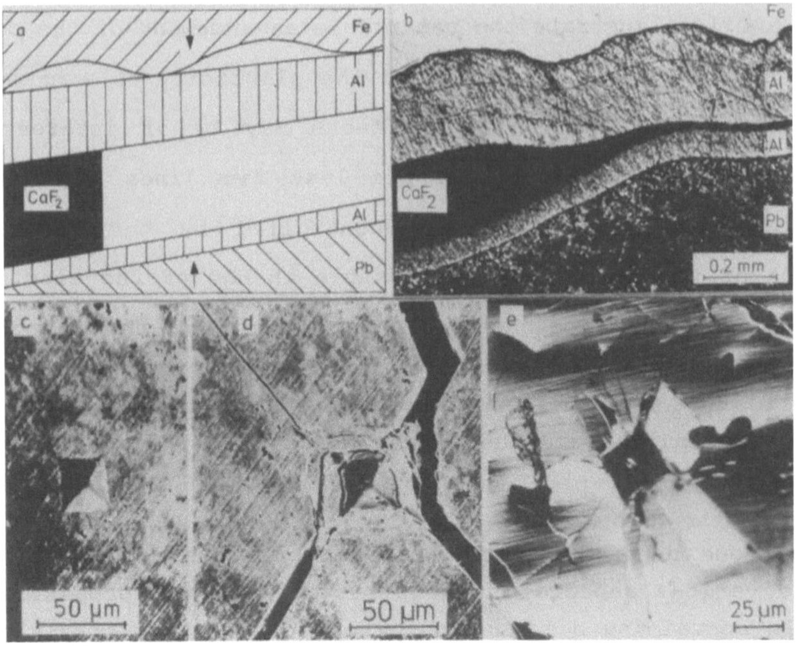

FIG. 10 Illustration of the mechanical properties of nanocrystalline materials. (a) Schematic drawing indicating the deformation procedure applied. (b) Plastic deformation of nanocrystalline CaF_2 by an arrangement as shown in Fig. 10 a. The deformation was carried out at 353 K and the deformation time was about 1 s. (c) Indentation of nanocrystalline TiO_2 at 293 K (load 2 N, deformation time 10 s). The hardness of the material (deduced from the size of the indent) was 250 HV. (d) Brittle behaviour of nanocrystalline TiO_2 (293 K, load 2 N). The indentation was generated by reducing the loading time from 10 s (Fig. 10 c) to 1 s. (e) Indentation of a conventional TiO_2 polycrystal under conditions similar to those in Fig. 10 c (293 K, load 5 N). The hardness deduced was about 800 HV. The different brightness of the fractured areas results from the tilting of these areas relative to incident light (43).

data available is beyond the limits of this paper. In order to keep this manuscript within reasonable lenght, we refer to existing review articles concerning the present state of the art (7) - (11), (40), (41). We shall now turn to the second part of the paper dealing with recent developments and future perspectives.

RECENT DEVELOPMENTS AND PERSPEKTIVES

In this section we shall point out some recent developments which may become the nuclei for new areas in the future. Naturally, the selection is bound to be subjective. Therefore, the author apologizes to all colleagues whose work will not be discussed.

ALLOYS WITH NANOMETER-SIZED MICROSTRUCTURES

Alloys made up of nanometer-sized crystals with different chemical compositions have recently been synthesized (44) and seem to open the way to generate alloys of chemically different components that are insoluble in bulk coarse-grained polycrystals. The preparation of nanometer-sized alloys was performed by means of an UHV evaporator (Fig. 4) filled with pure helium (≈0.1 kPa pressure). Inside of the evaporator, several evaporation (or sputtering) units were arranged for the substances A, B, C ... which were evaporated (or sputtered) simultaneously. The small crystals of A, B, C ... formed in the He atmospere were accumulated on the rotating cold finger as a loose powder. Under suitable evaporation (sputtering) conditions, the powder is a random mixture of small crystals of A, B, C ... . Subsequent consolidation of the stripped-off powder leads to a nanometer-sized polycrystal the individual crystallites of which consisted of A, B, C ... . This product was called a "nanocrystalline alloy". Compared to single component nanocrystalline solids, nanocrystalline alloys contain a high density of interphase boundaries as well as grain boundaries[2]. The

---

[2]Preparation of nanocrystalline materials and alloys under UHV conditions seems crucial to guarantee the formation of uncontaminated interfaces. In fact, nanometer-sized materials synthesized by using crystals with contaminated surfaces usually do not exhibit the features discussed in this paper. Of course deliberate adsorption or deposition of specific atoms or molecules at the surfaces of the small crystals prior to compaction may be attractive because it permits to incorporate certain atoms (molecules) into the boundaries. As a matter of fact, this method has been used to dope the boundaries of nanocrystalline materials e.g. doping of the grain boundaries in nanocrystalline Fe or Cu by Bi or of Ir with Cu.

atomic structure of nanocrystalline Ag-Fe and Cu-Fe alloys were studied recently (44). Ag-Fe alloys were selected for two reasons. Ag and Fe are insoluble in the solid and liquid state. Furthermore, the atomic structure may be investigated conveniently by X-ray dif-

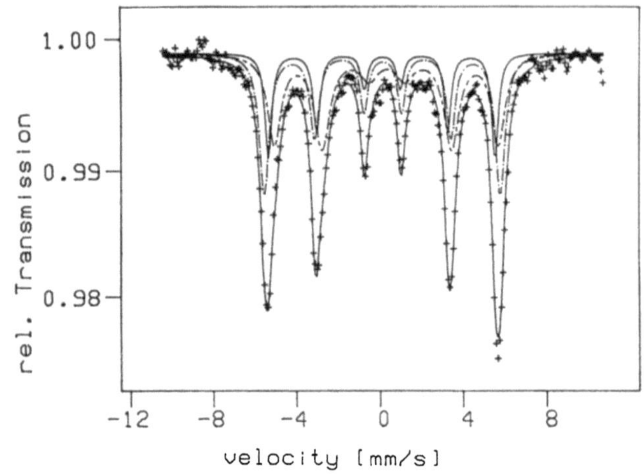

FIG. 11 Mössbauer spectrum of a nanocrystalline Fe-Ag alloy (30at% Fe, 10 K, average crystal size 8 nm). The spectrum consists of the following three components. (i) α-Fe (—), (ii) Fe atoms dissolved in Ag (-.-) and (iii) Ag atoms dissolved in Fe (---) (44).

fraction as well as Mössbauer spectroscopy. Fig. 11 displays the Mössbauer spectrum of a nanocrystalline Ag-30 at% Fe alloy with an average crystal size of 8 nm. The spectrum consists of three components of comparable volume fractions.

One component corresponds to body centered cubic (bcc) α-Fe crystals, the two other components originate from bcc α-Fe crystals with incorporated Ag atoms and face centered cubic (fcc) Ag crystals with incorporated Fe atoms. In addition, the X-ray scattering data indicate the presence Ag crystals. The atomic structure of the nanocrystalline Ag-30at%Fe alloy deduced from these observations is shown in Fig. 12.

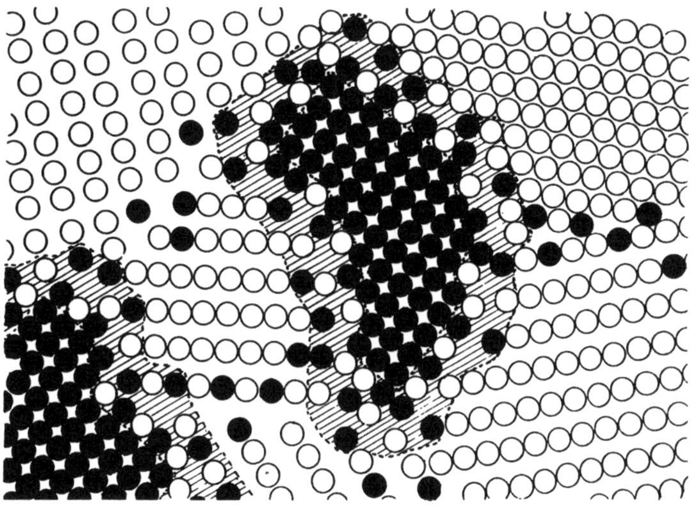

FIG. 12 Schematic model of a nanocrystalline Ag-Fe alloy according to the data of Mössbauer spectroscopy (Fig. 11). The alloy consists of a mixture of nanometer-sized Ag and Fe crystals. In the (strained) interfacial region between Ag and Fe crystals solid solutions of Fe-atoms in Ag and Ag-atoms in Fe are formed although both components are insoluble in the bulk liquid and/or solid state. Similar effects may occur in the grain boundaries between adjacent Fe and Ag crystals (44).

Apart from pure bcc α-Fe and pure fcc Ag crystals, solid solutions are proposed to be formed in the strained lattice regions in the vicinity of the Ag/Fe interphase boundaries and/or at the grain boundaries between adjacent Fe or Ag crystals, respectively. An enhancement of the solute solubility due to elastic strains has been demonstrated in the past theoretically (45), (46) and experimentally (47) for various alloys in terms of elastic strains in the vicinity of interfaces. The solute solubility in nanocrystalline alloys may not only be affected by elastic strains, but also by entropy and electronic effects (58). In fact, in order to reduce the Gibbs free energy of a nanocrystalline alloy with a structure as shown in Fig. 12, the enhanced solute concentration extends from the boundary cores into the adjacent crystal lattices so that a structure of higher entropy (and thus lower Gibbs free energy) results. If two crystals with different chemical composition form an interphase boundary, their different Fermi energies are equalized at the interface similar to a p/n junction and hence a region of locally enhanced or reduced electron density in the vicinity of the boundary results. The width of this region depends on the material and ranges from about one lattice constant (in metals) to microns (in certain ceramics and semiconductors). The local charge variation affects the solute solubility and other properties as well. For example the type and concentration of lattices defects in ceramics close to the boundary differs from the perfect lattice. And hence not only the solute solubility but also other properties e.g. transport properties (conductivitiy, diffusion) vary. Due to the small grain size of nanocrystalline alloys, the regions of local charge enhancement or depletion represent a large volume fraction of the material and thus the properties of nanocrystalline alloys are expected to differ from those of conventional polycrystals with the same chemical composition. The alternative interpretation of the data reported (Figs. 11 and 12) in terms of Fe of Ag atoms which become trapped in Ag or Fe crystals during the evaporation and/or crystallization process was ruled out by separating the Ag and Fe vapour sources. The formation of Ag/Fe solid solutions in Ag-Fe nanocrystalline alloys seems remarkable because Fe and Ag are insoluble in the solid and liquid state. If solid solutions are formed in such an extremely insoluble system, it seems conceivable that nanocrystalline alloys may open the way to the formation of alloys in systems in which alloying has not yet been possible. Alloying of Ag-Fe by co-evaporation on cold substrates is also possible. However, it is limited to thin films.

NANOGLASSES

As was pointed out in the section dealing with preparation, a frequently used procedure to synthesize nanocrystalline solids is by consolidation of small crystals. By analogy, non-crystalline solids may be generated by consolidating nanometer-sized glassy spheres (48) - (50). The resulting solids were called "nanoglasses". The preparation follows basically the same route as was discussed for nanocrystalline alloys (Fig. 4). However, instead of evaporating a material that crystallizes upon cooling in the He-atmosphere, one now evaporates (or sputters) a material that solidifies in the form of small glassy spheres. These spheres are subsequently consolidated into a nanoglass. The atomic structure proposed (as a working hypothesis) for nanoglasses in the as-consolidated state is indicated schematically in Fig. 13.

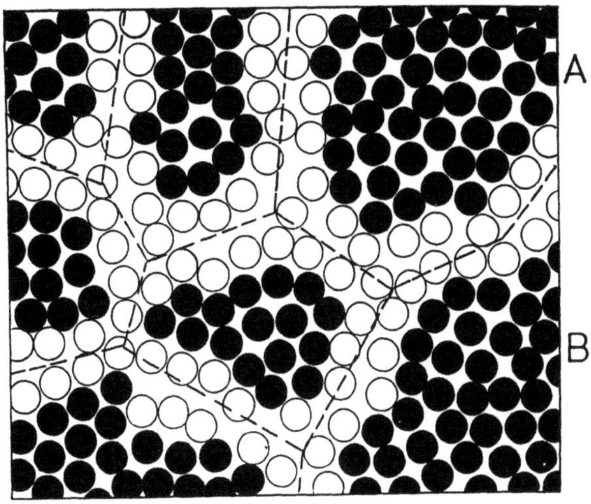

FIG. 13 Schematic cross sections through a two-dimensional nanoglass. The atoms are represented by circles. The material consist of small regions. In the central part of these regions (full circles), the interatomic spacings are similar to a bulk glass. In the interfacial areas (broken lines, open circles), a reduced density and a broad spectrum of interatomic spacings may exist.

The regions of contact between adjacent droplets differ structurally and/or chemically from the atomic structure in the center of the droplets for the following reason. The atoms at the surface of an isolated glassy droplet in vacuum form a glassy (short range ordered) atomic arrangement with the atoms in the interior so that certain interatomic spacings are preferred between the surface atoms and the atoms in the interior. If two (originally isolated) glassy droplets are brought into contact (e.g., in the region of contact between A and B in Fig. 13), the interatomic spacings between the surface atoms, belonging originally to the droplets A and B, deviate from the interatomic spacings between the atoms in the interior of A and B. Although some relaxational motion of the atoms in the region of contact may occur, the density and interatomic spacings in the interfacial regions are likely to differ from those of a bulk glass with the same chemical composition. In fact, by analogy to the structure of nanocrystalline materials, the atomic structure of the interfacial regions between the droplets was proposed to exhibit a broad distribution of interatomic spacings if one considers the average over all interfacial regions of a nanoglassy specimen (Fig. 13). This speculation agrees with recent measurements by Mössbauer spectroscopy (48). The observed Mössbauer spectra of a $Pd_{70}Si_{27}Fe_3$ nanoglass and the corresponding quadrupole splitting distribution are compared in Fig. 14b with the Mössbauer spectrum of a chemically identical glass prepared by melt spinning (Fig. 14a).

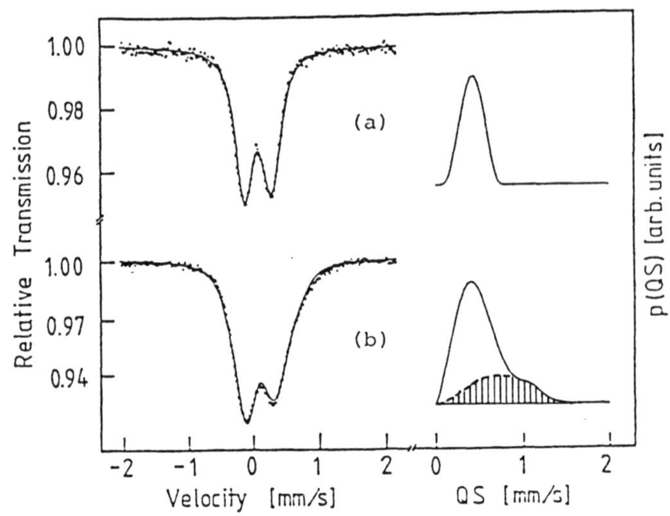

FIG. 14a Mössbauer spectrum and distribution, p(QS), of the quadrupole splitting (QS) of a melt-spun $Pd_{70}Si_{27}Fe_3$ metallic glass (48).

FIG. 14b Mössbauer spectrum and distribution, p(QS), of the quadrupole splitting (QS) of a nanoglass with the same chemical composition as in Fig. 14a. The diameter of the droplets used to generate the glass was about 4 nm. The quadrupole splitting consists of the following two components. A narrow one (solid line) similar to the quadrupole splitting of Fig. 14a and a broad component (broken line, hatched area) which is observed in the nanoglass only (48).

Obviously both materials exhibit different spectra and different distributions of the quadrupole splitting. In fact, the quadrupole splitting of the nanoglass may be considered as a superposition of the following two components. One (narrow) component (solid line in Figs. 14a and 14b) which corresponds to a melt-spun glass and a second (broad) component (broken line in Fig. 14b) which is observed in the nanoglass only. This result may be understood as follows. The quadrupole splitting depends on the nearest neighbour configuration around the iron atoms. Hence, the nanoglass consists structurally of two components. One component in which the atoms are arranged similarly to a melt-spun Pd-Si-glass. In Fig. 13, this component is formed by the regions consisting of the atoms represented by the full circles. In the second component, the distribution of the interatomic configurations between nearest neighbours is much wider than in the melt-spun glass as suggested by the broad distributions of the quadrupole splitting. This broad distribution was suggested to originate from the interdroplet regions (as was indicated in Fig. 13 by the open circles). Hence the structure of nanoglasses may be manipulated in a controlled way as follows. The volume fraction of the interdroplet regions deceases if the droplets become larger. As a consequence, by varying the diameter of the consolidated glassy droplets, the relative volume fractions of the glassy and the broad second component can be manipulated. In other words, by varying the size of the glassy droplets, one may be able to vary (at constant chemical composition) the atomic structure of nanoglasses continuously between a melt-spun glass and a glassy structure with a broad distribution of interatomic spacings. As a matter of fact, this speculations agrees with the experimental observation reported for Pd-Si-nanoglasses with different droplet diameters (48). If the chemical composition of the surface regions of an isolated glass droplet differs from the interior of the droplet (e.g., due to surface segregation effects), the interfacial regions deviate chemically as well as structurally from the center. Although the atomic structure of nanoglasses has been discussed in similar terms as was done for nanocrystalline materials, there seems to be the following difference between both materials. Nanocrystalline materials preserve the low energy structure in the interior of the crystallites at the expense of the boundary regions where the misfit between adjacent crystal lattices with different orientations is concentrated so that a structure far away from equilibrium (with a large free volume and a broad distribution of interatomic spacings) is formed. In the case of nanoglasses, the glassy interior of the droplets (full circles in Fig. 13) is not a structure of lowest free energy. Hence, the atomic misfit of the boundary region between adjacent glassy droplets (e.g., between A and B in Fig. 13) in the as-consolidated state may not remain in this position if the nanoglass is held for some time at ambient temperature. In fact, the diffusional rearrangement of individual atoms allows the system to redistribute the misfit and free volume from the interfacial regions in the as-consolidated morphology to other regions within the structure in order to reduce the total free energy of the systems. Hence, some time after consolidation, a nanoglass may have lost its "structural memory" to the as-consolidated morphology in the sense that the free volume distribution of the as-consolidated state is replaced by a new free volume distribution of lower free energy. In fact, such rearrangements may have been observed recently (51) by means of small angle scattering experiments on Au-Si-nanoglasses. The proposed rearrangement also agrees with the result of computer simulations of glassy structures into which "vacancies" or "dislocations" have been introduced by removing an atom or by local shear. In other words, nanoglasses may exhibit structural (microstructure and atomic structure) features which differ from the ones of nanocrystalline materials as well as from glasses prepared by conventional methods.

SUBMICRON MATERIALS

In several recent papers (52) - (56) evidence has been presented suggesting that polycrystals (called submicron-grained materials) with grain sizes between 100 and 1000 nm exhibit properties deviating from those of coarse-grained polycrystals with the same chemical composition. For example, the Curie temperature of Ni seems to decrease by 40K if the grain size was reduced to 70 nm. Similarly, the diffusivity in Ni and Al was noticed to be enhanced by a factor of about 10^3 or more for grain sizes between 70 and 300 nm. These effects were interpreted by postulating the grain boundary regions to have different properties than the interior of the crystals. In fact, experimental evidence supporting this idea comes from studies by Mössbauer spectroscopy of Fe with an average grain size of about 100 nm (56). The Mössbauer spectrum consisted of two components, one of which agreed with the spectrum of bcc α-Fe, whereas the second component exhibited a reduced hyperfine field and an enhanced isomer shift. From the intensities of both components at low temperatures it was concluded that in polycrystalline Fe the "boundary region" is about 16 nm wide. In terms of the presently existing theories of the structure of grain boundaries (57), it seems difficult to understand how a grain boundary can affect a region as wide as 16 nm. For example, measurements of the short circuit diffusion in isolated boundaries suggest a width between 0.5 to 1 nm for the region in which the diffusivity is enhanced (59). For practical applications submicron-grained materials seem to be attractive because they can be produced economically in large quantities by rapidly straining a material under pressure until a true strain of about 6 is achieved (53).

PLASTICITY OF CERAMICS AND INTERMETALLICS

Conventionally brittle ceramics have been observed to become ductile, permitting large (e.g., 100%) plastic deformations at low temperature (e.g., 293 K), if the ceramic is generated in the nanocrystalline form (43). The possibility to utilize this plasticity for net-shape forming processes has been demonstrated recently at low temperatures where little or no grain growth occurs (60). If the plastic deformation is performed at temperatures of about 50% of the absolute melting point of the ceramic (e.g. for TiO_2 deformation at about 800°C), total strains as high as 0.6 (without crack formation) were obtained in a period of about 15h (61) (Fig. 15). At 800°C, the grain size increased from about 50 nm to approximately 1 μm within 15h. During the deformation, the grains were noticed to remain equiaxed. With the exception of a few residual pores at grain triple junctions no porosity was detected. Strain rates as high as $8 \cdot 10^{-5} s^{-1}$ were observed at true stresses of 52 MPa. These rates are quite sufficient for technological applications. The increasing grain size during deformation does not seem to pose a basic problem, in utilizing the ductility technologically. Doping of nanocrystalline ceramics (e.g. TiO_2 by Y of Al) has been demonstrated (61) to reduce grain growth dramatically. Grain boundary sliding, grain rotation and grain shape accomodation by diffusional processes seem to play a crucial role in the deformation of nanocrystalline ceramics, i.e. processes which are typical for superplasticity (62). In fact, superplastic flow in ceramics has been reported previously, however, at temperatures that were considerably higher and technologically more

FIG. 15 Nanocrystalline TiO_2-sample before and after compression at 810°C for 15h with a stress of 40 MPa. The initial crystal size was about 50 nm (61).

difficult to achieve. Creep with high strains was reported in Al_2O_3 at 1750 - 1950 °C (63), Al_2O_3 doped with Cr_2O_3 and Y_2O_3 at 1500 °C (64) and MgO doped Al_2O_3 at 1420 °C (65). Only in ceramics with liquid phases at the interfaces was superplastic flow found at lower temperatures (64). Clearly, if the ductility of nanocrystalline ceramics is based on boundary sliding, grain rotation and accomodation, brittle intermetallic compounds may also be expected to exhibit ductility in the nanocrystalline form. If this would be so, it may open the way to utilize intermetallic compounds technologically. Their technological application has been hampered so far by brittleness. Ductile nanocrystalline ceramics may be used as a new type of ceramic material in their own right. However, the plasticity may also be utilized for processing by extrusion or rolling. Subsequently, the material may be fully or partially converted back into a conventional ceramic. Partial conversion (e.g., by surface laser annealing) results in a material which exhibits at the surface the properties (e.g., hardness and chemical resitivity) of a conventional ceramic. However the interior of such a material would still be ductile.

The nanocrystalline approach deviates from the conventional strategy to defeat the brittleness of ceramics by improvements in processing and/or composition. Macrodefekt-free cement, chemically modified silicon nitrides and transformation toughend zirconia (66), (67), represent succesful improvements based on these two strategies. The last of these was originally proposes (68) with the provocative title "Ceramic steel?", because doped zirconia, like high-tensile steel, depends on a partial phase transformation to temper strength with shock resistance: an advancing crack triggers a local transformation which hinders further propagation of the crack. None of these improvements has gone far enough. The technological, economical and organizational problems, that dimmed the promise of silicon nitride, which has now been the ceramic of the decade for three decades are described in ref. 68. Obviously, there is still a scope for new approaches.

MAGNETIC PROPERTIES

Measurements of the saturation magnetization (M_s) of nanocrystalline iron (6nm crystal size) revealed a reduction of M_s by up to 40 % relativ to the saturation magnetization of bulk α -iron (41). For comparison, in metallic iron glasses (extrapolated to pure iron), M_s is reduced by about 2 % relativ to α -Fe. The remarkable reduction of M_s by up to 40 % was attributed to the deviation of the interatomic spacings in the interfacial regions as compared to the α-iron structure. This interpretation is supported by Mössbauer spectroscopy measurements (Fig. 7) of the quadrupole splitting of nanocrystalline, antiferromagnetic FeF_2 (34) indicating a wide distribution of different atomic configuration in the boundaries. If this so, then one might speculate that other materials that are not ferromagnetic in the crystalline state, may become ferromagnetic in the nanocrystalline form. Ferromagnetism in the boundary regions may not only result from local structural variations but also from an enhanced solute concentration. In fact, the solute solubility of various nanocrystalline systems such as H dissolved in Pd, Bi in Cu, Bi in Fe, Ag in Fe has been shown (44), (69), (70) to be enhanced by one or several orders of magnitude by incorporation the solute atoms in the cores of the boundaries or in the adjacent regions. Hence in nanocrystalline alloys, the boundary regions differ structurally as well as chemically from the crystalline state which may result in ferromagnetism in the boundary regions of nanocrystalline alloys, even if the crystalline state is not ferromagnetic. Recently attractive magnetic properties have been reported for nanocrystalline Fe-alloys (75), (81) generated by partially crystallizing an Fe-Si-B-Cu glass. The resulting material exhibited core losses as low as 200 mW/cm^3 and effective permeabilities of 10^5 or more. Some observations (81) suggest that the ferromagnetic properties of the boundary regions play a critical role for these remarkable properties. Another feature which may be of relevance for future application of nanocrystalline substances as storage materials, is their magnetic microstructure. The magnetic microstructure of nanocrystalline Fe was found to differ from the one of crystalline and amorphous Fe and Fe alloys (71). The magnetic microstructure of crystalline and amorphous Fe consists of ferromagnetic domains separated by domain walls. In nanocrystalline Fe no domain structure was revealed by Bitter technique, Kerr microscopy and/or Lorentz electron microscopy (71). In fact, every crystallite of a nanocrystalline Fe specimen seems to be a single ferromagnetic domain. The orientation of the magnetization of neighbouring crystallites is controlled by two factors: the crystal anisotropy (which tries to align the magnetization of every crystallite in one of the easy directions), and the magnetic interaction between neighbouring crystallites (which tries to aling the magnetization of adjacent crystals into a common direction). If the crystallites are crystallographically oriented at random, a magnetic microstructure results in which the magnetization of the crystallites is correlated over to a few crystal diameters (e.g. a few nanometers). Long range correlations resulting in domain formation are prevented by the random crystal orientation and the crystal anisotropy. The local magnetic structure can be changed by local magnetization on a nanometer scale suggesting the application as storage materials.

PERSPECTIVES

One of the basic problems that has to be solved in order to permit the technological application of materials with nanometer-sized grains is the availability of economical methods to produce large quantities of such substances. The present situation seems to be somewhat analogous to metallic glasses. Shortly after the discovery of metallic glasses, several studies revealed technoligically attractive properties of these materials. Nevertheless it was the development of the melt spinning method (72), (73), which opened the way to the technological applications because it allowed to produce economically large quantaties. Similarly, the present methods to generate nanocrystalline materials seem too expensive for large scale production. In fact, there are many chemical methods known for synthesizing economically large quantities of small crystals. Several promising attempts have been reported at a recent conference (79) to utilize some of these method for generating economically large quantaties of nanocrystalline materials. A second area which seems to deserve attention is the thermal stability of nanometer-sized microstructures. The high energy stored in these structures provides driving forces for recovery and recrystallization processes. However nanometer-sized microstructures can be thermally stabilized in the form of multiphase materials consisting of crystals with different chemical compositions. If these crystals are mutually insoluble and if the various crystals are arranged in a non-perculative manner, thermally stable, nanometer-sized microstructures may be generated. So far, nanometer-sized microstructures formed by incorporating a high density of grain boundaries have been studied predominantly. Obviously, other nanometer-sized microstructures generated by incorporating a high density of other defects such as d i s l o c a t i o n s , s t a c k i n g f a u l t s etc. are likely to exhibit interesting properties as well.

Acknowledgement

This work has been supported by the BMFT under contract No. 523 - 400303 M 23 4 and the Deutsche Forschungsgemeinschaft (G. W. Leibniz-Programm). The co-operation with colleagues from the Universities of München, Saarbrücken and Stuttgart as well as from DESY Hamburg and ILL Grenoble, France, is appreciated.

REFERENCES

1 H. Gleiter in: Second Riso Internat. Symposium on Metallurgy and Materials Science, N. Hansen, A. Horsewell, T. Leffers and H. Lilholt (eds.) Risø Nat. Laboratory Roskilde, Denmark, p. 15 (1981).
P. Marquardt and H. Gleiter, Verhandl. d. Deutsch. Physikal. Gesellsch. (DPG) (VI) 15, 328 (1980).
2. K.L. Merkle, J.F. Reddy, C.L. Wiley and D.J. Smith, Phys. Rev. Lett. 59, 2887 (1987).
3. E. Taglauer, physica status solidi 29, 259 (1968).
42, 2639 (1971).
4. J. Friedel, Dislocations, Pergamon Press Oxford, p. 25 (1964).
5. M. R. Fitzsimmons and S.L. Sass, Jnl. de Phys. 49, C5-71 (1987).
6. H. Gleiter and P. Marquardt, Zeitschrift für Metallkunde 75, 263 (1984).
7. R.W. Siegel and H. Hahn, Current Trends in Physics of Materials, Ed. M. Yusouff, World Scientific Publ. Co., Singapore, p. 403 (1987).
8. R. Birringer and H. Gleiter, Advances in Materials Science, Suppl. Vol. 1 Encyclopedia of Mat. Sci. and Eng., R.W. Cahn (ed.),Pergamon Press, Oxford, p.339 (1988).
9. H.E. Schaeffer, R. Würschum and R. Birringer, J. Less-Common Metals 140, 161 (1988).
10. C. Suryanarayana and F.H. Froes, Proc. Physical Chemistry of Powder Metal Production and Processing, W. Murray and D.G.C. Robertson (eds), TMS Publ. Warndale P.A., p. 269 (1989).
11. R. Birringer, Materials Sci. and Eng. A 117, 33 (1989).
12. P. Haasen, Nachr. d. Akademie der Wissenschaften, Göttingen II. Mathem. Phys.: Klasse 6, 1 (1970)
13. G. Wasserman in: Proc. 4th Int. Conference on Strength of Metals and Alloys,Vol. 3, Nancy, p. 1343 (1976).
14. G. Frommeyer and G. Wassermann, Z.f. Werkstofftechnik, 2, 136 and 154 (1976).
15. J.D. Embury in: Strengthening Methods in Crystals, A. Kelly and R.B. Nicholson (eds.), Applied Science Publ. Ltd, London, p. 331 (1971).
16. R.B. Nicholson ibid p. 535.
17. R.A. Roy and R. Roy, Mater.Res. Bull. 19, 169 (1984).

18. D. Hoffman, R. Roy and S. Komarneni, Mater.Lett., 2, 245 (1984).
19. D. Hoffman, S. Komarneni and R. Roy, J. Mater. Sci.Lett. 3, 439 (1984).
20. R. Roy, S. Komarneni and D.M. Roy in: Better Ceramics Through Chemistry, C.J. Brinker (ed.),. Elsevier North Holland, New York, p. 345 (1984).
21. R. Roy, R.A. Roy and D.M. Roy, Mater. Lett. 4, 384 (1986).
22. R. Roy, Mater. Sci. Res. 21, 25 (1987).
23. R.P. Feynman, There's plenty of room at the bottom, in: H.D. Gilbert (ed.), Miniaturization, Reinhold, New York, p. 282 (1961).
24. L. Esaki und R. Tsu, IBM Research Note RC-2418 (1969).
25. L. Esaki und R. Tsu, IBM J. Res. Develop. 14, 61 (1970).
26. E.E. Mendez and K.v. Klitzing, Physics and Applications of Quantum Wells and Superlattices , Plenum Press, N.Y. (1988).
27. R.S. Lewis, T. Ming, J.F. Wacker, E. Anders and E. Steel, Nature 326, 160 (1987).
28. J.F. Kerridge and M.S. Matthews, Meteorites and the Early Solar System, Univ. Arizona Press (1988).
29. E. Jorra, H. Franz, J. Preisl, G. Wallner, W. Petry, R. Birringer, H. Gleiter and T. Haubold, Philos.Mag. B 60, 159 (1989).
30. M.R. Fitzsimmons, E. Burkel and J. Peisl, Verhandl. d. Dtsch. Physikal. Gesellsch. DPG (VI) 25, 625-DS 8.4 (1990).
31. G. Wallner, E. Jorra, H. Franz, J. Peisl, R. Birringer, H. Gleiter, T. Haubold, W. Petry, Mat. Res. Soc. Symp. Proc. 132, 149 (1989).
32. J.E. Epperson, R.W. Siegel, J.W. White, T.E. Klippert, A. Narayanasamy, J.A. Eastman and F. Trouw, Mat. Res. Soc. Symp. Proc. 132, 15 (1989).
33. T. Haubold, R. Birringer, B. Lengeler and H. Gleiter, Phys. Lett. 135, 461 (1989).
34. S. Ramasamy, J. Jiang, R. Birringer, U. Gonser and H. Gleiter, Solid State Commun. 74, 851 (1990).
35. X. Zhu, R. Birringer, U. Herr and H. Gleiter, Phys. Rev. B 35, 9085 (1987).
36. J. Jiang, S. Ramasamy, R. Birringer, U. Gonsor und H. Gleiter, to be published.
37. D. Korn, A. Morsch, R. Birringer, W. Arnold and H. Gleiter, J. de Physique 49, C5-769 (1988).
38. J. Rupp and R. Birringer, Phys. Rev. B36, 7888 (1987).
39. E. Hellstern, H.J. Fecht, Z. Fu and W.L. Johnson, J.Appl. Phys 65, 305 (1989).
40. H. Gleiter, Nanocrystalline Materials, Progress Materials Science, J. Christian, P. Haasen, T.B. Massalski (eds.), Pergamon Press, N.Y. 33, 224 (1990).
41. R. Birringer, U. Herr and H. Gleiter, Suppl. Trans. Jpn. Inst. Met. 27, 43 (1986).
42. S. Schumacher, R. Birringer, R. Strauss and H. Gleiter, Acta Metallurgica 37, 2485 (1989).
43. J. Karch, R. Birringer and H. Gleiter, Nature 330, 556 (1987).
44. U. Herr, J. Jing, U. Gonser and H. Gleiter, Solid State Commun. in press.
45. W.C. Johnson and J.I.D. Alexander, J. Appl. Phys. 59, 2735 (1986).
46. J.W. Cahn, and F.C. Larche, Acta Metallurgica 32, 1915 (1984).

47. R. Kelly and R.B. Nicholson Progress Materials Science, B. Chalmers (ed.) Pergamon Press N.Y. 10, 276 (1963).
48. J. Jing, A. Krämer, R. Birringer, H. Gleiter and U. Gonser J. of Non-Crystalline Solid 113, 167 (1989).
49. J. Weissmüller, R. Birringer and H. Gleiter, Phys. Lett. A145, 130 (1990).
50. R.S. Averback, H. Hahn, H.J. Hoefler and R.C. Logas, Appl. Phys. Lett. in press.
51. J. Weissmüller: Ph.D. Thesis Univ. of Saarbrücken 1990.
52. L.N. Paritskaya, V.I. Novikov and V.S. Kruzhanov, Soviet Powder Metallurgy and Metal Ceramics 21, 554 (1982).
53. R.Z. Valiev, R.R. Mulyukov, Kh.Ya. Mulyukov, V.I. Novikov and L.I. Trusov, Zh. tekh. Fiz. 15, 78 (1989).
54. R.Z. Valiev, Y.D. Vishnyakov, R.R. Mulyukov and G.S. Fainstein, physica status solidi (a) 117, 549 (1990).
55. R.Z. Valiev, R.Sh. Musalinov and N.K. Tsenev, physica status solidi (b) in press (1990).
56. R.Z. Valiev, R.R. Mulyukov and V.V. Ovchinnikov, Philos. Mag. Letters in press (1990).
57. H.F. Fischmeister, Materials Science, Research 21, 1 (1987), (Ceramic Microstructures 86; Role of Interfaces, J.A. Pask and G. Evans, eds.).
58. H. Gleiter, Proc. 1990 MRS Fall Metting, Boston, Symposium G, in press.
59. I. Kaur and W. Gust Handbook of Grain and Interphase Boundary Diffusion Data Vol. 1 and 2, Ziegler Press Stuttgart (1989).
60. J. Karch and R. Birringer, Ceramics International, in press.
61. H. Hahn, J. Logas, H.J. Hoefler and R.S. Averback, Mat. Res. Soc. Symp. (1990) in press.
62. J.W. Edington, K.S. Melton, and C.P. Cutler, Progress in Materials Science 21, 61 (1976).
63. A. H. Heuer, D.J. Sellers and W.H. Rhodes, J. Am. Ceram. Soc. 52, 468 (1969).
64. J.D. Fridez, C. Carry, A. Mocellin, Advances in Ceramics, Vol. 10, W.D. Kingery (ed.), Am. Ceram. Soc., Columbus, OH (1984).
65. K.R. Venkatachari and R. Raj, J. Am. Ceram. Soc. 69,135 (1986).
66. N. Claussen and A.H. Heuer, in Encyclopedia of Materials, Science and Engineering, M.B. Bever (ed.), Pergamon and MIT p. 5129 (1986).
67. F.F. Lange, J. Mat. Sci. 17, 225 (1982).
68. R.C. Garvie, R.H. Hannink and R.T. Pascoe, Nature 258, 703 (1975).
69. T. Mütschele and R. Kirchheim, Scripta Metall. 21,135 (1987).
70. H. Hahn, H.J. Hoefler and R.S. Averback, DIMETA-88, International Conference on Diffusion in Metals and Alloys, Balatonfüred, Hungary, September 5-9, 1988, to be published in Materials Science Forum.
71. W. Geibel, Diploma Thesis, Univ. of Saarbrücken 1989 to be published.
72. R. Pond and R. Maddin, Materials Sci. and Eng. 23, 87 (1969).
73. H.S. Chen and C.E. Miller Rev. Sci. Instr. 41, 1237 (1970). Materials Science Forum.
74. M.J. Weins, H. Gleiter and B. Chalmers, J. of Applied Physics 42, 2639 (1971).
75. Y. Yoshizawa, K. Yamauchi and S. Oguma, European Patent 0 271 657-A2, 22.06.1988.

76. U. Herr, H. Jing, R. Birringer, U. Gonsor and H. Gleiter, Appl. Phys.Lett. 50, 472 (1987).
77. Y. Ishida and T. Ozawa, Scripta Metall. 9, 1103 (1975).
78. D. Wolf, MRS Bulletin 15, 42 (1990).
79. B.H. Kear and R.W. Siegel (eds), Proc. Acta Metallurgica Conference on Materials with Ulrafeine Microstructures, 1-5 October 1990, Atlantic City, Acta Metall. in press.
80. S.R. Philipot, D. Wolf and S. Yip, MRS Bulletin Oct. 1990, 38.
81. G. Herzer and H. Warlimont, Proc. Acta Metallurgica Conference on Materials with Ultrafine Microstructures, 1-5 October 1990, Atlantic City, Acta Metall. in press.

Ceramics for Structural Purposes

R.J. BROOK
Max-Planck-Institut für Metallforschung
Pulvermetallurgisches Laboratorium
Stuttgart, Germany

Summary

The development of ceramic materials for high performance structural applications is reviewed. Although many attractive properties of these materials are intrinsic features of the chemical bonding, reliable performance depends on the mastery of microstructurally dependent properties and hence on the successful design and fabrication of microstructures. Progress in these two areas is reviewed and trends in current work are described. The ability of analytical and modelling methods to deal with increasing complexity and the tendency of refined processing methods to yield more controlled products has led to the point where improved interaction between modelling and experiment can be expected. The economic matching of these products to applications remains challenging particularly where materials substitution is involved.

Introduction

Ceramics have long been a favoured group of materials for structural purposes where compressive loads have been involved. The properties of hardness, stiffness, strength and corrosion resistance when coupled with a relatively low cost arising from availability and ease of manufacture have made them a natural choice in building construction over centuries. In the context of discussions of advanced structural and functional materials, however, the requirements have been very considerably widened and have taken ceramics to the limits of their structural stability. With the wish to extend technologies into new ranges of temperature, stress and chemical severity, ceramics are no longer proposed for their convenience but rather because they are the only materials offering even a possibility of survival under the required conditions.

The ambition to bring ceramics into advanced structural components is already some fifty years old dating from attempts to substitute alumina as blade materials in gas turbines with a view to the raising of engine temperature. The long history of the ensuing developments has resulted in part from the scale of the technological jump from traditional ceramic components (bricks, pottery, pipework) to components at the limit of current engineering development, in part from the need to guarantee reliability in the intended application if the engineering community is to be won over by the innovation and in part from the need to compete on a competitive economic footing with other materials already at an advanced stage in their evolution. A major factor has, however, been the peculiarly complex nature of the link between processing, microstructure and properties which, in ceramics as in other materials, underlies performance enhancement.

In the following sections, the three aspects of properties, microstructure and processing will be considered in terms of certain conceptual polarities which can be seen to exist, in terms of certain features which can be recognised in the work which has taken place and in terms of certain trends which are seen as promising for future developments. The extent of progress and understanding has been dramatic but the penalties of the advances in cost terms are not always surmountable. The precise patterns by which ceramics will come to the large scale, high technology markets which their properties promise are still matters of active debate.

Properties

A number of virtues that have attracted attention to ceramics in the first place and which allow favourable comparison with the metals can be termed intrinsic in that they are to a large extent determined by the strength and nature of the interatomic bonding. Thus the high stiffness, the relatively low crystal density, the high melting or decomposition temperatures and the hardness are lattice properties which are conferred without detriment to the finished component. They are matters of record and are not the subject of major development programmes. For oxide materials the same can be said in respect to atmospheric corrosion resistance.

It is in respect to the extrinsic properties, namely those that are dramatically influenced by the microstructure of the component and which are therefore determined by our ability to manufacture microstructures according to requirements, that the problems have arisen and that have therefore been at the basis of most research activity. The dependence of the strength, through the relation $\sigma_F \sim K/\sqrt{c}$, where K is the toughness and c the size of the flaw from which brittle fracture originates places immediate emphasis on processing skills through the need for flaw avoidance. The dependence of the diffusional creep rate through the relation $\dot{\epsilon} \sim G^{-n}$ where G is the grain size and n a mechanism-dependent constant emphasizes the impact of grain size; the presence of continuous viscous films at grain boundaries has an even more dramatic influence on creep. The dependence of toughness on the presence of microstructural elements able to induce hysteresis in the stress/strain cycle (1) indicates that microstructural influence does not simply lie in the avoidance of imperfections such as pores or impurity phases but rather that the imaginative and deliberate incorporation of structural features into a microstructure can play the critical role.

The patter is therefore one where nature has provided a class of materials with attractive intrinsic properties (the range is even wider when functional ceramics are considered) but where these properties cannot be accessed without the meeting of demanding microstructural targets and the development of the associated processing methods.

In looking over the work that has been done in resolving this issue, three aspects can be highlighted. The first relates to the need to incorporate statistical procedures into property evaluations where these properties are themselves determined, as in the case of strength, by flaws having a distribution of sizes within the components. Measures are needed first to describe the anticipated spread of property values - here the Weibull modulus, m, which can be related to a distribution of the flaw size, a, of the form $g(a) \sim a^{-r}$ where $r = (m/2) + 1$, has become the accepted parameter (2) reflecting the quality of reproducibility in processing and the narrowness of the flaw distribution - and secondly to allow guaranteed performance in those applications where no failure can be accepted - as in proof testing. Although such procedures have become established and allow accurate predictions of probable strength behaviour, the very need for this concept of probability in using ceramics has proved unattractive to potential users. The assumptions that underlie the statistical analysis, for example, that of a given flaw size distribution, can readily be changed by changes in the distribution in larger samples or in samples fabricated by modified processing methods and the statistical data base must be constantly revised.

A second aspect concerns the range of properties that must be considered in assessing the suitability of a material for an application. Much of the materials development has been steered by attention to a narrow set of properties. The initial choice of materials - Al_2O_3, Si_3N_4, SiC, ZrO_2, AlN... - of course makes reference to a wider property set - thermal expansion, conductivity, high temperature stability... - which has been used in materials selection. The progress in developing the material is, however, often linked to some single parameter such as strength which can then be used as a measure for improvements in processing skills. With the progress that has been made and with closer approach to applications, it is clear that a wider and indeed full set of load-related properties must be considered.

The relatively late recognition of cyclic fatigue crack growth in ceramics is an example of this tendency. There are now a group of applications areas, notably when both structural and functional requirements must be met as in fuel cell structures or in piezoelectric actuator elements, where the component is subjected to an array of mechanical, thermal, chemical and electrical loadings. Lifetime prediction for such uses will require the widest range of properties to be taken into quantitative consideration.

A third aspect arises generally but in particular for those applications where substitution of ceramics for a preexisting component in another material is to be attempted. Here one must recognise the cost context within which the development is to take place - extremely exacting for large-scale consumer applications, more forgiving for aerospace technologies - and recognise also the limits that are thereby set on the fabrication technologies that can be contemplated. Among the more significant of the properties of a ceramic are its fabricability and, in association, its cost.

Faced with such problems it has been important to develop methodologies which can succeed in integrating the array of requirements or property data into a form that is practicable for guiding development. Two such methods are the growing dependence on finite element procedures and the growing use of a variety of mapping methods. The use of finite elements to evaluate loads and material responses in components subjected to complex mechanical and thermal environments and the linking of such analyses with statistical data for strength is an established method (3) for the optimisation of component design. The extension of such methods to comprehend a wider range of loads and materials responses including for example mixed mechanical and electrical demands and indeed the full set of requirements to which the component is subjected is to be expected. This is a flexible and potentially rapid means of macrostructural design optimisation and it may be expected to find its place routinely in programmes of materials development.

Mapping methods (4) share the advantage of phase diagrams in providing a summary of the state of knowledge of a given subject at a given state of development and of allowing, on the basis of the summary, predictions of behaviour. They are perhaps most developed in ceramics for properties such as creep or deformation during hot pressing or hipping but the range of phenomena that can be covered is potentially wide. The ability to set out on a field of experimental parameters such as applied stress and temperature the quantitative response of the material, coupled with the ability of the method to present accumulated experimental data in a framework provided by theoretical models, is a powerful response to the complexities presented by these microstructurally sensitive - materials and to the vastness of the associated literature.

Microstructures

Two contrasting paths can be recognised in the design of microstructure, namely in the choice of the groupings of microstructural elements - grains, pores, second phases, interfaces - that compromise the ceramic. The first is the path of structural simplification where the objective has been to eliminate inhomogeneities or accidents in the structure which could be responsible for variations in properties and performance. The need to avoid such developments as abnormal grain growth in Al_2O_3 or in SiC or in $BaTiO_3$ has long been recognised; the approach has now been extended to the search for narrow pore size distributions and the avoidance of density variation, in short to the ambition that each small region within the material should be a faithful and identical contribution to the total homogeneous structure. Much of current processing science is directed to this objective.

A second trend is to greater microstructural differentiation as highly specific local structures - particular demands for grain shape or for structure within interfaces, complex phase assemblies as in whisker or platelet reinforced structures - are made with the ambition of exploiting or inducing highly specific microstructural responses to applied loads. The two approaches are not incompatible in that the finished component will only be able to demonstrate the benefits of the deliberate local microstructural engineering if the local complexity is then homogeneously repeated throughout the component. An interplay between the two trends does, however, arise in that the task of homogeneous processing can be made more demanding if the local structures become ever more elaborate and focused.

Selecting again three aspects of microstructural development which have been featured in recent work, recognition should be given first to the extent and success of the quest for toughening mechanisms. The most striking advance has arisen in the zirconia toughened materials (5) where the incorporation of fine ($\sim$ 0.5μm) zirconia grains can at ambient temperature bring about dramatic strength or toughness enhancement. Depending on details of the system - inclusion grain size, presence of dopants, character of the matrix -, different mechanisms are involved in establishing the energy barrier to crack extension; a common element is the martensitic phase transformation that occurs with lattice expansion as zirconia converts to the stable monoclinic form. The transformation can be responsible for microcrack development, with energy demand for the new surfaces, or for strain fields around the irreversibly transformed particles, with energy demand for the associated lattice strains. Toughness and strength values, if independently optimised, have reached 20 MPa$\sqrt{m}$ and 2 GPa respectively which are almost tenfold improvements on conventional ceramic practice. The microstructural design that underlies these advances in terms of inclusion type, size, form and location has become highly sophisticated.

The success of this development has prompted the search for alternatives and in particular for systems operative at higher temperature. A difficulty with the transforming systems is that the use of higher transformation temperatures (the mechanisms become less effective as the equilibrium bulk transformation temperature is approached) runs the risk of destructive instability should the material be brought to room temperature. Alternative concepts have involved the incorporation of metal inclusions (6) to take advantage of the metallic plasticity but the penalties of higher density and corrosion susceptibility have been recognised. The general feature, however, is that local microstructural engineering has proved a most fruitful avenue of ceramics development and the trend can be expected to continue.

A second aspect which is in part related to the first lies in the attention now being given to interface character. The role of thin films of residual amorphous material between grains of Si_3N_4 in determining the creep behaviour has long been known and a continuing issue is the extent to which such films can be crystallised or retracted from the boundaries to form concentrations at grain corners. The use of more elaborate microstructures, however, has placed growing emphasis on the relative strength of interfaces in determining the path that a growing crack will take through the microstructure. Value has been seen for example in increasing the aspect ratio of grains in Si_3N_4 because the nature of the boundary phase predisposes to intergranular fracture and the crack must follow a tortuous path around the elongated grains. Deliberate engineering of the interface strength in fibre and whisker reinforced systems by way of the use of coatings has been undertaken to achieve the same result (7). In a different example, the ability to exploit the plasticity of metal inclusions requires a critical degree of cohesion between inclusion and matrix and this again must be regulated by attention to interface character.

A third feature in microstructural studies has been the growing attention to the kinetics of the elemental steps in microstructural development. Global measurements of grain growth or densification are of practical relevance but it has become clear that the quantitative linking of processing method and microstructural result will require more detailed understanding of the contributing processes. It is difficult to predict the possibilities for grain shape modification in Si_3N_4 for example if the kinetic factors underlying grain development and its dependence on phase type, matrix properties and temperature are relatively unknown. Quantitative evaluation of the ability of additives to influence grain boundary mobility is similarly an essential element in the deliberate realisation of a desired microstructure. The role of segregation and of such influences as size difference, charge difference and chemical character on the extent of such segregation is now being systematically studied (8). This understanding of microstructural dynamics will be a deciding factor in determining the extent to which deliberate, predictive assembly of microstructures can be attempted.

In supporting the design and understanding of microstructures, two techniques have won a particular influence. The most important is certainly that of transmission electron microscopy: the ability to observe physical and chemical events with high precision was one of the foundations on which the understanding of the ZrO_2-toughened systems was built; the detailed examination of grain boundary structures in Si_3N_4 and related ceramics promises similarly to provide a comprehensive and systematic guide to additive selection. The trend to microstructural design on ever finer levels - nanostructures, fibre/matrix interfaces - and the need to evaluate microstructural responses such as crack development at the point of initiation - confirm the ongoing role of the method. It is, however, in the growing ability to observe such responses in situ that one of the most dramatic contributions will lie.

The second technique is that of computer simulation and modelling. The ability to build the conventional assumptions for such processes as grain growth into computer models and then to compare predictions with experimental observation has proved a valuable test of the assumptions which have not invariably proved valid. The growing powder of such modelling now allows the modelled systems to approach more closely those that occur in practice and the resonance between the two can be informative for both (9). This trend where experimental systems have become more reliable and controlled owing to the availability of cleaner systems and closer characterisation and where modelling procedures have become capable of tackling more complex geometries and more elaborate process interactions has led to the point where an effective synergy can be expected. The consequences in terms of accelerated understanding make this a most promising research sector.

<u>Processing</u>

The spotlight thrown onto processing by the flaw sensitivity of many properties and by the growing wish-list of those concerned with microstructural development has had two consequences. In the first place there has been a pragmatic response as people have tried to refine existing processing methods to the point where reliable results can be expected. The use of clean rooms to avoid the less frequent but seriously damaging occurrence of foreign particulates in the material, the relatively empirical optimisation of drying, granulation and forming procedures to avoid the frequent and still damaging occurrence of agglomerate residues in the material, these and other process refinements can be pursued with some sensitive parameter such as final component strength as the criterion for progress. This production engineering of ceramics will be a vital and necessary part in translating laboratory promise into commercial reality; it is complex and expensive and misguidedly open to charges of trial and error optimisation but its contribution to progress will not be the least significant.

A second consequence is that it has become important to evaluate any fundamental limits that may exist to the process mechanisms that are exploited in ceramics manufacture. A topical example is that of the densification of composites where it is known that the incorporation of film or whisker inclusions into a matrix can severely limit its ability to sinter. It is known that such inclusions induce stresses in the system with a hydrostatically compressive contribution in the inclusion as the matrix seeks to densify around it and with a compensating hydrostatic tension in the matrix. Although such stresses are now recognised to be small, the damage during processing can be severe. It is further known that the damage is limited to matrices of a polycrystalline nature as opposed to glass. Attempts to analyse such problems, as in the present instance where the modelling of structural development in crystal particle chains subjected to tensile loading has shown the tendency for chain division and hence for structural damage (10), are important in establishing the limits of heterogeneity that can be tolerated in a microstructure that is to be processed in a given way say by pressureless sintering. In a related example, it is important to assess the extent to which drying procedures (11) necessarily lead to agglomeration and heterogeneity; the need for colloidal processing as opposed to die pressing will depend upon such considerations.

In reviewing processing developments, three specific contributions should be noted. The first relates to the dramatic improvements in ceramic manufacture that have resulted from the availability of chemically prepared powders (12). A major impetus in this direction was the work of the MIT group in advocating monodisperse powders which would then in fact resemble the powder qualities treated in the earlier sintering modelling. This requirement for a single size has been relaxed in the interests of reacting a greater level of homogeneity in the powder assemblies (avoidance of regions of local ordering), but the target of fine, equiaxed, narrow size distribution has now been reached and, as in ZrO_2, converted to commercial practice. The same trend as earlier identified for microstructures namely that of removing accidental complexity and building in deliberate purposeful complexity can now be seen in the case of powders. The use of coatings to achieve target rheologies or to yield an optimum distribution of sintering additives is one example. The use of sub-assemblies of the target microstructure as in the use of whiskers coated with a soft precursor for the matrix powder to insure minimum damage in the pre-sintered component (13) is an indication of the progress that can be made in this direction.

A second aspect is the recognition of the chemical variable as offering perhaps the greatest potential for process improvement. The use of densification aids which as in the sialons and Si_3N_4 can be treated to yield specific boundary phase structures on the basis of understanding of the relevant phase equilibria or which, as in AlN, can be removed subsequent to densification by annealing treatments (14) are examples of such chemical selection. Grain growth suppressants in oxide systems offer further examples. With the understanding of the roles of particle size, size distribution, temperature, firing time and applied pressure (as in hot pressing), the chemical variable and its control both through additives and through atmosphere regulation become the main factors where substantial process improvements can be expected.

A third factor and one related to the pragmatism of process refinement noted earlier is that difficulties with densification as in the instance of the composite systems can often be circumvented by increasing the driving force as in hot isostatic pressing. The promise of this technique for the fabrication of complex microstructures and its ability to produce desired macrostructures between phases which would become too reactive at conventional sintering temperatures ensure its continuing importance. The fact that it lends itself to process modelling in the form of hipping maps and to finite element prediction of final component shape is also a benefit. The extent to which hipping will be needed to counter the more persistent deficiencies of classical process methods is still very much a matter for debate.

The demands that have been placed on processing and the increasing recognition of the subtleties of the mechanisms that are involved has led to two trends that can be identified in recent work. The first is to look for dramatic alternatives to the conventional methods and to explore the various powder-free routes to ceramic fabrication. The direct conversion of defect-free liquids to ceramic components as in sol-gel processing (15) has long been proposed. The present position is that the great shrinkages associated with the change to a dense product, illustrated by the development of cracking during drying of the gels, together with the necessary occurrence of heterogeneity associated with the nucleation and growth of new phases during crystallisation, make the process difficult for bulk ceramic manufacture. Its value seems more to lie in the preparation of low dimensional structures - coatings, films - or in the fabrication of totally novel systems as with aerogels.

Similar remarks apply to the use of organic polymer precursors (16) for the preparation of non-oxide ceramics such as SiC and Si_3N_4. The shrinkages arise from the decomposition of the polymer during pyrolysis with a similar tendency for the formation of flaws. The clear research need is for better understanding of the intermediate steps between polymer and product and of the interplay between the molecular structure in the one and the microstructure in the other. As with sol-gel methods, the promise of this route lies perhaps less in the direct preparation of ceramic products and more in the development of hybrid processes where the precursor can be used in conjunction with powders or dense structural elements in providing a better control of product microstructure.

A second trend is the increased recognition given to cost factors and to the development of processes which are designed to yield great cost benefits even if the product is not fully state of the art in respect to a property such as strength. Two methods capable in principle of yielding near net shape structures, i.e. with negligible dimensional change during consolidation, should be mentioned. The first, that of reaction bonded aluminium oxide (17) uses mixtures of alumina and aluminium metal in the green structure, the oxidation and expansion of the aluminium offsetting the shrinkage as the material sinters. A second proposal combines organic polymer precursors with metal inclusions, the reaction of carbide formation between the metal and organic species on decomposition of the polymer again offsetting the shrinkage that would otherwise occur (18). These processes are attractive not least because of the recognition they give to cost factors from the outset; they point the way to one of the surest methods of bringing an innovation to potential markets and offer an example which will no doubt be more frequently followed in future developments.

References

1. Evans, A.G.: Perspective on the development of high-toughness ceramics. J.Am.Ceram.Soc. 73 (1990) 187-206.

2. Danzer, R.: Performance prediction for ceramics. In Concise Encyclopaedia of Advanced Ceramic Materials, R.J. Brook (ed.) Oxford, Pergamon Press (1991).

3. Thümmler, F.: Engineering ceramics. J.Eur.Ceram.Soc. 6 (1990) 139-151.

4. Ashby, M.F.: On the engineering properties of materials. Acta Metall. 37 (1989) 1273-1293.

5. Green, D.J.; Hannink, R.H.J.; Swain, M.V.: Transformation toughening of ceramics. Florida, CRC Press (1989).

6. Flinn, B.; Rühle, M; Evans, A.G.: Toughening in composites of Al_2O_3 reinforced with Al. Acta Metall. 37 (1989) 3001-2.

7. Cao, H.C.; Bischoff, E.; Sbaizero, O.; Rühle, M.; Evans, A.G.; Marshall, D.B.; Brennan, J.J.: Effect of interfaces on the properties of fibre-reinforced ceramics. J.Am.Ceram.Soc. 73 (1990) 1691-99.

8. Hwang, S.-L.; Chen, I.-W.: Grain size control of tetragonal zirconia polycrystals using the space-charge concept. J.Am.Ceram.Soc. 73 (1990) 3269-77.

9. Hassold, G.N.; Chen, I.-W.; Srolovitz, D.J.: Computer simulation of final stage sintering. J.Am.Ceram.Soc. 73 (1990) 2857-64.

10. Cannon, R.W.; Carter, W.C.: Interplay of sintering microstructures, driving forces, and mass transport mechanisms. J.Am.Ceram.Soc. 72 (1989) 1550-55.

11. Scherer, G.W.: Theory of drying. J.Am.Ceram.Soc. 73 (1990) 3-14.

12. Segal, D.: Chemical synthesis of advanced ceramic materials. Cambridge, Cambridge University Press (1989).

13. Kapolnek, D.; De Jonghe, L.C.: Particulate composites from coated powders. J.Eur.Ceram.Soc. 7 (1991).

14. Udagawa, E.; Makihara, H.; Kamehara, N.; Niwa, K.: Influence of firing-gas pressure on the microstructure and thermal conductivity of AlN ceramics. J.Mat.Sci.Lett. 9 (1990) 116.

15. Klein, L.C. (ed.): Sol-gel technology for thin films, fibers, preforms, electronics and specialty shapes. New Jersey, Noyes Publications (1988).

16. Riedel, R.: Metallorganic compounds as ceramic precursor. In Concise Encyclopaedia of Advanced Ceramic Materials, R.J. Brook (ed.). Oxford, Pergamon Press (1991) 299.

17. Claussen, N.; Le, T.-Y.; Wu, S.-X.: Low shrinkage, reaction-bonded alumina. J.Eur.Ceram.Soc. 5 (1989) 29-35.

18. Seibold, M.; Greil, P.: AFCOP - ein neues Verfahren zur Herstellung keramischer Verbundwerkstoffe

Thermostructural Ceramic Matrix Composites: An Overview

R. NASLAIN
Laboratory for Thermostructural Composites
UMR-47 CNRS-SEP-UB1, University of Bordeaux
Europarc, 1, Av. Léonard de Vinci, F-33600 Pessac

ABSTRACT

Ceramic matrix composites (CMCs) consist of a ceramic matrix reinforced with ceramic fibers. They have been designed to be used in severe environments. With respect to monolithic ceramics, CMCs are characterized by a non-linear stress/strain mechanical behavior, a high resistance to crack propagation, a non-catastrophic failure and thus an improved reliability. CMCs are processed according to either a gas phase route (the matrix being deposited chemically from a gaseous precursor in the pores of a fiber preform) or a liquid phase route (by hot-pressing fiber tapes impregnated with a liquid precursor of the matrix). The most important CMCs are those made from SiC-based fibers embedded in glass-ceramic or SiC matrices. CMCs are inverse composites with the result that under loading the matrix fails first. They exhibit a tough behavior only when the fibers are weakly bonded to the matrix : the fiber/matrix interfaces arresting or deflecting the matrix microcracks preventing thus the early failure of the fibers and a catastrophic propagation of a macrocrack. The control of the fiber/matrix bonding is achieved through the use of an interphase material (e.g. a layer of carbon or BN) and a proper choice of the respective thermal expansion coefficients. Since long exposures to oxidizing atmospheres at high temperatures may have a detrimental effect (e.g. a tough/brittle transition due to the oxidation of the interphase), most CMCs receive a protective surface treatment, however, some of them exhibit a self-healing character.

INTRODUCTION

The main drawbacks of monolithic ceramics are their **sensitivity to defects** (even of very small size), their **low toughness** and **lack of reliability**. Conversely, ceramics have some unique properties which are attractive from an engineering point of view : (i) a high refractoriness, with melting points often higher than 2000°C, (ii) a high stiffness, with Young moduli often higher than 400 GPa, (iii) they are hard materials, (iv) most of them have a low density, ranging from 2.5 to 3.2 $g.cm^{-3}$ for covalent ceramics (whereas those of superalloys are of the order of

8-9 g.cm^{-3}), (v) they exhibit a wide range of thermal properties and finally (vi) some of them are very resistant to oxidation (i.e. oxide ceramics as well as silicon-containing ceramics such as SiC or Si_3N_4). A tremendous effort has been done during the last two decades in order to improve the toughness and reliability of monolithic ceramics through the use of : (i) **improved processing techniques** with a view to reduce the size of the defects and to increase the composition and grain size homogeneity, and (ii) **new materials concepts** such as phase transformation toughening or in an other field, superplasticity [1, 2]. However, the use of monolithic ceramics for the elaboration of large static structures or high speed rotating parts, heavily loaded and intended to be used at high temperatures, has been very limited up to now.

Twenty years ago, a new concept, that of **ceramics reinforced with ceramic fibers**, was at the origin of a dramatic break-through in the quest of tough ceramic materials. The idea was borrowed from the neighboring fields of the fiber-reinforced polymeric or metallic materials. However, the aims were very different. In fiber-reinforced plastics (FRP) and metal matrix composites (MMC), the objective was mainly to increase : (i) the stiffness and failure stress of the polymeric or metallic matrices at room temperature or/and at high temperatures and (ii) their resistance to fatigue. Those composites, in which the failure strain of the matrix ε_m^R is usually much higher than that of the fiber ε_f^R (typically : 1%) and the stiffness of the matrix E_m lower than that of the fiber E_f, are referred to as **regular composites** : as long as the fiber volume fraction is high enough, the failure of both FRP and MMC is controlled by that of the fibers.

In **ceramic matrix composites**, the situation is reverse inasmuch as ε_m^R (typically 0.1% or less for a ceramic material) is lower than ε_f^R. Moreover, depending on the respective nature of the matrix and the fibers : $E_f > E_m$ (e.g. in SiC/glass-ceramic composites) or $E_f < E_m$ (e.g. in SiC/SiC composites)(*). Thus, in CMCs, referred to as an **inverse composites**, the **matrix always fails first** (as long as the condition $\varepsilon_m^R < \varepsilon_f^R$ is fulfilled) [3]. The result, in terms of reinforcing effect and, more interestingly of toughness improvement, depends dramatically on the strength of the **fiber-matrix coupling**. When the fibers are **too strongly bonded** to the matrix (owing to high thermomechanical residual stress resulting from a thermal expansion mismatch or/and strong chemical bonding), the microcracks arising in the matrix when it fails yield stress-concentrations and notch effects at

(*) by convention, the nature of the reinforcement will be always written first, e.g. SiC/glass-ceramic is a composite consisting of SiC fibers embedded in a glass-ceramic matrix the fibers surface with a premature failure of the fibers. Under such conditions : (i) failure occurs at a low strain and the composite remains brittle, (ii) the reinforcing capability of the fibers is not fully used and (iii) toughness and reliability are not improved. It should be emphasized that this situation has remained for a long time that actually observed, owing to the formation of strong chemical bonds between fibers and matrix during processing (which always involved a high temperature step). When conversely, the fibers are only **weakly bonded** to the fibers, the matrix microcracks are either stopped or deflected at the fiber/matrix interface, owing to the ability of that interface to be easily debonded, with the following consequences : (i) the matrix failure no longer results in the premature failure of the fibers, (ii) the matrix is progressively damaged up to a saturation state, (iii) the fibers can be further loaded up to their intrinsic failure stress (the composite failure now takes place at ε_f^R (with $\varepsilon_f^R > \varepsilon_m^R$)) and (iv) the work of fracture of the composite as well as its toughness are dramatically improved owing to the **microdamaging mechanisms** (mainly : matrix microcracking, fiber/matrix debonding, friction of the debonded fiber against the matrix and pull out). An example of these two opposite mechanical behaviors is schematically shown in fig. 1.

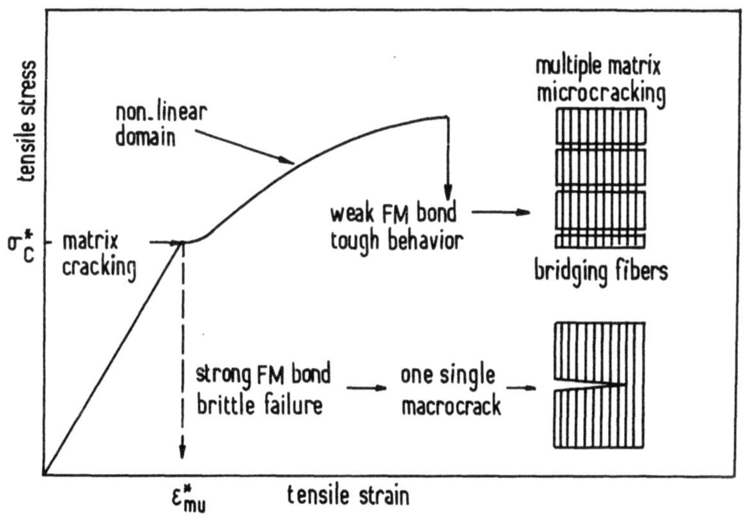

Fig. 1 : Mechanical behavior in tension of 1D-CMCs reinforced with continuous fibers in which the fiber/matrix bonding is weak or strong

The extension of the composite concept to ceramics, implies that some important features should be taken into account, namely : (i) the fibers should be stable at high temperatures, (ii) their reinforcing capabality should not be lowered during processing, (iii) the fiber/matrix coupling has to be controlled with care during processing, usually through the growth of a third component, the **interphase**, between the matrix and the fibers, (iv) the mechanical behavior is usually **non-linear** as the result of the damaging mechanisms and its characterization might need new concepts and (v) regarding the inhomogeneity of CMC, the effect of **environmental parameters** on the mechanical behavior appears to be of prime importance.

The aim of the present contribution is to address the specific requirements and features of CMCs with a view to understand how these materials have to be elaborated and how they behave particularly in extreme environments.

2- FIBERS FOR CMC IN EXTREME ENVIRONMENTS

Up to a recent past, the fibers which have been used in CMCs were those previously designed for the reinforcement of either polymeric or metallic matrices, i.e. mainly carbon, SiC-based and alumina fibers. As a result, they did not fulfil necessarily the requirements of the high temperature processing and use of CMCs.

Carbon fibers are often considered as the best reinforcements for high temperature composites owing to their refractoriness, microstructure stability, mechanical properties, low density, low cost and availability. Unfortunately, carbon exhibits a very poor resistance to oxidation and reacts with many ceramic matrices (e.g. most oxides, when temperature is high enough). Thus, attempts have been made to coat the carbon fibers with refractory diffusion barriers, such as SiC or B_4C (known to form protective oxide scales) by e.g. chemical vapor deposition (CVD) (fig. 2) [4].

From the early work of Yajima, various **silicon-based fibers** have been elaborated according to a general process involving : (i) the spinning (usually in the molten state) of an organosilicon polymer, e.g. a polycarbosilane (PCS), (ii) a curing step, necessary to render the green fiber infusible and generally performed by oxidation and (iii) a pyrolysis treatment in an inert atmosphere at about 1200-1300°C [5, 6].

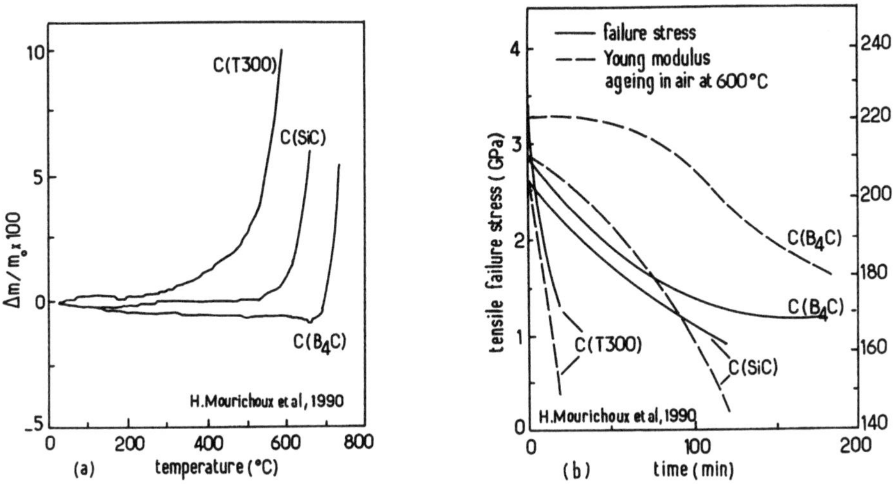

Fig. 2 : Coated-carbon (ex-PAN) fibers : (a) weight variations during ageing in air (heating rate : 300°C per hour), (b) effect of ageing on mechanical properties at 25°C [4]

The resulting **Si-C-O**[(1)] **and Si-C-Ti-O**[(2)] fibers are either amorphous or microcrystalline and they exhibit good mechanical properties up to about 1200°C, a feature which is compatible with the reinforcement of light alloys but which might be a limitation for that of ceramics [7, 8]. Such fibers have the advantage of a high resistance to oxidation (at least in the so-called passive oxidation regime) and are extensively used for the reinforcement of inorganic matrices. Unfortunately, they still contain after pyrolysis a **high percentage of oxygen** (≈ 10 wt.%) responsible for a decomposition process which takes place at high temperatures (i.e. beyond 1200°C) with : (i) an evolution of gaseous species (mainly SiO and CO) and a diameter shrinkage, (ii) the growth of β-SiC crystals and the formation of free carbon and (iii) a drop in the tensile failure stress and stiffness [9]. Adding **nitrogen** to the chemical composition of the fibers has some impedding effect on their decomposition/crystallization process but gaseous oxides are still formed [9]. Thus, the most effective improvement was achieved recently, through the use of novel curing agents avoiding any oxygen contamination (e.g. UV or electron beam curing under vacuum or inert

---

(1) Nicalon fibers from Nippon Carbon, (2) Tyranno fibers from Ube

atmosphere, chemical curing without oxygen). The resulting Si-C [10] or Si-C-N [9, 11] fibers[(1)] , which are still experimental, were reported to exhibit a higher microstructural stability when aged at high temperatures and consequently improved mechanical properties, as shown in fig. 3.

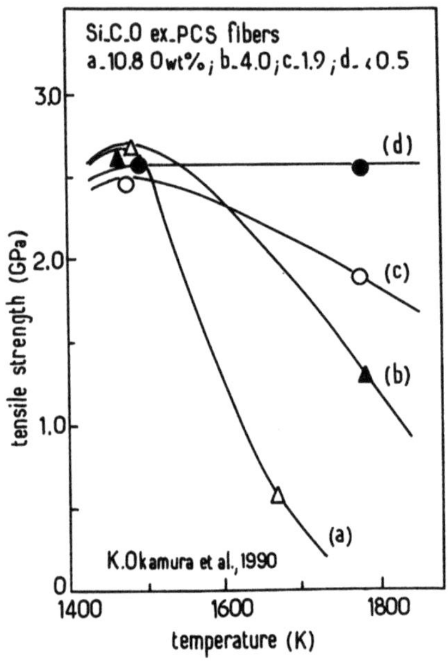

Fig. 3 : Failure strength at 25°C of Si-C-O (ex-PCS) fibers as a function of processing temperature [10]

Refractory oxide fibers, such as mullite, alumina and even stabilized zirconia fibers, have been also considered as potential reinforcements for CMC. A variety of **alumina**-based fibers are presently available including : amorphous alumina-silica fibers with a composition close to that of mullite, transition-alumina fibers (which contain less silica) [(2)] and polycrystalline corundum fibers (pure corundum or α-Al_2O_3 + ZrO_2 (Y_2O_3)[(3)] mixture). Unfortunately none of them has been specifically designed for CMCs : (i) all undergo a structural /microstruc -tural change when aged at high temperatures (i.e. recrystallization, phase transformation or grain growth) which lowers the tensile failure stress and (ii) the corundum-based fibers do not exhibit enough flexibility since their diameter is too large [12-15]. Zirconia fibers undergo also microstructural and morphological changes when aged at high temperatures [16]. Assuming that the tensile failure strength of a polycrystalline fiber is directly related to its grain size, it appears that one of the main limiting factors, in refractory oxides, is the microstructural change taking place at high temperatures. Finally, little is known on the creep resistance of these fibers which might be another limitation for their use in CMCs.

SiC-whiskers either prepared from the pyrolysis of rice hull [17] or according to the so-called VLS catalytic process [18], have also been envisaged as

---

(1) HPZ fibers from Dow Corning, (2) Saffil from ICI or Altex from Sumitomo Chemical,
(3) FP and PRD-166 fibers from du Pont

reinforcements in CMCs. Their main advantage lies in their outstanding tensile failure stress, which is at least one order of magnitude higher than those of the continuous fibers and close to that of the theoretical ≈ E/10 value (for SiC_W which have a Young modulus of the order of 500 GPa, one could expect a failure stress of about 50 GPa, i.e. ten times that of an excellent carbon fiber). Moreover, since whiskers are single crystals, they are expected to be morphologically more stable, when heated at high temperatures within a ceramic matrix, than amorphous or polycrystalline fibers. However, their size (≈ 1 μm or less in diameter and a few 10 or 100 μm in length) raises handling and health problems.

3- COMPOSITE ELABORATION

3.1- General features

Compared to the processing of FRP and MMC, that of CMC has to take into account several features which are specific to ceramic matrices and inverse composites : (i) in a CMC, the matrix has a very **high melting point** (often higher than 2000°) and almost **no plastic deformation** capability at medium temperatures and (ii) the fiber-matrix coupling should be **low enough** in order that the fiber-matrix interfaces could be easily debonded and used to arrest or deflect the matrix microcracks.

It clearly appears from (i) that brittle ceramic fibers cannot be embedded in a ceramic matrix in the molten state or by plastic deformation in the solid state since in both cases the required temperatures would be too high and thus too detrimental to the fiber strength (most ceramic fibers presently available undergo microstructural change at 1200 - 1400°C). Thus and if one excepts the particular case of glass-matrices (which are thermoplastic beyond Tg), the reinforcements in most CMCs are first infiltrated (or impregnated) with a **gaseous or liquid precursor** of the matrix which is then converted chemically into the refractory matrix at moderate temperatures (e.g. 900-1200°C). Under such conditions, it is often appropriate to arrange the fibers as a **preform** using the textile (continuous fibers) or paper (short fibers) industry techniques, solving thus two important problems, i.e. the fiber orientation and the fiber volume fraction V_f in the composite. Finally, hot-pressing processes should be regarded as detrimental to the fibers (due to surface abrasion or/and fiber breaking) unless the fibers are very short (whiskers) or already embedded in a viscous phase (e.g. a glass).

The second requirement, i.e. that of a low enough **fiber-matrix coupling**, raises the question of the physical and chemical compatibility between the two components during processing. In order to avoid a strong fiber-matrix clamping effect, due to differential thermal expansion when the composite will be heated to or cooled from the processing temperature, the two components should have similar coefficients of thermal expansion (CTE), an important feature which has to be taken into account in the choice of the fiber-matrix couple. During processing, a chemical bonding is usually formed between the fiber surface and the matrix. When it is expected that this bonding will be strong, a thin layer (a few 10 or 100 nm) of a third material, the **interphase**, has to be inserted between the matrix and the fiber. Two interphase materials i.e. **carbon and BN**, have been successfully tested [19-24]. Both can be deposited on the fiber surface, prior to the matrix infiltration, e.g. according to the following equations :

$$CH_{4\,(g)} \rightarrow C_{(s)} + 2H_{2\,(g)} \quad (1)$$

$$BX_{3\,(g)} + NH_{3\,(g)} \rightarrow BN_{(s)} + 3HX_{(g)} \text{ with } X = F, Cl \quad (2)$$

They can as well result from a chemical reaction taking place in-situ at the fiber/matrix interface during the high temperature processing step of the composite. As an example, the carbon interphase observed in SiC (ex-PCS)/glass-ceramic CMC processed by hot pressing at 1200-1300°C is the result of a fiber/matrix reaction or/and a surface decomposition of the Si-C-O ex-PCS fiber (known to yield free carbon) [9, 25-33].

It is not yet fully understood what precise requirements an interphase material has to fulfil, in terms of chemical composition, crystal structure, microstructure, thickness and mechanical properties, in order to arrest or/and deflect matrix microcracks. Both pyrocarbon and pyro-BN have **turbostratic layer structures** in which the layers are only weakly bonded to each other. Moreover, when deposited from a gaseous precursor, these materials were reported to have a tendency to grow with their layers parallel to the substrate (i.e. the fibers in the CVD/CVI process) at least in the vicinity of the substrate [21, 24]. Therefore, owing to the weak interlayer bonding in such materials, that part of the interphase located near the fiber surface could be the weakest link in the interfacial sequence. Such an assumption is supported by TEM data on microcracked SiC (ex-PCS)/BN/SiC (CVI) composites [34] : the microcracks propagate across the BN interphase (almost perfectly in mode I) and are deflected parallel to the fiber surface (i.e. in mode II) very near the fiber surface (consisting of a double carbon/silica layer thought to be the result of some surface decomposition of the ex-PCS fiber during the CVI-processing of the composite) [9, 35]. In an independent TEM study of SiC (ex-PCS)/C/SiC (CVI) composites, processed according to similar conditions, it has been reported recently that high failure strains correlated with the occurrence of carbon layers parallel to the fiber surface in the interphase [24]. There might be an optimal interphase thickness for each given fiber/matrix couple : interphase thickness as low as a few 10 nm and of the order of a few 100 nm have been reported for SiC (ex-PCS)/C/glass-ceramic [25-28] and SiC (ex-PCS)/C or BN/SiC (CVI) [24, 34] composites, respectively. Finally, the interphase has to be compatible with the atmosphere in which the CMC will be used. Both carbon and BN react with oxygen, the advantage of the latter lying in the fact that the chemical reaction starts at a higher temperature (i.e. about 900°C for BN vs 500°C for carbon) and yields a condensed oxide B_2O_3 in a rather large temperature range (whereas the oxides CO and CO_2 are gaseous).

3.2- The chemical vapor infiltration process

In the CVI-process , the ceramic matrix is deposited, within the pore network of the fiber preform maintained at a moderate temperature (e.g. 800-1100°C for most common ceramics), as a result of a chemical reaction involving gaseous species. The process can be used to infiltrate non-oxide or oxide ceramics according to the overall equations (1) to (8) :

$$CH_3SiCl_{3\,(g)} \xrightarrow{H_2} SiC_{(s)} + 3HCl_{(g)} \quad (3)$$

$$4BCl_{3\,(g)} + CH_{4\,(g)} + 4H_{2\,(g)} \rightarrow B_4C_{(s)} + 12\,HCl_{(g)} \quad (4)$$

$$TiCl_{4\,(g)} + CH_{4\,(g)} \xrightarrow{H_2} TiC_{(s)} + 4HCl_{(g)} \quad (5)$$

$$3SiCl_{4\,(g)} + 4NH_{3\,(g)} \rightarrow Si_3N_{4\,(s)} + 12HCl_{(g)} \quad (6)$$

$$2AlCl_{3\,(g)} + 3CO_{2\,(g)} + 3H_{2\,(g)} \rightarrow Al_2O_{3\,(s)} + 3CO_{(g)} + 6HCl_{(g)} \quad (7)$$

$$ZrCl_{4\,(g)} + 2CO_{2\,(g)} + 2H_{2\,(g)} \rightarrow ZrO_{2\,(s)} + 2CO_{(g)} + 4HCl_{(g)} \quad (8)$$

In CVI, two important mechanisms which can be each rate-limiting are involved : (i) the **chemical reaction** which takes place both on the external surface of the fiber preform and on the pore walls and (ii) the **mass transfers** of the gaseous reactants and products along the pores and across a boundary layer surrounding the preform (fig. 4).

In the **isothermal/isobaric** process (referred to as ICVI), which is presently used to produce CMCs, the preforms are set inside a hot-wall isothermal infiltration chamber swept by the reactant flow. Under such conditions, mass transfers across the boundary layer and along the pores occur mainly by **diffusion** (Fick or/and Knudsen diffusions). In order : (i) to maintain the pore entrances open during the whole infiltration process and (ii) to achieve a densification profile as homogeneous as possible across the preform thickness, surface reaction kinetics should be low with respect to mass transfers [36-38]. This requirement

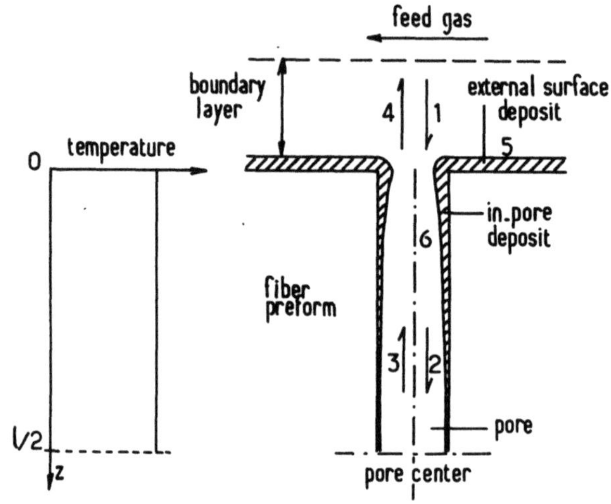

Fig. 4 : Densification of a pore by ICVI : mass transfers of reactants and products across the boundary layer (1,4) and along the pore (2, 3), deposition of the ceramic matrix on the preform surface (5) and pore wall (6)

means that ICVI has to be performed at low temperature and pressure with the result that : (i) the deposition rate is rather low with respect to those commonly reported for conventional CVD and consequently (ii) rather long durations are necessary to fill a high fraction of the initial porosity of the fiber preform. On the other hand, ICVI has a **high flexibility** related to its isothermal/isobaric character. It can be applied to the infiltration of the interphase (i.e. carbon or BN), to that of the ceramic matrix (non-oxide or oxide matrices) and finally to protective coatings deposited at the end of the infiltration procedure on the external surface of the composite. Furthermore, a large number of preforms different in size and shape can be infiltrated simultaneously with no individual tooling as long as each fiber preform has been properly consolidated. Finally, owing to the nature of in-pore mass transfer, the ICVI-process is well suited to thin fiber preforms corresponding indeed to most structural parts.

It has been suggested to apply a **temperature or/and a pressure gradients** along the fiber preform thickness in order to increase the infiltration rate or/and to improve the densification homogeneity. In the so-called **forced CVI process** (FCVI), the reactants are injected under a high pressure through one face of the preform maintained at a rather low temperature and the products and gas carrier are pumped at a low pressure through the opposite (or/and lateral) face heated at a higher temperature [39-40]. Under such conditions, mass transfers along the pores are mainly by forced convection. Moreover, the deposition rate in that part of the pores depleted in source species (i.e. far from the pore entrances) might be increased since its temperature is much higher. The main advantage of FCVI lies in the fact that the infiltration duration necessary to achieve a given state of densification is reduced by **one order of magnitude** with respect to ICVI. On the other hand, it requires one tooling per preform (to generate the temperature/pressure gradients) which can be complex if the preform has itself a complex shape and which should be maintained during the whole infiltration process. It thus appears that FCVI might be limited to preforms of simple shapes (e.g. cylinders of uniform diameters, tubes or thick plates).

Finally, **pulsed-CVI** (PCVI), in which the reactants are injected into the fiber preform and, alternately, the products are pumped out of the preform according to high frequency cycles (the period of a cycle being of the order of 1 s), has been suggested as another way to improve the mass transfer in the pore network [41, 42]. Although PCVI has been shown to be effective at the scale of a laboratory apparatus (actual volume : a few cm^3), it is doubtful that it could be easily and economically transferred to the scale of a plant (actual volume : several m^3).

CVI-processes require preforms with **specific pore networks**, i.e. consisting of pores with large enough diameters (in the range 10-500 μm) and connected to each other. For example, they are well suited to 2D-preforms (made of a stack of fiber fabrics) but poorly suited to 1D-preforms (in which all the fibers are parallel to each other). ICVI is used to produce C/SiC and SiC/SiC composites on an industrial basis both in Europe and America(1) . Its feasibility has been established at the laboratory scale for a number of other matrices (including B_4C, TiC, BN, Si_3N_4, Al_2O_3 and ZrO_2). The CVI-processed CMCs are characterized by : (i) a matrix of high quality (with a volume fraction of the order of 30% for a 2D-preform), (ii) almost undamaged fibers (since ICVI is a pressureless medium temperature process), (iii) well controlled interphases but (iv) some residual porosity (of the order of 10%).

---

(1) SEP in France and du Pont in the USA

3.3- The liquid phase processing route

CMCs can be processed according to a second route in which the matrix or more properly one of its precursors, is infiltrated into a fiber yarn or a fiber preform as **a liquid phase** (e.g. a molten element, an organometallic precursor, a sol or a slurry).

3.3.1- Impregnation of a fiber preform by a molten element

A few ceramic matrices are compounds of low melting point elements (e.g. alumina (mp = 2050°C) for aluminum (mp = 650°C) or SiC (dp = 2700°C) for silicon (mp = 1410°C)). In such cases, the related CMC can be processed according to a rather simple two-step procedure, namely : (i) the infiltration of a fiber preform by the molten element and (ii) a chemical reaction taking place either between the liquid and the preform or between the liquid and the gas phase, according to the following equations :

$$C\ (\text{preform}) + Si_{(l)} \rightarrow SiC_{(s)} \qquad (9)$$

$$2Al\ (l) + \frac{3}{2}O_{2\,(g)} \rightarrow Al_2O_{3\,(s)} \qquad (10)$$

The chemical reaction corresponding to (9) has been used to elaborate SiC-based CMCs(1) from a carbon preform, which consist mainly of SiC-fibers embedded in a silicon matrix [43] whereas that corresponding to (10) is used in a new processing technique(2) of alumina-matrix composites which seems promising and could be extended to other CMCs (fig. 5) [44, 45]. The main advantage of such processes lies in their rapidity. On the other hand, they raise the important question of the proper control of the fiber-matrix interactions.

3.2.2- Impregnation of a fibrous reinforcement with ex-glass liquids

Silica-based glasses give undercooled metastable liquids (when they are heated beyond Tg with Tg ranging from ≈ 600°C to ≈ 1150°C), whose viscosity decreases as temperature is raised. At a high enough temperature, they can be used to impregnate fiber preforms yielding after cooling ceramic fiber/glass matrix

---

(1) SILCOMP from General Electric, (2) Dimox from Lanxide

composites. Furthermore, the glass-matrix can be partly or totally crystallized (or ceramed), at least for specific glass compositions (table 1), by performing after the impregnation/densification step, an ageing treatment (e.g. 24 hrs at about 1050°C for the LAS glass) yielding after cooling ceramic fiber/**glass-ceramic** matrix composites [46-52].

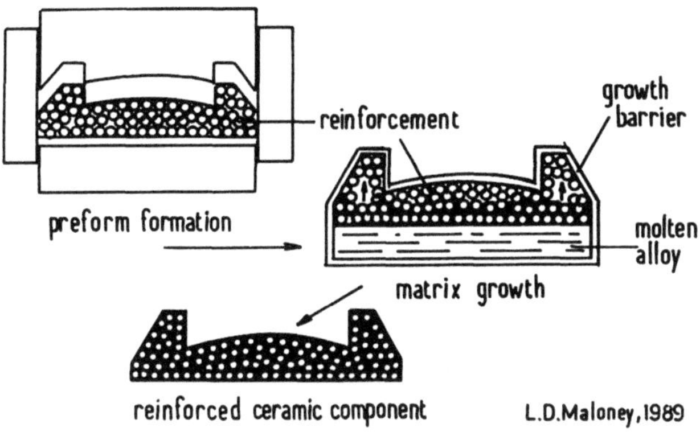

Fig. 5 : Processing of a CMC from a liquid element reacting with a gaseous atmosphere [44]

Matrix type	Major constituents	Major crystalline phases	Max. use Temper. (°C)
Glasses			
borosilicate	B_2O_3, SiO_2		600
aluminosilicate	Al_2O_3, MgO, CaO, SiO_2		700
high silica	SiO_2		1150
Glass-ceramic			
LAS-I	Li_2O, Al_2O_3, MgO, SiO_2	β-spodumene or	1000
LAS-II	Li_2O, Al_2O_3, MgO, SiO_2,Nb_2O_5		1100
LAS-III	Li_2O, Al_2O_3, MgO, SiO_2,Nb_2O_5	β-quartz sol.sol.	1200
MAS	MgO, Al_2O_3, SiO_2	cordierite	1200
BMAS	BaO, MgO, Al_2O_3, SiO_2	baryum osumilite	1250
Ternary mullite	BaO, Al_2O_3, SiO_2	mullite	≈ 1500
Hexacelsian	BaO, Al_2O_3, SiO_2	hexacelsian	≈ 1700

Table I : Glass and glass-ceramic matrices of interest for CMCs according to [49]

The starting material is usually a **slurry**, i.e. a stable suspension of the glass powder (mean grain size 8-12 μm) in a liquid (water or an alcohol) containing a binder [46-49]. However, sols or liquid organometallic precursors can also be used [50-52]. In order to avoid the problems related to the impregnation of complex shape preforms by slurries, a **pregreg route** is often preferred. The reinforcement, i.e. a continuous tow or tape of Si-C-O ex-PCS fibers is passed through an agitated slurry, as shown in fig. 6, and then wound on a reel of large size. After drying and cutting, 1D-prepreg plies are stacked according to a given orientation sequence and the binder is burnt out at about 600°C. The laminated structure is then **hot pressed** under an inert atmosphere (or vaccum) for a few minutes (15-30 min) at 1200-1300°C and under a pressure of 5-15 MPa. Under such conditions, the viscosity of the ex-glass liquid is low enough to allow all the individual filaments of the tow or tape to be properly embedded in the liquid, on the one hand, and to let the liquid matrix fill all the pores of the laminate, on the other hand, resulting in a composite with a **very low residual porosity** (typically a few percent). After densification, the composite can be formed to any desired shape as long as the matrix is not totally ceramed (i.e. as long as there remains enough viscous phase). Finally the degree of ceramization of the matrix is adjusted through an ageing treatment.

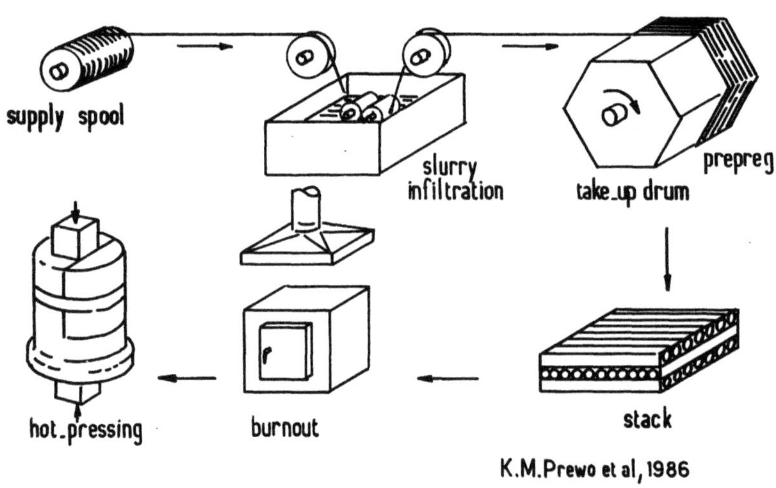

Fig. 6 : Steps in tape lay-up processing of glass-matrix composites [49]

The opimization of the glass composition and processing parameters has to take into account several important considerations. The undercooled liquid matrix should not crystallize too early and too fast in order to permit a full densification of the composite (the reverse situation being observed for e.g. exsol/gel matrices which are very reactive). The CTE of the matrix in the ceramed state should be of the same order (or lower) than the radial CTE of the fiber in order to avoid a too strong mechanical fiber/matrix bond during cooling. Since the Si-C-O ex-PCS fibers have already a low CTE (i.e. ≈ $3.10^{-6}K^{-1}$), glass-ceramic compositions with very low CTE are preferred. Probably still more important is the control of the chemical reactions taking place near the fiber/matrix interface during processing. It is now well established that the non-brittle mechanical behavior of Si-C-O/glass-ceramic composite is directly related to the occurence of a very **thin layer of carbon** at the fiber/matrix interface [25-33]. It is worthy of note that the carbon interphase is observed in materials processed at high temperatures, its thickness being eventually increased when the materials are further aged at a high enough temperature. On the other hand, it has not been observed for low processing temperatures (e.g. those permitted in the sol/gel approach), for composites reinforced with pure SiC (e.g. SiC-whiskers) and after ageing treatments performed in air on unprotected materials. It seems that the carbon interphase could be the result of two complementary phenomena : (i) a surface decomposition of the Si-C-O ex-PCS fibers (known to occur beyond 1100-1200°C and to yield free carbon), and (ii) a chemical reaction between the fiber and the matrix (presumably an oxidation of the fiber surface by some of the components of the matrix, the kinetics of formation of the carbon interphase being thus dependent on the matrix composition, e.g. on the occurrence of specific doping oxides) [28]. The occurrence of fiber/matrix reactions is clearly supported by the fact that in Nb_2O_5 modified glass-ceramic matrix composites, the interphase consists of a NbC/free carbon thin film, the weakest link in the interfacial sequence being the interface between the NbC crystals and the carbon sublayer [26, 29-32]. Since interfacial reactions in the condensed states are controlled by diffusion, it is understandable that both the matrix composition (which governs the diffusion driving force) and the processing parameters (namely temperature, duration and to a less extent pressure) play a key role in the achievement of high mechanical performances. Another way to control the fiber/matrix coupling through chemistry, i.e. the use of **precoated fibers,** has been studied to some extent. As an example, the use of **BN-interphases**, more resistant to oxidation than carbon, has been mentioned by several authors. BN is deposited from boron and nitrogen sources according to e.g. equation (2) (but borane and borazine have been also used) [19-23] or results from a chemical reaction between NH_3 and a boron containing fiber (e.g. the Al_2O_3-SiO_2-B_2O_3 fibers) [53]. Finally, another important criterion in the choice of the matrix composition is the use temperature of the composite. Although most glass-ceramic matrix composites have been designed for a use temperature of 800-1000°C, specific matrix compositions are known that could be used to extend the use temperature up to almost 1700°C (table 1).

3.3.3. Other liquid phase processing routes

SiC, Si_3N_4 and mullite-matrix composites have been elaborated by **hot pressing** a stack of plies (1D-tapes or 2D-fabrics) previously impregnated with an organometallic liquid precursor or a slurry. Such a process was reported to yield CMCs with high mechanical performances (e.g. a fluxural strength of 830 MPa and a toughness of 17 $MPam^{1/2}$ at room temperature for a 1D-C/mullite composite) [54,55]. However, it raises some important problems. **Organometallic** precursors (e.g. PS, PCS or PCSZ) undergo an important shrinkage during the organic/inorganic conversion due to the evolution of gaseous species. This drawback can be overcome by using an inert filler (e.g. a powder of BN) [56]. The **slurry route** [54] appears also to have some limitation : (i) the temperatures necessary to achieve a high degree of sintering are high, e.g. 1600-1800°C for SiC, Si_3N_4 or mullite, a feature which precludes the use of the Si-C-O ex-PCS fibers and limits the process to the less oxidation-resistant carbon fibers and (ii) it seems difficult to avoid an important fiber damage during the hot pressing step owing to the lack of plasticity of most ceramics within this temperature range (unless a glassy phase is formed).

3.4 - The solid state processing route

CMCs can be processed according to solid state routes derived from powder metallurgy. Although some attempts to elaborate composites with continuous reinforcements have been reported [54,55], the solid state processing route has been mainly used for **whisker-reinforced composites** (the most common reinforcement being SiC-whiskers) [57-59] in as much as : (i) long fibers are broken into small fragments during pressing and (ii) continuous fibers (with the exception of carbon fibers) are not stable, from a chemical and microstructural standpoint, at the sintering temperatures.

The solid state processing of whisker-reinforced composites raises a number of important problems : (i) the whiskers, owing to their morphology (high aspect ratio), have a tendency to form nest-like agglomerates (that would result in defects of large size), thus a particular care has to be taken when preparing the whisker-matrix powder mixture (e.g. use of very dilute suspensions), (ii) the whiskers impedde to some extent the sintering of the matrix, (iii) chemical reactions might occur between the whiskers and the matrix owing to the high temperatures required for sintering (however, good chemical compatibility has been reported for a few systems including SiCw/Al_2O_3) and (iv) as a result, the chemical bonding between the two components is usually strong. Finally, whiskers of very small diameters (i.e. less than 1 μm) raise health problem during processing, machining and use.

Despite the problems mentioned above, SiCw-reinforced composites exhibit attractive mechanical properties (in terms of toughness, failure stress and wear resistance) with respect to the related unreinforced ceramic matrices. The facts that they can be processed in the solid state according to well-established and cost effective techniques explain why they have been considered as potential materials for the volume production of small size ceramic items (e.g. cutting or grinding tool inserts).

4 - MECHANICAL BEHAVIOR

4.1 - The fiber/matrix mechanical coupling

As already mentioned, the mechanical behavior of a CMC strongly depends upon the strength of the fiber-matrix coupling which is governed itself by two parameters : (i) the occurence of **residual stress** which might result from differential thermal expansion between the fibers and matrix during cooling from processing to room temperatures and (ii) **chemical bonds**. The best results, in terms of tensile failure strain, work of fracture and toughness, are achieved when the **fiber-matrix bonding is not too strong** (when it is too weak, the load transfer from the matrix to the fibers is poor and when it is too strong, the fiber/matrix interfaces are no longer able to arrest or/and deflect the matrix microcracks). It is thus important to characterize **mechanically**, through simple tests, the fiber/matrix bonding in CMCs . In this field, suitable tests such as pull-out, push-out or fragmentation tests performed on monofilaments, have been already used for polymeric (and more recently metallic) matrix composites [60-63]. However, their extension to CMCs is not straightforward owing to the fact that CMCs are inverse composites (i.e. in the so-called fragmentation test, it is no longer the fiber but the matrix which undergoes multiple cracking) and to the intensity of the bonding itself (which can be one or two order(s) of magnitude higher than that commonly observed in polymeric matrix composites).

The most common test used to characterize the fiber/matrix bonding in CMCs is the **indentation test** in which a load is applied with a micro or even nano indenter (e.g. a Vickers hardness pyramid) onto the end of a fiber in a polished cross-section of the composite, measuring continuously the applied force F and the resulting displacement u [64-66]. If this loading causes debonding at the fiber/matrix interface with frictional sliding, the variations of u as a function of F can be expressed according to the following equation :

$$u = F^2/(4\pi^2 R_f^3 \tau E_f) - 2\,\Gamma/\tau \quad (11)$$

where R_f is the fiber radius, τ **the frictional force** that resists sliding at the interface, E_f the fiber Young modulus and Γ **the debond fracture energy**. The variations of u as a function of F^2, for a 1D-SiC (ex-PCS)/glass ceramic composite, are shown in fig. 7. As predicted from equation (11), the data obey a linear relation, the two interface parameters τ and Γ being calculated from the slope of the straight line $1/4\pi^2R^3\tau E_f$ and intercept $2\ \Gamma/\tau$, respectively. In this particular example, the values of τ and Γ are very small (τ = 3.5 MPa and $\Gamma \approx 0$), suggesting that the bonding between the two components is very weak, a well-known feature of these materials [64]. Futhermore, unloading and reloading cycles can be applied to the fiber : in fig. 7, the measured value of u after unloading is smaller than that predicted suggesting that the sliding resistance during reverse slip is lower. Finally, the $u = f(F^2)$ graph is changed when **residual stress** is present at individual fibers. Although the absolute values of the parameters derived from the indentation test, for a given CMC, can be questioned as discussed by e.g. Piquenot et al. [67], this test has two important advantages : (i) it yields quantitative data on the interfacial bonding and (ii) it can be performed directly on actual composites (requiring only classical metallographic sample preparation).

The indentation test has been used extensively to study the influence of the processing conditions, material parameters and environment on the strength of the fiber/matrix bonding in glass-ceramic matrix composites. A good example of such studies is that reported by Grande et al. in which the interfacial shear stress τ_{DEB}, was assessed according to an approach combining indentation test

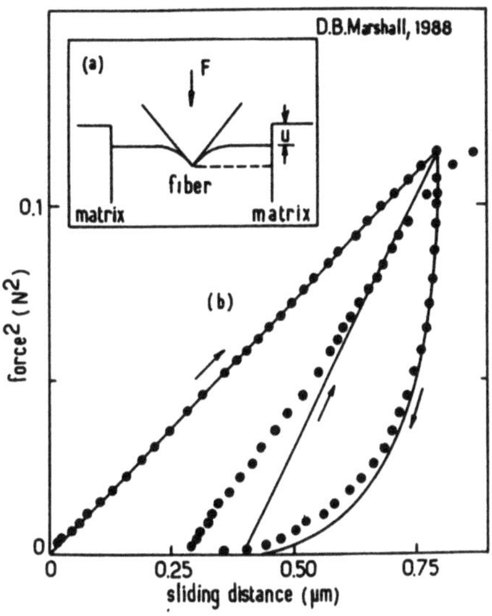

Fig. 7 : Indentation on a 1D-SiC/glass ceramic composite (curves represent theoretical predictions) [65]

data and results from a finite element analysis of an axisymmetric (cylinder) model which includes the fiber (Si-C-O ex-PCS fiber), surrounding matrix (glass or glass-ceramics) and average composite properties beyond the matrix [66]. As shown in fig. 8, the **processing temperature** Tp is a key parameter for achieving a high flexural strength : (i) for low Tp values (Tp < 900°C), the fiber/matrix bonding is strong and thus the flexural strength is low whereas (ii) for 1100 < Tp < 1300°C, the interfacial shear stress is much lower ($\tau_{DEB} \approx 200$ MPa), owing to the formation of a carbon-based interphase with the result that the flexural strength of the composite is increased by one order of magnitude. Another important parameter is the **matrix composition** : depending on the value of the CTE of the matrix, α_m with respect to that of the fiber, α_f^r, a state of radial residual stress (either compressive, when $\alpha_m > \alpha_f^r$, or tensile when $\alpha_m < \alpha_f^r$) is generated at the interface which strongly changes the value of the interfacial shear stress τ_{DEB}, as shown in fig. 9.

Fiber **push-out tests** have been also used to characterize the fiber/matrix bond strength. However their use was limited to large diameter filaments (e.g. 145 µm SiC CVD-filaments or 200 µm single crystal alumina fiber) [23, 68]. Two examples of push-out tool/sample slice configurations are shown schematically in fig. 10 and 11. In that shown in fig. 10, the filament is pushed out of the composite slice with a conical diamond indentor (machined to have a flat end ≈ 100 µm in diameter) and the loads F1 and F2 necessary to move the fiber in one direction (first push) and then in the opposite direction (second push) after having turned the sample over, are measured. The shear stresses corresponding to bonding, τ_b, and friction, τ_f, are calculated according to the following equations [23] :

$$\tau_1 = \tau_b + \tau_f = F_1 / (\pi d_f L) \quad (12) \qquad\qquad \tau_2 = \tau_f = F_2 / (\pi d_f L) \quad (12')$$

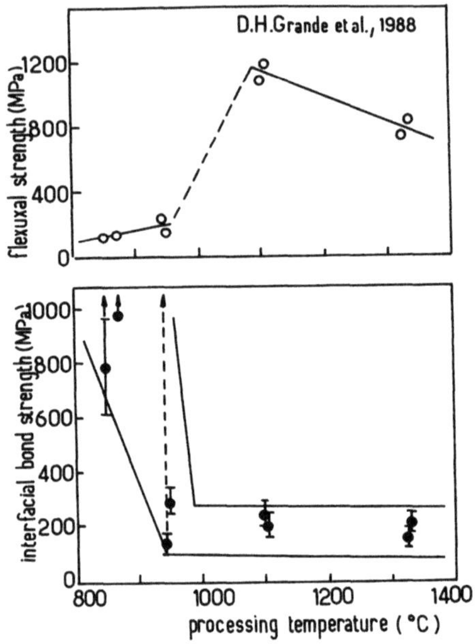

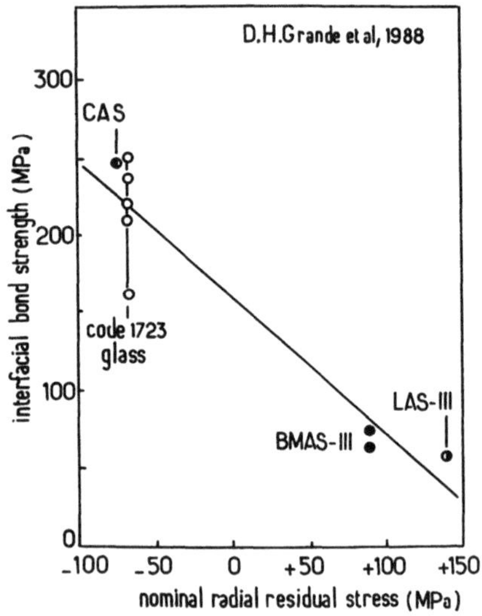

Fig. 8 : Bond strength τ and composite flexural strength for 1D-SiC/1723 composites processed at various temperatures [66]

Fig. 9 : Bond strength against residual interface stress in radial direction for ex-PCS fiber reinforced CMCs [66]

The test has been used to study the influence of a **BN-interphase** (deposited by LPCVD on the filament surface) and **CTE mismatch** (by varying matrix composition) on the fiber/matrix bonding. It clearly appears from table II and fig. 10 that : (i) the values of τ_1 and τ_2 are very low when $\alpha_f^r - \alpha_m > 0$ (namely for the glassy matrices with low CTEs) with almost no effect of the BN-coating (as could be expected from the fact that the SiC CVD -filaments(*) have already received a complex SiC/C interphase) whereas (ii) they are much higher when $\alpha_f^r - \alpha_m < 0$, i.e. when there is a clamping effect of the matrix upon the filament during cooling from processing to room temperatures. Under the latter condition, the BN-interphase yields a significant lowering of both τ_1 and τ_2 (the effect being more pronounced when the BN-coating is thicker) and seems to act as a buffer.

(*) SCS-6 from AVCO

In the configuration shown in fig. 11, the filament is pushed out with a WC-Co tool and the applied load F is recorded vs time for different sample slice thicknesses L [68]. As shown in fig. 11, the F = f(t) curve : (i) consists of a linear stage up to $F = F_D$ where the load is observed to drop instantaneously to a lower value owing to a partial debonding of the interface and (ii) exhibits a maximum, F_{max}, when the load is further increased which corresponds to the stage where the entire length of the embedded fiber is debonded and the fiber begins to exit from the opposite face. On the basis of a shear-lag model, the interface shear strength τ_b and the interfacial friction stress $\tau_f = \mu\sigma_o$ (where σ_o is a residual compressive stress and μ a coefficient of friction) are derived from the measured values of F_D and F_{max}, according to the following equations :

$$F_D = \frac{\pi d_f \tau_b}{\alpha} \tanh \alpha L \qquad (13)$$

$$F_{max} = \frac{\pi d_f^2 \sigma_o}{4 K} [\exp (\frac{4 \mu K L}{d} - 1)] \qquad (14)$$

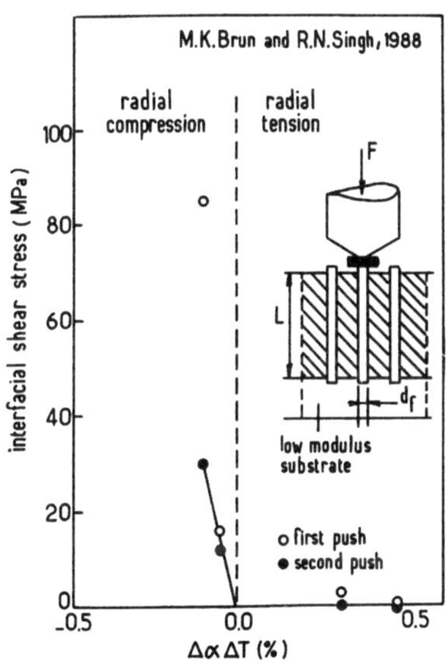

Fig. 10 : Push-out test on 1D-SiC (CVD)/BN/ceramic matrix composites [23]

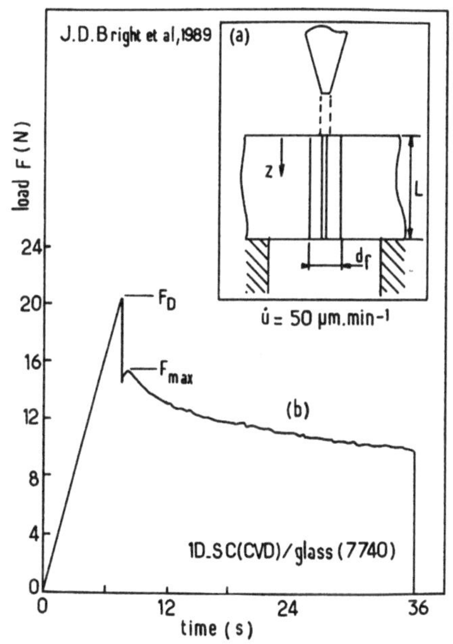

Fig. 11 : Push-out test on 1D-SiC (CVD)/borosilicate glass composite [68]

where α is a shear-lag parameter and K a non-dimensional parameter related to the Young's moduli and Poisson's ratios of the fiber and matrix. An iterative regression procedure applied to the $F_D = f(L)$ and $F_{max} = f(L)$ experimental data was used to derive the values of τ_b, σ_o and μ corresponding to the best theoretical fit (Table III). Push-out tests have an important advantage : the length of fiber which is debonded from the matrix is known accurately (i.e. L) whereas it has to be calculated in the indentation test. Furthermore, L can be varied easily. Conversely, push-out tests cannot be performed on small diameter fibers. Finally, from the values of the interfacial shear stress, it clearly appears that high strength CMCs correspond to low (or rather low) fiber/matrix bonding.

Matrix	τ_1 (MPa)	τ_2 (MPa)	$\alpha_f - \alpha_m$ (10^{-6}°C^{-1})
Silica	0.3	0	4.5
Silica (*)	0.3	0	4.5
Glass	2.9	0.06	2.7
Glass (*)	2.9	0.06	2.7
Si_3N_4 (*)	16	12	- 0.3
Mullite	>200	-	- 0.8
Mullite(*)	85	30	- 0.8
Mullite(**)	50	25	- 0.8

(*) 1 μm BN (**) 5 μm BN

Table II : Push-out tests performed at 25°C on CMCs reinforced with SCS-6 CVD SiC filaments (d_f = 140 mm) [23]

Matrix	τ_D (MPa)	τ_o (MPa)	μ	τ_f (MPa)
RBSN (1)	19.1	55	0.27	14.7
Borosilisate glass (2)	17.5	40	0.29	11.6
Borosilisate glass (3)	18.2	11	0.94	10.3

(1) unidirectional (2) unidirectional (7740 glass) (3) 0/90 cross-ply (7050 glass)

Table III : Push-out tests at 25°C on CMCs reinforced with SCS-6 CVD filaments (d_f 140 μm) [68]

4. 2 - The non-linear mechanical behavior of CMC and related damaging phenomena

Most (if not all) CMCs are **inverse composites**, i.e. the failure strain of the matrix (usually of the order of 0.1 %) is much lower than that of the fibers (typically 1 %). Thus, under loading e.g. in the fiber direction, the **ceramic matrix fails first**. The cracking of the matrix may -or may not- lead to a premature failure of the fibers depending on whether the fiber/matrix bonding is strong or weak. If the interfacial bonding is weak enough, **debonding** rather than fiber failure occurs at the crack front and further in the crack wake, as shown schematically in fig. 12. The extent of debonding is determined by the friction coefficient μ, the morphology of the debonded interface and can be extensive when the interface is in residual tension. Under such conditions, which are fulfilled when a layered structure interphase material (such as pyrocarbon or hex-BN) is present at the interface and $\alpha_f^r - \alpha_m \approx 0$ (or even slightly positive), the crack extends indefinitely in the matrix without breaking the fibers in its wake, at constant applied stress. Ultimately, it appears as bridged by the unbroken fibers.

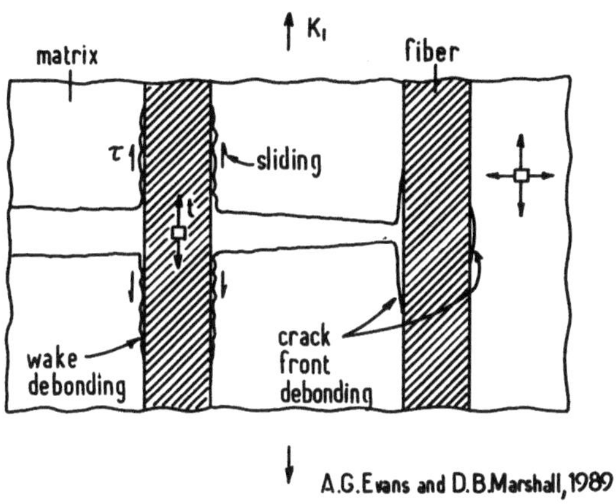

Fig. 12 : Initial debonding of fibers at the crack front and debonding in the crack wake [69]

Since the early work of Aveston, Cooper and Kelly (ACK-model), numerous studies have been devoted to the mechanical behavior of 1D-CMC (considered as model materials) in tension loading along the fiber direction [3, 49, 69, 70-72]. The main features of the stress-strain curve are shown in fig. 1 and 13. The material exhibits a linear elastic behavior up to $\varepsilon = \varepsilon^*_{mu}$ where the first crack occurs in the matrix. From that point, the material exhibits a **non-linear behavior** related to the extension of different damaging mechanisms. Further loading causes formation of **periodic matrix cracks**, which have been observed for both glass (or glass-ceramic) and SiC matrix composites [73,74], up to a **saturation stage** where the crack spacing and thus the size of the resulting matrix blocks reach a limit (related to the interfacial shear stress τ which controls the fiber/matrix load transfer). When load is increased beyond this saturation stage, frictional interactions occur between the fibers and the matrix blocks. Finally, failure takes place when $\varepsilon = \varepsilon_{fu}$.

The occurence of extensive **damaging phenomena** before failure, namely matrix microcracking, fiber/matrix debonding, sliding and fiber pull-out, is a key to understanting the mechanical behavior of the material as well as the effect of the environmental parameters. As the material is progressively damaged, its stiffness decreases (as apparent from the unloading-reloading cycles which can be performed for $\varepsilon^*_{mu} < \varepsilon < \varepsilon_{fu}$) thus the E/Eo ratio is often used to characterize the damage level (fig. 13)[73]. The **damaging phenomena** are also responsible for the high values of the failure strain and work of fracture (with respect to those of the unreinforced ceramic matrix). Finally, the occurrence of extensive fiber pull-out gives a brush-like appearance to the failure surface of tough CMC contrasting with that of the brittle monolithic ceramics (which is usually smooth and flat) [47,75]. Conversely when the interfacial requirements(in terms of interfacial shear stress τ and residual stress) are not fulfilled, the composite behaves as a brittle material and fails when matrix cracking begins to take place (fig.1).

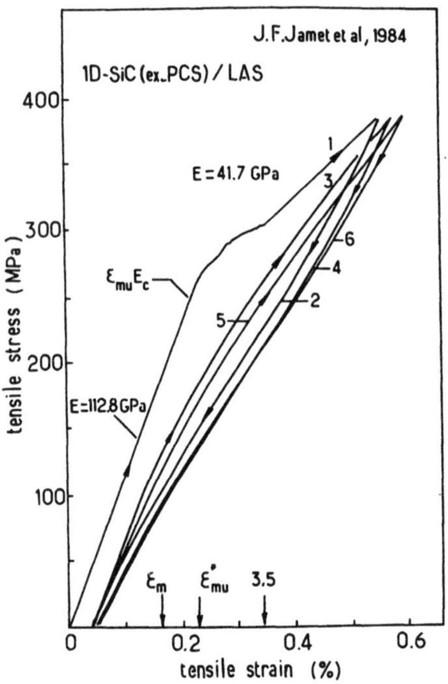

Fig. 13 : Mechanical behavior at 25°C of a 1D-SiC (ex-PCS)/glass-ceramic composite [71]

Tension tests provide two new ways for measuring **the interfacial shear stress** τ in 1D-CMCs. If one assumes that matrix cracking saturation is achieved, the interfacial shear stress τ is related to the mean crack spacing δ, according to the following equation [74] :

$$\tau = \lambda \, [(1 - V_f)^2 \, \gamma_m E_f E_m \, R_f^2 \, / \, V_f \, E \, \delta^3]^{1/2} \tag{15}$$

where $\lambda = 1.34$, γ_m is the matrix fracture energy and E the Young modulus of the composite. The mean crack spacing at saturation δ can be measured either from replicas of the tensile test specimen surface under loading or post mortem from polished and etched longitudinal sections of the specimen. Futhermore, assuming that the fibers are only weakly bonded to the matrix, the ACK model shows that the critical stress σ_c^* at which the matrix undergoes microcracking is directly related to the interfacial shear stress τ and the fiber volume fraction V_f, according to the following equations [71,72] :

$$\sigma_c^* = \varepsilon_{mu}^* \cdot E \quad (16)$$

$$\varepsilon_{mu}^* = \left[\frac{24\,\tau\,\gamma_m\,E_f\,V_f^2}{E\,E_m^2\,d_f\,(1-V_f)}\right]^{1/3} \quad (17)$$

The variations of σ_c^* as a function of both V_f and τ, are shown in fig. 14 for a 1D-SiC (ex-PCS)/glass-ceramic composite [71]. If V_f is known (it can be measured by e.g. image analysis on polished cross-sections) and σ_c^* measured on the stress-strain curve, then τ can be calculated. The use of the two methods for measuring τ is rather straightforward for glass or glass-ceramic 1D-CMC inasmuch as the matrix crack array is very regular, the carbon interphase very thin (i.e. less than 0.1 µm) and the matrix almost free of any residual porosity (Vp < 2-3 %). Conversely, their use for characterizing the fiber/matrix bonding in 1D-SiC (ex-PCS) /SiC(CVI) composites should be done with care owing to the thickness of the carbon interphase (0.1 < e < 1 µm), the large residual porosity (Vp ≈ 15 %) and the dispersion of the δ-values [73].

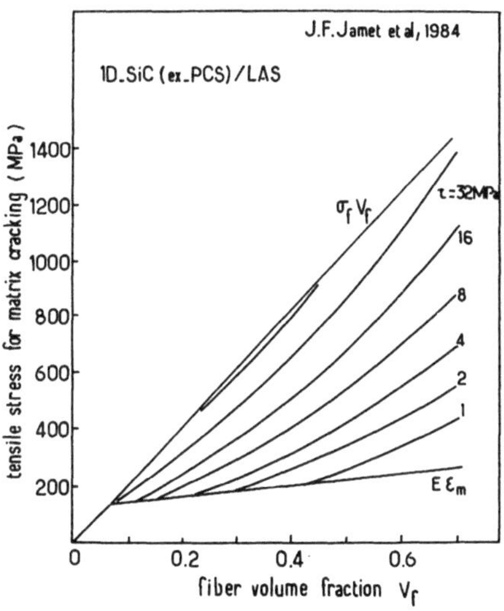

Fig. 14 : Critical stress σ_c^* as a function of fiber volume fraction and interfacial shear stress τ [71]

The mechanical behavior of the **2D-SiC(ex-PCS)/C/SiC(CVI)** composites, under tension loading at room temperature, exhibits the same general features as that reported for the 1D-CMC (fig. 15) : (i) an initial stress-strain linear elastic domain and (ii) beyond a critical strain value corresponding to the beginning of the matrix microcracking, a large domain of non-linearity, related to various damaging phenomena, and which is responsible for : (i) a progressive decrease in the material stiffness as clearly apparent from the unloading/reloading cycles (in first approximation, the composite behaves as an elastic damageable material), (ii) a high failure strain (up to 0.8 %) and (iii) a high work of fracture [76]. 2D-C/SiC and 2D-SiC/SiCcomposites exhibit an extremely **high resistance to crack propagation** as shown in fig. 16, related to the crack deflection mechanism by weak fiber/matrix interfaces depicted for 1D - CMC which results here in the formation of numerous small cracks extending on each side of the main crack (generating thus a processed zone). Therefore, these CMCs are non-brittle, damage - tolerant materials with a high reliability [77].

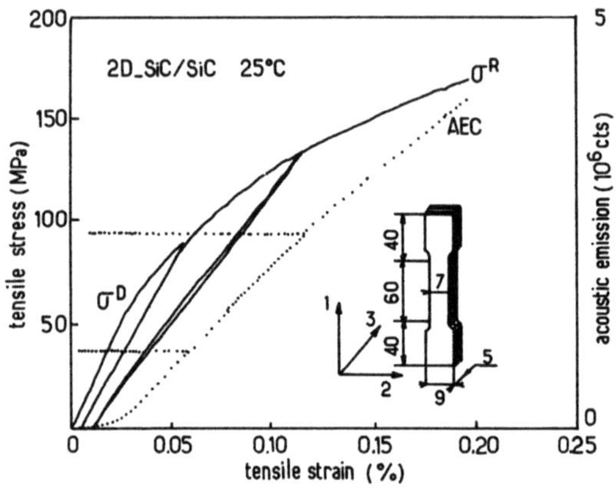

Fig. 15 : Mechanical behavior in tension of 2D-SiC (ex-PCS)/SiC (CVI) composite at 25°C [76]

5 - EFFECT OF ENVIRONMENTAL PARAMETERS

Most CMCs have been designed for use in severe conditions e.g. high temperatures, mechanical loading and corrosive atmospheres thus, the effect of environmental parameters on their mechanical behavior is an important subject. In order to decouple the respective effects of temperature, on the one hand, and nature of the atmosphere, on the other hand, high temperature mechanical tests or/and mechanical tests at room temperature after ageing treatment, are usally performed in both inert and oxidizing atmospheres.

5.1 Effect of temperature and time

The effect of ageing treatments (100 hours at 900 or 1400°C) performed in helium on the mechanical behavior under tension loading at 25°C of 1D-SiC

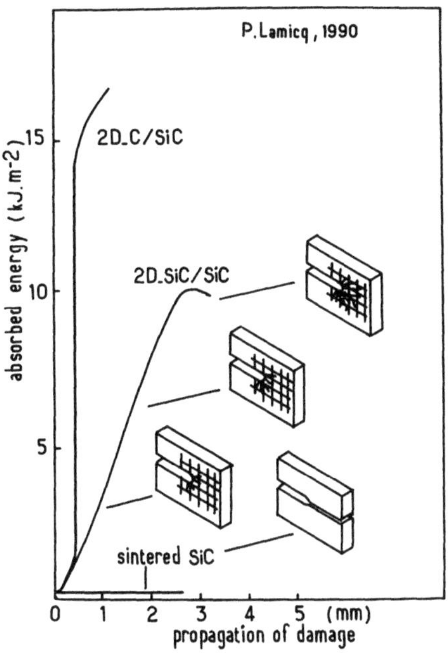

Fig. 16 : Resistance to crack propagation of 2D-C/SiC and 2D-SiC/SiC composites [84]

(ex-PCS)/C/SiC(CVI) with two different carbon interphase thicknesses (0.1 and 1 µm), has been studied by Filipuzzi et al. [78]. The treatment performed at 900°C was reported to improve slightly the ultimate tensile stress σ_R and the strain to failure ε_R, on the one hand, and to lower slightly the stress at first damage σ_D (equivalent to σ_c^*) whereas that performed at 1400°C resulted in a decrease of all these parameters. Futhermore, the failure surface exhibited limited fiber pull-out in as received materials and in materials aged at 1400°C. These results are consistent with the fact that ex-PCS fibers undergo a decomposition-/crystallization process at about 1100-1200°C, as discussed in section 2 [9, 35].

High temperature flexural and tensile tests have been performed in argon by Prewo et al. on SiC(ex-PCS)/glass-ceramic composites (0 and $0(0/90)_2$ fiber orientations) [79]. As shown in fig. 17, the tensile strength regularly increases as the test temperature is raised, even in the 1100-1300°C temperature range where ex-PCS fibers are no longer stable. Futhermore, the failure surface exhibited extensive fiber pull-out whatever the test temperature.

Generally speaking and assuming that the matrix is intrinsically stable within the temperature range considered, a treatment at high temperatures may have two effects on a CMC : (i) a decrease in the tensile strength and stiffness of the reinforcement related to the occurrence of crystallization/grain growth or/and decomposition processes and (ii) a change in the fiber/matrix interfacial zone resulting from the decomposition of the fiber or/and chemical reactions between the two constituents. As long as the treatment is performed at a high enough temperature but remains not too severe (in terms of time and temperature), its effect on the mechanical behavior of the CMC may be **beneficial** as suggested by the experimental data presented above : this is particularly true for composites elaborated from the metastable ex-PCS fibers.

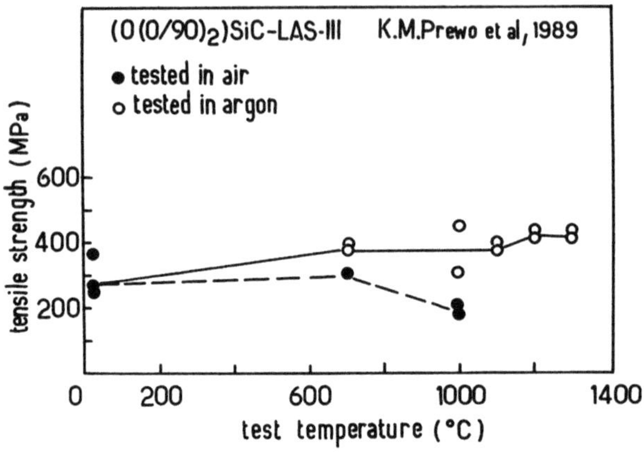

Fig. 17 : Tensile strength of [0 $(0/90)_2$] SiC (ex-PCS)/LAS-III composites in air and argon [79]

For glass-ceramic matrix composites, a treatment at a high enough temperature in an inert atmosphere (ageing treatment or high temperature mechanical test) can be considered from a chemical standpoint, as the continuation of the high temperature step (i.e. the hot-pressing step) of the processing procedure (which is responsible for the in-situ formation of the carbon-based interphase). Thus, its main effect (as long as the treament is not too severe to yield a significant lowering of the fiber strength and stiffness) will be to increase the interphase thickness and probably to lower the interfacial shear stress τ. Cooper and Chyung have reported the occurrence of a carbon interphase of 0.02 μm in SiC(ex-PCS)/CAS (calcium aluminosilicate) composites hot-pressed at 1200°C and an interphase of 0.06 μm after a post-processing ageing treatment of 2 hours at 1200°C [28]. Since in these composites the carbon interphase is very thin, it is anticipated that the effect of a post-processing ageing treatment on the value of τ might be more significant than for SiC/SiC composites where the carbon interphase thickness is already much thicker in as-prepared conditions (thus, the additional amount of carbon which might result from the decomposition of the ex-PCS fibers would be weak in this case as long as temperature and duration are not too high).

The fact that the high temperature tensile strength of SiC(ex-PCS)/glass ceramic composites still increases when the test temperature is raised beyond 1100°C could be considered as surprising since ex-PCS fibers are known to undergo a decomposition process at T > 1100°C (fig. 3). As a matter of fact, this decomposition process releases gaseous species CO and SiO and thus its kinetics are expected to depend upon the CO and SiO partial pressures in the gas phase. When it takes place in a fiber **encapsulated** in a compact matrix (there is almost no residual porosity in hot-pressed glass-ceramic matrix composites), the decomposition kinetics are expected to be significantly slowed down explaining thus that no strength loss was observed during the short time needed for a tensile test.

5.2 - Effect of an oxidizing atmosphere

5.2.1. Oxidation of 1D-SiC/C/SiC composites

The oxidation of 1D-SiC(ex-PCS)/C/SiC (CVI) model composites (with a carbon interphase of 0.1 or 1 μm) has been studied by Filipuzzi et al. [80,81]. The oxidation process involves two main phenomena : (i) chemical reactions between oxygen and the three components of the composite and (ii) diffusion mass transfers of the gaseous reactant (oxygen) and products (CO or/and CO_2) along the pore resulting from the consumption of the carbon interphase (fig. 18). The chemical reactions which take place, assuming that the test conditions remain those of the so-called **passive oxidation** regime (i.e. formation of a protective silica scale), correspond to the following equations :

$$C + O_2 \rightarrow CO_2 \qquad (18)$$
$$2C + O_2 \rightarrow 2CO \qquad (18')$$

for the carbon interphase,

$$SiC + 2O_2 \rightarrow SiO_2 + CO_2 \qquad (19)$$
$$SiC + \frac{3}{2}O_2 \rightarrow SiO_2 + CO \qquad (19')$$

for the SiC CVI-matrix and,

$$Si_{0.38}C_{0.47}O_{0.15} + 0.77\,O_2 \rightarrow 0.38\,SiO_2 + 0.47\,CO_2 \qquad (20)$$
$$Si_{0.38}C_{0.47}O_{0.15} + 0.54\,O_2 \rightarrow 0.38\,SiO_2 + 0.47\,CO \qquad (20')$$

for the Si-C-O ex-PCS fibers [82]. The reactions depicted by equations (18) and (18') yield only gaseous products and thus occur with a weight loss whereas those corresponding to equations (19)-(20') result in the formation of both a condensed phase, silica, and gaseous products and take place with a weight increase. The mass transfers along the annular pore occur by Fick and Knudsen diffusions (depending on the thickness of the carbon interphase and residual width of the pore). There is along the pore, a competition between the oxidation reactions taking place with the pore wall (equations (19) to (20') and the bottom of the pore (equations (18) and (18')), on the one hand, and the mass transfers of the reactant/product in the gas phase, on the other hand. As long as the aspect ratio of the annular pore is small, i.e. at the beginning of the oxidation test, the oxidation process is rate-controlled by the oxidation kinetics of the carbon interphase and a weight loss is observed (fig. 19). Conversely, when the aspect ratio of the annular pore is high enough (i.e. beyond a critical time), the mass transfers along the pore become slow and rate-controlling. Under such conditions, a weight increase (corresponding to equations (19) to (20')) is observed [80, 81].

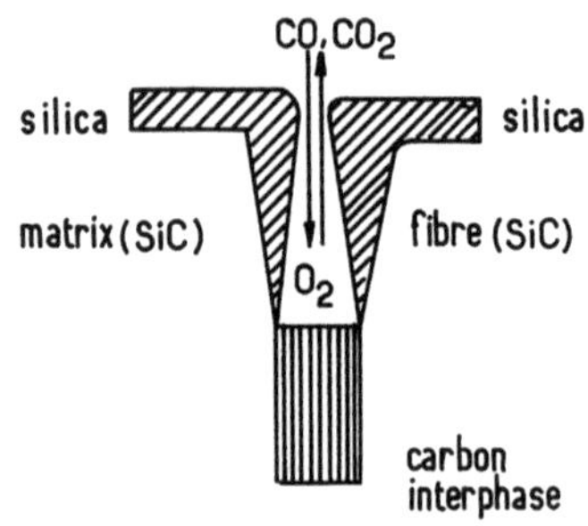

Fig. 18 : Mechanisms involved in the oxidation of uncoated 1D-SiC (ex-PCS)/C/SiC (CVI) model composites [81]

Two parameters play an important role in the oxidation process : the test temperature and the thickness of the carbon interphase. Low temperatures (e.g. T < 1000°C) and thick interphases (e.g. e = 1 μm) favor in-depth oxidation inasmuch as under such conditions the consumption of carbon is fast with respect to the formation of the silica scale, on the one hand, and the width of the annular pore is large, on the other hand. Conversely, the oxidation is limited to **near the surface** when it takes place at high temperature in a composite with a thin carbon interphase (e.g. e = 0.1 μm) owing to an early sealing of the annular pores by silica. Under such conditions, the material exhibits a **self-healing behavior** (fig. 19).

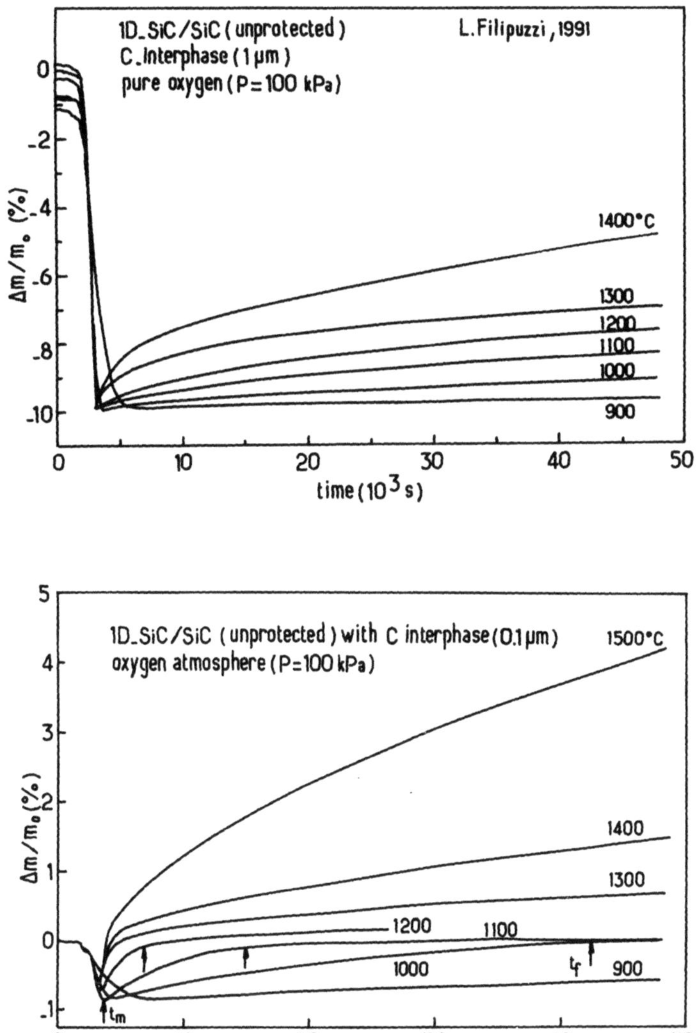

Fig. 19 : Variations of the relative mass $\Delta m/m_0$ as a function of time during oxidation tests performed on 1D-SiC (ex-PCS)/C/SiC (CVI) model composites in flowing oxygen [80-81]

As shown in fig. 20, ageing in air at high temperatures causes a progressive **weakening** of 1D-SiC/C/SiC composites at room temperature and in some cases an **embrittlement**. This change in the mechanical behavior is partly due to the metastability of the ex-PCS fibers at high temperatures (e.g. 1400°C) as discussed

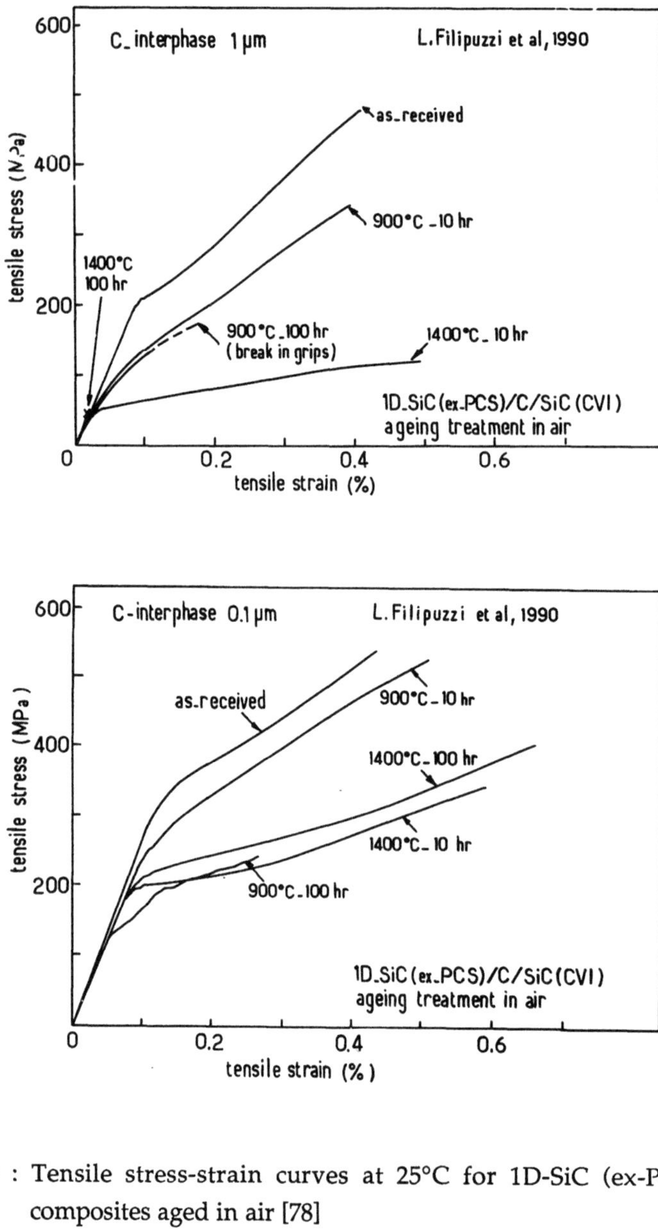

Fig. 20 : Tensile stress-strain curves at 25°C for 1D-SiC (ex-PCS)/C/SiC composites aged in air [78]

in section 5.1. It is also related to the consumption of the carbon interphase and the subsequent growth of interfacial silica layers which can form a strong bond between the fibers and the matrix (thought to be responsible for the embrittlement effect). It is noteworthy that the mechanical behavior at room temperature of the composite with the **thin interphase** is more effectively preserved after an ageing treatment at high temperatures (e.g. 1400°C) in air than that of the material with the thick interphase, owing to the self-healing character of the former. Additionally, the 100 hr-ageing treatment in air at 1400°C is less detrimental to the mechanical behavior of the composite (e = 0.1 mm) at 25°C than that performed at 900°C, a feature which is also consistent with the conclusions of the oxidation kinetics study [78].

5.2.2- Oxidation of 2D-SiC/C/SiC composites

As shown in fig. 21, the main effect of ageing treatments performed in air at 800 < T < 1550°C on the mechanical behavior of 2D-SiC/C/SiC industrial composites

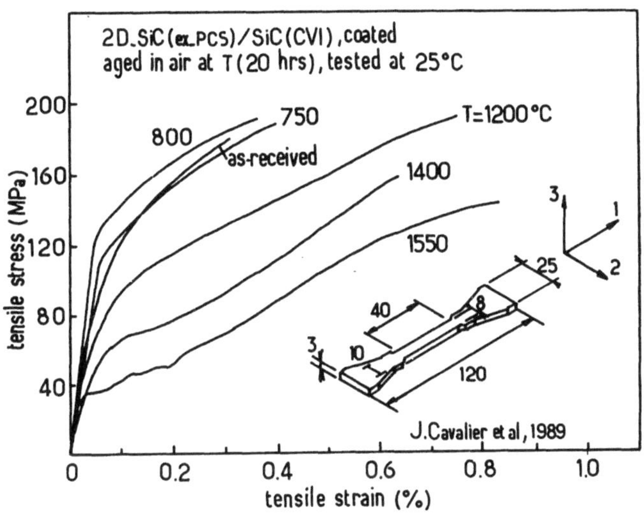

Fig. 21 : Tensile stress-strain curves at 25°C of coated 2D-SiC (ex-PCS)/C/SiC (CVI) composites aged 20 hours in air at various temperatures [83]

(with a protective coating) is similar to that reported for the model 1D-SiC/C/SiC composites (with no protective coating) [83]. Up to T = 800°C, the ageing treatment has no effect (owing to the coating which prevents the oxidation of the carbon interphase) whereas the ageing treatment performed at 1200°C is observed to **improve** the failure strain by more than 130% (with no failure stress loss but with some decrease in σ_D). It is only when the treatment is performed above about 1400°C that a noticeable strength loss occurs (presumably due to the metastability of the ex-PCS fibers), the material remaining tough (as evident from the high value of ε^R).

5.2.3- Oxidation of SiC (ex-PCS)/glass-ceramic composites

The effect of temperature, time and atmosphere on the mechanical behavior of SiC (ex-PCS)/glass-ceramic composites, has been studied extensively and correlated with the change occurring in the fiber/matrix interfacial bonding [28, 30, 31, 49, 64, 66, 70, 79].

As shown in fig. 17 the strength in air of SiC (ex-PCS)/glass ceramic composites is substantially reduced at 700-1000°C and the material is observed to be, at least partly, embrittled [70, 79]. The origin of this embrittlement has been associated with a transition in the fiber/matrix interfacial bond resulting from oxidation. In an oxidizing atmosphere, the thin carbon interphase (of the order of 50-150 nm in thickness) is rapidly consumed with an evolution of carbon oxides and the formation of an annular pore, as already discussed in section 5.2 (the oxidation rate of the carbon interphase being accelerated by the catalytic effect of trace amounts of heteroatoms picked up from the matrix during the in-situ formation of the interphase). In a second step, a silica layer grows in the annular pore as a result of a chemical reaction between oxygen and the SiC (ex-PCS) fibers. Thus, the oxidation of SiC/glass ceramic composites is mainly a diffusion-controlled process which starts at the specimen surface and moves inward at a rate which depends on temperature, p_{O2} and material parameters (e.g. interphase initial thickness, matrix composition). When the in-pore silica layer is thick enough (i.e. when its thickness is equal to that of the initial carbon interphase), it seals the pore giving rise to a strong bond between the fiber and the matrix as clearly evident from indentation measurements of the interfacial bond strength (fig. 22). As already discussed for SiC/SiC composites, the effect of an oxidizing atmosphere on the mechanical properties of SiC/glass-ceramic composites is more dramatic at low temperatures, e.g. 700°C, where an in-depth oxidation takes place, than at high temperatures, e.g. 1100°C, where the annular pore entrances are rapidly sealed by silica.

Generally speaking, it thus appears that the following phenomena play a key role in the understanding of the effect of the environmental parameters on the

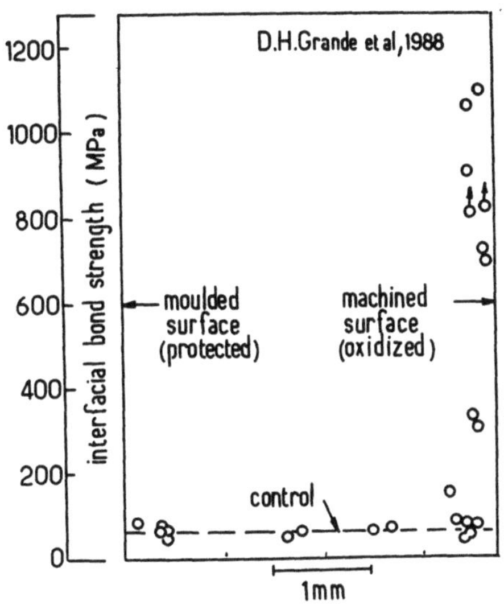

Fig. 22 : Interfacial shear stress as derived from indentation tests in cross-section of a SiC (ex-PCS)/BMAS composite aged in air (1000°C ; 12 hr.) prior to sectioning [66]

mechanical behavior of CMCs : (i) the metastability of ex-PCS fibers beyond about 1200°C, (ii) the vulnerability of the carbon interphase in oxidizing atmospheres particularly when the stress level exceeds that resulting in matrix microcracking and (iii) the oxidation of Si-based fibers and ceramics which yields a silica layer in the passive oxidation regime. This latter phenomenon has, on the one hand, a beneficial effect inasmuch as it is responsible for the self-healing character reported for some materials and use conditions but, on the other hand, a detrimental effect since the growth of a silica layer at the fiber/matrix interface dramatically increases the interfacial bond and embrittles the materials.

6- CONCLUSION

The research which has been performed on CMCs during the last two decades has clearly demonstrated the high potential of these composites as thermostructural materials. With respect to the related monolithic ceramics, CMCs exhibit the following advantages : a non-linear stress-strain behavior in tension due to the occurrence of diffuse damaging phenomena which are responsible for a high failue strain and work of fracture, a high resistance to crack propagation, a non-catastrophic failure mode with extensive fiber pull-out and, finally, an improved reliability. However and as it is often the case for new materials, a number of important points should be the subject of further research and development in the future and more particularly the following topics :

(i)- the high temperature performances of the available **ceramic fibers** are not satisfactory. The oxidation resistance of carbon fibers could be improved possibly through some chemical manipulation of the precursors (in which heteroatoms could be introduced) or/and the use of suitable coatings. More stable ceramic fibers could be designed from new organosilicon precursors or/and prepared from new curing/pyrolysis processing techniques, in order to raise their use temperature up to 1400-1600°C (this target does not seem too irrealistic from today knowledge). Finally, oxide fibers have obviously a high potential which has not been fully exploited yet.

(ii) the **interphase materials** (or more generally, the interphase concepts) are presently too limited and unsuitable for use in oxidizing atmospheres. New materials, which should be stable with respect to the most promising ceramic fibers/matrices **and** oxidizing atmospheres are needed : any new idea in this field should be strongly encouraged. Additionally, **reliable** methods for characterizing mechanically the fiber/matrix bond, the interphase material itself (particularly in terms of fracture energy), the crack deflection at the fiber/matrix interface, have to be worked out, properly tested on model materials and compared to each other.

(iii) serious progress has still to be made in the **processing** of CMCs, the most promising methods appearing to be the gas phase and the liquid phase routes, at least for composites consisting of continuous yarn fibers. The former could be improved in terms of infiltration rate and residual porosity whereas the latter, which has been up to now almost limited to glass and glass-ceramic matrices could be extended further to organo-metallic and gel precursors. Finally, processing techniques involving one (and even several) liquid element(s) **and** a chemical reaction with the gas phase, should be the subject of more research. Generally speaking, the effort in this field has to be oriented towards fast and low-cost processing techniques in order to enlarge the domain of application of CMCs.

(iv) the **mechanical behavior** at room/high temperatures of actual CMCs is insufficiently characterized. Up to now, most studies have been performed on model composites (e.g. 1D-composites) : they should be extended to materials that could be used by industry, e.g. cross-ply laminates, 2D-composites and multidirectional composites. **Constitutive laws** (including the non-linear domain and requiring more work on the extension of the damaging phenomena) are needed for the design of actual CMC-parts as well as **standard tests**. The lack of knowledge and well established testing procedures in this field, undoubtedly slows down the development of CMCs in industry. Finally, since CMCs have been designed to be used at high temperatures, the mechanical testing should not be limited to room temperature, as it is often the case presently.

(v) the effect of the **environmental parameters** and more particularly that of the atmosphere, on the mechanical behavior of CMCs at high temperatures has to be studied thoroughly. Static tests as well as temperature cycling tests in actual atmospheres (e.g. combustion gases) should be performed and surface/bulk treatments suggested in order to enlarge the lifetime of the materials.

Taking the examples of the superalloys and monolithic ceramics, the study of the different points which have been listed above will require a continuous and long effort of research and development before CMCs could be used **safely** in severe environments. However, CMCs could well be the most suitable materials for the elaboration of the hot structures of the next generation aerospace engines.

ACKNOWLEDGEMENTS

The author acknowledges the support received from SEP and the contribution of J. Forget and C. Dupouy to the preparation of the manuscript.

REFERENCES

1- F. Wakai, Br. Ceram. Trans. J., 88 (1989) 205-208
2-F. Wakai, Y. Kodoma, S. Sakaguchi, N. Murayama, K. Izaki and K. Niihara Nature, 344 [6265], (1990) 421-423
3- J. Aveston, G.A. Cooper and A. Kelly, in "The Properties of Fiber Composites" NPL Conf. Proc., Nov. 16, 1971, pp. 15-26, IPC, London, 1971
4- H. Mourichoux, J. Bouix, B. Bonnetot, C. Vincent and H. Vincent, in "Composite Materials for Applications at High Temperatures" (R. Naslain et al., eds), pp. 55-66, AMAC-CODEMAC, Bordeaux and Paris, 1990
5- S. Yajima, Y. Hasegawa, J. Hayashi and M. Iimura, J. Mater. Sci., 13 (1978) 2569-2576
6- Y. Hasegawa, M. Iimura and S. Yajima, J. Mater. Sci., 15 (1980) 720-728
7- R. Chaim, A.H. Heuer and R.T. Chen, 71 [11] (1988) 960-969
8- A.R. Bunsell and G. Simon, Composites Science and Technology, 27 (1986) 157-171
9- D. Mocaer, Thesis n° 585, Univ. Bordeaux, March 28, 1991
10- K. Okamura, M. Sato, T. Segushi and S. Kawanishi, in "Controlled Interphases in Composite Materials" (H. Ishida, ed.), pp. 209-218, Elsevier Science Publ. , 1990

11- K. Okamura, in Proc. 7th CIMTEC World Ceram. Congr. and Satellite Symp., Montecatini Terme, June 24-30, 1990 (in press)
12- J.D. Birchall, Trans. J. Br. Ceram. Soc., 82 (1983) 143-145
13- Y. Abe, S. Horikiri, K. Fujimura and E. Ichiki, in "Progress in Science and Engineering of Composites" (T. Hayashi et al., eds.), Proc. ICCM-IV, Tokyo, 1982, pp. 1427-1433
14- J.C. Romine, Ceram. Eng. Sci. Proc., [7-8] (1987) 755-765
15- V. Lavaste, M.H. Berger and A.R. Bunsell, in "Developments in the Science and Technology of Composite Mater." (J. Füller et al., eds.) Proc. EACM-4, Stuttgart, Sept. 25-28, 1990, pp. 561-566, Elsevier Applied Science, London and New York, 1990
16- D.B. Marshall, F.F. Lange and P.D. Morgan, J. Amer. Ceram. Soc., 70 [8] (1987) C187-C-188
17- S.R. Nutt, J. Amer. Ceram. Soc., 71 [3] (1988) 149-156
18- J.V. Milewski, F.D. Gac, J.J. Petrovic and S.R. Skaggs, J. Mater. Sci., (1985) 1160-1177
19- R.W. Rice, US Pat. 4,642,271 ; Feb 10, 1987
20- R.A. Lowden and K.L. More, Mater. Res. Soc. Proc., 170 (1990) 205-214
21- O. Dugne, S. Prouhet, A. Guette, R. Naslain, R. Fourmeaux, K. Hssein, J. Sévely, C. Guimon, D. Gonbeau and G. Pfister-Guillouzo, J. Physique, 50/C5 (1989) 333-341
22- R.N. Singh and M.K. Brun, Ceram. Eng. Sci. Proc., 8 [7-8] (1987) 636-643
23-M.K. Brun and R.N. Singh, Adv. Ceram. Mater., 3 [5] (1988) 506-509
24- D. Cojean, M. Monthioux and A. Oberlin, Proc. JNC-7 (G. Fantozzi and P. Fleischmann, eds.), Nov 6-8, 1990, Lyon, pp. 381-390, AMAC, Paris
25- J.J. Brennan, Mater. Sci. Research, 20 (1986) 549-560, Plenum Press, New York
26- J.J. Brennan, J. de Physique, Colloque C5 (Suppl. n° 10), 49 (1988) 781-809
27- J.J. Brennan, Mater. Sci. Research, 21 (1986) 387-399, Plenum Press, New York
28- R.F. Cooper and K. Chyung, J. Mater. Sci., 22 (1987) 3148-3160
29- R. Chaim and A.H. Heuer, Adv. Ceram. Mater., 2 [2] (1987) 154-158
30- T.W. Coyle, H.M. Chan and U.V. Deshmukh, in "Interfaces in Polymer, Ceramic and Metal Matrix Composites" (H. Ishida, ed.), Proc. ICCI-II, June 13-17, Cleveland, 1988, pp. 489-501, Elsevier
31- E. Bischoff, M. Rühle, O. Sbaizero and A.G. Evans, J. Amer. Ceram. Soc. 72 [5] (1989) 741-745
32a- M.Y. Chen, J.M. Battison and T. Mah, J. Mater. Sci., 24 (1989) 3213-3220
33b- J. Homeny, J.R. Van Valzah and M.A. Kelly, J. Amer. Ceram. Soc., 73 [7] (1990) 2054-2059
33- L.A. Bonney and R.F. Cooper, J. Amer. Ceram. Soc., 73 [10] (1990) 2916-2921
34- O. Dugne, Thesis n° 375, Univ. Bordeaux, Oct. 17, 1989
35- E. Bouillon, Thesis n° 307, Univ. Bordeaux, April 20, 1989
36- R. Naslain, J.Y. Rossignol, P. Hagenmuller, F. Christin, L. Heraud and J.J. Choury, Rev. Chimie Minérale, 18 (1981) 544-564
37- R. Naslain and F. Langlais, Mater. Sci. Research, 20 (1986) 145-164, Plenum Press, New York
38- R. Naslain, F. Langlais and R. Fedou, J. de Physique, 50 (1989) C5- 333- C5-341
39- D.P. Stinton, A.J. Caputo and R.A. Lowden, Ceram. Bull., 65/2 (1986) 347-350
40- T.M. Besmann, R.A. Lowden, D.P. Stinton and T.L. Starr, J. de Physique, Colloque C5 (Suppl. n° 10), 50 (1989) C5-229 - C5-239

41- R.L. Beatty and D.V. Kiplinger, Nuclear Applications and Technology, 8 (1970) 488-495
42- K. Sugiyama and E. Yamamoto, J. Mater. Sci., 24 (1989) 3756-3762
43- W.B. Hillig, R.L. Mehan, C.R. Morelock, V.J. de Carlo and W. Laskow, Amer. Ceram. Soc. Bull., 54 [12] (1975) 1054-1056
44- A.S. Fareed, B. Sonuparlak, C.T. Lee, A.J. Fortini and G. Schiroky, Ceram. Eng. Sci. Proc. 11 [7-8] (1990) 782-794
45- L.D. Maloney, Design News, March 13, 1989
46- R.A.J. Sambell,A. Briggs, D.C. Phillips and D.H. Bowen, J. Mater. Sci., 7 (1972) 676-681
47- J.J. Brennan and K.M. Prewo, J. Mater. Sci., 17 (1982) 2371-2383
48- K.M. Prewo, J. Mater. Sci., 17 (1982) 3549-3563
49- K.M. Prewo, J.J. Brennan and G.K. Layden, Amer. Ceram. Soc. Bull., 65 [2] (1986) 305-322
50- D.M. Dawson, R.F. Preston and A. Purser, Ceram. Eng. Sci. Proc., 7-8 (1987) 815-821
51- W. Pannhorst, M. Spallek, R. Brückner, H. Hegeler, C. Reich, G. Grathwohl, B. Meier and D. Spelmann, Ceram. Eng. Sci., Proc., 11 [7-8] (1990) 947-963
52- E. Fitzer and R. Gadow, Mater. Sci. Research, 20 (1986) 571-607
53- D.P. Partlow, Adv. Ceram. Mater., 3 [6] (1988) 553-556
54- K. Nakano, A. Kamiya, M. Iwata and K. Oshima, in "Developments in the Science and Technology of Composite Materials", Proc. ECCM-3, March 20-23, Bordeaux (A.R. Bunsell, P. Lamicq and A. Massiah, eds.), pp. 381-387, Elsevier Applied Science, London and New York
55- J.K. Guo, Z.Q. Mao, C.D. Bao, R.H. Wang and D.S. Yan, J. Mater. Sci., 17 (1982) 3611-3616
56- J. Jamet, J.R. Spann, R.W. Rice, D. Lewis and W.S. Coblenz, Ceram. Eng. Sci. Proc., 7-8 (1984) 677-694
57- R. Lundberg, L. Kahlman, R. Pompe, R. Carlsson and R. Warren, Amer. Ceram. Soc. Bull., 66 (1987) 330-333
58- S.T. Buljan, J.G. Baldoni and M.L. Huckabee, Amer. Ceram. Soc. Bull., 66 (1987) 347-352
59- P.D. Shalek, J.P. Petrovic, G.F. Hurley and F.D. Gac, Amer. Ceram. Soc. Bull., 65 (1986) 351-356
60-Y. Le Peticorps, R. Pailler and R. Naslain, Composites Sci. Technology, 35 (1989) 207-214
61- U. Gaur and B. Miller, Composites Sci. Technology, 34 (1989)35-51
62- W.A. Fraser, F.H. Ancker, A.T. Dibenedetto and B. Elbirli, Polymer Composites, 4 [4] (1983) 238-248
63- J.P. Favre, G. Desarmot, V. Orlionnet and F. Saint-Antonin, Proc. JNC-5 (C. Bathias and D. Menkes, eds.), Sept. 9-11, 1986, Paris, pp. 209-223, Pluralis, Paris
64- D.B. Marshall and W.C. Oliver, J. Amer. Ceram. Soc., 70 [8] (1987) 542-548
65- D.B. Marshall, J. de Physique, Colloque C5 (Suppl. n° 10), 49 (1988) C5-25 - C5-33
66- D.H. Grande, J.F. Mandell and K.C.C. Hong, J. Mater. Sci., 23 (1988) 311-328
67- N. Piquenot, V. Arnault and D. Rouby, in "Composite Materials for Applications at High Temperatures" (R. Naslain, J. Lamalle and J.L. Zulian, eds.), Proc. AMAC/CODEMAC Coll., March 29-30, 1990, Bordeaux, pp. 81-92, AMAC/CODEMAC, Paris and Bordeaux

68- J.D. Bright, D.K. Shetty, C.W. Griffin and S.Y. Limaye, J. Amer. Ceram. Soc., 72 [10] (1989) 1891-1898
69- A.G. Evans and D.B. Marshall, Acta Metall., 37 [10] (1989) 2567-2583
70- J.F. Mandell, D. H. Grande and J. Jacobs, J. Eng. Gas Turbines and Power, 109 (1987) 267-273
71- J.F. Jamet, D. Lewis and E.Y. Luh, Ceram. Eng. Sci. Proc., 7-8 (1984) 625-641
72- J.F. Jamet, D. Abbe and M.H. Guyot, in Proc. ICCM-V, July 30/Aug. 1, 1985, San Diego (W.C. Harrigan et al., eds), pp. 569-585, TMS-AIME, Warrendale (Penn., USA), 1985
73- P. Pluvinage, Thesis n° 584, Univ. Bordeaux, April 12, 1991
74- H.C. Cao, E. Bischoff, O. Sbaizero, M. Rühle and A. Evans, J. Amer. Ceram. Soc., 73 [6] (1990) 1691-1699
75- K.M. Prewo and J.J. Brennan, J. Mater. Sci., 17 (1982) 1201-1206
76- M. Bouquet, J.M. Birbis, J.M. Quenisset and R. Naslain, Proc. ICCM-6/ECCM-2 (F.L. Mathews et al., eds.), 2 (1987) 48-59, Elsevier Applied Science, London and New York
77- P.J. Lamicq, G.A. Bernhart, M.M. Dauchier and J.G. Mace, Amer. Ceram. Soc. Bull., 65 [2] (1986) 336-338
78- L. Filipuzzi, G. Camus, J. Thebault and R. Naslain, Proc. 11th Risø Int. Symp. Metallurgy and Mater. Sci., "Structural Ceramics-Processing, Microstructures and Properties" (J.J. Bentzen et al., eds.), pp. 283-289, Risø Nat. Lab., Roskilde, DK, 1990
79- K.M. Prewo, B. Johnson and S. Starrett, J. Mater. Sci., 24 (1989) 1373-1379
80- L. Filipuzzi, R. Naslain and J. Thebault, in "Composite Materials for Applications at High Temperatures" (R. Naslain, J. Lamalle and J.L. Zulian, eds.), Proc. AMAC/CODEMAC Coll., March 29-30, 1990, Bordeaux, pp. 289-301, AMAC-CODEMAC, Paris and Bordeaux, 1990
81- L. Filipuzzi, Thesis n° 593, Univ. Bordeaux, March 29, 1991
82- L. Filipuzzi and R. Naslain, Proc. 7th CIMTEC World Ceram. Congr. and Satellite Symp., Montecatini Terme, June 24-30, 1990 (in press)
83- J.C. Cavalier, A. Lacombe and J.M. Rouges, in "Developments in the Science and Technology of Composite Materials", (A.R. Bunsell, P. Lamicq and A. Massiah, eds.), Proc. ECCM-3, March 20-23, 1989, Bordeaux, pp. 99-110, Elsevier Applied Science, London and New York, 1989
84- P. Lamicq, in "High Temperature Materials for Power Engineering" (E. Bachelet et al., eds.), Proc. Conf., Liège, Sept. 24-27, 1990, Part 2, pp. 1559-1574, Kluwer Academic Publishers, Dordrecht, 1990

Intermetallics - Fundamentals and Prospects

H.-J. Engell, A. v. Keitz, and G. Sauthoff
Max-Planck-Institut für Eisenforschung GmbH
D - 4000 Düsseldorf

Abstract

An overview is given on intermetallics with respect to the possibilities for the development of new structural materials for high-temperature applications. In the first main section the fundamentals of intermetallics are discussed, i.e. the type and strength of bonding and its correlation with materials properties, in particular with elasticity and diffusion. In the second main section present materials developments are overviewed which are based on titanium aluminides, nickel aluminides and less-common phases. Characteristic data are presented, and the problems and prospects are discussed.

---

1. Introduction

Intermetallic phases are compounds of metals with long-range ordered lattice structures which are different from those of the components. These structures form because there is a very strong bonding of the unlike atoms, and from this strong bonding particular physical and mechanical properties result. Intermetallic phases had been in use for various purposes since many centuries because of their comparitively high hardness (Table 1 a), whereas in modern times their particular physical properties have been of primary interest (Table 1 b).

Table 1: Some applications of intermetallic phases-

a) in past centuries (1):

phase	application	since about
Cu_3As	coating of bronze (Egypt)	2500 B.C.
$Cu_{31}Sn_8$	mirrors (China)	0
Sn_8Hg	mirror surfaces (Venice)	1500
$Ag_2Hg_3 + Sn_6Hg$	dental amalgam (China)	600
Cu_4Hg_3	dental amalgam (Germany)	1500
SbSn	Type metal	1540

b) in this century:

phase	application	since about
Ni_3Fe	high permeability magnetic alloy (Permalloy)	1920
FeCo(-2V)	soft magnetic alloy (Permendur)	1930
$Fe_3(Si,Al)$	magnetic head material (Sendust)	1935
$SmCo_5$	permanent magnets	1970
Nb_3Sn	A15-superconductors	1965
CuZnAl CuNiAl	shape-memory alloys	1960
NiTi	shape-memory alloys (Nitinol)	1965
$MoSi_2$	heater elements (Mosilit, Super-Kanthal)	1955
NiAl,CoAl	protective coatings	1965

The high hardness is correlated with high strength at high temperatures, and thus intermetallic phases have been regarded as promising with respect to the development of new structural materials for high-temperature applications. This was realized already more than 30 years ago, and materials developments were initiated (2). However, because of the brittleness of the intermetallic phases and experimental difficulties these first initiatives were not continued.

Since several years there is a renewed and more pronounced interest in intermetallic phases with respect to materials developments for high-temperature applications all over the world - in particular in the U.S.A., Japan and Germany (3). The present situation is to be reviewed in this paper. In the first main part the basic physical principles and mechanisms from which the formation, the crystal structure and the particular physical and mechanical properties of the IPs result are overviewed. In the second part the present materials developments and their prospects are summarized.

2. Fundamentals

In the first part of this section the formation of intermetallic phases is related to the various types of atomic bonding. In crystals the bonding strength is characterized by the bonding energy which is correlated with the macroscopic enthalpy of phase formation. In the second part the correlation of the formation enthalpy with some basic properties of intermetallics is discussed in order to get guidelines for the selection of candidate phases for high-temperature applications.

2.1 Bonding in crystalline solids

The different types of bonding are (4,5):
a) Covalent bonding: symmetric orbital overlap effected by common electron pairs, e.g. silicon.
b) Metallic bonding: free electrons in a partially filled valence band common to the crystal, e.g. natrium.
c) Ionic bonding: electrostatic interaction between oppositely charged atoms i.e. ions, e.g. NaCl.
d) Other weaker types of bonding, e.g. van der Waals bonding: the bonding atoms mutually induce a dipole moment, e.g. rare gas crystals.
Figure 1 shows the bonding energy for crystalline solids with various types of bonding.

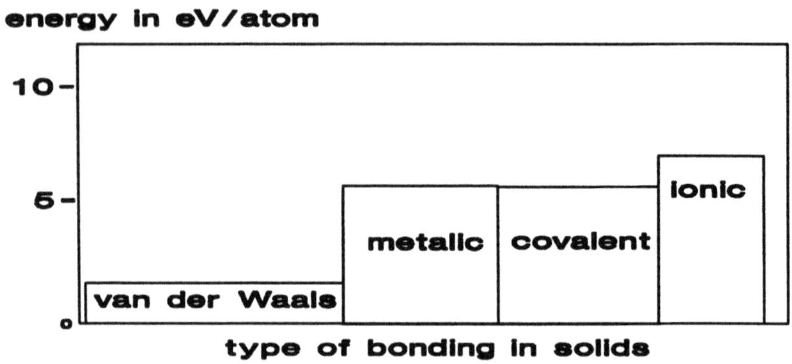

Figure 1: Bonding energies for different types of bonding (6).

2.2 Types of intermetallics

For the present discussion intermetallics are defined as chemical compounds of two ore more metals with limited range of stoichiometry, the crystal structures of which exhibit atomic long range order and are different from those of the constituents (7).

The type of bonding in intermetallics is determined by the electron configuration of the compounds. The electron configuration of an atom is reflected by the position of the element in the periodic system of the elements (PSE) (8,9). The electron configuration is described by the valence state which is characterized by:

- period number n (n = 1, 2 ...)
- type of valence orbital a (a = s, p, d, f)
- number of electrons in the valence orbital m (m = 1, 2 ...)

The valence state is expressed as $n\,a^{m}$. The valence states of the metals are shown in Figure 2.

$2s^m$ Li Be			
$3s^m$ Na Mg			$3p^m$ Al Si
$4s^m$ K Ca	$3d^m$ Sc Ti V Cr Mn Fe Co Ni	$3d^{10}3s^m$ Cu Zn	$4p^m$ Ga Ge As
$5s^m$ Rb Sr	$4d^m$ Y Zr Nb Mo Tc Ru Rh Pd	$4d^{10}4s^m$ Ag Cd	$5p^m$ In Sn Sb
$6s^m$ Cs Ba	$5d^m$ La Hf Ta W Re Os Ir Pt	$5d^{10}5s^m$ Au Hg	$6p^m$ Tl Pb Bi
$7s^m$ Fr Ra	$6d^m$ Ac		
m = 1-2	m = 1-8	m = 1-2	m = 1-3
s-metals	transition-metals	B-subgroup-metals	

Figure 2: Valence states of the metals.

The crystal structure of intermetallic phases can be classified with respect to the valence states of the constituents. For binary intermetallics the correlation between phase type, valence state of the constituents and crystal structure is summarized in Table 2. According to Hume-Rothery the electron compounds in Table 2a are compounds which crystallize in a particular structure if the number of valence electrons per atom corresponds to 21/14, 21/13 or 21/12 (11). The size factor compounds comprise the Laves phases for which the ratio of the atomic radii is described by $R_A/R_B \sim 1{,}225$ and the sigma phases the structure of which results from a particular stacking of atoms (Kagomé nets) (12). The wurtzite and zincblende type structures are similar to the diamond structure or stacking variations of it (13). The Zintl phases are compounds with a total sum of valence electrons equalling 8. The coarse classification of phases -by various different criteria- in Table 2 according to valence state leads to rather complex groups (a) - (c). This indicates the complexity of the intermetallic phases and the limited significance of simple classification criteria.

Table 2: Phase type, atom valence state and structure type of some binary intermetallics A_xB_y (10):

valence state		example	
A_x	B_y	phase	structure type

a) Electron compounds (Hume-Rothery phases) and size-factor compounds (e.g. Laves phases, Sigma phases)

nd^{1-8}	ns^{1-2}	$MgNi_2$	hexagonal (C36)
nd^{1-8}	nd^{1-8}	$TaFe_2$	hexagonal (C14)
nd^{1-8}	nd^{10}	Ni_3Pt	cubic ($L1_2$)
nd^{1-8}	np^{1-3}	NiAl	cubic (B2)
nd^{10}	nd^{10}	Cu_3Au	cubic ($L1_2$)
nd^{10}	ns^{1-2}	$MgZn_2$	hexagonal (C14)
ns^{1-2}	ns^{1-2}	KNa_2	hexagonal (C14)

b) Phases with crystal structures derived from wurtzite-type and zincblende type structure

nd^{10}	np^{1-3}	$CuSe_2$	orthorhombic (C18)
		Ag_3Al	cubic (A13)

c) Zintl phases

ns^{1-2}	np^{1-3}	NaTl	cubic (B32)
		Mg_2Si	tetragonal (C1)
np^{1-6}	np^{1-3}	AlB_2	cubic (C32)

2.3 Bonding in intermetallics

A first group of intermetallics shows physical properties which are similar to those of the metals. This indicates prevalent metallic bonding, and an example is given by the Laves phase KNa_2: bonding results from delocalized bonding electrons which interact with the atomic core (14).

A second group of intermetallics has localized bonding electrons, i.e. strong ionic bonding. This is exemplified by the Zintl-phases (15). The group of Zintl phases form two subgroups:

- intermetallics with 8 valence electrons, i.e. complete octets. The ion with np^{4-5} valence is supposed to be isolated in the crystal. (Interactions between these ions are only weak because of their large distances). The crystal structures of these intermetallics correspond to those of typical salts. The acting bonding forces are Coulomb forces with a very small contribution of covalent bonding. An example is Mg_2Si with the anti-CaF_2-type structure (C1).

- intermetallics without complete electron octets which form anion sublattices. The small atom separation in these sublattices is supposed to indicate a strong covalent bonding between anion atoms. Between the anions and cations there is the strong ionic bonding.

A third group of intermetallics has partially localized bonding electrons, i.e. there is a contribution of covalent bonding (16). These partially localized electrons still have a high mobility because they can be promoted into non-bonding states corresponding to the conduction band in metals. This bonding concept is similar to that of pure transition metals of valence state nd^{1-8}, e.g. the fcc metal nickel with $3d^8$.The bonding in intermetallics is effected by the electron orbitals which overlap and form bands (17,18,19). The various types of overlapping orbitals lead to complex band structures (20,21). In particular, there are bonding electron states which enhance the overall atom bonding, and anti-bonding states which weaken the overall bonding. This is illustrated by Figure 3 which shows the bonding energy in the crystal for the pure $3d^{1-10}$ metals. Clearly visible is the change from bonding states to anti-bonding states when the number of valence electrons increases from 5 to 6 (V→Cr) and the inverse change for the transition from Mn to Fe.

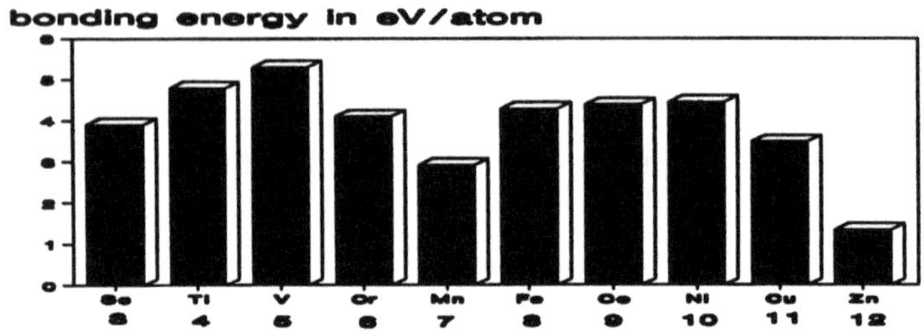

Figure 3: Bonding energy and number of valence electrons for the transition metals (6).

In the intermetallic NiAl the bonding is determined by the 3d orbitals as in pure nickel (22,23). In NiAl there are interactions between the $3d^8$-transition metal nickel and the $3p^1$ B-subgroup metal aluminium. The main contribution to the bonding energy results from the splitting into a bonding 3d-3p band and an antibonding $3d^*$-$3p^*$ band. It may be supposed that because of symmetry the 3d orbital contributes only partially to d-p bonding, i.e., there is still some 3d-3d bonding as in pure nickel. The d-d bonding is weaker than the p-d bonding. The s-orbitals (low energy orbitals) can be regarded as non-bonding in first approximation which means, that these orbitals don't contribute directly to the bonding character in NiAl. From this, the resulting band structure can be derived as shown schematically in Figure 4.

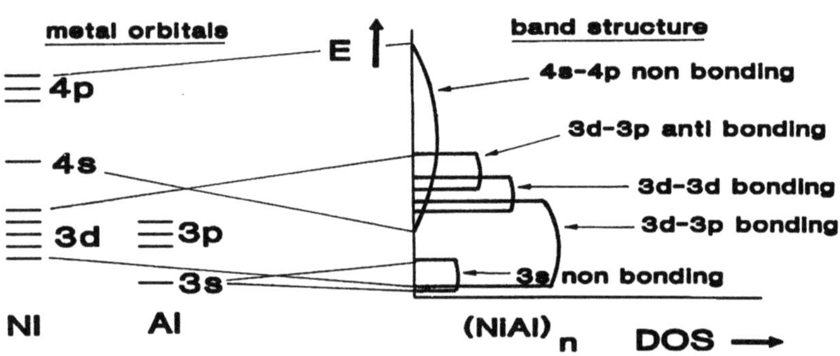

Figure 4: Density of states as a function of energy (schematic) (24).

The band structures can be determined experimentally by various methods, in particular by XPS (X-ray photon spectroscopy) and BIS (Bremsstrahlung isochromatic spectroscopy). The XPS and BIS spectra for NiAl were measured and a sufficient agreement with the model was found (22).

2.4 Correlation between bonding energy and thermodynamic quantities

The macroscopic equivalent to the bonding energy is the so-called internal energy U of the intermetallic which is related to the enthalpy of formation (25). This enthalpy can be calculated in an approximative way by Miedema's semi-empirical approach (26). The basic assumption is that the crystal can be subdivided into atomic cells (Wigner-Seitz Cells) (27), which contain the valence electrons. Mixing of metals ,e.g. Ni and Al, leads to cell size changes from which a negative contribution to the formation enthalpy results. Hybridization effects which result from the bonding between nd^{1-8} metals and sp metals are considered by a particular R-term which is obtained by fitting to experimental data (28). The calculated enthalpy ΔH is shown in Figure 5 for the Ni-Al system.

Δ H can also be calculated by means of a quantum mechanical model which is based on the density function theory of Kohn and Sham and on the energy-band theory by Andersen (29). The model considers the particular elements, their concentrations and the particular crystal structures. The model was applied to the Ni-Al system (23), and Figure 5 shows the calculated heat of formation in comparison to Miedema's estimate and experimental values by Hultgren (30).

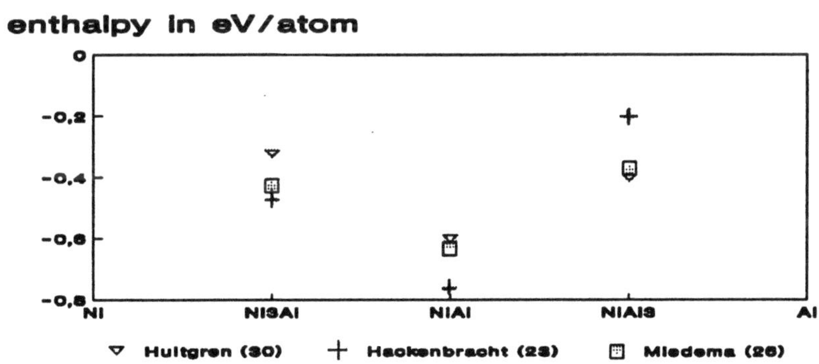

Figure 5: Heat of alloy formation in the system Ni-Al.

Miedema's simple approach gives a sufficient agreement with experimental values and with the quantum mechanical model, and thus it may be used for estimating the formation enthalpy. In the following the formation enthalpy is used for characterizing the bonding energy and the stability of the regarded intermetallic phases. It was found that the maximum formation enthalpy for phases in a binary system is correlated with the number of phases formed, i.e., the higher the formation enthalpy, the higher the number of phases in the system (31). This correlation is shown in Figure 6. In the regarded Ni-Al system with an enthalpy maximum of about 48 kJoule/mol atom (stoichiometric NiAl) four stable intermetallics exist.

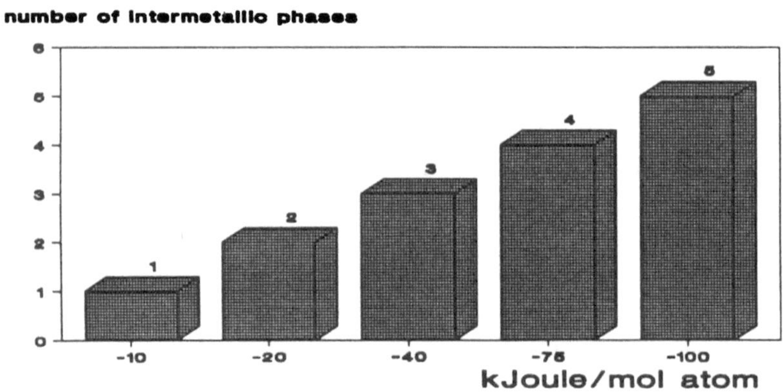

Figure 6: Relation between enthalpy and number of theoretically stable intermetallic phases in a binary system (31).

Figure 7 shows the correlation between formation enthalpy ΔH and melting temperature T_m for the 3d-metal aluminides. A close correlation between ΔH and T_m is visible only for the phases FeAl, NiAl, CoAl which have the same crystal structure B2. Obviously the relation between ΔH and T_m is different for other phases with different structures. The reason for this behaviour is that Tm is determind by the differences in enthalpy and entropy between the liquid state and solid state, i.e. there is only an indirect correlation between ΔH and T_m.

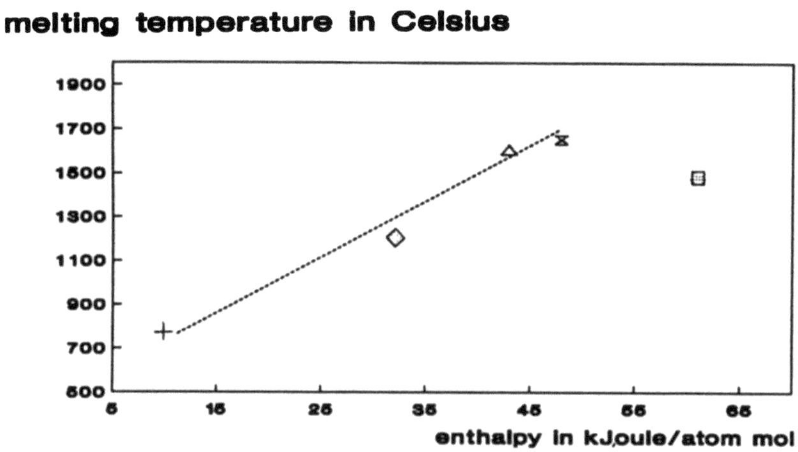

Figure 7: Correlation between measured formation enthalpy and measured melting temperature (30) in the binary $3d^{1-8}$ valence metal aluminides which have the crystal structures B2: ◇ FeAl, △ CoAl, ⧗ NiAl, + CuZn, and $L1_0$:□ TiAl.

2.5 Properties of intermetallic phases

In view of applications as structural materials the behaviour during deformation, i.e. mechanical behaviour of intermetallics, is of primary interest. Deformation is either elastic or plastic. In the simplest case the elastic deformation of a specimen is described by Hooke's law (32):

$$\sigma = E * \varepsilon \qquad \text{Equ. 1}$$

with σ = stress, ε = strain and E = Young's modulus. E is a material parameter which characterizes the material elasticity and depends directly on the atom bonding strength (33). Therefore E should be related to ΔH. Figure 8 shows the correlation for the Ni-Al system. There is an appreciable scatter of the experimental E values which may be due to differences in preparation and testing methods. However, it can be seen that E and ΔH vary with phase composition in similar ways if the single deviating value for Ni_3Al from ref.(34) is neglected.

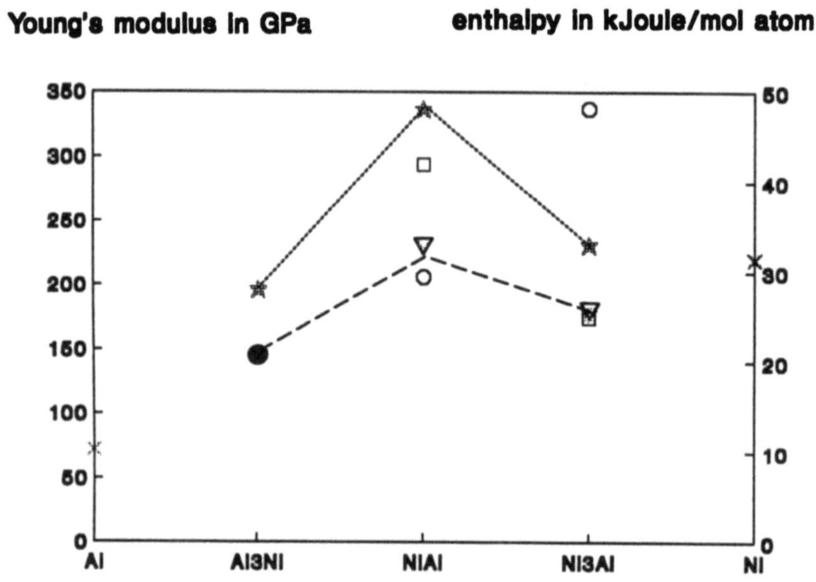

Figure 8: Measured Young's modulus (GPa)(a) and measured formation enthalpy (kJoule/mol atom)(b) in the Ni-Al system:
a): ○ Munroe/Baker (34), ▽ Destefani(35), □ Marmouche/Stoloff (36) and ● v.Keitz/Sauthoff (37) × Landolt/Börnstein (38).
b): ✱Hultgren (30).

Plastic deformation at low temperatures is due to the motion of dislocations on crystallograghic slip planes. The dislocation is characterized by its Burgers vector which gives the slip direction in the slip plane. In B2 structures there are various possible slip systems depending on the bonding character and ratio of atomic radii (39). According to the observed active slip systems B2 compounds form three groups: phases with ionic bonding, phases with metallic-covalent bonding and phases which can be described by pure metallic bonding. The observed active slip systems in B2 compounds with different types of bonding are shown in Table 3.

Table 3: Observed active slip systems in differently bonded B2-structures (R_A, R_B = atomic radi).

bonding character	active slip system	example	R_A/R_B	Ref.
ionic	<001>{110}	TlCl	0,833	39
metallic-covalent	<001>{110}	MgTl	0,780	39
	<001>{100}	NiAl	0,847	39
	<001>{110}			
	<001>{110}	FeAl	0,945	40
	<111>{1$\bar{1}$0}			
metallic	<111>{1$\bar{1}$0}	CuZn	0,913	39

In intermetallics with metallic-covalent bonding the type of observed Burgers vectors in {110} plane is correlated with the ratio of atomic radii which depends on the respective valence states. In long-range ordered structures the dislocations can split into partials with two-dimensional antiphase boundaries (APBs) or stacking faults (SFs) in between to form superdislocations. The energy of the superdislocation is determined primarily by the APB energy which has been calculated by Hong and Freemann for 1/2<111>{110} slip systems for stoichiometric NiAl and for several NiAl alloys where Ni or Al has been substituted partially by other metal atoms (41) (see Table 4). In this calculation a quantum mechanical approach was used which considers the directional metallic-covalent bonding between the nickel and aluminium atoms and the effect of solute atom. The production - and to some extent the mobility - of dislocations depends on the dislocation energy, i.e. on the APB-energy. Therefore, partial substitution of the constituent elements gives the possibility to control the deformation behaviour.

Table 4: Calculated energies for the 1/2<111>{110} APB for stoichiometric NiAl with various solute metals (41).

composition	APB energies (mJoule/m^2)
$Al_6\ Ni_6$	1000
$Al_6Ni_4Mn_2$	740
$Al_6Ni_4Cr_2$	510
$Al_6Ni_4V_2$	250
$Al_4V_2Ni_6$	550
$Al_4Cr_2Ni_6$	250

Plastic deformation at high temperatures is controlled by non- conservative dislocation movements which are controlled by local diffusion mechanisms. The temperature dependence of diffusion is described by an Arrhenius-type equation:

$$D = D_0 * \exp(-Q_D/k*T) \qquad \text{Equ. 2}$$

with D_0 = frequency factor and Q_D = activation energy. The effective activation energy Q_D consists of two terms correlated with the formation and the motion of point defects. Although these prozesses may be very complicated in ordered lattice structures they are correlated with the bonding energy in the lattice and therefore with the enthalpy of formation. Tracer diffusion data are given in Table 5 for the Ni-Al system and compared with data for CuZn which has the same crystal structure as NiAl, but its bonding energy is appreciably lower. The comparison between NiAl with metallic-covalent bonding and CuZn with typical metallic bonding shows a good correlation between bonding energy and activation energy. The smaller difference in activation energy and formation enthalpy between NiAl and Ni_3Al implies similar diffusion mechanisms and bonding types in both intermetallics. However, this necessitates a more detailed discussion which has to consider the particular diffusion mechanisms and the particular point defects which are involved.

Table 5: Tracer diffusion data and formation enthalpies for NiAl, Ni_3Al and CuZn.

intermetallic	D_0 (cm^2/s)	Q_D (eV)	ΔH (kJoule/mol*atom)	Ref.
NiAl (Ni^{63})	4.5	3.19	-48	42
Ni_3Al (Ni^{63})	146	3.00	-33	43
CuZn (Zn^{65})	80	1.56	-10	44

The deformation behaviour of ordered intermetallics, e.g. NiAl, is affected by constitutional defects which are present in off-stoichiometric phases. At high temperatures the stoichiometric alloy is the most creep-resistant one (Fig.9), because the constitutional defects in off-stoichiometric phases are mobile and result in increased diffusion rates, i.e. increased creep rates as can be seen in

Fig.10. At lower temperatures creep resistance is minimum at the stoichiometric composition, since the constitutional defects then are immobile and impede dislocation movements (45).

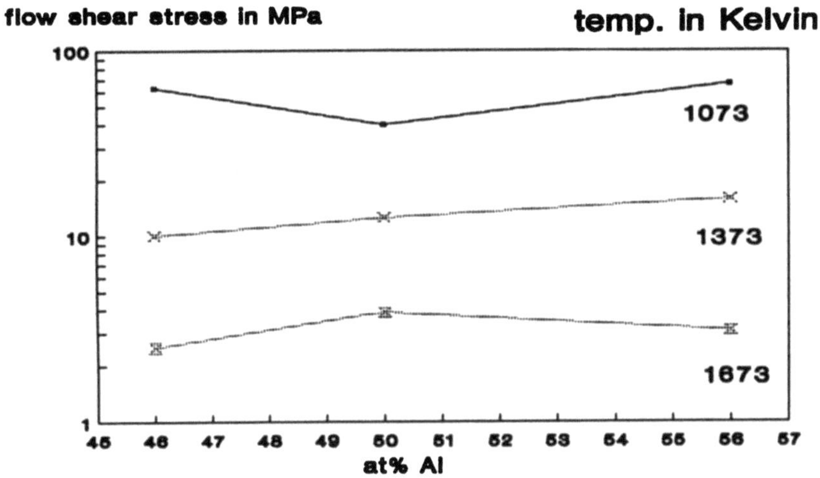

Figure 9: Creep resistance at $10^{-4}s^{-1}$ deformation rate as a function of composition for NiAl for various temperatures (45).

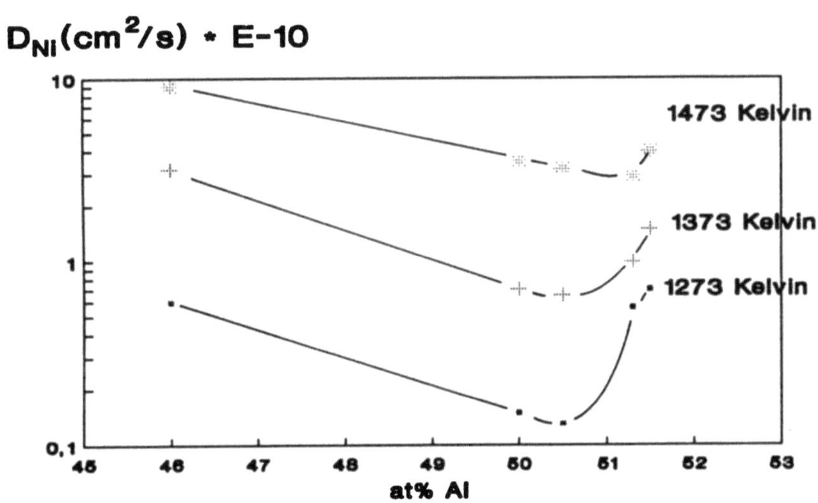

Figure 10: Composition dependence of diffusion coefficient D_{Ni} in NiAl (46).

The pronounced bittlenes of most intermetallics is due to the reduced number of slip systems and the high dislocation energy, which prevent the relaxation of stress at crack tips by plastic deformation. Furthermore in many intermetallics cracking is intergranula, i.e. the grain boundaries are the preferred crack path. This can be correlated to the fact that in addition to the lattice distortion at the grain boundary there exists a reduced number of bonding A-B atom pairs. In any case, a correlation between formation energy and brittlenes is observed as Table 6 shows.

Table 6: Correlation between fracture strain and heat of formation for various polycrystalline $L1_2$ phases (47).

$L1_2$ phases	fracture strain %	ΔH (kJoule/mol)
Ni_3Al	5	-28
Co_3Ti	13	-25
Ni_3Fe	22	- 1
Ni_3Mn	23	- 8

3. Materials Developments

A high potential of development is attributed to the intermetallic phases because of their outstanding physical and chemical properties, in particular after the successful ductilization of some phases (47-50). Therefore, intermetallic phases have become subject of fundamental research, materials developments on the basis of intermetallic phases are under way, and new developments are started as is documented by the proceedings of international conferences on intermetallic phases. (e.g. (51-53)).

In West Germany only few isolated studies were done before 1985 (3). After 1985 various research programs were established for promoting the fundamental research on intermetallic phases and for initiating new materials developments(54).

The expectations are illustrated by Fig. 11. The use of superalloys with a good combination of strength and toughness is presently limited to temperatures below 1100°C. The modern engineering ceramics can be used at much higher temperatures, but they are inherently brittle because of the covalent atomic bonding. The intermetallics are expected to fill the gap between the superalloys and the ceramics not only with respect to the service temperature range, but also with respect to the properties: higher strength than the superalloys because of the strong atomic bonding and less brittleness than the ceramics because metallic bonding predominates.

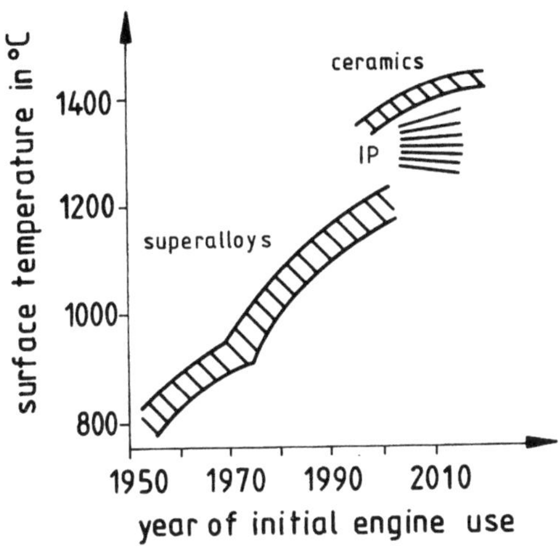

Figure 11: Developments of superalloys and ceramics for gas turbine applications(3) and desired range of the intermetallic phases (IP)(55).

3.1 Criteria for Phase Selection

First of all, a phase for high-temperature applications must have a sufficient strength at service temperature which means a sufficient creep resistance. The creep resistance scales with the diffusion coefficient and with the shear modulus (56), and both parameters scale with the melting temperature T_m(57). Thus the phase should have a sufficiently high T_m. For conventional high-temperature materials the maximum service temperature is usually about 3/4 T_m, i.e. $T_m \geq$

1600°C for service temperatures of 1100°C or higher.

Density is an additional parameter of high importance. For moving parts the ratio of rupture strength σ_f and specific weight ρ is decisive. This ratio is called rupture length and is about 15 km for advanced superalloys. Phases which contain light elements - e.g. Ti, Al, Si, Mg - may have such a low density that they compare favourably with conventional alloys in spite of lower strength or restricted service temperature range.

The major problem of the strong intermetallics is their brittleness which, however, should be less severe than that of the ceramics because of the different type of atomic bonding. A closer look indeed shows that the deformability of intermetallics is between those of metals and ceramics as is exemplified by Fig. 12. The intermetallic NiAl which is a candidate phase for high-temperature applications softens with rising temperature at about half the melting temperature T_m which is the range of the covalent Si and is higher than that of the ceramic Al_2O_3. However, the brittle-to-ductile transition of NiAl is at about 0.4 T_m, whereas this transition is only above 0.8 T_m for Si and Al_2O_3. Thus NiAl indeed bridges the gap between metals and ceramics.

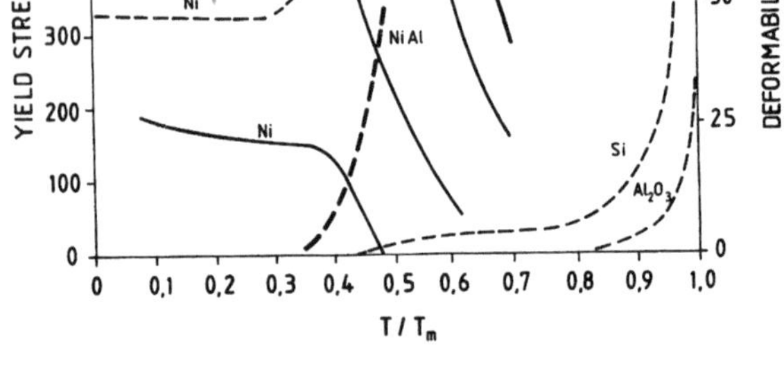

Figure 12. Yield stress and deformability for various materials as a function of homologous temperature T/T_m (T_m = melting temperature)(58).

Plastic deformation is more difficult in intermetallics than in metals because of the stronger atom bonding and the ordered atom distribution, i.e., the less symmetric crystal lattices(59). Experience indeed shows that the brittleness of intermetallic phases increases with decreasing lattice symmetry and increasing unit cell size(60). Thus intermetallic phases with high crystal symmetry - possibly cubic phases - and small unit cells are preferred for developments of structural materials. Figure 13 shows some simple crystal structures with ordered atom distributions which are derived from the bcc and fcc lattice, respectively. The $L1_0$ structure and the DO_{22} structure are slightly tetragonal whereas the others are cubic.

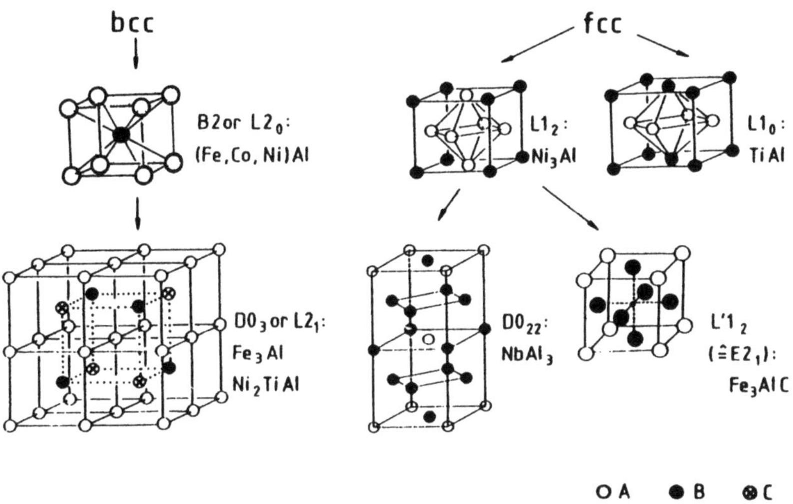

Figure 13: Some simple intermetallic crystal structures which are derived from the bcc and fcc structure, respectively, with typical examples(55).

Intermetallics for high-temperature applications must be oxidation-resistant in most cases. Oxidation resistance is provided by the presence of elements which can form protective oxide layers. Such elements are first of all Cr, Al and Si (61). However, chromium oxide and silica are volatile at low oxygen pressures above 1000°C, and silicon oxide may form low-melting silicates. Thus aluminides are strongly favoured for high-temperature applications. In cases of insufficient oxidation resistance protective coatings may be applied (62-64). However, the stability of such coatings decreases with increasing temperature, and therefore phases for applications at 1100°C and higher should be oxidation-resistant inherently.

Finally the preparation of components from the intermetallic in question must be possible with reasonable quality, i.e. with a sufficiently low content of inherent macroscopic defects. The processing of the high-strength intermetallics is difficult because of their brittleness, and thus the development of the necessary processing techniques is a very demanding task. Inherent defects increase the apparent brittleness and decrease the fracture strength, and thus they may preclude any application.

In the following survey which is far from being complete the intermetallic phases are discussed which today are subject of materials developments or which are attractive with respect to such developments. The first group of "classic" intermetallics is only small and the most important representatives are the two titanium aluminides Ti_3Al and TiAl and the $L1_2$ nickel aluminide Ni_3Al. Interest is concentrated on these phases, and they are subject of most of the present research. Other phases which are attractive according to the above criteria have received little or no interest up to now and thus may be called less-common phases.

3.2 Titanium Aluminides

3.2.1 Ti_3Al and TiAl

The low density of the titanium aluminides made them attractive for the development of lightweight materials, and already in 1956 TiAl was considered as a base for high-temperature alloys(65). Materials developments were started in the USA about 16 years ago and have reached a very high level in comparison with other intermetallics(66). The phases on which materials developments have been based are Ti_3Al with DO_{19} structure (ordered close-packed hexagonal) and TiAl with $L1_0$ structure (tetragonally distorted ordered fcc).

On the basis of Ti_3Al the so-called α_2 alloys have been developed in the USA which are now being commercialized (67). The "ductilization" of these alloys has been achieved by alloying with Nb and thermomechanical processing. The alloys contain 23 - 25 at. % Al, 8 - 18 at. % Nb and few % Mo, V, Ta, and the resulting fine-disperse multiphase structures primarily consist of α_2-Ti_3Al and disordered ß-Ti-Al-Nb with the additional orthorhombic O-phase Ti_2AlNb. Besides thermomechanical processing hydrogen treatments are used for refining the microstructure.

On the basis of TiAl the γ alloys are developed which are near commercialization, too, in the USA(67). The γ alloys contain 46 - 52 at. % Al and in addition up to 10 at. % V, Mn which improve ductility, and Nb, Ta, W, Mo which improve strength and oxidation resistance. The γ alloys are single-phase, or they contain α_2-Ti_3Al as additional phase, and they can be prepared by powder metallurgy or ingot metallurgy. Optimum properties are obtained by thermomechanical treatments(67). The best ratio of strength and ductility is shown by two-phase γ alloys with minimum oxygen content. In contrast to α_2-Ti_3Al the hydrogen solubility in γ-TiAl is low, but there are indications of hydrogen embrittlement, too(67).

With respect to applications in the USA, the "National Aero-Space Plane (NASP)" program is of particular importance(68). For this program the α_2 alloys have been selected besides other non-metallic materials because of their advanced development stage. However, a higher potential is attributed to the γ alloys because of higher elastic modulus, lower density, and better high-temperature capability(69). For improving the strength the α_2 alloys as well as the γ alloys are considered as matrix alloys for composite materials with SiC or Al_2O_3 fibres (68-71). With respect to present jet engines, various components have been fabricated from α_2 alloys and γ alloys, and they have been tested successfully. However, up to now there are not yet any flying applications (67).

In Japan a TiAl-based turbocharger rotor has been developed (71). The highest strength and ductility at room temperature is shown by an alloy with 47 at. % Al which contains about 20 vol. % Ti_3Al. The specific strength (i.e., divided by density) surpasses that of the superalloy Inconel 713C. The oxidation resistance of the rotor is achieved by coating. Such rotors are tested under practical conditions and are near commercialization (72).

In Germany various projects are under way or being started in cooperation with industry to develop TiAl-based materials for specific applications. The present state is illustrated by Figs. 14 a - b which show the temperature dependence of strength and fracture strain for a cast TiAl alloy with 45 at. % Al (73-74). Obviously the aging treatments of the alloy which is two-phase with Ti_3Al as second phase reduce both the strength and the deformability, i.e. ductility and toughness. Of particular interest is the anomalous temperature dependence of strength (75-77), i.e. the strength increases with increasing temperature, which is analogous to that of Ni_3Al (see next section). The further work aims at increasing the strength by varying the composition, alloying with further elements and thermomechanical treatments. In particular the alloying with Cr not only increases the hardness, but also the oxidation resistance (73-74).

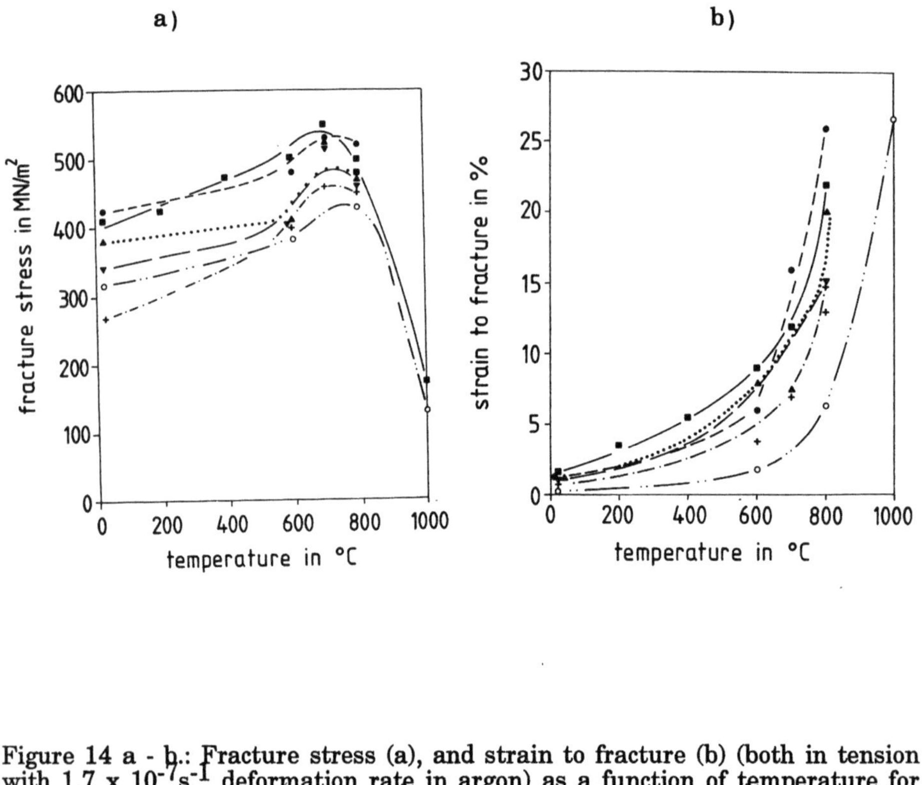

Figure 14 a - b.: Fracture stress (a), and strain to fracture (b) (both in tension with 1.7 x $10^{-7} s^{-1}$ deformation rate in argon) as a function of temperature for cast TiAl with 45 at. % Al after various heat treatments (73,74): ■ as cast, ▲ 168 h at 600°C in air, ▼ 336 h at 650°C in air, ✚ 168 h at 700°C in air, • 3 h at 1000°C in argon, ○ 168 h at 1000°C in argon.

Other work on TiAl is directed at studying the effects of processing on the microstructure and the resulting mechanical behaviour in order to obtain a basis for alloy optimization (78). For the preparation of alloys both ingot metallurgy and powder metallurgy was used. Additions of third elements affect the microstructure in a sensitive way, and besides Nb Mn can be used for improving the ductility (79). Hot working of TiAl induces dynamic recrystallization by which the microstructure - including texture - and the mechanical behaviour can be controlled (80).

3.2.2 Other Titanium Aluminides

The Al-rich titanium aluminide $TiAl_3$ (with tetragonal DO_{22} structure) exhibits only a low strength in spite of its brittleness (81-82). The low ductility of $TiAl_3$ can be improved by alloying with Zr, Hf, Li, B or by substituting Ti by V (83), but up to now $TiAl_3$ has not been subject of materials developments.

The alloying of $TiAl_3$ with Cr, Mn, Fe, Ni, Co, Cu or Pd leads to ternary phases with $L1_2$ structure, and it is expected that these $L1_2$ phases can be "ductilized" as Ni_3Al (see next section) (84-89). Up to now only brittle materials were obtained which may be due to insufficient material quality. The brittleness is lower than that of $TiAl_3$ with DO_{22} structure and an anomalous temperature dependence of strength has been observed (88).

A different development has been started in China on the basis of $TiNbAl_3$ the constitution and crystal structure of which is not yet clear (90). These alloys are regarded as promising for high-temperature applications because of their high strengths and oxidation resistances.

3.3 Nickel Aluminides and Iron Aluminides

3.3.1 Ni_3Al and $L1_2$ Phases

The most familiar and best investigated intermetallic phase is Ni_3Al with $L1_2$ structure (Fig. 13) which is the strengthening γ' phase in the superalloys. The flow stress of Ni_3Al and other $L1_2$ phases exhibits an anomalous temperature dependence as is exemplified by Fig. 15. The 0.2 % flow stress of Ni_3Al increases with increasing temperature up to a maximum at about 650°C, and only at higher temperatures the material softens with rising temperature. This anomaly results from the anisotropy of the energy and mobility of the superlattice screw dislocations which govern the plastic deformation of Ni_3Al (92). The screw dislocations can split on {111} planes and on {010} planes. The splitting on {010} is favoured energetically because the energy of the antiphase boundaries between the partials is lower on {010} (93). However, this superlattice dislocation is sessile because the cores of the partials spread outside the plane of the antiphase boundary. On the other hand, the superdislocation on {111} with higher energy is glissile because the core spreading is confined to the slip plane (92). This glissile state is metastable since the partials must first be coalesced into a single dislocation before cross slip to {010} can occur. Hence dislocations are generated primarily in {111} planes on loading, and slip is confined to {111} at low temperatures. With rising temperature cross slip to {010} becomes possible by thermal activation by which the dislocations are immobilized (Kear-Wilsdorf mechanism). The sessile dislocations act as obstacles and give rise to an increased flow stress. Only at much higher temperatures the glissile dislocations are mobilized again by the enhanced thermal activation which leads to the observed softening above 650°C.

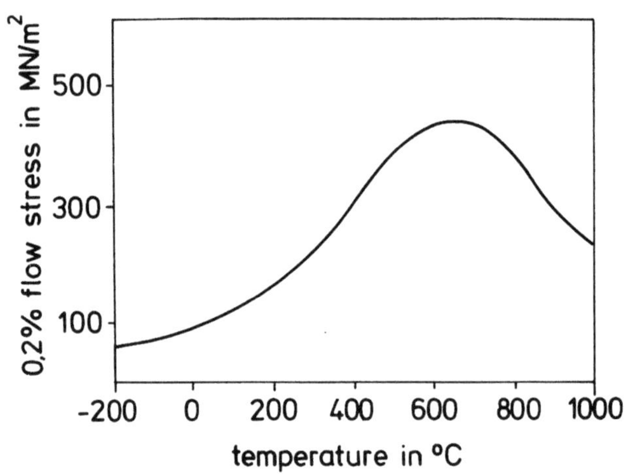

Figure 15: Temperature dependence of the 0.2 % flow stress of polycrystalline Ni_3Al (after (91)).

From this discussion it is clear that the anomalous temperature dependence of the strength of $L1_2$ phases is a function of the bonding forces between the atoms, and indeed the anomaly is exhibited by various $L1_2$ phases to different extents being closely correlated with the stability of $L1_2$ with respect to other possible structures(94). In Ni_3Ge the anomaly is still more pronounced than in Ni_3Al, it is less pronounced in Co_3Ti, Zr_3Al, Fe_3Ga and Cu_3Au, and there is a continuous transition from a positive temperature dependence of the strength to the familiar negative one in $(Ni,Fe)_3Ge$ from Ni_3Ge to Fe_3Ge(95). It has to be noted that analogous effects have been observed in other phases with different structures, too, - e.g. FeCo-2V (B2), CuZn (B2), TiAl ($L1_0$), Fe_3Al (DO_3), Mg_3Cd (DO_{19}), Ni_3V (DO_{22}), Fe_2B (C16) - and various mechanisms have been proposed for these phases (58,60,96-99).

Ni_3Al is not inherently brittle: single crystals are ductile, and only polycrystalline Ni_3Al is brittle. However, the polycrystalline Ni_3Al can be "ductilized". Additions of 200 to 500 ppm boron together with a slight Ni excess and a special casting technique (drop casting) result in a material with about 50 % strain to fracture at room temperature which can be hot-rolled, cold-rolled and deep-drawn, and the high-temperature properties are as good as can be expected for an ordered phase (47,48,100-102). A complete physical understanding of this effect has not yet been reached, and therefore predictions with respect to other phases can not be made (50,96,101-107). For example, the boron effect has not been found with NiAl because either the effect is non-existent or the necessary conditions have not been met(50).

The discovery of the ductilization of Ni_3Al by boron in 1978 by Aoki and Izumi(101) in Japan initiated extensive materials developments in the USA from which the Advanced Aluminides resulted in particular(47). These aluminides are Ni_3Al base materials with Al contents below 25 at. % and with additions of further elements - primarily B, Hf, Cr - for high strength, oxidation resistance and ductility. The Advanced Aluminides are stronger and lighter than the superalloys and are now in the process of commercialization(108). Potential applications include gas, steam and water turbines, aircraft fasteners, heating elements, automotive turbochargers, pistons and valves (109-110).

In spite of this successful materials development and in spite of the fact that Ni_3Al is the most studied intermetallic much more fundamental research is necessary to reach a sufficient understanding of the effects of composition and microstructure on ductility and strength. In particular, the long-term behaviour of the Advanced Aluminides is far from satisfactory with respect to oxidation resistance, ductility and strength. In contrast to the high flow stress of the Advanced Aluminides (47,108) the creep resistance is comparatively low (111), even lower than that of TiAl (112), and both are much lower than that of the advanced ODS superalloy MA 6000(113) as is illustrated by Fig. 16. If, however, the comparison is made with respect to specific creep resistance, i.e. creep resistance per unit density, the situation is more favourable for the aluminides.

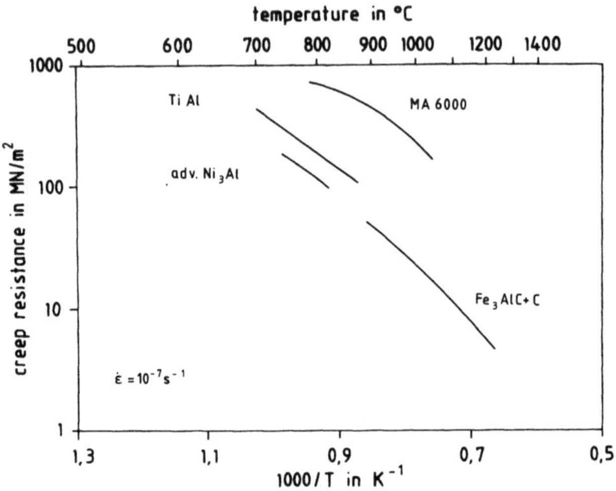

Figure 16: Creep resistance (at a steady state creep rate of $10^{-7}s^{-1}$) as a function of temperature for the Advanced Aluminide Ni_3Al (111), for TiAl (112), for the perovskite-type phase Fe_3AlC with graphite (114) and for the advanced superalloy MA 6000 (113).

3.3.2 NiAl

Ni_3Al-base materials are expected to be applicable in the temperature range of the superalloys, i.e. up to about 1100°C which is already about 80 % of the melting point temperature of Ni_3Al. Thus it cannot be expected to extend this temperature range for such materials. A higher melting point is shown by the other nickel aluminide NiAl with B2 structure which has been regarded as promising for high-temperature applications since more than 30 years (115). However, the two major problems withstanding applications - low-temperature brittleness and inadequate creep resistance at temperatures of interest - are still unsolved (106,117). The creep resistance of NiAl can be increased appreciably by adding third elements(118,119). A significant increase in creep resistance is achieved by alloying NiAl with the B2 aluminides CoAl and FeAl as is illustrated by Figure 17. In particular, it is to be noted that even alloying with the softest aluminide FeAl still strengthens the harder aluminides NiAl and CoAl.

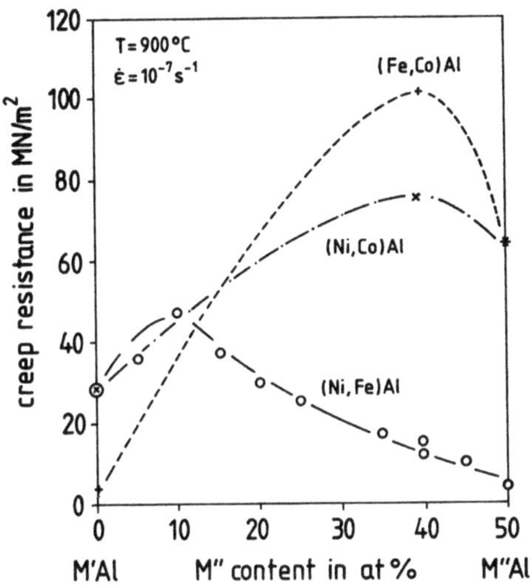

Figure 17 Creep resistance (at $10^{-7}s^{-1}$ secondary rate) as a function of composition for the ternary B2 aluminides (Ni,Fe) Al, (Ni,Co) Al and (Fe,Co) Al at 900°C (56,120).

Apart from this kind of solid solution strengthening the creep resistance can be increased by adding second phases (118,119). The second phase has not to be harder than the matrix. Indeed, two-phase NiAl with precipitated disordered α-Fe particles shows a higher creep resistance than single-phase NiAl of the matrix-phase composition (121). However, in such two-phase materials the compositions of matrix and precipitate are controlled by the two-phase equilibrium and cannot be chosen independently. This leads to an off-stoichiometric matrix in the mentioned case which means a loss in creep resistance. For improving the strength of NiAl harder phases are dispersed or precipitated in the intermetallic matrix, and examples are Ni_2TiAl and TiB_2(30),(122-124). Composites on the basis of such systems are regarded as most promising for applications at temperatures above 1000°C (116).

Apart from this, NiAl forms eutectics with the refractary metals Cr, Mo, W which may be used to produce composite materials by directional solidification (125), and indeed the NiAl-Cr composite shows promising high-temperature strength (126). Such alloys - in particular NiAl-Cr - are now studied intensively in order to develop high-temperature high-strength alloys (127,128). As an example, Figure 18 shows compressive strength data for PM NiAl-Cr alloys according to which the strength indeed increases significantly with increasing Cr content, i.e. increasing volume fraction of the α-Cr phase in these two-phase alloys. An even higher strength is obtained with cast alloys with much coarser grain structure. Fig. 19 demonstrates the drastic effect of different ways of processing on ductility and toughness of such an eutectic alloy. This means that the processing of intermetallic alloys must be controlled carefully for a good and reproducible materials quality. On the other hand, there are good chances for increasing the low toughness by optimizing the alloy with respect to composition, phase distribution and microstructure.

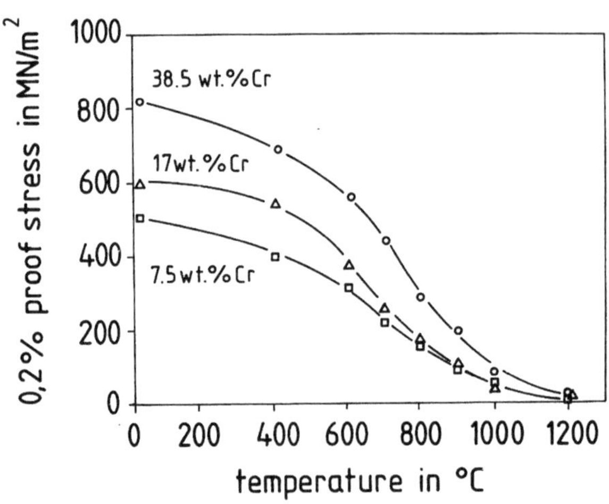

Figure 18: Temperature dependence of strength (in compression) for two-phase PM NiAl-Cr alloys with 7.5 wt. % Cr, 17 wt. % Cr, and 38.5 wt. % Cr (eutectic composition), respectively, as HIPed(127,128).

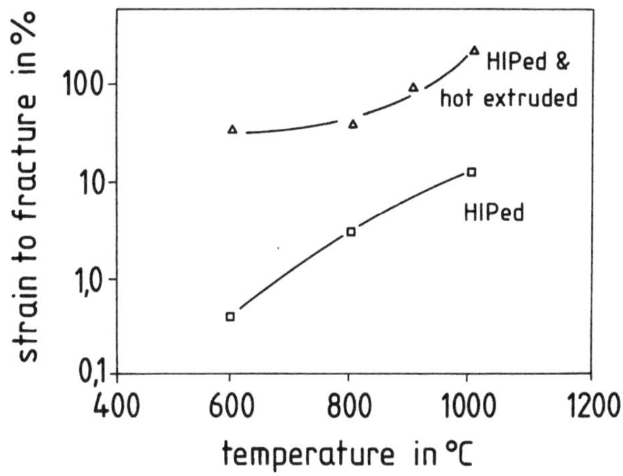

Figure 19: Temperature dependence of strain to fracture (in tension) for a PM NiAl-Cr alloy with 38.5 wt. % Cr (eutectic composition) after HIP and after HIP and hot extrusion, respectively (127,128).

3.3.3 Iron Aluminides

The iron aluminides FeAl with the B2 structure of NiAl and Fe_3Al with cubic DO_3 structure have been selected for materials developments in spite of their comparatively low strength at high temperatures (117,129). Intermetallic iron alloys with 25 - 35 at. % Al exhibit a sufficient room temperature ductility and a good corrosion resistance with respect to oxidation and sulfidation. Strength can be increased by alloying with further elements to such an extent that the strengths of heat resistant steels are reached. Further strength improvements are possible by precipitating or dispersing other phases (130,131). It is to be noted that the room temperature ductility depends not only on alloy composition, but also on the surrounding atmosphere in a sensitive way(132).

A particular case is the phase Fe_3AlC with $L'1_2$ structure (perovskite-type) which contains about 0.66 atoms C per unit cell depending on temperature (114). The $L'1_2$ unit cell is that of the $L1_2$ structure with an additional interstitial site for the C atom in the unit cell centre (Fig. 14), i.e. Fe_3AlC is an ordered fcc phase which is stabilized by interstitial C whereas Fe_3Al without C is an ordered bcc phase with DO_3 structure. Thus Fe_3AlC offers the possibility of exploring the potential of Fe-base $L1_2$ phases. Fe_3AlC is presently studied, and the first results with respect to strength and oxidation resistance are encouraging(114). In particular, the creep resistance corresponds to that of Ni_3Al as is shown in Fig. 16.

3.4 Less-Common Phases

Some less-common phases have become subject of materials developments primarily because of their low density. For example, the silicide Mg_2Si with cubic C1 structure has a density of 1.95 g/cm^3 and is used for a materials development which aims at applications in conventional combustion engines (133). An outstanding combination of high strength and low density is shown by the silicide Ti_5Si_3 with the hexagonal $D8_8$ structure, a density of 4.15 g/cm^3 and a melting point at 2130°C (134,135). Ti_5Si_3 is very brittle but there are good prospects for developing two-phase materials with Ti_5Si_3 and a second phase - e.g. Ti_3Al - with a tolerable brittleness (134-139).

The materials developments which have been discussed so far do not allow application temperatures above 1100°C which is the maximum temperature for superalloys. For increasing the efficiency of gas turbines higher service temperatures are necessary. Therefore other "exotic" - i.e. unknown - high-strength intermetallics are tested in application-oriented programs in spite of brittleness and lacking oxidation resistance of the respective phases or high prices of the constituents (140-147).

Al_3Nb is an intermetallic with a melting temperature of 1680°C (148) which is much higher than that of Ni_3Al and even higher than that of NiAl. The crystal structure of Al_3Nb - DO_{22} - is a tetragonal variant of the $L1_2$ structure of Ni_3Al. Thus Al_3Nb should be significantly stronger than both nickel aluminides NiAl and Ni_3Al. However, it is rather brittle in spite of indications of plasticity at room temperature, and in the past it was regarded as not useful for high-temperature applications (149). Recently the deformation behaviour of Al_3Nb and Al_3Nb-base alloys has been studied in detail, and the findings show that there are possibilities for materials developments(150,151). On one hand, Al_3Nb can be alloyed with the lightweight DO_{22} phase Al_3Ti to form single phase ternary alloys $Al_3(Nb,Ti)$. On the other hand, Al_3Nb can be alloyed with the B2 phase NiAl to form two-phase Al_3Nb-NiAl alloys. Figures 20-22 illustrate the mechanical behaviour of the resulting alloys, always compared with the respective binary phases. It can be seen that the binary Al_3Nb shows the highest strength at temperatures above 1000°C, but there is practically no ductility at lower temperatures. The alloying with the other phases extends the range of ductility to significantly lower temperatures whereas the high-temperature strength is affected only to a small degree. The most pronounced effect has been observed for the multiphase alloys where the amount of softening increases with increasing volume fraction of the less brittle phase NiAl (150,154). Such a behaviour which corresponds to a rule of mixtures is expected since the phase distributions in such alloys were very coarse. In a separate study of NiAl-Fe alloys it could be shown that strength and in particular creep strength can be described with a rule of mixtures as long as the phase distribution is sufficiently coarse whereas very fine phase distributions result in an extra threshold stress for creep (153,154). It is noted that the oxidation resistance of these Al_3Nb based alloys has been studied, too, in some detail, and it was found that the good oxidation resistance of the binary NiAl can nearly be reached by proper adjustment of the volume fractions of the respective phases (155).

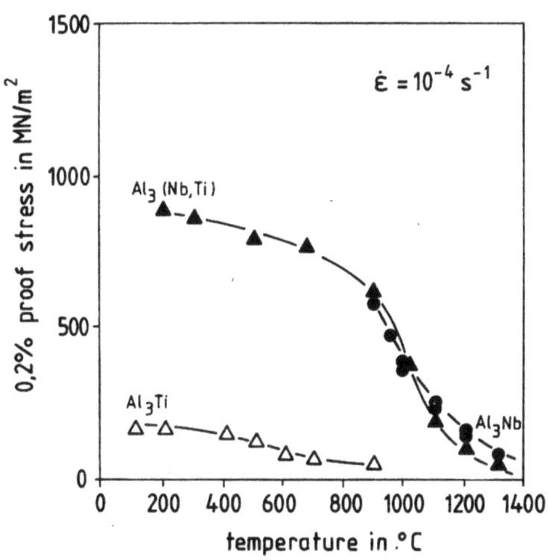

Figure 20: Temperature dependence of the 0.2 % proof stress ($10^{-4}s^{-1}$ strain rate in compression) for stoichiometric Al_3Nb, $Al_3(Nb_{0.75}Ti_{0.25})$ (150,151), and Al_3Ti (152), all with DO_{22} structure.

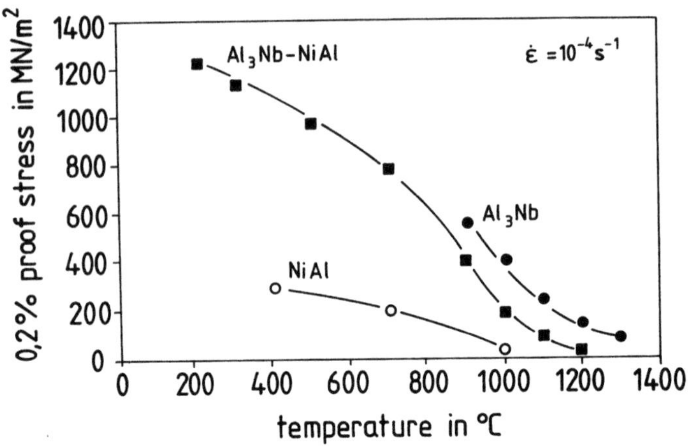

Figure 21: Temperature dependence of the 0.2 % proof stress in compression (strain rate $10^{-4}s^{-1}$) for stoichiometric B2-NiAl, DO_{22}-Al_3Nb and three-phase Al_3Nb-NiAl with 22 vol. % NiAl and 4 vol. % C14-NbNiAl (150,151).

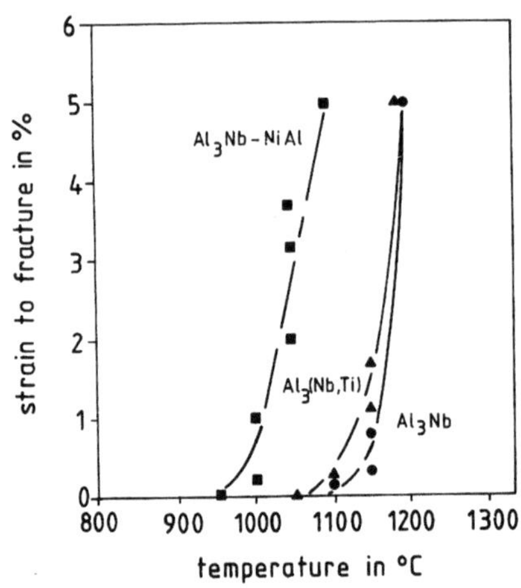

Figure 22: Temperature dependence of strain to fracture in bending (strain rate $10^{-4}s^{-1}$) for various Al_3Nb alloys as given in Figures 20 and 21 (150,151).

Combining NiAl with Al_3Nb increases the high-temperature strength appreciably, but the resulting strengths only reach or slightly surpass those of the superalloys. For entering the high-temperature strength range above that of the superalloys, one has to look for still stronger intermetallic phases. NbNiAl is one of the Al containing ternary Laves phases with the hexagonal C14 structure (156). It is significantly stronger than Al_3Nb, but also more brittle, and crack-free specimens can be produced only by powder metallurgy (157). However, as in the case of Al_3Nb, the Laves phase forms a stable equilibrium with the B2 phase NiAl, and the resulting NbNiAl-NiAl alloys can be produced by vacuum induction melting. The brittleness of these two-phase alloys is reduced to such an extent that they can be handled safely (157).

Figure 23 shows flow stress data of NbNiAl-NiAl alloys at 1200°C as a function of the volume fraction of the strengthening Laves phase. The compressive strength follows a rule of mixtures for the larger volume fractions, whereas for lower volume fractions it is higher. The reasons for this behaviour are presently studied. The bending strength does not increase with the NbNiAl volume fraction as the compressive strength at the higher volume fractions. This is supposed to be due to alloy defects - pores and microcracks, and thus the further work has to concentrate on improving the materials quality. Up to now the brittle-to-ductile transition temperature is between about 500°C and 700°C depending on composition and preparation method. The analysis of the microstructures of the deformed alloys has shown that in the transition-temperature range cracks in the Laves phase are stopped at the Laves phase-NiAl interface by producing a plastic zone in the NiAl phase ahead of the crack (157,158).

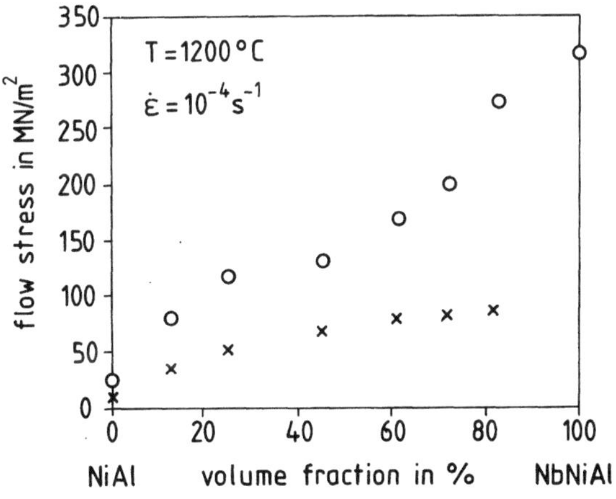

Figure 23: Flow stress of two-phase NbNiAl-NiAl as a function of the volume fraction of C14-NbNiAl at 1200°C in compression (o: 0.2 % proof stress) and in bending (x)(157).

Other and still harder ternary Laves phases with Al are NbFeAl, TaFeAl, TaNiAl, and again these phases form stable equilibria with the respective B2 aluminides FeAl and NiAl. Fig. 24 shows strength data of such phases and of respective two-phase alloys. It can be seen that it is indeed possible to enter the strength-temperature range above that of superalloys by making use of such intermetallic alloys. The creep strength of such NiAl-base intermetallic alloys is distinctly lower than the flow stress because of the strain-rate dependence of the thermally activated deformation processes, but the observed creep resistance of the TaNiAl-NiAl alloy is regarded as sufficiently for eventual applications above 1100°C (Figure 25).

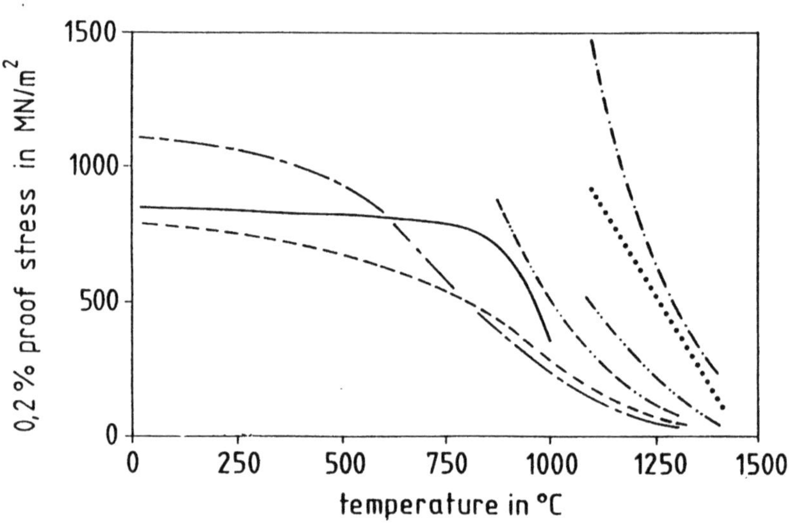

Figure 24: Temperature dependence of the 0.2 % proof stress in compression (strain rate $10^{-4}s^{-1}$) for the Laves phases (single phase) TaFeAl (·—·—·—), TaNiAl (······), NbNiAl (—··—) and the alloys $Nb_{17}Ni_{42}Al_{41}$ (-·-·--) (NbNiAl-39 vol. % NiAl), $Nb_4Ni_{49}Al_{47}$ (—— — ——) (NbNiAl-87 vol. % NiAl), $Ta_3Ni_{49}Al_{48}$ (- - - - -) (NiAl-12 vol. % TaNiAl) compared with the superalloy IN 100 (——) (tension data)(157).

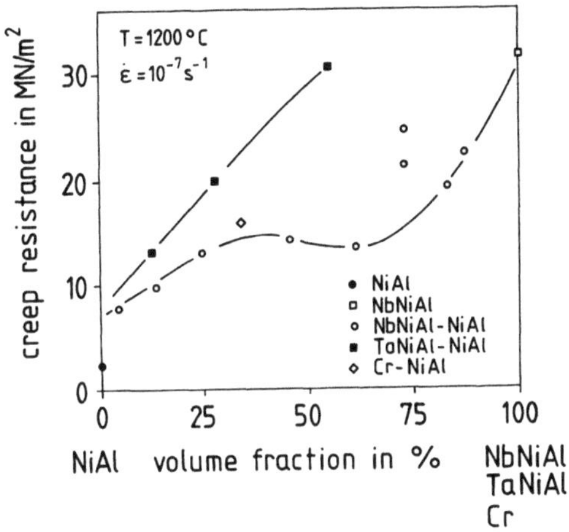

Figure 25: Creep resistance (stress for $10^{-7}s^{-1}$ secondary creep rate) at 1200°C as a function of volume fraction for NiAl-NbNiAl, NiAl-TaNiAl and NiAl-Cr (157).

4. Conclusions and Prospects

Any summarizing assessment of the prospects of intermetallic phases has to consider that the intermetallics do not represent a homogeneous group of materials. There is a manifold and divergent spectrum of intermetallic phases with large differences in properties. Thus intermetallic phases which are considered as candidate phases for high-temperature applications require specific materials developments from which specific applications may result.

This is demonstrated by the advanced materials developments on the basis of the classic phases Ti_3Al, TiAl, and Ni_3Al which are at the beginning of commercialization. These classic phases allow applications only up to about 1000°C whereas materials developments for application temperatures above those of the superalloys have to rely on less-common phases with sufficiently high stability and strength. In spite of the various materials developments the understanding of the physical mechanisms which control the properties - in particular ductility, toughness, corrosion resistance - is not yet sufficient and more fundamental research is necessary.

Literature References

(1) J.H. Westbrook, Metall. Trans. 8A (1977) 1327.

(2) J.H. Westbrook, J. El. chem. Soc. 103 (1956) 54.

(3) G. Sauthoff, Z. Metallkde. 77 (1986) 654.

(4) L. Pauling, Die Natur der chemischen Bindung, Verlag Chemie Weinheim 1964, Chap. 1.

(5) C. Kittel, Introduction To Solid State Physics, New York 1956, Chap.3.

(6) L. Brewer, LBL-Report, 2/1975.

(7) R.W. Cahn, P. Haasen, Physical Metallurgy, North Holland 1983, Chap.5.

(8) K. Schubert, in: Westbrook, Intermetallic Compounds, New York 1967, Chap. 6.

(9) D.G. Pettifor, Structure Maps for Pseudo-binary and Ternary Phases, Mat.Sc. Tech., 4 (1988), 675.

(10) P. Villars, L.D. Calvert, Pearson's Handbook of Crystallographic Data for Intermetallic Phases Volume 1- 3, American Society for Metals, 1985.

(11) W. Hume-Rothery, The Structure of Metals and Alloys, New York 1962, 198, 230.

(12) F. Laves, Crystal Structure and Atomic Size in Theory of Alloy Phases, Cleveland OH, 1956, 124.

(13) E. Parthè, Chrystal Chemistry of Tetrahedral Structures, Gordon and Breach Science Publisher, New York 1964.

(14) J. Hafner, From Hamiltonians to Phase Diagrams, Springer Verlag Berlin, 1987, 231.

(15) H. Schäfer, B. Eisenmann and W. Müller, Angew. Chemie 85, 1987, 871.

(16) R. Hoffmann, Angew. Chemie 99, 1987, 871.

(17) J.B. Goodenough, Magnetism and the Chemical Bond, Krieger, New York 1966, 28.

(18) F. Fukui, Accts. Chem. Res. 4, 1971, 57.

(19) R. Hoffmann, J. Am. Chem. Soc. 106, 1984, 2006.

(20) T.A. Albright, J.K. Burdett, W.H. Whangbo, Orbital Interactions in Chemistry, Wiley-Interscience, New York 1985.

(21) T. Hughbanks, R. Hoffmann, J.Am.Soc. 105, 1983, 3528.

(22) D. Sarma, J.Phys. Condes Matter 1, 1989, 9131.

(23) D. Hackenbracht, J.Physic F: Metal Physic 10, (1980), 427.

(24) A.v. Keitz, G. Sauthoff, unpublished 1990.

(25) W.P. Pearson, The Crystal Chemistry and Physics of Metals and Alloys, Milley, New York 1972, Cap. 3.

(26) A.R. Miedema, Cohesion in Metals - Transition Metals Alloys, North Holland Publ., 1988, Cap. 1,2.

(27) F. Seitz, The Modern Theory of Solids, New York 1940, 329.

(28) A.R. Miedema, J. Less-Common Met. 41, (1975), 283, 46, 1976, 67.

(29) K. Anderson, Phys. Rev. B 12 , (1975), 3060.

(30) R. Hultgren, Selected Values of Thermodynamic Properties of Metals and Alloys, John Willey, New York 1963.

(31) A.R. Miedema, Philips Technical Review 36, (1976, 217.

(32) J.F. Bell, The Experimental Fundations of Solid Mechanics, Springer Verlag Berlin 1973.

(33) P. Paufler and G.E.R. Schulze, Physikalische Grundlagen mechanischer Festkörpereigenschaften, Vieweg, Braunschweig 1978, Chap. 2.

(34) P.R. Munroe, in: Metals and Materials, 1989, 27.

(35) D. Destefani, in: Advanced Materials and Processes, (1989), 27.

(36) N.S. Stoloff, Physical and Mechanical Metallurgy of Ni_3Al and its Alloys, in : International Materials Reviews, 34, (1989), 153.

(37) Landolt/-Börnstein, Vol.4., part 2, Springer Verlag Berlin 1965, 53, 307.

(38) A.v.Keitz, G. Sauthoff, unpublished, 1990.

(39) E.P. Lautenschlager, T. Hughes, J.O. Brittan, Acta Metall., 15, (1967), 1347.

(40) M.G. Mendiratta, C.C. Law, Materials Science 22, (1987), 607.

(41) T. Hong, A.J. Freemann, Electronic structure and mechanical properties of intermetallics: APB-energies in NiAl-based systems, in: High-Temperature ordered intermetallic alloys (III), (1989), 75.

(42) G.F. Hancock, McDonnell, phys. stat. sol. (a) 4, (1971), 143.

(43) H. Hoshino, S.J. Rothman, R.S. Averbach, Acta metall. 36 (1988) 1271.

(44) A.B. Kuper, D. Lazurus, J.R. Manning and C.T. Tomizuka, Phys. Rev. 104 (156), 1536.

(45) R.R. Vandervoort, A.K. Mukherje, J.E. Dorn, Trans. ASM, 59, (1966),.930.

(46) O.Izumi, T. Takasugi, Mechanismus of ducility improvement in $L1_2$ compounds, J.Mater.Res.3, 1988, 426.

(47) C.T. Liu and C.L. White, in: Symposium High-Temperature Ordered Intermetallic Alloys (MRS Symp. Proc. Vol. 39), C.C. Koch, C.T. Liu, and N.S. Stoloff (eds.), Mat. Res. Soc., Pittsburgh (1985), pp. 365.

(48) C.T. Liu and J.O. Stiegler, Science 226 (1984) 636.

(49) C.T. Liu, Int. Met. Rev. 29 (1984) 168.

(50) N.S. Stoloff, in: High-Temperature Ordered Intermetallic Alloys (MRS Symp. Proc. Vol. 39), C.C. Koch, C.T. Liu, and N.S. Stoloff (eds.), Mat. Res. Soc., Pittsburgh (1985), 3.

(51) High-Temperature Ordered Intermetallic Alloys, C.C. Koch, C.T. Liu, and N.S. Stoloff (eds.), Mat. Res. Soc., Pittsburgh (1985).

(52) High-Temperature Ordered Intermetallic Alloys II, N.S. Stoloff, C.C. Koch, C.T. Liu, and O. Izumi (eds.), Mat. Res. Soc., Pittsburgh (1987).

(53) High-Temperature Ordered Intermetallic Alloys III, C.T. Liu, A.I. Taub, N.S. Stoloff, and C.C. Koch (eds.), Mat. Res. Soc., Pittsburgh (1989).

(54) G. Sauthoff, Z. Metallkde. 81 (1990), 855.

(55) G. Sauthoff, Z. Metallkde. 80 (1989) 337.

(56) I. Jung, M. Rudy, and G. Sauthoff, in High-Temperature Ordered Intermetallic Alloys II, N.S. Stoloff, C.C. Koch, C.T. Liu and O. Izumi (eds.), MRS Pittsburgh (1987) 263.

(57) H.J. Frost and M.F. Ashby, Deformation Mechanism Maps, Pergamon Press, Oxford (1982) chapter 18.

(58) J.H. Westbrook, in High-Strength Materials, V.F. Zackay (ed.), J. Wiley & Sons, New York (1965) 724, especially 725.

(59) P. Paufler, Z. Kristallographie 170 (1985) 144.

(60) P. Paufler, in Intermetallische Phasen, authors collective, VEB Deutscher Verlag für Grundstoffindustrie, Leipzig (1977) 165-.

(61) H. Hindam and D.P. Whittle, Oxid. Met. 18 (1982) 245- 284.

(62) F.S. Pettit and G.W. Goward, in Coatings for High- Temperature Applications, E. Lang (ed.), Appl. Science Publ., London (1983) 341.

(63) H.W. Grünling, S. Leistikow, A. Rahmel, and F. Schubert, in Aufbau von Oxidschichten auf Hochtemperatur- werkstoffen und ihre technische Bedeutung, A. Rahmel (ed.), Deutsche Gesellschaft für Metallkunde, Oberursel (1983) 7.

(64) R. de Planque and L. Singheiser, in Metalle und nicht- metallische Werkstoffe und ihre Verarbeitungsverfahren im Vergleich, Teil IV, VDI-Berichte 600.4, VDI Verlag, Düsseldorf (1988) 211.

(65) J.B. Mc Andrew and H.D. Kessler, J. Metals 8 (1956) 1348.

(66) H.A. Lipsitt, in Advanced High-Temperature Alloys: Processing and Properties, S.M. Allen, R.M. Pelloux, R. Widmer (eds.), ASM, Metals Park (1986) 157.

(67) Y.-W. Kim and F.H. Froes, in: High-Temperature Aluminides and Intermetallics, S.H. Whang, C.T. Liu, D.P. Pope and J.D. Stiegler (eds.), TMS, Warrendale 1990, p. 465.

(68) T.M.F. Ronald, Adv. Materials & Processes 5/89 (1989) 29.

(69) J.M. Larsen, K.A. Williams, S.J. Balsone and M.A. Stucke, in: High-Temperature Aluminides and Intermetallics, S.H. Whang, C.T. Liu, D.P. Pope and J.D. Stiegler (eds.), TMS, Warrendale 1990, 521.

(70) R. Bowman and R. Noebe, Adv. Materials & Processes 8/89 (1989) 35.

(71) J.R. Stephens, Adv. Materials & Processes 4/90 (1990) 35.

(72) Y. Nishiyama, T. Miyashita, S. Isobe and T. Noda, in: High-Temperature Aluminides and Intermetallics, S.H. Whang, C.T. Liu, D.P. Pope and J.D. Stiegler (eds.), TMS, Warrendale 1990, 557.

(73) T. Kremser, G. Frommeyer, and K.-H. Matucha, Mechanische Eigenschaften schmelzmetallurgisch hergestellter TiAl-Basislegierungen, poster presented at the DGM 1989 Annual Meeting in Karlsruhe.

(74) W. Wunderlich, T. Kremser, and G. Frommeyer, Z. Metallkde. 81 (1990) 802.

(75) T. Kawabata, T. Kanai and O. Izumi, Acta Metall., 33 (1985) 1355.

(76) G. Hug, A. Loiseau and P. Veyssiere, Phil. Mag., A57 (1988) 429.

(77) B.A. Greenberg, Scripta Metall., 23 (1989) 631.

(78) C. Hartig, S. Chen, P.A. Beaven, and H. Fukutomi, in: Proc. Sixth World Conf. on Titanium, Part II, P. Lsacombe, R. Tricot, and G. Beranger (eds.), Les Editions de Physique, Les Ulis Cedex, France, 1989, pp. 1021.

(79) B. Dogan, R. Wagner, and P.A. Beaven, presentation at the 1989 TMS Fall Meeting in Indianapolis, to be published in Scipta Met. 24 (1990).

(80) H. Fukutomi, C. Hartig, and H. Mecking, Z. Metallkde. 81 (1990) 272.

(81) M. Yamaguchi, Y. Umakoshi and T. Yamane, Phil. Mag. A 51 (1987) 301.

(82) M. Yamaguchi, S.R. Nishitani and Y. Shirai, in: High- Temperature Aluminides and Intermetallics, S.H. Whang, C.T. Liu, D.P. Pope and J.D. Stiegler (eds.), TMS, Warrendale 1990, 63.

(83) M. Yamaguchi, Y. Shirai and Y. Umakoshi, in: Dispersion Strengthened Aluminium Alloys, Y.-W. Kim and W.M. Griffith (eds.), TMS, Warrendale 1988, 721.

(84) S. Mazdiyasni, D.B. Miracle, D.M. Dimiduk, M.G. Mendiratta and P.R. Subramanian, Scripta Met. 23 (1989) 327.

(85) J.P. Nic, S. Zhang and D.E. Mikkola, Scripta Met. Mat. 24 (1990) 1099.

(86) C.D. Turner, W.O. Powers and J.A. Wert, Acta metall. 37 (1989) 2635.

(87) H. Mabuchi, K. Hirukawa, H. Tauda and Y. Nakayama, Scripta Met. Mat. 24 (1990) 505.

(88) K.S. Kumar and J.R. Pickens, in: Dispersion Strengthened Aluminium Alloys, Y.-W. Kim and W.M. Griffith (eds.), TMS, Warrendale 1988, 763.

(89) W.D. Porter, K. Hisatsune, C.J. Sparks, W.C. Oliver and A. Dhere, in: High-Temperature Ordered Intermetallic Alloys III, C.T. Liu, A.I. Taub, N.S. Stoloff and C.C. Koch (eds.), Mat. Res. Soc., Pittsburgh 1989, 657.

(90) G. Chen, Z. Sun, X. Xie, Y. Ren, Y. Xu, K. Yao, X. Zhou, L. Sha, S. Fu and W. Yang, in: Intermetallic Compounds and High Temperature Materials (Proc. C-MRS International '90), Peking 1991, im Druck.

(91) N.S. Stoloff and R.G. Davies, Progr. Mater. Sci. 13 (1966) 3.

(92) V. Vitek and D.P. Pope, in: Symposium High-Temperature Ordered Intermetallic Alloys (MRS Symp. Proc. Vol. 39), C.C. Koch, C.T. Liu, and N.S. Stoloff (eds.), Mat. Res. Soc., Pittsburgh (1985), 205.

(93) J.A. Horton and C.T. Liu, Acta Metall. 33 (1985) 2191.

(94) Y. Mishima, Y. Oya, and T. Suzuki, in: Symposium High- Temperature Ordered Intermetallic Alloys (MRS Symp. Proc. Vol., C.C. Koch, C.T. Liu, and N.S. Stoloff (eds.), Mat. Res. Soc., Pittsburgh (1985), 263.

(95) T. Suzuki, Y. Oya, and D.M. Wee, Acta Metall. 28 (1980) 301.

(96) N.S. Stoloff, Int. Met. Rev. 29 (1984) 123.

(97) N.S. Stoloff and R.G. Davies, Acta Metall. 12 (1964) 473.

(98) P. Haasen, in Ref.(7), 1341.

(99) T. Kawabata, T. Kanai and O. Izumi, Acta Metall. 33 (1985) 1355.

(100) C.T. Liu and C.C. Koch, in Technical Aspects of Critical Materials Use by the Steel Industry, Vol. IIB, U.S. Department of Commerce, National Bureau of Standards (1983) paper 42.

(101) K. Aoki and O. Izumi, J. Japan Inst. Metals 43 (1979) 1190.

(102) C.T. Liu, C.L. White and J.A. Horton, Acta Metall. 33 (1985) 213.

(103) C.C. Koch, C.L. White, R.A. Padgett, and C.T. Liu, Scripta Metall. 19 (1985) 963.

(104) C.T. Liu, C.L. White and E.H. Lee, Scripta Metall. 19 (1985) 1247.

(105) T. Takasugi, O. Izumi and N. Masahashi, Acta Metall. 33 (1985) 1247; 1259.

(106) T. Takasugi, E.P. George, D.P. Pope and O. Izumi, Scripta Metall. 19 (1985) 551.

(107) T. Ogura, S. Hanada, T. Masumoto, and O. Izumi, Metall. Trans. 16A (1985) 441.

(108) V.K. Sikka, in Interdisciplinary Issues in Materials Processing and Manufacturing, Vol. II. ASM, Metals Park (1988) 451.

(109) V.K. Sikka, Materials & Manufact. Processes 4 (1989) 1.

(110) J.W. Patten, in: High-Temperature Aluminides and Intermetallics, S.H. Whang, C.T. Liu, D.P. Pope and J.D. Stiegler (eds.), TMS, Warrendale 1990, p. 493.

(111) J.H. Schneibel, G.F. Petersen, and C.T. Liu, J. Mat. Res. 1 (1986) 68.

(112) P.L. Martin, M.G. Mendiratta, and H.A. Lipsitt, Metall. Trans. 14A (1983) 2170.

(113) T.E. Howson, D.A. Mervyn, and J.K. Tien, Metall. Trans. 11A (1980) 1609.

(114) I. Jung and G. Sauthoff, Z. Metallkde. 80 (1989), in print.

(115) E. Fitzer and P. Gerasimoff, Z. Metallkde. 50 (1959) 185.

(116) J.D. Destefani, Adv. Materials & Processes 2/89 (1989) 37.

(117) P.R. Munroe and I.Baker, Metals & Materials 4 (1988), 435.

(118) K. Vedula and J.R. Stephens, in Horizons of Powder Metallurgy, Part II, Verlag Schmid, Freiburg (1986) 1313.

(119) K. Vedula, V. Pathare, I. Aslanidis, and R.H. Titran, in High-Temperature Ordered Intermetallic Alloys, C.C. Koch, C.T. Liu, N.S. Stoloff (eds.), Mat. Res. Soc., Pittsburgh (1985) 411.

(120) G. Sauthoff, in Symposium Materialforschung 1988, BMFT- PLR, Jülich (1988) 399.

(121) I. Jung and G. Sauthoff, Z. Metallkde. 80 (1989), in print.

(122) J.D. Rigney, P.S. Khadkikar, J.J. Lewandowski and K. Vedula, in: High-Temperature Ordered Intermetallic Alloys III, C.T. Liu, A.I. Taub, N.S. Stoloff and C.C. Koch (eds.), Mat. Res. Soc., Pittsburgh 1989, 603.

(123) J.D. Whittenberger, R.K. Viswanadham, S.K. Mannan and K.S. Kumar, in: High-Temperature Ordered Intermetallic Alloys III, C.T. Liu, A.I. Taub, N.S. Stoloff and C.C. Koch (eds.), Mat. Res. Soc., Pittsburgh 1989, 621.

(124) J.D. Whittenberger, D.J. Gaydosh and K.S. Kumar, J. Materials Sci. 25 (1990), 2771.

(125) H.E. Cline, J.L. Walter, E. Lifshin, and R.R. Russel, Metall. Trans. 2 (1971) 189.

(126) J.L. Walter and H.E. Cline, Metall. Trans. 1 (1970) 1221.

(127) G. Frommeyer, W. Kowalski, and C. Lüdecke, Plastizität und Festigkeit intermetallischer NiAl-Legierungen, poster presented at the DGM 1989 Annual Meeting in Karlsruhe.

(128) W. Kowalski and G. Frommeyer, to be published in Z. Metallkde 2 (1991).

(129) K. Vedula and J.R. Stephens: B2 Aluminides for High Temperature Applications, in: High-Temperature Ordered Intermetallic Alloys II, N.S. Stoloff, C.C. Koch, C.T. Liu and O. Izumi (eds.), Mat. Res. Soc., Pittsburgh (1987), 381.

(130) M.G. Mendiratta, S.K. Ehlers, D.M. Dimiduk, W.R. Kerr, S. Mazdiyasni and H.A. Lipsitt, in: High-Temperature Ordered Intermetallic Alloys II, N.S. Stoloff, C.C. Koch, C.T. Liu and O. Izumi (eds.), Mat. Res. Soc., Pittsburgh 1987, 393.

(131) C.G. Mc Kamey, J.A. Horton and C.T. Liu, in: High- Temperature Ordered Intermetallic Alloys II, N.S. Stoloff, C.C. Koch, C.T. Liu and O. Izumi (eds.), Mat. Res. Soc., Pittsburgh 1987, 321.

(132) C.T. Liu and C.G. Mc Kamey, in: High-Temperature Aluminides and Intermetallics, S.H. Whang, C.T. Liu, D.P. Pope and J.D. Stiegler (eds.), TMS, Warrendale 1990, 133.

(133) E.E. Schmid, K. von Oldenburg, and G. Frommeyer, Z. Metallkde. 81 (1990) 809.

(134) G. Frommeyer, R. Rosenkranz, and C. Lüdecke, Z. Metallkde. 81 (1990) 307-313.

(135) R. Wagner, in Symposium Materialforschung 1988, BMFT- PLR, Jülich (1988) 26.

(136) O. Schob, H. Nowotny, and F. Benesovsky, Planseeberichte f. Pulvermetallurgie 10 (1962) 65.

(137) J.S. Wu, P.A. Beaven, R. Wagner, C. Hartig, and J. Seeger, in ref. 6 (Boston III), 761.

(138) P.A. Beaven, J.S. Wu, B. Dogan, C. Hartig, J. Seeger and R. Wagner, in: GKSS - Jahresbericht 1989, 49.

(139) J.S. Wu, D. Chen, P.A. Beaven, R. Wagner, and J. Seeger, in: Proc. C-MRS International '90, Beijing (1991), in print.

(140) D.L. Anton and D.M. Shah, in: High-Temperature Ordered Intermetallic Alloys III, C.T. Liu, A.I. Taub, N.S. Stoloff and C.C. Koch (eds.), Mat. Res. Soc., Pittsburgh 1989, 361.

(141) R.L. Fleischer, in: High-Temperature Ordered Intermetallic Alloys III, C.T. Liu, A.I. Taub, N.S. Stoloff and C.C. Koch (eds.), Mat. Res. Soc., Pittsburgh 1989, 305.

(142) R.L. Fleischer, D.M. Dimiduk and H.A. Lipsitt, Ann. Rev. Mater. Sci. 19 (1989) 263.

(143) A.I. Taub and R.L. Fleischer, Science 243 (1989) 616.

(144) R.L. Fleischer and R.J. Zabala, Metall. Trans. 21A (1990) 1951.

(145) R.L. Fleischer and R.J. Zabala, Metall. Trans. 21A (1990), in print.

(146) S.R. Schuon and V. Rehzets, in: High-Temperature Ordered Intermetallic Alloys III, C.T. Liu, A.I. Taub, N.S. Stoloff and C.C. Koch (eds.), Mat. Res. Soc., Pittsburgh 1989, p. 647.

(147) H.P. Kao, C.R. Brooks, and M. Sanganeria, in: High- Temperature Ordered Intermetallic Alloys III, C.T. Liu, A.I. Taub, N.S. Stoloff and C.C. Koch (eds.), Mat. Res. Soc., Pittsburgh 1989, p. 749.

(148) R.L. Fleischer, J. Mat. Sci. 22 (1987) 2281.

(149) D. Shechtman and L.A. Jacobson, Metall. Trans. 6A (1975) 1325.

(150) G. Sauthoff, Mechanical Properties of Intermetallics at High Temperatures, in: High Temperature Aluminides and Intermetallics, C.T. Liu et al. (eds.), ASM/TMS, Metals Park (1990), 329.

(151) C.-P. Reip and G. Sauthoff, Verformungsverhalten der intermetallischen Phase Al_3Nb, poster presented at the DGM 1989 Annual Meeting in Karlsruhe.

(152) M. Yamaguchi, Y. Umakoshi and T. Yamane, Phil. Mag. A55 (1987) 301.

(153) J. Klöwer, Untersuchung des Kriechverhaltens gerichtet erstarrter, lamellarer Eisen-Nickel-Aluminium- Legierungen, Dr. rer. nat. thesis, Rheinisch- Westfälische Technische Hochschule Aachen, 1989, 45.

(154) J. Klöwer and G. Sauthoff, Z. Metallkde. 82 (1991), in print.

(155) M. Steinhorst, Oxidationsverhalten des Aluminids $NbAl_3$ und mehrphasiger Legierungen aus dem System Nb-Ni-Al, Dr.-Ing. thesis, Universität Dortmund 1989.

(156) P. Villars and L.D. Calvert, Pearson's Handbook of Crystallographic Data for Intermetallic Phases, ASM, Metals Park (1985).

(157) L. Machon and G. Sauthoff, Verformungsverhalten zweiphasiger intermetallischer Legierungen der Laves- Phase NbNiAl und der B2- Phase NiAl, poster presented at the DGM 1989 Annual Meeting in Karlsruhe.

(158) W. Wunderlich, L. Machon and G. Sauthoff, Versetzungsanalyse an Rißspitzen an der Phasengrenzfläche NiAl/NbNiAl in intermetallischen Ni-Al-Nb-Legierungen, presentation at the DPG 1990 spring meeting in Regensburg.

Shape Memory Alloys

E. Hornbogen, Ruhr–Universität Bochum, Institut für Werkstoffe, D–4630 Bochum 1

Abstract

Shape memory is the ability of a material to remember its original shape, either after mechanical deformation (one–way effect) or by cooling and heating (two–way effect). This phenonemon is based on a structural phase transformation, which is diffusionless and connected with large amounts of homogeneous shear. It is known as martensitic transformation. Particularities of many bulk and microstructural properties are found in the environment of the transformation temperatures. It is explained how the alloys acquire their final shape, how they learn the two–way effect, and how they fatigue. Finally, a systematic survey is given on present and perspective applications in engineering and medicine.

1. Introduction

1.1 Thermal and mechanical properties

Shape memory alloys (SMA) show an anomalous behavior if they are exposed to a change in temperature and/or mechanical stress. The normal behavior of most materials is characterized in a stress (σ)–, strain (ϵ)–, temperature (T)–diagram. Time (t) can be neglected in this context, because diffusionless (time–independent) reactions are responsible for shape memory. If, however, heat treatments for the primary production of the microstructures, or ageing effects are discussed the duration of the experiment may come into play.[1-6]

In this field we have to deal with temperature– or stress–induced changes in shape and volume. They are described by the bulk properties: coefficients of thermal expansion α, elastic moduli E, and descrete values like yield strength σ_y and fracture strength $\sigma_f \geq \sigma_y$ (Fig. 1):

$$\epsilon_e = \alpha\Delta T \tag{1a}$$

$$\epsilon_e = E^{-1}\sigma \tag{1b}$$

$$\epsilon_p = A(\sigma-\sigma_y)^n \tag{1c}$$

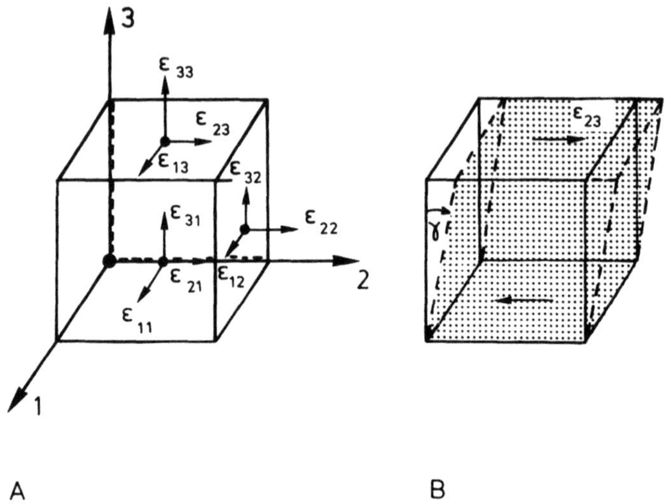

Fig. 1: a) Strains in bulk body
b) Shear strain which may transform crystal structures

These equations are valid for isotropic materials only. Anisotropy may stem from the structural levels of phase (crystals, texture) or microstructure (fibre reinforced composites). Thermal expansion and shape change require the strain tensor ϵ_{ij} for their description where i = 1,2,3, j = 1,2,3 are the three principle directions in space:

$$\epsilon_{ij} = \alpha_{ij}\Delta T \tag{2a}$$

$$\epsilon_{ij} = \begin{pmatrix} \epsilon_{11} & \epsilon_{12} & \epsilon_{13} \\ \epsilon_{21} & \epsilon_{22} & \epsilon_{23} \\ \epsilon_{31} & \epsilon_{32} & \epsilon_{33} \end{pmatrix}$$

$$\alpha_{ij} = \begin{pmatrix} \alpha_{11} & \alpha_{12} & \alpha_{13} \\ \alpha_{21} & \alpha_{22} & \alpha_{23} \\ \alpha_{31} & \alpha_{32} & \alpha_{33} \end{pmatrix}$$

A change in volume is described by $\alpha_{11}, \alpha_{22}, \alpha_{33}$:

$$v_T = \epsilon_{11} + \epsilon_{22} + \epsilon_{33} \tag{2b}$$

For shear strains ϵ_{ij} $i \neq j$, $\epsilon_{ij} = \epsilon_{ji}$ is valid. Consequently, they provide three additional components to the tensor. A total of six strains are necessary to describe volume and shape changes. For SMA $\epsilon_{11} = \epsilon_{22} = \epsilon_{33} = 0$, and $\epsilon_{12}, \epsilon_{13}, \epsilon_{23} \rightarrow$ max. is aspired.

1.2 Phase, microstructure, transformations

Shape memory is connected with solid, mostly crystalline phases. Only quite recently a crystal $\leftrightarrow$ amorphous transformation has been found in $AlPO_4$ (Aluminiumphosphate) which occurs at 150 kbar pressure. We deal with transformations between crystalline phases. In shape memory alloys they may fulfill certain crystallographic and thermodynamic prerequisites:

1. Compatibility: one crystalline phase α must be able to form inside another one β with maximum coherency, in any case without breaking. This implies similarity of the lattices as for

$$\begin{array}{lcl} \text{fcc} & \leftrightarrow & \text{hcp} \\ \text{fcc} & \leftrightarrow & \text{bcc} \\ \text{hcp} & \leftrightarrow & \text{bcc} \end{array} \tag{3a}$$

   and minimum change in specific volume (equ. 2b).

2. Lack of diffusion: the transformation temperatures $T_{\beta\alpha}$ must be situated in a temperature range in which individual motion of atoms by diffusion is negligible.

$$\text{one year} < t_0 = \frac{b^2}{D} \tag{3b}$$

$$T_{\beta\alpha} \leq \frac{1}{3} T_{cl} \tag{3c}$$

   For substitutional alloys this amounts to 150 °C for β–CuZn–, 200 °C for β–NiTi–, and 300 °C for γ–Fe–base alloys.

3. Reversibility: The transformation reaction $\beta \rightarrow \alpha$ which takes place during cooling must be reverted in such a way that the original structure is completely restored, i.e. no defects must be left behind. If the transformation takes place by a large amount of shear ($\epsilon_{12} > 0.01$) the reaction is of first order and named martensitic. For small amounts of strain the reaction may be of second order and a typical tweed–like pattern is formed. The term spinodal shear is sometimes used in analogy to spinodal decomposition by diffusion.

It is a principle feature of all b.c.c. (or β–) crystals to show a tendency to transform into a structure of higher coordination number (c.n.8 → c.n.12) at low temperatures. This is due to entropy (of lattice vibration) terms which become dominant at elevated temperatures. For alloys of iron the same is valid. However the less closely packed b.c.c. (α) phase reoccurs due to contribution of magnetic ordering to phase stability.

Related to the phase transformation is a change in microstructure. In rare cases a single crystal of β is transformed into a single crystal of α. More often crystallites of the high temperature phase (β,γ) desintegrate into a large number of orientational variants of α–crystals. A wide variety of martensitic morphologies can originate in this way (Fig. 2).

1.3 Description of three memory effects

The occurrence of a $\beta \leftrightarrow \alpha$, or $\gamma \leftrightarrow \alpha$ martensitic transformation is the prerequisite for the anomalies shown in the T–, σ–, ϵ–diagrams (Fig. 3). For the one–way (1w) behavior the material evidently shows considerable plastic deformation ($\epsilon_{1w} \lesssim 0.10$) at a yield stress $\sigma_{\beta\alpha}$. Straining should be discontinued below a second yield stress $\sigma_{y\beta}$ and the fracture stress σ_f. Heating this strained state will lead to a reversion of the shape change. Consequently, the term pseudo–plastic may be used for this type of deformation.

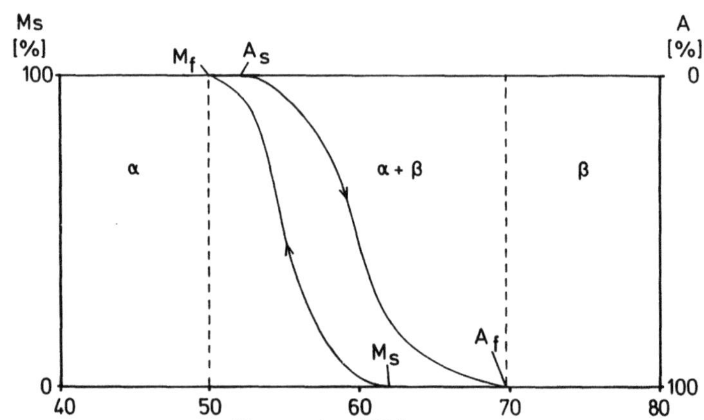

Fig. 2: Transformation cycle in an Ni–48.6 At.% Ti alloy ($\beta \equiv$ austenite, $\alpha \equiv$ martensite)

A T–, ϵ–diagram is required to describe the two–way effect. By heating and cooling only the material performs a shape change. A special training treatment is required to induce any desired deformation which may amount up to $\epsilon_{2w} \leq 0.06$.

Finally, a rubber like pseudo–elastic behavior is shown in the σ–, ϵ–diagram (Fig. 3c). The material starts to deform similarly to the 1w–condition. However, if stress is re–

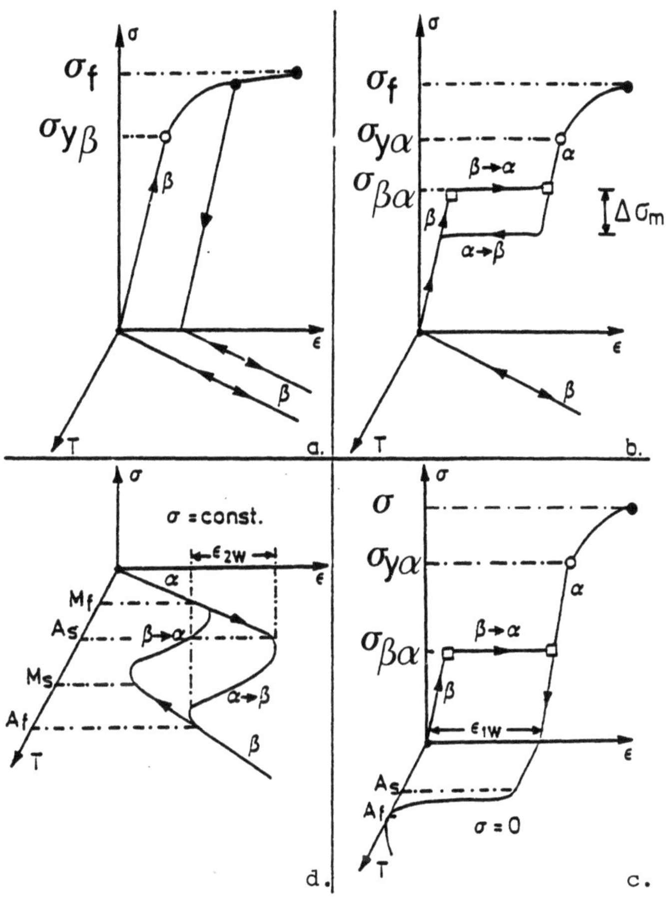

Fig. 3: Temperature T–, Stress σ–, strain ϵ–diagrams.
a) normal behavior
b) pseudo–elasticity
c) one–way effect
d) two–way effect or pseudo–plasticity

laxed strain does not persist. With a certain hysteresis it is reverted during lowering of stress. All effects are caused either by temperature– or by stress–induced martensitic transformation. Which effect is found depends (besides temperature and mechanical stress) on the chemical composition and microstructure of the particular type of transforming alloy.

1.4 New materials, new properties

Shape memory alloys provide an example for a special, rather rare case of materials' development (Fig. 4). More common is that a well–known type of material such as aluminum, is modified in order to provide improved, but known properties, such as a combination of high strength and low weight with new Al–Li or Mg–Li base alloys. Completely new properties are utilized for example if the old material glass fibres is used for the new application transmission of information. Shape memory alloys are materials which provide new not only improved properties. In this way they are similar to semi–conductors in microelectronics. For their introduction into safe practical use both a large amount of fundamental and applied research is required and in addition much missionary work to inform design engineers and convince them about their usefulness.

2. Particular physical properties of SMA

2.1 Geometry: Non fractal and fractal microstructures

The anomalous properties of SMA are based on a particular behavior of physical properties. Some of them are mentioned here. During cooling the transformation starts at a temperature M_s (Fig. 5a), but it requires additional undercooling to continue. Martensite crystals form with lenticular or lath shape. In polycrystals they originate at

Fig. 4: Ways of progress in materials		Properties	
		old + improved	new
materials	old + improved	Ni–base superalloys	optical glass fibers for transmission of information
	new	Al+Li–, Mg–Li– superlight alloys	shape memory alloys

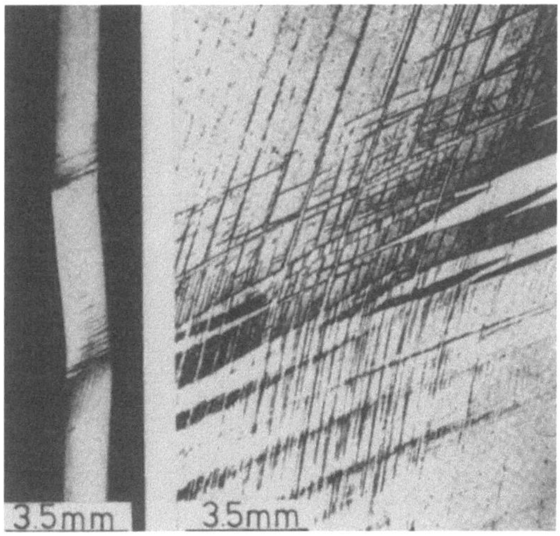

Fig. 5: Microstructural aspects of SMA.
a) Transformation of a β–single crystal into a single crystal of α by shear stress, Cu–39 % Zn

b) Martensitic microstructure in a CuZnAl alloy with $M_f > 20\,^{\circ}C$

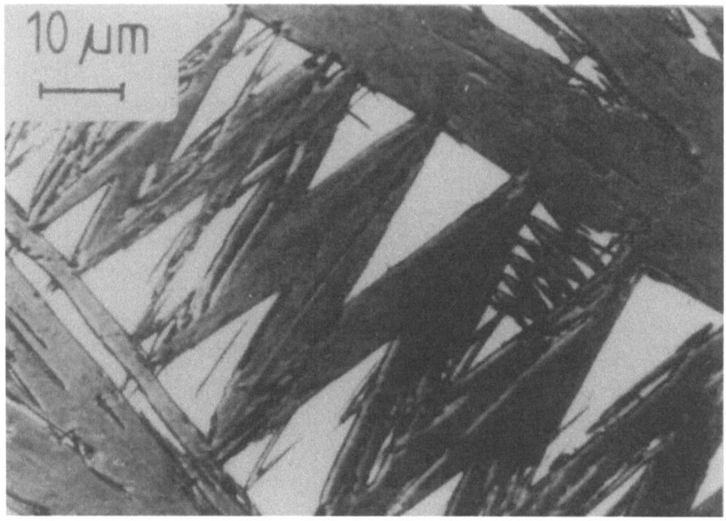

c) Fractal martensite α and residual austenite γ in an FeNiAl–alloy

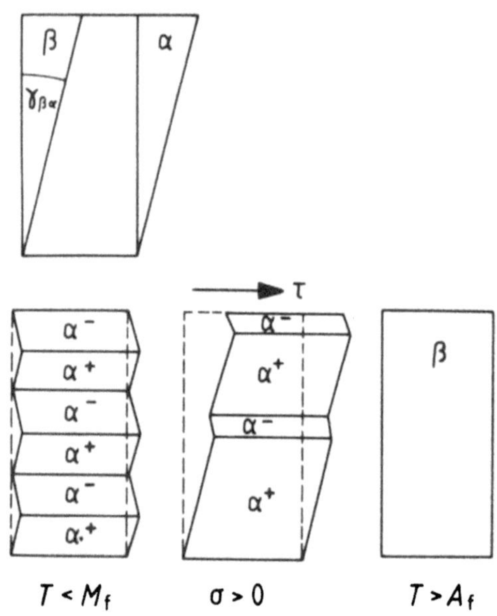

d) Shape change and domain structure (α^+, α^-) in martensite. The original shape is always restored by transformation into β

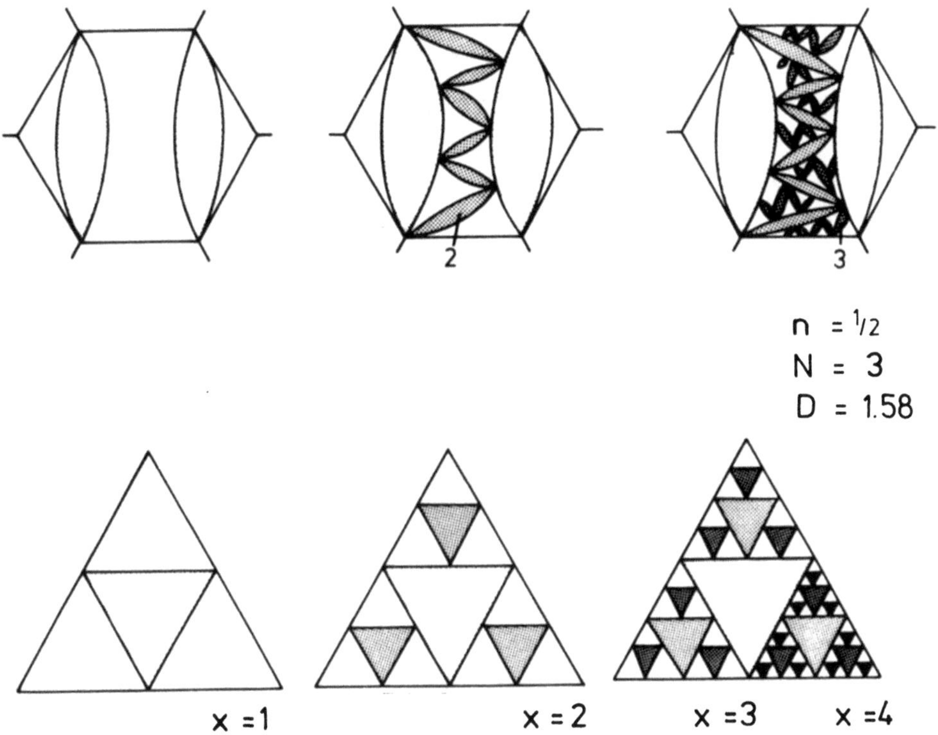

e) Fragmentation of an austenite crystallite by martensitic transformation, schematic.

grain boundary sites, and expand across the crystallite. The reaction may continue by sidewise growth and the formation of self–accommodating groups (Fig. 5b). An alternative is the formation of lenticular orientational variants of martensite by nucleation at the β/α–interface of new lenticular crystals which grow into the untransformed spaces (Fig. 5c).[7] This process may be repeated several times, and a fractal microstructure originates. The bulk course (Fig. 5a) can be composed of these descrete iterations. If the transformation takes place under stress, or is stress–induced, the number of orientational variants is reduced by interaction of the external shear stress τ with the transformation strain $\epsilon_{\beta\alpha}$ (Fig. 6). A single crystal of martensite (α) may form from a single crystal of the high temperature phase β (Fig. 5d). The transformation temperature M_s and the bulk course of the $\beta \rightarrow \alpha$ transformation are important properties of a shape memory alloy. The reverse transformation $\alpha \rightarrow \beta$ requires an overheating $\Delta T = A_s - M_f$ to start. The last martensite crystal which has formed disappears first taking exactly the reverse path of its formation so that the perfect β–crystal is restored. This process continues until the first generation of martensite crystals last disappears at A_f.

Type of alloy	Designation of crystal structures	Δv
Fe–Ni–	$\gamma \rightarrow \alpha$	+ 0.03
Cu–Zn–	$\beta \rightarrow \alpha$	~ 0
Ni–Ti–	$\beta \rightarrow \alpha$	~ 0
Fe–Mn–	$\gamma \rightarrow \epsilon$	– 0.015

Fig. 6: Crystallographic aspects of martensitic transformation.
a) Volume change during martensitic transformation

In deformed alloys and alloys of iron the transformation to martensite α never becomes complete. The residual austenite shows a fractal site distribution. A high degree of completion of the reaction is desirable for SMA because it facilitates a maximum shape change. Incomplete transformation is related to a volume change $v_{\beta\alpha} \neq 0$ and consequent strains (equ. 2b, Fig. 6).

$$v_\beta - v_\alpha = v_{\beta\alpha} \lesseqgtr 0 \qquad (2c)$$

2.2 Crystallography: complex stacking sequences

Some new crystalline phases can be formed without diffusion by homogeneous shear (Fig. 6). In other cases slight reshuffling is required i.e. motions of atoms which amount to $x << b$. Large amounts of shear strain $\gamma_{\beta\alpha} \approx 2\epsilon_{\beta\alpha} \approx 0.2$ are required for the transformations listed in equ. 3a. The diffusionless nature and the fact that a martensitic reaction takes place inside a matrix crystal (β, austenite) implies that large strains and consequently energies would be required for the formation of α (martensite). A lattice invariant internal plastic deformation permits to minimize these strains. The nature of this deformation is important for shape memory behavior. It produces the following types of defects:

- dislocations,
- twin boundaries,
- stacking faults

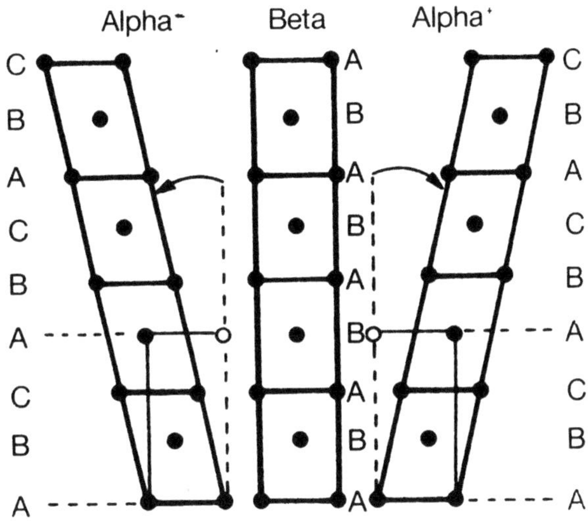

b) Change in stacking sequence by shear (compare fig. 1b), formation of two variants (compare fig. 5d)

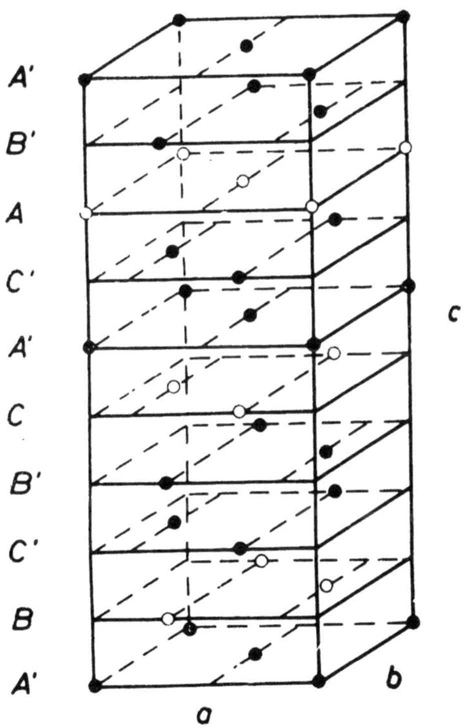

c) 9R–structure of α'–CuZn

in addition to elastic stresses. The energy required for their formation must be less than the energy saved. Their nature depends on the particular type of martensitic phase. It deforms in a similar way as it would for normal plastic deformation. In close–packed structures (f.c.c., h.c.p.: c.n.12) it depends on the stacking fault energy (SFE) which is low in martensitic CuZn– and NiTi–base alloys. Deformation leads to the formation of stacking faults or twins depending on the chemical composition. Less close packed (b.c.c., b.c.t.: c.n.8) martensites have a high SFE, and form undissociated dislocations, but deformation twins, if the transformation temperature is low. The formation of dislocation modes in b.c.c. martensite by:

$$\frac{a}{2}[110]_\alpha + \frac{a}{2}[\bar{1}10]_\alpha \rightarrow a[100]_\alpha \qquad (4a)$$

leads to structures which cannot be reverted. Reversion is possible for twinned and faulted structures. Irreversible are all alloys of iron, except Fe–Mn–base alloys with their f.c.c. $\leftrightarrow$ h.c.p. transformation.

Of special importance for current SMA is the structure of the martensites that form from the b.c.c. structures (A2 ≡ b.c.c. disordered, B2 ≡ AB ordered, DO_3 ≡ A_3B). The six close packed $\{110\}_\beta$ planes in b.c.c. austenite become basal planes by martensitic shear in <110> directions. Together with the sign of shear 24 combinations result which can originate from a β–single crystal. If the martensite structures are f.c.c. or h.c.p. the basal planes are $(111)_\alpha$ and $(0001)_{\alpha'}$. In reality, often complex stacking sequences are found which require up to 18 close packed layers for an elementary cell: designations are: 2H, 3R, 9R, 18R. Some are shown in fig. 6. They are in fact rather similar. 9R originates from B2–, 18R from DO_3–β structure.[18]

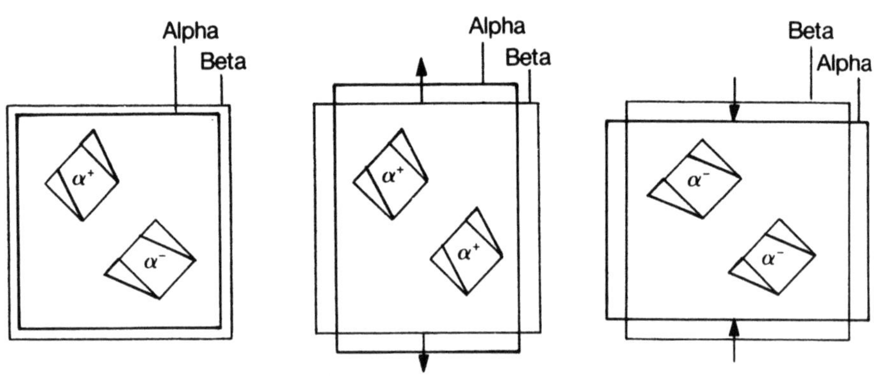

d) Shape change due to selection of variants by external stress.

$$B2 \rightarrow 9R \equiv ...ABCBCACAB... \qquad (4b)$$
$$DO_3 \rightarrow 18R \equiv ...AB'CB'CA'CA'BA'BC'BC'AC'AB'... \qquad (4c)$$

A description in conventional crystallographic terms leads to large, low symmetry elementary cells.

A final note has to be made on the effect of order. Lattice–variant shear does not change the nearest neighbour relations. Consequently, metastable ordered structures may have to be formed from stable ones. An example is provided by alloys of iron with ordered (B1) austenite γ'

$$\gamma' \rightarrow \alpha' \qquad (4a)$$

α' cannot assume a DO_3–structure except by diffusion and therefore a metastable phase is formed with a strong tendency for reversion into γ' (Fig. 6b).

2.3 Thermodynamics: Equivalence of undercooling ΔT and shear stress τ

The existance of the phase transformation requires the intersection of the free energy functions of the phases β and α. As no change in chemical composition is permitted, an alloy can be treated as one–component system (Fig. 7). T_0 is the temperature of equilibrium of the two phases, but transformation always takes place at $M_s < T_0$:[9]

$$T_0 - \Delta T_{\beta\alpha} = M_s \qquad (5a)$$

The undercooling $\Delta T_{\beta\alpha}$ is proportional to a free energy density $G_{\beta\alpha}$ which is required to nucleate and/or drive the reaction.

$$G_{\beta\alpha} = S_{\beta\alpha} \Delta T_{\beta\alpha} \qquad (5b)$$

In the temperature range $M_s < \Delta T < T_0$ the material shows a variety of anomalies, due to its tendency to transform: tetragonal distortion of the cubic crystal structure of β, spinodal shear waves (similar to spinodal decomposition), soft phonons and shear moduli in the $<110>_\beta$–directions. In this range an external stress σ, with its component in a direction of lattice variant shear $\tau_{\beta\alpha}$ is able to rise the transformation temperature M_s. The Clausius–Chapeiron equation

$$\frac{d\Delta T_{\beta\alpha}}{dp} = v_{\beta\alpha} S_{\beta\alpha} \qquad (5c)$$

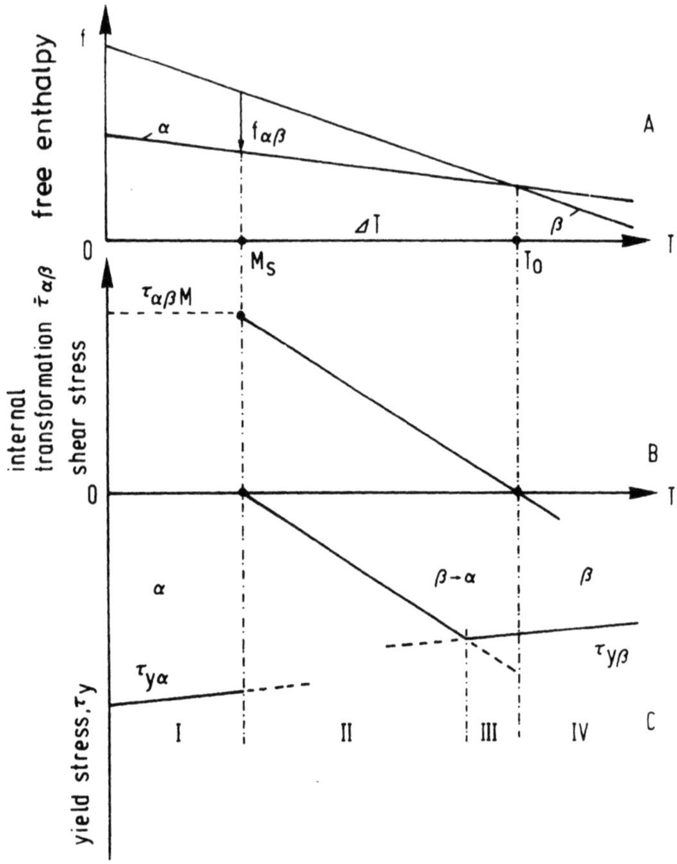

Fig. 7: Schematic representation of thermo–dynamical and mechanical properties of an undercooled high temperature phase β.

a) Temperature dependence of free energy of the phases β and α
b) At $T < T_0$ an internal shear stress $\tau_{\alpha\beta}$ arises which induces spontaneous transformation at $\tau_{\alpha\beta M}$
c) The critical resolved shear stress of $\tau_{y\beta}$ is reduced by $\tau_{\alpha\beta}$ because of stress induced transformation. The material deforms differently in the temperature ranges I to IV: I – Plastic deformation of martensite α_M. II – Stress–induced martensite in undeformed β. III – Stress–induced martensite in deformed β. IV – Plastic deformation of β.

is modified to consider the interaction of the shear stress τ with the transforming phase instead of the hydrostatic stress p and the volume change $v_{\beta\alpha}$:

$$\frac{dM_s}{d\tau} = \frac{\epsilon_{\beta\alpha}}{S_{\beta\alpha}} \tag{5d}$$

The microstructure not the crystal structure of stress induced martensite differs from that of thermally induced martensite. The stress selects favorably orientated varients. This, in turn, leads to a shape change which is limited by the lattice variant deformation $\epsilon_{\beta\alpha} \leq 0.1$. The upper temperature limit for purely stress–induced transformation is defined by the condition

$$\tau_{\beta\alpha} = \tau_{y\beta} \tag{5e}$$

For $\tau_{\beta\alpha} > \tau_{y\beta}$ yielding of the β–phase precedes the transformation. This implies a permanent change in structure by defects such as dislocations and antiphase boundaries. These, in turn, modify start and course of the thermal transformation.[9 10)]

2.4 Mechanical properties: Interrelation between yield strength and transformation temperatures

Equation 5c implies that an SMA yields, above at M_s, by phase transformation. Above T_0 yielding due to true plastic deformation, i.e. motion of dislocations, the β phase takes over. However, in the temperature range $M_s < \Delta T < T_0$ two yield stresses $\tau_{\alpha\beta}$ and $\tau_{y\beta}$ can be defined. The latter defines the load carrying ability of an SMA (Fig. 3). This becomes significant if the alloys are used in a constrained system. The boundary condition may vary between free motion

$$\sigma = 0 \;, \quad \epsilon = \max \tag{6a}$$

and full constraint

$$\sigma = \max \;, \quad \epsilon = 0. \tag{6b}$$

In this context a high yield stress $\tau_{y\beta}$ should be aspired. This is provided by the elementary hardening mechanisms, i. e. defects which act as obstacles to the motion of dislocations in the β–phase.

A second thought has to be given to the effect of such measures on M_s and the course of the transformation (Fig. 8). For a quantitative analysis of M_s the factors affecting the thermodynamical equilibrium T_0 have to be separated from those (such as nucleation and growth) that affect $\Delta T_{\beta\alpha}$ (equ. 5a). This may be demonstrated by an equation which describes the course of M_s during a precipitation (dc/dt) hardening sequence (duration t). Precipitation of atoms from the matrix changes T_0. Hardening $\Delta\tau_y/dt$ raises $\Delta T_{\beta\alpha}$, if it impedes the propagation of the $\beta\alpha$ transformation interface.

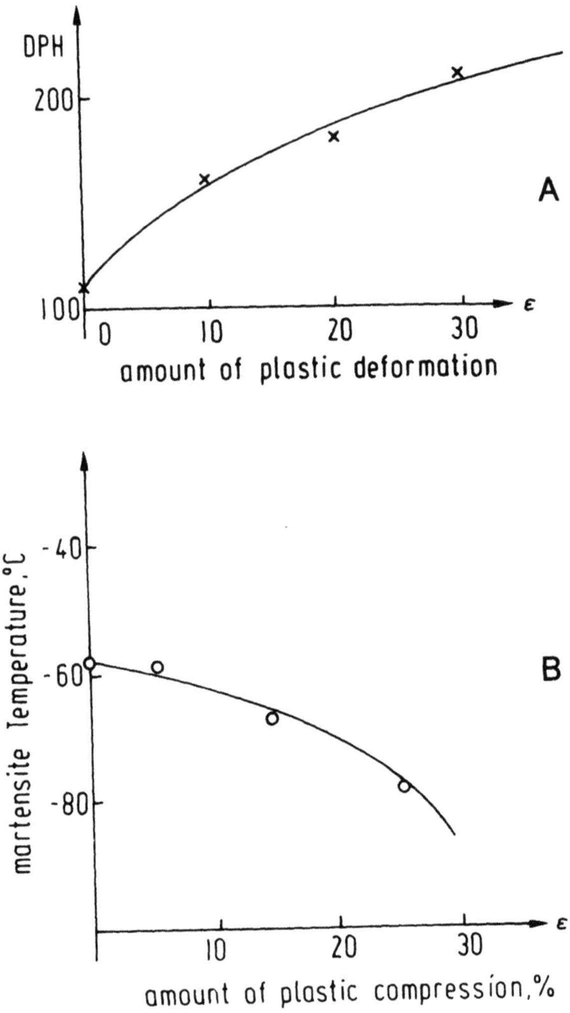

Fig. 8: The Effect of strength in austenite on the transformation.
a,b) Lowering of M_s (b) by hardening (a), β–Cu + 39.2 wt% Zn

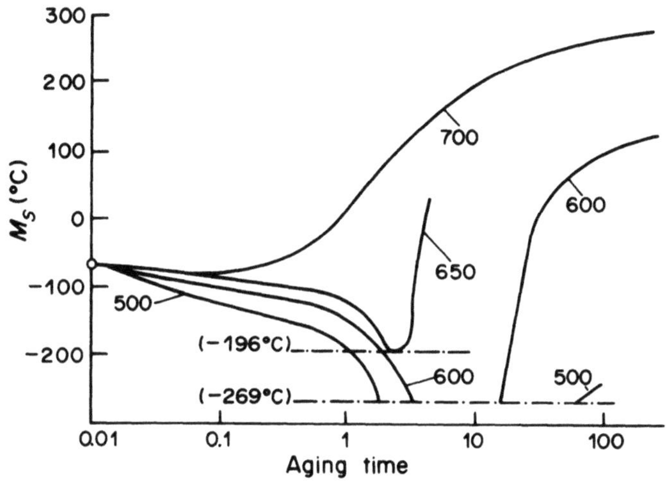

c) The effect of ausaging on M_S, Fe + 27 at.% Ni + 12 at.% Al, different ageing temperatures are indicated, time: hours

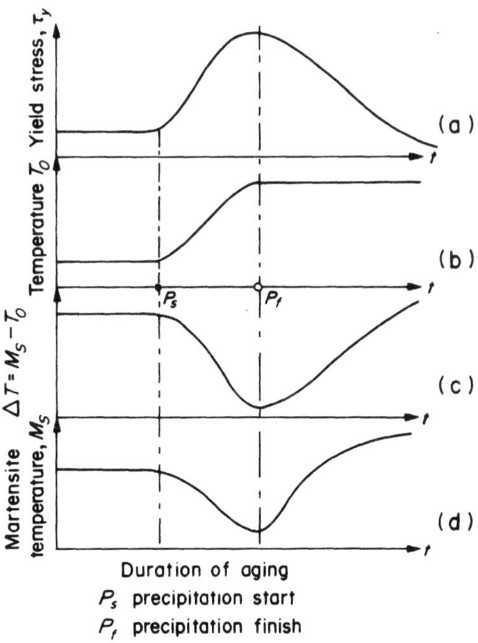

d) Interpretation of the course of the curves shown in fig. 8c, by separation T_0 and ΔT, schematic. (a) Precipitation hardening, (b) Change of T_0 due to decomposition, (c) Change of ΔT due to hardening, and (d) Change of $M_S = T_0 - \Delta T$.

$$\frac{dM_s}{dt} = \frac{dT_0}{dc}\frac{dc}{dt} + \frac{d\Delta T_{\beta\alpha}}{d\Delta\tau_y}\frac{d\Delta\tau_y}{dt} \tag{7}$$

The particularities of the mechanical properties of SMA can be summarized as follows. An external (and internal) shear stress τ favors the formation of martensite i.e. raises the transformation temperature M_s. Strengthening of the parent structure β (austenite) lowers it by increasing $\Delta T_{\beta\alpha}$. This effect can be compensated if the chemical composition or degree of order in the β–solid solution provides for an increase in T_0. Thus thermodynamic and mechanical properties are required for an understanding of the transformation temperature and consequent tailor making of SMA.

Mechanical properties also play a role for the understanding of the course of the transformation, hysteresis, and course of the reverse transformation. Many fundamental aspects of the transformation cycles of SMA are analogous to magnetization loops in ferromagnetic materials. A cross–fertilization can be expected from an interaction of the two fields.

3. Preparing SMA for applications

3.1 Betatizing

This term has been invented for b.c.c. SMA in analogy to austenitization of steels. The purpose of it is the formation of a β–structure which is able to perform the desired phase transformation at lower temperatures. In addition microstructural properties such as grain size, texture, degree of order, and coherent precipitation may have to be provided by the betatizing treatment. Least problematic is the treatment for binary TiNi–alloys. This phase is stable and ordered up to the melting temperature. Usually a recrystallization treatment from a cold worked state is applied to produce a fine grained microstructure. Copper based alloys are betatized from a few degrees below the melting temperature (~ 900 °C). They have to be rapidly cooled to avoid either the massive transformation into the α–phase or unwanted precipitation of an intermetallic compound. In some copper alloys eutectoid decomposition can occur which can be regarded as a combination of massive transformation and precipitation:

massive transformation:	$\beta \rightarrow \alpha$	(8a)
precipitation:	$\beta \rightarrow \beta' + \gamma$	(8b)
eutectoid transformation:	$\beta \rightarrow \alpha + \gamma$.	(8c)

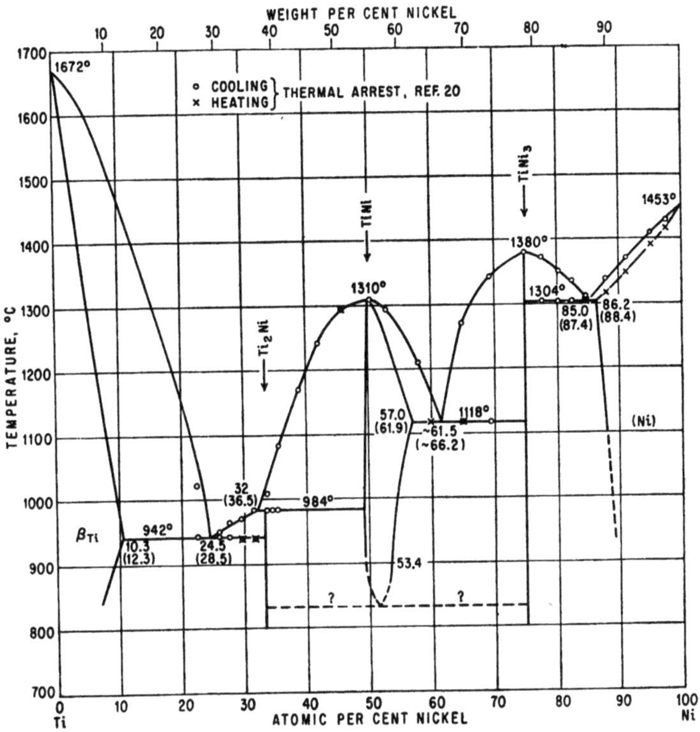

Ni-Ti.

Fig. 9a: Ni–Ti phase diagram showing the stability of β–NiTi between melting temperature and A_f.

It is often difficult to avoid these reactions during cooling from the homogeneous β—range of the phase diagrams.[11] The time–temperature–transformation diagram provides a quantitative description. The required cooling rates can be read from such diagrams. The maximum cross sections that can be successfully betatized follow from the critical cooling rates in a similar way as it is well known from hardenability of tool steels (Fig. 9b).

A homogeneous β–phase can also be obtained by rapid cooling from a liquid. Liquid—solid segregation and dendrite formation have to be avoided. This method is appropriate if thin ribbons of SMA are to be obtained. In addition, it can provide a columnar grain structure with a solidification texture which is favorable for maximum shape change in bending.

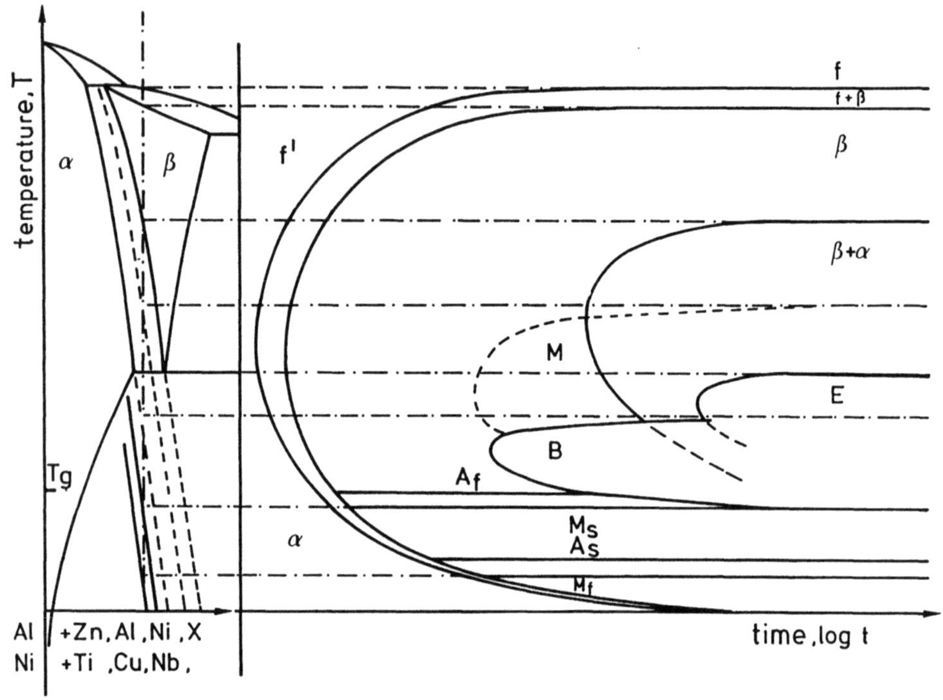

b) T–c, and T–t–transformation diagram of CuZn–alloys. Betatization is possible by rapid quenching from the liquid or the homogeneous β–Cu–Zn solid solution to avoid unwanted solid state reactions.

3.2 Production of the desired final shape (one way)

After heating above A_f a specimen of arbitrary shape should assume a desired shape. This is achieved by micro–creep in the β–state. The procedure is shown in figure 10. A die has to be used into which the alloy is bent in the martensitic state. It appears soft due to mobile boundaries in the microstructure: twin boundaries, variant boundaries. During heating to $T > A_f$ the original β–crystal is reformed. A stress is expected after reverse transformation (equ. 6b). The self induced stress σ surmounts the value of the yield stress $\tau_{y\beta}$ at the particular annealing temperature $T >> A_f$ (Fig. 7).[12]

Local plastic flow $d\epsilon/dt$ is controlled by climb of dislocation in the β–crystals and therefore by the activation energy of diffusion of the atoms Q.

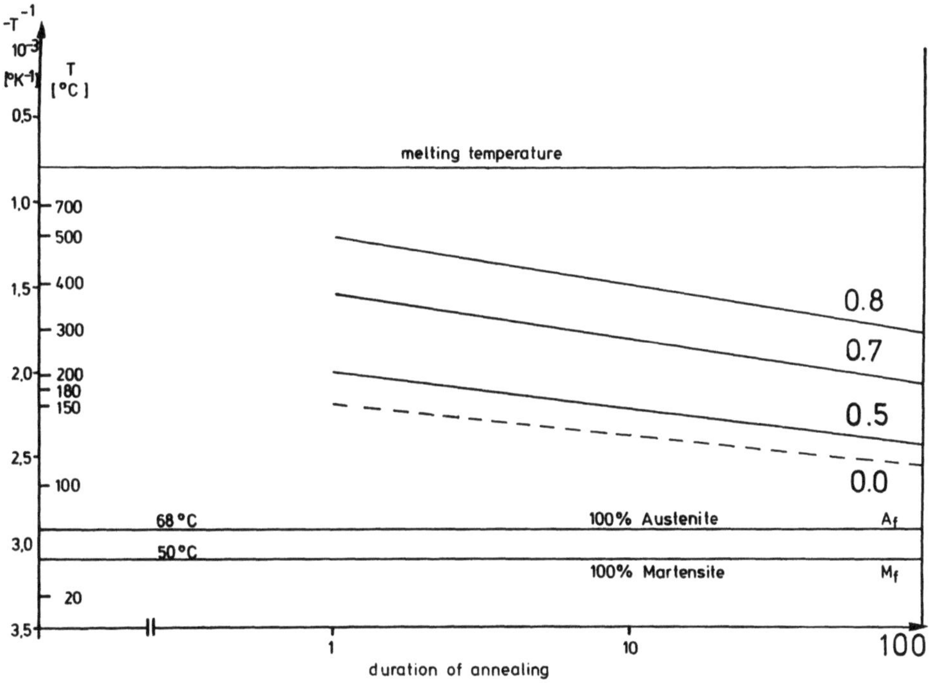

Fig. 10: Temperature–time diagram for producing a certain shape in the β–state for the one–way effect ($\varphi/3\pi$): fraction of the final shape determined by the constraint in a die

$$\frac{d\epsilon_p}{dt} = \dot{\epsilon}_p = A\sigma^m \exp\left(-\frac{Q}{RT}\right) \tag{9a}$$

The plastic strain leads to local relaxation of the transformation induced stress and therefore to a shape change. If a certain strain ϵ_p is required to produce this desired shape, the duration t of an annealing temperature at a temperature T can be estimated by a linear approximation

$$t = \frac{\epsilon_p}{\dot{\epsilon}_p} \tag{9b}$$

Figure 10 shows for NiTi as an example the results of heat treatments on final deformation ϵ_p.

It is evident from the results that this heat treatment must not disturb the structure produced by the betatizing treatment by inducing coarse grain size, embrittlement etc. For CuZn–base alloys shape treatment and betatizing treatment have to be combined.

3.3 Training for the two–way effect

This treatment starts with a primary shape which is produced as described in 3.2. The two way effect is due to a selection of martensitic variants without any external stress (Fig. 3).[12,13] Training must introduce microstructural features which affect nucleation or growth of martensite in such a way that the shape change takes place during a transformation cycle between M_f and A_f (Fig. 2). All possible structural features of two way (2w) memory are summarized in figure 11. Most frequently internal stresses caused by dislocation groupings are used as structural base of 2w–memory. They are introduced by repeated application of small amounts of local plastic deformation – for example by bending in the direction of the desired shape. The applied stress should not be much higher than the yield stress $\tau_{y\beta}$. The temperature for this treatment must surmount A_f. It should however be less than that required for considerable creep (equ. 9a). Figure 11b shows the effect of strain ϵ, number of cycles n, and temperature on the resulting two–way effect ϵ_{2w}. An optimum treatment can be found in this way. However, ϵ_{2w} is always somewhat smaller than ϵ_{1w}, because external stresses can be higher than inter–

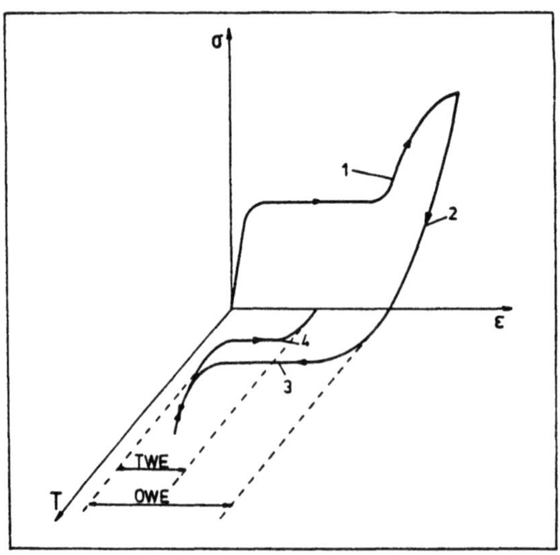

Fig. 11: Training for the two–way effect.
a) T–, σ–, ϵ–training cycle, schematic

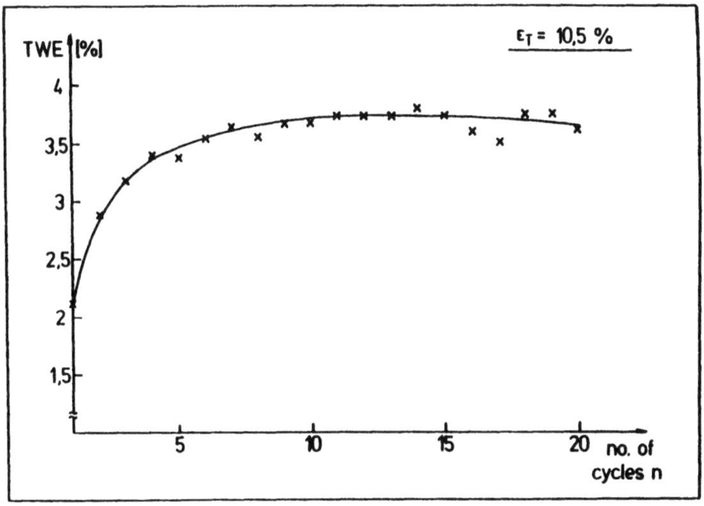

b) Amount of 2w–effect in an NiTi–alloy deformed 10.5 % in the α–state as a function of the number of training cycles.

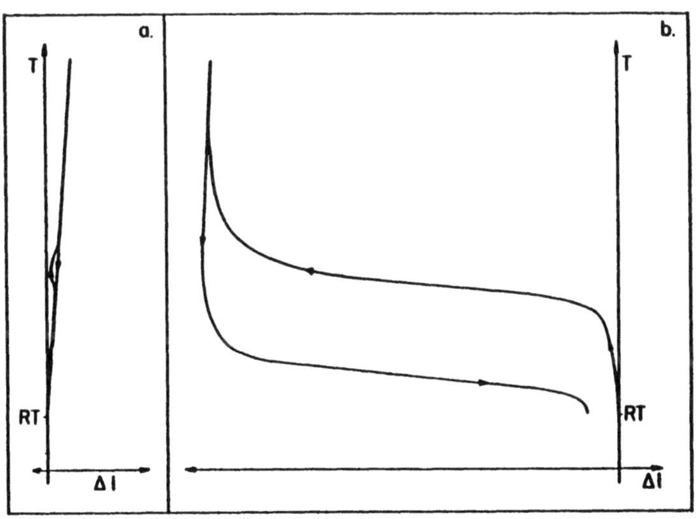

c) Behavior of unstrained (left) and trained alloy, schematic

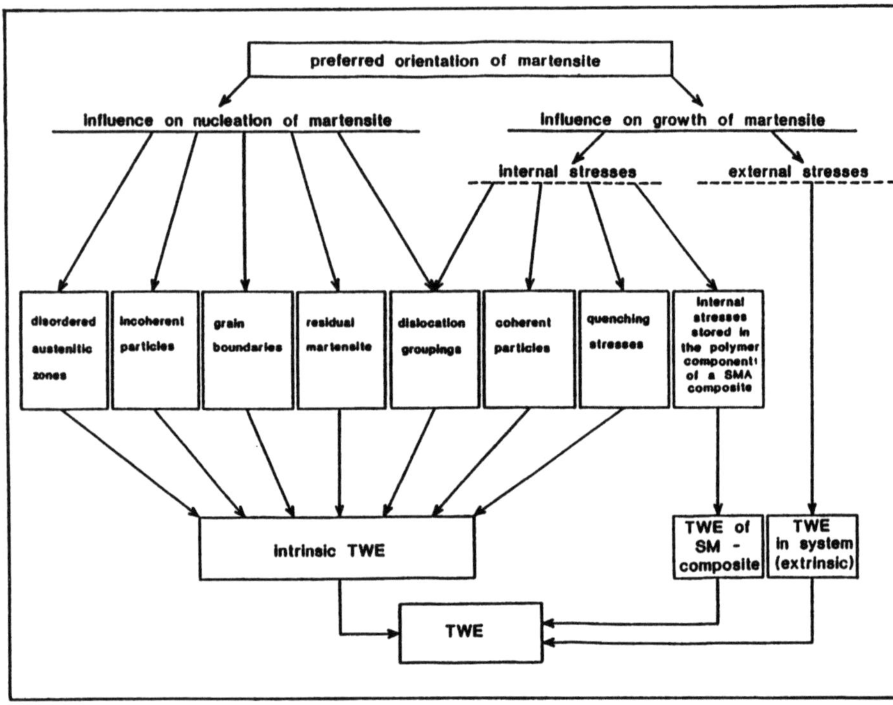

d) Survey of microstructural and macroscopic features that affect the two way behavior.

nal stresses. If the dislocation arrangements which cause internal stresses are not stable, the material may "forget" what has learned during training after repeated thermal cycles.

3.4 Thermal and mechanical fatigue

Fatigue is due to the accumulation of lattice defects during repeated application of thermal or mechanical cycles. For fatigue in the one–way mode, mechanical and thermal cycles are applied alternatively. All the types of fatigue which can be expected for the three effects (Fig. 3) and the two boundary conditions (equ. 6) have been summarized in fig. 12a. It has to be noted that SMA show two particularities in fatigue:[14,15]

1. Not only mechanical but also thermal cycling can induce considerable fatigue, because the passage of the transformation interface may leave behind more and other defects than non–transforming alloys.

T range / mode of loading	$T > A_f \approx T_0$ stable β	$A_f > T > M_s$ pre–martensitic	$M_s > T > M_s$ transform–ing β	$T < M_f$ martensite
mechanical	fatigue in austenite	pseudo–elastic	stress in–duced in $\beta + \alpha$ mixture	fatigue in martensite
thermal	cycling in the range of $\Delta T = T_1 - T_2$; $T_2 > A_f$; $T_1 < M_f$			

Fig. 12: Fatigue of shape memory alloys.
a) Loading conditions form which fatigue may arise

2. Besides classical fatigue (cyclic work hardening → crack formation → crack propagation → rupture) there is "shape memory fatigue". The fatigue–induced defects cause shifts in transformation temperature, hysteresis and course of the transformation cycle. Figure 12b,c show results of measurements of fatigue in the pseudoelastic

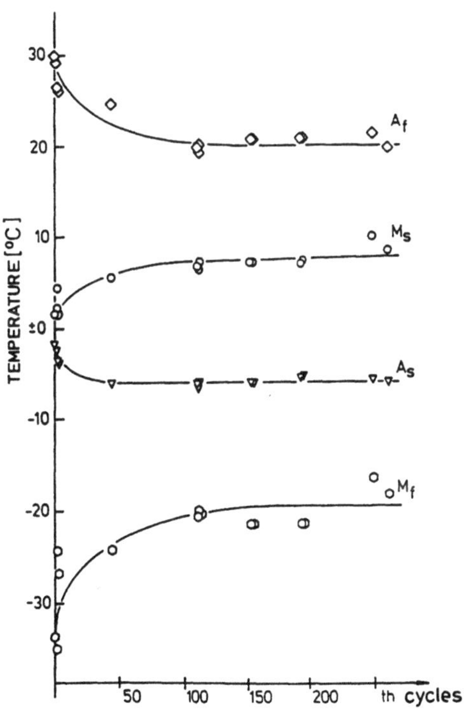

b) Thermal cycling between $T_1 < M_f$ and $T_2 > A_f$ leads to change in transformal be–haviour (compare fig. 2, Cu 26 % Zn, 4 %Al, M_s = +8° C).

c) Formation of fatigue cracks at martensite variant boundary in mechanically cycled CuZnAl–alloy.

state and of shape memory fatigue in CuZn–base alloys. Microstructural investigations provide evidence for easy nucleation of cracks at the boundaries (twins, variants) in the martensitic state. Crack growth is slow due to multiple crack–branching. Stress induced pseudoelastic transformation at a crack tip may retard its growth. A high yield stress $\tau_{y\beta}$ is the most important requirement for a good fatigue strength of shape memory alloys. This is an important property, because many present and prospective applications such as control systems or heat engines require multiple transformation and loading cycles.

4. Applications

There is a wide field for possible applications of shape memory alloys in engineering and medicine. Probably only a small portion has been explored. They can be subdivided according to application of the three effects (Figure 13). The boundary conditions for most applications are intermediate between free motion and full constraint (equ. 6): A robot hand has to move before it grips. A number of fields in engineering have been listed, in many of them application of SMA is still in the state of exploratory development. These materials still offer solutions for which problems are sought.

SM Effect / Technology	One–way	Two–way	Pseudo–elasticity
connecting, fastening	pipe connections expanding rivets		seals, spectacle frames
control		valve control, temperature protection, greenhouse opener	
motor vehicle		opening of fog–lights high noise absorption	
data processing	plug connections for circuits	head/disc system magn. data memory colour changes for opt. storage	
power/ engines		rotating, solar battery control	
automation	manipulators	robot limbs,	
medicine	implants (osteosynthesis)	endoscope control	dental braces
clothing	fabric reinforcement		rubber substitute

Fig. 13: Application of SMA
a) Survey of application of the three effects in different fields of engineering and medicine.

b) 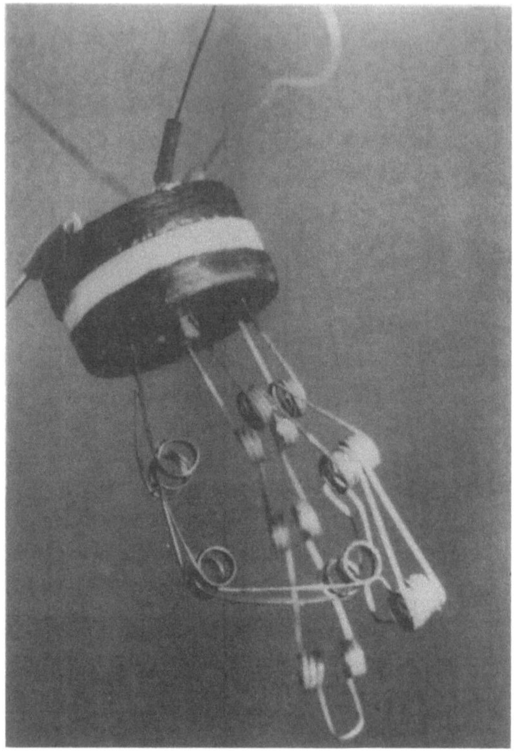

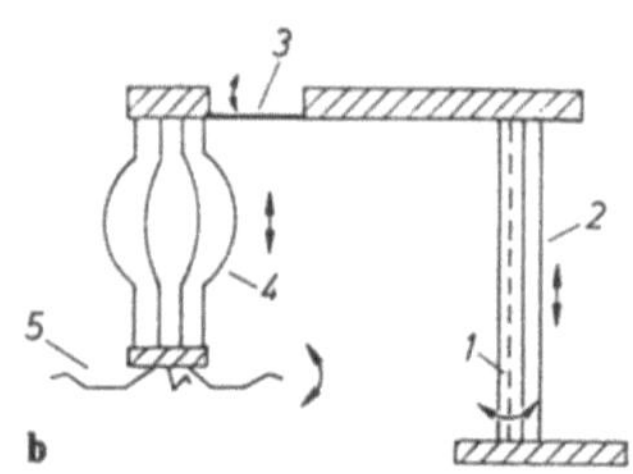

c) 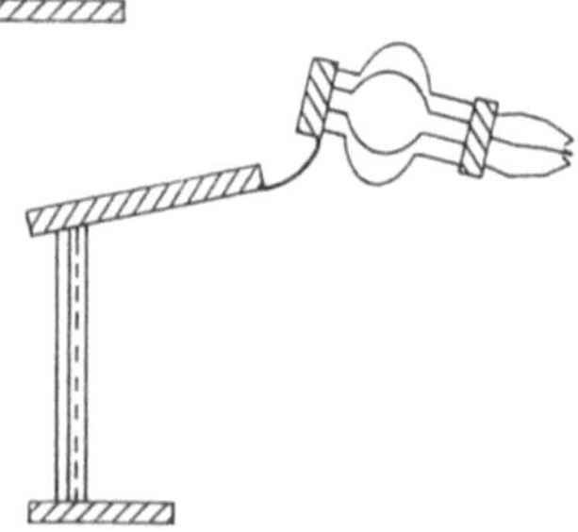

b,c) Prototype of an SMA micro–robot which is electrically heated and operates without joints and bearings

Acknowledgement

Thanks are due to the German Federal Ministry of Technology and Research who supported this work (BMFT 03M 5006 B5) and to Mr K. Escher and Mr N. Jost for providing some of the results as well as some of the pictures.

References

1. C. M. Wayman, Introduction to the crystallography of martensitic transformations, Macmillan, New York 1964.
2. J. Perkins, ed., Shape memory effects in alloys, Plenum Press, New York 1975.
3. Z. Nishiyama, Martensitic transformations, Academic Press, New York 1979.
4. E. Hornbogen, M. Thumann, eds., Die martensitische Phasenumwandlung, DGM Informationsgesellschaft, Oberursel 1986.
5. D. Stöckel, ed., Legierungen mit Formgedächtnis, expert Verlag, Böblingen 1988.
6. E. Hornbogen, N. Jost, eds., The Martensitic Transformation in Science and Technology, DGM Informationsgesellschaft, Oberursel 1989.
7. E. Hornbogen, On fractal aspects of martensitic microstructure, Z. Metallkde. 78 (1987) 352.
8. T. Saburi, C. M. Wayman, Crystallographic Similarities in Shape–Memory Martensite, Acta Met. 27 (1979) 979.
9. E. Hornbogen, On the Martensite Start Temperature M_s, Z. Metallkde. 75 (1984) 741.
10. Li Lu et al., Simulation of transformation hysteresis, Z. Metallkde. 81 (1990) 613.
11. M. Meuris, E. Hornbogen, Heat treatments of β–bronces and shape memory alloys, Sonderbd. der Prakt. Metallogr. 21 (1990) 299.
12. M. Meuris, E. Hornbogen, Das Eintrainieren des Ein– und Zweiweggedächtnisses von NiTi–Formgedächtnislegierungen, Sonderbd. der Prakt. Metallographie 22 (1991) (in print)
13. K. Escher et al., The Two–way Shape Memory Effect for Mechanical Hands, Proc. 31st Structures, Structural Dynamics and Materials Conf., AIAA, Washington D.C. (1990) 29.
14. M. Thumann, E. Hornbogen, Thermal and Mechanical Fatigue in Cu–base SMA, Z. Metallkde. 79 (1988) 119.
15. M. Sade, J. Kumpfert, E. Hornbogen, Thermo–mechanical and pseudo–elastic fatigue of a polycrystallization CuZn24Al3alloy, Z. Metallkde. 29 (1988) 678.

Symbol	unit	
A_f	K	austenite finish temperature
A_s	K	austenite start temperature
α, α_{ij}	K^{-1}	coefficient of thermal expansion
b	m	atomic spacing
c	1	chemical composition of solid solution
c.n.	1	coordination number of crystal structures
D	m^2s^{-1}	diffusion coefficient
E	Pa	elastic modulus
ϵ	1	any strain
ϵ_e	1	reversible strain
ϵ_p	1	plastic strain
$\epsilon_{\beta\alpha}$	1	transformation strain
ϵ_{1w}	1	one–way effect
ϵ_{2w}	1	two–way effect
$\dot{\epsilon}_p$	1	$d\epsilon_p/dt$, creep rate
$\gamma_{\beta\alpha}$	1	$2\gamma_{\beta\alpha} = \epsilon_{\beta\alpha}$, transformation shear
$G_{\beta\alpha}$	$Jmol^{-1}$	free enthalpy of transformation
m	1	creep exponent
M_f	K	martensite finish temperature
M_s	k	martensite start temperature
n	1	work hardening exponent
p	Pa	hydrostatic pressure
Q	$Jmol^{-1}$	activation energy for diffusion
R	$Jmol^{-1}K^{-1}$	gas constant
$S_{\beta\alpha}$	$Jmol^{-1}K^{-1}$	entropy of transformation
σ	Pa	any stress
σ_y	Pa	yield stress
σ_f	Pa	fracture stress
$\sigma_{y\beta}$	Pa	yield stress of β–phase
$\sigma_{\beta\alpha}$	Pa	stress required to induce transformation
T	K	any temperature
ΔT	K	temperature difference
$\Delta T_{\beta\alpha}$	K	$\Delta T_{\beta\alpha} = T_0 - M_s$, required for transformation
T_{cl}	K	melting temperature (crystal → liquid)
T_0	K	equilibrium temperature
t	s	any period time for heat treatment

t_0	s	time for one atomic hop
τ	Pa	any shear stress
$\tau_{\beta\alpha}$	Pa	transformation shear stress
$\tau_{\beta y}$	Pa	critical shear stress in β–phase
$\Delta\tau_y$	Pa	increase in yield stress by hardening mechanism
v_T	1	specific volume at T
$v_{\beta\alpha}$	1	change in volume by transformation

Self-Reinforcing Polymer Composites

Gerhard Wegner, Max–Planck–Institut für Polymerforschung
Ackermannweg 10, D–6500 Mainz/Germany

Introduction

Fibre reinforced polymers play an enormous role as structural materials in present day technology. The same is true for highly filled polymers, where the filler consists of isotropic particles which, however, provide impact strength, thermal conductivity and rigidity. The use of mixtures of form anisotropic (fibres) or form isotropic (powders) particles and polymers is linked to a number of serious problems in processing technology. Moreover, the presence of internal surfaces and phase boundaries in the composite materials is at the origin of the failure, ageing and other misbehaviors which limit the technical performance of these otherwise interesting materials.

In the recent years attempts have been made to create novel classes of polymers which cana be processed in the same way or similarly as conventional thermoplastics which, however, solidify in a structural form which resembles a reinforced polymer. In other words high strength, high rigidity polymers may be obtained which show directional properties as if they would contain fibres embedded in a continuous matrix. Such polymers usually belong to the class of so called main–chain liquid crystal polymers. They are processed from either a liquid crystal melt, which crystallizes after processing retaining the molecular orientation of the liquid–crystal state, or they are brought into the final form going through a series of phase transitions involving various homogeneous, liquid crystal and crystal phases.

Strength and Thermal Stability

Heating of a polymer sample or part causes an increase of the mobility of the segments of the macromolecules and triggers both phase transitions and chemical changes.

Increased exposure to high temperatures may result in irreversible changes. The following changes have to be considered:

- Changes of state not involving chemical reactions, e.g. glass transition, melting, phase transitions involving different liquid crystalline phases etc.
- Evaporation of low molecular weight components of the material, e.g. plastisizers, stabilizers.
- Depolymerization with complete degradation.
- Thermally induced splitting of components of polymer bound residues, however, with retention of the chain structure; a typical example is the imidization of poly(imide) precurcors.
- Chain scission with or without subsequent crosslinking,
- Oxidation (burning off) of volatile fragmentation products with formation of coal as the residue.

In the context of this paper we are only interested in high temperature and high strength materials, that is materials which may be exposed over prolonged times to temperatures well above 150°C.

The general principles which govern the relations between the details of the chemical structure, phase formation, strength and thermal stability have been amply described in the literature. A standard text is contained in ref. 1 (1). Among the longest known high–temperature (HT)–polymers are the fluorine containing polymers, e.g. PTFE (poly(tetrafluoroethylene)) and the silicones. However, their rather limited strength and creeps–behavior does not allow their use as structural materials notwithstanding the fact, that they are excellent in terms of insulation materials for high–tech-applications. A list of structural elements contained in polymers specifically designed to be used at high temperatures is given in Table 1.

Among the newer HT–thermoplastic materials are the poly(sulfones) (PSU) showing a heat stability (WFT) of 174°C, poly(ethersulfones) (PES) with WFT of 215°C, copolymers like PPSU exposing a WFT of 200–280°C, depending on the actual composition. Poly(phenylensulfide) (PPS) (WFT = 240°C), poly(phenyleneoxide) (PPO) (WFT = 220°C), copolyesters of hydroquinone with iso– or terephthalic acid (WFT ≃ 240°C) or aromatic copolyesters based on bisphenol A (WFT ≃ 300°C) have also become popular. Poly(amidimides) (PAI) (WFT ≃ 274°C) are – in a formal sense – intermediate between classical poly(amides) and duroplastic poly(imides).

Table 1.1

Structural elements of HT–polymers; the elements marked a,b,c,d–1, d–2, e–1, and e–2 are units to be found in various ratios in HT–thermoplastic polyesters, some of which show lc–phases.

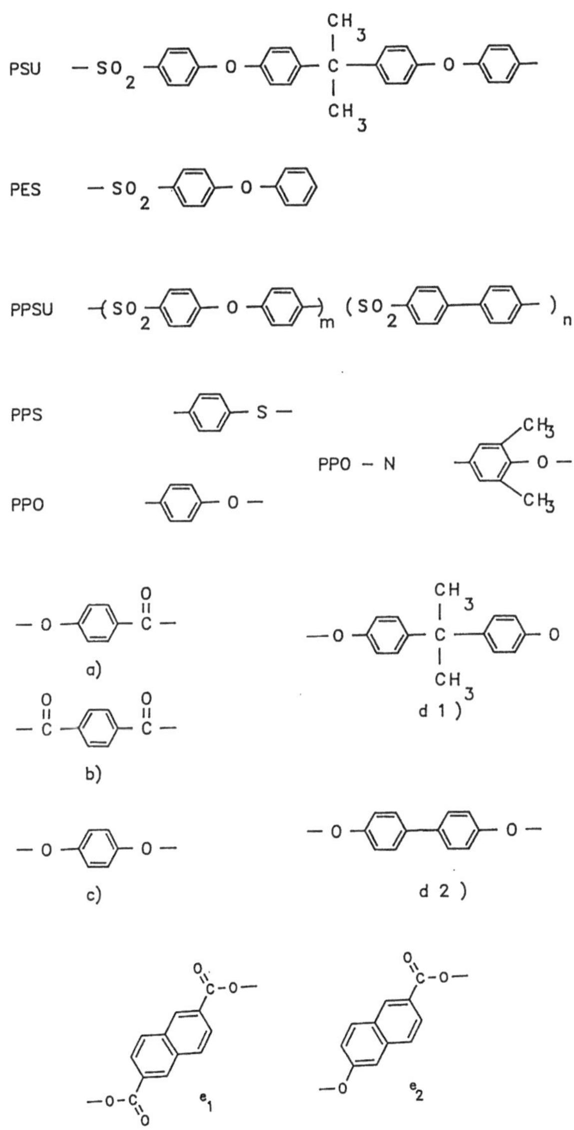

Table 1.2

Further examples of HT–polymers, containing all–aromatic structural elements.

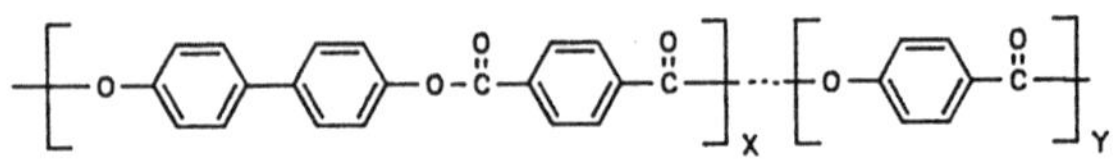
Xydar

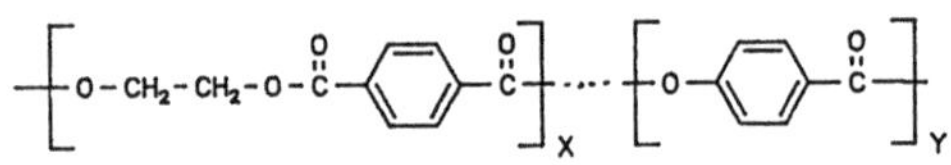
LCP–K

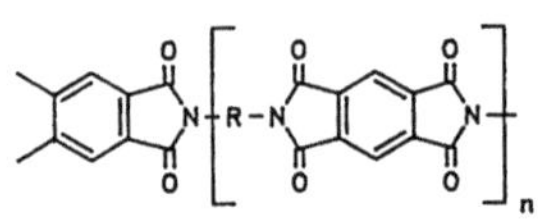
Polyimide

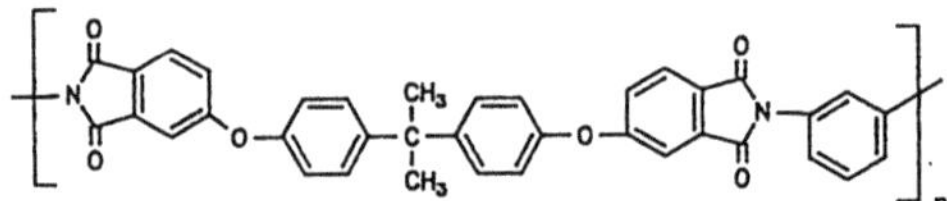
Ultem

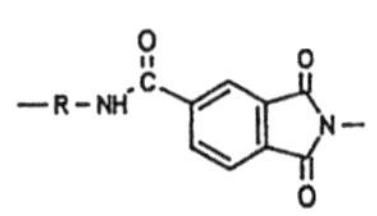
PAI

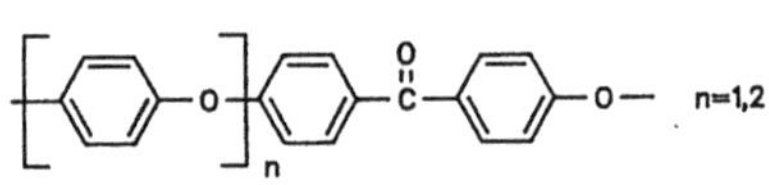

PEK
PEEK

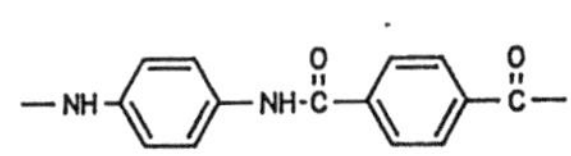
PAA

Poly(imides) are extremely high–temperature stable duroplastic materials which perform over very long time at temperatures in the range of 500°C. Since they are made by chemical reaction (imidization) starting from suitable, soluble or meltable precursor polymers with volatilization of low molecular weight reaction products, they are only available as thin films or coatings and are not used in terms of structural materials.

The most common HT–materials together with values characterizing their strength (at room temperature) and thermal stability are summarized in Table 2. This table contains data of some liquid–crystalline materials which constitute a class of polymers attracting great interest currently (2,3).

Table 2

Comparison of the properties of some "Engineering Plastics" showing temperature stability (WFT); materials to be processed from liquid–crystalline phase (LCP) are indicated.

Polymer type	Name Producer	Bending modulus (Nmm^{-2})	Tensile strength (Nmm^{-2})	WFT (°C at 1.85 Nmm^{-2})
APE	Ekkcel (LCCP) (Carborundum)	2500	100	295
APE	Vectra (LCP) (Hoechst–Celanese)	23000	200	180
APE	Xydar (LCP) (Dartco)	14000	140	355
PAI	Torlon (Amoco)	4800	190	275
PES	Ultrason (BASF)	2600	85	210
PEEK	Victrex (ICI)	3800	95	150
PPS	Ryton (Philipps)	3400	76	135
PPO	Noryl	2500	95	130

Liquid Crystal Polymers

In the context of structural materials the so-called mainchain liquid-crystal (lc) polymers receive increased attention. Traditional polymers, when melted, consist of an isotropic entanglement-network of the chains. The entanglement network depicted schematically in Figure 1a is at the origin of the well-known visous behavior of polymer melts. In contrast to these conventional materials lc-polymers consist of rather rigid rod-like or at least worm-like chains which form spontaneously anisotropic structures of the type sketched in Figure 1b. The lack of an entanglement network

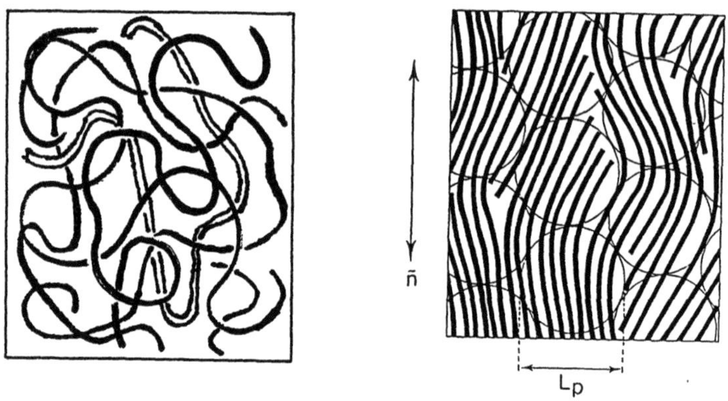

Figure 1

a) Schematic structure of a polymer melt of conventional macromolecules

b) Schematic description of a lc-polymer in the nematic melt

provides for the very low viscosity of the melts of such material. The ease, by which the chains or rather the domains (compare Figure 1b) of many chains can be oriented in a stress field provides for the exceptional mechanical strength observed in these materials when tested after quenching to the solid state. Due to the anisotropic nature of the bulk material, the strength is very high in chain direction but very low in transverse direction. This causes the materials to exhibit low compressional strength, if there was no care taken to randomize the domain structure prior to quenching. Thus, a series of problems exist in the processing of these novel materials. Processing problems also exist if one tries to spin fibres of liquid crystal polymers either from the thermotropic or lyotropic phase. A better known example are the fibres of poly(aramides) which are obtained by spinning a lc–lyotropic solution of poly-terephthalamide in conc. sulfuric acid into water. Fibres of exceptional strength in fibre direction (E ≃ 150 GPa) and considerable heat–stability are obtained which, however, suffer from lack of compressional strength. In addition, these fibres exhibit a rather complicated morphology and it is by no means clear, to what extent this morphology influences the overall mechanical and thermomechanical performance including properties of composite materials where such fibres are used to reinforce a matrix resin. A picture describing some of the more important morphological details seen in such fibres is depicted in Figure 2 as an illustration of the complications which one encounters in the processing of such high–strength materials.

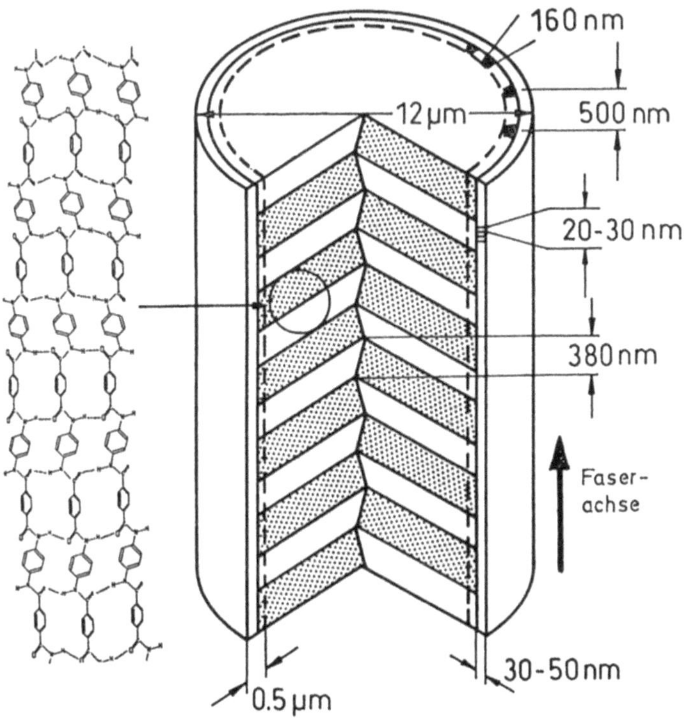

Figure 2

Morphology of a fibre of poly(terephthalamide) as obtained by spinning from lc–solution in conc. sufuric acid.

Table 3

Rigid macromolecules with flexible side chains showing self–reinforcing behavior in terms of formation of nanocomposites (4).

Polyester

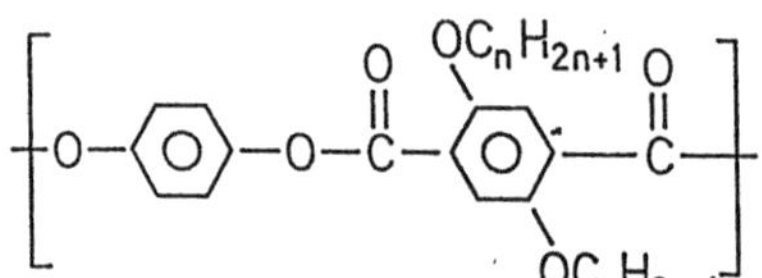

1

Polyamide

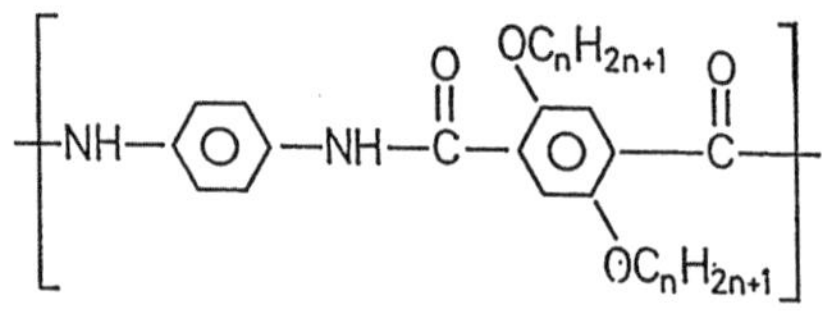

2

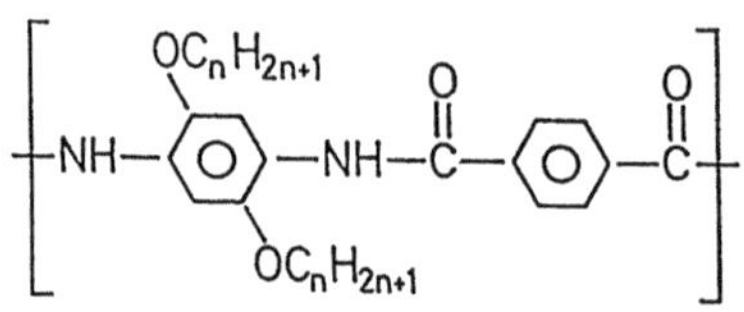

3

Polyimide

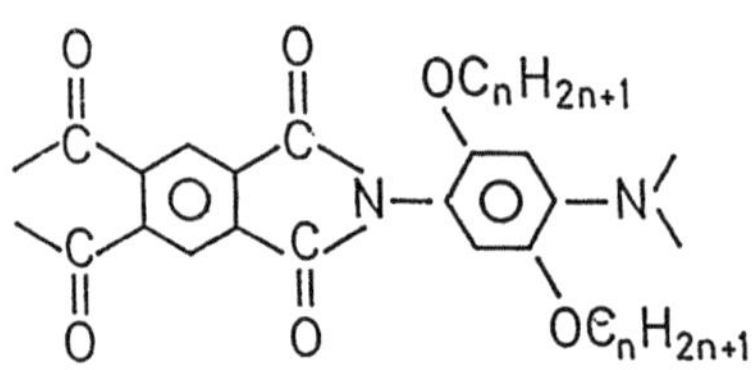

4

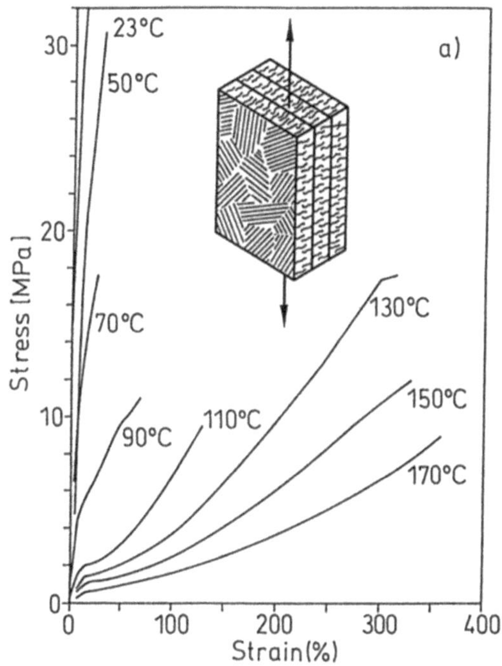

Figure 3

Stress–chain curves at various temperatures of a polyester of type 1 (compare Table 3) with C–16 side–chains performing in a manner described as "nanocomposite".

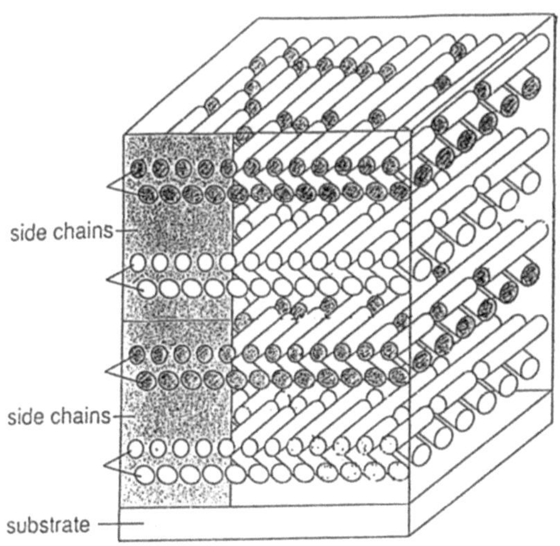

Figure 4
Nanocomposite of alternating double–layers of rod–like macromolecules embedded in a continous matrix of their own side chains as obtained by application of the Langmuir-Blodgett method (5,6).

One of the larger projects undertaken in the Max–Planck–Institut für Polymerforschung in the recent years was concerned with lc–polymers which combine rigidity of the main chains and the matrix properties of composites that is elasticity and relaxational properties in the same molecule (4). Thus, rigid–rod polymers containing short and flexible side chains as part of the constitutional units were designed and tested. Some of the typical structures are collected in Table 3. The mechanical behavior of a polyester of type 1 (Table 3) with C–16–side chains is shown in Figure 3 together with a sketch of the molecular architecture of the material to be tested. The short and flexible side chains form a continous matrix in which the main chains are embedded as reinforcing elements: thus, the overall behavior of a reinforced material is the result although it consists of a homopolymer only. The name "nanocomposite" is, therfore, a proper designation of these novel materials. An even more sophisticated approach to nanocomposites becomes possible using the Langmuir–Blodgett–method to construct architectures of rod–like macromolecules decorated by flexible side chains (5). Macromolecules of elongated cylindrical shape may be spread to the surface of water and compressed to a two–dimensional nematic solid, in other words, rafts as of logs swimming on a lake, are formed and subsequently transferred layer by layer to planar substrates. Ultrathin layers of nanocomposites as depicted in Figure 4 are obtained. They exhibit exceptionally interesting properties both as model materials and as components in electrical, optical and electrooptical devices presently tested in our Institute (6).

References

1. "Polymere Werkstoffe", Bd. 1, H. Batr, Hrsgb. G. Thieme Verlag, Stuttgart, 1985

2. "High Modulus Polymers – Approaches to Design and Development", A.E. Zachariades, R.S. Porter, Hrsgb., Marcel Dekker, Inc., New York, 1988

3. "Structure and Properties of Oriented Polymers", I.M. Ward, Hrsgb., Appl. Sci. Publ., London 1975

4. a) M. Ballauff, Makromol. Chem. Macromol. Symp. 26, 57, 1989
   b) M. Ballauff, Angew. Chem. 101, 261–276, 1989
   c) U. Falk, H.W. Spiess, Makromol. Chem. Rapaid Commun. 10, 149, 1989
   d) J.M. Rodriguez–Parada, R. Duran, G. Wegner, Macromolecules 22, 2507, 1989

5. F. Embs et al., Adv. Materials 3, 25, 1991

6. W. Hickel et al., Langmuir 6, 1403, 1990

Functional Molecular Materials Derived from Organic Conducting Polymers

Francis GARNIER, Laboratoire des Matériaux Moléculaires, CNRS, 2 rue Dunant 94320 THIAIS FRANCE.

ABSTRACT

Organic conjugated polymers form a new class of materials in which each macromolecular chain presents a high electrical conductivity. This unique property allows to represent these materials as a tridimensional network of molecular wires, able to carry informations. These wires can be chemically functionalized by the covalent binding of various prosthetic groups, presenting specific interactions whether with the chemical environment or with external physical quantities. Molecular transducers can thus be developed, and examples already exist on the molecular recognition exerted by functionalized conducting polymers toward ions or toward optically active species in solution. These structures open the promising field of molecular electronics, although the development of this new research area first requires the demonstration that the charge transport properties of these materials are efficient enough for them to be used as active layers in electronic devices. This step has been recently achieved, by the development of a new class of organic conjugated materials, conjugated oligomers, which present shorter conjugation length but much lower content of structural defects. The semiconducting properties of these oligomers have been determined through the characterization of field-effect transistors fabricated from these organic materials. A new all-organic thin film architecture has been proposed for these transistors, which shows comparable characteristics to those of a-Si:H based devices. These results confirm that organic conjugated polymers and oligomers can be used as active materials for the construction of electronic devices, and that their further functionalization will allow the development of new types of molecular transducers.

INTRODUCTION

The last decade has seen the emergence of a new research field, at the frontier between physics and chemistry, which concerns organic conducting polymers. Until the late seventies, electrical conductivity and semiconductivity were the priviledged area of inorganic materials, which formed the heart of electrical and electronic devices. The use of organic compounds was mainly limited to the construction process of devices, for instance as resists for microlithography, and also for ensuring mechanical strength and protection against the environment. Electronic devices have been constructed at the image of biological functions, underdoing the basic operations of gate, memory and amplification, and various sensors have been also proposed for performing recognition operations. These developments have been based on a macroscopic approach, in the sense that large scale materials and structures have been involved in the construction of the devices. The trend toward integration led to important technological improvements, which allowed electronics to come to the area of Very and now Ultra Large Scale Integration, with an actual limit of resolution at about 0.2 micron, each step raising technological problems of exponentially increasing complexity. On the other hand, we know from biology that such basic operations can be performed at a molecular level, but also that the various attempts to isolate the biological units associated with these processes have failed, due to our incapacity to maintain a long term activity to these biological functions, when deconnected from their ensemble. Devices based on biological materials, although very promising, have been thus limited to the use of some enzymes of short term stability. It appears that the approaches followed by solid physicists and by biologists toward the replication of molecular type devices and sensors are runing into extremely difficult problems. Organic chemists on the other hand know since very long how to engineer molecules and how to design various kinds of functions, R, which give a selective response whether to the chemical environment or to external physical quantities. More recently, they have also learned how to assembly these molecules in an ordered way into materials, and how to control supramolecular properties in the corresponding molecular assemblies. A step was however still missing for bridging the gap between the area of functionalized molecular materials and that of electronics, which concerned the development of conducting or semiconducting properties which appear necessary for the transfer of the various informations. As a matter of fact, the classical structures, used by chemists as supports for chemical functions, involve insulating molecules or macromolecules, whereas the transfer of the informations collected from the functions R requires the existence of charge transfer properties in the molecular backbone supporting these prosthetic functions R. Ideally, molecular sized wires would correspond to the optimized internal organisation of such a functionalized molecular material. Although the principle of electrical conductivity in organic compounds was already discovered in the early sixties, when H. Naarmann (1) showed that conjugated polymers could become electrically conducting upon doping, the development of the field of organic conducting polymers started only at the end of the seventies. The first step in this research field was devoted to the understanding of the fundamental processes involved in the transfer of charge in these organic materials, and also to the experimental control of the physical and chemical properties of organic conducting polymers. The obtained results allowed then to develop a new generation of organic conducting polymers, obtained by their functionalization with various chemical groups, and chemists know actually how to engineer these molecular materials in such a way that basic operations of electronics and sensors can now be created.

ORGANIC CONDUCTING AND SEMICONDUCTING POLYMERS

The field of organic conducting polymers has been mainly developed since A.J. Heeger et al. observed that polyacetylene showed, upon doping, a fairly high electrical conductivity, of the order of $\sigma = 10^3$ S cm^{-1}. This result raised a large interest and the many works conducted in this area allowed a better understanding of the physical and chemical properties of these organic materials. The macromolecular chains of these polymers are made of a regular alternation of single and double bonds, as exemplified in polyacetylene, which is the prototype of these conducting polymers. Other classes of conducting polymers, such as polypyrroles(3), polythiophenes (4) and polyanilines (5), have appeared since in the literature. These polymeric chains present a cross section which has the geometrical characteristics of their corresponding monomer, with typical dimensions of the order of some angströms. Their length is controlled by the polymerization conditions, and can exceed hundreds of microns. The chains are not individually dispersed in the three dimensional polymeric material, as it has been generally observed that they agreggate together to form fibrils, of some few thousand of angströms thickness, as shown in Figure 1.

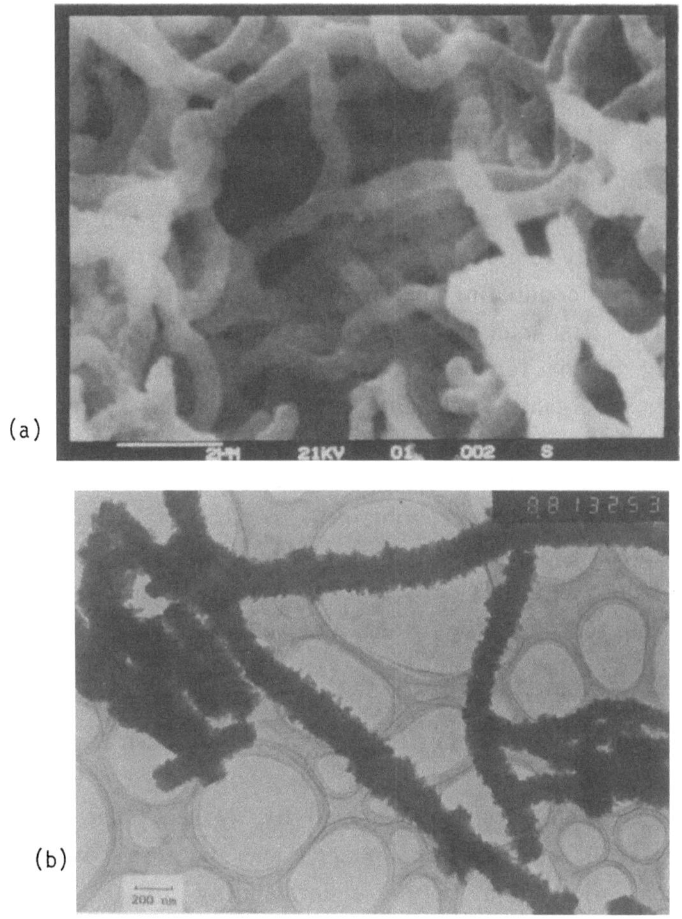

Figure 1. (a) Scanning electron microscope picture of the fibrilar structure of poly(3-methyl thiophene), and (b) transmission electron microscope picture of individual fibers of the same conducting polymer.

Although the distribution of the individual polymer chains inside these fibers is not yet precisely known, the consideration of the molecular structure of the monomers allows to discuss the conformation of these chains.

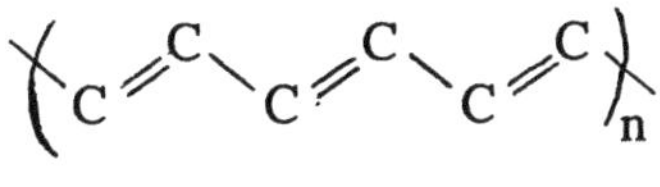

Polyacetylene

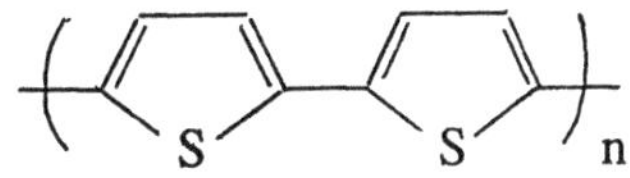

Polythiophene

The conducting properties of these macromolecular chains originate from the conjugation of the π−electron system, which creates a molecular orbital extending throughout the polymer chains. Using in a first approximation the concepts derived for the one-electron model, these low dimensional materials have been described by a fully occupied valence band and an empty conduction band, formed respectively by the conjugated π and π* systems. In this state, these polymeric chains show semiconducting behavior associated to a band gap of about 2 eV. Holes or electrons can then be injected in these chains, by p-type or n-type doping respectively, which leads to the appearance of electrical conduction, a phenomenon which has been described by the creation of solitonic or polaronic states in the gap, using thus the concepts of solid physics. This doping is experimentally performed by the inclusion of electron acceptors , such as halogens, or donors such as alkaline metals, which intercalate between the conjugated polymer chains and lead to their oxidation or reduction respectively. It must be pointed out that the term of doping is used here in a chemical sense, and not in the commonly accepted one in solid physics.

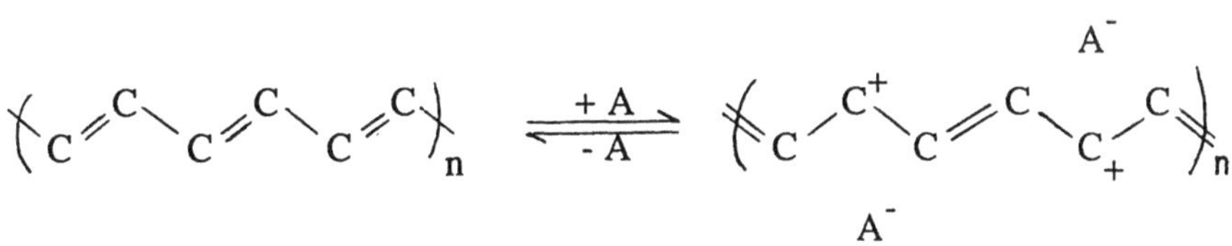

Figure 2. Reversible doping of polyacetylene with an electron acceptor species A.

The obtained conductivity values depend on the amount of charges, electrons or holes, created in the polymer chains, which is determined by the dopant concentration. Thus, very high concentrations of dopants appear necessary for reaching high conductivities, up to one dopant molecule for every two or three monomeric units in the polymeric chains. Due to an uncomplete control of the polymerization reaction, structural defects exist along the polymeric chains, which interrupt the conjugation and impose interchain hopping for the overall charge transfer to proceed. The low efficiency of this charge hopping mechanism has appeared as the main limitation for the macroscopic conduction observed in these molecular materials, as confirmed by the recent work of H. Naarmann who has been able, by a controlled improvement of conjugation length in polacetylene, to reach a conductivity value of $2x10^{5}$ S. cm^{-1}, which is equivalent to that of a metal such as iron (6). It must be however noted that a real metallic state has not yet been observed for these organic conducting materials, as shown for instance by their conducting behavior in function of temperature (2).

An important property, which has been observed for these organic materials concerns the reversibility of the doping process, which can be also electrochemically driven by the use of an electrode for injecting electrons or holes. This reversible doping and undoping mechanism confers to these organic polymers interesting electroactive properties and it has been shown that even for very thick polymer films, up to some millimeters, almost fully reversible electrochemical systems can be experimentally obtained, by which every two or three monomeric units of the polymer are associated with a dopant molecule. This result, which indicates that even in a large macroscopic material the very molecular monomeric units are involved in the doping process, means that electroactivity is in this case a molecular process. A large concentration of dopant molecules can thus be reversibly intercalated in such an organic electrode, up to 30 to 40 % per monomeric units present in the polymer film. This result has been the starting point for numerous studies on the so-called "organic batteries", mainly based on polypyrrole and polyaniline, which already appeared on the market, and also for some proposals of electrochromic devices. In these applications however, the electroactive conducting polymer chains have been only considered as playing a passive role of charge carriers for the corresponding reversible insertion of a large amount of dopant molecules or ions, and their potential has been largely underevaluated. As a matter of fact, this backbone of polymeric chains, which possess individually a high electronic conductivity, form a real network of molecular wires, i.e. wires which consist of the sequence of just one type of molecule, such as ethylene, thiophene, pyrrole or aniline. Furthermore, we know from their electroactive behavior toward dopant molecules that almost each molecular unit is involved in the conducting and electroactive processes, and can thus be adressed from an electrode. This unique molecular property confirms the potential of these molecular materials to be used as a tridimensional conducting molecular network for the development of new sensors and devices operating at the molecular level.

MOLECULAR SENSORS AND TRANSDUCERS.

These molecular wires, organized in a tridimensional conducting network, form priviledged routes for the reversible transfer of electronic information between an electrode, onto which these polymers chains are grafted, and various prosthetic groups R which can be covalently bonded as side groups all along the conducting chains. This polymer functionalization can be underdone by the initial chemical modification of the monomer unit, prior its polymerization, assuming that the polymerization reaction is controlled in such a way that the presence of substituents R onto the monomer units is still compatible with the formation of long conjugated conducting polymer chains. By varying the nature of R, many types of interactions can be envisionned, from molecular recognition, which will allow the development of molecular transducers, to logic functions or to memories, which will open the new research area of molecular electronics. A second generation of organic conducting polymers has thus appeared in the recent years, obtained by their functionalization with various prosthetic groups R. All classes of polymers have been the object of such chemical modifications, the most studied ones concerning the polyheterocycles derived from pyrrole and thiophene.

It must be however pointed out that this functionalization appears as a difficult experimental task, since the electronic and steric properties of most of the potentially interesting substituents R on a given monomer are not compatible with the synthesis of a long conjugated conducting polymer.

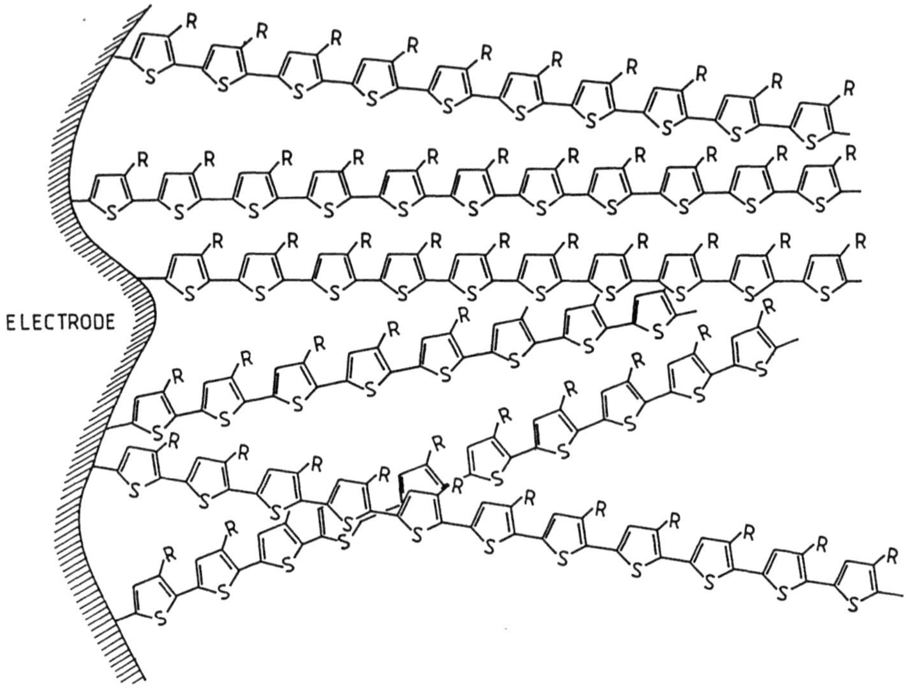

Figure 3. Three-dimensional network of molecular wires, formed by conducting polythiophene macromolecular chains, bearing functional groups on each monomeric thiophene unit.

As previously mentionned, the conducting and electroactive properties of these conjugated polymers are highly dependant on the stereoregularity of the single and double bond alternation, and also on the absence of any trace of chemical impurities such as polymerization catalyst. The most critical point concerns the steric effects presented by higly encumbered functionalized substituents, which hinder the chemical coupling of the corresponding substituted monomers and lead to very short chain lengthes, or which can also provoke important torsions between adjacent monomeric units in the polymer chain, reducing thus largely the conjugation. A large number of functionalized polymers have been thus described in the literature, such as poly(N-substituted pyrroles), possessing conductivity values as low as 10^{-4} S. cm^{-1}, which clearly shows that the concepts of long conjugated chains and of molecular wires cannot anymore be applied to these materials (7). A recent carefull analysis of the steric and electronic factors which, for a given substituent R on a monomer, control the conducting and electroactive properties of the corresponding polymer, has led to the development of a new concept of "functionalized environment" , which allows to overcome these problems and to realize functionalized polymers being at the same time conducting, electroactive and stable (8).

Prosthetic substituents are now grafted along the conducting chains of conjugated polymers, leading to molecular materials consisting of a three dimensional distribution of functionalized centers, linked together by a network of conducting molecular wires. This distribution is mostly isotropic, corresponding to the amorphous state generally observed for organic conducting polymers, but it has to be underlined that long range order and even crystallinity are also known for some conjugated polymers, which may lead to preferential orientation for the molecular wires and their pendant substituents (9) Furthermore, various techniques have also been developed which allow to control the spatial ordering of the elemental units of these molecular materials. Poling under a high electric field has already been used for the spatial alignement of molecular units possessing a high dipole moment, and the well known Langmuir Blodgett technique, which allows the construction of a material, almost molecule by molecule, will play a major role in the development of this field. Molecular engineering, performed at the level of the elementary bricks (functionalized monomers) and during the controlled assembly of these bricks into a material, will afford thus new type of transducers, presenting a large variety of specific interactions with external physical or chemical phenomenon. Some examples can be given of the possible interactions that can be developed in functionalized conducting polymers, some of which have been already achieved, and some others being within the hands of chemists.

Photon sensing can be realized by the use of a photoactive center as substituent R, such as the well known transition metal complexes (RuL_3). The N-substitution of pyrrole with the complexing moiety bipyridine, L = bipyridine, allows to realize a polymer in which the bipyridine substituents, bpy, coordinate to ruthenium, leading to the existence of pendant $Ru(bpy)_3{}^{2+}$ groups along polypyrrole chains.

Although this functionalized polymer shows a very low conductivity, due to the high steric constraints from the bulky substituents, the appearence of photocurrents under illumination has been demonstrated, associated with the transfer of charge from the photoexcited states of the ruthenium complexes to the conducting polypyrrole chains (10), as presented in the Figure 4.

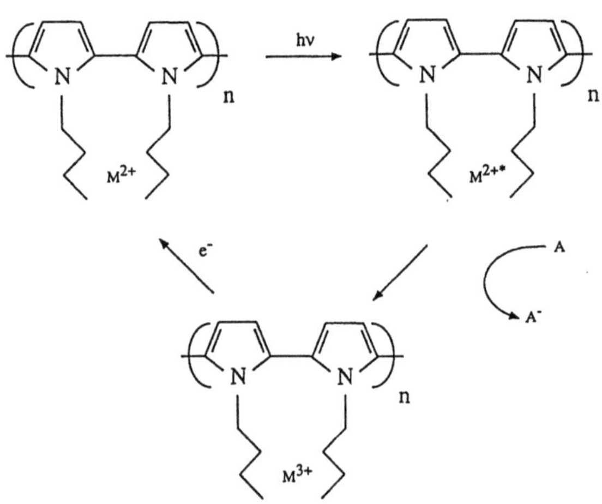

Figure 4. Reaction pathway converting an incoming photon, absorbed by the photoactive Ru^{2+} complex, into an electron which is transfered to the conducting polypyrrole chain, evidenced by the observation of a current.

More elaborated systems have been constructed, involving a layer of polypyrrole functionalized with a photosensitizer center, $Ru(bpy)_3{}^{2+}$, and a layer of polypyrrole substituted with an electron acceptor such as viologene. A charge transfer between this two-layer structure has been experimentally demonstrated, which confirms the principle of this sensing process (11) Specific interaction with other physical quantities, such as electric field, is also possible, although such transducers have not yet been experimentally described in the literature. Thus, if using a highly polarizable substituent, the application of an external electric field will lead to the appearance of an excess of charge in the vicinity of the conjugated chain, and hence to the variation of its conjugation and conducting properties. Sensitivity to pressure has also been described and devices have already been marketed which show a very high sensitivity (12).

Selective interactions with external chemical species such as ions have also been developped by the appropriate functionalization of conducting polymers. These interactions are based on the electroactive properties of the conjugated chains, which can be reversibly doped and undoped with various ions, and also on the grafting of polyether groups along the conjugated polymer chains, which are known to be very efficient and selective complexing agents toward metallic cations. When using these modified electrodes in presence of the appropriate cations corresponding to the cage size created by the polyether groups, new features appear on the electrochemical voltammograms, associated to the selective interaction of the metallic cations with these functionalized polymer chains.

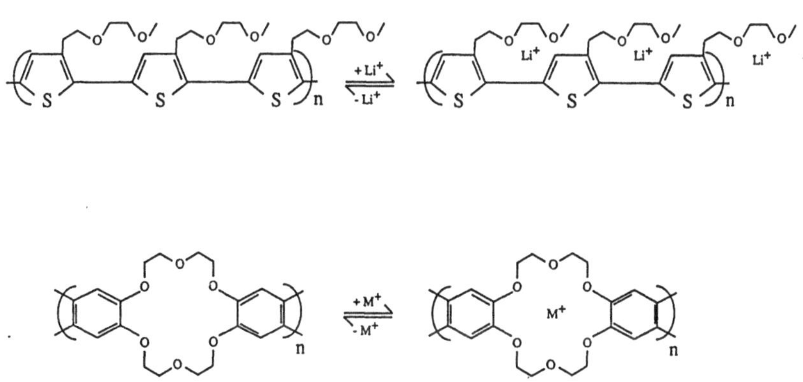

Figure 5 (a) Poly (3,6-dioxaheptyl)thiophene, and (b) poly(dibenzo crown ethers) , showing reversible specific complexation with cations.

Various examples have been already described in the literature, which operate according to this principle. Thus ionic selectivity has been demonstrated in the case of poly(dibenzo crown ethers), shown in Figure 5b, which contain crown ethers of variable size and which can thus be adapted to the specific complexation of various cations (13). Short polyether chains have also been grafted on pyrrole and on thiophene monomers, and their subsequent polymerization has led to materials which show specific interaction with cations. Thus a thiophene monomer has been functionalized with a (3,6-dioxaheptyl) group, which is known to be a complexing group for lithium cations. The corresponding functionalized conducting polythiophene, shown on Figure 5a, has been electrochemically characterized in electrolytic media containing various cations. The electrochemical voltammograms associated to the doping-undoping process show that whereas the majority of cations lead to the same shape of the voltammogram, with a maximum at 0.8 V /SCE, the use of the lithium cation Li^+ is accompanied with the appearance of a new peak located at 0.5 V/ SCE, as shown on Figure 6, (14).

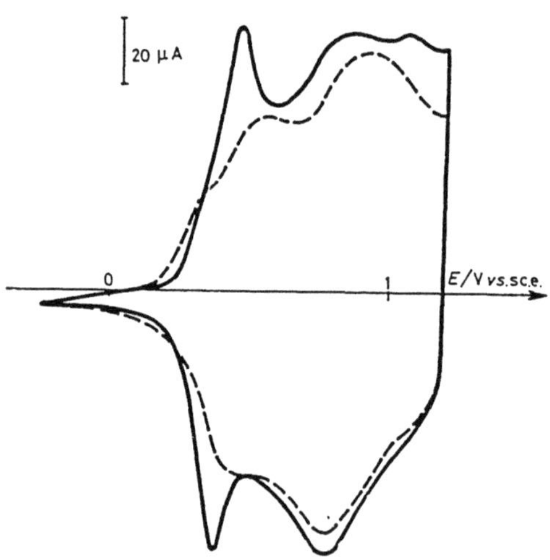

Figure 6. Cyclic voltammogram of an electroactive polythiophene bearing diether substituent, R = $(CH_2)_2O(CH_2)_2OCH_3$, in presence of various cations. —— : Li^+ ; ---- : $N\ Bu_4^+$ or $N\ Et_4^+$, in CH_3CN ; E vs. saturated calomel electrode

A specific interaction with lithium cations, characterized by a definite electronic signal, has thus been created in these electroactive conducting polythiophenes, by their functionalization with polyether groups, and it must be pointed out that this recognition toward cations is exerted at the level of every thiophene units in all of the macromolecular chains forming the material on the electrode. This recognition, which is thus operating at the molecular level of the material, can be furthermore designed for various types of cations, by the adaptation of the polyether cage size to the cation size, as already largely demonstrated in the literature (15). Among ionoselective conjugated polymers, thiophene based ones possess a high electrical conductivity, of the order of 10^2 S cm^{-1}, and the simultaneous control of electrical and ionic conductivity opens up a large potential in the realization of ion selective, electrochemically controlled, membranes (14).

Besides ionoselectivity, the most challenging recognition property concerns enantioselectivity, i.e. the ability to discriminate between optically active substrates, at the image of some enzymatic processes for instance. We have recently succeeded in achieving this ultimate recognition property, by the controlled functionalization of polythiophenes. Thiophene monomers bearing large chiral groups, such as R* = (2S)-(+)-$(CH_2)_3OCH(Et)C_6H_5$ were first synthesized, and then polymerized as thin films on an electrode.

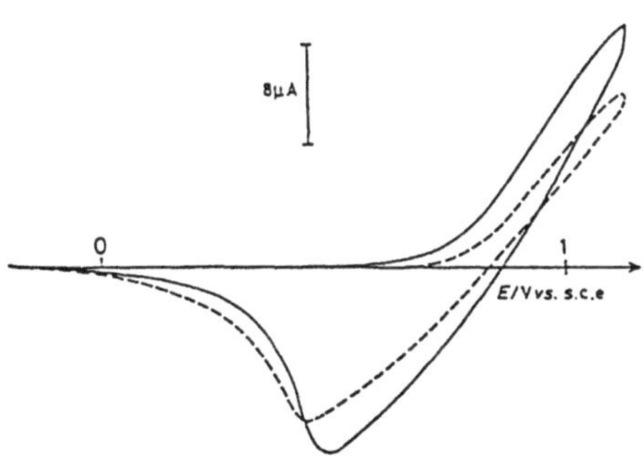

Figure 7. Electrochemical cyclic voltammogram of a chiral polythiophene electrode, using an optically active anion as doping agent : (——) , (-----) : (+)- and (-)-10-camphorsulfonate respectively. E versus calomel electrode.

The characterization of the resulting polymer has confirmed that it is optically active (specific rotation = + 3000°), conducting (σ = 12 Scm^{-1}) and electroactive. When used as an organic electrode in an electrolytic medium containing for instance ClO_4^- anions, the electroactivity is characterized by a reversible doping-undoping process, corresponding to the intercalation of these anions inbetween the conducting polymer chains. If using an optically active anion, such as (1S)-(+)-10- or (1R)-(-)- 10-camphorsulfonate anion instead of ClO_4^-, this chiral electrode shows a different electrochemical response, depending on the camphorsulfonate enantiomer used (16). A ratio of about 2 has been observed for the doping level in the chiral polymer matrix, which shows that this polymer has exerted a molecular recognition toward optically active species in solution, a phenomenon which mimics biological processes, as presented in Figure 8.

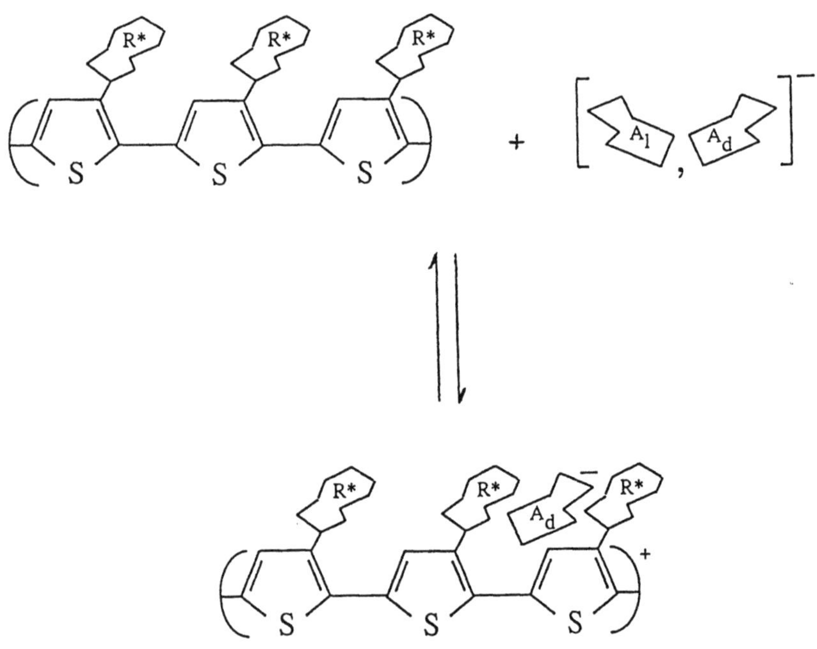

Figure 8. Molecular recognition exerted by a chiral substituted conducting polymer toward optically active anion A^- in solution, during the electrochemically driven doping process.

This effect has been interpreted on the basis of the chiral environment created by the bulky R* substituents around the polythiophene chains. During the doping process, the anions have to percolate through this chiral environment in order to reach the thiophene rings and ensure the electronic doping interaction.

These iono and enantioselectivies can be used for the development of new types of sensors. As a matter of fact, microelectrochemical field-effect type, FET, transistors based on organic conducting polymers have been already described by M. WRIGHTON, in which a thin film of conjugated polymer is used as active layer (17). Two microelectrodes imbeded in this film form the source and drain contacts, and this device, immersed in an electrolytic medium, is operated like an organic electrode with a counter electrode forming the gate. As a sharp potential rise is applied between source and counter gate electrode, the conjugated polymer is switched from an insulating to a conducting state by an electrochemical doping process, which leads to the appearance of a drain current between source and drain electrodes onto which a source-drain voltage has been applied. An interesting amplification ratio of about 10^3 has been demonstrated , but the lack of selectivity shown by these organic microelectrochemical devices toward the doping anions involved in the doping process limits their potential interest. The use of functionalized electroactive polymers instead of the conventional ones will bring the needed selectivity, whether toward cations or toward enantiomers, and microelectrochemical FET type transistors are now achievable which will show molecular recognition toward ions or optically active species in solution. This new type of transducers appear of large interest, both on the fundamental point of vue and also for their possible applications.

MOLECULAR MATERIALS FOR ELECTRONICS

The concept of molecular wires, which can be used for describing the properties of organic conducting polymers, appears also extremely attractive for the development of a new generation of electronic devices. The ultimate step could be expressed by the term "molecular electronics", although much care has to be taken for the real significance of this new research area (18). Thus the functions of switching, gating and amplification result from the possibility to modulate the charge transfer properties of these molecular wires, and memory effects will be obtained through the control of bistability. Many ways exist for the control of these phenomena, by a proper chemical engineering of these molecular wires. For example, donor (D)-acceptor (A) systems are known to undergo an internal charge transfer, which can be photochemically induced, and which leads to the system D^{+}---A^{-} where a large separation of charges exist. If the donor and acceptor molecules are grafted as side groups along a conjugated polymer chain, such as depicted in Figure 9, it appears that the conjugation along the polymer chain, and thus its conductivity, will be highly dependant on the degree of charge transfer in the D---A system.

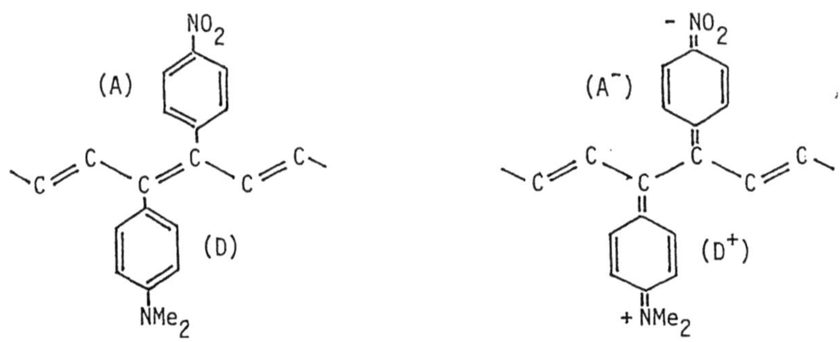

Figure 9 (a) Conjugated chain of polyacetylene substituted with a donor group D and an acceptor group A. The internal charge transfer from donor to acceptor (b) leads to interruption of conjugation.

The conducting properties of this functionalized molecular wire will be controlled and modulated through any external physical quantity, such as photons or electrical field, which is able to modify this degree of charge transfer, leading thus to promising molecular gate properties. The expected switching time of this molecular gate should be ultrafast, as Franck Condon excited states are involved, which are reached in the femtosecond scale. Although already proposed some ten years ago, no clear experimental evidence has yet been given for these systems, due the very difficult synthetic task for reaching these functionalized conjugated polymers. On the other hand, interesting results have appeared recently in the literature, which describe the principle of a molecular memory working on the principle of charge transfer variation in a molecular donor acceptor system. An acceptor center A and a donor center D are covalently bonded at the periphery of a particular conjugated system which undergoes, under light excitation, a conformational twist which modifies its conjugation properties. These molecular systems, operating in a process known as Twisted Internal Charge Transfer, present the potential interest of bistability, and bring the demonstration that molecular memories can be realized. Although the characterization of these molecular devices is possible by the use of a nanoprocessor such as a tunneling microscope, the technological aspect of adressing such molecular systems, i.e. writing and reading at the level of a molecule, remains at a far distance, and this consideration brings one of the major limitations to the actual significance of the area of molecular electronics. Other bistable phenomena have been also described in the literature for molecular systems, such as the antiferroelectric to ferroelectric transition induced by the application of an external electric field (19), or the ferrimagnetic to ferromagnetic transition in heterobimetallic complexes under the application of a magnetic field (20), or on the well known modification of optical properties occuring in photochromic materials following a light pulse.

As pointed out, if many research groups have already experimentally demonstrated on various examples that the charge transport properties of molecular systems can be controlled and modulated by external physical or chemical quantities, our state of art in technology does not give reasonable hope for even mid term access to molecular scale devices. But, on the way to molecular electronics, these results open up a very promising field which concerns the use of molecular functionalized materials for new devices, in an area which can be depicted by the term Electronics with Molecular Materials. Owing to the preceedingly described possibilities of molecular engineering, chemists can realize new types of transducers of large potential interest. This application of organic materials to electronics sets however other kind of problems, which concern the material properties generally required for building electronic devices. Many attempts have thus been described in the literature for the realization of organic based devices, such as Schottky junctions, homo- and hetero junctions, organic on inorganic junctions and field-effect transistors, FET. Conjugated polymers, polyacetylene, polypyrrole, polythiophene have been used for the fabrication of FET, (21), such as presented in Figure 10a : a highly doped n-type silicon wafer, on which a back contact is realized for building the gate electrode, is thermally treated for growing a thin SiO_2 insulating layer. Two small gold contacts are evaporated, which form source and drain electrodes, and the organic semiconductor is further deposited by using various techniques, such as spin-coating in the case of soluble polyalkylthiophene, direct polymerization in the case of polyacetylene or polythiophene, or vacuum evaporation for metallophthalocyanines.

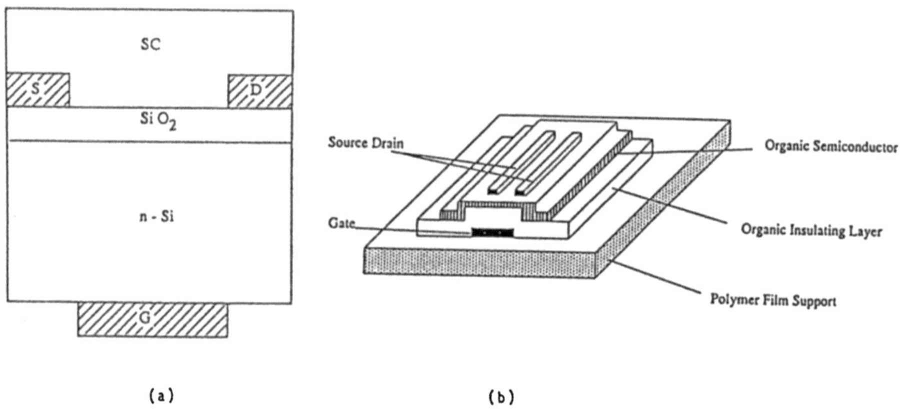

Figure 10. Schematic views of (a) an organic-on inorganic field-effect transistor, FET, in which an organic semiconductor, SC, is deposited on an inorganic insulating layer, such as SiO_2, and (b) an all-organic thin film transistor, TFT, in which a thin layer of organic semiconductor, SC, is deposited on an organic insulating polymer, IP, using a polymer substrate P. Gate, source and drain are presented by G, S and D.

The first amplification characteristics obtained with these organic-based FETs confirmed the principle of using molecular materials for the electronics, but the experimental results evidenced that the electronic properties of these organic materials were far from meeting the requirements for them to be used as active layers in devices. Thus the current output was very small, of the order of some nanoamperes, and a large ohmic current was flowing in the FET channel in the "off" position, which were respectively due to low mobility and high conductivity values in these organic semiconductors. As a matter of fact, most of reported organic semiconductors present large concentrations of impurities and structural defects which create efficient traps and recombination centers, and which impose interchain hopping of charge in the general charge transport process. Thus the carrier mobility value μ, one of the most critical parameter for a semiconductor, was very low for these organic conjugated materials, rarely exceeding the value of $10^{-4}\ cm^2V^{-1}s^{-1}$ whereas that of amorphous silicon is of the order of 10^{-1} to $1\ cm^2V^{-1}s^{-1}$, and that of monocrystalline Si is larger than $10^3\ cm^2V^{-1}s^{-1}$. It was not clear whether this low mobility value had to be attributed to the low efficiency of the interchain hopping of charge, or to the high efficiency of traps. In the first case, a longer conjugation length would have been needed for increasing the mobility of charges, and in the second case a higher structural control of these conjugated materials would have been required. Most of research groups have assumed that conjugation length was the critical parameter, and have used various conjugated organic polymers and organometallic materials for the construction of FET. The obtained characteristics however were deceiving, and owing to the experimental difficulties encountered for increasing significantly the conjugation lenth in these conjugated polymers, various chemical and physical tretaments, such as annealing under oxygen, have been proposed for increasing the mobility. However, none of these treatments led to a real improvement of the charge transport efficiency, as confirmed by the mobility μ which still remained around the value of $10^{-4}\ cm^2V^{-1}s^{-1}$. On the other hand, following the hypothesis that these low mobilities were due to the existence of high density of traps present in these conjugated polymers, we developed a new class of semiconducting conjugated organic material, in which the concentration of structural defects could be largely if not completely suppressed. Thus series of oligomers have been proposed, which consist in shorter conjugated chains of the parent polymers polythiophene,n Figure 11, and polyacene, and which can be chemically synthesized as pure and well defined molecule (22). The conjugation length can also be precisely controlled in this new series of semiconductors, which allows to undertake the study of relationship between conjugation and charge transport properties.

As a matter of fact, when compared to their parent conjugated polymers, these oligomers show much shorter conjugation length, which require thus a higher contribution of interchain charge transfer, but a much lower content of structural defects and chemical impurities.

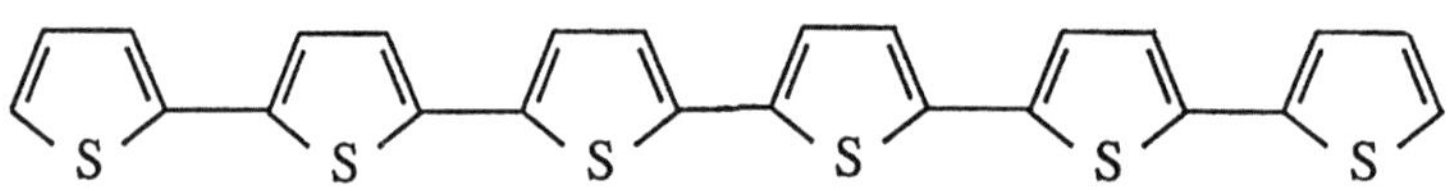

Figure 11. Chemical structure of sexithienyl, a short conjugated oligomer derived from thiophene.

Oligomers of variable chain length derived from thiophene and from acene have been synthesized, and the analysis of the charge transport properties of these organic semiconductors has been underdone through the characterization of field-effect transistors (23). A first analysis of the mode of operation of these organic-based field-effect transistors has shown that these devices operate through the formation of an accumulation layer, in which an ohmic current is flowing in the bulk of the semiconductor in parallel with the channel amplification current, and that the best adapted device architecture requires a very thin layer of organic semiconductor, corresponding thus to a Thin Film Transistor structure, TFT (24). The characteristics of these TFT being controlled by the interface between insulator and semiconductor, we developed an "all-organic" thin film transistor (25), in which the insulating layer is formed by an insulating polymer layer, as shown on Figure 10 b. The amplification characteristics of such an all-organic thin film transistor, Figure 12, based on the thiophene oligomer sexithienyl as active semiconductor, have been determined, and the results showed a dramatic increase of the mobility μ, which rises to values higher than $10^{-1} cm^2 V^{-1} s^{-1}$ (25). One important point also concerns the conductivity which remains at a low value, of the order of 10^{-7} S cm^{-1}, leading thus to a negligeable contribution of ohmic current in the total observed amplification current (26)

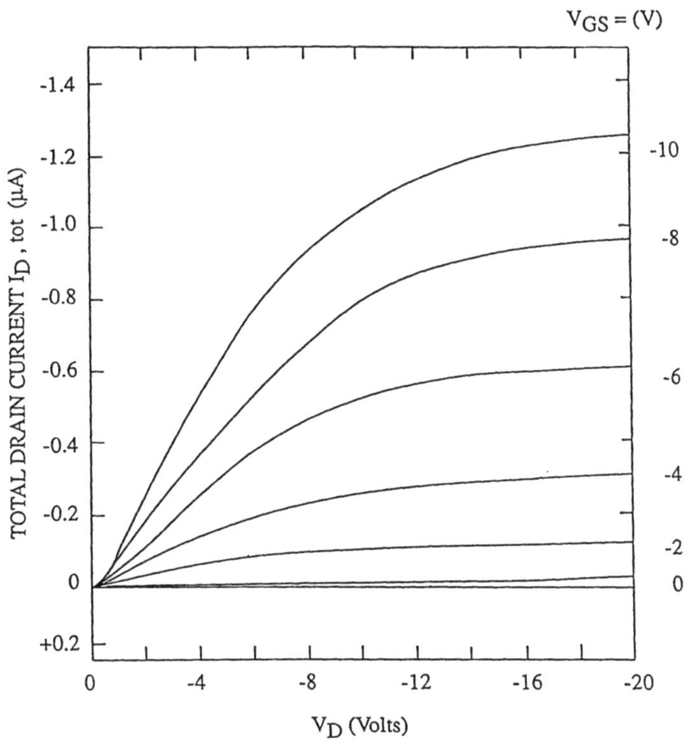

Figure 12. Amplification characteristics of an all-organic thin film transistor based on sexithienyl oligomer as organic semiconductor, and on neoprene as insulating polymer. The thickness of the semiconducting layer is 14 nm, and that of the insulator is 3 μm.

This increase in semiconducting properties of organic semiconducting materials is not limited to the case of thiophene oligomers, as shown by the study of another class of materials, oligoacenes, which also show a large increase of carrier mobility. Short conjugated oligomers appear thus very close to amorphous hydrogenated silicon, and even if the dynamic of amplification of these all-organic devices is still lower than that of inorganic-based ones, such as MOSFETs, they appear very promising.

Beyond the potential interest presented by these results for applications in fields such as display devices, the important conclusion of this work concerns the demonstration that organic molecular materials can meet the requirements for being used as active semiconducting layers in electronic devices, which gives a clear answer to the pertinence of these materials for the electronics. In addition, this new class of oligomers presents fundamental differences when compared to classical inorganic materials. As a matter of fact, the semiconducting properties of inorganic semiconductors such as silicon are linked to the existence of a macroscopic material, in which atoms are held together by strong covalent bonds, and any disruption of this assembly by a foreign atom or by an impurity has to be rigorously controlled for mastering its semiconducting properties. In the case of organic materials on the other hand, few modifications appear when passing from the oligomer molecule to the molecular material, in which the individual molecules are only held together by weak Van der Waal forces. The properties of such molecular assembly reflect the properties of the individual molecules, and this consideration has two important consequences. First, the concept of impurity become different in the case of these molecular materials, as no cohesion of a macromolecular assembly is required for the semiconducting properties to appear. The construction of electronic devices from such organic materials will be much less sensitive to purity conditions, as has been experimentally observed in the realization of the described all-organic devices. Secondly, the molecular nature of these semiconductors will be much more compatible with the scaling down of devices, being of course aware that the main problem for this move toward larger scale integration will concern the available adressing technology. As a main result, this work shows that conjugated molecular systems possess the charge transport properties needed for building electronic devices, and that the way is now open for their further functionalization which will allow the development of new type of sensors, transducers and memories operating at the molecular level.

In conclusion, organic conjugated polymers and oligomers form a new class of molecular materials, which form a three dimensional network of macromolecular chains possessing intrinsic charge transport properties. A concept of molecular wires can thus be developed, which can be used for the transfer of electronic information. The further functionalization of these molecular wires with prosthetic groups presenting specific interaction with external physical or chemical quantities allow to build a new generation of transducers and devices which operate at a molecular level., opening thus an exciting field toward molecular electronics.

REFERENCES

(1) H. Naarmann, F. Beck, E.G. Kastning, BASF AG Ludwigshafen, *DB Patent* 1195497 01 February 1963.

(2) T.A. Skotheim (Ed.) : *Handbook of Conducting Polymers*, Marcel Dekker, New York 1966.

(3) A.F Diaz, K.K. Kanazawa, J.P. Gardini, *J. Chem. Soc. Chem. Comm.* 1979, 635.

(4) G. Tourillon, F. Garnier, *J. Electroanal. Chem. 135* (1982) 173.

(5) A.G. MacDiarmid, J.C. Chiang, M. Halpern, W.S. Huang, S.L. Mu, N.D.L. Somasiri, W. Wu, S. Yaniger, *Mol. Cryst. Liq. Cryst. 121* (1985) 173.

(6) H. Naarmann, N. Theophilou, *Synth. Met. 22* (1988) 1.

(7) J. Bargon, S. Mohmand, R.J. Waltman, *IBM J. Res. Dev. 27* (1983) 330.

(8) J. Roncali, R. Garreau, D. Delabouglise, F. Garnier, M. Lemaire, Synth. *Met. 28* (1989) 341.

(9) M. Lemaire, R. Garreau, F. Garnier, J. Roncali *New J. Chem. 11* (1987) 703.

(10) G. Bidan, A. Deronzier, J.C. Moutet, *Nouv. J. Chim. 8* (1984) 501.

(11) A.J. Downard, N.A. Surridge, T.J. Meyer, S. Cosnier, A. Deronzier, J.C. Moutet, *J. Electroanal. Chem. 246* (1988) 321.

(12) S. Yaniger, personal communication.

(13) V. Leberre, L. Angely, N. Simonet-Gueguen, J. Simonet, *J. Electroanal. Chem. 206* (1986) 115.

(14) J. Roncali, R. Garreau, D. Delabouglise, F. Garnier, M. Lemaire, *J. Chem. Soc., Chem. Comm.* 1989, 679;

(15) J.M. Lehn, *Angew. Chem. Int. Ed. Engl. 27* (1988) 89.

(16) M. Lemaire, D. Delabouglise, R. Garreau, A. Guy, J. Roncali, *J. Chem. Soc. Chem. Comm.* 1988, 658.. J. Roncali, R. Garreau, D. Delabouglise, F. Garnier, M. Lemaire, *Makromol. Chem. Makromol. Symp. 20* (1988) 601.

(17)J.W. Thackerey, H.S. White, M.S. Wrighton, *J. Phys. Chem. 89* (1985) 5133.

(18) F. Carter (Ed) : *Molecular Electronic Devices*, Marcel Dekker, New York, 1982.

(19) J. Malthete, A. Collet, *J. Amer. Chem. Soc. 109* (21987) 6887

(20) O. Kahn, Y. Pei, M. Verdaguer, J.P. Renard, J. Slettern, *J. Amer. Chem. Soc. 110* (1988) 782.

(21) J.H. Burroughes, C.A. Jones, R.H. Friend, *Nature, 335* (1988) 3255. A. Tsumura, H. Koezuka, T. Ando, *Synth. Met. 25* (1988) 11. G. Guillaud, M. Al Sadoun, M. Maitrot, J. Simon, M. Bouvet, *Chem. Phys. Lett. 167* (1990) 503.

(22) D. Fichou, G. Horowitz, B. Xu, F. Garnier, in H. Kuzmani, M. Mehring, S. Roth (Edts) *Electronic Properties of Conjugated Polymers III*, Springer Verlag Series in Solid State Sciences, Berlin, 1989, 386.

(23) G. Horowitz, D. Fichou, X.Z. Peng, F. Garnier, *Solid State Comm. 72* (1989) 381.

(24) G. Horowitz, X.Z. Peng, D. Fichou, F. Garnier, *J. Appl. Phys. 67* (1990) 528.

(25) X.Z. Peng, Horowitz, D. Fichou, F. Garnier, *Appl. Phys.Lett; 57* (1990) 2013; F. Garnier, G. Horowitz, X.Z. Peng, D. Fichou, *Adv. Mat. 2* (1990) 592.

(26) F. Garnier, X.Z. Peng, G. Horowitz, D. Fichou, *Molec. Engin.* under press.

Piezoelectric Composites

Helmut THOMANN

Siemens AG
Corporate Research and Development
Material Science and Electronics
Applied Materials Research
Otto-Hahn-Ring 6, D 8000 Munich 83
Fed. Rep. of Germany

Abstract

Piezoelectric composites have been developed to meet the requirements for the design of special electromechanic and electroacustic transducers. The main incentives are electrical and acoustic impedance matching. The composites are fabricated by very different preparation routes, the most important ones being tape casting and laminating. The properties of piezocomposites can be derived from the properties of ferroelectric ceramics and the connectivity of the composite material.

1. Piezoelectricity - One of the Fundamental Principles of Electrotechnology.

Certain materials display the physical effect of piezoelectricity. Their elastic and dielectric properties are linked together by the piezoelectric coefficient. This property is described by the basic piezoelectric equations

$$S = s^E \cdot T + d \cdot E \qquad (1)$$

$$D = \epsilon^T \cdot E + d \cdot T \qquad (2)$$

where S is the elastic strain, T the elastic stress, and s^E the compliance of the material for the boundary condition of constant electric field. D is the dielectric displacement, E the electric field strength, and ϵ^T the permittivity for the boundary condition of constant stress. d is the piezoelectric modulus of the material. For non-piezoelectric materials one has d=0. Equation (1) then describes the normal elastic behavior, equation (2) the dielectric polarizability of the material. As a matter of fact, equations (1) and (2) are tensor equations. A detailed discussion is given in [1].

Piezoelectric materials have the property of transforming mechanical energy partially into electrical energy and vice versa. Every change of strain or stress will cause a change of the electric field or of the dielectric displacement, depending on the external boundary condition. On the other hand, a change of the field or displacement will bring forth a change of the strain or of the stress. Therefore, piezoelectrics are used widely as electromechanical transducer materials.

Mutual transformation of mechanical and electrical energy can be achieved also by the application of electromagnetic forces. Thus, piezoelectricity has to compete with the electrodynamic principle. But, in general, the fields of applications of these two principles are very different, and competition is restricted to certain areas. The power densities in piezoelectric bodies are limited. Accordingly all high power applications, such as generators, power transformers, and electric motors, are the domain of electromagnetic machines. Piezoeletric bodies, however, are capable of vibrating at well defined high frequencies, and ultrasonic transducers are one of the main applications of the piezoelectric effect. They are used in ultrasonic radar systems, in air as well as in water, to measure distances or to detect the presence and position of objects, or they are applied in medical diagnostics. In the audio frequency range, there is really a competition to electrodynamic transducers. Microphones, earphones, and alarms in telephone sets are often made from piezoelectric devices. In the high frequency range, from about 10 MHz up to some GHz, frequency filters, delay lines, and more sophisticated devices such as signal correlators are designed by piezoelectric surface acoustic wave technology (SAW), which includes the fabrication of devices by microphotolithography.

On the other hand, quasi static transducers can be designed on the basis of piezoelectricity, too. They offer the advantage of a capacitive characteristic, without galvanic losses. Piezoelectric relays, or switches, however, have not yet been introduced into mass production because the piezoelectric effect in classical materials is too small to fulfil the requirements of modern technology. This is one of the starting points to the development of piezocomposites. We shall come back to that in chapter 3.1. But at first we should have a glance on the homogeneous piezoelectric materials established in electrotechnology.

2. Homogeneous Piezoelectric Materials

2.1 Piezoelectric Single Crystals

Piezoelectricity is a property of certain crystals which have a non-centrosymmetric structure. Additionally, piezoelectricity is restricted to insulating materials where atoms are combined by heteropolar chemical bonds. Therefore, the list of materials with good piezoelectrical properties, and with proper physical and chemical stability, which are available in good quality at low cost, is rather short.

Quartz crystals are used as piezoelectric resonators, e.g. in electronic watches. Natural quartz crystals are extremely stable with respect to their elastic properties, and are therefore used as time normals of highest accuracy. Synthetic quartz crystals can be fabricated by the hydrothermal process, and are available in large scale at moderate cost. But the piezoelectric effect is rather small, and the application of quartz is confined mainly to piezoelectric resonators.

Lithiumniobate $LiNbO_3$ and Lithiumtantalate $LiTaO_3$ are synthetic crystals displaying a considerably larger piezoelectric effect. They are the most important basic materials for high frequency devices, e.g. SAW frequency filters and delay lines.

The main problem with all piezoelectric single crystals is impedance matching. This holds for the electric impedance and for the acoustic impedance as well.

The electric impedance of a piezoelectric transducer in the case of non-resonant condition is

$$|Z_e| = \frac{1}{2\pi\ f \cdot C} \qquad (3)$$

where f is the working frequency and C is the capacity of the transducer. In most cases, this impedance is too high for an efficient energy transfer from the biasing circuit into the transducer or vice versa.
The acoustic impedance of a material is

$$Z_a = \varrho \cdot v \qquad (4)$$

where ϱ describes the density and v the sound velocity. A good flow of acoustic energy from the transducer to an adjacent medium is possible only if the acoustic impedances of both materials are of the same order of magnitude. Piezoelectric crystals have high density, they are very stiff and therefore display a high sound velocity. Thus, a good coupling to other solid bodies is possible. The energy transfer to water is poor, the transfer to air is even worse.

2.2 Polycrystalline Piezoelectric Ceramics

In general, polycrystalline bodies are not piezoelectric. Even if the crystal structure is asymmetric and polar, no macroscopic effect will be observed because of the random distribution of the orientation of the piezoelectric axes of the crystallites. Ferroelectricity is the key to polycrystalline piezoelectrics [2], and consequently to piezoelectric composites. A ferroelectric is a polar crystal capable of reversal of the polar axis by the impact of an external electric field. Thus, in a ferroelectric ceramic, ordering of the axes can be achieved. Fig. 1 shows the dielectric behavior of a ferroelectric ceramic. In the as-fired state, the total polarization of the material is zero. With a strong d.c. field applied, the polarization increases rapidly due to ordering of the polar axes in the individual crystallites. When this process comes to saturation, the slope of the P/E-curve becomes flat, approaching the small signal permittivity of the material. If now the external field is reduced, the polarization drops according to the permittivity and arrives at a remanent polarization with zero field. This state is stable as long as no strong external field is applied, and the ceramic body now displays piezoelectricity. With strong fields exceeding the coercive field in opposite direction, of course, the polarization ordering will be reversed. Thus the material displays a hysteresis loop.

In a ferroelectric ceramic, the crystallites are devided into domains. Fig. 2 is a schematic representation of the structure of a ferroelectric ceramic in the as-fired state and after poling.

With some complex oxide compositions which crystallize in perowskite structure, e.g. modified $Pb(Ti, Zr)O_3$, extremely high piezoelectric moduli can be achieved [3]. These ceramics exhibit very high permittivities, up to 2000; therefore the electric impedances of transducers made from these ceramics are considerably lower as compared to single crystal transducers. A great variety of transducers for different applications has been developed based on piezoelectric ceramics [4]. But for operating them with semiconductor devices, the electric impedances are still high. On the other hand, solid bodies containing high concentrations of lead oxide exhibit very high acoustic impedances.

2.3 Piezoelectric Polymers

Polyvinylidenefluoride (PVF_2) is a polymer which displays properties similar to ferroelectricity. The acoustic impedance of this material is low, and would be suitable for operation in water. But the piezoelectric coupling is rather small, the permittivity is low, and dielectric losses are high. PVF_2 is available as large-area foils with good properties. Unfortunately, compact bodies show poor piezoelectric properties. Therefore, PVF_2 is used only for special applications, e.g. large area ultrasonic detectors.

3. Challenges to Piezoelectric Composites

Piezoelectric composites were developed to solve special technical problems. Indeed, the objective of this work was to design new devices. The new materials were results of device engineering. This will be demonstrated now with four different piezoelectric composite systems:

- multilayer piezoelectrics, connectivity [5] 2-2 piezoceramic-metal
- porous piezoceramics, connectivity 3-3 or 3-1, piezoceramic-polymer
- hydrophones, connectivity 1-3, piezoceramic-polymer
- ultrasonic antenna arrays, connectivity 1-3, piezoceramic-polymer.

The figures of connectivity give the dimensions of coherence of the different phases, the first figure corresponds to the piezoelectric phase. Thus, a 3-1-connective piezocomposite consists of a piezoelectric phase which is connective in all directions, with a second phase in one-dimensional holes; a 1-3 connective piezocomposite, in contrast, consists of one-dimensional piezoelectric bars in a 3-connective polymer matrix (Fig. 3).

3.1 Multilayer Piezoelectrics

Multilayer piezoelectric ceramics are 2-2-connective layered bodies consisting of ceramic and metal. They were designed to overcome the problem of electrical impedance matching, especially with low frequency transducers and quasistatic positioning elements [6,7]. From equation (3) we can see that the impedance increases with decreasing frequency. This can be compensated for by increasing the capacity of the transducer, which is achieved without changing the shape of the piezoelectric body by laminating it, with thin metal electrodes between the laminae. The optimum lamination procedure was developed for ceramic capacitors. Multilayer piezoelectrics are fabricated in the following way (Fig. 4). At first a ceramic powder is prepared by conventional techniques. This powder is mixed with the aquous or non-aquous solution of a binder, e.g. polyvinylalcohol. From the slurry, a tape is formed (e.g. 100 µm thick) by tape casting or doctorblading, using a polymer foil as a carrier. After drying again, the carrier is removed. The flexible "green" ceramic-metal-tapes are stacked and pressed. They are cut to proper size, and fired like a bulk ceramic. Afterwards, both ends of the composite bodies are covered by silver paste, and are fired again at lower temperature to form the outer electrodes (Fig. 5). Finally, a d.c. voltage is applied to the electrodes sufficent to polarize the thin ceramic layers. A piezoelectric body consisting of n layers displays n^2 times the capacity of a bulk capacitor of equal outside dimensions.

Piezoelectric ceramics must be fired in an oxidizing atmosphere. Otherwise the lead oxide, one of the main constituents of the ceramic, would be reduced to metallic lead. The internal electrodes of the composite therefore have to consist of a metal which is stable under the firing conditions, e.g. 1100°C in air. Only noble metals, such as platinum and palladium, and some alloys like palladium-silver are suitable for this purpose. Multilayer piezoelectrics are therefore pretty expensive.

A technology has been developed to find a substitute for noble metals, as e.g. low-cost lead [8]. Screen printing on the green ceramic tape is not done with a noble metal paste but with a paste consisting mainly of carbon and an organic binder admixed with a small amount of powdered ceramic. After drying, pressing, cutting, and firing as described above, open-structured areas appear between the ceramic layers. (Fig. 6) These are evacuated and then impregnated with molten lead under pressure. Multilayers of this type, however, are not yet produced on large scale.

3.2 Porous Piezoceramics

Porous piezoelectric ceramics have been developed to improve the acoustic impedance which is too high for many applications, as discussed above. The acoustic impedance is diminished when the density is lowered. The density, of course, decreases with increasing porosity. This is a simple way to adjusting the acoustic impedance as long as the size of the pores is small in comparison to the acoustic wavelength. At a frequency of 100 kHz, the wavelength is in the order of 30 mm; thus this requirement can easily be met. But with a ferroelectric, an other point must be considered. Ferroelectricity is an effect of long range interaction in the crystal lattice. The polarization is brought forth by a lattice distortion. The positively charged metal ions are shifted in relation to the negatively charged oxygen ions. Thereby the lattice is distorted. The lattice parameter increases in the direction of the polarization by a few per cent, and decreases perpendicular to it. A minimum size of the crystallites of more than 1 µm is necessary for the development of the ferroelectric polarization. Thus we need big pores of equal size, e.g. 20 µm, in a matrix of fully dense ceramic without smaller pores, where big grains can grow during the firing procedure after shrinking and densification. Special techniques have been developed to obtain such a porous material of well defined porosity. We shall discuss two different preparation routes which are of technical importance. One of them results in randomly distributed equisized pores. The other one leads to a variety of pore shapes and pore distributions and renders new combinations of material properties possible.

3.2.1 Randomly Distributed Defined Pores

Piezoelectric ceramics of this type are made by conventional techniques. A powder is prepared e.g. by the mixed oxide method. When it is mixed with a binder before pressing a combustible material of proper shape is added, e.g. a polymer

with a particle size of 100 µm. During firing, cavities of that size, or slightly smaller due to shrinking, are formed in a dense, well crystallized matrix [9, 10]. With small fractions of polymer, closed cavities are produced, and the result is a "3-0-composite" of piezoceramic and air. With higher fractions of polymer open pores are produced (Fig. 7). In this case it is advantageous to fill an epoxy into the pores to avoid spurious surface effects. A 3-3-composite of piezoceramic and polymer is the result.
The sound velocity is lower in porous ceramics. Together with the lowered density, the acoustic impedance can be lowered drastically, corresponding to equation (4). Porosity up to 80 per cent is possible, thus lowering the acoustic impedance by a factor of ten. The epoxy filler has only little effect, because of its low density and high elastic compliance. As a disadvantage, the permittivity is lowered, too. But this is not important for high frequency applications in the Megahertz range, e.g. in medical ultrasonic imaging.

3.2.2 Ordered Porosity

Piezoelectric ceramics with an ordered porosity are fabricated by a multilayer technology. It starts in the same way as multilayer ceramics are made. A ceramic powder of proper formulation is prepared by the mixed oxide method, or by chemical powder preparation methods. It is admixed to an aquous solution of a binder to form a slurry. A tape is made by tape casting or doctor blading.
Now we continue by laminating the green ceramic sheet with a photoresist (Fig. 8). The resist is structured by exposing and development, as well known from microtechnology. In Fig. 8, long straight strips are shown, but many other geometries are possible, of course. The tapes are then stacked and pressed, and cut to proper size. During firing, the photoresist structures burn out, and voids of the same geometry are formed. In our example, long pinholes in a regular arrangement are the result (Fig. 9). Electrodes can now be applied to the porous

ceramic body using burnt-in silver, or by an evaporation or sputter process. The voids are filled with an expoxy, to avoid spurious effects of moisture. Finally, the ferroelectric ceramic is polarized in a d.c. field.
By this procedure, a large variety of pore structures can be achieved [11], e.g. closed cavities in arbitrary arrangements (3-0-connective), or parallel straight long pinholes (3-1-connective). With those structures, all macroscopic material parameters become anisotropic, opening a wide field for the design of piezoelectric transducers. The properties of these composites can be calculated from the well known properties of the bulk materials, and the geometry of the composites [11, 12].

3.3 Hydrophones

Hydrophones are non-resonant broadband sensors for acoustic vibrations in water. If a hydrostatic pressure acts on a piezo-electric body, two effects are brought forth simultaneously (Fig. 10):

- the direct effect of pressure on the piezoelectric axis, and
- the transverse effect of pressure perpendicular to the piezoelectric axis.

In general, both effects have opposite signs and compensate partially. For the most efficient piezoelectric ceramics holds $d_{31} \approx -0,4\ d_{33}$, where d_{33} is the direct piezoelectric modulus, and d_{31} the transverse modulus. The hydrostatic piezoelectric modulus is $d_h = d_{33} + 2\ d_{31}$, and is rather small.

The hydrostatic piezoelectric effect is enhanced drastically in ceramic-polymer composites [13, 14, 15]. In the simplest case, slim ceramic bars are imbedded in a polymer matrix of high elastic compliance. All bars are in parallel (1-3-connectivity) (Fig. 11). If there is a tight coupling of ceramic and polymer, this will act as a stress transformer. The hydrostatic piezoelectric sensitivity of transducers is increased drastically by several different effects. There is a better ratio of d_{33}/d_{31}, but additionally a better fit of acoustic impedance to water. An analysis of the very complex behavior of such hydrostatic low-frequency transducers cannot be given here, and should be read from the literature [16]. They are produced on a large scale, mainly for military applications.

3.4 Ultrasonic Array Materials

Ultrasonic imaging technologies have been developed to a very high standard for material testing and even more for medical diagnosis. When an ultrasonic transducer emits a short pulse into the human body, echoes are generated by all structures which display a change of acoustic properties of the tissue. The echo returns to the transducer and is detected as an electric signal. The propagation time is a measure of the distance. Mostly, the echoes are very weak. Therefore, a good acoustic coupling of the transducer to the human body is necessary, i.e. a good acoustic impedance matching, as discussed above.

With a single transducer, an approximately spherical wave front is generated, and no angular information is gained.

A linear wave front is generated, according to Huygen's principle, if a linear or preferentially a two-dimensional array consisting of small transducers is applied [17]. Such an array gives information mainly in one direction. In order to gain a picture, it is necessary to rotate the array. In direct contact to the human body, a rotation only by a small angle is possible. Otherwise, a water bag is necessary as a coupling medium.

A better means to gain angular information is the phased array. It consists of many single transducers, similar to a linear array; but the lateral dimensions of the transducers are very small compared to the acoustic wave length. If there is a phase shift from each of the transducers to the next one, a linear wave is emitted at an angle (Fig. 12). The phase shift can be modified by electronic means, and thus the complete information on distance and angle is gained.

For a good spatial resolution of the image, however, the acoustic wave length should be distinctly below 1 mm. Therefore, frequencies up to 10 MHz are required. The lateral dimension of the individual transducers must be distinctly below the wave length, and we arrive at e.g. 20 µm. Ultrasonic arrays of this type cannot be fabricated by mounting small individual transducer elements on a support. Sash sawing is a technology to make fine structures down to 100 µm [18]. Even fine structure are produced by the "lost wax" process (Fig. 13). At first, a structure similar to a honeycomb consisting of a photoresist material is made by a photolithografic process. A ceramic powder is prepared, and mixed with a binder solution, to form a slurry. Now the voids of the honeycomb are filled with slurry, and some slurry is poured in excess on the structure to form a continuous layer of green ceramic (Fig. 13, above). After drying, the honeycomb is burnt away. The ceramic is then fired to density. Now we have a ceramic plate with deep trenches. These are filled with a polymer. The plate is ground to the desired thickness. Thereby the continuous layer of the ceramic may be removed (not shown in Fig. 13), and a 1-3-connective composite is the result. Electrodes are now applied and the ferroelectric ceramic is poled by the proper electric field. The possibility of making very different "lost wax" structures is an advantage of this technology. Concentric rings of ceramics combined by the polymer can be fabricated for example. Such an "annular array" is capable of focussing an ultrasonic beam on an optional distance by the proper phase shift applied to the individual rings.

Another way of fabrication is laser ablation. This process starts with a solid ceramic body fired to maturity, cut to a plate and polished. Deep trenches (Fig. 14) are produced by the irradiation of ultraviolet light coming from a high power eximer laser. With every laser pulse, all atoms within a thin layer of the ceramic are evaporated simultaneously, because of the extremely high photon density of the laser pulse and the very good UV absorption of the material [19]. This process offers the advantage of very exact control of the dimensions, because the structure is developed after firing and shrinking.

4. Conclusions

Piezoelectric composites are object of interdisciplinary research. They are fabricated by advanced processing technologies. The piezoelectric effect in most composites occurs only after ordering of the ferroelectric polarisation. The macroscopic properties can be calculated from the well-known properties of the components, as can be read from the literature [10, 11, 12, 14]. Research and development of piezoelectric composites is a very stimulating field, and new applications of the piezoelectric effect will be opened.

References

[1] W.G. Cady; Piezoelectricity, Vol. 1, Dover, New York, 1964

[2] B. Jaffe, W.R. Cook, H. Jaffe; Piezoelectric Ceramics, Academic, London, New York, 1971

[3] W. Heywang, H. Thomann; Ann. Rev. Mater. Sci. 14, 27-47, 1984

[4] J.A. Gallego-Juárez; J. Phys. E:Sci. Instrum 22, 804-816 (1989)

[5] R.E. Newnham, D.P. Skinner, L.E. Cross; Mater. Res. Bull. 13, 525-536 (1978)

[6] W. Wersing, M. Schnöller, H. Wahl; Ferroelectrics 68, 145-156, (1986)

[7] W. Wersing, H. Wahl, M. Schnöller; Ferroelectrics 87, 271-294, (1988)

[8] T.C. Rutt, US Patent 3679950 (1972), US Patent 3965552 (1976)

[9] T.R. Shrout, W.A. Schulze, J.V. Biggers; Mat. Res. Bull. 14, 1553, (1979)

[10] W. Wersing, K. Lubitz, J. Mohaupt; Ferroelectrics 68, 77-97 (1986)

[11] W. Wersing, Proc. Sixth IEEE Internat. Sympos. Appl. Ferroelectrics, Bethlehem, USA, 212-213, (1986)

[12] U. Bast, W. Wersing; Ferroelectrics 94, 229-242, (1989)

[13] R.E. Newnham, A. Safari, E. Sa-gong, J. Ginicwicz, Proc. IEEE Ultrasonics Sympos., 501-506, (1984)

[14] R.Y. Ting, Ferroelectrics 67, 143-157, (1986)

[15] R.E. Newnham, L.J. Bowen, K.A. Klicker, L.E. Cross; Materials in Engineering 2, 93, (1980)

[16] T.R. Gurmaja, W.A. Schulze, L.E. Cross, R.E. Newnham, B.A. Auld, Y.J. Wang; IEEE Trans. on Soncis and Ultrasonics SU 32, 481-498 and 499-513, (1985)

[17] W.A. Smith, Proc. Sixth IEEE Internat. Symp. Applic. Ferroelectrics, Bethlehem, USA, 249-256, (1986)

[18] C. Nakaya, H. Takeuchi, K. Katakura, A. Sakamoto, Proc. IEEE Ultrasonics Sympos., San Franciso, USA, 634-636, (1985)

[19] M.Eyett, D. Bäuerle, W.Wersing, H. Thomann; J. Appl. Phys. 62, 1511-1514 (1987)

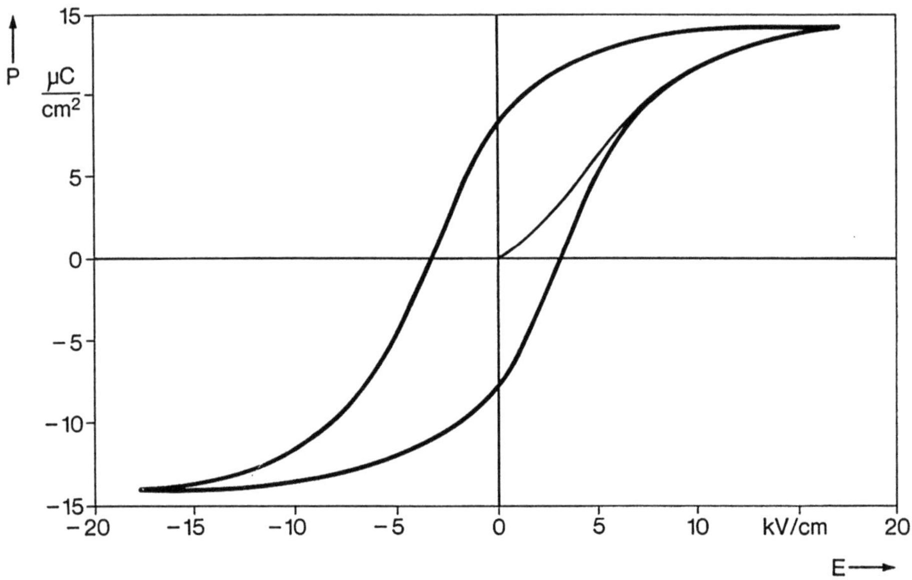

Fig. 1 Ferroelectric Hysteresis loop (polarization versus electric field)

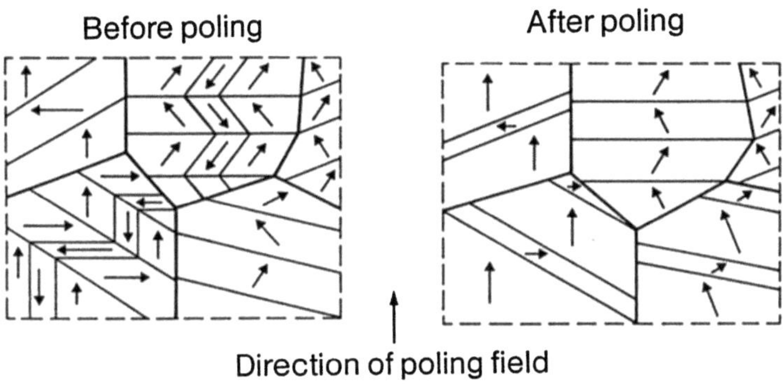

Fig. 2 Ferroelectric domains in piezoceramics: a before poling, b after poling

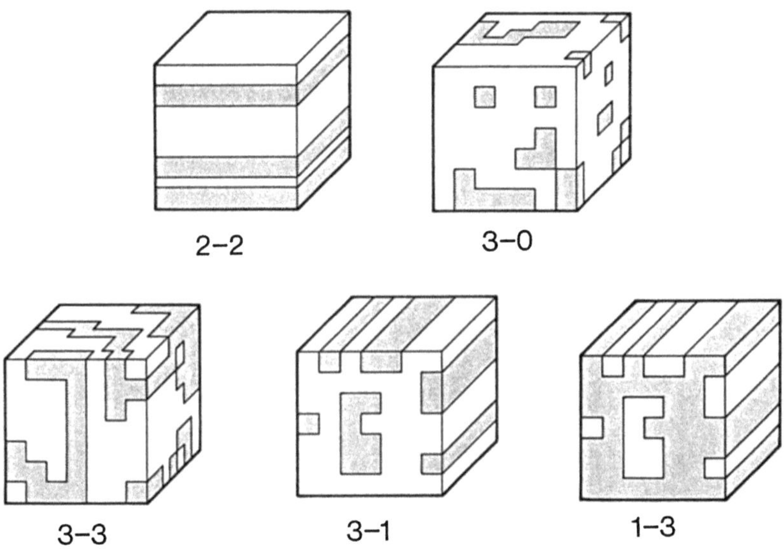

Fig. 3 Connectivities of piezoelectric composites

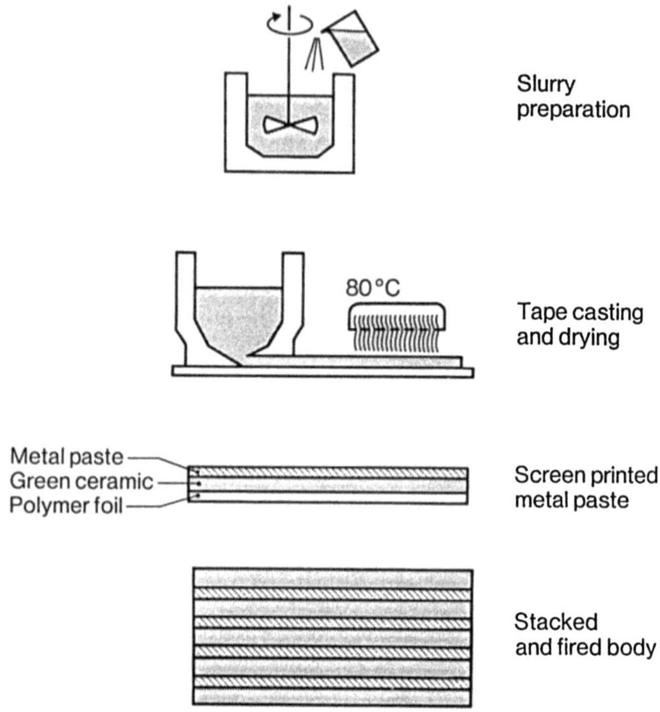

Fig. 4 Preparation route of multilayer ceramics (2-2-connective)

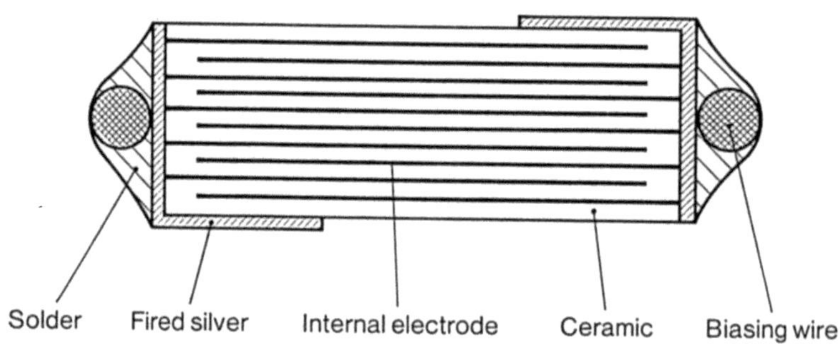

Fig. 5 Cross section of a multilayer ceramic component

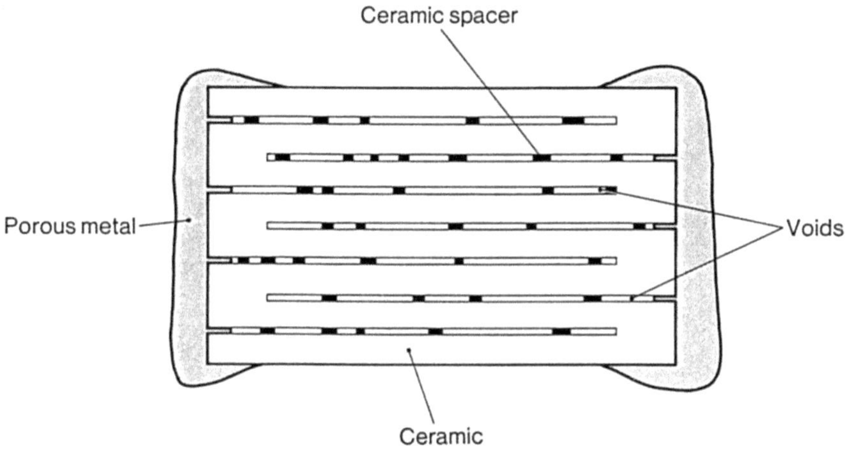

Fig. 6 Ceramic body prepared for lead injection

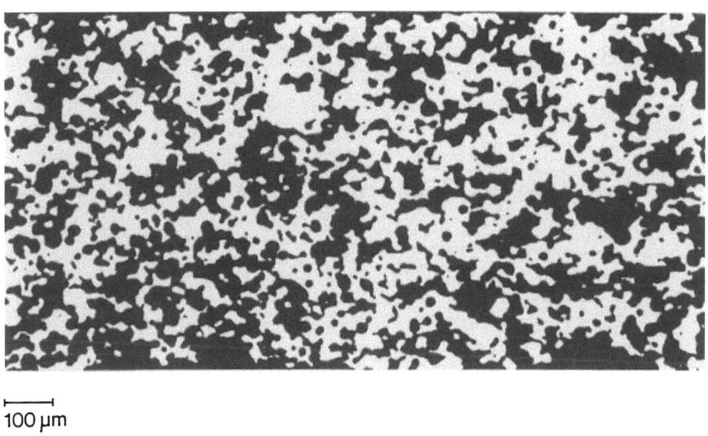

Fig. 7 Piezoelectric ceramic with randomly distributed equisized pores

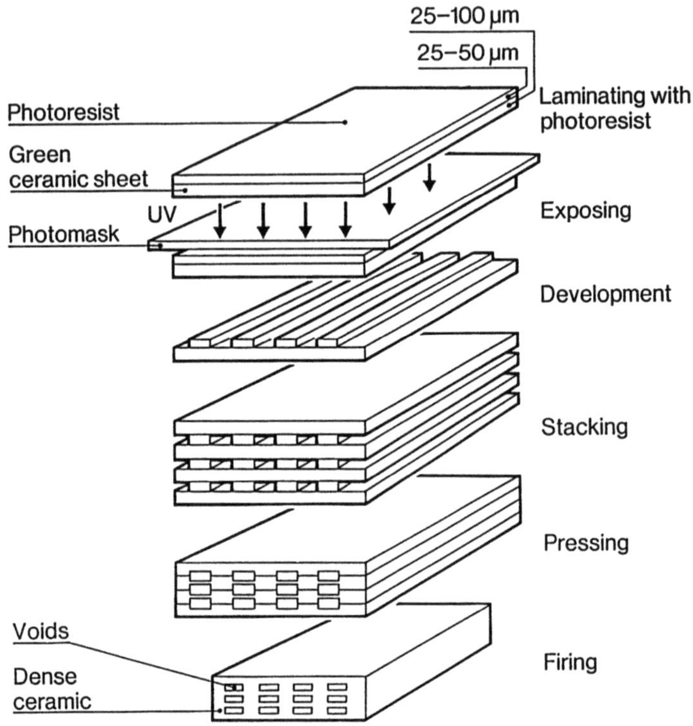

Fig. 8 Preparation route of 3-1-connective piezoceramics

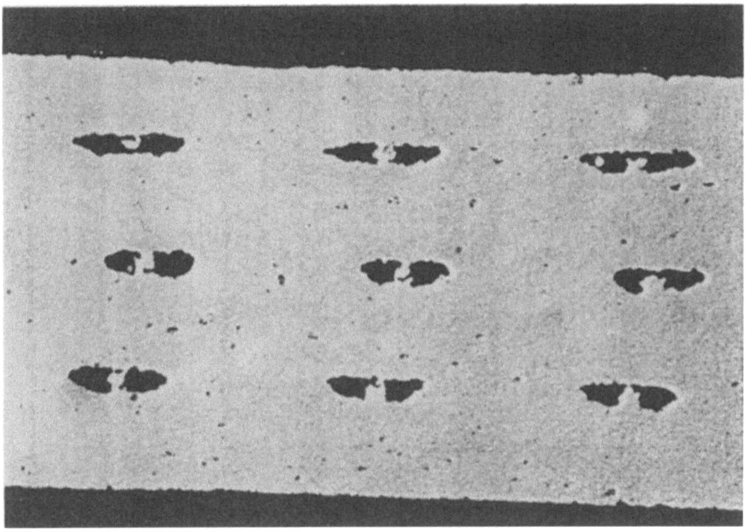

Fig. 9 Cross section of 3-1-connective piezoceramics

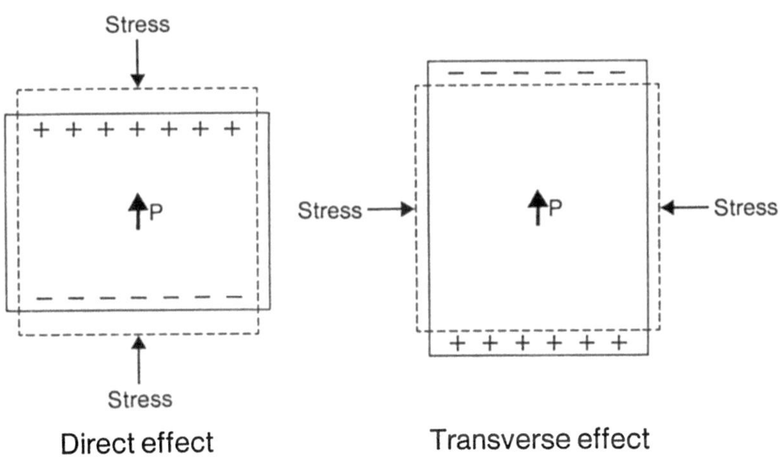

Fig. 10 Direct and transverse piezoelectric effect. With hydrostation pressure, both effects appear simultaneously

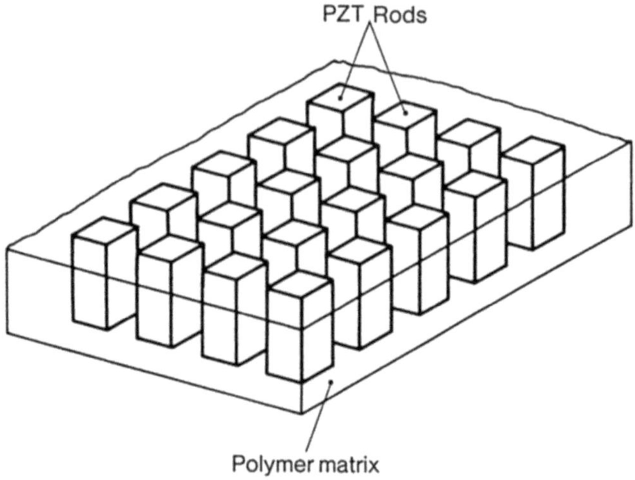

Fig. 11 Hydrophone structure

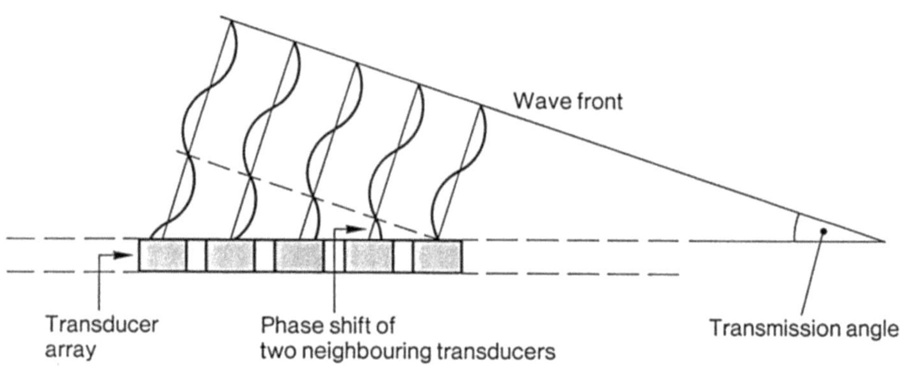

Fig. 12 Principle of a phased array

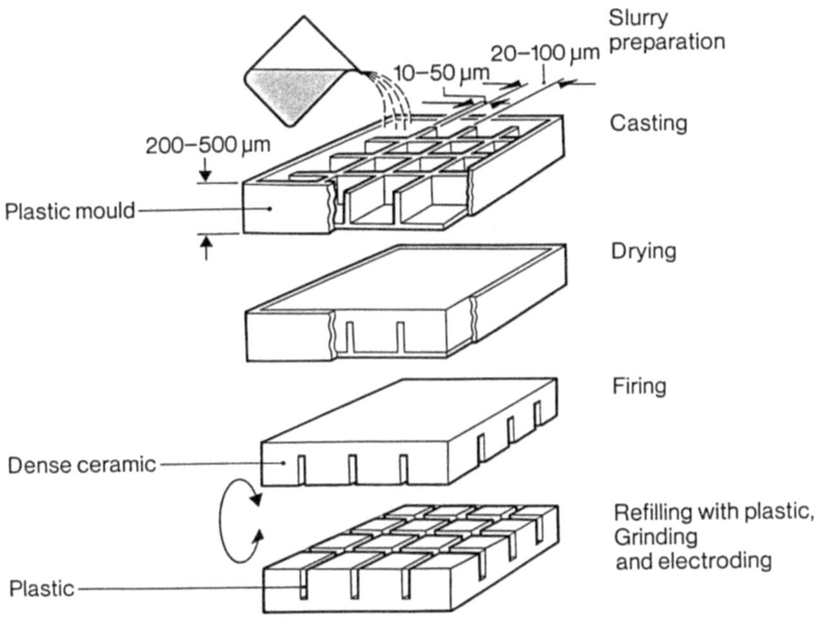

Fig. 13 Preparation route of 1-3-connective array structures ("lost wax" process)

Fig. 14 Array structure fabricated by laser ablation

MIX
Papier aus verantwortungsvollen Quellen
Paper from responsible sources
FSC® C105338

If you have any concerns about our products,
you can contact us on
ProductSafety@springernature.com

In case Publisher is established outside the EU,
the EU authorized representative is:
Springer Nature Customer Service Center GmbH
Europaplatz 3, 69115 Heidelberg, Germany

Printed by Libri Plureos GmbH
in Hamburg, Germany